P9-CBK-722

Harper & Row's
# Complete Field Guide to
# North American Wildlife

# Consultants

# Harper & Row's
# Complete Field Guide to North American Wildlife

## Eastern Edition

Covering 1500 species of birds, mammals, reptiles, amphibians, food and game fishes of both fresh and salt waters, mollusks, and the principal marine invertebrates occurring in North America east of the 100th meridian from the 55th parallel to Florida north of the Keys

Assembled by Henry Hill Collins, Jr.

Illustrations by Paul Donahue, Jennifer Emry-Perrott, Nancy Lou Gahan, John Hamberger, Michel Kleinbaum, Klarie Phipps, Stephen Quinn, Guy Tudor, Nina L. Williams, John Cameron Yrizarry

Harper & Row, Publishers

New York, Hagerstown, Philadelphia, San Francisco
Cambridge, London, Mexico City, São Paulo, Sydney

1817

Portions of this work originally appeared in somewhat different form in *Complete Field Guide to American Wildlife: East, Central and North.* Copyright © 1959 by Henry Hill Collins, Jr. Published by Harper & Row, Publishers, Inc.

HARPER & ROW'S COMPLETE FIELD GUIDE TO NORTH AMERICAN WILDLIFE— *Eastern Edition.* Copyright © 1981 by Harper & Row, Publishers, Inc. All rights reserved. Printed in the United States of America. No part of this book may be used or reproduced in any manner whatsoever without written permission except in the case of brief quotations embodied in critical articles and reviews. For information address Harper & Row, Publishers, Inc., 10 East 53rd Street, New York, N.Y. 10022. Published simultaneously in Canada by Fitzhenry & Whiteside Limited, Toronto.

FIRST EDITION

Library of Congress Cataloging in Publication Data

Collins, Henry Hill, 1905–1961.
  Harper & Row's complete field guide to North American wildlife, Eastern edition.
  Includes index.
  1. Zoology—North America. 2. Zoology—Atlantic States. 3. Animals—Identification. I. Title.
QL151.C62     591.974     80-8198
ISBN 0-06-181163-7

81 82 83 84 85 10 9 8 7 6 5 4 3 2 1

# Contents

# Harper & Row's
# Complete Field Guide to North American Wildlife

**Sponsoring Editor**
Nahum Waxman

**Project Editor**
Marta Hallett

**Production Editor**
Bernard Skydell

**Project Design Director**
Abigail Moseley

**Manufacturing Direction**
Thomas Malloy

**Copy Chief**
Dolores Simon

**Manufacturing Coordination**
Pat Slesarchik

**Copy Editing**
Carole Berglie
Duane Berry
Bernie Borok
Virginia Ehrlich
Janet Field
Rebecca Finnell

**Design Assistants**
Sarah Hartman
Barbara Knight

**Indexer**
Sydney Wolfe Cohen

Composition: TriStar Graphics
Color Separation: Sterling Regal Incorporated
Printing and Binding: R. R. Donnelley & Sons Co.
Jacket Printing: Longacre Press, Inc.

# Introduction

*Harper & Row's Complete Field Guide to North American Wildlife* takes its inspiration from *Complete Field Guide to American Wildlife,* compiled by the late Henry Hill Collins, Jr., and published by Harper & Row in 1959. This important book, covering the Eastern United States and Canada only, was the first single-volume field guide covering all of the major families of larger wildlife—birds, mammals, reptiles, amphibians, fishes, mollusks, and other marine invertebrates.

In the mid-1960s a decision was made at Harper & Row to carry out important revisions in Mr. Collins's field guide and to add a second volume covering the Western United States and Canada. To this end, Jay Ellis Ransom was engaged to prepare a text for the Western Edition, and, in due course, boards of consultants were assembled to review and edit in detail the texts of both volumes.

Working with Mr. Collins's and Mr. Ransom's compilations, the two boards, one for the eastern volume and one for the western, reviewed exhaustively each and every entry, revising, rewriting, and bringing the content into conformity with scientifically attested descriptive information and the latest taxonomic thinking.

In addition, the eastern volume, which did not include species of the southeastern United States, was expanded to cover that region.

Fifteen of the country's outstanding wildlife artists were brought in to work on the projects. All but a dozen of the plates from the original eastern volume were scrapped and replaced with totally new art. The western volume contains only new art, done especially for this work.

Now, after more than fifteen years in the making, this remarkable collaboration is complete and we are pleased to present it for the use of both professional and amateur field natural scientists.

The purpose of the Field Guides is field identification. Included are more than 3300 species, described by appearance, habitat, life zones, behavior, reproduction, and other important aspects. Covered are all species as follows:

*In the East*
East of the 100th meridian, north through temperate Canada (approximately the 55th parallel) and south to southern Florida (the 26th parallel, taking in the peninsula but not the Keys). Included are food and game fish species to a depth of 25 fathoms and mollusks and other marine invertebrates to a depth of 10 fathoms.

*In the West*
West of the 100th meridian, north through temperate Canada (approximately the 55th parallel) and south to the Texas-Mexico border, including all major Texas species but not necessarily specialized coverage of all desert species). Included are all food and game fish species to a depth of 25 fathoms and other marine invertebrates to 10 fathoms.

### Names and Classification
The common and scientific names used in these books represent the viewpoint of the consultants concerned with each section. In general the consultants follow the prevailing scientific checklists in their field, and in the instructions to their sections cite the

authority they have used. In many individual cases these scientists have developed their own conclusions on certain points, and in these cases the terminology or the order of the materials may not conform to the checklists.

## How to Use This Book

In trying to identify any species, note (and jot down, if possible) all the visible characteristics before the animal disappears from sight. If it is a bird, in addition to general color and size, try to determine the color of its beak, legs, feet, and undertail feathers. Notice whether it has wing bars, an eye-ring, white on its tail; stripes on its body; a patch on its cheek or a line over its eyes. Check the shape of its bill. It is important to remember that quick observation is an essential part of field identification, for frequently a species darts in and out of sight very rapdily. Thus, it is necessary for the observer to note visual characteristics quickly, and this can be achieved only if the observer knows what to look for in a particular group of animals.

For mammals, particularly important to note are color patterns—whether they be on the sides, back, or underparts. Color patterns and scales are the key to identification of snakes. So too the number and character of stripes for a lizard. In a fish, the color markings are of great import, but observe them, to the extent possible, while the fish is still alive, for, as with the marine invertebrates, once the animal dies, its colors will change and fade. Naturally, size, habitats, habits, and voice, if any, can be important clues to the identity of any species.

When you turn to the appropriate section of the book, check the ranges and eliminate all species not within your range and season. Check also habitat data, and eliminate those species which appear in a habitat not normally observed. Then turn to the illustrations, which combined with the text, will provide all necessary information for making an identification.

Be sure to note the following points in using these guides:

1. Entries are by common name, with the scientific (Latin) name below. Many marine invertebrate species, however, have no common names and are listed by scientific names only.

2. To the right, opposite each entry, is a set of boldface numbers providing illustration references. Numbers preceded by the designation *Fig.* refer to figures directly in the text. Such figures are generally no more than two or three pages away from the text entry they illustrate. Numbers not preceded by the designation *Fig.* and taking the form of two numerals separated by a colon (e.g., **32:5**) refer to an illustration on one of the plates, with the first half of the number indicating the plate and the second half indicating a specific figure on the plate (thus, **32:5** refers to figure 5 on Plate 32).

3. Plate captions include page references back to the text entries for the species illustrated, so that throughout this book cross-referencing in both directions is complete.

Some species may show color variations because of age, season, sex, environment, or molt—and, as a result, an animal seen in the field may not be colored exactly as the one illustrated. To the extent possible, this factor has been considered, so that in many cases the most prevalent color variations are illustrated. However, if this is not the case with the species sighted, read once again the text description to learn if the species seems to conform in every other respect. It is essential to use both the illustrations and the text for accurate identification. If there is doubt about the identity of a species, don't mark it as having been observed. Scrupulous accuracy in field identification is necessary for any field person.

## Life List

This book can be used to keep a life list of species seen and identified. First, there is included in a section, just before the index, an alphabetical list of all the species in the book by category. Second, the reader may make notes in the index or in the margins of the book, right beside the text descriptions of each species identified. Some people may find it useful to specify the place and date of observation. In this way the guide can become a personal lifetime natural history diary of each new sighting.

## Acknowledgments

For their aid in providing reference materials for these volumes, thanks go to Mildred Bobrovich, Assistant Librarian for Reference Services at The American Museum of Natural History; Jill Fairchild at the Sea Library in Santa Monica, California; and Mary Ann Nelson at the Duke University Marine Laboratory in Beaufort, North Carolina.

To the consulting editors particular thanks are due, for their exemplary efforts in guiding the plan of each chapter, nurturing and editing the text descriptions, and examining each and every piece of artwork.

And, of course, grateful acknowledgment to the artists, whose fine work and whose enthusiasm for the project have done so much to help make these volumes the accomplishment we believe they are. Special thanks to Guy Tudor, who contributed so much more than his paintings.

The Editors

# Birds

**Consulting Editor**
Lester L. Short
Curator of Ornithology
American Museum of Natural History

**Illustrations**
Water Birds and Game Birds, Plates 1–7, 10–11, 17, 24–28
Guy Tudor

Flying Ducks, Plates 8–9 Michel Kleinbaum
Birds of Prey, Plates 12–16 Paul Donahue
Owls and Nightjars, Plates 29–30 Stephen Quinn

Shorebirds, Woodpeckers, Perching Birds, Plates 18–23, 31–48
John Cameron Yrizarry

Text Illustrations, Michel Kleinbaum, Guy Tudor,
John Cameron Yrizarry

# Birds
Class Aves

A bird is an animal with feathers; most, but not all, birds fly. It is generally presumed that those that do not fly are descended from ancestors that did. For this ability, the bodies of birds are specially adapted. The bones are light but strong, and some contain air sacs connected with the lungs. Wings—adaptations of the vertebrate forelimb—are designed to propel the bird through the air. In flightless forms, they balance the bird as it runs, as in the ostrich, or propel it through the water, as in the penguin.

The tail is made entirely of feathers, with a bony base, in contrast to the mammals in which the bones run down the middle. The tail may be either long or short, and is sometimes forked in strong flying species. What looks like, and is generally called, the leg is more or less the equivalent of the human foot (tarsometatarsus), with toes at the end. What looks like a reversed knee is really a joint somewhat the equivalent of the human ankle, the true knee usually being hidden by feathers next to the bird's body. Most birds have four toes, three forward and one, the hallux, behind. Some birds, like the woodpeckers, parrots, and cuckoos, have two toes forward and two behind. In many birds the toes lock automatically around the perch when a resting bird lowers its body, allowing it to sleep without falling.

Birds have no teeth, so a bill adapted for the bird's preferred diet is used to tear, crush, or seize the flesh, seeds, or insects upon which a particular species subsists. Food is sometimes temporarily stored in a crop, an enlargement of the esophagus. The gizzard, or main stomach, is a tough organ, effectively replacing the teeth and heavy jaw muscles of other animals, and the process of digestion is often aided by stones or gravel, which the bird swallows for that purpose. The products of the digestive, excretory, and reproductive systems are all discharged through a common opening, the cloaca. Birds are warm-blooded, with a body temperature varying between 98°F and 112°F (36.7°–44.4°C).

Birds show a complex mixture of unmodifiable ("instinctive"), somewhat modifiable, and highly modifiable (or "learned") behavior. Among the most remarkable of the essentially unlearned patterns is the migration flight of some young shorebirds that leave their summer homes independently of their parents and follow, for thousands of miles to their winter homes, a route they have never before traveled.

### Evolution
Birds arose from the reptiles; the feather is, in essence, a modified scale such as still occurs on their legs and feet. Birds are seldom preserved as fossils, and their record in the rocks is still scant compared with other groups, such as the mammals. It is known, however, that the gull-like *Ichthyornis* and the flightless, somewhat loonlike *Hesperornis* had already appeared by the Cretaceous period, and that the giant flightless land predator, the seven-foot (2.1-m) tall *Diatryma*, inhabited what is now Wyoming in the Eocene Epoch.

In the Pleistocene, in what is now Los Angeles, the Giant Condor, *Teratornis,* was occasionally engulfed in the La Brea tar pits. Almost into historic times the flightless Elephant Bird, *Aepyornis,* persisted on the island of Madagascar. One of the flightless moas of New Zealand, which reached a height of over ten feet (3.1 m), was contemporaneous with early humans, who exterminated the last of them.

### Adaptation

Within the stringent limitations imposed by flight, birds have adapted in diverse ways to varied environments. Birds in their thousandfold varieties are the unchallenged champions of the air. Arctic Terns may fly 22,000 miles (35,405 km) a year in their extended migrations. Some swifts reach speeds of up to 200 miles per hour (322 km/hr). Hawks, vultures, and albatrosses can soar for long periods, hardly flapping a wing. And, because they are warm-blooded, birds can survive in climates from the Poles to the Equator.

### Conservation

Birds have long served humans for game, food, and feathers, as well as in their predatory capacity as destroyers of insects and rodents. Because of the prodigious abundance of wildlife in the early days, the possibility of the permanent disappearance of any species seemed inconceivable. As a result, conservation came too late to save the Great Auk, Labrador Duck, Passenger Pigeon, and Carolina Parakeet, although the Passenger Pigeon was once probably the single most numerous bird species in the entire world.

In the 1890s the public was finally shocked into action at the disappearance of so great a number of birds. The National Audubon Society was formed and set out to save the egrets, which the hunters of plumes for women's hats had almost totally destroyed. Long, hard campaigns resulted in laws that protect virtually all birds. A recent step forward is the Federal Rare and Endangered Species Program.

Urbanization, industrialization, the draining of wetlands, and unwise agricultural practices are continually eliminating the habitats and food supplies of many species of birds, thus causing the greatest modern threat to birds. Also, the widespread use of pesticides has been implicated in the reduction in numbers of certain birds.

### Habitat

Birds can be found almost anywhere—even in city parks and backyards. The best places, however, are where different types of habitat meet, such as the edges of woods, shorelines, and marshes. National, state, and county parks, wildlife refuges, and nature sanctuaries are good places to find birds. Gardens, parks, and open suburbs also may have a variety of species, particularly where sufficient undergrowth and cover have been left, and where dead trees and dead limbs (for nesting holes and exposed perches) have not all been cleared away. On the other hand, although deep woods, desert, mountains above timberline, and certain other places of uniform habitat are not rich in birds, the species found there include some not likely to be seen elsewhere.

The best time of day for bird-watching is the early morning, when birds start singing and actively feeding. During the heat of the day most land birds become inactive and silent, but some will start singing and feeding again toward evening. There may be considerable bird activity throughout the day when it is cloudy or raining lightly. Spring, when migrating birds in breeding plumage are hurrying north, is the best time of the year to see birds; however, the drawn-out fall migration, with the birds in nonbreeding or immature plumage, offers a greater challenge.

*Fig. 1*
Parts of a Typical Bird

See Glossary for:
axillars
carpal
speculum

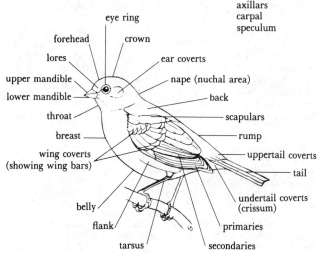

## Voice

It is important to be able to identify birds by sound, especially if the bird in question is hidden by foliage or silhouetted against the sky. In some cases, it is the best means of distinguishing between species that visually may appear almost identical.

For each species discussed in the text, such data on voice are given as may be useful for identification. Whenever possible, the song or call has been rendered into English words, or into syllables that can be pronounced. Inadequate as any graphic attempt to portray sound must be, the use of English phrases is the most useful memory aid.

## Food

The food a bird eats is correlated closely with the kind of bill, or beak, it has. A sparrow with a stout, conical beak is adapted to a diet of hard-coated seeds; a hawk with a sharp, hooked beak, to the tearing and shearing of flesh. Knowledge of a bird's diet is always of interest and may be of help in identification. Hence, brief descriptions of the principal food of each species are given.

## Reproduction

All birds are hatched from eggs laid by the female, usually in a nest, sometimes on the bare ground. The eggs are incubated by the female or male, or by both alternately. In North America, the cowbirds are a notable exception to this practice; they lay their eggs in the nests of other birds.

Incubation takes from eleven to sixty-three days. The young of some species, such as ducks, grouse, and shorebirds, are able to run about almost as soon as they are hatched; these are termed *precocial*. But the *altricial* young of the majority of species must be fed in the nest for several days or weeks until they are fledged.

Mating is customarily preceded by a more or less elaborate courtship performance. When their plumages differ, males of the species, except in a few cases, are generally more brightly colored than the females. However, in many groups, such as herons, shorebirds, and sparrows, the sexes are alike in plumage. In the birds of prey the female is often larger than her mate.

### Nest and Eggs

With a little practice, the nests and eggs of many species can be identified in the field. The concise descriptions given in the text will be of assistance, but the advanced student will want to use one of the several available guides to birds' nests and eggs. There is often much variation in the color and markings of birds' eggs, so the descriptions given are of typical examples only. When found in the field, a bird's nest should not be disturbed, nor robbed of its eggs.

### Plumage

Most birds have more than one plumage; that is, covering of feathers. The variations that occur in the coloration of certain birds usually correspond to specific periods of the year (seasonal), sexual differentiation (male or female), or to stages of development (adult and immature).

There is almost always a difference between the immature, those incapable of breeding, and adult plumages. This is especially prolonged in the water birds and birds of prey. In some groups, gulls for example, there are distinctions between the juvenals and immatures as well. During the juvenal period, usually when the bird is first out of its nest in summer, a bird may exhibit one plumage; this may distinctly change to another plumage when the bird subsequently grows into its immature stage.

Changes that relate to a period of time in the life of the bird may also appear in adults, birds which are capable of reproduction. Generally, the terms *spring, fall, winter,* and *summer,* in relation to plumage, apply to the differences in adult breeding and nonbreeding plumages, rather than to the season of the year. *Spring* and *summer* usually apply to breeding plumages; *fall* and *winter* to nonbreeding. A spring bird, for example, is generally considered to be an adult in its bright breeding plumage. In this context, the term does not apply strictly to the spring season as we think of it (March to June), although it is loosely related to this time of year. The bird may be in this plumage somewhat before, and for a time after, the season as we know it. By the same definition, the "fall" plumage, as used herein, may be acquired by midsummer in a group such as the shorebirds.

The most common plumages of the individual species are discussed within the text when they are a distinct factor in identifying a species.

### Range

The area of coverage for this chapter includes species found in eastern North America north of the 25th parallel, and west to the 100th meridian. The central prairie states along the axis of the 100th meridian constitute the "twilight zone," where the ranges of eastern and western birds converge and overlap, to some extent. Ranges apply to the species as a whole only within the geographic coverage of the chapter.

Most birds within the given range that are not permanent residents breed in the north in summer and migrate south for the winter.

Exceptions are a few petrels and shearwaters, which breed in the Southern Hemisphere and then cross the Equator to "winter" in North American waters during our summer. Both the normal breeding and wintering ranges of each species are given in the text.

Range boundaries are seldom clearly defined in the wild and tend to change slightly as time passes; hence the ranges given should be understood to be approximations only. Species may occasionally be found somewhat outside, or be absent from some space inside, the indicated range. Within a range area, of course, a species will normally occur only in its usual habitat.

### Nomenclature

Except in a few cases in which recent studies indicate a change, the English and scientific names and the taxonomy of birds used in this chapter are those set forth in the 1957 *Check-list of North American Birds* by the American Ornithologists' Union, and supplements since published in *The Auk*. The families, genera, and species are arranged more or less in phylogenetic order; that is, the sequence employed by taxonomists to reflect their understandings of avian evolution, from more primitive to more advanced types. Because of gaps in knowledge and difficulties posed by any linear sequence, the taxonomic arrangement is subject to constant reinterpretation and change. To aid in identifying, the most specific characteristic appears in italics, preceded by a bullet (•), under the description heading for most species.

### Illustrations

Virtually every bird that is discussed in this chapter is illustrated and, where appropriate, varying plumages are shown. All the birds are adults unless otherwise specified in the caption. If there is no gender symbol shown, it is because the sexes are identical, or nearly so.

A page of illustrations divided by a rule indicates that there is a change in scale. Otherwise, all the species shown are proportionately sized on the page. The sizes given are the average known sizes.

## USEFUL REFERENCES

### Books

American Ornithologists' Union. 1957. *Check-list of North American Birds*. 5th ed. Baltimore: American Ornithologists' Union.

Austin, O. L., Jr., and Singer, A. 1961. *Birds of the World*. New York: Golden Press.

Baker, J. H. 1941. *The Audubon Guide to Attracting Birds*. New York: Doubleday & Co.

Barton, R. 1955. *How to Watch Birds*. New York: McGraw-Hill.

Bent, A. C. 1919–68. *Life Histories of North American Birds*. 23 vols. Washington, D.C.: U.S. National Museum.

Berger, A. J. 1971. *Bird Study*. Reprint. New York: Dover Publications.

Bull, J. 1964. *Birds of the New York Area*. New York: Harper & Row.

Bull, J. *Birds of New York State*. New York: Harper & Row.

Bull, J., and Farrand, J. 1977. *The Audubon Society Field Guide to North American Birds, Eastern Region*. New York: Knopf.

Darling, L., and Darling, L. 1962. *Bird*. Boston: Houghton Mifflin.

Gilliard, E. T. 1958. *Living Birds of the World*. New York: Doubleday & Co.

Godfrey, W. E. 1966. *The Birds of Canada*. Bulletin 203. Ottawa: National Museum of Canada.

Harrison, H. H. 1975. *A Field Guide to Birds' Nests*. Boston: Houghton Mifflin.

Hoffmann, Ralph. 1927. *Birds of the Pacific States*. Boston: Houghton Mifflin.

Lanyon, W. E. 1963. *Biology of Birds*. Garden City: Natural History Press.

Leck, C. 1975. *Birds of New Jersey: Their Habits and Habitats*. New Brunswick: Rutgers University Press.

Ovington, R. 1976. *Birds of Prey in Florida*. St. Petersburg, Fla.: Great Outdoors.

Palmer, R., ed. 1962– . *Handbook of North American Birds*. Vols. 1, 2, 3. New Haven: Yale University Press.

Peterson, R. T. 1947. *A Field Guide to the Birds*. 2nd ed. Boston: Houghton Mifflin.

———. 1961. *A Field Guide to Western Birds*. Boston: Houghton Mifflin. 60 pl.

———. 1963. *The Birds*. New York: Time-Life Books.

Pettingill, O. S., Jr. 1953. *A Guide to Bird Finding West of the Mississippi*. New York: Oxford University Press.

Pough, R. H. 1946. *Audubon Land Bird Guide*. New York: Doubleday & Co. 48 col. pl.

———. 1951. *Audubon Water Bird Guide*. New York: Doubleday & Co.

Reed, C. A. 1965. *North American Birds' Eggs*. New York: Dover Publications. Illus.

Reilly, E. M., Jr. 1968. *The Audubon Illustrated Handbook of American Birds*. New York: McGraw-Hill.

Robbins, C. S.; Brunn, B.; Zim, H. S.; and Singer, A. 1966. *A Guide to Field Identification: Birds of North America*. New York: Golden Press.

Snyder, D. E., and Griscom, L. 1955. *Birds of Massachusetts*. Salem: Peabody Museum.

Stefferud, A., ed. 1966. *Birds in Our Lives*. Washington, D.C.: U.S. Department of Interior.

Van Tyne, J., and Berger, A. J. 1976. *Fundamentals of Ornithology*. New York: John Wiley.

Welty, J. C. 1975. *The Life of Birds*. New York: W. B. Saunders.

Wing, L. W. 1956. *Natural History of Birds*. New York: Ronald. Lists many state and local books and periodicals.

**Periodicals**

*American Birds*. Bimonthly. National Audubon Society, 950 Third Avenue, New York, New York.

*Audubon Magazine.* Bimonthly. National Audubon Society, 950 Third Avenue, New York, New York.

*The Auk.* Quarterly. American Ornithologists' Union, Allen Press, Inc., Lawrence, Kansas.

*Birding.* Bimonthly. American Birding Association, Inc., Box 4335, Austin, Texas.

*The Condor.* Quarterly. Cooper Ornithological Society, Allen Press, Inc., Lawrence, Kansas.

*The Wilson Bulletin.* Quarterly. Wilson Ornithological Society, Allen Press, Inc., Lawrence, Kansas.

### Records and Tape Cassettes

Kellogg, P. P., and Allen, A. A. *A Field Guide to Bird Songs.* Boston: Houghton Mifflin. More than 300 eastern species, to accompany Peterson's *A Field Guide to the Birds.*

# GLOSSARY

♂—Male
♀—Female

*Axillar* (*pl.* axillars) Feathers of the underwing, between the wing lining and the body ("armpits").

*Breeding* An adult bird in the plumage in which it reproduces.

*Carpal* Relating to the wrist (carpus).

*Displaying* A male bird in breeding display pose.

*Form* A distinct plumage type, such as of a subspecies, or color phase.

*Immature* A bird in its first year, usually the first fall and winter following juvenal plumage; occasionally in large birds may extend to second or third year.

*Juvenal* A bird in its first plumage after natal down during summer; followed by immature plumage.

*Molt* The loss of one feather plumage and acquiring of another.

*Nonbreeding* An adult bird that is in a plumage other than that in which it reproduces.

*Phase* A coloration other than the normal plumage, irrespective of sex, age, or season.

*Race* A subgroup (or subspecies) within a species.

*Resident* A species that breeds and winters in the same general latitude.

*Sibling Species* Closely related, look-alike species that coexist.

*Speculum* A patch of color, usually iridescent, on the secondaries; mainly waterfowl.

*Sub-Adult* A plumage between immature and adult; sometimes or regularly breeds in this stage.

# Water and Game Birds and Birds of Prey

## Loons
Order Gaviiformes

## LOONS
Family Gaviidae

Loons are large water birds with strong, pointed bills and thick necks, midway in size between ducks and geese. They are dark above and white below, with white spots on the back in summer. Their wings are slender and pointed. They fly with rapid, fairly deep wingbeats (but slower than any duck), with necks extended and drooping, and with feet trailing. They dive rapidly, and can swim long distances under water. Loons are northern birds that feed on fish and are seen in most of the east only on migration and in winter.

### COMMON LOON
*Gavia immer*                                                    1:1

**Description**
Size, 28–36 in. (71.1–91.4 cm). Heavy bill held straight. Summer: head dark, • *back evenly spotted with white squares*, but spots not visible at great distance. Winter: hindneck and back unspotted gray; throat, cheek, and underparts white.
**Similarities**
Arctic and Red-throated have thinner necks; Common's back is darker than smaller Red-throated.
**Habitat**
Prefers large, deep freshwater lakes for breeding; at other seasons more often found in fresh water than other loons; commonest loon in most of the east.
**Habits**
Swims low in water, sometimes with only neck and head visible; patters on water into wind to take flight; ungainly on land.
**Voice**
Wild and quavering, often compared to insane laughter.
**Eggs**
2; olive-brown with spots; 3.5 x 2.2 in. (8.9 x 5.6 cm).
**Range**
Breeds throughout N., from n. Canada, s. to N.H. and e. from Mass. to Minn.; winters s. to Gulf Coast.

### ARCTIC LOON
*Gavia arctica*                                                 1:2

**Description**
Size, 23–29 in. (58.4–73.7 cm). • *Bill straight* or downcurved. Summer: • *head gray*; white spots above only on scapulars, • *2 patches* on each side. Winter (difficult identification): • *head and neck grayish*, back black (often with scaly appearance), body small and thin, bill thin. Breeding: throat sometimes turns black first

along lower edge, showing black line across lower throat (helpful but not diagnostic).

**Similarities**
In general, Arctic is similar to, but smaller than, Common; bill of Arctic is appreciably smaller; in winter, Arctic's body is smaller, thinner, and with thinner bill. Red-throated has upturned bill and a speckled brownish-gray back.

**Habitat**
Large lakes, ponds, tundra waters; in winter, most maritime of eastern loons.

**Habits**
Similar to Common.

**Voice**
Loud squalls and screams; seldom heard in winter.

**Eggs**
2; olive-buff with dark spots; 3.0 x 1.8 in. (7.6 x 4.6 cm).

**Range**
Breeds in Arctic, s. to Hudson Bay; winters mainly in W., but e. to Man., Ont., Que.; rarely farther s. and e.


## RED-THROATED LOON
*Gavia stellata* 1:3

**Description**
Size, 24–27 in. (61.0–68.6 cm). Head held at upward tilt when at rest. In eastern range, only loon with • *upturned bill.* Summer: head gray; nape striped; throat rusty-red; back brownish-black,• *speckled with white.* Winter: head and neck gray above, white below; back brownish-gray, more conspicuously speckled with white.

**Similarities**
Other loons are darker-colored and heavier with straighter bill.

**Habitat**
In summer, small ponds and other bodies of water, fresh or salt; other seasons, usually salt water.

**Habits**
Springs into the air without pattering on water; will often fly rather than dive; walks, but not with ease.

**Voice**
Not as loud or as wild as other loons; more complaining.

**Eggs**
2; brownish-olive, sparingly spotted; 2.8 x 1.8 in. (7.1 x 4.6 cm). Nest is muddy platform at edge of tundra lake or grassy pool, or on islet.

**Range**
Breeds on tundra along n. Atlantic Coast, s. to Newfoundland and w. to n. Man.; winters s. to Fla., w. to Gulf Coast.


# Grebes
Order Podicipediformes


# GREBES
Family Podicipedidae

These ducklike birds sit low in the water and have thin necks, sharp bills, and no apparent tails; their feet are very much to the rear. They are almost helpless on land, but are excellent swimmers and divers. They hold their heads erect when swimming and run along on the surface of the water before taking flight. They fly with neck inclined downward. Most eat aquatic invertebrates and fish.

## RED-NECKED GREBE
*Podiceps grisegena*                                                1:4

**Description**
Size, 18–22½ in. (45.7–57.2 cm). Short body; long neck; straight,
• *yellow bill*. Summer: • *cheeks white*, neck red-brown. Winter: cap
dark, neck red, body gray; • *2 white patches* on each wing and
(except in some first-year birds) white crescent on cheek; dark
upperparts and light underparts blend without sharp line of
contrast. Neck thicker, head and bill heavier than in other grebes.
Most loonlike of all grebes (but no loon has white wing patches).

**Similarities**
Common Loon has larger, longer body, shorter neck. Horned Grebe
is smaller, neck white, bill dark. Red-throated (winter) Loon has
whitish throat and upraised pale bill. Red-breasted Merganser in
flight shows only 1 white wing patch, has faster wingbeat, and holds
neck horizontally.

**Habitat**
Freshwater ponds in summer; fresh and salt water in winter.

**Voice**
Various loonlike wails; a single bray.

**Eggs**
4–5; bluish-white, usually heavily nest-stained; 2.3 x 1.4 in. (5.8 x
3.6 cm).

**Range**
Breeds in n. Canada, s. to N.H., w. to Minn.; winters s. to N.Y.,
coastally to Fla.

## HORNED GREBE
*Podiceps auritus*                                                1:6

**Description**
Size, 12–15¼ in. (30.5–38.7 cm). Summer: crown and cheeks dark
green, ear tufts orange-buff, neck and flanks dark chestnut. Winter:
gray above; cheeks, foreneck, and underparts clear white. In flight,
shows white wing patch.

**Similarities**
See Eared Grebe.

**Habitat**
In summer, freshwater ponds and lakes; in winter, fresh and salt
water.

**Eggs**
4–5; bluish- or olive-white, usually heavily nest-stained; 1.7 x 1.2
in. (4.3 x 3.0 cm).

**Range**
Breeds in n. Canada, s. to Wis.; winters along Atlantic Coast, s. to
Gulf Coast.

## EARED GREBE
*Podiceps nigricollis*                                            1:7

**Description**
Size, 12–14 in. (30.5–35.6 cm). Summer: • *crest,* neck, and back
• *black*; ear tufts golden; cheeks orange; flanks dark chestnut.
Winter: gray above, light below; lower cheek and spot behind ear
are whitish; • *sides of neck grayish. Bill appears upturned.*

**Similarities**
Horned Grebe shows entire cheek bright white and contrasting
gray and white neck in winter; has wing patches.

**Habitat**
Fresh water; breeds in colonies in shallow marshy parts of lakes; many winter on salt water.
**Voice**
*"Poo-eep, poo-eep; hick-rick-up, hick-rick-up"* (Dawson).
**Eggs**
4–5; indistinguishable from those of Horned; 1.7 x 1.2 in. (4.3 x 3.0 cm).
**Range**
Breeds in W.; winters casually to Atlantic Coast.

## WESTERN GREBE
*Aechmophorus occidentalis*                                              1:5

**Description**
Size, 22–29 in. (55.9–73.7 cm). Large, long neck; head held high above water; dark gray above, white below, giving a two-toned,
• *black-and-white appearance.* Crown and hindneck black; back gray; cheek, foreneck, wing patch, and underparts white; bill yellow, slightly upturned; eyes red; neck long, slender.
**Similarities**
Red-necked Grebe in winter is smaller, bill duskier yellow, plumage dingy gray, neck especially gray. Loons have shorter neck, no wing patches.
**Habitat**
Lakes and ponds.
**Habits**
Neck droops in middle in flight.
**Voice**
"A loud, double-toned, whistled *c-r-r-ee-ee-r-r-ee*" (Chapman).
**Eggs**
3–6; dull, bluish-white to olive-buff, usually nest-stained; 2.3 x 1.5 in. (5.8 x 3.8 cm).
**Range**
Breeds in W.; winters casually to Great Lakes and se. coastline.

## PIED-BILLED GREBE
*Podilymbus podiceps*                                                    1:8

**Description**
Size, 12–15 in. (30.5–38.1 cm). Brown above, light below; bill short, heavy, whitish; no white wing patches; undertail white. Summer only: • *black throat and ring around bill.*
**Habitat**
In spring and summer on ponds and marshes; in winter on fresh and salt water.
**Habits**
On water, carries tail cocked, showing white below; can dive or submerge by sinking body until only head shows.
**Voice**
Series of harsh *cow-cow-cow* notes on 1 pitch, somewhat cuckoolike.
**Eggs**
5–7; indistinguishable from those of Horned Grebe; 1.7 x 1.2 in. (4.3 x 3.0 cm).
**Range**
Breeds throughout Canada and U.S.; winters s. from New England.

# Fulmars, Shearwaters, and Petrels
Order Procellariiformes

These web-footed, powerfully winged water birds have nostrils in a tube on top of the bill. Their home is the open sea, but they breed on islands.

## FULMARS AND SHEARWATERS
Family Procellariidae

For long periods, these birds glide on stiff wings close to the surface of the sea. They swim well. Frequently they collect in numbers around fishing boats for scraps and waste. Shearwaters are open-ocean birds. They are dark above and, except for the Sooty, white below. They alternate deep flaps with wheeling glides just over the waves. In gliding, the long slender wings are held straight out and stiff. In the East, they are summer or fall visitors, generally well offshore. Their food consists of small fish, crustaceans, and plankton.

### NORTHERN FULMAR
*Fulmarus glacialis*                                         2:7

**Description**
Size, 17–20 in. (43.2–50.8 cm). Light phase: back and mantle gray; head, tail, and underparts white; bill stubby, yellow (sometimes darker). Dark phase: dark gray all over, bill dusky-brown. Intermediates occur frequently. Tube-nosed bill and manner of flight distinguish it from gulls. Habitat is open ocean.
**Similarities**
Shearwaters do not flap as much; dark phases of Sooty similar to Fulmar, but darker.
**Habits**
Ship follower and scavenger; flaps more than most shearwaters.
**Voice**
Grunts, chuckles, cackles.
**Eggs**
1; white; 2.9 x 2.0 in. (7.4 x 5.1 cm); little or no nest on rock ledge.
**Range**
Breeds along coastline, s. to Newfoundland; winters at sea, s. to N.J.

### CORY'S SHEARWATER
*Puffinus diomedea*                                          2:8

**Description**
Size, 21 in. (53.3 cm). Largest eastern shearwater and the only one with • *yellow bill.* Brownish-gray above, white below; bill long, thick, yellow; • *cap same color as back* and blends into white of throat, sometimes shows indistinct white V at base of tail.
**Similarities**
Greater is darker, has conspicuous white V at base of tail, and black cap makes sharp line of contrast with white cheek.
**Eggs**
1; 3.0 x 2.0 in (7.6 x 5.1 cm).
**Range**
Breeds in N. Hemisphere, e. of N. America; appears during our summer off Atlantic Coast, s. to S.C.

## GREATER SHEARWATER
*Puffinus gravis*                                          2:5

**Description**
Size, 18 in. (45.7 cm). Black cap contrasts with brown back. Above, sooty brown; underparts, half collar, and conspicuous white V at base of tail; • *black cap contrasts sharply with white cheek.*

**Similarities**
Bill of Greater is thin, black, and shorter than Cory's.

**Remarks**
This species reverses the normal pattern of migration. During our winter it breeds in the Tristan da Cunha Islands in the Atlantic, then spends its winters in the North Atlantic during our summer.

**Range**
Breeds in S. Hemisphere; appears offshore in N. Atlantic waters during late spring to summer.

## SOOTY SHEARWATER
*Puffinus griseus*                                         2:4

**Description**
Size, 16–18 in. (40.6–45.7 cm). All-dark, sooty brown, appearing black at a distance; underwings sometimes lighter, but difficult to see at a distance. Looks like a small, all-black gull that glides (not flaps) on stiff wings over the waves.

**Voice**
"Low, guttural *wok-wok-wok* when much excited" (Rich).

**Range**
Breeds in subantarctic; appears offshore in N. Atlantic waters during summer.

## MANX SHEARWATER
*Puffinus puffinus*                                        2:10

**Description**
Size, 12½–15 in. (31.8–38.1 cm). Bill, cap, and • *upperparts black; underparts white;* white undertail coverts; feet pinkish.

**Similarities**
Audubon's is slightly smaller, and has gray undertail coverts.

**Habits**
Flight resembles that of Greater. Manx glides more, flaps less than Audubon's; wingbeats slower, flight less fluttery.

**Voice**
*"Cuck-cuck-oo,* repeated several times" (Saunders).

**Range**
Breeds in e. Atlantic; appears rarely offshore along Atlantic Coast.

## AUDUBON'S SHEARWATER
*Puffinus lherminieri*

**Description**
Size, 12 in. (30.5 cm). Feet slaty; white-bellied, with short, gray undertail coverts.

**Habits**
Flaps more, glides less than Manx.

**Voice**
Weird cacklings on breeding grounds: *"whither did he go"* (hence the local name, Wedrigo).

**Range**
Breeds in tropical seas; appears along coastline in summer, n. to N.Y.

# STORM-PETRELS
Family Hydrobatidae

These small, dark seabirds have white rumps and fly close to the surface of the sea, with their feet hanging down. Their habitat is the open ocean, occasionally large bays, and they feed on small invertebrates.

## LEACH'S STORM-PETREL
*Oceanodroma leucorhoa*                                     **2:6**

**Description**
Size, 7½–9 in. (19.1–22.9 cm). Dusky black; • *white rump;* wings long, top of wing with pale band; legs short, dark, and in flight do not show beyond end of tail; tail forked.

**Habits**
Flight like that of Common Nighthawk; wingbeats slower, more irregular than Wilson's, which has rounded tail; glides with wings curved down; swims in water with wings uplifted, then springs into air; not a ship follower.

**Voice**
On breeding grounds, 7 to 8 low cooing notes; also twitterings, screams, trills.

**Eggs**
1; white; 1.3 x 0.9 in. (3.3 x 2.4 cm).

**Range**
Breeds along Atlantic Coast, from Labr., s. to Maine; winters at sea.

## WILSON'S STORM-PETREL
*Oceanites oceanicus*                                       **2:9**

**Description**
Size, 7 in. (17.8 cm). Dusky black; wings with narrow white line; legs long, in flight show beyond end of tail; yellow feet; square tail.

**Similarities**
Leach's appears lighter with longer wings and rounded tail.

**Habitat**
Open ocean. More common than Leach's.

**Habits**
Flight is swallowlike. A ship follower.

**Voice**
A peeping, when feeding at sea.

**Eggs**
1; 1.3 x 0.9 in. (3.3 x 2.4 cm).

**Range**
Breeds in Antarctic; appears offshore along Atlantic and Gulf coasts in summer.

# Pelicans and Allies
Order Pelecaniformes

# PELICANS
Family Pelecanidae

These are large, fish-eating water birds, with oversized pouched beaks and long wings. They fly, with necks drawn in, in follow-the-leader or V-shaped lines, alternately flapping and gliding in unison.

# AMERICAN WHITE PELICAN
*Pelecanus erythrorhynchos*                                    *Fig. 2*

**Description**
Size, 54–70 in. (137.2–177.8 cm). • *White*; wing tips black, bill yellow to orange. The only white pelican in the east.

**Habitat**
Breeds on western freshwater lakes, migrates along river valleys, winters on saltwater bays and bayous.

**Habits**
Frequently soars.

**Voice**
Low croaks on breeding ground.

**Eggs**
2; white; 3.5 x 2.2 in. (8.9 x 5.6 cm). Nest is of vegetation, stones, and debris with a raised rim; on ground or flattened vegetation, in colonies on islands in shallow water

**Range**
Breeds from nw. Canada, s. to w. Ont.; winters from Fla. along Gulf Coast, southward.

*Fig. 2*

summer

winter

Brown Pelican
adults

American White Pelican

# BROWN PELICAN
*Pelecanus occidentalis*

**3:4**
*Fig. 2*

**Description**
Size, 44–55 in. (111.8–139.7 cm). • *Brown;* the only brown pelican in east. Strictly coastal.

**Habitat**
Resident with some movement on south Atlantic and Gulf coasts, in small numbers.

**Habits**
Fishes from surface or dives from low height.

**Voice**
Usually silent, rarely croaks.

**Eggs**
2–3; dusky white; 2.8 x 1.8 in. (7.3 x 4.6 cm). Nest is variably bare ground to stick nest in low trees; on islands offshore.

**Range**
Resident along coastline, from N.C. to Fla., southward.

# GANNETS
Family Sulidae

These are large, big-billed seabirds that fly on stiff, long, pointed wings. They eat fish, which they capture by diving, often deeply and from a considerable height.

## GANNET
*Morus bassanus*                                                3:3

**Description**
Size, 35–40 in. (88.9–101.6 cm). Larger than any gull. Adult: white; head sometimes yellowish; • *wing tips black*; bill long, sharp, carried point down; tail pointed; brilliant white of adults visible and recognizable from a great distance; silhouette when soaring makes an almost-perfect cross. Immature: brown above, flecked with white; light below. Birds in changing plumage often mixed black and white, especially on mantle.

**Habitat**
Open coastal sea, sometimes large bays.

**Habits**
Often seen along eastern coasts in spring and fall, diving offshore; flies, soars, dives well; alternately flaps and sails low over waves; wingbeats relatively rapid. Breeds in great colonies.

**Voice**
Guttural croaks and grunts on breeding grounds, *kur-r-r-ruk, kur-r-ruk*.

**Eggs**
1; bluish-white; 3.0 x 1.8 in. (7.6 x 4.6 cm). Nest is large mound of vegetation on top or on the ledge of an island cliff.

**Remarks**
The cliffs of Bonaventure Island off the Gaspé Peninsula, in Quebec, contain one of the largest Gannet colonies.

**Range**
Breeds on islands in Gulf of St. Lawrence; winters along coast from New England to Fla.

# CORMORANTS AND ANHINGAS
Family Phalacrocoracidae

These are black, long-necked water birds about the size of geese. They have long, hook-tipped bills and are darker and longer-tailed than any goose or loon. They swim well, sometimes with only the head, neck, and upturned bill showing. They fly with heronlike flaps and intermittent glides, with their necks slightly raised. Their diet is fish. They nest in colonies on the ground on rocky islands or ledges, sometimes in trees. The nest is of seaweed, sticks, twigs, and grasses. The three to four pale blue eggs have a chalky covering. Although they swim and dive, cormorants and anhingas seem incompletely adapted to the water, for their feathers are not waterproof like a duck's and, when perched, they habitually spreadeagle their wings to dry them.

## GREAT CORMORANT
*Phalacrocorax carbo*                                           3:6

**Description**
Size, 30–40 in. (76.2–101.6 cm). One quarter larger than Double-crested. In breeding plumage (March–June), only cormorant in the east with white flank patches. Adult: black; throat pouch white, chin pouch yellow; except in breeding plumage, flanks dark.

Immature: dark brown, underparts whitish (generally), as far back as undertail coverts; bill heavy; neck stocky.

**Similarities**
Can be confused with Double-crested, but area of regular overlap is small: Long Island and New Jersey in winter, Gulf of St. Lawrence, and Newfoundland in breeding season.

**Habitat**
Ocean and large bays; prefers rocky shores.

**Habits**
As for family; wingbeats slower than Double-crested's; flies in lines, wedges, or irregular bunches; can soar.

**Voice**
Deep croaks.

**Eggs**
3–4; 2.5 x 1.6 in. (6.4 x 4.1 cm).

**Range**
Breeds in Canadian Maritime Provinces, s. to N.S.; winters along coast s. to N.J.

# DOUBLE-CRESTED CORMORANT
*Phalacrocorax auritus* 3:5

**Description**
Size, 30–36 in. (76.2–91.4 cm). Smaller, slimmer than Great, especially in the neck. Adult: black; • *throat pouch orange*; never any white on flanks or chin. Immature: brown; breast and forebelly white with brownish cast; hindbelly black.

**Similarities**
Care is necessary to distinguish this from Great; size is a poor characteristic, unless direct comparison is possible. Some adult Double-cresteds have bleached throat pouches, which look yellowish; some immatures may have unusually light bellies.

**Habitat**
Lakes, rivers, bays, ocean.

**Habits**
Wingbeats more rapid than Great.

**Voice**
Normally silent; a rare croak when alarmed.

**Eggs**
3–4; pale blue; 2.4 x 1.5 in. (6.1 x 3.8 cm).

**Range**
Breeds in U.S. and Canada; winters s. from N.Y. and n.–cen. U.S.

# NEOTROPICAL (OLIVACEOUS) CORMORANT
*Phalacrocorax olivaceus* 3:7

**Description**
Size, 23–29 in. (58.4–73.7 cm). Adult: black; • *orange throat pouch has narrow white border*, visible at close range.

**Similarities**
Double-crested is larger, heavier.

**Habitat**
Salt, fresh, and brackish water.

**Voice**
Grunts or croaks.

**Eggs**
2–6; chalky-bluish; 2.2 x 1.4 in. (5.6 x 3.6 cm). Nest is in bush, tree, or on ground; of small twigs.

**Range**
Resident in s. U.S., along coast of La. and Tex., southward.

## ANHINGA
*Anhinga anhinga*                                                    **3:2**

**Description**
Size, 32–36 in. (81.3–91.4 cm). Bill long, pointed; thin, snakelike
neck. Mainly black, female with brownish foreparts; long, fanlike,
white-tipped tail.
**Habitat**
Quiet southern marshes and swamps in fresh or salt water.
**Habits**
Frequently swims with only head and bill above water; often
spreads wings to dry.
**Voice**
Long, chattering call.
**Eggs**
2–5; 2.0 x 1.4 in. (5.1 x 3.6 cm). Nest is singly or in colonies in
trees.
**Range**
Breeds along Atlantic and Gulf coasts, from N.C., s., and in
Mississippi Valley from Ark., southward; winters from Fla. and
Gulf Coast southward.

## FRIGATEBIRDS
Family Fregatidae

These are long-winged, gliding and soaring birds of the tropical
seas. They feed at the water's surface on jellyfish, fish, squid, and
sewage items, or by robbing other fish-eating birds. None breed in
east.

## MAGNIFICENT FRIGATEBIRD
*Fregata magnificens*                                                **3:1**

**Description**
Size, 37–41 in. (94.0–104.1 cm); wingspread to 8 ft. (2.4 m). Male:
glossy black. Female: browner with breast and sides of abdomen
white. Immature: white-headed are most often seen in eastern
waters. Long, forked tail and narrow, crooked wings with a long,
hooked bill are diagnostic. Feeds over sea, but in sight of land.
**Range**
Breeds s. of U.S.; appears along coasts of Fla., La., and Tex.

# Long-Legged Waders
Order Ciconiiformes

These usually large wading birds of marshes, mud flats, and shores
have long bills, necks, and legs and short tails.

## HERONS AND BITTERNS
Family Ardeidae

Members of this narrow-bodied family have sharp, pointed bills;
slender necks; and large, rounded wings. Many flock to large roosts
in evening. They fly with necks drawn in and legs extended; their
wingbeats are slow and deliberate. Their food consists of crayfish,
small fishes, and other invertebrates.

## GREAT BLUE HERON
*Ardea herodias*                                         5:1

**Description**
Size, 48 in. (121.9 cm). Adult: head white with 2 black plumes,
neck brownish-gray, shoulder patches black, rest of plumage
grayish-blue; largest heron in eastern range. Immature: duller,
with dark cap and no plumes.

**Similarities**
Sandhill Crane is larger, more uniform gray in color, and flies
with neck extended. In far southern Florida, an all-white form, the
"Great White Heron," is larger than Great Egret and has yellow,
not black, legs.

**Habitat**
Shallow water anywhere, but rarely oceanfront. Most common
heron in east.

**Habits**
Fishes by standing motionless in water for protracted periods; flies
with slow, steady, heavy wingbeats; can soar well.

**Voice**
Guttural squawks; a honk reminiscent of Canada Goose but flatter,
harsher, longer.

**Eggs**
4; greenish-white; 2.5 x 1.8 in. (6.4 x 4.6 cm). Nest is a platform
of sticks in swamp trees, on rocky islets, in marshes; loosely
colonial.

**Range**
Breeds throughout s. Canada and U.S., from N.S. and Que. s. to
Mexico; winters from New England and Great Lakes southward.

## GREEN HERON
*Butorides striatus*                                     4:9

**Description**
Size, 16–22 in. (40.6–55.9 cm). A small, dark heron. Adult: neck
maroon, front white, upperparts blue-green (often looks more blue
than green, and may appear dark at a distance), • *legs orange.*
Immature: browner; throat streaked, legs greenish.

**Similarities**
Least Bittern is only heron that is smaller.

**Habitat**
Any shallow water, even tiny ditches and pools, usually near trees.

**Habits**
Stretches neck and raises crest when alarmed; often alights in trees.
In flight, shows downward bend to wing tips.

**Voice**
Various squawks and grunts; a loud *skyau.*

**Eggs**
4–5; greenish-white; 1.5 x 1.2 in. (3.8 x 3.0 cm). Nest is of sticks;
in tree, shrub, or grass clump, not necessarily near water; nests
singly or in small colonies.

**Range**
Breeds from N.B. s. to Minn. and throughout U.S.; winters from
S.C. and Gulf Coast southward.

## GREAT EGRET
*Casmerodius albus*                                 4:2, 5:2

**Description**
Size, 37–41 in. (94.0–104.1 cm). Largest white heron in the east;
only one with a • *yellow bill.* • *Legs, feet, and extreme tip of bill*

*black,* bill largely yellow; up to 50 long plumes (aigrettes) on its back in breeding season.

**Similarities**
See Habits, below.

**Habitat**
Streams, ponds, marshes; fresh or salt water.

**Habits**
Flocks to evening roosts, waits motionless for prey (does not rush around like Snowy); wingbeats slower than smaller Snowy's.

**Voice**
Low, heavy croak.

**Eggs**
3–4; pale bluish-green; 2.2 x 1.6 in. (5.6 x 4.1 cm). Nest is of sticks; in trees near water; in colonies.

**Former name**
Common or American Egret.

**Range**
Breeds from n. U.S. s. throughout rest of U.S.; winters from S.C. and Gulf Coast southward.

## LITTLE BLUE HERON
*Egretta caerulea*                                                    4:6

**Description**
Size, 20–24 in. (50.8–61.0 cm). • *Bill bluish with black tip;* • *feet and legs greenish.* Adult: completely dark; head and neck deep maroon (no white in front); rest of plumage slaty-blue; looks black at a distance. Immature: all-white. Changing birds show dark and white patches.

**Similarities**
Snowy Egret has black bill and black legs, yellow feet. Green Heron is smaller, short-necked.

**Habitat**
Fresh- and saltwater marshes.

**Habits**
Gathers at communal roosts at sunset; moves about actively when feeding (but motions slower than Snowy's); in flight, wingbeats more rapid than Great Blue or Great Egret.

**Voice**
Croaks, and screams, *"tell you what, tell you what"* (Chapman).

**Eggs**
4–5; bluish-green; 1.7 x 1.3 in. (4.3 x 3.3 cm). Nest is of sticks, in bushes and trees near water.

**Range**
Breeds mainly along Atlantic Coast, from New England s. and from Ark. and Tenn. s.; winters in s. states, southward.

## SNOWY EGRET
*Egretta thula*                                                       4:5

**Description**
Size, 20–27 in. (50.8–68.6 cm). Adult: all-white with many beautiful plumes (aigrettes) on its back in the breeding season; bill narrow and black with yellow or red skin near base; • *legs black, feet yellow.* Immature: has yellow line up back of legs.

**Similarities**
Little Blue Heron's bill is shorter, heavier, bluish with black tip.

**Habitat**
Marshes and shallow water, fresh or salt.

**Habits**
Shuffles legs and dashes about erratically when feeding (Little Blue Heron and other egrets do not); wingbeats more rapid than Great Egret's.
**Voice**
Harsh hiss.
**Eggs.**
4–5; pale bluish-green; 1.7 x 1.3 in. (4.3 x 3.3 cm). Nest is of sticks; in trees, shrubs, or marsh grass; in colonies.
**Range**
Breeds in Atlantic coastal states from Maine s. and along Gulf Coast; winters from S.C. s. and along Gulf Coast.

## LOUISIANA HERON
*Egretta tricolor*                                                            **4:3**

**Description**
Size, 26 in. (66.0 cm). Dark, very slim, white rump and belly. Adult: upperparts slaty-blue with some maroon. Immatures: head, neck, and part of wings brown; rest of upperparts grayish.
**Similarities**
Little Blue; changing from white to dark plumage, sometimes shows white belly with dark upperparts, but there would also be some white feathers above.
**Habitat**
Coastal and larger inland marshes.
**Habits**
Fairly active feeder; flocks to roost in evening; engages in aerial games over breeding pond.
**Voice**
Low groans and grunts.
**Eggs**
4–5; pale bluish-green; 1.7 x 1.3 in (4.3 x 3.3 cm). Nest is of sticks (with a soft lining); in low trees and bushes; in colonies.
**Range**
Breeds along Atlantic and Gulf coasts from Fla. w.; after breeding, some wander n. to Mass.

## REDDISH EGRET
*Egretta rufescens*                                                           **4:4**

**Description**
Size, 27–32 in. (68.6-81.3 cm). More common dark phase: gray-backed with rusty brown head and neck. White phase: all-white.
• *Bill pink with black tip.* Legs bluish in both phases.
**Habitat**
Feeds along coast and in adjacent brackish and fresh water.
**Habits**
Usually dashes about, actively hunting.
**Voice**
Croaks, honking noise.
**Eggs**
3–4; bluish-gray; 2.0 x 1.5 in. (5.1 x 3.8 cm). Nest is on ground or in bush.
**Range**
Resident along Tex. Gulf Coast and in s. Fla.

## CATTLE EGRET
*Bubulcus ibis*                                               **4:1**

**Description**
Size, 20 in. (50 cm). Adult: white, with buff patches on head, breast, and back (buff paler in nonbreeding season); reddish bill and legs. Immature: all-white; bill yellow; legs dark.
**Similarities**
Stockier than other white-colored herons; bill shorter, eye red.
**Habitat**
Wet fields.
**Habits**
Associates with cattle, flocks to roost in evening.
**Voice**
Various croaks when breeding.
**Eggs**
4–5; bluish-white; 1.8 x 1.3 in. (4.6 x 3.3 cm). Nest is of sticks; in trees; in colonies.
**Range**
Breeds from s. New England s. and w. to Great Lakes; winters throughout se. states from Va. and w. to Ark.

## BLACK-CROWNED NIGHT HERON
*Nycticorax nycticorax*                                       **4:8**

**Description**
Size, 23–28 in. (58.4–71.1 cm). The stockiest heron; neck and legs short, legs yellow (red in breeding season). Adult: • *crown and back black*, wings gray, underparts and 2 long plumes from nape white. Immature: brown above with white spots, white below with brown streaks.
**Similarities**
See American Bittern and Yellow-crowned Night Heron.
**Habitat**
Ponds, lakes, rivers, bays, and marshes, both fresh and salt water; also built-up areas.
**Habits**
Most active at night, often roosts in small groups by day; stands with neck pulled in, with a hunched-up appearance. In flight, looks short-necked; flies with slow flaps; sometimes glides with wings bent down.
**Voice**
Loud *quawk*.
**Eggs**
3–6; pale bluish-green; 2.0 x 1.5 in. (5.1 x 3.8 cm). Nest is of sticks; in trees and shrubs; in colonies, not necessarily near water.
**Range**
Breeds from N.B. to Sask. s. throughout U.S.; winters along Atlantic Coast from N.J. s. and in s. states to Gulf Coast.

## YELLOW-CROWNED NIGHT HERON
*Nycticorax violacea*                                         **4:7**

**Description**
Size, 24 in. (61.0 cm). Adult: blue-gray with black markings on back and wings, • *head black*, crown pale yellow or white, cheek stripe white. Immature: gray-brown above with white dots, white with brown streaks below.

**Similarities**
Immature Yellow-crowned is grayer than Black-crowned, dots smaller. Stocky, but slimmer than Black-crowned; bill blunter, neck and legs longer, legs yellower.

**Habitat**
Swamps and marshes, fresh or salt water.

**Habits**
Not as nocturnal as Black-crowned. Feet and part of legs extend beyond tail in flight.

**Voice**
Similar to, but higher pitched and more pleasant than, Black-crowned's.

**Eggs**
3–4; greenish-blue; 2.0 x 1.5 in. (5.1 x 3.8 cm). Nest is of sticks (with lining); in trees, bushes, or on the ground; in small colonies.

**Range**
Breeds along Atlantic Coast, from Mass. to Fla. w. to Mississippi R. and Tex. with some to Kans., Ky.; winters mainly along Gulf Coast.

## LEAST BITTERN
*Ixobrychus exilis*                                        4:10

**Description**
Size, 11–14 in. (27.9–35.6 cm). A tiny heron. Male: crown, back, primaries, and tail greenish-black; cheeks and neck bright chestnut-buff; white line down either side of back; underparts buffy; • *buff wing patches*. Female: black replaced by dark brown, neck streaked. Young: like female, but paler. Very rare and local dark phase: buff replaced by dark red-brown.

**Habitat**
Marshes, especially with cattails.

**Habits**
Climbs along plant stems; slips through reeds like a rail; "freezes" like American Bittern when alarmed; flushes with fluttering wings, legs dangling.

**Voice**
Dovelike or cuckoolike *coo-coo-coo-coo*, somewhat similar to Pied-billed Grebe.

**Eggs**
4–5; bluish-white; 1.2 x 0.9 in. (3.0 x 2.3 cm). Nest is of twigs and grasses; in thick vegetation over water.

**Range**
Breeds from s. Canada and throughout U.S. s. to W. Indies; winters from Fla. and Gulf Coast southward.

## AMERICAN BITTERN
*Botaurus lentiginosus*                                    4:11

**Description**
Size, 27 in. (68.6 cm). Above rich brown; below tan, heavily marked with brown; bill yellow; throat white; legs green; primaries black.

**Similarities**
Immature Black-crowned Night Heron is grayer, has yellow legs, black bill, and no black on wings or neck. Immature Yellow-crowned is grayer still.

**Habitat**
Marshes, fresh and salt water.

**Habits**
When alarmed, stands still, bill pointing straight up. During flight, wings are less curved and wingbeat is faster than other herons; wingbeat has "snap" to downstroke.

**Voice**
"Pumping sound, *plum pudd'n*" (Collins); a noise as of driving a stake; when flushed, a croaking, *kok-kok-kok*.

**Eggs**
4–5; buffy-brown; 1.9 x 1.4 in. (4.8 x 3.6 cm). Nest is of vegetation; just above water.

**Range**
Breeds from Newfoundland, s. to Fla. and w. from Man. to Gulf Coast; winters from Del. and Ohio, southward.

# STORKS
Family Ciconiidae

## WOOD STORK (WOOD IBIS)
*Mycteria americana*                                      **5:6**

**Description**
Size, 34–47 in. (86.4–119.4 cm). Long-legged white bird with dark, naked head. Bill decurved, head bare, tail black; rear of white wings black.

**Similarities**
See White Pelican, White Ibis. Note also that white herons lack black in wings and retract neck in flight. Rare Whooping Crane is very similar, with straight bill.

**Habitat**
Marshes, swamps of Deep South.

**Habits**
Feeds quietly from water on shore. Flies with legs and neck extended.

**Voice**
Usually silent, but occasionally a hoarse croak.

**Eggs**
3–4; white; 2.7 x 1.7 in. (7.0 x 4.4 cm). Nest is of sticks in trees; in colonies.

**Range**
Resident from S.C. to Fla. and w. to Tex., southward; some wander n. in summer.

# IBISES AND SPOONBILLS
Family Threskiornithidae

These are long-legged marsh birds with decurved or spatulate bills; they fly with legs and bills extended. They feed on crustaceans and insects.

## GLOSSY IBIS
*Plegadis falcinellus*

**5:8**
*Fig. 3*

**Description**
Size, 22 in. (55.9 cm). A dark ibis, very common to the East Coast. Adult: dark, glossy chestnut with some iridescent green; looks black at a distance; narrow white border on skin between bill and eye in breeding condition; legs grayish. Immature: grayer, neck streaked with white.

**Similarities**
In flight, silhouette resembles Whimbrel, but is bigger, heavier-billed.
**Habitat**
Ponds, wet fields, mud flats, and marshes, both fresh and salt.
**Habits**
Flies in long, often undulating lines or ranks; flocks to roost in evening.
**Voice**
Combination grunt and bleat.
**Eggs**
3–4; bluish-white; 2.0 x 1.5 in. (5.1 x 3.8 cm). Nest is of sticks; in trees, bushes, or reeds over water; in colonies.
**Range**
Breeds along Atlantic Coast, from Maine to Fla. and along Gulf Coast to Tex.; also inland to Great Lakes; winters in Fla. and along Gulf Coast.

Note: The **WHITE-FACED IBIS,** *Plegadis chihi (Fig. 3)*, size, 23 in. (58.4 cm), on the western Gulf Coast and east to Louisiana. A close relative of the Glossy Ibis, it is exceedingly similar. Adults in breeding condition may be differentiated from Glossy with care under favorable circumstances by a broader area of white feathering extending back behind eye.

Glossy Ibis                          White-faced Ibis

*Fig. 3*

breeding adults

## WHITE IBIS
*Eudocimus albus*                                  5:5

**Description**
Size, 22–28 in. (55.9–71.1 cm). Adult: all-white except for some black on wing tips; • *decurved bill* and • *red legs*. Immature: streaky brown with white belly, dark legs, and bill.
**Similarities**
Wood Stork has dark, naked head.
**Habitat**
Coastal fresh- and saltwater areas.
**Habits**
Gregarious.
**Voice**
Squawking notes.
**Range**
Resident from Ga. s. along Atlantic Coast and w. along Gulf Coast to Tex.

## ROSEATE SPOONBILL
*Ajaia ajaja*                                                    **5:9**

**Description**
Size, 30–34 in. (76.2–86.4 cm). • *Pinkish* with bare head (adults)
or white (immature) with long, spatulate bill.
**Habitat**
Coastal salt- and freshwater marshes and lagoons.
**Habits**
In small groups, moves rapidly through water sweeping head and
bill from side to side.
**Eggs**
1–4; white with brown spots; 2.5 x 1.75 in. (6.4 x 4.4 cm). Bulky
nest of sticks in island trees.
**Range**
Breeds locally from Fla., southward, and along Gulf Coast; winters
s. from s. Fla. and s. Tex.

# Swans, Geese, and Ducks
Order Anseriformes

## SWANS, GEESE, AND DUCKS
Family Anatidae

These are medium to large birds of fresh or salt water, with
webbed feet and bills that have tiny, toothlike projections along the
edges. Their flight is swift and sustained; but, when the flight
feathers are molting, most species for a time are unable to fly.
Except when specifically noted otherwise, all make ample-sized
nests of vegetation on the ground near water. This family includes
some of our choicest game birds. The major groups of waterfowl
are:

*Swans:* Very large, long-necked birds, usually all-white when
adult. They feed on aquatic vegetation from the bottom by
extending their necks deep under water.

*Geese:* Smaller, shorter-necked than swans and larger than ducks,
geese have blunt, triangular bills. The sexes are alike. They feed in
stubble and grain fields or grassy marshes, consuming seeds,
aquatic plants, and grasses.

*Whistling-Ducks:* Also called Tree Ducks, these are small,
gooselike ducks that perch upright, may feed away from water on
seeds and grass, and they nest in trees.

*Surface-feeding or Pond Ducks* (Genus *Anas* and Allies): These
largely freshwater ducks spring almost vertically into the air when
taking flight. They feed on aquatic plants and seeds by "tipping
up," with heads below the surface and tails pointing skyward.
Males are often highly patterned about the head. Females are
predominantly brown.

*Diving Ducks* (Pochards [Genus *Aythya*], Goldeneyes, Eiders,
Scoters): The legs of diving ducks are set farther back than those of
the surface-feeders, and the hind toe is free and lobed. Confusingly
similar females are often best identified by their accompanying
males. They dive for aquatic plants, snails, and insects, and patter
along the water's surface before they fly.

*Mergansers*: These fish-eating ducks have crests and long, saw-toothed, slightly hooked bills, with some white in the wing and on the secondaries. They feed on fish, crayfish, and some amphibians.

*Stiff-tailed Ducks*: These are small diving ducks that often erect the tail as they swim. They feed on water plants, insects, and other small aquatic creatures.

## MUTE SWAN
*Cygnus olor*                                                     *Fig. 4*

**Description**
Size, 60 in. (152.4 cm). The only swan with a • *black knob at the base of the orange bill*. Knob is less conspicuous in female and almost absent in young. Immature: bill pink or dusky, blackish at base.
**Similarities**
Bill of immature Mute is darker than young Whistling's.
**Habitat**
Freshwater marshes, ponds; coastal waters in winter.
**Habits**
Sits higher in water than Whistling; carries neck gently arched, bill pointed down; wing secondaries often raised over back aggressively (no other swan does this); wingbeats make a singing, throbbing sound audible from afar; very aggressive in defense of nest.
**Voice**
Usually silent, but can hiss.
**Eggs**
4–8; bluish-white; 4.5 x 2.9 in. (11.4 x 7.4 cm).
**Range**
Resident s. New England to Md.; introduced in U.S. from Europe and established along middle Atlantic Coast.

Mute Swan

*Fig. 4*

Whistling Swan

adults

## WHISTLING SWAN                                    **6:4**
Cygnus columbianus                                      *Fig. 4*

**Description**
Size, 52 in. (132.1 cm). Adult: bill black with yellow spot at base. Immature: bill pinkish, tip dusky, lacks yellow spot at base.
**Similarities**
Smaller than Mute Swan. Immature is difficult to distinguish from immature Mute.
**Habitat**
Lakes and ponds.
**Habits**
Holds neck erect, bill horizontal, wings not arched above back; spends more time feeding with head under water than does Mute.

**Voice**
Musical whooping whistle and soft trumpeting; *wow-wow-ou.*
**Eggs**
4–5; creamy-white; 4.2 x 2.7 in. (10.7 x 6.9 cm).
**Range**
Breeds in far n. Canada from Baffin Is. w.; winters coastally from Md. w. to Tex., occasionally w. to Great Lakes.

## CANADA GOOSE
*Branta canadensis*                                                      **6:5**

**Description**
Size, 22–43 in. (55.9–109.2 cm). Quite variable, with several different and disputed races. Above grayish-brown; • *head, neck, and tail black;* • *cheek patches white;* base of black "stocking" neck clearly defined against whitish underparts; bill, legs black. Smaller races very stubby-billed; western races often darker-breasted than eastern races.
**Habitat**
Ponds, rivers, bays, fields.
**Habits**
Usually flies in V formation. In flight, black neck stretched out and slightly downcurved; usually very vocal. (Cormorants are silent and darker.)
**Voice**
Loud, from 2-syllabled honking to high-pitched yelping, depending on race.
**Eggs**
4–6; white; 2.3–3.0 x 1.5–2.5 in. (5.8–7.6 x 3.8–6.4 cm). Nest usually on ground near water, well guarded by adults.
**Range**
Breeds s. from n. Canada to N.C.; winters along Atlantic Coast s. from N.J. and interior from Nebr. to Gulf Coast.

## BRANT
*Branta bernicla*                                                        **6:3**

**Description**
Size, 24 in. (61.0 cm). A small goose with a black breast clearly set off from the pale underparts and an all-black head. Above dark brown; head, neck, breast, and tail black; white patch on each side of throat; upper- and undertail coverts white, making white V over tail in flight; • *underparts light brownish-gray;* on water, • *sides look white.* Immature lacks throat patch.
**Similarities**
Canada Goose has white on breast above waterline and white cheek patches.
**Habitat**
Sheltered salt water.
**Habits**
Sits high in water like a gull; flight somewhat undulating, hence hunters' name of "wavy"; flight rapid for a goose, in irregular flocks, rarely V-shaped; wings pointed.
**Voice**
Deep, loud, grunting honks.
**Eggs**
3–5; creamy; 2.8 x 1.9 in. (7.1 x 4.8 cm).
**Range**
Breeds in Arctic; winters along Atlantic Coast to S.C.

## WHITE-FRONTED GOOSE
*Anser albifrons*                                        6:2

**Description**
Size, 29 in. (73.7 cm). Adult: gray-brown; • *area around bill* and
upper- and undertail coverts *white;* black speckles on belly (hence
hunters' name, Specklebelly); bill pink; • *legs yellow or orange.*
Immature: similar, but bill yellow; no white around bill or black
speckles on belly.
**Similarities**
Immature White-fronted resembles immature blue form of Snow,
but has yellow bill and legs.
**Habitat**
Ponds, lakes, rivers, bays; likes shallow water.
**Habits**
Flies rapidly in V formation.
**Voice**
High-pitched, double-noted *wah-wah, wah-wah,* like human
laughter.
**Eggs**
5–7; white; 3.1 x 2.1 in. (7.9 x 5.3 cm).
**Range**
Breeds in Greenland and nw. Canada; winters s. from Ill. to Gulf
Coast; uncommon on Atlantic Coast.

## SNOW GOOSE
*Chen caerulescens*                                      6:1

**Description**
Size, 23–31 in. (58.4–78.1 cm). Adult: • *body white,* variegated, or
entirely gray-brown; legs pink; head white (may be rust-stained);
• *wings always with black tips;* bill pink with dark basal streak.
Immature: similar but duskier; darker bill and feet.
**Similarities**
Swans have longer necks, no black in wings.
**Habitat**
Ponds, lakes, rivers, bays, marshes, and fields.
**Habits**
Flies in loose V's.
**Voice**
Single nasal honk.
**Eggs**
5–7; white; 3.2 x 2.1 in. (8.1 x 5.3 cm).
**Remarks**
"Blue" and white phases now considered same species. Birds
wintering in Atlantic Coast are mainly white, while both blue and
white phases occur on the Gulf Coast.
**Range**
Breeds in Arctic; winters along Atlantic Coast from N.J. to Fla.
and w. along Gulf Coast to Tex.; casual to interior.

## FULVOUS WHISTLING-DUCK
*Dendrocygna bicolor*                                    7:2

**Description**
Size, 18–21 in. (45.7–53.3 cm). Upright posture, buff and rusty
color, and • *white rump.*
**Similarities**
Cinnamon Teal is smaller, deeper color.

31

**Habitat**
Pastures and rice fields, and adjacent waters.
**Habits**
Active by night and day, feeds in fields as well as in water, dives regularly.
**Voice**
High-pitched double whistle.
**Eggs**
12–17; white; 2.2 x 1.5 in. (5.6 x 3.8 cm). Nest is in dense ground vegetation in wet areas.
**Range**
Resident along Gulf Coast from La. to s. Tex.

## MALLARD
*Anas platyrhynchos*                                        **7:10, 8:2**

**Description**
Size, 23 in. (58.4 cm). • *Speculum bluish-purple with white borders,* wing linings white. Male: breast chestnut; body gray; tail white; some black tail coverts curl forward; • *head green, narrow white neckband,* bill yellow. Female: brown, bill mottled orangish, tail with only some white. Southeast coastal (Florida and Gulf Coast) Mottled Duck **(7:10)** is a subspecies in which males are colored as females, but both sexes are darker than female Mallards, with a clear orange or yellow bill; wing speculum is bordered with black (very little white is evident).
**Similarities**
Northern Shoveler has uncrested green head, but breast is white, sides chestnut, and bill huge. Black Duck is darker than female Mallard and lacks white borders to speculum. Female Pintail is longer and neck more slender than female Mallard, has only 1 white rear border on speculum, wing linings not white, bill gray, tail more pointed. Female American Wigeon has more white in wing.
**Habitat**
Almost any water.
**Habits**
Flight is like Black Duck, wingbeat not very deep.
**Voice**
Male, a low, reedy *kwek-kwek-kwek;* female, the familiar barnyard quack.
**Eggs**
8–10; olive-green; 2.3 x 1.6 in. (5.8 x 4.1 cm). Nest is not always near water.
**Remarks**
This species is the ancestor of most of our domestic ducks. It frequently hybridizes with the Black Duck in the northeast.
**Range**
Breeds throughout Canada and n. U.S., from Greenland to Alaska, s. to Va. and Tex.; winters s. from s. Canada, throughout cen. and s. states.

## BLACK DUCK
*Anas rubripes*                                             **7:9, 8:1**

**Description**
Size, 23 in. (58.4 cm). Sexes similar. • *Very dark brown,* throat and cheeks paler; speculum purple bordered with black; paler head contrasting with very dark body; bill yellow to greenish; legs dusky

or red. Female Mallard is lighter, shows less head and body contrast, and has an orangish bill.

**Similarities**
Scoters have bigger heads, shorter necks, and are found along the coast.

**Habitat**
Ponds, lakes, often in woodland.

**Habits**
Similar to Mallard. In flight, shows all-silver wing linings, flashing against a nearly black body.

**Voice**
Similar to Mallard.

**Eggs**
6–12; creamy to greenish; 2.3 x 1.7 in. (5.8 x 4.3 cm). Nest is usually on ground, occasionally in an abandoned crow or hawk nest in a tree.

**Range**
Breeds throughout e. and cen. North America, from Labr. and Man. s. to Va., Ill., and Wis.; winters from coastal Newfoundland and s.-cen. Canada to Fla. and Tex.

# GADWALL
*Anas strepera*                                                  7:8, 8:3

**Description**
Size, 20 in. (50.8 cm). The only surface-feeding duck with a • *white speculum,* conspicuous in flight, not easily seen on the water. Male: • *gray;* head and neck light brown, shoulder red-brown; • *rump and undertail coverts black* (conspicuous in flight), bill gray. Female: light brown, bill yellow-brown. In both sexes, belly white, feet yellow.

**Similarities**
White wing patch of American Wigeon more conspicuous and on forepart, not hindpart, of wings; on water, Wigeon shows white on flanks. Female Wigeon is ruddy-flanked, gray-headed. Female Mallard is larger. Female Pintail has blue-gray bill.

**Habitat**
Ponds, lakes, streams, marshes; fresh water and brackish.

**Habits**
Wingbeat faster than Wigeon's; flight swift, direct, in small compact flocks; wings long, pointed.

**Voice**
Male whistles and trills, female quacks.

**Eggs**
7–13; creamy; 2.2 x 1.6 in. (5.6 x 4.1 cm). Nest is not necessarily near water.

**Remarks**
No other duck has so wide a distribution. Of the larger regions of the world, it is missing only from South America and Australia.

**Range**
Breeds from s. Canada, s. to N.C. along Atlantic Coast and interior to Gulf Coast; winters from s. New England southward.

# NORTHERN PINTAIL
*Anas acuta*                                                  7:11, 8:4

**Description**
Size, 22–27 in. (55.9–68.6 cm). Male: A noticeably long, pointed tail; distinctive white line up side of neck. Female: light brown, somewhat darker above; lacks very long central tail feathers. Both

sexes, white below, speculum brown; bill and feet blue-gray. In flight, long-necked and long-tailed, slender, white underparts and white line on trailing edge of wing visible from afar.

**Similarities**
Female Mallard is larger, with shorter neck and tail, orangish bill, and speculum with 2 white borders. Female Gadwall has white speculum. The only other long-tailed duck, the Oldsquaw, is a diving species.

**Habitat**
Ponds, lakes, marshes, rivers; in winter, saltwater bays.

**Habits**
A fast flier, sits high on water.

**Voice**
Male, in flight, a loud *kwa, kwa;* a Teal-like whistle; female, a hoarse quack.

**Eggs**
6–12; olive-buff; 2.1 x 1.5 in. (5.3 x. 3.8 cm). Nest can be as much as a mile from water.

**Range**
Breeds from n. Canada, s. to w. Pa. and w. to Nebr. and casually farther e. in U.S.; winters s. from s. New England and s. Ill.

## COMMON TEAL
*Anas crecca*                                              **7:5, 8:5**

**Description**
Size, 14 in. (35.6 cm). Male: spotted tan breast, gray sides, • *chestnut-colored head;* • *green ear patch.* American form (Green-winged Teal) has • *white vertical crescent behind breast;* European form, which is rare in our area, lacks vertical bar but has horizontal white stripe on back above wing. Female: grayish-brown, speckled below; European form indistinguishable from American form.

**Similarities**
Shorter-necked, shorter-bodied than Blue-winged, and without blue on wing. From below, in flight, male Common shows light belly; male Blue-winged shows dark. Female resembles a pint-sized female Mallard. If a small duck without conspicuous wing colors springs up from a marsh, it is probably this species.

**Habitat**
Marshes, ponds, lakes, streams; brackish and salt water.

**Habits**
Sometimes walks about in mud feeding like shorebird; flight fast, buzzy, and erratic, in compact flocks wheeling like pigeons; wings whistle in flight.

**Voice**
Male, piping whistles; female, high-pitched quack.

**Eggs**
10–12; pale buff; 1.8 x 1.4 in. (4.6 x 3.6 cm). Nest is occasionally found some distance from water.

**Range**
Breeds from Que. to N.Y. and from Man. to Nebr.; winters s. from New England and from Mo. southward.

## BLUE-WINGED TEAL
*Anas discors*                                            **7:7, 8:6**

**Description**
Size, 15 in. (38.1 cm). Bill relatively large. A very small duck, with • *chalky-blue shoulder patches.* Male: grayish above, tan spotted

with dark below; white patch on rear of flanks; large • *white face crescent.* Female: brownish-gray above, pale gray marked with darker below.

**Similarities**
Northern Shoveler bill is shaped similar to Blue-winged. Female Northern Shoveler is larger and much heavier-billed; female Common Teal has no blue on wing, bill is smaller.

**Habitat**
Freshwater marshes, ponds; rarely salt water.

**Habits**
Flight erratic, blue wing patch may look white in poor light; dabbles in mud when feeding.

**Voice**
Male, whistling peep; female, quack similar to Common Teal's.

**Eggs**
6–12; white; 1.8 x 1.3 in. (4.6 x 3.3 cm).

**Range**
Breeds from Newfoundland, Que., and Man. s. to N.C. and Gulf Coast; winters from Md. and s. Ill. southward.

## EURASIAN WIGEON
*Anas penelope*                                                          7:3

**Description**
Size, 19 in. (48.3 cm). Dusky axillars. Male: gray upperparts and flanks; pinkish breast; white wing coverts, rump, rear flanks, and belly; black primaries and tail coverts; green speculum, • *red-brown head, creamy-colored crown.* General impression is of a gray duck with a red-brown head.

**Similarities**
Male American Wigeon lacks brown on the head; female American looks grayer, has gray head, and white axillars. Redhead has black breast.

**Habitat, Habits, Eggs**
Similar to American Wigeon.

**Voice**
Shrill whistling *whee-you.*

**Remarks**
Common in Europe, this bird is regular, but quite rare, in this country.

**Range**
Breeds in Eurasia; regularly appears in small numbers along East Coast.

## AMERICAN WIGEON
*Anas americana*                                                    7:4, 8:8

**Description**
Size, 20 in. (50.8 cm). Head and neck gray; speculum green; wing coverts and belly white; bill blue, black-tipped. Male: • *shiny white crown,* ear patch green (looks black at a distance); back brown; breast and sides pink; rump, rear of flanks, and large patch on front of wing white; primaries and undertail coverts black. In flight, white front wing patches and black undertail coverts conspicuous. Female: head gray, wing patches grayish-white, upper parts ruddy-brown, breast and sides tan, undertail coverts white.

**Similarities**
Gadwall lacks white crown and flank patches, sits lower on water; blue wing patches of Northern Shoveler and Blue-winged Teal may sometimes look white. See Eurasian Wigeon. Female Mallard is larger, darker, and has less white on wing.

**Habitat**
Lakes, streams, rivers, marshes, bays; prefers fresh water but also seen on salt.
**Habits**
Sits high on water, pivots as it feeds; flies in compact, irregular flocks.
**Voice**
Male: "a wild and musical note, *whew, whew, whew.*" Female: "a *qua-awk, qua-awk*" (Kortright).
**Eggs**
6–12; creamy; 2.1 x 1.5 in. (5.3 x 3.8 cm). Nest is often at a distance from water.
**Other name**
Baldpate.
**Range**
Breeds in Nw., s. to Nebr. and Minn.; rare farther e.; winters along coast from N.J. southward, and in s. states, southward.

## NORTHERN SHOVELER
*Anas clypeata*                                            **7:6, 8:7**

**Description**
Size, to 19 in. (48.3 cm). Has a large, flat, spoon-shaped bill. Male: head green (looks black at a distance); body and tail white; • *sides rufous*; on water shows most white of any surface-feeding duck. In flight, from head to tail shows alternating pattern of dark-light-dark-light-dark. Female: brownish above, paler below. Both sexes have chalky-blue wing coverts, green speculum.
**Similarities**
Shoveler is larger and bigger-billed than Blue-winged Teal.
**Habitat**
Marshes and shallow water, fresh or brackish.
**Habits**
Flies and swims with bill pointing down at an angle, wings seem set far back on body; flight slower, more hesitating than teal's; sits low on water, feeds from surface, using bill as a strainer.
**Voice**
Male: *"woh, woh, woh"* (Collins); female: a weak quack.
**Eggs**
6–14; pale olive-green or buffy; 2.1 x 1.5 in. (5.3 x 3.8 cm).
**Range**
Breeds from N.S. and Man. s. to Del., N.C. and w. to Nebr.; winters s. from Ga.; some s. from New England.

## WOOD DUCK
*Aix sponsa*                                              **7:1, 8:9**

**Description**
Size, 18 in. (45.7 cm). Has a conspicuous crest. Male: boldly patterned with iridescent maroon, green, purple, and white; eclipse plumage much duller and with little or no crest. Female: head gray, crested; eye-ring, throat, and underparts white; back gray-brown, speculum blue. In flight, head held above level of body, bill pointed down at an angle; short neck and long square tail conspicuous; also dark breast, wings, and tail; white belly and white trailing edge of wing.
**Similarities**
See American Wigeon.
**Habitat**
Freshwater marshes, swamps, and creeks.

**Habits**
Often perches in trees within swamp; sits high on water; flight swift and direct.
**Voice**
Distinctive whistle in flight.
**Eggs**
10–15; dull white; 2.0 x 1.6 in. (5.1 x 4.1 cm). Nest is in hole in tree, often some distance from water.
**Range**
Breeds from N.S. and s.-cen. Canada s. to Fla. and Gulf Coast, w. to Tex.; winters from N.C. southward.

# REDHEAD
*Aythya americana*                                        9:1, 11:1

**Description**
Size, 20 in. (50.8 cm). Male: • *rounded, brownish-red head*, gray back; blue bill with black tip. Female: brown; belly and area around bill white; head rounded.
**Similarities**
Canvasback has blackish bill, long sloping forehead, rustier head, whiter body. Female Common Goldeneye has brown head, white collar. Female scaup have broad white circle around base of bill. Female Ring-necked Duck is smaller, darker, and has white eye-ring and ring around bill. In flight, Redhead is shorter and darker than Canvasback, wingbeats more rapid, and flight more erratic; wing stripe of Redhead is long and gray, not white as in scaup.
**Habitat**
Ponds, lakes, rivers, bays.
**Habits**
Gathers in big rafts on large lakes and bays in winter.
**Voice**
Male: a catlike *me-ow*. Female: "a growl *r-r-r-rwha, r-r-r-r-wha*" (Griscom).
**Eggs**
10–15; pale olive-buff; 2.4 x 1.7 in. (6.1 x 4.3 cm).
**Range**
Breeds in nw. Canada, s. and e. to Wis. and Nebr., and occasionally farther e.; winters from Great Lakes s., and from s. New England southward.

# RING-NECKED DUCK
*Aythya collaris*                                             11:3

**Description**
Size, 17 in. (43.2 cm). Head rather triangular in shape; bill with white ring, speculum bluish-gray. Male: head, breast, and back black; head with purple iridescence, sides gray (sometimes seem white), chestnut ring on neck seldom visible; on the water, shows a • *vertical white crescent* on its side behind the breast. Female: brown with white eye-ring and indistinct white area near bill. In flight, black back and forewing in male (dark brown in female) and gray wing stripe.
**Similarities**
Male scaups have lighter backs; female Redhead is larger and has less white on cheeks. Female scaup have more distinct white area around base of bill. In flight, male scaup show broad white wing stripe.
**Habitat**
Ponds, wooded lakes, streams; in South, saltwater bays.

**Habits**
Travels in small groups, alights without circling.
**Voice**
Similar to Lesser Scaup, seldom heard.
**Eggs**
6–12; dark olive-buff; 2.3 x 1.6 in. (5.8 x 4.1 cm).
**Range**
Breeds from Newfoundland and Man. s. to Maine and Iowa; winters from New England and Ill. southward.

## GREATER SCAUP
*Aythya marila*                                   9:3, 11:5

**Description**
Size, 18 in. (45.7 cm). Black with a white saddle, this is only eastern duck with a broad white wing stripe extending almost to wing tip; bill bluish. Male: black head, breast, primaries, and tail; gray back; white sides; • *head glossed with green* and quite rounded. Female: dark brown with sharply defined white patch around bill.
**Similarities**
Lesser Scaup has a thinner neck, shorter wing stripe, and purple gloss on the somewhat angular head of the male.
**Habitat**
Lakes, ponds, rivers, bays, often ocean.
**Habits**
Large rafts collect on bays in winter; in east, more inclined to salt water and larger bodies of fresh water than Lesser Scaup.
**Voice**
"A loud, discordant *scaup, scaup*" (Kortright).
**Eggs**
8–9; buffy-olive; 2.5 x 1.7 in. (6.4 x 4.3 cm).
**Remarks**
The extreme similarity of the 2 scaup is noteworthy. Presumably they represent formerly separated populations that evolved into separate species, the Greater in Eurasia, the Lesser in North America. The Greater is the more northerly and westerly, and probably has recently entered North America from Asia.
**Range**
Breeds in nw. Canada, s. and e. to Mich.; winters along Atlantic and Gulf coasts from N.B., southward, and in Great Lakes.

## LESSER SCAUP
*Aythya affinis*                                   9:4, 11:4

**Description**
Size, 17 in. (43.2 cm). Male: • *head iridescent purple* (but iridescence may vary with angle of light), flanks grayish, head somewhat angular. Female: dark brown, bill clearly "masked" with white at base.
**Similarities**
Smaller than Greater, especially in apparent bulk, and white stripe on wing noticeably shorter. Ring-necked Duck has a black back; female Redhead and Canvasback are larger; female Ring-necked has less distinct white around bill but has white eye-ring.
**Habitat**
In the east, in smaller bodies of fresh water than the Greater and less often in salt water.
**Habits**
Flight swift, erratic; often in closely bunched large flocks.

**Voice**
Male: a coarse *scaup;* in flight, a repeated *pppr-pppr.*
**Eggs**
6–15; buffy-olive; 2.3 x 1.6 in. (5.8 x 4.1 cm).
**Other names**
Bluebill, Broadhull.
**Remarks**
If a male scaup has a purple head sheen, it is a Lesser, but this is usually difficult to determine. On the water, watch for one to flap its wings, showing the length of the white wing stripe.
**Range**
Breeds in nw. Canada, se. to Nebr. and Iowa; occasionally farther e.; winters from Mass. and Ill. s. to Gulf Coast.

## CANVASBACK
*Aythya valisineria*                                           9:2, 11:2
**Description**
Size, 21 in. (53.3 cm). • *Bill long, blackish, merges into forehead* in a long continuous slope. Male: • *rusty head,* white back. Female: head and breast light brown, forehead sloping, back grayish.
**Similarities**
Male Redhead is gray above, bill bluish, head rounded. Similar females all have rounded foreheads.
**Habitat**
Ponds, lakes, rivers, bays.
**Habits**
In flight, long bill, head, and neck carried slightly down; pointed wings appear set far back on body. Flies in lines or V's; collects in great rafts on the water.
**Voice**
Male grunts, female quacks.
**Eggs**
7–9; grayish-olive; 2.5 x 1.8 in. (6.4 x 4.6 cm).
**Range**
Breeds in nw. Canada, s. and e. to Minn. and Nebr.; winters from Mass. and Great Lakes s. to Gulf Coast.

## COMMON GOLDENEYE
*Bucephala clangula*                                           9:5, 10:3
**Description**
Size, 18 in. (45.7 cm). Chunky body, high-domed head, short neck, golden eye, white wing patch; in flight, looks large-headed, short-necked. Male: white-sided duck with black head and • *round white jowl spot;* head, upperparts, and bill black (head with greenish gloss), rest of body white; great white squares in wings and blackish lining to wings from below. Female: brown head; gray back and sides; collar, breast, belly, and divided wing patch white; outer third of bill yellow in spring.
**Similarities**
Barrow's has shorter bill; winter females indistinguishable in field. Scaups have a black breast; mergansers are long-necked and have a rakish, not a stocky, look.
**Habitat**
Lakes, rivers, bays, and ocean.
**Habits**
Wings whistle in flight; dives frequently when feeding; flock rises all at once, does not string out in flight; female sits lower in water than male.

**Voice**
Male: "a penetrating *spear, spear*" (Collins). Female: a low quack.
**Eggs**
8–15; pale green; 2.4 x 1.7 in. (6.1 x 4.3 cm).
**Range**
Breeds from Labr., s. to n. New England, w. to Minn.; winters in n. waters, s. to Fla. and Gulf Coast.

# BARROW'S GOLDENEYE
*Bucephala islandica*                                            10:1
**Description**
Size, 18 in. (45.7 cm). Male: • *head has purplish gloss* and bulges fore and aft; crown has a snoodlike droop at rear (not high-domed, as in Common); has a white • *crescent-shaped jowl spot*. On water, sides and wings show much more black than does Common; black wedge points down at shoulder, separating sides from breast; row of white spots on black scapulars; black line divides white wing patch. Female: in spring breeding plumage, bill all-yellow, head darker than female Common; forehead more abrupt; in winter, indistinguishable in field.
**Similarities**
Similar to the more abundant Common Goldeneye, but bill shorter and thicker at base, head shaped differently.
**Habitat**
Lakes, bays, ocean.
**Habits**
In winter, remains farther north than Common.
**Voice**
Croakings.
**Eggs**
6–14; pale greenish; 2.4 x 1.7 in. (6.1 x 4.3 cm).
**Range**
Breeds along Labr. coast; winters from Maritime Provinces s. to N.Y.

# BUFFLEHEAD
*Bucephala albeola*                                       9:6, 10:2
**Description**
Size, 14 in. (35.6 cm). Small, puffy head on chunky body, short-necked; bill blue-gray, stubby; white wing patches are conspicuous in flight. In flight, Bufflehead resembles goldeneye, but flies faster and nearer water. Male: on water, whitest eastern duck; suggests a small goldeneye, with a • *large, white head patch* extending over the rear of crown. Female: upperparts dusky, head with a slanting white cheek patch; at a distance, suggests female goldeneye.
**Similarities**
Male Hooded Merganser looks dark, has long head and body, long, thin black bill, black neck; its white head patch has a black border.
**Habitat**
Lakes, rivers, bays, ocean.
**Habits**
Occurs in small groups, can dive from wing, and can fly directly into air from under water.
**Voice**
Seldom heard; male whistles, female quacks.
**Eggs**
6–14; buffy; 1.9 x 1.4 in. (4.8 x 3.6 cm).

**Other names**
Buffleheads are locally known as Butterball, Cock-dipper,
Didapper, Dopper, Marionette, Scotchman, Shotbag, Spirit Duck,
and Woolhead.
**Range**
Breeds from Ont. to Man.; winters along Atlantic Coast from N.B.
to Fla. and from Great Lakes to Gulf Coast.

## OLDSQUAW
*Clangula hyemalis*                                        **9:9, 10:5**

**Description**
Size, male, 21 in. (53.3 cm); female, 16 in. (40.0 cm). Only duck in
this area with • *all-dark wings* and white on the body. On water
looks chunky, bill small. Male: in summer, dusky with • *white on
face and belly;* in winter, white with dark cheeks, back, and breast.
From below, in flight, white with conspicuous black wings and
breast band. In winter has an all-white crown and at all seasons
has long, pointed tail. Female: largely brown, including crown and
ear/cheek patch, but plumage lighter in winter than in summer;
tail short. Immature: in fall often quite dark.
**Similarities**
Head pattern of young female resembles changing Harlequin.
**Habitat**
Large lakes, bays, ocean.
**Habits**
In flight, wings low, curved, pointing to rear; wingbeats rapid;
flight erratic, buzzy, low over water, in small flocks veering (like
shorebirds) and flashing black and white; often alights with great
splash. Dives for shellfish.
**Voice**
A "musical, gabbling *south, south-southerly* or *how doodle do*"
(Collins).
**Eggs**
6–10; light grayish-olive; 2.1 x 1.6 in. (5.3 x 4.1 cm).
**Range**
Breeds in Arctic, s. to Hudson Bay; winters along Atlantic Coast s.
to S.C.

## HARLEQUIN DUCK
*Histrionicus histrionicus*                                **10:4**

**Description**
Size, 17 in. (43.2 cm). On water, small and chunky; in flight,
suggests goldeneye but is dark; bill small. Male: slate-blue body,
white markings, and rusty flanks are unique, but looks black at a
distance; conspicuous white marks on head, neck, sides, wings;
changing male presents bizarre pattern. Female: brown with 3
white spots on head, no wing patch.
**Similarities**
Female has shape of Bufflehead and pattern of female scoters.
Female Surf and White-winged scoters are bigger, heavier, and
have 2 white head spots. Female Bufflehead has 1, and a white
wing patch; female Oldsquaw has whiter head and neck.
**Habitat**
In winter, the ocean, especially near rocky shores; in summer,
mountain streams.
**Habits**
Floats high in water in close formation, often with tail cocked;

feeds around rocks; flies fast, low to water, in compact flocks; can dive from wing.

**Voice**
Whistle or squeak.

**Eggs**
5–10; buffy; 2.3 x 1.6 in. (5.8 x 4.1 cm). Nest is in rock crevices, logs, or holes in trees.

**Range**
Breeds in Arctic, s. to Labr.; winters along Atlantic Coast s. to N.Y.

# COMMON EIDER
*Somateria mollissima*                                        *Fig. 5*

**Description**
Size, 24 in. (61.0 cm). Head wedge-shaped, bill and forehead form a continuous slope. Male: • *black underparts*, crown, and tail, white head and breast; black wings with white coverts. Female: reddish-brown barred with black. First-year male: often with a white collar, is intermediate between female and adult male, or all-dark with whitish breast and wing patch (like female goldeneye with a slanting forehead).

**Similarities**
Slope of forehead resembles Canvasback. Female Common Eider is similar to female King Eider, but top of bare upper bill slopes almost to eye.

**Habitat**
Rocky seacoasts, islands; the ocean over shoals.

**Habits**
Flies low, sluggishly with neck drooping; heavy wingbeats alternate with gliding; looks very large in flight; swims with neck in and bill down.

**Voice**
Male gives a low, plaintive *he-ho-ha-ho*, female quacks.

**Eggs**
5–10; dull green; 3.1 x 2.1 in. (7.9 x 5.3 cm).

**Range**
Breeds from Greenland and Arctic to Maine; winters along Atlantic Coast s. to N.Y.

*Fig. 5*

immature ♂

♂ adult

♀ adult

Common Eider          King Eider

# KING EIDER
*Somateria spectabilis*                          **9:10, 10:9**
                                                 *Fig. 5*

**Description**
Size, 22 in. (55.9 cm). Male: distinguished by his large orange forehead shield; pearly crown; black back, wings, and belly; white

breast, wing patch, and flank spot near tail; face white with greenish cast. Forehead slopes up abruptly from bill, back appears shorter than in Common; in flight, shows large white wing patches; at a distance looks white in front, black to rear. Female: difficult to distinguish from female Common Eider, but forehead more abrupt, top of bare upper bill much shorter, plumage lighter, more rusty. Immature male: dusky; abrupt forehead; dark brown head with indication of adult facial pattern; light breast.

**Similarities**
Female Common Eider very similar; see above.

**Habitat**
Large lakes, rocky ocean shorelines, offshore reefs.

**Habits**
In flight, similar to Common, but beak droops less.

**Voice**
Male moans, female quacks.

**Eggs**
4–7; dull green; 2.0 x 2.0 in. (5.1 x 5.1 cm).

**Range**
Breeds in Arctic from Greenland to Hudson Bay; winters along Atlantic Coast s. to Mass.; casually to great Lakes.

## WHITE-WINGED SCOTER
*Melanitta fusca*                                    9:11, 10:8

**Description**
Size, 21 in. (53.3 cm). Both male and female have • *white wing patches*. These patches are not always evident, when the bird is in the water, until it flaps its wings. Male: white streak under eye; bill orange with a black knob. Female: 2 whitish cheek patches prominent in younger birds, sometimes obscure in adult females.

**Similarities**
See Black Guillemot.

**Habitat**
Ocean, bays, large lakes.

**Habits**
Travels in large flocks.

**Voice**
In flight, a low, bell-like whistle; also a croak.

**Eggs**
9–14; pinkish-buff; 2.6 x 1.8 in. (6.6 x 4.6 cm).

**Range**
Breeds from Man. s. to N.Dak.; winters s. to S.C. and La.

## SURF SCOTER
*Melanitta perspicillata*                            9:12, 10:7

**Description**
Size, 19 in. (48.3 cm). No white wing patches; bill forms an even slanting line with forehead. Male: white forehead and nape; bill varicolored red, white, orange, and black. Female: face has 2 light patches in younger birds, more obscure in older females (similar to female White-winged, but without wing patch); sometimes shows whitish patch on nape.

**Similarities**
In flight, resembles Black Scoter, but lacks silvery sheen under flight feathers.

**Habitat**
Oceans, bays, large lakes.

**Habits**
Similar to White-winged, but alights with wings held upward; wings hum in flight.
**Voice**
Croak.
**Eggs**
5–9; creamy; 2.4 x 1.7 in. (6.1 x 4.3 cm).
**Range**
Breeds in n. Canada s. to n. Que.; winters on Atlantic and Gulf coasts.

## BLACK SCOTER
*Melanitta nigra*                                                    10:6
**Description**
Size, 19 in. (48.3 cm). Smallest-looking, most ducklike scoter, primaries with silvery sheen beneath. Male: all-black with an orange bill knob. Female: brown with black crown and whitish cheeks.
**Similarities**
Female White-winged and Surf Scoters have 2 light patches on cheeks; winter Ruddy is smaller, paler, and with white chest.
**Habitat**
Coast and tundra, rare inland.
**Habits**
Travels in large flocks, flight more clean-cut than other scoters; on water, often cocks pointed tail, rides high with head high and bill horizontal or uptilted (other scoters carry bill pointed down).
**Voice**
Musical *cour-loo.*
**Eggs**
6–10; pale ivory-yellow; 2.5 x 1.7 in. (6.4 x 4.3 cm).
**Former name**
Common or American Scoter.
**Range**
Breeds in Labr. and Newfoundland; winters along Atlantic Coast s. to S.C.; rarely in interior.

## HOODED MERGANSER
*Mergus cucullatus*                                              9:7, 11:6
**Description**
Size, 18 in. (45.7 cm). The smallest, slimmest, merganser, male unlike others in pattern; bill small, thin. Male: • *black-bordered white crest* is unique among eastern water birds; head, upperparts, and 2 vertical lower breast stripes black; underparts and wing patch white, sides brown; in flight, crest shows as a white streak on head, which is held lowered. Female: crest buffy; smooth-edged • *brown head*; gray-brown upperparts, flanks, and breast; white wing patch and belly.
**Similarities**
See Bufflehead.
**Habitat**
Ponds, streams, lakes, rivers.
**Habits**
On water, sometimes cocks tail, can rise into air with great speed; usually found in pairs or small groups; male frequently raises and lowers crest.
**Voice**
Grunting *crew, crew.*

**Eggs**
6–12; white; 2.1 x 1.8 in. (5.3 x 4.6 cm). Nest is in hole in tree or stump near water.

**Range**
Breeds from N.S. and Man. s. to Tenn. and Nebr.; occasionally seen in se.; winters from New England and Nebr. s. to Gulf Coast.

## COMMON MERGANSER
*Mergus merganser*                                                  **11:8**

**Description**
Size, 24 in. (61.0 cm). Large, loonlike, long bill. Male: crest not noticeable; back, primaries, and tail black; • *head green-black*, bill red; rest of bird is white, breast with rosy blush. In flight, shows more • *white on body* and wings than any other duck. Female: moderate crest; head bright red-brown, sharply contrasting with white throat and neck; upperparts gray, wing-patch white.

**Similarities**
Red-breasted male has red breast, conspicuous crest; female is more crested, and red-brown of head blends into white of throat and breast; female goldeneyes are shorter and stockier, with shorter bill.

**Habitat**
Lakes, ponds, rivers.

**Habits**
Submerges by jumping, then diving, or by gradually sinking; flies in string formation, low, loonlike, horizontal; flight shape rakish.

**Voice**
Usually silent; occasionally "an unmelodious squawk" (Swarth).

**Eggs**
6–17; creamy; 2.5 x 1.8 in. (6.4 x 4.6 cm). Nest is in hollow tree or stump, sometimes on ground.

**Former name**
American Merganser.

**Range**
Breeds from Newfoundland and Man. s. to New England and Mich.; winters s. to Gulf Coast.

## RED-BREASTED MERGANSER
*Mergus serrator*                                              **9:8, 11:7**

**Description**
Size, 22 in. (55.9 cm). Often has a • *double-pointed crest* and red bill. Male: black head and back, head with a greenish gloss and an untidy crest; sides gray; collar, wing patch, and • *underparts white*; breast red-brown, dark patch with white spots near shoulder. On water, looks slim and dark; in flight, red breastband conspicuous between white neck and belly; white wing patch appears framed. Female: see Common Merganser.

**Similarities**
Uncrested male Common Merganser is mostly white, and has more white on wing patch; female Common has a brighter head, more of a crest, and sharp line of demarcation between throat and breast. See also Red-necked Grebe.

**Habitat**
Lakes, ponds, rivers; in winter, bays and ocean. (Common prefers fresh water.)

**Habits**
Flight is swift, noiseless, and direct, with head, neck, and body horizontal; flattens crest before diving; flies in string formation.

**Voice**
Guttural croaks when alarmed.
**Eggs**
6–12; creamy buff; 2.5 x 1.8 in. (6.4 x 4.6 cm).
**Range**
Breeds in Arctic from Labr. s. to Maine, Mich.; winters along
Arctic Coast s. to Gulf Coast.

## RUDDY DUCK
*Oxyura jamaicensis* **11:9**

**Description**
Size, 15 in. (38.1 cm). Thick-necked, chunky; bill broad, and
upturned; • *cheeks white*; wings entirely brown. Male: in summer,
mostly rich red-brown, cap black, bill bright blue, cheeks white; in
winter, red is replaced by gray and brown, cap becomes dark
brown, and bill much duller. Female: similar to winter male, but
with black streak on cheek and bill dusky.
**Similarities**
Winter male resembles female Black Scoter but is smaller, whiter-
cheeked, and seldom seen on ocean.
**Habitat**
Ponds, lakes, streams, rivers, sheltered bays.
**Habits**
Swims buoyantly with tail often cocked; can sink slowly under
water or dive abruptly; needs long run for takeoff into air; flight
buzzy.
**Voice**
Silent, except for weak cluck when courting.
**Eggs**
5–15; pale buff; 2.5 x 1.8 in. (6.4 x 4.6 cm).
**Range**
Breeds in cen. Canada from Que. s. to Gulf Coast, southward;
winters s. from Mass. and Ill.

# Vultures, Hawks, and Falcons
Order Falconiformes

These birds have hooked beaks and, except for the vultures, strong
toes and long claws. Females are generally larger than males.
(Where two sizes are given, the smaller is the average size for the
male, the larger the average for the female.) These birds hunt by
day. Vultures feed on carrion, the others usually on freshly killed
animal life. Many are powerful fliers and soarers.

# AMERICAN VULTURES
Family Cathartidae

The vultures in east are black and have naked heads. They have
excellent sight and are magnificent soarers. Recent studies show
that they may locate their food—carrion—by smell. Their voice is
a hiss. They nest in hollow logs, crevices in rocks, or other
hideaways on the ground, and they feed their young by
regurgitation.

## TURKEY VULTURE
*Cathartes aura*      **12:1**

**Description**
Size, 29 in. (73.7 cm); wingspread, 6 ft. (1.8 m). All-black bird
that soars on long wings held at an angle. Head small, red (dark in
immatures); flight feathers lighter than coverts, giving wings a 2-
toned look from below; outer flight feathers spread like fingers; tail
long, narrow, rounded.

**Similarities**
Black Vulture has a shorter, squared tail and a white spot under
end of wings; it is clumsier, and flaps more. Northern Harrier also
holds wings at an angle, but is smaller, slimmer, and white-
rumped. Eagles soar on horizontal wings.

**Habitat**
Usually seen soaring over mountains, woods, farm country, but
rare over unbroken forest.

**Habits**
Perches with shoulders hunched, roosts in groups; soars effortlessly,
flaps infrequently.

**Eggs**
1–3; white, marked with brown and purple; 2.8 x 1.9 in. (7.1 x 4.8
cm).

**Range**
Breeds throughout most of e. U.S., s. to Gulf Coast; winters from
s. New England s. and w. to Nebr.

## BLACK VULTURE
*Coragyps atratus*      **12:2**

**Description**
Size, 25 in. (63.5 cm); wingspread, 5 ft. (1.5 m). An all-black bird
with whitish patches near the ends of its wings. Head small, black;
tail short, square (legs often protrude in flight); underwings silvery.

**Similarities**
Black's wings are shorter, broader than Turkey Vulture's. Eagles
and dark phases of large hawks have large heads.

**Habitat**
Open sky, beaches, garbage dumps, slaughterhouses.

**Habits**
Gregarious; in air, appearance more compactly built, heavier than
Turkey Vulture; wings held almost horizontal, tail often fanned;
flaps wings more often, flight more labored than Turkey's.

**Eggs**
1–3; bluish-white, marked with brown; 3.1 x 2.0 in. (7.9 x 5.1 cm).

**Range**
Resident from Pa., Ind., and Kans. southward.

# KITES, HAWKS, EAGLES, AND HARRIERS
Family Accipitridae

In the past, the predatory habits and seeming love of independence
of these birds made them suitable symbols for coats of arms. Today
hawks are recognized as of great value to agriculture because of the
millions of rodents, grasshoppers, and other insects they annually
destroy.

Kites are graceful, falcon-shaped hawks with pointed wings. They
often have forked tails, weak bills, and feet adapted for capturing
insects and small reptiles and mammals.

## KITES, HAWKS, EAGLES, HARRIERS

The accipiters, or bird hawks (Genus *Accipiter*), have short, rounded wings, are finely barred below, and have long tails barred with narrow black bands. Adults are slate-gray above with dark caps and barred below. They frequent woods.

The buteos (Genus *Buteo*) are medium-sized to large hawks with broad wings and generally short, fan-shaped tails. All are dark above. They hunt by soaring and circling, then dropping down to seize prey, mainly rodents.

The eagles are essentially very large hawks. They have fully feathered heads, large bills, and long tails and wings that are held more or less horizontally in flight. They make huge nests on trees or ledges.

Harriers are slender birds of prey that hunt small birds or rodents in open rangeland. They have slim wings and long tails and fly lazily at low altitude, on dihedral wings.

## WHITE-TAILED KITE
*Elanus leucurus*                                                      **16:4**
**Description**
Size, 15–16 in. (38.1–40.6 cm); wingspread, 3⅓ ft. (1.0 m). A gull-like gray-and-white hawk with black shoulders. Adult: • *square tail* and • *whitish head, black* of *wing bend* visible in flight. Immature: similar but with rusty streaking above and below.
**Habitat**
Meadows and marshes.
**Habits**
Circles low over meadows, hovering in search of prey, then pounces.
**Voice**
Whistled call.
**Food**
Small mammals.
**Eggs**
3–5; blotchy; 1.6 x 1.2 in. (4.2 x 3.2 cm). Nest is of twigs in a tree near open country.
**Range**
Resident s. Tex., southward.

## SNAIL KITE
*Rostrhamus sociabilis*                                         **15:3, 16:7**
**Description**
Size, 15–17 in. (38.1–43.2 cm). Black or brown, white-rumped hawk with a long needle-hook on the bill and square tail. Male: black with white base of tail and rump, reddish legs and base of bill. Females heavily blotched, immatures streaked with brown on head and underparts; whitish in tail and wings, orange legs.
**Similarities**
Northern Harrier is distinguished from female Snail Kite by white on its tail.
**Habitat**
Freshwater lakes and marshes.
**Habits**
Glides low over marshes, dropping to the ground for food, then perching to eat.
**Voice**
Shrieking note.

**Food**
Freshwater snails.
**Eggs**
3–4; brown, marked white; 1.7 x 1.4 in. (4.3 x 3.6 cm).
**Former name**
Everglade Kite.
**Remarks**
This bird is endangered in the United States and is limited to a small area of Florida, where at most a few score remain.
**Range**
Resident in s. Fla., chiefly Lakes Okeechobee and Loxahatchee.

## MISSISSIPPI KITE
*Ictinia mississippiensis*         **16:3**

**Description**
Size, 14 in. (35.6 cm); wingspread, 3 ft. (0.9 m). A • *black-tailed*, falcon-shaped hawk. Adult: bluish-gray above; • *pale head, with unmarked gray below;* long, pointed wings; dark primaries, light secondaries; tail notched or squared; eyes and legs reddish. Immature: streaked above, spotted below; tail banded.
**Similarities**
Graceful flight and gray color of adult distinguish this from hawks of similar size.
**Habitat**
Open country and woods.
**Habits**
Flight buoyant and gull-like; often tilts, showing 2-toned upper surface of gray wings; gathers in flocks in migration.
**Voice**
"*Phee-phew*" (Sutton).
**Food**
Insects.
**Eggs**
2–3; blue-green; 1.6 x 1.3 in. (4.1 x 3.3 cm). Lined nest of twigs in treetop.
**Range**
Breeds in se. and s.-cen. U.S.; winters from Fla. and Tex. southward.

Note: In southern Florida observers may see an occasional **SWALLOW-TAILED KITE**, *Elanoides forficatus* **(15:4)**, size, 22–24 in. (55.9–61.0 cm). A beautiful, fork-tailed bird of prey, this is often seen in small groups over swampy forests and adjacent marshes. It is considered an occasional visitor to the Gulf Coast. White head and underparts, with blackish wings and tail characterize this distinctive bird.

## NORTHERN GOSHAWK
*Accipiter gentilis*         **13:1**

**Description**
Size, 20–26 in. (53.3–63.5 cm); wingspread, 4 ft. (1.2 m). A large gray hawk with short, rounded wings and a long, nearly squared tail. Adult: • *white line over eye*, dark cap and ear patch, and finely gray-barred pearly breast; paler above than Cooper's. Immature: brown above, streaked below, distinct white eye line.
**Similarities**
In flight, proportionately heavier about head and neck than smaller

Cooper's, and tail barring less conspicuous. Gyrfalcon has long, pointed wings.

**Habitat**
Forests, especially coniferous.

**Habits**
Silently glides and swoops through woods, dropping onto prey.

**Voice**
Silent except when breeding, then various cackles.

**Food**
Large birds, rabbits, other mammals.

**Eggs**
2–5; bluish-white; 2.3 x 1.8 in. (5.8 x 4.6 cm). Nest of sticks is in tree.

**Range**
Breeds throughout Canada from Newfoundland and n. Que. s. to New England and Mich.; seen along Appalachians s. to Md.; winters s. to Va., Tenn., and Tex.

## COOPER'S HAWK
*Accipiter cooperii*                                    **13:2**

**Description**
Size, 14–20 in. (35.6–50.8 cm); wingspread, 3 ft. (0.9 m). A medium-sized hawk with short, rounded wings and a • *long, rounded tail*; size about that of a crow. Adult: breast barred with red brown. Immature: usually more finely streaked below than young Sharp-shinned.

**Similarities**
Broad-winged Hawk has longer wings, which are light and unbarred below, and a fan-shaped tail; wings of Merlin are long and pointed; difference between tails of Cooper's and Sharp-shinned is best seen when tail is folded.

**Habitat**
Deciduous forest.

**Habits**
Wingbeats slower than Sharp-shinned's; circles and soars more.

**Voice**
Noisier than Sharp-shinned; "a shrill *quick, quick, quick*" (May), sometimes suggestive of a Common Flicker.

**Food**
Birds, small mammals.

**Eggs**
3–5; bluish-white, often with brown spots; 1.9 x 1.5 in. (4.8 x 3.8 cm).

**Range**
Breeds in s. Canada, s. to Fla.; winters from Maine and s. Ont. southward.

## SHARP-SHINNED HAWK
*Accipiter striatus*                                    **13:3**

**Description**
Size, 10–14 in. (25.4–35.6 cm); wingspread, 2 ft. (0.6 m). A small hawk with • *short, rounded wings* and a • *long, square tail*, sometimes slightly notched. Adult: slate-gray above, breast barred with red-brown. Immature: above brown; below white with brown streaks; more reddish in female, darker in male. Tail may look slightly rounded if spread like fan. When perched, Sharp-shinned's wings reach to lower one-third or one-half of tail.

**Similarities**
Merlin's wings when perched reach almost to tip of tail. Large female Sharp-shinned may be as large as a small male Cooper's, but Cooper's tail is rounded; American Kestrel and Merlin have long, pointed wings.

**Habitat**
Woodland and edges.

**Habits**
Rarely soars, except in migration, but occasionally ascends in tight circles with much flapping.

**Voice**
*Cack, cack, cack.*

**Food**
Mainly small birds.

**Eggs**
3–5; bluish-white, beautifully spotted with brown and lilac; 1.5 x 1.2 in. (3.8 x 3.0 cm).

**Remarks**
Hawk Mountain Sanctuary, Pa., is famous for fall migrations of this and other hawks.

**Range**
Breeds in cen. Canada, from Newfoundland s. to Fla. and Tex.; winters s. from New England southward.

# RED-TAILED HAWK
*Buteo jamaicensis*                                    **13:7, 14:4**

**Description**
Size, 19–25 in. (48.3–63.5 cm); wingspread, 4½ ft. (1.4 m). A large hawk with a red tail; chunkier than Red-shouldered and with wider wings, shorter tail. Adult: from below, pale, white area on breast set off by dark area of throat and dark band on lower breast; often with black wrist mark. Individuals vary to almost wholly dark below, but adults normally have a red tail. Immature: underparts streaked with brown; no wrist mark; tail brown above, barred low. Red tail is not always obvious from below; wait until the soaring hawk wheels, showing upper surface of tail. Adult of northern Plains race is very pale, with whitish head and pinkish tail. Western race typically rufescent below, but dark phase occurs. Far northwestern race is very dark, tail mottled without red; winters in small numbers in southern Plains to Louisiana and Arkansas. All races may mix on Southern Plains in winter.

**Similarities**
Hard to distinguish Red-tailed from young Red-shouldered, although latter is usually uniformly streaked below, without tendency to zoning, as in Red-tailed; immature Red-taileds are sometimes whitish at base of tail; compare with Swainson's, Rough-legged, Ferruginous.

**Habitat**
Common in wooded or open areas, marshes.

**Habits**
Soars and circles for long periods, often twisting tail at angle to body (wings shorter than Turkey Vulture and held horizontally), occasionally hovers.

**Voice**
Squealing *kee-a-a-a-r-r-r*, suggesting escaping steam; longer and higher than *"blue-kay"* note of Red-shouldered.

**Food**
Rodents and other small mammals.

**Eggs**
2–4; white, sparingly spotted with brown; 2.3 x 1.9 in. (5.8 x 4.8 cm).

**Former name**
The dark northwestern form of this hawk was known as Harlan's Hawk.

**Range**
Breeds throughout s. Canada and U.S.; winters s. from Maritime Provinces southward.

## RED-SHOULDERED HAWK
*Buteo lineatus*                                    13:5, 14:1

**Description**
Size, 17–24 in. (43.2–61.0 cm); wingspread, 4 ft. (1.2 m). Noted for the narrow white bands on its black tail; intermediate in size between larger Red-tailed and smaller Broad-winged. Wings and tail longer and slimmer than other buteos; white "translucent" spots near wing tip are set off by dark reddish-brown underwing coverts; upperwings barred with black and white. Adult: underparts cross-barred with bright rufous. Immature: duller, often with pale band on upper wing near tip; underparts uniformly streaked, tail bands usually brown and reddish. Races from southern Florida and the Keys are fairly to very pale about the head and chest.

**Similarities**
Immature Red-tailed's streaks tend to be in bands; Broad-winged has broad white bands on tail.

**Habitat**
Wooded or open areas, often found in moister terrain and smaller woodlands than Red-tailed.

**Habits**
Perches less conspicuously, soars less, and has less buoyant flight than Red-tailed.

**Voice**
Clear whistled *kee-yer*; never wheezy, like Red-tailed; often imitated by Blue Jay.

**Food**
Rodents and other small mammals, reptiles, and amphibians.

**Eggs**
2–4; white, marked with brown and buff; 2.2 x 1.7 in. (5.6 x 4.3 cm).

**Range**
Breeds in s. Canada s. to Gulf Coast; winters from s. New England and s. Minn. southward.

## BROAD-WINGED HAWK
*Buteo platypterus*                                 13:4, 14:3

**Description**
Size, 15–17 in. (38.1–43.2 cm); wingspread, 3¾ ft. (.8 m). A buteo with • *2 or 3 broad, white bands on its black tail;* chunky, like a little Red-tailed; the broadest-winged, broadest-tailed, and smallest buteo. Throat and eyebrow white; wings from below mainly white with dark tips and black wrist marks, without the translucent spots of the Red-shouldered. Adult: cross-barred with rusty below. Immature: underparts buff, streaked with brown; dark tail bands more numerous, white bands narrower and less distinct than in adult (more like Red-shouldered, but wings relatively broader, tail shorter—sometimes a difficult identification).

**Similarities**
Broad-winged is duller than Red-shouldered, and without zoning of Red-tailed.

**Habitat**
Deciduous woods.

**Habits**
Sluggish, tame; much given to soaring; when hunting, sometimes hovers.

**Voice**
Drawn-out, high-pitched whistle.

**Food**
Insects, mammals, reptiles, amphibians.

**Eggs**
2–4; whitish, marked with reddish-brown; 1.9 x 1.6 in. (4.8 x 4.1 cm).

**Range**
Breeds from s. Canada s. throughout e. U.S.; winters s. of U.S.

# SWAINSON'S HAWK
*Buteo swainsoni*                                              **13:8, 14:2**

**Description**
Size, 19–22 in. (48.3–55.9 cm); wingspread, 4½ ft. (1.4 m). In light plumage, with • *dark breast* and • *2-toned wing linings*; wings long and somewhat pointed for a buteo; tail long. Adult: light phase shows throat and belly white; tail dark gray above, lighter at base, with 9–12 indistinct dark bands below; dark phase shows nearly entire plumage sooty brown, sometimes with some white underwing or ashy bars on tail. Immature: buff, streaked with dark brown below (usually darker than other young buteos); breast usually darker; wings and tail indistinctly barred with brown. Many variations among these 3 plumages occur.

**Similarities**
Red-tailed has white breast, streaks at rear; dark phase of Red-tailed has reddish tail above. Dark phase of Rough-legged has clear white underwing flight feathers.

**Habitat**
Open areas.

**Habits**
Sluggish, tame; wingbeats faster than Red-tailed's; hunts by cruising low over prairie with wings in open V, as does Northern Harrier or Rough-legged, not by soaring high as does Red-tailed; gregarious, especially in migration.

**Voice**
Long "plaintive whistle, *kree-e-e-e*" (Bent), suggesting Broad-winged.

**Food**
Grasshoppers, rodents.

**Eggs**
2–4; dull white, with umber spots; 2.3 x 1.8 in. (5.8 x 4.6 cm).

**Range**
Breeds chiefly in w. North America; occasionally e. to Minn., Mo., Tex.; winters in S. America, small numbers in Fla.

Note: The **SHORT-TAILED HAWK,** *Buteo brachyurus* **(14:5)**, size, 17 in. (43.2 cm), is a buteo the size of a crow. It is found in swampy forests of South Florida and the Keys. It has 2 phases: the adult white phase has a white belly and underwing coverts, with a conspicuous white patch on its forehead; the adult black phase is black on the belly and has underwing coverts; tails of both show indistinct barrings.

## ROUGH-LEGGED HAWK
*Buteo lagopus*                                   **13:9, 14:6**

**Description**
Size, 20–23 in. (50.8–58.4 cm); wingspread, 4½ ft. (1.4 m). Large
and dark, with rather long and pointed wings for a buteo and
longish tail; • *rump and base of tail white*; feathered legs sometimes
visible as feet drop before swoop. Adult: light phase shows head
and upper breast buffy streaked with brown, with black patch at
bend of light underwing; • *belly, broad end of white tail*, wing tips,
and rest of plumage *blackish*; in dark phase shows all-dark except
for white on underwing and base of tail, occasionally even lacks
white base of tail. Immature: similar to light adult, but dark belly
band more pronounced. Birds with intermediate plumages not
uncommon. In most of our area only in winter.

**Similarities**
Dark phase of Rough-legged may resemble young Golden Eagle.
Dark phase of Ferruginous usually has some rufous mixed with
the black, and has more white at base of primaries. Dark phase of
Red-tailed usually has some rufous in tail.

**Habitat**
Open areas.

**Habits**
Sluggish, perches on observation post by meadow, hovers in air like
Kestrel, or hangs motionless on updraft. Quarters low over
meadows like a Northern Harrier, but is larger, with broader
wings, and may light on a tree, which Harrier does not. Soars with
wings "fingered," and sometimes spirals upward; at other times
alternately flaps heavily and sails.

**Voice**
Silent in winter; a squealing *hurry-up* when breeding.

**Food**
Mice, lemmings.

**Eggs**
2–5; whitish, blotched with brown; 2.2 x 1.8 in. (5.6 x 4.6 cm).

**Range**
Breeds in Arctic s. to Newfoundland and Man.; winters s. to Va.

## FERRUGINOUS HAWK
*Buteo regalis*

**Description**
Size, 22½–25 in. (57.2–63.5 cm); wingspread, 4¾ ft. (1.5 m). The
largest buteo; the reddish-brown legs make a distinctive dark V
against light underparts. Light area in extended primaries in all
plumages; more slender than Rough-legged and with longer tail
and larger bill. Adult: light phase shows shoulders, rump, and
thighs rusty; • *head and tail whitish;* underwings and underparts
also whitish, lightly streaked and barred. Dark phase (rare) shows
all-dark with some rufous mixed in, and with tail light or with
several narrow white bands. Immature: "underparts and tail whiter
than in other hawks" (Bent).

**Similarities**
Western race of Red-tailed lacks reddish-brown upperparts of
Ferruginous. The darker Rough-legged has a broad black band on
its tail. Ferruginous is smaller than young Golden Eagle, with
wings more pointed and tail proportionately longer. Swainson's is
smaller, usually with finely barred tail and dark rear edge of wing.

**Habitat**
Open areas, prairies, badlands.

**Habits**
Perches on 1 leg on observation post; takeoff is slow and heavy, but flight is swifter than Rough-legged's; quarters like a Northern Harrier alternately flapping and sailing; also soars like a Red-tailed.

**Voice**
Various squeals; also a gull-like *kaah*.

**Food**
Rodents, especially ground squirrels.

**Eggs**
3–5; white marked with brown; 2.4 x 1.9 in. (6.1 x 4.8 cm).

**Range**
Breeds primarily in W., from Canadian prairies, s. to Ariz. and Okla.; winters in sw. U.S.

## GOLDEN EAGLE
*Aquila chrysaetos*                                                    **12:6**

**Description**
Size, 31–40 in. (78.7–101.6 cm); wingspread, 7½ ft. (2.3 m). The only eagle that is all-dark below, including the wing linings. Adult: above dark brown, with golden-brown nape (visible only at close range); base of undertail lighter. Immature: base of tail white both above and below, giving • *"ring-tailed"* effect; base of primaries white, rest of body dark brown; amount of white diminishes with age.

**Similarities**
Wings and tail narrower in Bald Eagle; immature Bald has white on coverts, not flight feathers, and never has sharply ringed tail.

**Habitat**
Woods, mountains, badlands (rare).

**Habits**
In flight, tips of flight feathers are outspread and upcurved; soars high, then dives for prey; in straight flight, alternately flaps and glides; wings beat faster than Bald Eagle's; occasionally hovers.

**Voice**
"A shrill whistled *kee-kee-kee*" (Bendire).

**Food**
Mainly smaller mammals, a few birds.

**Eggs**
1–3; white, blotched with red-brown; 3.0 x 2.3 in. (7.6 x 5.8 cm).

**Range**
Breeds in n. Canada, s. to N.Y. and Nebr.; also occurs in Appalachians s. to N.C.; winters casually s. to Gulf states.

Note: The Golden Eagle of the mountains also nests near the sea abroad. The **WHITE-TAILED** or **GRAY SEA EAGLE**, *Haliaeetus albicilla*, size, 27–36 in. (69–91 cm), resembles the Golden Eagle. It breeds in western Greenland and possibly winters into the eastern Canadian Arctic. It is a coastal eagle, like the Bald Eagle, but with a brown head and fully white tail.

## BALD EAGLE
*Haliaeetus leucocephalus*                                            **12:3**

**Description**
Size, 32–40 in. (81.3–101.6 cm); wingspread, 7½ ft. (2.3 m). Adult: large, brown; • *head and tail white*; bill and legs yellow. Immature: dark brown with mottled white on wing linings; bill dusky; tail and head gradually assume full white plumage with age.

**Similarities**
Dark phase of Rough-legged and Ferruginous hawks have shorter wings. See also Golden Eagle.
**Habitat**
Near water; rare.
**Habits**
Wings held horizontal in flight; soars like a Red-tailed; steals fish from Ospreys.
**Voice**
Loud cackles.
**Food**
Dead fish, small animals, rarely birds.
**Eggs**
2–3; white; 2.8 x 2.3 in. (7.1 x 58.4 cm). Nest of sticks is in trees.
**Range**
Breeds in n. and e. Canada, n. U.S., and Fla.; winters coastally and along large rivers in interior.

## NORTHERN HARRIER
*Circus cyaneus*                                                  **13:6, 15:5**

**Description**
Size, 18–22 in. (45.7–55.9 cm); wingspread, 4½ ft. (1.4 m). A hawk with long, slim wings and tail and a white rump. Adult male: • *pale gray* above, whitish below; black wing tips. Female: dark brown above, lighter below. Immature: like female but underparts rufescent with less streaking.
**Similarities**
Rough-legged is larger; has broader wings and tail; and has white mainly on upper tail, not on rump.
**Habitat**
Marshes, fields.
**Habits**
Flight buoyant, gull-like; flies low over grass with wings angled upward; flaps and glides low over grass; in courtship or migration, may soar high and circle on level wings like a buteo.
**Voice**
Low *chu-chu-chu*; a weak nasal *pee-pee*.
**Food**
Small mammals, birds.
**Eggs**
4–6; whitish; 1.8 x 1.4 in (4.6 x 3.6 cm). Nest is of grass and reeds, usually in marsh.
**Former name**
Marsh Hawk.
**Range**
Breeds in cen. Canada, from Newfoundland to Va.; winters from s. Ont. and n. U.S. southward.

# OSPREYS
Family Pandionidae

# OSPREY
*Pandion haliaetus*                                               **12:5, 15:2**

**Description**
Size, 21–24½ in. (53.3–62.2 cm); wingspread, 6 ft. (1.8 m). The only hawk that dives into the water. Wings long, narrow; dark

brown above, with narrow, blackish bands on tail; • *white crown, throat, and underparts;* • *black eye patch* and wrist mark on underwings; immature has dark crown. In flight, wings characteristically angled, not flat or upturned.

**Habitat**
Near water.

**Habits**
Flaps slowly; sometimes sails, often hovers before plunging; dives from up to 100 ft. (30.5 m) in the air.

**Voice**
Whistled *you-you-you*; a "complaining *shriek, shriek, shriek*" (Cruickshank); also chickenlike peeps.

**Food**
Fish.

**Eggs**
2–4; white to rusty, blotched with deep brown; 2.4 x 1.8 in. (6.1 x 4.6 cm). Nest is of sticks on tree or other bare pole.

**Range**
Breeds in cen. Canada, s. to Fla. and Gulf Coast; winters from Gulf Coast, southward.

# CARACARAS AND FALCONS
Family Falconidae

This family contains fast-flying, long-legged birds of prey having large heads; long, narrow, pointed wings; and long tails. The sexes are alike in color, but females are larger. The various species primarily eat small birds and animals or carrion. The major group of these species is the true falcons (Genus *Falco*).

## CRESTED CARACARA
*Polyborus plancus*                                    12:4, 15:1

**Description**
Size, 22–24 in. (55.9–61.0 cm); wingspread, 4 ft. (1.2 m). Adult: dark above; long-legged, long-necked; black crest and red face; throat and breast white; belly black; tail white, tipped with black, presenting an alternating flight pattern of light and dark viewed from below; conspicuous whitish patches near wing tips. Immature: duskier; breast streaked, not barred.

**Habitat**
Open rangeland, prairies.

**Habits**
Often observed on fence posts or feeding with vultures.

**Voice**
Rattling cackle.

**Eggs**
2–3; whitish, blotched; 2.3 x 1.8 in. (5.9 x 4.6 cm). Nest is reeds or sticks; atop tree, yucca, or cactus.

**Range**
Resident in s. Ariz., s. Tex.; disjunct population in prairies of cen. Fla.

## GYRFALCON
*Falco rusticolus*                                           16:5

**Description**
Size, 20–25 in. (50.8–63.5 cm); wingspread, 4 ft. (1.2 m). Gull-sized; white phase is all-white with some black markings; gray

phase is gray above, paler below; dark phase is almost solid black with some white markings; various difficult-to-identify intermediates occur—dark above, lighter below, with underparts barred or spotted in adult, streaked in immature.

**Similarities**
Peregrine is smaller, slimmer; darker above, lighter below; with black crown and mustache; relatively faster wingbeats. See Goshawk, Rough-legged.

**Habitat**
Northern coasts and tundra.

**Habits**
Flight swift but heavier than Peregrine's; rapid wingbeats alternate with short sails; may hover before stooping for prey.

**Voice**
Chattering and screaming.

**Eggs**
3–4; white to rusty, marked with brown; 2.3 x 1.8 in. (5.8 x 4.6 cm).

**Range**
Breeds from Greenland and n. Canada s. to cen. Canada; winters in breeding range; occasionally wandering s. to n. U.S.

## PEREGRINE FALCON
*Falco peregrinus* **16:6**

**Description**
Size, 15–21 in. (38.1–53.3 cm); wingspread, 3¾ ft. (1.4 m). Sturdy-bodied; dark above with a black cap, light below. Adult: slaty above, tail lightly banded, throat and upper breast white, underparts barred. Immature: brown above, streaked below; throat buffy.

**Similarities**
Color of Gyrfalcon is more uniform, less contrasting; Merlin is smaller, more streaked below, tail more banded, less conspicuous mustache.

**Habitat**
Cliffs, coasts, high buildings; many migrate along seashore. Rare.

**Habits**
Wingbeat rapid, flight swift, resembling that of a pigeon; captures prey by pursuit or by stooping (plunging); observation perch a crag or dead branch. In pursuit one of fastest of all birds.

**Voice**
Loud cackles, *kak, kak, kak,* and other cries.

**Eggs**
2–4; creamy, blotched with chocolate; 2.1 x 1.7 in. (5.3 x 4.3 cm). Nest is on ledge.

**Range**
Breeds in Greenland; winters from Mass. southward.

## MERLIN
*Falco columbarius* **16:2**

**Description**
Size, 10–13½ in. (25.4–34.3 cm); wingspread, 2 ft. (0.6 m). A medium-sized falcon with a heavily banded tail. Male: slaty above, buffy steaked with brown below. Female, immature: gray upperparts replaced by dark brown.

**Similarities**
Sharp-shinned has short, rounded, not long pointed, wings; see American Kestrel.

**Habitat**
Woodlands; also frequents seashore in migration and open country in winter.

**Habits**
Flies low and fast, somewhat like a pigeon, or alternates beats with glides; perching outlook is a post, knoll, or dead branch; like Kestrel, sometimes hovers and pumps tail.

**Voice**
Various cries and cackles.

**Eggs**
3–6; white to maroon, with dark blotches of brown; 1.6 x 1.2 in. (4.1 x 3.0 cm). Nest is in evergreens or on ledge.

**Former name**
Pigeon Hawk.

**Range**
Breeds from Labr. s. to N.S., and Mich.; winters from Newfoundland southward.

## AMERICAN KESTREL
*Falco sparverius*                                                16:1

**Description**
Size, 9–12 in. (22.9–30.5 cm); wingspread, 2 ft. (0.6 m). Back and tail rufous; head multicolored with black head markings; tail with black band near tip; underparts buffy. Male: wings blue-gray, tail unbarred. Female: duller; wings rusty, rufous tail barred.

**Similarities**
Merlin seems heavier and with shorter wings. Sharp-shinned has short, rounded, not long pointed, wings; very dark immature Kestrel sometimes suggests Merlin.

**Habitat**
Open country, roadsides, cities.

**Habits**
Flight usually unhurried, frequently hovers, pumps tail when perched; keeps lookout from telephone wires, poles, dead branches.

**Eggs**
3–5; white to reddish-white, spotted with brown; 1.4 x 1.2 in. (3.6 x 3.0 cm). Nest is in holes, boles, nooks.

**Former name**
Sparrow Hawk.

**Range**
Breeds from Newfoundland s. to S. America; winters from New England and Ill. southward.

# Fowl-like Birds
Order Galliformes

These are small-headed, full-bodied birds with short, rounded wings and strong legs and feet. They usually live on the ground, migrate little, and are good scratchers and runners. Their wings make a loud whirring sound in flight. Their food is largely vegetation. The young run about almost as soon as hatched; all are considered game birds.

# GROUSE, PARTRIDGES, PHEASANTS, AND TURKEYS
Family Phasianidae

Grouse are ground-dwelling birds with feathered legs that feed on seeds, buds, berries, and some insects. Ptarmigan differ from other grouse in having feathered toes as well as feathered legs, an adaptation to their often snowy environment.

Partridges and pheasants are similar in many ways to grouse, differing mainly in that their legs are not fully feathered. They nest in concealed, lined hollows on the ground, and eat insects, grains, and berries.

The Turkey is the largest upland game bird. It is the species from which all domestic turkeys are descended. Its food is acorns and other nuts, berries, plants, seeds, and insects.

## SPRUCE GROUSE
*Canachites canadensis*                                               16:8

**Description**
Size, 16 in. (40.6 cm). Male: grayish; breast, throat, and tail black; red over eye. Female and immature: brown barred with black above, lighter below.
**Similarities**
Ruffed Grouse is larger and lighter brown; spotted, not barred, above; with black band at end of longer tail.
**Habitat**
Northern coniferous forest, especially spruce.
**Habits**
Solitary, often remarkably tame, flushes to small tree and sits.
**Voice**
Cluckings; drums with wings.
**Eggs**
6–15; brown spots on buff; 1.6 x 1.2 in. (4.3 x 3.1 cm).
**Range**
Resident in cen. Canada from N.S. to Man., s. to New England, n. N.Y., and Mich.

## RUFFED GROUSE
*Bonasa umbellus*                                                    16:9

**Description**
Size, 18 in. (45.7 cm). Has a conspicuous black tail band. Sexes similar; with a slight crest; above • *red-brown* or • *gray-brown* (2 phases); tail rufous or gray, white-tipped, fan-shaped, bare red area over eye; black ruff on neck; underparts white barred with dark brown. Female: duller, lacks crest and ruff.
**Similarities**
Female Ring-necked Pheasant has longer, pointed tail; slower wingbeats; less noisy flight; and is found in more open country.
**Habitat**
Woodlands with openings, second growth.
**Habits**
Tame in wilderness, wary near man; seeks safety by lying close, flying behind tree, or running away and flying at a distance; flushes with an explosive whir.

**Voice**
Nervous *quit-quit*; various clucks and coos; male makes a characteristic drumming with its wings in the breeding season.
**Eggs**
6–15; buffy to olive, spotted with brown; 1.4 x 1.1 in. (3.8 x 2.9 cm). Nest is on the ground.
**Range**
Resident in n. Canada s. to S.C. and S.Dak.; also in Appalachians s. to Ga.

## WILLOW PTARMIGAN
*Lagopus lagopus*                                                17:2

**Description**
Size, 16 in. (40.6 cm). Tail black. Spring male: white; head, neck, breast, and front of back red-brown. Summer male: brownish, darkest above; wings white. Spring and summer female: black and buff; scaly above, barred below; usually found with mate. Much variation; individuals in molt are mottled brown and white. Winter: pure white; bill and tail black.
**Similarities**
Bill heavier and more arched than in Rock Ptarmigan; female Rock in summer almost indistinguishable; male and some female Rocks in winter have black patch between bill and eye.
**Habitat**
In summer, low tundra, upland valleys; in winter, willow bottoms.
**Habits**
Gregarious after breeding season; roosts in snow; numbers subject to periodic fluctuations.
**Voice**
Crows, cackles.
**Eggs**
5–10; blotchy-red; 1.7 x 1.2 in. (4.4 x 3.1 cm). Nest is tundra depression lined with grass and feathers.
**Remarks**
Ptarmigans differ from other grouse in having feathered toes as well as feathered legs, an adaptation to their snowy environment.
**Range**
Resident in n. Canada from Baffin I. and Greenland s. to Newfoundland, cen. Que., and Man.

## ROCK PTARMIGAN
*Lagopus mutus*                                                 17:1

**Description**
Size, 13 in. (33.0 cm). Similar to Willow Ptarmigan but somewhat smaller; female only pigeon-sized; bill smaller and all-black. Male: in summer gray-brown, not red-brown, below. Female: in fall, looks more like male Rock. Winter: all males and 60% of females have black between bill and eye; the other females, except for the bill, look like small Willow Ptarmigan.
**Similarities**
Willow is more reddish in summer than male Rock. Female Willow is more buffy than female Rock.
**Habitat**
In summer, the most barren rocky areas; in winter, more sheltered situations.
**Habits, Voice**
Similar to Willow Ptarmigan.

**Eggs**
6–15; buffy-olive, spotted with brown; 1.7 x 1.2 in. (4.4 x 3.2 cm).
Nest is on ground.
**Range**
Resident in Arctic from Greenland and Ellesmere I. s. to
Newfoundland and n. Que.

## GREATER PRAIRIE CHICKEN
*Tympanuchus cupido*                                        17:3

**Description**
Size, 18 in. (45.7 cm). Brown and buff with light scales above,
• *dark bars* below; • *tail square, black* in male, barred in female.
**Similarities**
Female pheasant has long, pointed tail; Sharp-tailed Grouse has
white on sides of pointed tail; Ruffed Grouse has fan-shaped tail.
**Habitat**
Prairie grasslands.
**Habits**
In flight, alternately flaps and sails; in courtship male raises
elongate neck feathers and tail.
**Voice**
Clucks and cackles; male in courtship utters hollow, booming notes.
**Eggs**
7–17; olive, spotted; 1.7 x 1.2 in. (4.4 x 3.2 cm). Nest is grassy
depression in prairie.
**Range**
Resident locally from Man. s. through Great Plains to Okla., and
in Ill., Mich., and Wis.; also along e. Tex. coastal prairies.

Note: The **LESSER PRAIRIE CHICKEN**, *Tympanuchus
pallidicinctus*, size, 16 in. (40.6 cm), occurs in southeastern
Colorado and southwestern Kansas. It is smaller and paler than
the Greater. Some consider them a single species.

## SHARP-TAILED GROUSE
*Tympanuchus phasianellus*                                 17:4

**Description**
Size, 15–20 in. (38.1–50.8 cm). Brown above, marked with white;
below white with dark V-shaped marks on breast; • *tail with white
sides, short and pointed*; neck sacs of male purple; in flight, white
spots on wings conspicuous; in winter, more rufous and buffier.
**Similarities**
Tail of female Ring-necked Pheasant is long and pointed; of
prairie chickens, short, dark, and rounded; of Ruffed Grouse, fan-
shaped and banded.
**Habitat**
Open woodlands and brush.
**Habits**
Flight speedy, straight; rapid wingbeats alternate with glides on
downcurved wings; in winter, perches in trees, roosts in snow; has
elaborate courtship similar to that of Greater Prairie Chicken.
**Voice**
Clucking *"whucker, whucker, whucker"* (Bent).
**Eggs**
7–13; brownish-green, spotted; 1.7 x 1.2 in. (4.4 x 3.2 cm). Nest is
grassy depression under bush or in tall grass.
**Range**
Resident from Que. and n. Man. s. to Mich. and Nebr.

## BOBWHITE
*Colinus virginianus*                                      **16:10**

**Description**
Size, 10 in. (25.4 cm). Body plump, brownish, and with much
white below; throat and line over eye white in male, buffy in
female; tail short and dark.

**Similarities**
Much smaller than Ruffed Grouse; lacks white outer tail feathers
of meadowlarks.

**Habitat**
Pastures, brush.

**Habits**
When flushed, explodes into flight, then scales quickly down into
cover; group or covey roosts on ground in circle facing out.

**Voice**
Clear whistled *bob-white* or *poor-bob-white*; also covey call, *quoi-
lee* (suggests Sora).

**Eggs**
10–20; white; 1.2 x 0.9 in. (3.0 x 2.3 cm).

**Range**
Resident from Mass. and s. Ont., s. to Fla. and Gulf Coast.

---

## GRAY PARTRIDGE
*Perdix perdix*                                           **16:11**

**Description**
Size, 13 in. (33.0 cm). Brownish-gray above, tan face, gray breast
with chestnut V, barred flanks, • *red-brown tail* (conspicuous in
flight).

**Similarities**
Bobwhite is smaller and not as gray, with whitish face marks.

**Habitat**
Fields, brushy edges.

**Habits**
Flight swift, direct, noisy, followed by scaling down to cover; covey
sleeps on ground in circle facing out.

**Voice**
Loud *kee-hah*.

**Eggs**
15–17; olive; 1.4 x 1.0 in. (3.6 x 2.5 cm).

**Range**
Locally in s. Canada and n. U.S., from N.S., N.B., and Ont., s. to
Ohio, Ind., s. Mich., Iowa, and Minn., w.; introduced from
Eurasia, but generally unsuccessful in becoming established.

---

## RING-NECKED PHEASANT
*Phasianus colchicus*                                      **17:5**

**Description**
Size, male, 30–36 in. (76.2–91.4 cm); female, 21–25 in. (53.3–63.5
cm). Usually with a white neck ring, this pheasant is easily
distinguished by brilliant plumage. Male: • *tail very long*, tan
barred with black. Female: brownish, tail similar but shorter.

**Similarities**
Ruffed Grouse has shorter, fan-shaped tail; Sharp-tailed Grouse
has white on sides of shorter tail.

**Habitat**
Brushy edges, pasture.

**Habits**
Flushes with loud whir, flies fast, scales quickly into cover; enjoys dust baths, roosts on ground or in trees, flocks in winter.
**Voice**
Harsh *c-a-a-a-a*, like antique automobile horn.
**Eggs**
6–14; greenish-brown; 1.6 x 1.3 in. (4.1 x 3.3 cm).
**Range**
Resident from Maritime Provinces to Ont., s. to N.J., Md., and Okla.; introduced from Asia and Europe.

## TURKEY
*Meleagris gallopavo* Fig. 6

**Description**
Size, male, 48 in. (121.9 cm); female, 36 in. (91.4 cm); wingspread, 5 ft. (1.5 m). Identical to domestic bird but • *tail tip brown.*
**Similarities**
None.
**Habitat**
Extensive open woodlands.
**Habits**
Good flier, but prefers to escape by running; roosts in trees.
**Voice**
Gobbles like barnyard bird.
**Eggs**
8–15; buff spotted with gray; 2.7 x 1.8 in. (6.9 x 4.6 cm).
**Range**
Resident locally from n. New England to Ill., s. to Gulf Coast.

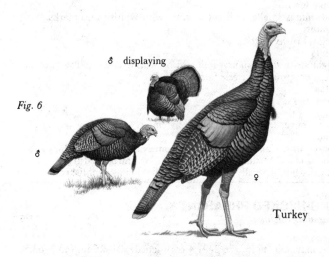

♂ displaying

*Fig. 6*

♂

♀

Turkey

# Cranes and Rails
Order Gruiformes

These are a diverse group of marsh and open-country birds that nest on the ground. The young can run about almost immediately after they leave the egg.

# CRANES
Family Gruidae

The long-necked cranes, our tallest birds, are stockier than herons but their straight bills are shorter and less pointed. The long, curved tertials create a "tufted" effect over the tail. They fly with their necks and legs stretched straight out. They are waders and walkers, and never perch in trees. Their food is reptiles, amphibians, insects, grains, aquatic plants.

## WHOOPING CRANE
*Grus americana*                                                      5:4

**Description**
Size, 50 in. (127.0 cm); wingspread, 7½ ft. (2.3 m). A tall, long-necked white bird with black primaries. Adult: head with naked red area, bill dark with yellow base, legs black. Immature: tan above, no naked red area, bill all-dark.

**Similarities**
Snow Goose has much shorter bill, neck, and legs; White Pelican is heavy, not tall and slender; egrets lack black wing tips; White Ibis is much smaller and has curved bill.

**Habitat**
Marshes; very rare.

**Habits**
Very wary; walks about marsh; in flight, wingbeats slow, fly in single file; sometimes spirals to great height and performs aerial revolutions.

**Voice**
Loud vibrating trumpeting.

**Eggs**
2; buff, blotched with brown; 3.9 x 2.5 in. (9.9 x 6.4 cm).

**Range**
Breeds in nw. Canada (Wood Buffalo Park); winters on Tex. Gulf Coast (Aransas National Wildlife Refuge).

## SANDHILL CRANE
*Grus canadensis*                                                    5:3

**Description**
Size, 34–48 in. (86.4–121.9 cm); wingspread, 7 ft. (2.1 m). Uniformly gray or brown. Adult: gray; bill and legs dark; forehead red. Immature: all-brown, no red.

**Similarities**
Great Blue Heron is varicolored and flies with neck drawn in.

**Habitat**
Marshes and open country.

**Habits**
Gregarious; walks much; in flight, neck extended, wingbeats slow and with flick on upstroke; at times spirals high in air.

**Voice**
Ringing trumpetings, *"garoo-oo-oo-oo, garoo-oo-oo"* (Laing).

**Eggs**
2; drab, sparingly spotted with brown; 3.6 x 2.3 in. (9.1 x 5.8 cm).

**Range**
Breeds in Arctic and nw. Canada, s. to Mich. and Minn.; also along Gulf Coast from Fla. to Tex.; winters s. of U.S. and in Tex.

# LIMPKINS
Family Aramidae

## LIMPKIN
*Aramus guarauna*                                                 **5:7**

**Description**
Size, 24–28 in. (61.0–71.1 cm). A long-legged brown marsh bird
with • *white spots on head, neck, and back,* and a long, slightly
decurved bill; often flies with dangling legs.

**Similarities**
Immature night herons have straight bills, shorter, heavier necks.

**Habitat**
Marshes.

**Habits**
Feeds mainly at night.

**Voice**
Wailing *rr-ow*, piercing the night air.

**Food**
Snails.

**Eggs**
5–6; buff; 2.2 x 1.7 in. (5.7 x 4.4 cm). Nest of marsh plants is
fastened to vegetation or platform in bush or tree.

**Range**
Resident locally in S., from s. Ga. to Fla., from Okefenokee
Swamp to Everglades, and s. to S. America.

# RAILS, GALLINULES, AND COOTS
Family Rallidae

Members of this family are somewhat chickenlike birds of wet
areas. They seldom fly far, except on migration. They are often
heard, especially at dawn, but less often seen. Their nest is a
platform in the marsh. The young chicks are downy black. Rails
and crakes are thin skulkers amid marsh reeds and grasses. The
more ducklike gallinules and coots often swim about in open fresh
water. Rails (Genus *Rallus*) have slightly curved bills and
conspicuously barred flanks; crakes are small and short-billed with
barred flanks. Rails prefer running to swimming, and prefer
swimming to flying; if flushed, they usually flutter only a few
yards with feet dangling.

## KING RAIL
*Rallus elegans*                                                 **18:5**

**Description**
Size, 17 in. (43.2 cm). A red-brown rail of the freshwater marsh,
this is the largest. Above brown streaked with black, cheeks and
underparts cinnamon.

**Similarities**
Clapper replaces cinnamon with buff and gray. Virginia Rail is
similar to King but much smaller, and with gray cheeks.

**Habitat**
Freshwater marshes, sometimes salt marshes in winter.

**Habits**
Skulks.

**Voice**
Grunting *umph-umph-umph-umph* on same pitch (call of Virginia
Rail descends scale).

**Food**
Insects, aquatic animal life, seeds.
**Eggs**
6–12; buffy, blotched with red-brown; 1.6 x 1.1 in. (4.1 x 2.8 cm).
**Range**
Breeds from Mass. to Minn., s. to Fla. and Tex.; winters from Gulf Coast southward.

## CLAPPER RAIL
*Rallus longirostris*                                                    18:4

**Description**
Size, 15 in. (38.1 cm). A large, gray-brown rail of salt marshes. Some buff below, throat and undertail coverts white.
**Similarities**
Freshwater King Rail (by some recognized as part of same species) has cinnamon instead of gray, blacker streaks above, and rusty on wings; the browner young Clapper resembles young King.
**Habitat**
Salt marshes.
**Habits**
Swims well and dives; often seen swimming and walking about salt marsh during high or flood tides; head bobs in swimming; tail often raised to reveal white undertail coverts.
**Voice**
Frequently heard and oft-repeated staccato *kek-kek-kek-kek.*
**Food**
Crustaceans, mollusks, worms.
**Eggs**
5–14; buffy, marked with red-brown; 1.7 x 1.2 in. (4.3 x 3.1 cm).
**Range**
Breeds along Atlantic Coast, from Mass. to S. America.; also along Gulf Coast to Tex.; winters from N.J., southward.

## VIRGINIA RAIL
*Rallus limicola*                                                        18:6

**Description**
Size, 10 in. (25.4 cm). A small, red-brown rail. Above olive streaked with dusky; eye red, cheeks gray; below; red-brown, forewing reddish (conspicuous in flight); immatures have much black but with long bill.
**Similarities**
King Rail is considerably larger and has red-brown cheeks. Sora has short, yellow bill.
**Habitat**
Freshwater marshes, rarely brackish.
**Habits**
Can climb reeds and vines.
**Voice**
Harsh *kid-ik, kid-ik, kid-ik*; other "kicking," squealing, clucking, and vaguely chickenlike noises.
**Food**
Aquatic animal life, seeds, berries, insects.
**Eggs**
5–12; similar to King's; 1.3 x 0.9 in. (3.3 x 2.3 cm).
**Range**
Breeds s. Canada, from Newfoundland to Minn. s. to C. America; winters from Va. southward, occasionally farther north.

## SORA
*Porzana carolina*                                                    **18:7**

**Description**
Size, 9 in. (22.9 cm). Adult: chunky shape, dark brownish above, black face and throat, gray below, • *short yellow bill*. Immature: buffy below, bill duller.
**Similarities**
Buffy Yellow Rail is smaller, rarer, and has white wing patches.
**Habitat**
Freshwater marshes; salt marshes in migration.
**Habits**
Swims, dives well; migrates at night, often striking obstructions.
**Voice**
High-pitched, horselike whinny; a whistled *cur-wee* (like covey call of Bobwhite); a single *keek*; other rail-like noises.
**Food**
Insects; wild rice and other seeds.
**Eggs**
6–15; buffy, marked with dull brown; 1.2 x 0.9 in. (3.0 x 2.2 cm).
**Range**
Breeds in cen. Canada, from Newfoundland to Ont., s. to Pa. and Okla.; winters from S.C., southward.

## YELLOW RAIL
*Coturnicops noveboracensis*                                         **18:9**

**Description**
Size, 7 in. (17.8 cm). Buffy yellow; back streaking has "checkerboard" effect; • *white wing patch*; bill very short, yellow.
**Similarities**
Immature Sora is larger, with no white in wing.
**Habitat**
Wet meadows, higher parts of fresh and salt marshes; extremely rare.
**Habits**
Most secretive, almost never flies except at night on migration; can conceal itself in extremely short, sparse grass.
**Voice**
High-pitched clicking notes in series. *"kik-kik-kik-kik-queeah"* (Ames).
**Food**
Little known; includes snails.
**Eggs**
7–10; buff with small, red-brown dots; 1.1 x 0.8 in. (2.8 x 2.0 cm).
**Range**
Breeds in cen. Canada and n. U.S., from N.B. and Que. s. to Maine and N.Dak.; winters from S.C., Fla., and Gulf Coast southward.

## BLACK RAIL
*Laterallus jamaicensis*                                             **18:8**

**Description**
Size, 6 in. (15.2 cm). A tiny black rail, the size of a sparrow, with a maroon-brown nape. Above black with white dots; below dark gray; • *bill black*.
**Similarities**
Downy young chicks of all rails are black, but they lack the Black's nape patch and white dots; they cannot fly.

**Habitat**
Fine grass salt marshes; grassy edges of freshwater marshes.
**Habits**
Very hard to flush; runs like a mouse, with head down and neck out.
**Voice**
Female: *"croo-croo-croo-o* . . . like the commencement of the song of the Yellow-billed cuckoo" (Wayne). Male: *"kik, kik, kik, kik* or even *kuk, kuk, kuk, kuk"* (Wayne). *"Did-ee-dunk, did-ee-dunk, did-ee-dunk"* (McMullen).
**Food**
Isopods (small crustaceans).
**Eggs**
4–9; white, marked with red-brown; 1.0 x 0.8 in. (2.5 x 2.0 cm).
**Range**
Breeds along Atlantic and Gulf coasts from N.Y. to S. America and locally inland; winters from Gulf Coast southward.

## PURPLE GALLINULE
*Porphyrula martinica*                                            18:1
**Description**
Size, 13 in. (33.0 cm). • *Blue forehead shield* and a purple breast.
Adult: back and wings glossy green; head, neck, and underparts glowing purple and blue (looks black in poor light); undertail coverts white; bill red with a yellow tip; • *legs bright yellow.*
Immature: above dull brown with some greenish reflections; below, dingy buff and white; bill dull.
**Similarities**
Common Gallinule has a red frontal shield and white stripe along sides; immature much grayer below.
**Habitat**
Freshwater marshes, especially those with small ponds.
**Habits**
Runs or walks on lily pads with dovelike nods, flirting white undertail; clambers about marsh reeds, climbs vines; perches in shrubs and trees; bobs head as it swims; but less aquatic than Common; flight weak and with legs dangling.
**Voice**
Loud, chickenlike; in flight, *kek, kek, kek, kek.*
**Food**
Frogs, shellfish, aquatic insects, seeds.
**Eggs**
6–10; white, lightly spotted with brown; 1.6 x 1.1 in. (4.1 x 2.8 cm).
**Range**
Breeds from S.C. and Tenn. s. to Fla. and Tex.; winters from Fla. and Tex. s. to S. America

## COMMON GALLINULE
*Gallinula chloropus*                                            18:3
**Description**
Size, 13 in. (33.0 cm). Head, neck, and underparts gray; back olive; divided undertail coverts white. • *White stripe along flanks;*
• *red bill* and frontal shield with yellow tip; green legs. Immature: paler below; bill dusky.

**Similarities**
American Coot is plumper, shorter-necked, uniform slate gray,
larger headed with a white bill and shield. Immature Purple is
buffier-white below.
**Habitat**
Freshwater marshes, occasionally salt.
**Habits**
Runs and walks over lily pads jerking tail; tips up like a pond
duck to feed in shallow water; also feeds on land; short flights
weak with legs dangling; long flights with neck and legs extended;
swims stern high, pumping head (but Coot swims more); can dive.
**Voice**
Loud, chickenlike, extremely varied; *cac, cac, cac,* etc.
**Food**
Vegetation, snails, insects.
**Eggs**
6–12; buffy, marked with brown; 1.7 x 1.2 in. (4.3 x 3.0 cm).
**Other name**
Moorhen.
**Range**
Breeds throughout U.S., from s. Canada to Fla. and Gulf Coast;
winters from Ga. and Gulf Coast s. to S. America.

---

# AMERICAN COOT
*Fulica americana*                                              **18:2**

**Description**
Size, 15 in. (38.1 cm). A • *slaty-black* ducklike water bird with a
conspicuous, short • *white bill.* Adult: head black, bill short,
undertail coverts white, legs greenish, toes with gray lobes.
Immature: paler below, and with a duller bill.
**Habitat**
Freshwater marshes, ponds, rivers; at times, salt bays.
**Habits**
Gregarious; swims, dives, and tips up; floats high on water with
back level (not stern high, as do gallinules); pumps head when
swimming; patters along surface before taking to air. In flight,
shows white on trailing edge of wing; neck and legs are extended;
feet protrude behind tail. Where protected, becomes quite tame,
often approaching people to be fed.
**Voice**
"Coughing sounds, froglike plunks, and a rough sawing or filing
*kuk-kawk-kuk, kuk-kawk-kuk,* as if tree saw were dull and stuck
. . . [and] a grating *kuk kuk kuk kuk kuk*" (Bailey); and many other
noises.
**Food**
Aquatic plants, grass, grain.
**Eggs**
8–12; whitish, with pinhead-sized dark brown dots; 1.9 x 1.3 in.
(4.8 x 3.3 cm).
**Range**
Breeds from s. Canada s. to Gulf Coast and Fla.; winters s. from
Md. and Ohio Valley to n. S. America.

# Shorebirds, Gulls, Auks, and Allies
Order Charadriiformes

## OYSTERCATCHERS
Family Haematopodidae

### COMMON OYSTERCATCHER
*Haematopus palliatus*

19:16
Fig. 7

**Description**
Size, 19 in. (48.3 cm). A shorebird with a long, • *heavy red bill.*
Head and neck black; back, wings, and tail dark brown;
underparts, rump, and wing stripe white; eye red, feet pink; bill
brownish in young; conspicuous bold black-and-white appearance
in flight.
**Habitat**
Sand beaches, rocky coasts.
**Habits**
Wary; walks sedately, wades to belly; can swim and dive, prods in
wet sand with bill; flight strong, swift.
**Voice**
Penetrating *wheep, wheep.*
**Food**
Shellfish, crustaceans, marine worms.
**Eggs**
2–3; buff, blotched with dark brown; 2.2 x 1.5 in. (5.6 x 3.8 cm).
Nest is in slight hollow in beach.
**Range**
Breeds along Atlantic and Gulf coasts, from Mass., s. to Fla. and
S. America; winters from N.C., southward.

Marbled Godwit, p. 87

Whimbrel, p. 76

*Fig.7*

fall

fall

Willet, p. 80

Hudsonian Godwit, p. 87

Common Oystercatcher

**Large Shorebirds in Flight**

# PLOVERS
Family Charadriidae

These small to medium-sized plump, round-headed, short-necked shorebirds are runners rather than waders. They have pigeonlike bills that are shorter than their heads, and shorter and thicker than those of sandpipers. They have a stop-and-go manner of feeding, rather than the continuous probing of sandpipers. When standing still, many bob the body forward and back. The bulk of their food is shellfish, other marine invertebrates, and worms. Plovers all nest in a depression in the sand or on the ground; the eggs are pear-shaped. The young can run about almost as soon as they are hatched. The parent simulates a broken wing to lure intruders away from the nest or young. The Ringed Plovers (Genus *Charadrius*) are usually small, dark brown or whitish above and white below, with one or two black breastbands and some white on the tail; females and young are paler.

## SEMIPALMATED PLOVER
*Charadrius semipalmatus*                                    19:2, 23:1

**Description**
Size, 7 in. (17.8 cm). Spring: above color of wet sand; • *black band across upper breast* and black line through eye to bill; white forehead, throat, underparts, and sides of tail; yellow-orange bill with black tip, yellow-orange legs. Fall and immature: black areas browner, more white on forehead, legs paler, bill more dusky.
**Similarities**
Piping Plover is lighter, with fewer face markings. Killdeer has 2 breastbands.
**Habitat**
Mud flats, beaches, shorelines, wet fields.
**Habits**
Often flies in compact flocks wheeling in unison; on alighting, birds spread out to feed, stoop intermittently to pick up food.
**Voice**
Plaintive rising *cheer-wee*.
**Eggs**
3–4; buff, marked with blackish-brown; 1.3 x 0.9 in. (3.3 x 2.3 cm).
**Range**
Breeds in Arctic, s. to N.S., Que., and n. Man.; winters from Gulf Coast southward.

## PIPING PLOVER
*Charadrius melodus*                                         19:1, 23:2

**Description**
Size, 7 in. (17.8 cm). The only pale-backed plover on East Coast and Great Lakes. Above color of dry sand. Spring: tip of bill, line over forehead, and end of tail black; single black band across upper breast, sometimes broken; base of bill and • *legs yellow-orange*; underparts, rump, and wing stripe white. Fall and immature: black on head and breast lost or replaced by brown, bill darker.
**Habitat**
Sandy beaches, mud flats, and fill.
**Habits**
Similar to Semipalmated but prefers drier sand of higher beach.
**Voice**
Soft, whistled descending *peep-lo*.
**Eggs**
3–4; creamy, finely dotted with chocolate; 1.2 x 1.0 in. (3.0 x 2.5 cm). Nests amid dunes.

### Range
Breeds along Atlantic Coast from Newfoundland, s. to Va. and along shores of Great Lakes; winters along Atlantic and Gulf coasts, from S.C., southward.

Note: The **SNOWY PLOVER,** *Charadrius alexandrinus* **(19:4)**, size, 6 in. (15.2 cm), is a Piping Plover-like bird with a paler back, broken breastband, thinner black bill and slaty legs; it occurs in eastern Colorado and southwest Kansas, and along entire Gulf Coast.

## WILSON'S PLOVER
*Charadrius wilsonia*                                                **19:5**

### Description
Size, 8 in. (20.3 cm). • *Bill thick and long, black* at all seasons. Spring male: above color of wet sand (head and nape of male sometimes rusty); very broad black breastband; legs pinkish. Female and fall plumages: black areas are brown.

### Similarities
Semipalmated Plover is smaller, slightly darker, with a much smaller bill, narrower breastband, and yellow legs.

### Habitat
Sandy beaches, mud flats, shores of salt marshes.

### Habits
Movements quite deliberate; not as shy or nimble as Piping Plover; often active after dark.

### Voice
Loud whistled *wheat;* on wing, a *tut-tut* (like alarm note of Wood Thrush, but lower); near nest, a high-pitched *"queet, queet, quit it, quit it"* (Pennock).

### Eggs
3; olive or drab, marked with dark brown; 1.3 x 1.0 in. (3.3 x 2.5 cm).

### Range
Breeds along Atlantic and Gulf coasts from s. N.J. to Fla. and w. to Tex.; winters along Gulf Coast s. to n. S. America.

## KILLDEER
*Charadrius vociferus*                                         **19:3, 23:3**

### Description
Size, 10 in. (25.4 cm). Distinguished by • *2 black breastbands;* earth-brown above, with white forehead, throat, collar (continuing around neck), underparts, wing stripes, and edges of tail; • *rump and upper part of rather long tail orange-brown;* black bill; pinkish legs; young, paler, and with grayish breastband.

### Similarities
Other ringed plovers are smaller with only 1 band.

### Habitat
Mud flats, fields, parks, open areas, and fill; often near water.

### Habits
Swift runner and flier, often active after dark and on moonlit nights; has spectacular nuptial flight; if nest is approached, circles in air screaming above intruder, or gives "broken-wing" display on ground.

### Voice
Loud *kill-dee, kill-dee;* a noisy and persistent trilling.

### Eggs
4; buff, blotched with brown; 1.4 x 1.1 in. (3.6 x 2.8 cm).

### Range
Breeds throughout most of U.S., from Newfoundland s. to West Indies; winters from N.J. and Ohio, southward.

## AMERICAN GOLDEN PLOVER
*Pluvialis dominica*                    **19:7, 21:11, 23:6**

**Description**
Size, 9½–11 in. (24.1–27.9 cm). Spring: speckled gold above, all-black below; forehead and sides of neck white; underparts black to tail. Fall: brownish or yellowish-brown, darker above, tail always dark, bill black, legs slate, no hind toe. In flight, wings gray below.

**Similarities**
Black-bellied is slightly larger, less slender, and has shorter wings; underparts black only to thigh. In winter, Black-bellied is grayer, has thicker bill and neck; in flight, has white wing stripe and rump, some contrast between gray belly and white undertail coverts, and black wing pits.

**Habitat**
Marshes, fields, mud flats, beaches, open areas.

**Habits**
Gregarious, flight swifter, more buoyant than Black-bellied; on ground, more aggressive; raises wings on alighting, often bobs head.

**Voice**
Quavering *quee-i-i-a* or a harsh *queedle*.

**Eggs**
4; buffy-olive, marked with brown and black; 2.0 x 1.3 in. (5.1 x 3.3 cm).

**Range**
Breeds in Arctic and nw. Canada; winters in S. America. Fall migration is from Ne. coast across Atlantic; spring migration is through cen. states and Canadian provinces to tundra. The Golden Plover's migration covers 2400 miles from N.S. to S. America.

## BLACK-BELLIED PLOVER
*Pluvialis squatarola*                    **19:6, 21:12, 23:8**

**Description**
Size, 11 in. (27.9 cm). Stout head and bill; hind bill and legs black; toe present; wing stripe, • *black wing pits*, rump, black-barred tail, and undertail coverts white. Spring: pale speckled gray above, black (to thighs) below; forehead and sides of neck white. Fall: grayish, lighter below.

**Similarities**
White area of Golden is smaller; for other comparisons, see American Golden Plover.

**Habitat**
Beaches, mud flats, salt-marsh meadows.

**Habits**
Sedate, somewhat stolid shoreline figure of erect carriage; white forehead is conspicuous in birds on ground amid grass.

**Voice**
Wild, plaintive, somewhat Bluebird-like *toor-a-wee*.

**Eggs**
4; buffy-olive, marked with brown and black; 2.1 x 1.4 in. (5.3 x 3.6 cm).

**Range**
Breeds in Arctic and s. Baffin Is.; winters along coast from Va. to S. America.

# WOODCOCK, SNIPE, SANDPIPERS, AND ALLIES
Family Scolopacidae

These shorebirds have thin, relatively long bills; long, pointed wings (except Woodcock); rather long legs; and short tails. Most are waders and prefer moist areas or shallow water and shorelines; most can swim and dive. Almost all have a distinctive courtship song, occasionally heard in migration. With one exception (Solitary Sandpiper), all nest in a depression upon the ground, often lined with grasses, and usually near water. Parents may simulate a broken wing to lure intruders from the nest. The eggs are usually four and pear-shaped; the young can run about almost immediately after leaving the egg. Their diet consists primarily of shellfish, crustaceans, insects, and seeds.

## AMERICAN WOODCOCK
*Scolopax minor*                                               **20:17, 23:9**

**Description**
Size, 11 in. (27.9 cm). Chunky; • *long bill*, short neck, big eyes; upperparts mottled like dead leaves, underparts orange-tan; crown black with narrow, tan crossbands; legs pinkish, • *wings rounded*; no seasonal change.
**Similarities**
Common Snipe is thinner, darker above, rises in a zigzag on thin, pointed, noiseless wings.
**Habitat**
Swamps, wet woods, and thickets.
**Habits**
Solitary, secretive, largely nocturnal; probes deep into mud when feeding, leaving visible borings; rises straight, with round body, long bill pointing down, and on short whistling wings.
**Voice**
*Peent* somewhat like Common Nighthawk.
**Eggs**
3–4; buff, marked with brown; 1.5 x 1.1 in. (3.8 x 2.8 cm).
**Range**
Breeds in e. N. America from s. Canada to Gulf Coast; winters from Va. and Mo. southward.

## COMMON SNIPE
*Gallinago gallinago*                                          **20:14, 23:7**

**Description**
Size, 11 in. (27.9 cm). Wings rather pointed; • *bill long*; head striped; streaked brown above; breast spotted, belly white; • *tail orange*, tail corners pale; legs greenish; no seasonal change.
**Similarities**
Dowitchers have conspicuous white stripe on lower back and frequent seashore.
**Habitat**
Meadows and marshes, fresh or salt.
**Habits**
Solitary, secretive; hides by squatting; most active at dawn, dusk, and on cloudy days; usually rises in zigzags, uttering nasal escape note; can swim and dive.
**Voice**
A "rasping *escape, escape*; on breeding grounds, a melodious *wheat wheat*" (Collins).

**Eggs**
3–4; dull olive, boldly blotched with brown; 1.5 x 1.1 in. (3.8 x 2.8 cm).
**Range**
Breeds from Labr. s. to Mass., Ind. and some farther s.; winters from Va. southward.

## RUDDY TURNSTONE
*Arenaria interpres*                    19:8, 21:5, 23:10

**Description**
Size, 9 in. (22.9 cm). Ruddy upperparts of spring replaced by brown in fall; in flight at all seasons, calico pattern with 5 prominent white stripes on back and wings; black bill may be slightly upturned; legs short, orange.
**Habitat**
Primarily coastal; beaches, shorelines, mud flats, rocks, jetties.
**Habits**
Pokes bill under pebbles, shells; digs shallow holes in sand; pugnacious; sometimes perches off ground; can swim; large migrating flocks separate into small groups on ground.
**Voice**
Harsh *chut-chut*; a melodious *quit-tock*.
**Eggs**
4; cream, splashed with brown; 1.5 x 1.1 in. (3.8 x 2.8 cm).
**Range**
Breeds in Arctic; winters from S.C. s. to S. America; some wander farther n.

## WHIMBREL                              19:13
*Numenius phaeopus*                         Fig. 7

**Description**
Size, 15–18 in. (38.1–45.7 cm); bill long and decurved, 2¾–4 in. (7.0–10.2 cm). A common curlew, the only one with • *prominent stripes on the head*. Grayish brown above, barred pinkish-buff underwings, grayish-white below, blue-gray legs.
**Similarities**
Much smaller, grayish not buffy, and bill shorter than Long-billed.
**Habitat**
Salt marshes, tidal flats, shorelines, river bars, tundra.
**Habits**
Flight steady; high over land in flocks like ducks, or in long lines low over water; often scales on set wings.
**Voice**
Soft musical *cur-lew*; a series of harsh *ku-ku-ku* notes on same pitch.
**Eggs**
4; buff, marked with brown; 2.4 x 1.6 in. (6.1 x 4.1 cm).
**Former name**
Hudsonian Curlew.
**Range**
Breeds in Arctic and nw. Canada; winters from Gulf Coast s. to S. America.

## LONG-BILLED CURLEW
*Numenius americanus*                                       **19:14**

**Description**
Size, 20–26 in. (50.8–66.0 cm); bill, 5–7 in. (12.7–17.8 cm). An impressive shorebird distinguished by its very long, downcurved bill. Buffy; head unstriped; underwings and • *wing pits bright cinnamon*.

**Similarities**
Whimbrel has center crown stripe, no cinnamon underwings.

**Habitat**
Plains, prairies, open areas near water; lagoons and mud flats along coast in winter.

**Habits**
Wary, flies in V-shaped flocks; can swim.

**Voice**
Loud whistled *curlew-curlew*; a rapid *"wheety, wheety, wheety"* and a loud, rattling *"que-he-he-he-he-he"* (Bent).

**Eggs**
4; buff, marked with brown and lavender; 2.6 x 1.9 in. (6.6 x 4.8 cm).

**Range**
Breeds in w. N. America; winters from La. to C. America and from S.C. to Fla.

## UPLAND SANDPIPER
*Bartramia longicauda*                                  **19:9, 23:5**

**Description**
Size, 12 in. (30.5 cm). A medium-sized upland species with a straight bill and white outer tail feathers. Streaked brownish above, lighter below; small head, slender bill, rather long neck and tail; line over eye and outer tips of tail white; underwing black and white; dark rump, yellowish legs.

**Similarities**
Buff-breasted Sandpiper is smaller and with an unmarked breast.

**Habitat**
Grassy inland fields, prairies.

**Habits**
Flies brief distances with short strokes of downcurved wings like Spotted Sandpiper; extended flight is swift, buoyant; often perches on fence posts.

**Voice**
Song is a mournful, mellow whistle, *"wh-e-e-e-e-e-e-e-e-o-o-o-o-o-o-o-o-o-o . . . . [alarm note] quitty-quit-it-it"* (Knight).

**Eggs**
4; buff, finely spotted with brown; 1.8 x 1.3 in. (4.6 x 3.3 cm).

**Former name**
Upland Plover.

**Range**
Breeds in cen. Canada, s. to cen. U.S.; winters in S. America.

## SPOTTED SANDPIPER
*Actitis macularia*                                     **21:18, 22:6**

**Description**
Size, 8 in. (20.3 cm). Olive-brown above; line over eye, wing stripe, and edges and tip of tail white. Spring: distinctly • *spotted with dark below*. Fall: spots lacking; dusky sides of neck separated from wing by white mark.

**Similarities**
Fall Solitary is darker above, has streaked breast, lacks white wing stripe; bobs, does not teeter.
**Habitat**
Shorelines.
**Habits**
Usually solitary, constantly teeters tail up and down as it walks; flight highly characteristic, somewhat like a meadowlark, short rapid beats of downcurved wings alternate with glides low over water; also has a seldom-seen full, free flight like a yellowlegs; can swim and dive from water or wing; perches on posts, wires, and branches.
**Voice**
Low, distinct *peet-weet*.
**Eggs**
4; cream, spotted with chocolate and gray; 1.3 x 0.9 in. (3.3 x 2.3 cm). Nest is grassy scrape near water or in brush.
**Range**
Breeds from n. Canada to s. U.S.; winters from S.C. s. to S. America.

## SOLITARY SANDPIPER
*Tringa solitaria*       21:16, 22:9

**Description**
Size, 7½–9 in. (19.1–22.9 cm). No seasonal change. Upperparts and underwing dark brown; eye-ring white; • *tail appears white* with dark center; breast streaked, underparts white; bill dark, slender; legs olive, no wing stripe. Young more white-dotted above.
**Similarities**
Spotted is paler above, has black spots below in spring, and white wing stripe all seasons. Yellowlegs are larger, lighter, taller, and with white rump.
**Habitat**
Freshwater edges, streams, pools in marshes and woods, ditches.
**Habits**
Rather solitary; flight light, airy, often zigzagging; wings have good upstroke (unlike Spotted); engages in extensive aerial revolutions; however, short-distance flight is jerky with wings only partially spread; drops abruptly to a landing, raises wings on alighting; bobs head, as if hiccuping (does not teeter); can swim, dive.
**Voice**
One or more *peet* notes, higher-pitched than Spotted's.
**Eggs**
4–5; pale greenish, with red-brown spots and blotches; 1.4 x 1.0 in. (3.6 x 2.5 cm). Nests in the deserted tree nests of other birds.
**Range**
Breeds in n. forest fringe at edge of tundra in Canada; winters from cen. U.S. and Gulf Coast s. to S. America.

## GREATER YELLOWLEGS
*Tringa melanoleuca*       21:21

**Description**
Size, 12½–15 in. (31.8–38.1 cm). • *Bright yellow legs* and • *white rump* distinguish the 2 yellowlegs. Slender, above grayish, no wing stripe; • *tail whitish;* below, white with streaks on breast; long legs. No seasonal change.
**Similarities**
Lesser is smaller, with smaller, straighter bill and different call.

**Habitat**
Marshes, mud flats, rain pools, streams, muskeg.
**Habits**
Not so gregarious as Lesser; bobs up and down, wades deeply, occasionally swims; very noisy; its loud cries warn of intruders; responds to imitations of its note.
**Voice**
Clear whistled *yew,* usually repeated 3 or 4 times; "a rolling *toowhee, toowhee*" (Nichols).
**Eggs**
4; grayish, splashed with brown and lilac; 1.8 x 1.3 in. (4.6 x 3.3 cm). Nest is a muskeg hollow.
**Range**
Breeds in s. and cen. Canada from Newfoundland to James Bay; winters from S.C. and cen. U.S. s. to S. America.

## LESSER YELLOWLEGS
*Tringa flavipes*                                        **21:20, 22:15**

**Description**
Size, 11 in. (27.9 cm). • *Bright yellow legs,* a white rump, a • *straight, thin bill* and its call identify this smaller yellowlegs. Similar to Greater, but smaller, and with a shorter, more slender bill. Other differences slight; size alone deceptive, but direct comparison in field often possible.
**Similarities**
Stilt Sandpiper in fall is smaller, with greenish legs and bill slightly drooped at tip. Fall Wilson's Phalarope is smaller and has more needlelike bill, lacks yellow legs, and usually swims in groups.
**Habitat**
Same as Greater.
**Habits**
More gregarious than Greater, less suspicious; flight "more buoyant and hence not so suggestive of momentum" (Brewster); goes north later than Greater and south earlier.
**Voice**
Usually only 1 or 2 *yew* notes, similar to Greater, somewhat flatter and less forceful; rolling *toowhee* and several other notes similar to those of larger bird.
**Eggs**
4; 1.7 x 1.1 in. (4.3 x 2.8 cm); in other respects similar to Greater. Nest is ground hollow in open, often far from water.
**Range**
Breeds from n. Man. to cen. Que.; winters from S.C. and cen. U.S. s. to Gulf Coast and S. America.

Note: Fall-plumaged **RUFFS (20:11)** (males) and **REEVES** (females), *Philomachus pugnax*, size, 10–12½ in. (25.4–31.8 cm), occasionally occur in company with Lesser Yellowlegs, which they somewhat resemble. Fall: distinguished by generally browner plumage; narrow wing stripe; shorter, stouter bill; shorter legs, often yellow; and 2 long, oval white patches on sides of dark rump and tail. Very rare in spring when male has large, distinctive ruff varying greatly in color and neck and ear tufts. Voice: *To-wit.* This is an Old World species, in which the sexes bear different names (smaller female called Reeve).

## WILLET

*Catoptrophorus semipalmatus*

19:12, 21:17
*Fig. 7*

**Description**
Size, 14–17 in. (35.6–43.2 cm). A plain-colored shorebird with a broad • *white stripe* on its • *black wings;* one of the larger species. On ground, rather nondescript; grayish above, white below with numerous black markings in spring, unmarked in fall; wing pattern occasionally visible; bill thick, black, and with a bluish base; legs bluish; white rump and pale tail conspicuous in flight.

**Similarities**
Greater Yellowlegs is less robust, has plain wings, and bright yellow legs.

**Habitat**
Marshes, beaches, mud flats, sloughs.

**Habits**
Bobs less than yellowlegs; flight is strong, direct, ducklike; wingbeat is flat; occasionally scales, may hold wings aloft for several seconds after alighting; often perches on trees, fences, or posts; wades up to belly, swims well.

**Voice**
Whistling *pill-will-willet;* "a loud vehement *wek, wek, wek* or *kerwek, kerwek, kerwek*" (Bent).

**Eggs**
Variable; often grayish-olive spotted with brown; 1.9 x 1.5 in. (4.8 x 3.8 cm). Nest is grassy cup in green.

**Range**
Breeds locally in s. Canada, N.S., and N.Y., s. along coast to Fla. and w. along Gulf Coast; winters from Va. and Gulf Coast s. to S. America.

## PURPLE SANDPIPER

*Calidris maritima*

20:10, 21:7, 22:4

**Description**
Size, 9 in. (22.9 cm). Plump. Fall and winter; • *slaty above;* gray breast, white underparts and wing stripe, bill orange with a black tip, • *legs yellow.* Spring: browner and with a more heavily streaked breast.

**Habitat**
Wave-washed rocks and jetties; with Sanderling and Dunlin the only regularly wintering sandpipers from New York north.

**Habits**
Usually seen in groups; stands low; can swim.

**Voice**
Like *kip* of Sanderling; also a double-noted *twit-twit.*

**Eggs**
3-4; olive-drab, boldly marked with brown; 1.5 x 1.0 in. (3.8 x 2.5 cm).

**Range**
Breeds in Arctic; winters along Atlantic Coast from Maritime Provinces s. to N.J.

## RED KNOT

*Calidris canutus*

20:12, 21:10, 22:11

**Description**
Size, 11 in. (27.9 cm). Legs and bill short. Spring: reddish-brown and gray with heavy black streaks above; breast is robin-red. Fall: gray above, white below, giving a washed-out appearance; wing stripe faint. Rump and tail whitish; legs greenish.

**Similarities**
Dowitcher has a long bill; Sanderling and Dunlin are smaller, less chunky, and dark rumped.
**Habitat**
Shorelines, flats, beaches, salt marshes, tundra.
**Habits**
Gregarious on ground, sluggish; in flight, occurs in large, tight, twisting flocks.
**Voice**
Rather quiet, a low harsh chut; a soft *wah-quoit;* "a soft *whit whit,* like a man whistling for a dog" (Hoffmann).
**Eggs**
3–4; light-greenish, marked with brown; 1.8 x 1.2 in. (4.6 x 3.0 cm).
**Range**
Breeds in Arctic; winters from Mass. and Gulf Coast s. to S. America.

## PECTORAL SANDPIPER
*Calidris melanotos*                         20:8, 22:3

**Description**
Size, 8–9½ in. (20.3–24.1 cm). Noted for its • *streaked, buffy breast* abruptly divided from a white belly; male larger than female. Above brownish striped with black; rump pied; legs greenish; wing stripe very thin and invisible at any distance; greenish bill sometimes droops at tip.
**Similarities**
All peeps are similar; all but Baird's have white wing stripe. Baird's has a scaly back pattern.
**Habitat**
Grassy edges of mud flats and marshes, both fresh and salt.
**Habits**
Lies close; if flushed, jumps suddenly with a harsh *kriek* and hurriedly zigzags away; then may circle high before pitching abruptly down; suggests Common Snipe in flight behavior; occurs in scattered flocks; can swim.
**Voice**
Reedy *kriek-kriek.*
**Eggs**
3–4; pale olive, spotted with umber; 1.5 x 1.0 in. (3.8 x 2.5 cm). Nest is grassy tundra hollow.
**Range**
Breeds in Arctic, s. to Hudson Bay; winters in S. America; migrates coastally and in interior.

## WHITE-RUMPED SANDPIPER
*Calidris fuscicollis*                     20:4, 21:6, 22:1

**Description**
Size, 7½ in. (19.1 cm). Above rusty in spring, gray in fall; • *breast streaked* and dark in front of wing; wing stripe white; prominent white line over eye in fall; bill straight, heavy for a peep, occasionally slightly drooped at tip; legs dark greenish; • *rump white.*
**Similarities**
Semipalmated has white in front of wing; when feeding in mixed flocks, it can be distinguished as larger than Least or Semipalmated. This species and next 6 are termed "peep."
**Habitat**
Mud flats, rocky beaches, shorelines, salt marshes; fresh marshes inland.

**Habits**
Tame; actions deliberate; feeds in water up to its belly.
**Voice**
"A squeaky mouselike *jeet*" (Nichols); like 2 marbles being knocked together, uttered on wing.
**Eggs**
3-4; olive, marked with dark brown; 1.3 x 0.9 in. (3.3 x 2.3 cm).
**Range**
Breeds in Arctic s. to n. Hudson Bay; winters in S. America. Spring migration mainly through interior; fall migration more commonly along Atlantic Coast.

## BAIRD'S SANDPIPER
*Calidris bairdii*                                                  **20:5**

**Description**
Size, 7½ in. (19.1 cm). A buffy-brown, long-winged peep with a scaly pattern on its back; no seasonal change. • *Head and breast buffy;* dark in front of wing, white stripe weak or missing; white throat; black legs and bill.
**Similarities**
Least and Semipalmated are smaller, slightly darker above; Semipalmated shows white in front of wing, and its breast is paler. Pectoral is browner with bill shorter and more slender; Pectoral and White-rumped are more streaked on crown and back. Sanderling has a prominent wing stripe. Western has a longer bill. Dunlin has a curved bill. Buff-breasted is a trifle larger and has the crown paler, breast and throat buffy. This identification requires special care.
**Habitat**
Grassy areas inland, shorelines, sometimes tidal flats, high-altitude lakes in migration. Rarest peep on coast.
**Habits**
Allows close approach, less active than other peeps; flies with them, but feeds by itself.
**Voice**
Similar to that of Semipalmated; a distinctive *creep* in flight.
**Eggs**
4; clay-colored, spotted with umber; 1.3 x 0.9 in. (3.3 and 2.3 cm).
**Remarks**
Like many of its relatives, this sandpiper migrates long distances. Many individuals that summer in the Arctic winter in Chile, sometimes in areas up to 13,000 feet (3962 m) in the Andes.
**Range**
Breeds in Greenland and Canadian Arctic; winters in S. America; migrates mainly through interior.

## LEAST SANDPIPER
*Calidris minutilla*                                         **20:2, 21:2**

**Description**
Size, 6 in. (15.2 cm). The smallest sandpiper, with • *yellowish or greenish legs,* thin bill.
**Similarities**
Resembles Semipalmated, but browner above, with neck and breast more streaked in spring and darker in fall; bill thinner, legs dusky yellow. Suggests a diminutive Pectoral.
**Habitat**
Shorelines, mud flats, wet fields; most common away from beach.

**Habits**
Perhaps the most common shorebird; quite tame; when flushed, zigzags off like Common Snipe.
**Voice**
A *peep*; also a short *kreep,* higher-pitched and more squeaky than Semipalmated's.
**Eggs**
3–4; buff, with rich, dark markings; 1.1 x 0.8 in. (2.8 x 2.0 cm). Nest is mossy depression in bog or tundra marsh.
**Range**
Breeds in Arctic, s. to James Bay and Newfoundland; winters from N.C. and Gulf Coast s. to S. America.

## CURLEW SANDPIPER
*Calidris ferruginea*                                    **20:9; 21:8**
**Description**
Size, 8½ in. (21.6 cm). Slender; • *bill noticeably downcurved;* rufous in spring; gray above, white below in fall; white wing stripe, • *rump white.*
**Similarities**
Dunlin has bill drooped at tip only and lacks white rump in fall. White-rumped has shorter, straight bill.
**Habitat**
Coastal. Very rare.
**Habits**
Usually seen as a solitary bird with Dunlins.
**Voice**
Soft, musical double-note *chirrup,* quite different from harsh ternlike call of Dunlin.
**Range**
Breeds in Asian Arctic; winters in Africa, Asia, and Australia; in migration, seen along Atlantic and Gulf coasts.

## DUNLIN
*Calidris alpina*                                    **20:7, 21:9, 22:5**
**Description**
Size, 8½ in. (21.6 cm). Short-legged, rather stocky; • *bill long, heavy, and with droop at tip;* legs black. Spring: bright reddish above; wing stripe, sides of rump, and underparts white; • *belly black.* Fall and winter: gray above, white below with grayish band across breast.
**Similarities**
Purple Sandpiper is darker, Sanderling lighter with straight bill.
**Habitat**
Beaches, shorelines, flats.
**Habits**
On ground, tame, sluggish; often looks hunched up.
**Voice**
Harsh, rather loud *cheer-ur;* "flushing note . . . a fine *chit-l-it*" (Nichols).
**Eggs**
4; buff, marked with chestnut brown; 1.5 x 1.0 in. (3.8 x 2.5 cm).
**Former name**
Red-backed Sandpiper.
**Range**
Breeds in Arctic s. to Hudson Bay; winters along Atlantic Coast from New England s. to Fla.

## SEMIPALMATED SANDPIPER
*Calidris pusilla*                                        **20:3, 21:1**

**Description**
Size, 5½–6¾ in. (14.0–17.1 cm). A small, common peep. Streaked gray-brown above, slightly browner in spring; breast ashy with light streaks; paler or unmarked in fall; black bill stout and straight. • *Slaty legs* sometimes appear greenish when slimy with mud.

**Similarities**
Least very similar, but browner and less chubby; breast more streaked in fall; legs yellow. Bill of Western is longer, thicker at base, and with slight droop at tip.

**Habitat**
Beaches, shorelines, flats; more common than Least near beach.

**Habits**
Often stands or hops on 1 leg, dashes about feeding with head down (unlike plovers); retreats and advances before waves, often in groups or flocks; at high tide, rests higher up on beach in groups behind some shelter or facing wind.

**Voice**
Flight call a grating *churp* or *check,* shorter than Least's; flushing note *ki-i-ip.*

**Eggs**
3–4; variable, buffy marked with brown; 1.2 x 0.8 in. (3.0 x 2.0 cm).

**Range**
Breeds in n. Canada s. to Hudson Bay; winters s. of U.S.

## WESTERN SANDPIPER
*Calidris mauri*                                    **20:1, 21:3, 22:2**

**Description**
Size, 6½ in. (16.5 cm). Similar to Semipalmated, but • *bill longer, thicker* at base, and with slight • *droop at tip.* Fall: almost indistinguishable, but "somewhat larger, rangier, paler, grayer" than Semipalmated and with "better developed white stripes over the eyes" (Nichols). Spring: more reddish above, giving a 2-toned rusty-and-gray effect; breast more streaked, often with dark V's running down the sides.

**Habitat**
Similar to Semipalmated's.

**Habits**
Carries bill pointed down more than does Semipalmated; usually feeds in deeper water.

**Voice**
Many notes like those of Semipalmated, but flight note *chee-rp (ee* as in *kreep* of Least Sandpiper).

**Eggs**
Similar to Semipalmated's.

**Remarks**
Only typical birds should be identified in the field.

**Range**
Breeds in Asia and Alaska; winters along coasts, from N.C. to S. America.

## SANDERLING
*Calidris alba*                                    **20:6, 21:4, 22:8**

**Description**
Size, 8 in. (20.3 cm). Somewhat larger and stockier than other peep. Conspicuous • *white wing stripe;* underparts and sides of

tail white; bill and legs black. Spring: head, upperparts, and breast bright red-brown. Fall: upperparts light gray, dark at bend of wing, forehead and underparts foam-white.

**Habitat**
Sea beaches, tidal flats, sandy edges of bays.

**Habits**
Follows edge of advancing and receding waves, often in small flocks; when flushed, flock semicircles out over breakers and lands farther up the beach; often stands on 1 leg; makes lines of little holes probing in sand.

**Voice**
Sharp and distinct *kit,* singly or in series.

**Eggs**
4; dull olive, marked with brown; 1.4 x 1.0 in. (3.6 x 2.5 cm).

**Range**
Breeds along Arctic coast s. to Hudson Bay; winters from s. New England s. along coast to Fla., Gulf Coast, and S. America.

## STILT SANDPIPER
*Micropalama himantopus*                    21:13, 22:14

**Description**
Size, 7½–9 in. (19.1–22.9 cm). Legs long, greenish; no wing stripe; white on uppertail coverts • *horseshoe-shaped,* not straight, across back; whitish tail; long bill, slightly drooped at tip. Spring: • *rusty ear streak,* dark back, and evenly barred underparts. Fall: gray above, line over eye and underparts white.

**Similarities**
Lesser Yellowlegs is larger, bill does not droop, and legs are longer and yellow. Dowitchers are heavier billed, chunkier, have a white wing stripe and back stripe. Stilt "looks and acts more like a short-billed, long-legged, pale-breasted dowitcher" (Pough).

**Habitat**
Shallow pools, fresh or salt.

**Habits**
Flight similar to Lesser Yellowlegs; quiet, sedate, frequently feeds in compact groups in pools up to its belly in water, often with head and neck under water and bill thrusting down perpendicularly, or swinging widely from side to side; does not bob head, as do yellowlegs.

**Voice**
Low *thu,* resembling note of Lesser Yellowlegs.

**Eggs**
3–4; dull white, marked with brown; 1.4 x 1.0 in. (3.6 x 2.5 cm).

**Range**
Breeds in n. Canada; winters in S. America; migrates w. of Mississippi except for a small number along Atlantic Coast.

## BUFF-BREASTED SANDPIPER
*Tryngites subruficollis*                    19:10, 23:4

**Description**
Size, 7½–8¾ in. (19.1–22.2 cm). No seasonal change. Bill short, head rounded; appears big-eyed because of conspicuous white eye-ring; head, neck, and • *underparts plain buff,* a few dark spots on the side; upperparts blackish-brown tinged with buff; wing stripe obscure, • *wing linings whitish;* tail dark; wings long, pointed, reaching at rest beyond end of tail; legs yellow.

**Similarities**
Remotely resembles a diminutive Upland Sandpiper, which is larger.

**Habitat**
Grassy open areas; rare along coast.
**Habits**
Tame; while on ground often raises wing and extends head; seeks safety by hiding; behavior often similar to that of Upland Sandpiper; flies in compact flocks.
**Voice**
Rather silent; occasionally a sharp, thin, clicking *tik;* also a Robin-like *chwup,* or a *pr-r-r-reet.*
**Eggs**
4; clay-colored; boldly marked with umber and slate; 1.5 x 1.0 in. (3.8 x 2.5 cm).
**Range**
Breeds in Arctic and nw. Canada; winters in S. America; migrates through Great Plains, rarely to Atlantic coast.

## SHORT-BILLED DOWITCHER
*Limnodromus griseus*                           20:15, 21:14, 22:12

**Description**
Size, 10½–12 in. (26.7–30.5 cm); bill, male, 2¼ in. (5.7 cm); female, 2½ in. (6.4 cm). Chunky, wing stripe on trailing edge, not in middle as in some other shorebirds; barred tail; greenish legs. Spring: brown above, rusty below. Fall: gray instead of red-brown and with a white eye line.
**Similarities**
See Long-billed Dowitcher, Red Knot, Common Snipe, yellowlegs. Fall Stilt Sandpiper is more slender, shorter-billed, and longer-legged.
**Habitat**
Marshes, mud flats, shorelines, inner beaches, still waters.
**Habits**
Flies in compact flocks with bill pointed partly down; feeds in deeper water than most shorebirds, probing like a sewing machine in the mud with its long bill; swims readily.
**Voice**
Flight call resembles that of Lesser Yellowlegs, but is faster, softer; it is a whistled, *too, too, too* or *dow-itch,* or *dow-itch-er.*
**Eggs**
4; grayish, with dark brown spots; 1.6 x 1.1 in. (4.1 x 2.8 cm).
**Remarks**
Dowitchers are the only shorebirds with long straight bills and white rumps, continuing as a wedge up the back. The bill of the Short-billed averages shorter than that of the Long-billed, but is not a reliable factor.
**Range**
Breeds in cen. to s. Canada, from Que. to ne. Man.; winters from S.C. to Gulf Coast and S. America.

## LONG-BILLED DOWITCHER
*Limnodromus scolopaceus*

**Description**
Size, 12 in. (30.5 cm); bill, male, 2½ in. (6.4 cm); female, 3 in. (7.6 cm). Spring: dark above, underparts red-brown to undertail coverts, which are barred, not spotted. Winter: very like Short-billed; bill longer (see Voice).
**Similarities**
Very difficult to distinguish from Short-billed except by voice in winter plumage; tends to be longer billed, in spring has rusty belly with bars.

**Habitat, Habits**
Similar to Short-billed's.
**Voice**
High *keek* or *keek, keek, keek*.
**Eggs**
4; olive, marked with brown; 1.7 x 1.2 in. (4.3 x 3.0 cm).
**Range**
Breeds in nw. Canada; winters from Fla. and Gulf Coast to C.
America.

## MARBLED GODWIT
*Limosa fedoa*

**19:11**
*Fig. 7*

**Description**
Size, 16–20 in. (40.6–50.8 cm); bill, 4¼ in. (10.8 cm). One of the
largest shorebirds. • *Upturned bill,* sometimes almost straight; buffy
brown with black markings; no white; legs blue-gray. In flight,
patch of cinnamon under wings; wings also show black outer
primaries.
**Similarities**
Hudsonian Godwit is smaller and has a white rump; Long-billed
Curlew has downcurved bill.
**Habitat**
Shorelines, flats, salt marshes, wet grasslands. Rare.
**Habits**
Flight swift, strong, direct, head somewhat drawn in; bill straight
forward, legs stretched out behind.
**Voice**
Noisy on breeding grounds, *god-WIT, god-WIT, god-WIT* or
*you're-crazy-crazy-crazy* and *cor-RECT, cor-RECT;* flight call is
*queep, queep, queep.*
**Eggs**
4; buff, spotted with brown; 2.2 x 1.5 in. (5.6 x 3.8 cm).
**Range**
Breeds in Prairie provinces, from s.-cen. Canada to Minn. and
S.Dak.; winters from S.C. to Fla. and Gulf Coast to C. America.

## HUDSONIAN GODWIT
*Limosa haemastica*

**20:13, 21:15**
*Fig. 7*

**Description**
Size, 14–16½ in. (35.6–41.9 cm). • *Bill slightly pink, upturned,
black at tip,* sometimes virtually straight; wings blackish with
narrow white wing stripe; underwing and wing pit black
(conspicuous in flight); tail black, with base and tip white; legs
blue-gray. Spring: below rich chestnut marked with black. Fall:
above dark gray; below whitish.
**Similarities**
Willet in flight has much more white on wing and tail is whitish,
not ringed; standing, Willet is sturdier and has a thicker, shorter
bill. Greater Yellowlegs is smaller, has yellow legs, and bobs.
Neither has an upcurved bill.
**Habitat**
Similar to Marbled Godwit, but breeds on tundra, not prairie.
**Habits**
Probes in mud and deep water, often up to belly and with head
submerged; flies in dark, undulating lines.

**Voice**
Low *qua-qua* when flushed; lower, less harsh than Marbled's; a low *ta-it* in flight.
**Eggs**
2–4; dull olive, marked with brown; 2.2 x 1.4 in. (5.6 x 3.6 cm).
**Range**
Breeds on Canadian tundra; winters in S. America; migrates through interior in spring and reaches Atlantic Coast in fall.

# PHALAROPES
Family Phalaropodidae

Phalaropes, a group (often merged into sandpiper family) of only three species, are small, long-necked, swimming sandpipers with pointed wings and lobed toes. The females are larger and more brilliantly plumaged than the males, and they take the lead in courtship. Phalaropes fly swiftly and swim buoyantly, often whirling. Their food is insects and small aquatic animals. The male builds the nest on the ground, incubates the eggs, and cares for the young.

## RED PHALAROPE
*Phalaropus fulicarius*                    20:18, 21:23, 22:10

**Description**
Size, 7½–9 in. (19.1–22.9 cm). Bill thick, wholly or partly yellow. Spring female: • *reddish below* and on neck, black crown, and white cheeks and wing stripe; dark-tipped yellow bill. Spring male much duller. Fall: pale, unstreaked gray above; head, neck, and underparts white; black patches on head; notably curved "phalarope mark" behind eye. Bill short, blunt, thicker at base.
**Similarities**
Rather than needle bill of other phalaropes, Red has shorter, stouter bill; see Wilson's and Northern Phalaropes. Whiter-backed Sanderling in fall seldom swims, and it lacks mark on head.
**Habitat**
Most maritime of the phalaropes; breeds on tundra, winters on open sea.
**Habits**
Active, unsuspicious; flies like a Sanderling, swims like a tiny gull, bobs head and dabs bill for food, or tips up and feeds; in feeding, makes twice as many dabs per second as Northern.
**Voice**
Similar to Sanderling, but thinner and higher-pitched.
**Food**
Jellyfish, fish.
**Eggs**
4; buffy or greenish, marked with brown; 1.2 x 0.8 in. (3.1 x 2.2 cm).
**Range**
Breeds in Arctic; winters in S. Hemisphere oceans; migrates along Atlantic Coast.

## WILSON'S PHALAROPE
*Phalaropus tricolor*                    20:16, 21:19, 22:13

**Description**
Size, 8½–10 in. (21.6–25.4 cm). Has the thinnest, most needlelike bill of any shorebird. White rump and tail, no white wing stripe.

Spring female: underparts and cheeks white; black band from eye turns • *chestnut* on back. Spring male: much duller. Fall: plain brownish-gray above, plain white below; legs turn greenish-yellow.

**Similarities**
Other phalaropes have wing stripes and dark marks through mid-rump and tail. Winter Sanderling has thicker, shorter bill and conspicuous wing stripe. Stilt Sandpiper has similar flight pattern, but is darker, bill is much heavier, forehead is dark. Lesser Yellowlegs has streaks on breast and lemon-yellow legs.

**Habitat**
Pools, shallow lakes inland; not oceanic.

**Habits**
Spins and runs nervously in shallow water, making abrupt twists and turns; probes with bill in mud and swings bill from side to side; feeds on foot in wet meadows and mud flats.

**Voice**
Soft nasal *oit-oit*, unlike other phalaropes.

**Food**
Crustaceans, insects.

**Eggs**
4; buffy or greenish, marked with brown; 1.2 x 0.9 in. (3.3 x 2.3 cm). Nest is grassy depression in either wet or dry meadow.

**Range**
Breeds from s.-cen. Canada, s. to Ind. and Mich.; winters in S. America; migrates through Mississippi Valley region and occasionally to Atlantic Coast.

# NORTHERN PHALAROPE
*Phalaropus lobatus*                         20:19, 21:22, 22:7

**Description**
Size, 6½–9 in. (16.5–20.3 cm). Small head, thin neck; needlelike black bill, white wing stripe, dark tail and legs. Spring female: gray above, darkest on head; • *sides of neck chestnut*; throat and belly white; back streaked; Spring male: much duller. Fall: above blackish streaked with gray; below white; black eye stripe.

**Similarities**
Fall Red Phalarope is larger, more heavily built, darker above, and has shorter, heavier bill. Fall Wilson's has white rump, no white wing stripe, and no face markings. Sanderling has a longer wing stripe, shallower wingbeat.

**Habitat**
Breeds on tundra, winters on ocean, sometimes seen on inland ponds.

**Habits**
Gregarious, tame; flight swift, erratic; often alights on floating seaweed and runs about; bathes with characteristic upjerk and ducking motion.

**Voice**
Similar to Sanderling's.

**Food**
Insects, crustaceans, plankton.

**Eggs**
4; buffy or greenish, marked with brown; 1.1 x 0.7 in. (2.9 x 2.0 cm).

**Range**
Breeds in Arctic, s. to Labr. and St. James Bay; winters in S. Hemisphere oceans; migrates off Atlantic Coast.

# AVOCETS AND STILTS
Family Recurvirostridae

These are very long-legged, thin-billed wading birds spectacularly patterned with black and white. They frequent shallow freshwater, alkaline or brackish ponds and coastal flats, feeding on insects, shellfish, and seeds.

## AMERICAN AVOCET
*Recurvirostra americana*                                    **19:17**

**Description**
Size, 15½–20 in. (39.4–50.8 cm). • *Bill upturned;* black-and-white pattern unique, both at rest and in flight; head and neck buffy in spring.
**Similarities**
Black-necked Stilt is more slender, has black on head and neck, and has red legs.
**Habitat**
Shallow lakes and marshy ponds inland.
**Habits**
Flight direct, strong, and fairly swift, with neck and legs extended. Feeds by walking quickly through water, working bill from side to side in the mud; or while swimming, by tipping up.
**Voice**
Loud yelp.
**Eggs**
3–4; olive, marked with brown; 2.0 x 1.3 in. (5.1 x 3.3 cm).
**Range**
Breeds in nw. Canada, s. to Tex.; winters from s. Tex. s. to C. America; casual to Atlantic Coast in migration.

## BLACK-NECKED STILT
*Himantopus mexicanus*                                       **19:15**

**Description**
Size, 13–17 in. (33.0–43.2 cm). A tall shorebird, black above and white below. Adult male: long, slender black bill; • *very long, thin red legs;* white rump and tail. In flight, appears to have a white body, black wings. Immature: browner, legs paler.
**Habitat**
Shallow water, primarily fresh.
**Habits**
Aggressive; feeds while wading, steps high in walking and often tosses head. In flight, neck somewhat drawn in, legs extended.
**Voice**
Noisy; loud, high-pitched, insistent yipping.
**Eggs**
3–5; buff, marked with dark brown; 1.7 x 1.2 in. (4.3 x 3.0 cm).
**Range**
Breeds along Atlantic Coast from Del. s. to Fla. and S. America; winters s. of U.S.; casual in migration along Atlantic Coast.

# SKUAS AND JAEGERS
Family Stercorariidae

This is a family of strong-flying, hook-billed, hawklike, predominantly darkish seabirds flashing white in their wings. Adult jaegers are generally dark above, with black caps and whitish collars, and light below. The amount of wing flash varies with the species and each has its characteristic lengthened tail feathers. The young have dark, barred breasts and lighter bellies barred on the sides. Their tails are not lengthened, thus making identification difficult and sometimes impossible. Members of this family breed in the Arctic, winter at sea, and only occasionally are seen from shore in migration. These predators feed by harassing other seabirds into dropping their food, which they then seize. Problems of identification of jaegers are enhanced by the occurrence of dark and light color phases.

## POMARINE JAEGER
*Stercorarius pomarinus*

2:2
*Fig. 8*

**Description**
Size, 20–23 in. (50.8–58.4 cm). Center • *tail feathers blunt and twisted* in adults. Typically dark-capped, light below, but also all-dark phase.
**Similarities**
• *Broader-winged*, larger than other jaegers. Young larger, bill heavier, and buffier below than other jaegers; more white in wings.
**Habitat**
Arctic coasts when breeding; offshore winter.
**Habits**
Aggressive.
**Voice**
"A sharp *which-yew*" (Rich).
**Food**
Mainly fish.
**Eggs**
2; olive-brown with dark brown spots; 2.2 x 1.6 in. (5.7 x 4.1 cm). Nest is in grass-lined depression on ground.
**Range**
Breeds in Greenland and Arctic Canada; winters at sea, s. from N.C.

## PARASITIC JAEGER
*Stercorarius parasiticus*

2:1

**Description**
Size, 16–21 in. (40.6–53.3 cm). Black legs, pale breastband, bill smaller than Pomarine's; center tail feathers pointed, project beyond others.
**Similarities**
Less contrast between cap and back in light phase than in Long-tailed; also yellow on head is paler, and white collar on hind neck is narrower.
**Habitat**
See Pomarine.
**Voice**
Various wails, shrieks, and cries.
**Food**
Small birds, lemmings in summer, fish at other times.
**Eggs**
2; olive-brown with dark brown spots; 2.2 x 1.6 in. (5.7 x 4.1 cm). Nest is tundra hollow.
**Range**
Breeds in Greenland and n. Canada, s. to cen. Canada; winters at sea, s. to S. America; migrates along coasts.

## LONG-TAILED JAEGER
*Stercorarius longicaudus*                                    **2:3**

**Description**
Size, 20–23 in. (50.8–58.4 cm), including tail. • *Bluish legs;* breast
white shading into gray on belly; tail pointed, over 4 in. (10.2 cm).
Light phase: pale, ashy back contrasts sharply with black cap,
cheeks bright yellow, white collar on hindneck broader than
Parasitic's and almost surrounds neck.
**Similarities**
Other jaegers are larger-bodied with shorter central tail feathers.
**Habitat**
Open sea when not breeding.
**Habits**
Less aggressive, harasses gulls less frequently than larger-bodied
jaegers.
**Voice**
"A shrill *pheu-pheu-pheu-pheo* . . . often followed by a harsh *qua*"
(Nelson).
**Food**
Lemmings, young birds in summer, fish, marine invertebrates
otherwise.
**Eggs**
2–3; olive, spotted with sepia; 2.1 x 1.5 in. (5.5 x 3.9 cm).
**Range**
Breeds in Arctic; winters at sea, s. of Va.; migrates along Atlantic
Coast and Great Lakes area.

## NORTHERN SKUA
*Catharacta skua*                                           *Fig. 8*

**Description**
Size, 21 in. (53.3 cm). Large, robust; • *wings broad with
conspicuous white patch.* Buteolike; sooty brown above, rusty
below; • *tail short,* slightly uptilted, faintly forked in young.
**Similarities**
Herring Gull lacks white wing patches.
**Habits**
Flight hawklike or eaglelike, and very rapid.
**Voice**
"*Skua*" (McGillivray).
**Food**
As jaegers; carrion.
**Eggs**
2–3; white blotched with brown; 2.7 x 1.9 in. (7.0 x 4.9 cm).
**Range**
Breeds in Iceland and n. Europe; winters at sea off Newfoundland
s. to New England.

*Fig. 8*

Northern Skua

Pomarine Jaeger, p. 91

dark phase immature

# GULLS AND TERNS
Family Laridae

# GULLS
Subfamily Larinae

Gulls are fairly large, long-winged, generally white water birds, many with black wing tips, often with a pearly "mantle" (back and top of wings), sometimes with a black head or back. Juveniles and immatures frequently are of varying degrees of dusky brown, as it sometimes takes several years to reach the adult plumage. Gulls fall into three plumage groups: (1) the white-wings: Glaucous, Iceland, and Ivory; (2) the hooded gulls: Laughing, Franklin's, Bonaparte's, Black-headed, Little, and Sabine's; (3) the pearly-backs: all the others except the Great Black-backed.

Gulls fly with the bill straight forward, soar for protracted periods, often follow boats, swim buoyantly, pick food from the surface, and, being scavengers, often collect in great numbers at sewer outlets, fish-processing areas, and garbage dumps. They perch on buoys, pilings, and roofs, and walk easily. Gulls sometimes frequent croplands and follow the farmer as he plows in order to eat the grubs in the furrow.

## GLAUCOUS GULL
*Larus hyperboreus*                                         **24:1**

**Description**
Size, 28 in. (71.1 cm). Almost size of Great Black-backed, the largest gull. Folded, white-tipped wing does not reach tip of tail. Adult: white with a pale gray mantle, bill heavy and yellow with red spot near tip, eye-ring and eye yellow in breeding season, feet pinkish, • *white wing tips*. First-winter immature: pale buffy throughout, primaries equally pale. Second-winter immature: mantle gradually changes to creamy white, then to all-white in third year; bill white, bill spot black.
**Similarities**
Adult Iceland is smaller, has a less robust head and bill, eye-ring sometimes red, and longer wings; immature has smaller, darker bill. Iceland is more buoyant; at rest, wings reach beyond tail.
**Habitat**
Coast, offshore.
**Habits**
Predatory; flight steady, soaring, somewhat hawklike.
**Voice**
Hoarse *ku-ku-ku, ku-lee-oo*.
**Food**
Carrion, refuse, birds, marine invertebrates.
**Eggs**
2–3; olive-buff, marked with dark brown; 3.0 x 2.1 in. (7.6 x 5.3 cm). Nest of seaweed, grasses, etc., up to 3 ft. (0.9 m) high, on cliffs and ledges; in colonies.
**Range**
Breeds in Arctic; winters in N. America to N.Y. and the Great Lakes.

## ICELAND GULL
*Larus glaucoides*                                                    **24:3**

**Description**
Size, 24 in. (61.0 cm). Folded wing reaches beyond tip of tail.
Adult: white with a pale gray mantle, yellow bill with red spot
near tip, eye ring purple, iris mottled to dark in breeding season,
wings white-tipped like Glaucous or white with gray spots. Feet
pinkish. Immature: plumages similar to immature Glaucous, but
smaller bill more extensively black-tipped and turns greenish in
second year. The subspecies called "Kumliens" **(24:3c)** has white
wing tips with gray spots.

**Similarities**
Glaucous is larger and has a proportionately heavier head and bill,
whereas Iceland's wings are longer, proportionately more pointed,
its flight more buoyant. However, size alone is confusing; large
Iceland may be as large as small Glaucous and vice versa. In
breeding season Glaucous and Herring Gulls have yellow eyes and
eye-rings, the Iceland and Thayer's Gulls have purple eye-rings
and mottled or dark eyes; see Thayer's Gull.

**Habitat**
Coasts, inland lakes, and rivers.

**Habits**
Less aggressive and predatory than Glaucous.

**Voice**
Similar to Herring's, but shriller.

**Food**
Fish, refuse, carrion, offal, crowberries.

**Eggs**
2–3; buff, often tinged with green and marked with brown; 2.8 x
1.7 in. (7.1 x 4.3 cm). Nest of vegetation on cliffs; in colonies.

**Range**
Breeds in Greenland, s. to Baffin Is.; winters in ne. U.S. waters s.
to N.J. and w. to Great Lakes.

## THAYER'S GULL
*Larus thayeri*

**Description**
Size, 22 in. (55.9 cm). Resembles Herring Gull and Iceland Gull
closely, but dark-eyed. Adult: pearly mantle; bill and legs as
Herring Gull; iris dark, surrounded by dark (violet) eye-ring; wing
tip variable, gray with white spots to gray-black with white spots
(between Herring and Iceland in color). Immature:
indistinguishable from Herring Gull.

**Similarities**
Herring Gull has yellow eyes and black-and-white wing tips.
Usually distinguishable from Iceland Gull by darker mantle;
darker wing tips; Kumlien's race of Iceland has paler gray spots in
wing tips. A very difficult identification.

**Habitat, Habits, Voice, Food**
Similar to Herring Gull's.

**Eggs**
Similar to Herring Gull's but nest is of vegetation on Arctic cliffs.

**Range**
Breeds in Greenland and n. Canada; winters along north Pacific
Coast; rarely to Atlantic Coast and Great Lakes.

# HERRING GULL
*Larus argentatus*                                    **24:4, 26:11**

**Description**
Size, 22½–26 in. (57.2–66.0 cm). Adult: white with pearly-gray
mantle, black wing tips with white spots; iris and eye-ring
yellowish in breeding season, bill yellow with a red spot near tip,
• *legs pinkish;* in winter, head and neck streaked with brown. First-
winter immature: almost uniform sooty. Second-winter immature:
dusky, broad diffuse dark band near end of tail; body becomes
lighter and uppertail coverts whiter with age; bill dark, sometimes
with black ring near end.

**Similarities**
Ring-billed Gull is smaller, has black ring around middle of bill
and yellowish legs, and has more buoyant flight. Immature Great
Black-backed is larger, darker above, lighter below. Immature
Ring-billed is lighter above and below, and has narrow blackish
band near end of white tail. See also Iceland and Thayer's Gulls.

**Habitat**
The commonest gull in most areas including shorelines, bodies of
water, rivers, garbage dumps.

**Habits**
Gregarious; a great ship-follower and soarer, often high in air;
"trades" back and forth between bodies of water morning and late
afternoon, at middle height; swims, drinks sea water; drops
mollusks from height on a hard surface to break shells.

**Voice**
*"Queeeeeeah-ah, kak, kak, kak"* (Collins).

**Food**
Carrion, garbage, refuse, marine animals, eggs, young birds.

**Eggs**
2–5; variable, often olive-buff spotted and blotched with dark
brown; 2.8 x 1.9 in. (7.1 x 4.8 cm).

**Range**
Breeds from Greenland to N.C.; winters from s. Canada, s. to W.
Indies and Mississippi Valley.

# RING–BILLED GULL
*Larus delawarensis*                                  **24:5, 26:10**

**Description**
Size, 18–21 in. (45.7–53.3 cm). Adult: white with pearly mantle
slightly paler than Herring's; black wing tips with white spots,
wing tips all-black below; • *yellowish legs;* black ring around bill.
First-winter immature: dusky. Second-winter immature: light
dusky above and whitish below, narrow blackish band near end of
white tail, bill and legs usually pinkish, bill with a black tip.

**Similarities**
Immature Herring Gull tail mainly brown; Ring-Billed has dark-
tipped white tail, body whiter.

**Habitat**
Coast and inland waters; garbage dumps.

**Habits**
Gregarious; carriage more sprightly than Herring's; follows plow
and ship; flight said to be lighter, more buoyant, more dovelike
than Herring's, but not easy to distinguish.

**Voice**
Notes higher-pitched than Herring's; alarm note a hawklike *cree-
cree.*

**Food**
Garbage, refuse, carrion, aquatic animals, rodents, insects.

### Eggs
2-4; similar to Herring Gull but smaller; 2.3 x 1.6 in. (5.8 x 4.1 cm).

### Range
Breeds from Labr. to Great Lakes; winters from s. New England s. to Cuba.

Note: The **MEW GULL**, *Larus canus*, size, 17 in. (43.2 cm), occurs in East only as a breeder east to W. Mackenzie, northeastern Alberta, northern Saskatchewan, and recently Wisconsin. It resembles a Ring-billed Gull but has a smaller unmarked yellow bill.

## GREAT BLACK-BACKED GULL
*Larus marinus*                                          24:2

### Description
Size, 29 in. (73.7 cm); white-headed, • *slate-colored back and wings*. Adult: bill yellow with red spot near tip, legs pinkish. First-winter immature: dusky streaked with dark band near end of tail; back darker, head and underparts lighter than Herring, older young approach adult appearance; bill blackish, feet whitish to pinkish. In flight, bigger head and neck protrude more than Herring.

### Similarities
Immature Glaucous Gull is lighter and lacks blackish primaries and dark band near tip of tail. Immature Herring has streakier head, less dark on mantle.

### Habitat
Coastal waters.

### Habits
Predatory, flight strong, wingbeats slow, impressive; soaring flight.

### Voice
*Keeaw,* lower than Herring's; also a quick *gaw-gaw-gaw;* a high *key-key-key;* and a hoarse laugh, *ha-ha-ha.*

### Food
Carrion, refuse, marine invertebrates, birds, eggs, rodents.

### Eggs
2-3; olive or buffy gray, blotched with brown; 3.2 x 2.1 in. (8.1 x 5.3 cm).

### Range
Breeds along Atlantic Coast from Greenland to N.J.; winters s. to Ga. and w. to Great Lakes.

Note: The accidental **LESSER BLACK-BACKED GULL**, *L. fuscus*, size, 23 in. (58 cm), of Europe, is smaller (size of Herring Gull), slimmer, lighter above, and has yellow legs.

## BLACK-HEADED GULL
*Larus ridibundus*                                       25:5

### Description
Size, 15 in. (38.1 cm). Summer adult: white with • *brown head, gray mantle, bill red to yellowish*. Winter adult: like winter Bonaparte's, but bill and underwing differ as above. Immature: As Bonaparte's, but younger birds browner above, bill and legs yellow to pinkish, yellow bill with black tip.

### Similarities
Bonaparte's is smaller, with smaller black bill.

### Habitat
Sewer outlets, estuaries, harbors, usually in winter.

**Habits**
Similar to Bonaparte's.
**Voice**
"A harsh *kwarr*, a short *kwuh*" (Mountfort).
**Range**
Breeds in Eurasia; winters s. of Europe to Africa and in s. Asia;
casual to Atlantic Coast in winter.

## LAUGHING GULL
*Larus atricilla*                                                25:1

**Description**
Size, 17 in. (43.2 cm). Summer adult: wing tips black, • *trailing
edge of wing white;* neck, tail, and underparts white; bill and legs
dark red; bill relatively long. Winter adult: head white with dark
smudge; bill and legs duller. First-winter immature: head and
breast dark, rump and belly white, tail with black band near white
tip, bill and legs blackish.
**Similarities**
The only black-headed gull lacking a distinctive black or white
wing-tip pattern.
**Habitat**
Shorelines, rivers; breeds in marshes.
**Habits**
Flight light, strong, but heavier than other black-headed gulls;
often soars, catches insects in air; follows plow.
**Voice**
Noisy; a characteristic laughter; *"hah-ha-ha-ha-ha-hah-hah-hah"*
(Langille).
**Food**
Aquatic animals, insects, carrion, offal, garbage, refuse.
**Eggs**
3; variable, often buff blotched with umber; 2.1 x 1.5 in. (5.3 x 3.8
cm).
**Range**
Breeds along Atlantic Coast from Maine to W. Indies and along
Gulf Coast to Tex.; winters s. from N.C.

## FRANKLIN'S GULL
*Larus pipixcan*                                                 25:2

**Description**
Size, 14 in. (35.6 cm). Summer adult: head black, • *white bar* or
"window" between gray primary base and white-tipped black ends,
bill and legs dark red, underparts with faint rosy bloom, bill
relatively long. Winter adult: white head with dark band behind
head from eye to eye. First-winter immature: forehead white, nape
black, wings dark all over, rump and underparts white, legs
blackish.
**Similarities**
Somewhat similar, Laughing Gull is strictly coastal, and in east
never occurs in the summer range of Franklin's.
**Habitat**
Prairies, marshes, fields, and bodies of water inland.
**Habits**
Gregarious; follows plow, sometimes soars upward in spirals,
captures insects in air, swims buoyantly, often flies in V-shaped
flocks.

**Voice**
Unlike Laughing; "a soft *krruk* . . . a louder and more plaintive . . . *pway* or *pwa-ay*" (Bent); when soaring, a "*weeh-a-weeh-a weeh-a po-lee po-lee po-lee po-lee*" (Miller).

**Food**
Grasshoppers, other insects, fish, frogs, mollusks.

**Eggs**
2–4; buffy, marked with brown; 2.0 x 1.4 in. (5.1 x 3.6 cm).

**Range**
Breeds in prairie marshes, s. from s. Canada to Minn. and S.Dak.; winters from La. and Tex., s. to S. America.

## BONAPARTE'S GULL
*Larus philadelphia*                                        25:3

**Description**
Size, 13 in. (33.0 cm). Summer adult: gray mantle, black-headed; long, white wedge down outer side of black-tipped primaries; wing tips white from below; bill is small, black; legs orange-red. Winter adult: white head with black spot behind ear coverts. First-winter immature: similar to winter adult but with narrow black band near tip of tail; upper wing with grayish band from wrist to back; bill dark with light base; legs pinkish.

**Similarities**
Laughing adult has much darker mantle; immature, neck and sides of breast dark and broader band on tail. Franklin's has much darker mantle in adult. Kittiwake young is larger; black bar on back of neck, yellowish legs.

**Habitat**
Coastal and interior waterways; breeds in coniferous forests.

**Habits**
Gregarious; flight somewhat ternlike, but seldom dives; points bill down in flight, which seems desultory but is rapid; swims buoyantly.

**Voice**
Shrill, nasal *peer;* "sparrowlike conversational notes" (Jones).

**Food**
Largely insects, some fish, crustaceans, worms.

**Eggs**
2–4; olive-buff, spotted with chocolate; 1.9 x 1.3 in. (4.8 x 3.3 cm). Nest of sticks in conifers, to height of 20 ft. (6.1 m).

**Range**
Breeds in nw. Canada; winters along Atlantic Coast from s. New England to Fla. and Gulf Coast.

## LITTLE GULL
*Larus minutus*                                        25:6

**Description**
Size, 11 in. (27.9 cm). This rare European visitor is the only gull with • *all-black wing linings*. Wings and tail somewhat rounded. Summer adult: black head; gray mantle and wings above; wings blackish below with white trailing edges; bill and legs red. Winter adult: head white with black ear patch, bill blackish. First-winter immature: wide, black open V on gray wing; black band near tip of tail; wing linings white; perched bird shows dark stripe on wing.

**Similarities**
Immature resembles larger small Kittiwake, but has black on back and distinct crown patch. Like immature Bonaparte's but more black on back and crown; primaries solid black without white bases.

**Habitat**
Ocean coasts and Great Lakes harbors; sewer outlets in winter.
**Habits**
Behavior in pursuit of insects suggests Black Tern; tame, fearless; feeds from surface of water in flight; often found with Bonaparte's.
**Voice**
"A rather low *kek-kek-kek*" (Mountfort).
**Food**
Fish, aquatic life, insects, refuse.
**Eggs**
3; olive-brown with dark spots; 1.6 x 1.2 in. (4.1 x 3.1 cm).
**Range**
Breeds in Europe and recently in Ont. and Wis.; winters in E., particular along Atlantic Coast from N.B. to N.J., and around Great Lakes.

## IVORY GULL
*Pagophila eburnea*

**Description**
Size, 17 in. (43.2 cm). Adult: white, plump; wings long, wedge-shaped; bill dark with yellow tip; • *legs black.* Immature: blackish spots above bill dusky.
**Habitat**
On Arctic coasts in summer, edge of pack ice in winter.
**Habits**
Not highly gregarious; restless, runs on ice like large plover; flight ternlike; perches on ice pinnacles, seldom swims.
**Voice**
"Shrill and not unlike that of the Arctic Tern" (Yarrell).
**Food**
Carrion, refuse, lemmings, marine organisms.
**Eggs**
2; buff-olive with dark blotches; 2.4 x 1.7 in. (6.1 x 4.4 cm).
**Range**
Breeds primarily in E. Hemisphere, some on Somerset Is. in Arctic; winters at sea in Arctic, sometimes s. to N.Y. and N.J.

## BLACK-LEGGED KITTIWAKE
*Rissa tridactyla*                                    **25:7, 26:9**

**Description**
Size, 17 in. (43.2 cm). Only eastern gull with a solid • *black triangular tip on a gray wing.* Wings long; tail long, broad, slightly forked. Adult: white head, gray mantle, yellow bill, black legs, nape gray in winter. First-winter immature: dark open V on each wing; bill, outer primaries, and end of slightly notched tail black; spot behind eyes; dusky bar on nape.
**Similarities**
Young Bonaparte's is smaller, lacks dark bar on back of neck. See also Little and Sabine's Gulls.
**Habitat**
Coasts, ocean.
**Habits**
Gregarious; drinks salt water, sleeps on waves; follows ships; flight buoyant, graceful, swallowlike, and distinctive at great distance; wingbeats rapid; only gull that dives from wing and swims under water.

**Voice**
"*Kitti-wake, ka-ake;* sharp and piercing *ki, ki, ki* . . . harsh rattling *kaa, kaa, kae, kae* and *kaak-kaak*" (Townsend).
**Food**
Fish, crustaceans, mollusks, refuse (but less of a scavenger than larger gulls).
**Eggs**
1–2; olive-buff, marked with brown; 2.2 x 1.6 in. (5.6 x 4.1 cm). Nest of seaweed, etc., on cliffs; in colonies.
**Range**
Breeds from Arctic Ocean to Gulf of St. Lawrence; winters off Atlantic Coast, s. to Bermuda.

Note: The **ROSS' GULL**, *Rhodostethia rosea,* size, 14 in. (35.6 cm), is a small, wedge-tailed gull with rosy-tinged underparts. It breeds in Siberia and winters in the Arctic seas. It is an accidental visitor to eastern coastal or Great Lakes areas.

## SABINE'S GULL
*Xema sabini*                                                   **25:4**
**Description**
Size, 13–14 in. (33.0–35.6 cm). Summer adult: • *tail forked;* wings with long, black triangular area at tip; long, white triangular area at rear of wings. Winter adult: head white, brownish smudge around back of head. First-winter immature: head gray above, forked tail black-tipped, wing pattern like adult.
**Habitat**
Tundra, coast, ocean.
**Habits**
Feeds ploverlike over flats at low tide; flight ternlike, seldom dives.
**Voice**
Cry similar to Arctic Tern's, but shorter, harsher.
**Food**
Aquatic life, insects.
**Eggs**
2–3; olive-buff, faintly spotted with brown; 1.7 x 1.3 in. (4.3 x 3.3 cm).
**Range**
Breeds along Arctic shoreline; winters at sea in N. Atlantic; seen in migration along Atlantic Coast from Labr. to N.Y.

# TERNS
Subfamily Sterninae

Terns are generally smaller, lighter, more streamlined than gulls; and have sharp, pointed bills, long pointed wings, and forked tails. They start changing into fall plumage in midsummer. These "sea swallows" often hover, usually fly with the bill pointing down, and dive readily for food (or at intruders around the nest), but swim and walk little. The young can run about soon after hatching, but, unlike shorebirds, they are dependent for a long time upon their parents for food. Members of the large Genus *Sterna* have thin bills, black caps, pearl-gray or black mantles, and forked tails.

# GULL-BILLED TERN
*Gelochelidon nilotica*                                    **26:5, 27:5**

**Description**
Size, 14 in. (35.6 cm). • *Body white, stocky*; feet black; • *bill black*.
Spring adult: black cap. Fall adult: head white with black ear
patch and mottled crown. Immature: similar but with brown band
on end of tail, mottling on back, and dark-tipped, lighter bill.

**Similarities**
Tail less forked than in Common or Forster's; immature
particularly gull-like, but has notched tail.

**Habitat**
Coastal islands, marshes.

**Habits**
Somewhat like Laughing or Franklin's Gull; flight characteristic,
wingbeats slower than *Sterna* terns, dives less, picks food off
surface more; hawks back and forth over marshes for insects;
follows plow.

**Voice**
*"Katydid, katydid*; a gull-like *ka-ka-ka"* (Collins).

**Food**
Insects, some crustaceans.

**Eggs**
2–3; variable, often buff marked with brown; 1.9 x 1.3 in. (4.8 x
3.3 cm). Nest is in depression in sand or of vegetation among grass.

**Range**
Breeds along Atlantic Coast from N.Y., s. to Gulf of Mexico and
W. Indies; winters from Gulf Coast to S. America.

# FORSTER'S TERN
*Sterna forsteri*                                          **26:4, 27:7**

**Description**
Size, 14–16½ in. (35.6–41.9 cm). • *Whitish primaries* in breeding
plumage. Spring adult: bill orange-red with a black tip; cap black;
tail gray, moderately forked, dark inner edges, light outer edges;
feet orange-red. Fall adult: black ear patch on white head, pale
nape, dusky bill, yellowish feet.

**Similarities**
Bill occasionally without black tip; Common is darker, with shorter
tail, paler, and tail has reverse color—dark outer and light inner
edges. Fall Common has a larger black area on head going across
nape and crown from eye to eye.

**Habitat**
Coast, marshes, inland waters.

**Habits**
Wingbeats have quicker, sharper snap than Common's.

**Voice**
Rasping *tsa-a-ap, zreep,* suggesting a Nighthawk; also a rapid
peeping *pip, pip, pip*.

**Food**
Insects, fish, frogs.

**Eggs**
3–4; similar to Common Tern's; 1.7 x 1.2 in. (4.3 x 3.0 cm). Nest
is usually of vegetation on ground or marsh; in colonies.

**Range**
Breeds along Atlantic Coast, from Md. to Gulf Coast, w. to Tex.,
and inland w. of Ill.; winters along coasts from S.C. to C. America.

## COMMON TERN
*Sterna hirundo*                                           **26:3, 27:9**

**Description**
Size, 13–15 in. (33.0–38.1 cm). A common tern, distinguished in
breeding plumage by its black-tipped red bill and dusky primaries.
Spring adult: black cap, pearl-gray mantle, moderately forked tail
white except for gray outer edge of outermost feathers, orange-red
legs. Fall adult: head white with black band around nape from eye
to eye, bill blackish, legs paler. At rest, wings extend beyond tip of
tail. Immature young: resembles fall adult; similar in color to
young Arctic or Roseate.

**Similarities**
Bill occasionally without black tip in late summer; see Roseate. Bill
and legs of Common are longer and tail shorter than in Arctic. See
also Forster's.

**Habitat**
Coast, ocean, shorelines, bodies of water inland.

**Habits**
Gregarious; flocks collect over schools of fish, thus guiding
fishermen to good fishing areas.

**Voice**
Descending *tee-arr*, a rapid *kik-kik-kik*.

**Food**
Fish, aquatic life, insects.

**Eggs**
2–4; variable, often pale brown blotched with gray or lilac; 1.6 x
1.2 in. (4.1 x 3.0 cm). Nest is depression in sand, of vegetation.

**Range**
Breeds throughout e. N. America from Labr. s. to W. Indies and
w. to Wis.; winters from Fla. s. to S. America.

## ARCTIC TERN
*Sterna paradisaea*                                        **26:6, 27:8**

**Description**
Size, 13–16 in. (33.0–40.6 cm). • *Bill blood-red to the tip*, legs
short. Spring adult: black cap with white border above grayish
face. Fall adult: bill black, legs blackish. At rest, body is closer to
ground than Common, wings and tail about equal in length. Bill
does, rarely, have a black tip; identifying this species requires care.

**Similarities**
Face of Common is all-white; Arctic is grayer above and below
than Common and tail is longer and more deeply forked. Young
Common as Roseate is similar in color to fall Arctic.

**Habitat**
Coast, ocean, interior lakes.

**Habits, Food**
Similar to Common.

**Voice**
Rising *key, key, key*; a rising *tee-arr* shriller than Common.

**Eggs**
2; similar to Common's; 1.6 x 1.2 in. (4.1 x 3.0 cm). Nest is
depression in sand in colonies.

**Range**
Breeds in n. Canada from Ellesmere Is. s. to Newfoundland,
Mass., and w. to Que. and Man.; winters at sea s. of Equator. The
22,000 mile-(35,405 km) a-year flight of Arctic Terns from the
Arctic to the Antarctic is the longest migration of any bird.

# ROSEATE TERN
*Sterna dougallii*                                    **26:7, 27:6**

**Description**
Size, 14–17 in. (35.6–43.2 cm). This has the longest, most deeply
forked tail of any tern. Pale mantle evenly colored throughout; at
rest, tail projects beyond wings. Spring adult: • *bill black*,
sometimes with red at base; black cap; faint rosy tinge to creamy
underparts; primaries light; tail pure white; feet red. Fall adult and
immature: crown and forehead white, nape dark.

**Similarities**
Similar to young Common or Arctic, but forward edge of inner
wing not dark. Bill dark and slimmer compared with mainly red
bill of Common; tail longer than wings at rest, while it is shorter
than wings in Common; and about the same as wings in Arctic.

**Habitat**
Coast.

**Habits**
Flight most graceful of all terns, "the greyhound of its tribe"
(Bent). Wingbeats a little slower, deeper than Common's; flies
lower over water than most terns.

**Voice**
Distinctive rasping *krack*, louder, lower than other terns, like
tearing a piece of strong cloth; a musical whistlelike *chee-bee*.

**Food**
Fish.

**Eggs**
2–3; buff, spotted with brown; 1.7 x 1.2 in. (4.3 x 3.0 cm). Nest is
depression in sand or crevice in rocks; in colonies.

**Range**
Breeds along Atlantic Coast from Gulf of St. Lawrence s. to S.
America; winters from Gulf Coast to S. America.

Note: The **SOOTY TERN,** *Sterna fuscata (Fig. 9)*, size, 16 in. (40.6
cm), of the Gulf of Mexico is occasionally blown north into our
area from West Indian seas by hurricanes. Adults are black above,
white below with a black cap; black bill and feet; and a forked,
mainly black tail. Immatures are almost all brown, flecked above
with white. The accidental **BRIDLED TERN,** *S. anaethetus*, a rarer
hurricane waif from the West Indies, is similar but paler gray, not
black, with a white collar between black cap and gray back.

*Fig. 9*

Sooty Tern

# LEAST TERN
*Sterna albifrons*                                    **26:1, 27:10**

**Description**
Size, 8½–9 in. (21.6–22.9 cm). A • *small tern*. Spring adult: white
forehead, underparts, and tail; black cap, eye line, and outer
primaries; yellow feet; tiny black tip to bill, • *bill yellow*. Fall
adult: head white, only nape and line to eye black, bill dusky, feet

fuller. Immature: similar, but with primaries and front edge of wing dusky.

**Similarities**
Immature Black has front edge of wing light, flanks grayish (not white), and tail dark.

**Habitat**
Sandy coasts, interior islands, and shorelines.

**Habits**
Very rapid and distinctive wingbeat.

**Voice**
A "rasping *zreeeep*; a rapid, high *kik kik kik*" (Collins).

**Food**
Small fish and aquatic life.

**Eggs**
2–3; variable, often olive-buff marked with drab; 1.2 x 1.0 in. (3.0 x 2.5 cm). Nest in depression in sand; in colonies.

**Range**
Breeds along Atlantic Coast from Maine s. to S. America and also in Mississippi Valley; winters from Gulf Coast southward.

---

## ROYAL TERN
*Sterna maxima*                                  **26:12, 27:1**

**Description**
Size, 18–21 in. (45.7–53.3 cm). A big, crested tern with a • *thick orange-red bill*. At rest, wing tips do not extend beyond end of tail. Spring adult: black cap, well-forked tail, underwings light, feet black. Fall adult: more white on forehead. Immature: sometimes with yellow legs.

**Similarities**
Bill thinner than Caspian's.

**Habitat**
Coast.

**Habits**
Flight heavier than Common's, lighter than Caspian's.

**Voice**
Shrill *keer*, a squawking *kowk*; a rolling, liquid whistle, *tourrreee*, suggesting Upland Sandpiper.

**Food**
Fish.

**Eggs**
1–2; whitish, evenly spotted with dark brown; 2.5 x 1.8 in. (6.4 x 4.6 cm). Nest is in depression in sand; in colonies.

**Range**
Breeds along Atlantic Coast from Va. to Fla. and along Gulf Coast to Tex.; winters from Gulf Coast, southward.

Note: The uncommon **SANDWICH TERN**, *Sterna sandvicensis* **(26:8, 27:4),** size, 15 in. (38.1 cm), breeds with Royal Terns along the Virginia coast, wintering to the south. It is small, gray-backed with a white, forked tail, and has a black cap with a slight crest and a black, yellow-tipped bill. Voice: *"Kirhitt, kirhitt"* (Yarrell). Nests with Royal Terns but the eggs are smaller, more variable—often buff with brown and lavender marks.

## CASPIAN TERN
*Sterna caspia*                                          **26:13, 27:2**

**Description**
Size, 19–23 in. (48.3–58.4 cm). A big, • *crested tern* with a • *thick
red bill*. Heavy body, broad wings; at rest, wing tips extend beyond
end of tail. Spring adult: black cap; primaries dusky above, darker
beneath; feet black. Fall adult: crown streaked, head appears dusky
above to below eyes. Immature: similar to fall adult, but marked
with black above; bill more orange.

**Similarities**
Bill is thicker, tail less forked than Royal's; build is stockier, flight
heavier, and has less crest than Royal.

**Habitat**
Coasts, large interior lakes, rivers.

**Habits**
Makes powerful Gannet-like dives; flight heavy; gull-like; flies low
when fishing, at a great height and with bill forward when
traveling; soars occasionally; often seen with Ring-billed Gulls.

**Voice**
"A hoarse, croaking *kraaa*" (Bent).

**Food**
Fish.

**Eggs**
1–3; buff, sparingly spotted with brown; 2.7 x 1.9 in. (6.9 x 4.8
cm). Nest is depression in sand; in colonies.

**Range**
Breeds from Newfoundland s. to Gulf Coast and w. to Great
Lakes; winters from N.C., southward.

## BLACK TERN
*Chlidonias nigra*                                       **26:2, 27:3**

**Description**
Size, 9–10¼ in. (22.9–26.0 cm). Mantle and tail slate-gray, tail
slightly forked, bill black. Spring adult: • *black head* and body, dark
red feet. Fall adult: nape black with point extending down behind
eye; forehead, collar, and underparts white; presents a pied
appearance when molting. Immature: mottled brown or mantle.

**Similarities**
In fall, gray color and slightly forked tail are helpful; immature
Least is paler, but with darker, more contrasting wings.

**Habitat**
Breeds in marshes and lakes, migrates along coast and marshes;
seen on North Atlantic Coast only in fall.

**Habits**
Flight in pursuit of insects like a Nighthawk's; usually picks food
from surface rather than by diving.

**Voice**
Short, shrill *crik* (longer, not as high-pitched as Least's) or longer
screaming *creek*.

**Food**
Insects, fish, small aquatic life.

**Eggs**
2–3; variable, often buffy-olive blotched with brown; 1.3 x 1.0 in.
(3.3 x 2.5 cm). Nests on ground or in marsh; in colonies.

**Range**
Breeds from N.S. s. to Pa. and w. to Mo.; winters in S. America.

# SKIMMERS
Family Rynchopidae

## BLACK SKIMMER
*Rynchops niger*                                        *Fig.10*

**Description**
Size, 18 in. (45.7 cm). Distinctive for lower bill, which is longer than upper, and very long wings. Adult: black above, white below; bill and legs red, bill with black tip. Immature: streaked brownish above, bill and legs paler.

**Habitat**
Sheltered bays on sandy coasts.

**Habits**
Flies low in flocks, cutting water surface with lower mandible, often flying back over same route; often active at night.

**Voice**
Like yelping of hounds.

**Food**
Fish, small aquatic organisms.

**Eggs**
4; pale buff, blotched with dark brown; 1.7 x 1.3 in. (4.3 x 3.3 cm). Nest is depression in sand; in colonies.

**Range**
Breeds along Atlantic and Gulf coasts from Mass. to Fla. and w. to Tex.; winters from S.C., southward.

*Fig. 10*

adult

immature

Black Skimmer

# AUKS, MURRES, AND PUFFINS
Family Alcidae

Alcids, as these birds are called, have short necks, short pointed wings, short tails, and webbed feet. Their legs are placed far back on the body. The plumage is generally black above and white below. They swim well, dive from the surface, and patter along the surface of the water to get aloft. They have a buzzy flight with rapid wingbeats. Their wings are used in swimming under water. These birds frequent rocky coasts, northern coastal cliffs, or rocky islands. The larger species feed mainly on fish; smaller alcids feed on crustaceans, mollusks, and other invertebrates. Dovekies are notable for their small size. They are compact seabirds with extremely short necks and white throats. Alcids are the Arctic ecological counterparts of the Antarctic penguins, which they superficially resemble.

## RAZORBILL
*Alca torda*                                              28:5

**Description**
Size, 17 in. (43.2 cm). Summer: heavy head and neck; black head
and upperparts, white below; • *white line from eye down side of
bill.* Winter: throat, lower face, and sides of neck white. Immature:
bill much smaller, but curved and blunt.
**Similarities**
Murres are less robust, have thinner, pointed bills.
**Habitat**
Coastal cliffs, offshore in water.
**Habits**
Swims lightly with head drawn in, bill tilted upward, and tail
cocked; looks heavy-headed, hump-backed in flight; migrates in
lines low over water; inquisitive.
**Voice**
Guttural croaks.
**Eggs**
1; often bluish-white spotted with brown; 3.0 x 2.0 in. (7.6 x 5.1
cm).
**Range**
Breeds along coast from Greenland to Maine; winters s. to N.J.

## COMMON MURRE
*Uria aalge*                                              28:1

**Description**
Size, 16–17 in. (40.6–43.2 cm). Bill thin, pointed; • *no white patch;*
white line on rear of wing. Summer: black head, throat, and
upperparts; white below. Ringed phase: eye-ring and streak behind
eye white. Winter: throat and side of head white, black line back
from eye.
**Similarities**
Head smaller, bill thinner, neck longer than Razorbill's; see also
Thick-billed Murre. Winter loons are larger and have spotted
backs and no white line on wing.
**Habits**
Flies with head and neck extended and drooping; looks long and
slender in flight (Razorbill looks short and compact; Thick-billed
Murre is intermediate); feet project beyond short tail.
**Voice**
Purring *mur-r-r-r-e.*
**Eggs**
1; light green, marked with brown and lilac; 3.2 x 2.0 in. (8.1 x
5.1 cm).
**Former name**
Atlantic Murre.
**Range**
Breeds in Arctic s. to Gulf of St. Lawrence; winters along Atlantic
Coast s. to N.J.

## THICK-BILLED MURRE
*Uria lomvia*                                             28:6

**Description**
Size, 18 in. (45.7 cm). Bill stout, pointed. Summer adult: black
above, white below, as Common Murre, but pale line along nape.
Winter: lower cheek only and throat white.
**Similarities**
Thicker, shorter bill than Common Murre; winter Common

107

Murre has more white on sides of head; Razorbill has thicker bill; winter loons lack white line on wing.

**Habits**
Flies with head and neck extended; less extended and less drooping than Common, less "heavy-headed" than Razorbill; can swim high or low in water.

**Voice**
Adults, a soft purr; young, an emphatic *beat-it, beat-it.*

**Eggs**
1–2; pear-shaped, variable, but usually indistinguishable from Common Murre's; 3.2 x 2.0 in. (8.1 x 5.1 cm).

**Former name**
Brünnich's Murre.

**Range**
Breeds in Arctic s. to Gulf of St. Lawrence; winters along Atlantic Coast s. to N.J.

# DOVEKIE
*Alle alle*                                                                     28:4

**Description**
Size, 8 in. (20.3 cm). Unmistakable by its small size, only that of a Starling. Bill stubby, neck short. Summer: head and upperparts black; streaks on back, line on wing, and underparts white. Winter: similar, but throat and sides of nape white.

**Habits**
Inquisitive; can swim high or low in water; flight suggests that of Chimney Swift, not as fast as guillemot or murre; drops off cliffs rapidly.

**Voice**
*Al-le.*

**Eggs**
1; pale greenish-blue; 1.7 x 1.2 in. (4.3 x 3.0 cm).

**Range**
Breeds in n. Arctic s. to Greenland; winters off Atlantic Coast from N.S. s. to N.J.

# BLACK GUILLEMOT
*Cepphus grylle*                                                                28:2

**Description**
Size, 13 in. (33.0 cm). Summer adult: black with • *white patch on wing* (this patch is rarely missing), • *inside of mouth and feet red.* Winter adult: foreback mottled gray, end of wing black, much of plumage white. Immature: dark above with white specks and much white on wing, barred with brown on sides, white below.

**Similarities**
The much larger, longer-billed White-winged Scoter is remotely similar in summer.

**Habits**
Can swim high or low in water, often dabbles nervously in water with bill, usually flies close to surface, takes flight readily.

**Voice**
"A high-pitched, complaining whine" (Snyder).

**Eggs**
2; whitish, sometimes spotted with brown; 2.4 x 1.5 in. (6.1 x 3.8 cm).

**Range**
Breeds in Arctic s. to Maine; winters off Atlantic Coast s. to N.Y.

## COMMON PUFFIN
*Fratercula arctica*                                                      28:3

**Description**
Size, 13 in. (33.0 cm). Bill high and brightly colored. Short-necked, chunky. Summer adult: black upperparts and collar, white underparts and cheek, • *large and varicolored bill,* red feet. Winter adult: cheek grayish, but head pattern similar; bill smaller. Immature: similar to winter adult, but bill small, yellow.

**Similarities**
Immature is smaller, floats higher in water, and looks chunkier than Razorbill or Thick-billed Murre.

**Habits**
Inquisitive, unsuspicious; stands high; flocks wheel in unison.

**Voice**
"A low, purring note, *purr-la-la-la*" (Townsend); a "deep nasal *hey-al*" (Cruickshank).

**Eggs**
1–2; dull white, occasionally freckled, usually earth-stained; 2.5 x 1.8 in. (6.4 x 4.6 cm). Nest is in burrows.

**Range**
Breeds along coast from Greenland s. to Maine; winters offshore in breeding range.

# Land Birds

## Pigeons and Doves
Order Columbiformes

## PIGEONS AND DOVES
Family Columbidae

This is a family of rapid-flying, small-headed birds with slender bills swollen at the base, long pointed wings, and short legs. They spend much time on the ground; bob their heads as they walk; eat grain, seeds, and nuts; and drink, unlike other birds, without raising their heads.

### ROCK PIGEON
*Columba livia*                                                    30:1

**Description**
Size, 13 in. (33.0 cm). This is the common street pigeon introduced from Europe and often gone wild. Tail squared; most wild birds bluish-gray with a white rump, 2 black bars on wing, and 1 on tail; eyes and legs red; feral birds may be highly variegated with shades of gray, brown, white, and black, but populations tend to revert to original wild plumage in time.
**Similarities**
Mourning Dove has pointed tail, tan color.
**Habitat**
Originally cliffs; now also adapted to cities, parks, bridges, freight yards, farmyards, beaches.
**Habits**
Gregarious; tame, may take food from the hand; flight strong.
**Voice**
Characteristic cooing.
**Eggs**
2; white; 1.5 x 1.1 in. (3.8 x 2.8 cm). Nest is of trash, on building ledge, cliff.
**Other names**
Domestic Pigeon, Rock Dove.
**Range**
Resident throughout much of N. America, from s. Canada, s. to Mexico.

Note: The **WHITE-CROWNED PIGEON,** *Columba leucocephala* **(30:2),** size, 13 in. (33.0 cm), occurs in southernmost Florida and the Keys. Looking much like an all-dark Rock Pigeon, it has a snowy white crown. It is a gregarious, colonial species feeding on fruits.

### MOURNING DOVE                                               30:3
*Zenaida macroura*                                             *Fig. 11*

**Description**
Size, 11–13 in. (27.9–33.0 cm). One of the most abundant and widespread North American birds. Wings long, pointed; • *tail long,* white-edged, • *pointed;* head brown; body buffy-gray with a • *bluish cast on wings.* Slimmer than Rock Pigeon and with pointed tail.

**Similarities**
American Kestrel in flight lacks sharp downstroke of dove, has
heavier head and shoulders, and squared tail.
**Habitat**
Open woodlands, suburban roadsides.
**Habits**
Feeds on ground, often takes dust baths or picks gravel from
roadside; flight direct, wings whir as it rises; in dry areas flies
daily to water; forms loose flocks in winter; frequently seen on wire
lines; may nest in cities on fire escapes.
**Voice**
Mournful *coo-ah, coo, coo.*
**Eggs**
2; white; 1.1 x 0.9 in. (2.8 x 2.3 cm); 2 or more broods in South.
Nest is usually in tree, loosely built of twigs, or on ground.
**Range**
Breeds from N.S. w. across Canada and s. through U.S. to C.
America; winters from n. U.S. southward.

*Fig. 11*

Mourning Dove

Common Ground Dove

# COMMON GROUND DOVE
*Columbina passerina*

30:4
*Fig. 11*

**Description**
Size, 6–7 in. (15.2–17.8 cm). A tiny dove; showing • *rusty patches
in wings* during flight. Spots on wings, tail short and black.
**Similarities**
Mourning Dove larger with pointed tail.
**Habitat**
Open areas, usually near coast.
**Habits**
Aptly named, found usually feeding on ground.
**Voice**
Repetitive *woo-oo, woo-oo,* often with rising intonation.
**Eggs**
2; white; 0.7 x 0.6 in. (1.9 x 1.7 cm). Nest is on ground, in tangle,
or tree, sometimes old nest of other bird, of twigs.
**Range**
Resident from s. U.S., s. to S. America.

# Parrots
Order Psittaciformes

## PARROTS
Family Psittacidae

This family includes the various parrots and parakeets. None of
these are illustrated here; the only native eastern parrot, the
Carolina Parakeet, *Conuropsis carolinensis,* is extinct. The only
parrot likely to be seen away from larger cities (where zoo and
aviary parrots of diverse species often escape and puzzle bird-
watchers) and southern Florida (where several "escaped" parrots
are feral and breeding) is the Monk Parakeet, *Myiopsitta
monachus,* a long-tailed, gray-headed, greenish parrot that is highly
social. It was introduced accidentally near New York City, and
now occurs in scattered eastern areas.

# Cuckoos and Allies
Order Cuculiformes

## CUCKOOS
Family Cuculidae

Members of this family are slender birds with long, slightly curved
bills; long, narrow, rounded tails; and zygodactyl feet (two toes
forward, two toes behind). North American cuckoos usually do not
lay their eggs in other birds' nests as do their European
counterparts. The family feeds on seeds, grasshoppers, and other
insects and small fruits.

### YELLOW-BILLED CUCKOO
*Coccyzus americanus*                                    **32:8**

**Description**
Size, 11–13½ in. (27.9–34.3 cm). Eyelids and • *lower bill yellow;*
olive-brown above, • *rufous on wing,* white underparts, • *large
white spots on tail.*
**Similarities**
Black-billed has small tail spots, no rufous on wings.
**Habitat**
Second growth, orchards, thicketed streamsides.
**Habits**
Secretive, moves noiselessly about upper foliage of small trees, sits
motionless; flight in open is direct, horizontal.
**Voice**
Wooden *kuk-kuk-kuk-kow-kow-kow* (or *kiaow-kiaow-kiaow* or
*kowlp-kowlp)*, the *kow, kiaow,* or last notes retarded; other *kek,
kow,* and *koo* notes, some indistinguishable from Black-billed's.
**Eggs**
2–6; bluish-green; 1.2 x 0.8 in. (3.0 x 2.0 cm). Nest is frail
platform of twigs 4–10 ft. (1.2–3.0 m) up in shrub or low tree.
**Range**
Breeds from Maine and Minn. s. to Mexico; winters in S.
America.

## BLACK-BILLED CUCKOO
*Coccyzus erythropthalmus*                                    **32:9**

**Description**
Size, 11–12½ in. (27.9–31.8 cm). Uniform olive-brown above,
white below; • *bill black;* • *eye-ring red.*

**Similarities**
Yellow-billed is not as slim. Distinguish from Yellow-billed by
absence of rufous on wing, large white spots on tail.

**Habitat**
Second growth.

**Habits**
Secretive, similar to Yellow-billed, but more active at night.

**Voice**
Notes evenly spaced, not retarded at end; a series of 2-syllable
notes, *kuk* or *coo* in groups. Yellow-billed may give similar calls.

**Eggs**
2–6; bluish-green; 1.3 x 0.8 in. (3.3 x 2.0 cm). Nest is a platform
of twigs, firmer than Yellow-billed's; 2–10 ft. (0.6–3.0 m) up in
shrub.

**Range**
Breeds from s. Canada s. to S.C. and w. to Kans.; winters in S.
America.

## MANGROVE CUCKOO
*Coccyzus minor*                                             **32:6**

**Description**
Size, 11–12 in. (27.9–30.5 cm). Resembles Yellow-billed, but has
buffy underparts, blackish ear coverts, and no rusty patch on the
wings.

**Similarities**
Resembles other cuckoos in form and movements but has black
mask and buffy underparts.

**Habitat**
Mangrove swamps.

**Voice**
Call resembles Yellow-billed's, but is slower, huskier.

**Range**
Resident only in s. Fla., s. to W. Indies and from Mexico to S.
America.

Note: In south Colorado, and central Kansas south to Texas, the
**ROADRUNNER,** *Geococcyx californianus (Fig. 12),* size, 23 in.
(58.4 cm), is a very large, ground-dwelling, long-tailed, streaky
cuckoo of open country. It is unmistakable as it runs about,
occasionally erecting its crest and eating lizards and insects. The
**SMOOTH-BILLED ANI,** *Crotophaga ani,* size, 12 in. (29.4 cm), is an
all-black, social cuckoo with a deeply arched bill, occurring locally
in open areas of southern Florida and the Keys.

*Fig. 12*

Roadrunner

# Owls
Order Strigiformes

Owls have big eyes that face forward, acute hearing, a broad head with a facial disk, a short neck, and soft, fluffy plumage that covers the base of the hooked bill and legs. They have sharp talons and are silent-flying, largely nocturnal birds of prey, although some are abroad by day. Owls feed extensively on rodents and are highly beneficial to the farmer. They usually perch upright, and periodically disgorge pellets of indigestible fur and bone. Owls are usually silent except in the mating season. Their eggs are white, almost round.

## BARN OWLS
Family Tytonidae

These owls are "monkey-faced" and have long legs extending beyond the tail in flight.

### BARN OWL
*Tyto alba*                                                   29:8

**Description**
Size, 14–20 in. (35.6–50.8 cm). Among owls, this species is unique for its "monkey-face" and pale breast. Long wings, dark eyes; buffy-brown above, pale below, relatively unmarked; • *face white, heart-shaped.* In flight, looks big-headed, slender-bodied, white below.

**Habitat**
Wood edges, farmland, haunts of man.

**Habits**
Quite nocturnal; frequents old belfries, water towers, barns, deserted buildings; flies with deep wingbeats, buoyantly, often reeling, legs trailing behind.

**Voice**
"(1) a discordant scream . . . (2) a snapping of the bill . . . (3) a flight call, resembling *ick-ick-ick-ick-ick*" (Potter and Gillespie); young in nest hiss.

**Food**
Rodents, insects.

**Eggs**
5–8; more oval than most owls; 1.8 x 1.3 in. (4.6 x 3.3 cm). Nest is in hollow trees, buildings, or cavities elsewhere.

**Remarks**
One of the few bird species of almost worldwide distribution, this owl in its various races is found on every continent, and even in islands as remote from each other as Curaçao and Madagascar.

**Range**
Resident throughout most of U.S. from s. Canada s. to S. America.

## TYPICAL OWLS
Family Strigidae

These owls have round or squarish faces and short legs that usually do not reach beyond the tail in flight.

## SCREECH OWL
*Otus asio*                                    **29:3**

**Description**
Size, 7–10 in. (17.8–25.4 cm). Breast streaked, eyes yellow; • *ear tufts*. Red phase: rufous; the reddest owl. Gray phase: brownish-gray; our grayest owl.

**Similarities**
Long-eared is taller, thinner, browner; ear tufts are set closer together.

**Habitat**
Orchards, open woods, suburbs, small towns.

**Habits**
Strictly nocturnal; often lives in hollow tree and sits in entrance hole by day; enjoys bathing.

**Voice**
Tremulous wail, descending, or all on 1 key; *"oh-o-o-o-o that I had never been bor-or-or-or-orn"* (Thoreau).

**Food**
Rodents, insects, other animal food.

**Eggs**
3–5; white; 1.4 x 1.3 in. (3.6 x 3.3 cm). Nest is in hole in tree, nook, or nesting box.

**Range**
Resident throughout most of U.S. from New England, w. across s. Canada and s. to Gulf Coast and Mexico.

## SNOWY OWL
*Nyctea scandiaca*                            **29:10**

**Description**
Size, 20–27 in. (50.8–68.6 cm). Eyes yellow. Adult male • *white* with some dark scaly barring, heavier on female and young.

**Similarities**
White Gyrfalcon is slenderer, has smaller head, longer neck, and pointed wings. Many young owls are whitish in downy plumage.

**Habitat**
Rolling tundra; in U.S., coastal marshes, farmland.

**Habits**
Forages mainly during the day; perches on dune, post, or other elevation for a lookout; flight strong, direct, but jerky, upbeat faster than down; often sails.

**Voice**
"A deep angry *krohgogogok,* almost like a raven" (Hantzsch).

**Food**
Small mammals (especially lemmings), ptarmigan.

**Eggs**
4–10; white; 2.3 x 1.8 in. (5.8 x 4.6 cm). Nest is depression in tundra.

**Range**
Breeds in Arctic and n. Canada; winters s. rarely to S.C.; populations migrate further s. when food is scarce in n.

## GREAT HORNED OWL
*Bubo virginianus*                            **29:11**

**Description**
Size, 18–25 in. (45.7–63.5 cm). • *A very large owl with ear tufts;* female larger. Wings long, broad; eyes yellow. Dark brown, heavily barred, and streaked with black; throat white; northern races paler. In flight, ear tufts flattened; dark, looks neckless.

**Similarities**
Long-eared is smaller; streaked, not barred below; has ear tufts closer together. Red-tailed Hawk is smaller, with proportionally smaller head and wings.
**Habitat**
Woodland edges, woods.
**Habits**
Most powerful of owls; sometimes hunts by day as well as by night, often sails on fixed wings; sometimes soars like a Red-tailed.
**Voice**
Bass; a deep *hoo-hoo-hoo-hoo, hoo-hoooo* (deeper than Barred); a *waugh-HOO;* less commonly, a blood-curdling shriek.
**Food**
Small mammals, birds.
**Eggs**
2; white; 2.2 x 1.8 in. (5.6 x 4.6 cm), often laid in February. Nest is often in deserted hawk's or crow's nest, usually in heavy timber.
**Range**
Resident throughout most of N. America from Arctic to S. America, excluding W. Indies.

---

# HAWK OWL
*Surnia ulula*                                                              29:5
**Description**
Size, 14½–17½ in. (36.8–44.5 cm). Appearance dark, plump; wings short, pointed; • *tail long,* graduated, banded. Gray-brown above, head with white spots and black sideburns.
**Similarities**
Boreal and Saw-whet lack dark bars and long tail.
**Habitat**
Natural openings in northern coniferous forest.
**Habits**
Diurnal; often perches hawklike on tops of trees with body bent forward; pumps, sometimes cocks tail; flight falconlike, direct, swift, close to ground, and with glides; hovers, drops on prey like a shrike; tame.
**Voice**
"A trilling whistle *tu-wita-wit, tuwita-tu-wita, wita, wita"* (Henderson).
**Food**
Rodents, other small mammals, grouse.
**Eggs**
3–7; white; 1.5 x 1.3 in. (3.8 x 3.3 cm). Nest is in hollow tree, top of stump, or deserted bird's nest.
**Range**
Breeds in n. Canada from Newfoundland w. to Que.; winters from breeding range s. to n. U.S.

---

# BURROWING OWL
*Athene cunicularia*                                                        29:1
**Description**
Size, 9–11 in. (22.9–28.9 cm). Lives on the ground. Legs long; tail short; eyes yellow; brown above, spotted with white; white below, barred with brown.
**Habitat**
Open places, even vacant lots in cities.
**Habits**
Lives in rodent or other holes, often standing near entrance or

perching in eminence on ground, post, wire, or shrub; bobs head and tail up and down in comical fashion on its long legs; often follows moving animals; flies little.

**Voice**
In flight, a chattering note; alarm note, *tsip-tsip.*

**Food**
Insects, small vertebrates.

**Eggs**
5–9; white; 1.3 x 1.1 in. (3.3 x 2.8 cm). Nest is grass-lined chamber at end of manure-bordered burrow in ground.

**Range**
Resident in Fla., W. Indies, and s. to S. America.

## BARRED OWL
*Strix varia*                                                      **29:9**

**Description**
Size, 17–24 in. (43.2–61.0 cm). A big, round-headed brownish owl without ear tufts. Bill yellow, eyes dark; gray-brown above; pale below, with • *dark crossbars on breast* and collar and dark streaks on belly.

**Similarities**
Great Horned and Long-eared have ear tufts and reddish facial disks; all other owls, except Barn, have yellow eyes.

**Habitat**
Moist woods.

**Habits**
Wingbeat slow, flight buoyant; inquisitive.

**Voice**
Hoots higher-pitched than Great Horned, commonly 8 hoots in 2 groups of 4; at distance may be confused with dog barking.

**Food**
Mice, other small animals, birds.

**Eggs**
2–3; white; 2.0 x 1.6 in. (5.1 x 4.1 cm). Nest is in hollow tree or deserted nest of hawk, crow, or squirrel.

**Range**
Resident from cen. Canada s. to Gulf of Mexico and w. to Rockies.

## GREAT GRAY OWL
*Strix nebulosa*                                                  **29:12**

**Description**
Size, 24–33 in. (61.0–83.8 cm). A big, round-headed gray owl with no ear tufts. • *Eyes yellow;* plumage loose, fluffy; gray, streaked below with black; wings and tail long and broad; face disks very large, black chin spot; no bars on breast.

**Similarities**
Barred Owl is smaller and browner, with bars on breast, smaller face disk, larger and darker eyes, and shorter tail. Great Horned Owl has ear tufts, is browner.

**Habitat**
Northern coniferous forest, rarely wandering south in winter.

**Habits**
Flies with slow flaps of broad, rounded wings; hunts by daylight in Arctic summer; tame.

**Voice**
"Several deep-pitched *whoos*" (Grinnell and Storer); also a wavering cry suggesting Screech Owl.

**Food**
Small mammals.

**Eggs**
2–5; white; 2.2 x 1.8 in (5.6 x 4.6 cm). Nest is of sticks from 20 ft. (6.1 m) up in evergreens.
**Range**
Resident throughout much of Canada and most of U.S., from Arctic to S. America.

## LONG-EARED OWL
*Asio otus*                                             **29:7**

**Description**
Size, 13–16 in. (33.0–40.6 cm). Long wings and tail; ear tufts close together; blackish-brown above; rusty face; • *streaked and lighter below;* yellow eyes. In flight, ear tufts not visible; brown spot shows on buff underwing.
**Similarities**
Great Horned is larger, bulkier; has ear tufts wider apart, a white throat, and more crossbars below; in flight, Short-eared looks lighter and occurs in more open country.
**Habitat**
Mixed woodlands or, preferably, evergreens.
**Habits**
Flight wavering; perches close to trunk of tree, stretches body very thin, can successfully hide in slight cover; may give distraction displays at nest, spreading wings and body, looking fierce, or feigning injury, leading one away from nest. Flocks in migration, and in winter often in groves of dense evergreens.
**Voice**
A "dove-like *hoo hoo hoo . . . a* slurred whistle, *WHEE-you"* (Griscom); snaps bill and makes great variety of noises at nest, when disturbed.
**Food**
Small mammals.
**Eggs**
3–8; white; 1.6 x 1.3 in. (4.1 x 3.3 cm). Nest is in conifer, of sticks.
**Range**
Breeds from s. Canada s. to s. U.S.; winters in s. parts of breeding range s. to Gulf states and Mexico.

## SHORT-EARED OWL
*Asio flammeus*                                        **29:6**

**Description**
Size, 13–17 in. (33.0–43.2 cm). A day-flying owl with its streaked breast, wavering flight, but no ear tufts. Bright buffy-brown, darker on back; head and pale breast streaked with black; facial disk dark, bordered with white; eyes yellow; wings with light patch above and black spot near bend below; appears neckless in flight.
**Similarities**
Buffier, and with longer wings than Long-eared. Owl head, short tail, and absence of white rump distinguish it in flight from Northern Harrier. Barn Owl is paler, lacks wing patches. Rough-legged Hawk also has black mark under wing, but has smaller head, more neck, white rump, and less wavering flight.
**Habitat**
Marshes, meadows, prairies, open country.
**Habits**
Often hunts by day, quarters low over ground like a Northern Harrier or sits on observation post; roosts on ground; has impressive aerial courtship flight.

**Voice**
*Toot-toot-toot* (to 20 times), higher than Great Horned; a sharp, catlike *ski-ow*.
**Food**
Mice.
**Eggs**
4–9; white; 1.5 x 1.3 in. (3.8 x 3.3 cm). Nest is on ground near clump of vegetation in marsh or meadow.
**Remarks**
This cosmopolitan species hunts by night the same marshes hunted by the Northern Harrier by day, but at times both may be abroad in daylight.
**Range**
Breeds locally from n. Canada s. to N.J.; winters from s. parts of breeding range s. to C. America.

## BOREAL OWL
*Aegolius funereus*                                              **29:2**

**Description**
Size, 8½–12 in. (21.6–30.5 cm). • *No ear tufts;* • *yellow bill and eyes;* brown above, with white spots on forehead and back, and • *black rim around face disk;* underparts pale, streaked with brown.
**Similarities**
Darker brown and with longer wings and tail than red phase of ear-tufted Screech-Owl. Saw-whet is smaller, streaked instead of spotted on forehead, with no black ring around face disk, and with a black bill.
**Habitat**
Coniferous forests.
**Habits**
Very tame; hunts by daylight in Arctic summer.
**Voice**
Notes like dripping water.
**Food**
Mice, birds, insects.
**Eggs**
4–7; white; 1.2 x 1.0 in. (3.0 x 2.5 cm). Nest is in holes in trees or deserted nests of other birds.
**Former name**
Richardson's Owl.
**Range**
Breeds from Labr. w. across Canada and s. to N.B., Ont., and s. Man.; winters from s. parts of breeding range s. as far as N.J.; Ill. (rare).

## SAW-WHET OWL
*Aegolius acadicus*                                             **29:4**

**Description**
Size, 7–8½ in. (17.8–21.6 cm). A very small owl with a black bill and • *no ear tufts*. Wings long and broad; eyes yellow; brown above, with white streaks on head; underparts white, heavily streaked with rufous. Young: chocolate brown with • *dusky face* and white V on forehead.
**Similarities**
Boreal is larger and has yellow bill, black rim around face, and white spots on forehead.
**Habitat**
Evergreen forests, swamps.

**Habits**
Strictly nocturnal, inquisitive; in taking off, bird often drops a little
before flying forward; flight suggests that of a woodcock.
**Voice**
In spring, like the filing or whetting of a saw with notes tapering
off at end; also a ventriloquial "soft *co-co-co-co-co-co*" (Eckstrom).
**Food**
Insects, small mammals.
**Eggs**
3–7; white; 1.0 x 0.9 in. (2.5 x 2.3 cm). Nest is in woods or
swamps in old woodpecker hole.
**Range**
Breeds from N.S. w. to Man. and s. to Md. and Kans.; winters
from s. parts of breeding range s. to Gulf Coast and C. America.

# Goatsuckers
Order Caprimulgiformes

# GOATSUCKERS
Family Caprimulgidae

Goatsuckers have huge, often bristle-bordered mouths; huge, flat
heads; big, dark eyes; long wings, and often long tails; and small,
weak feet. Their fluffy plumage is associated with their silent,
wavering flight. Their mottled colors afford protection when they
rest motionless by day on leaves or gravel, or lengthwise on the
branch of a tree. These birds are most active by night, have highly
distinctive voices, feed on the wing on insects, and lay their
protectively colored eggs, with no nest, on bare leaves or gravel.
The name goatsucker comes from ancient times and the erroneous
belief that with their wide, gaping mouths they sucked milk from
goats.

## CHUCK-WILL'S-WIDOW
*Caprimulgus carolinensis*                                      **30:7**

**Description**
Size, 12 in. (30.5 cm). Has bristles around its very wide mouth;
• *tail rounded,* longer than the rounded, owl-like wings; mottled
buffy-brown, no white spot on wing. Male: narrow white band
across lower throat, white tail corners. Female: white on throat and
tail replaced by buff. Whip-poor-will has more white on throat
and outer tail feathers.
**Habitat**
Deciduous woods, common in southern woodlands.
**Habits**
Starts calling at dusk; its eyes often shine in headlights along
country roadsides.
**Voice**
Emphatic *chuck-wills-WID-ow;* also a *quak* when flushed.
**Food**
Large insects such as moths, rarely small birds.
**Eggs**
2; white, mottled with gray and brown; 1.4 x 1.0 in. (3.6 x 2.5
cm). Laid on leaves or bare ground.
**Range**
Breeds from N.Y. w. to Kans. and s. to Gulf states; winters in W
Indies and from Mexico, s. to S. America.

## WHIP-POOR-WILL
*Caprimulgus vociferus*                                    **30:9**

**Description**
Size, 10 in. (25.4 cm). Bristles about mouth; • *tail rounded,* longer
than wings; mottled brown, sides unbarred, no white wing spot.
Male: fairly narrow white band low on black throat and much
white on outer tail feathers. Female: buff instead of white on
throat and dark outer tail feathers.
**Similarities**
Common Nighthawk is gray-brown, has white spot on wing, and
has forked tail.
**Habitat**
Woods, especially near fields.
**Habits**
Calls most near dusk, dawn, and on moonlit nights; feigns broken
wing to lure intruder from young; flight low, erratic; sails on
horizontal wings, tilting them sometimes almost vertically in
turning. Red eyes sometimes shine in headlights along roadside.
**Voice**
Cluck often precedes the *whip-poor-WILL,* more accurately
written *pur-ple RIB.*
**Food**
Various moths, beetles, flying insects.
**Eggs**
2; white, mottled with brown and gray; 1.2 x 0.8 in. (3.0 x 2.0
cm). Laid on leaves or bare ground.
**Range**
Breeds from s. Canada s. to s. U.S., and in mountains s. to
Mexico; winters from Gulf of Mexico s. to C. America.

## POOR-WILL
*Phalaenoptilus nuttallii*                                 **30:8**

**Description**
Size, 7–8½ in. (17.8–21.6 cm). Tail short, rounded; no white spot
on rounded wing; mottled with gray-brown; throat white, buffy in
young; breast very dark; extreme corners of tail white, narrower in
female.
**Similarities**
Suggests a small Whip-poor-will, has less white in tail.
**Habitat**
Dry, open country.
**Habits**
Pink eyes often shine in headlights along roadsides, bird flutters up
as car passes; rests by day on ground, sometimes in tree; flight low,
mothlike, in short sallies.
**Voice**
At a distance sounds like *poor-will* or, farther off, *p'will;* also a
clucking note in flight.
**Food**
Nocturnal flying insects.
**Eggs**
2; plain white; 1.1 x 0.8 in. (2.8 x 2.0 cm). Laid on bare ground.
**Remarks**
This is one of the few birds that ever become dormant. In winter,
torpid specimens with distinctly lowered body temperatures have
been found in rock crevices.
**Range**
Breeds in W., but e. to Nebr. and Iowa, s. to cen. Tex.; winters
from s. Tex. southward.

## COMMON NIGHTHAWK
*Chordeiles minor*                                                    **30:6**

**Description**
Size, 8½–10 in. (21.6–25.4 cm). No bristles about mouth; wings long, thin, extend beyond end of forked tail; mottled gray-brown; sides pale, barred; male has • *broad white band on throat* and near end of tail, throat band in female is buffy.

**Similarities**
Wings of Whip-poor-will are not as pointed.

**Habitat**
Open country, cities.

**Habits**
Often seen by day; flies high in bouncing, erratic flight, hawking insects in spectacular aerial revolutions; also swoops low over cities and near street lights; often seen in migration, especially along river valleys, in loose flocks of 20–100 birds. Defends young by attacking intruder, hissing, with mouth agape or by feigning injury or death; sometimes roosts on fence posts.

**Voice**
Harsh *peenk,* given in flight with 3 double-speed flips of wings; suggests ground call of courting male Woodcock; in courtship, a booming sound at end of dive.

**Food**
Diverse insects, including mosquitoes.

**Eggs**
2; grayish-white, blotched with brown; 1.2 x 0.9 in. (3.0 x 2.3 cm). On bare ground or flat roofs in towns and cities.

**Range**
Breeds throughout most of Canada and U.S. from cen. Canada s. to C. America; winters in S. America.

# Swifts and Hummingbirds
Order Apodiformes

# SWIFTS
Family Apodidae

Swifts have short bills; long, narrow curved wings; short heads and tails; and small, weak feet. The ends of the tail feathers are stiff. They are among the swiftest fliers. Their habitat is the sky, where they feed on flying insects, drink, mate, and collect twigs (without perching), which they cement into a nest with their saliva.

## CHIMNEY SWIFT
*Chaetura pelagica*                                                  **30:5**

**Description**
Size, 4¾–5½ in. (12.1–14.0 cm). An all-dark bird that looks like a flying cigar; the only swift in most of the East. Wings sicklelike, tail squared, plumage sooty throughout.

**Habitat**
The sky, usually near habitations.

**Habits**
Alternately flies and sails, often high up, on set wings that seem not to bend; 2 or 3 birds frequently fly together; roosts clinging upright in chimney, well, cave, or hollow tree, supported by its stiff tail; many collect for migration into a great flock, which spirals funnellike at evening into a selected chimney to roost.

**Voice**
Characteristic repeated twittering in flight.

**Eggs**
4–6; white; 0.7 x 0.5 in. (1.8 x 1.3 cm). Nest is inside chimney or hollow tree.

**Range**
Breeds throughout much of e. U.S. from N.S. w. to s. Man. and s. to Gulf of Mexico; winters in S. America.

Note: The **WHITE-THROATED SWIFT,** *Aeronautes saxatilis,* size, 7 in. (17.8 cm), may be seen in the mountains and valleys of the western fringes. It is larger than the Chimney Swift and is easily distinguished by its white throat.

# HUMMINGBIRDS
Family Trochilidae

This family includes the smallest birds—those with the fastest wingbeat and the only ones that can fly backward. "Hummers" have long, needlelike bills; extensible tongues; partly iridescent plumage; and small, weak feet.

## RUBY-THROATED HUMMINGBIRD
*Archilochus colubris*                                                    **34:15**

**Description**
Size, 3½ in. (8.9 cm). This is the only hummingbird in most of the East. Male: tail forked; bronze-green above, whitish below; with
• *ruby throat* (which appears black in some lights); green back.
Female: tail rounded, throat and tips of outer tail feathers white. Note absence of rufous edgings in tail feathers, and relatively silent flight; do not confuse with large hawkmoths.

**Habitat**
Gardens, woodlands, areas with flowers.

**Habits**
Flies 500 miles across Gulf of Mexico in migration; readily frequents feeding stations.

**Voice**
Shrill squeals, chirps, and chippering.

**Food**
Small insects, some nectar.

**Eggs**
2; white; 0.5 x 0.3 in. (1.3 x 0.9 cm).

**Range**
Breeds from s. Canada s. to Gulf Coast; winters from Mexico s. to C. America, with some n. to Gulf Coast.

Note: The **BROAD-TAILED HUMMINGBIRD,** *Selasphorus platycercus,* size, 4–4½ in. (10.2–11.4 cm), is a Rocky Mountain species frequenting the western fringe of the East, especially in migration. Males are very like the Ruby-throated, but have a paler, rosier throat, no strong fork in the tail, and shrill sound of whistling wings; females have rusty sides and rufous in the tail.

# Kingfishers and Allies
Order Coraciiformes

## KINGFISHERS
Family Alcedinidae

Kingfishers have strong, pointed bills, longer than the head; big heads; big eyes; short tails; and small legs and feet. The third and fourth toes are joined together. They fish from a perch above water or by hovering and diving headfirst, feeding on fish, some insects, and lizards. The scales and bones are ejected in pellets.

### BELTED KINGFISHER
*Ceryle alcyon*                                                          35:12

**Description**
Size, 13 in. (33.0 cm). Crested, looks top-heavy. Male: upperparts and breastband gray-blue, underparts and collar white. Female: with an additional band of chestnut on lower chest and flanks.
**Habitat**
Shorelines.
**Habits**
Solitary; each has own territory along watercourse with a series of observation posts; hovers before diving; flight straight, 2 regular wingbeats alternate with several faster ones.
**Voice**
Loud, wild rattle, often heard in flight.
**Eggs**
5–8; white; 1.3 x 1.1 in. (3.3 x 2.8 cm). Nest is chamber at end of 4–8 ft. (1.2–2.4 m) tunnel excavated in earth or gravel bank, often away from water.
**Range**
Breeds from n. Canada s. through most of U.S.; winters in C. America.

# Woodpeckers and Allies
Order Piciformes

## WOODPECKERS
Family Picidae

Woodpeckers are arboreal, medium-sized birds with strong skulls; strong, pointed bills; extensible tongues; stiff tails; and zygodactyl feet (two toes pointing forward and two pointing back). Many species are resident, but the Common Flicker, Red-headed Woodpecker and Yellow-bellied Sapsucker are partly migratory. Most feed largely by clinging to the bark of a tree with their feet, bracing with the tail, and chiseling with the beak in search of wood-boring insects. A woodpecker often signals its presence by the tattoo it plays on a dead limb. Woodpeckers have an undulating flight. They nest, and usually roost, in holes, which they excavate with their bills in dead timber.

## COMMON FLICKER
*Colaptes auratus*                                                    **31:2**

**Description**
Size, 13 in. (33.0 cm). Upperparts brown with barring; • *white
rump* (conspicuous in flight), • *yellow* (usually) *or reddish
underwings*, black crescent on breast, black spots below. Males
have black (or red) mustaches, lacking in female.

**Habitat**
Suburbs, farms, wood edges, woods.

**Habits**
Spends much time on ground, may loudly drum on dead limb or
tin roof; glides upward as it alights on tree; has noteworthy
displays in breeding season.

**Voice**
Long series of *wik* notes; alarm *pee-ah;* a repetitive *wick-a, wick-a*
or *wick-up, wick-up* during displays; many others.

**Food**
Ants, other insects, fruits, berries.

**Eggs**
6–8; white; 1.0 x 0.8 in. (2.5 x 2.0 cm).

**Former names**
Yellow-shafted Flicker, Red-shafted Flicker.

**Range**
Breeds from Newfoundland w. to Man. and s. to s. Fla. and Gulf
Coast; winters from N.Y. and s. Ont. s. to Gulf Coast and Mexico.


## PILEATED WOODPECKER
*Dryocopus pileatus*                                                  **31:1**

**Description**
Size, 16–19 in. (40.6–49.5 cm). A crow-sized woodpecker, with a
crest. Male: black, with • *red crest* and *mustache* and big, heavy
bill; white stripe on thin neck; big, white wing patch conspicuous
in flight. Female: less red on crest and no red mustache.

**Habitat**
Woods with large trees.

**Habits**
Strips off quantities of bark and chisels big rectangular holes in
dead or diseased trees; shy; flight vigorous, somewhat slow, either
straight and crowlike or in long undulations.

**Voice**
A *yuk-yuk-yuk* sounding hollow, louder, heavier, slower than
Common Flicker's; in flight, a slow *puck, puck;* "call, a loud *cack-
cack-cack*" (Bendire); drumming signals are higher, faster, and
louder in the middle, dying away at end.

**Food**
Wood-boring beetles and ants, berries, nuts.

**Eggs**
3–6; white; 1.4 x 1.0 in. (3.6 x 2.5 cm).

**Remarks**
This big woodpecker seems gradually adapting itself to well-
wooded suburbs, even nesting within sight of New York City.

**Range**
Resident from s. Canada to Gulf Coast and s. almost to Mexico.

Note: The probably extinct (except for a few in Cuba) **IVORY-
BILLED WOODPECKER,** *Campephilus principalis,* size, 11 in. (48.2
cm), has been reported but not proven to exist in certain
southeastern states (Florida, South Carolina, Louisiana, Texas).

Similar to the Pileated Woodpecker, it has a black throat, white patches in front and back of wings in flight, and conspicuous, large, white patch at rest.

## RED-BELLIED WOODPECKER

*Melanerpes carolinus*                                                31:5

**Description**
Size, 10 in. (25.4 cm). Distinctive • *red nape;* • *barred back and wings*, pale, unmarked face. Adult: white rump and wing patches conspicuous in flight; gray underparts; reddish wash on belly inconspicuous; entire crown red in male, gray in female. Immature: duller, head all-gray.

**Habitat**
Woods, edges, farms, suburban areas, swamps.

**Habits**
Comes to feeders, will also feed on ground; stores acorns, nuts in crevices for future use.

**Voice**
Noisy; *cha-cha-cha;* a *churr, churr,* like Red-headed's, but lower; a wide variety of other notes.

**Food**
Ants, other insects, fruits, acorns.

**Eggs**
4–5; white; 1.0 x 0.8 in. (2.5 x 2.0 cm).

**Range**
Resident throughout most of se. U.S. from Atlantic Coast w. to Tex., and from Conn. s. to Fla.

## RED-HEADED WOODPECKER

*Melanerpes erythrocephalus*                                         31:10

**Description**
Size, 9 in. (22.9 cm). Adult: above dark blue-black; • *head and neck all-red;* white rump and underparts; conspicuous white wing patch; in some light appears red, white, and blue. Immature: mottled brown head and body, but has white rump and wing patch like adult.

**Habitat**
Orchards, roadsides, farming country, open woods.

**Habits**
Social, quarrelsome; catches insects in air like a flycatcher; stores nuts in cracks and cavities; tends to drop from tree almost to ground and then fly low; many killed by automobiles; distribution spotty.

**Voice**
*Churr, churr,* higher-pitched than Red-bellied's; a trill like the Gray Tree Frog: *"ker-r-r-ruck, ker-ruck-ruck-ruck"* (Merriam).

**Food**
Beechnuts, acorns; insects, fruit; occasionally eats young birds.

**Eggs**
4–6; white; 1.0 x 0.8 in. (2.5 x 2.0 cm).

**Range**
Breeds from Que. and Man. s. to Fla., Gulf Coast, and La.; winters in S. America; scarce in ne. U.S.

Note: The **LEWIS' WOODPECKER,** *Melanerpes lewis,* size, 11 in. (27.9 cm), is found on the western fringe of the East. This dark woodpecker has a distinctive gray-white collar, pinkish underparts, and a red face. Its flight is direct, unwoodpeckerlike. Often found in rather open country.

## YELLOW-BELLIED SAPSUCKER
*Sphyrapicus varius*                                                31:8

**Description**
Size, 8 in. (20.3 cm). Conspicuous • *white wing stripe;* • *red forehead* and black upper breast; belly yellowish. Male: red throat. Female: white throat. Immature: very different, brownish.

**Similarities**
Immature Red-headed has differently placed wing patch, shows separation of brown upper breast and throat from whitish rest of underparts. (No such contrast in sapsuckers.)

**Habitat**
Forests, wood lots, orchards.

**Habits**
Perforates bark of trees with even rows of small holes, mainly to get sap, but also insects attracted to it; secretive in migration, perching quietly, often on side of tree away from the observer.

**Voice**
Noisy in spring, a querulous *wha-ee;* a ringing clear, 5-6 times; drums in a disconnected roll, individual taps often being well separated.

**Food**
Various insects in and on trees, also regularly drills holes and feeds on sap flowing from them.

**Eggs**
5-7; white; 0.9 x 0.7 in. (2.3 x 1.8 cm).

**Range**
Breeds from cen. Canada, s. to Va.; winters from N.J. and Ohio Valley s. to C. America W. Indies.

## HAIRY WOODPECKER
*Picoides villosus*                                                 31:4

**Description**
Size, 8½-10½ in. (21.6-26.7 cm). Size of Robin. Black-and-white pattern above and on wings, white down center of back and below; 3 outer tail feathers white. Male has • *red on back of head.*

**Similarities**
Bill much longer than Downy's, length twice the distance from base of bill to eye.

**Habitat**
Woods.

**Habits**
Woodland counterpart of the dooryard Downy; sometimes travels in mixed groups with Downies, nuthatches, and chickadees; comes closer to habitations in winter; shier, noisier, and more restless than smaller Downy.

**Voice**
Loud *peek*, louder and sharper than Downy's; a rattle, loud and on same pitch; drumming usually louder than Downy's, but often indistinguishable at a distance.

**Food**
Borers, caterpillars, other tree-bark insects.

**Eggs**
3-6; white; 0.9 x 0.7 in. (2.3 x 1.8 cm).

**Range**
Resident from n. Canada s. to s. U.S. and C. America.

## DOWNY WOODPECKER
*Picoides pubescens*                                                31:3

**Description**
Size, 6–7 in. (15.2–17.8 cm). Size of House Sparrow; commonest
small woodpecker. Similar to Hairy but smaller; 3 outer tail
feathers have black bars; bill shorter, length equals distance from
base of bill to eye.
**Similarities**
Hairy is larger, has no black bars in outer tail feathers.
**Habitat**
Open woodland, wood edges, trees about habitations, orchards.
**Habits**
Easily attracted to feeders, tame; often travels in mixed flocks with
chickadees, nuthatches, kinglets, and Brown Creepers.
**Voice**
Short, sharp *pik;* a rattle of 15 or so rapid staccato notes, like
Hairy's, but not so loud, normally descending in pitch; tattoo a
long roll.
**Food**
Borers, surface insects, some vegetation.
**Eggs**
4–7; white; 0.7 x 0.6 in. (1.8 x 1.5 cm).
**Range**
Resident from n. Canada, s. to s. U.S.

Note: The **LADDER-BACKED WOODPECKER**, *Picoides scalaris,* size,
6–7½ in. (15.2–19.1 cm), barely reaches the southwestern fringe of
the East. It has a striped face and barred back, and is a black-and-
white woodpecker of the desert and riverside timber.

## RED-COCKADED WOODPECKER
*Picoides borealis*                                                 31:6

**Description**
Size, 8 in. (20.3 cm). • *Ladder back, black crown;* large white cheek
patch. Male: has a small, often hidden red "cockade" at side of
crown behind eye. Immature: buffy below, instead of white, and
with red crown patch.
**Habitat**
Mature southern pine forests with old trees.
**Habits**
Somewhat social, noisy, active, quarrelsome, wary.
**Voice**
Call note suggests small woodwind instrument; also a nasal,
querulous, scolding note.
**Food**
Mainly wood-boring insects.
**Eggs**
3; white, 1.0 x 0.7 in. (2.5 x 1.8 cm).
**Remarks**
Its requirement for old pines with rotted centers has made it
vulnerable through pinewoods management and timber practices.
On the decrease, it is classed as an Endangered Species.
**Range**
Resident of se. U.S. from Md. and Ky. s. to Fla. and w. to e. Tex.

## BLACK-BACKED WOODPECKER
*Picoides arcticus*                                              31:7

**Description**
Size, 9–10 in. (22.9–25.4 cm). • *Black-backed* with barred sides.
Male has a • *yellow crown patch.*

**Similarities**
Three-toed is smaller, back barred or mainly white.

**Habitat**
Coniferous forests, often near water.

**Habits**
Solitary, unsuspicious, movements deliberate; makes periodic winter
incursions south of normal range; freshly barked tree trunks are
often a sign of this species.

**Voice**
Sharp, metallic *pik;* distinctive rattle call, a screaming, *wreeoh;*
noisy and aggressive about nest.

**Food**
Mostly subsurface arboreal insects, few nuts and fruits.

**Eggs**
4–5; white; 0.9 x 0.7 in. (2.3 x 1.8 cm).

**Former names**
Arctic or Black-backed Three-toed Woodpecker.

**Remarks**
Look for this species in boggy areas amid dead conifers, and about
clearings or burns.

**Range**
Resident from cen. Canada s. to n. U.S.

## THREE-TOED WOODPECKER
*Picoides tridactylus*                                          31:9

**Description**
Size, 8–9½ in. (20.3–24.1 cm). Noticeably smaller, browner, less
contrastingly black-and-white than Black-backed. Male has a
• *yellow crown patch;* in flight, • *barred back* is conspicuous and tail
flashes white.

**Similarities**
Some eastern birds show little white on back and may be mistaken
for Black-backeds—look for more broadly white face marks, black
marks in white of outer tail, and signs of white on back to
distinguish from Black-backed.

**Habitat**
Coniferous forest, especially spruce.

**Habits**
Solitary, unsuspicious, sedentary; often rarer, more local and is less
noisy, and flight less vigorous than Black-backed.

**Voice**
Call differs from Black-backed, a weak *pik* or *peet;* a rattle, like
Hairy's, but softer.

**Food**
Beetle larvae and other bark insects.

**Eggs**
4; white; 0.9 x 0.7 in. (2.3 x 1.8 cm).

**Former names**
American or Northern Three-toed Woodpecker.

**Range**
Resident from Arctic s. to n. U.S.

# Perching Birds
Order Passeriformes

This order includes far more species than any other, and has most
of the land birds. The feet have three toes forward and one behind,
making them well designed for grasping a perch. The young are
hatched naked, blind, and helpless and are cared for in the nest
until fledged.

# TYRANT FLYCATCHERS
Family Tyrannidae

Flycatchers have flattened bills with a little hook at the tip and
bristles about the broad base. They make their headquarters on a
perch, from which they sally forth to snap up passing flying
insects, their main food. Some species are quite pugnacious, whence
the family name. All, except the Eastern Phoebe, spend the winter
in southern regions.

The *Empidonaces* (Genus *Empidonax*) are flycatchers 5½ to 6
inches (14–15.2 cm) in size that look alike, so most species cannot
safely be identified by plumage alone. All have olive-grayish, or
brownish, backs and whitish or yellowish breasts. Most have
conspicuous light eye-rings, two wing bars, and pale lower
mandibles. However, their notes in the breeding season are
distinctive. The somewhat smaller wood pewees (Genus *Contopus*)
and Eastern Phoebe are larger and lack white eye-rings.

## EASTERN KINGBIRD
*Tyrannus tyrannus*                                        **33:7**

**Description**
Size, 8–9 in. (20.3–22.9 cm). Black above, white below; 2 very
narrow white wing bars; prominent white band at end of tail; male
with a concealed crimson crown patch.
**Similarities**
Western lacks white band at tail tip and has yellow in plumage.
**Habitat**
Roadsides, farms, orchards; woods, meadows, streams; shorelines.
**Habits**
Very pugnacious; attacks crows, hawks, vultures, sometimes
alighting on their backs; flies horizontally on quivering wings.
**Voice**
Incisive *tzee,* alone or rapidly repeated; an excited *kipper, kipper*
with descending inflection; a shrill twitter; also a dawn song.
**Eggs**
3–5; creamy, with brown spots; 0.9 x 0.7 in. (2.3 x 1.8 cm). Nest
of twigs and grasses; bulky; ragged without, neat within; 3–20 ft.
(0.9–6.1 m) up in bush, tree, stump, or structure.
**Range**
Breeds from cen. Canada s. to s. U.S.; winters in S. America.

## WESTERN KINGBIRD
*Tyrannus verticalis*                                      **33:6**

**Description**
Size, 8–9½ in. (20.3–24.1 cm). Olive above, gray head and upper
breast, • *black tail with white edges on outer feathers,* yellow belly;
black ear patch.

**Similarities**
Young sometimes lack white edges of outer tail feathers; Eastern
Kingbird is much darker above. Great Crested Flycatcher, a
woodland bird, has a rufous tail.
**Habitat**
Open country, ranches, towns, roadsides.
**Habits**
Flight less fluttery than Eastern's.
**Voice**
Noisier than Eastern, many similar notes; also "a single *kip* . . . a
*quer-ick*" (Stevens); and a *"ker-er-ich-ker-er-ich"* (Barnes).
**Eggs**
3–5; creamy, with brown spots; 0.9 x 0.7 in. (2.3 x 1.8 cm). Nest is
similar to Eastern's, but more often on man-made structures.
**Former name**
Arkansas Kingbird.
**Range**
Breeds from Man. s. to Kans., Okla., and Mexico; winters in C.
America.

Note: The **CASSIN'S KINGBIRD,** *Tyrannus vociferans* **(33:5)**, size,
8–9 in. (20.3–22.9 cm), occurs on the westernmost fringe of eastern
region. It resembles the Western Kingbird, but is darker, with a
white throat and a white-tipped tail, and it lacks the white edges of
the outer tail feathers of the western. The **GRAY KINGBIRD,**
*Tyrannus dominicensis* **(33:8)**, inhabits Florida, and the Florida
Keys especially, but sometimes wanders north to the Carolinas. It
is 7 in. (17.8 cm), very like the Eastern, but lacks white on the tail,
is pale gray above, and has a larger bill.

# GREAT CRESTED FLYCATCHER
*Myiarchus crinitus* **33:1**

**Description**
Size, 8–9 in. (20.3–22.9 cm). Crested; olive-brown above; gray
throat and breast, yellow belly; rufous on wings, 2 white wing
bars; tail rufous.
**Habitat**
Woods, orchards, farms.
**Habits**
Aggressive; glides from tree to tree on outspread wings; often feeds
from tops of tall forest trees.
**Voice**
Raucous *wheap;* a whistled *whit-whit, whit-whit;* a "rattling cry,
*creep* or *cr-r-r-reep*" (Allen).
**Eggs**
3–8; creamy, with black and brown lines and penlike scratches; 0.9
x 0.7 in. (2.3 x 1.8 cm). Nest is in cavity, 5–60 ft. (1.5–18.3 m)
high, of trash, often with discarded snakeskin.
**Range**
Breeds throughout much of e. U.S., from s. Canada to Gulf of
Mexico; winters from Mexico to S. America.

# SCISSOR-TAILED FLYCATCHER
*Muscivora forficata* **33:4**

**Description**
Size, 11–15 in. (27.9–38.1 cm). Conspicuous long streamer tail.
Gray above, flanks and underwings pink, underparts white, scarlet
patch on underwing and crown (concealed).

**Similarities**
Western Kingbird resembles immature with short tail, but is yellowish below instead of pinkish.
**Habitat**
Plains, prairies; roadsides, farms, ranches.
**Habits**
Often seen on wire, tail not spread when perched; in flight it opens and shuts; flight swift, graceful, low over ground.
**Voice**
"Twittering *psee, psee, psee;* a harsh *thish-thish*" (Collins); also a twilight song.
**Eggs**
3–6; creamy, spotted with brown; 0.9 x 0.7 in. (2.3 x 1.8 cm). Nest of twigs, 4–20 ft. (1.2–6.1 m) up on limb of a deciduous tree.
**Range**
Breeds in s.-cen. U.S. from Mo. and Nebr., s. to Mexico; winters from Mexico s. to C. America; also winters in s. Fla.

## EASTERN PHOEBE
*Sayornis phoebe* 33:3

**Description**
Size, 6¼–7¼ in (15.9–18.4 cm). • *Bill black,* tail notched. Olive-gray above, darker on head and tail; grayish on throat and breast, whitish below; • *no eye-ring or wing bars.* In fall, yellowish below, young with cinnamon wing bars.
**Similarities**
Eastern Wood Pewee has wing bars and does not pump its shorter tail.
**Habitat**
Farms, orchards, gardens, streamsides; about buildings.
**Habits**
Tame; an early migrant; gets some food from ground; head looks blackish in flight; tail pumping diagnostic.
**Voice**
Burred, often repeated *PHE-bee,* sometimes uttered in flight; a chip like a Swamp Sparrow's.
**Eggs**
4–5; white, rarely dotted; 0.7 x 0.5 in. (1.8 x 1.4 cm); 2 broods. Nest is of mud and moss, often under a bridge, eaves, cave, or man-made structure.
**Range**
Breeds from s. Canada s. to n. Gulf states; winters from Va. and Gulf Coast s. to Mexico.

## SAY'S PHOEBE
*Sayornis saya* 33:2

**Description**
Size, 7–8 in. (17.8–20.3 cm). Pale, gray-brown back; black tail; • *rusty below.*
**Habitat**
Open country, ranches, ravines.
**Habits**
Early migrant, very active, restless; flight is low, strong, zigzaggy; wingbeats deeper, slower than Eastern's; occasionally pumps tail.
**Voice**
Plaintive *phee-ur* (given with twitch of tail); "a swift *pit-tsee-ar*" (Hoffmann).

**Eggs**

4–5; white, rarely dotted; 0.7 x 0.5 in. (1.9 x 1.5 cm). Nest is similar to Eastern, with less mud (often wool-lined).

**Range**

Breeds chiefly in W., but e. to N.Dak., e. Nebr., and cen. Tex.; winters from s. Tex. southward.

## YELLOW-BELLIED FLYCATCHER

*Empidonax flaviventris*                                                      33:15

**Description**

Size, 5½ in. (14.0 cm). This *Empidonax* says *pu-ee*. Olive-green above, eye-ring yellow. This species in spring may be identified by its color, yellow below including throat, but in fall other *Empidonaces* may be yellowish below.

**Habitat**

Near streams in coniferous forests; cold bogs; alder swamps in migration.

**Habits**

Secretive; jerks head and tail and flutters wings as it sings.

**Voice**

Rising *pu-EE;* a sneezelike *pse-ek;* less musical than Acadian, which does not breed within its range; in migration, "a plaintive *pea*" (Maynard); flight song *pu-ee, killik, pu-ee, killik.*

**Eggs**

4; creamy, finely dotted with brown; 0.6 x 0.5 in. (1.7 x 1.3 cm).

**Range**

Breeds from Newfoundland and Man. s. to Maine, Wis., and N.Dak. and Pa. mountains; winters from Mexico to C. America.

## ACADIAN FLYCATCHER

*Empidonax virescens*                                                       33:14

**Description**

Size, 6 in. (15.2 cm). This *Empidonax* says *ka-ZEEK*. Somewhat greener than other *Empidonaces*. Throat white, pale yellowish below, brighter in fall.

**Habitat**

Woods and swamps near streams in southern woodlands; likes beeches.

**Habits**

Sings through heat of day, twitches tail as it sings, frequents the middle layer of branches.

**Voice**

Explosive *ka-ZEEK*, second syllable higher; call a sharp *peet.*

**Eggs**

2–4; pale cream, sparingly spotted with brown; 0.7 x 0.5 in. (1.8 x 1.4 cm). Nest of rootlets, etc., swung like a hammock between twigs near end of branch, 4–20 ft. (1.2–6.1 m) up, often over water.

**Range**

Breeds from s. New England, Pa., and Great Lakes s. to se. U.S.; winters from C. to S. America.

## WILLOW FLYCATCHER

*Empidonax traillii*                                                        33:12

**Description**

Size, 5½–6¾ in. (13.3–17.1 cm). This *Empidonax* says *fitz-BEW*. Greenish-brown above with whiter throat than most *Empidonaces*.

**Similarities**
Only by its song can this species be distinguished with certainty from the Alder.
**Habitat**
Usually near water, low bushes along streams, edges of marshes and meadows; occasionally in dry fields; usually in more open, grassy, less wooded areas than Alder.
**Habits**
Late migrant; active, restless; sings from tops of shrubs.
**Voice**
A *fitz-BEW,* buzzy, an even buzzier *fizz-BEW;* also "a buzzy *creet*" (R. C. Stein).
**Eggs**
3–4; whitish, with fine brown spots; about 0.7 x 0.5 in. (1.8 x 1.3 cm). Nest is bulky, of shreds, grasses, low in shrub, to 3 ft. (0.9 m) up.
**Remarks**
This and the Alder Flycatcher were formerly considered one species, called the Alder or Traill's Flycatcher.
**Range**
Breeds from Maine, N.Y., and Alta. s. to Va., Ark., and Okla.; winters in S. America.

## ALDER FLYCATCHER
*Empidonax alnorum*                              **33:12**

**Description**
Size, 6 in. (15.2 cm). This *Empidonax* says *fe-BE-O.* Tends to be grayer, less green above than Willow, with same whitish throat.
**Similarities**
Alder cannot be distinguished from Willow except by voice.
**Habitat**
Low bushes in wet areas, swamps, around marshes, streamsides, near woods.
**Habits**
As Willow Flycatcher.
**Voice**
A *fe-BE-O,* the *fe* rasping, the *BE-O* whistled.
**Eggs**
3–4; buffy, with large purplish-brown spots; about 0.7 x 0.5 in. (1.8 x 1.3 cm). Nest is compact, of cottony material and grasses; resembles that of Yellow Warbler; whitish, in shrubs 3 to 8 ft. (0.9–2.4 m) up.
**Range**
Breeds from cen. Que. and Man. s. to s. New England and Minn.; winters in C. and S. America.

## LEAST FLYCATCHER
*Empidonax minimus*                              **33:13**

**Description**
Size, 5¼ in. (13.3 cm). This *Empidonax* says *che-bec.* Grayest above, whitest below of *Empidonaces;* lower mandible fairly dark.
**Similarities**
Light lower mandible in other eastern *Empidonaces,* noticeable at close range.
**Habitat**
Orchards, woods, streamsides, rural countryside, haunts of man.
**Habits**
Jerks head and tail as it calls.

**Voice**
Noisy; an emphatic *che-BEC,* much repeated; call, a short *whit;*
also a flight song, *"che-BEC-treee-treo, che-BEC-treee-cheu"*
(Samuels).
**Eggs**
3–6; creamy; 0.6 x 0.4 in. (1.6 x 1.2 cm). Nest is of plant fibers
and grasses, in upright fork of tree, 8–40 ft. (2.4–12.2 cm) up.
**Range**
Breeds from s. Canada s. to cen. U.S., and in n. Ga. mountains;
winters in C. America.

Note: The western fringes of our region are reached in summer by
2 western species. The **DUSKY FLYCATCHER,** *Empidonax
oberholseri,* size, 5½ in. (14.0 cm), is a gray-olive species that
frequents low trees in the wooded foothills of the Rocky
Mountains. Its song is a *clip-whee-ZEE.* In the same region the
more yellow-greenish **WESTERN FLYCATCHER,** *Empidonax
difficilis,* size, 6 in. (15.2 cm), favors open places, especially near
streams within the forests. Its vocalizations include a wheezy *pee-
IST,* and its song is *ps-SEET, ptsick, sst.*

## EASTERN WOOD PEWEE
*Contopus virens*                                                33:9

**Description**
Size, 6–6¾ in. (15.2–17.1 cm). Wings longer than Eastern
Phoebe's. • *Lower mandible partly yellow;* dark olive-gray above,
with • *2 white wing bars* (buffy in young); whitish below, mid-
breast with light line dividing darker sides.
**Similarities**
Eastern Phoebe has all-black bill, no wing bars, pumps longer tail
and usually feeds near ground. *Empidonaces* are smaller, have
white eye-ring, and lack whitish line down breast. See also
Western Wood Pewee.
**Habitat**
Pinewoods, orchards, deciduous woods, shade trees.
**Habits**
Sits motionless, does not pump tail; not pugnacious, frequents
middle and upper layers of branches.
**Voice**
Sweet, plaintive *PEE-wee, PEE-a-WEE,* or *pee-AAA* with a
rising inflection, repeated; also a more elaborate song at dawn and
twilight; sings well into August.
**Eggs**
2–4; with ring of brown or purple spots; 0.7 x 0.5 in. (1.8 x 1.3
cm). Nest is shallow, constructed of fibers covered with lichens 6–
50 ft. (1.8–15.2 m) up on a horizontal branch.
**Range**
Breeds from se. Canada s. to Gulf of Mexico; winters from C. to S.
America.

## WESTERN WOOD PEWEE
*Contopus sordidulus*                                           33:10

**Description**
Size, 6–6½ in. (15.2–16.5 cm). Plumage similar to Eastern Wood
Pewee's; slightly grayer and less greenish above, darker below, but
a difficult identification without voice.
**Habitat**
Open woodlands.

**Habits**
Similar to Eastern.
**Voice**
Nasal *peeer,* suggesting note of Common Nighthawk, a sad *dear-me; "Tswee-tee-teet, TSWEE-tee-teet, bzew (bzew,* a downward slur)" (Miller); often sings at night.
**Eggs**
3–4; similar to Eastern's. Nest is like Eastern's; tends to be larger, deeper, lined with yellow grasses, and usually without lichens.
**Range**
Breeds chiefly in W., but from Man. s. through Nebr. and cen. Tex. to C. America; winters in C. and S. America.

## OLIVE-SIDED FLYCATCHER
*Contopus borealis*                                          **33:11**

**Description**
Size, 7–8 in. (17.8–20.3 cm). Large bill and head, sturdy body. Olive-gray above; • *gray sides and white stripe down mid-breast* suggest gray vest half open over white shirt; white tufts in side of rump visible at times.
**Similarities**
Wood pewees are smaller, with light wing bars, and lack contrast between sides and center of breast.
**Habitat**
Dead timber in coniferous forest, often near water.
**Habits**
Selects a particularly exposed perch.
**Voice**
Loud whistle, quick *THREE BEERS*; usually silent in migration.
**Eggs**
3; creamy, blotched with chestnut; 0.9 x 0.7 in. (2.3 x 1.8 cm). Nest is of twigs, 10–50 ft. (3.0–15.2 m) up, in crotch of horizontal branch on conifer.
**Range**
Breeds from s.-cen. Canada s. to Mich., Wis., and n. Minn.; winters in S. America.

# LARKS
Family Alaudidae

Largely terrestrial birds, larks are mostly brown streaked, have musical voices, are gregarious; the sexes are virtually alike. The hind claw is elongated, and nearly straight.

## HORNED LARK
*Eremophila alpestris*                                        **48:7**

**Description**
Size, 7–8 in. (17.8–20.3 cm). Long hind claw. Male: brownish above and on sides, unique face pattern with • *tiny black "horns"* and black sideburns, • *tail black,* outer tail feathers and underparts white, line over eye yellow or white; the "yellow-faced" northern form occurs in winter; the "white-faced" form breeds in our area. Female and immature: duller, except the juvenal, but all show suggestion of face pattern.
**Similarities**
Juvenal might be confused with molting longspur, but no longspur has Lark's tail pattern, and Lark's bill is thinner.

**Habitat**
Grasslands, golf courses, fields, parkway shoulders, beaches, dunes, marshes.
**Habits**
In winter, gregarious; walks (does not hop) or runs mouselike on ground; or "freezes" into invisibility. Flight is light, bounding, showing contrast of black tail and white underparts.
**Voice**
Common flight note, *"p-seet"* (Collins); song, thin, jumbled tinkling and bubbling often given in flight, or as it parachutes earthward from a height.
**Food**
Seeds, insects.
**Eggs**
3–5; grayish, marked with brown; 0.9 x 0.6 in. (2.3 x 1.5 cm). Nest is on ground, of grasses.
**Range**
Breeds from Arctic s. to N.C., Mo., and Tex. Gulf Coast; winters from s. Canada s. to Gulf Coast and southward.

# SWALLOWS
Family Hirundinidae

Swallows have short, flat bills and wide mouths; long, pointed wings; weak, small feet; and usually notched tails. They frequent open country and bodies of water, and spend much time catching insects on the wing. Mixed flocks often hawk for insects low over ponds prior to bad weather. Swallows are gregarious and migrate in flocks by day, often along the coasts. None, except some Tree Swallows, winter in our area.

## TREE SWALLOW
*Iridoprocne bicolor*                                          32:3

**Description**
Size, 5–6½ in. (12.7–15.9 cm). Adult: glossy greenish-blue above, including area around eyes; • *white below.* Immature: dusky-brown above, sometimes with faint, broken dark band across upper breast.
**Similarities**
Brown Rough-winged has dingy throat; resembles young Tree in late summer. Bank has complete breastband.
**Habitat**
Open areas, especially near water; often breeds in dead trees in open swamps.
**Habits**
Flight slightly flickering, wingbeats faster than Barn's; when sailing, seems to arch back and lower wing tips; wings look triangular in air, somewhat like Purple Martin's. An early migrant; collects in vast numbers on coastal marshes in fall migration, swarming over trees, wires, bushes, roads, beaches.
**Voice**
Chirrups, twitters, *silip, silip, silip* or *"tsip-prrup, tsip-prrup-prrup"* (Allen), much repeated.
**Eggs**
3–6; white; 0.7 x 0.5 in. (1.9 x 1.3 cm). Nest is single or in colonies, in woodpecker holes or boxes, often near water.
**Range**
Breeds from Newfoundland w. to n. Man. and s. to Md. and Nebr.; winters from N.C. southward.

Note: The **VIOLET-GREEN SWALLOW**, *Tachycineta thalassina*, size, 5¼ in. (13.3 cm), is found in the mountains and valleys from the Black Hills and Pine Ridge of Nebraska westward. It is colored somewhat like the Tree Swallow but with white around the eyes and a white patch on either side of the rump.

## BANK SWALLOW
*Riparia riparia*                                                    32:4

**Description**
Size, 4¾–5¼ in. (12.1–13.3 cm). • *Brown above;* head, wings, and tail darker; • *white below;* distinctive • *brown breastband.*
**Similarities**
Immature Tree is larger and whiter below, darker above.
**Habitat**
Meadows, ponds, other open areas, banks.
**Habits**
Flight low, erratic, somewhat fluttering and mothlike; keeps wings close to body when sailing; rows of holes in bank mark site of nesting colony.
**Voice**
Notes more gritty than other swallows; "soft abrupt *ffrrutt*" (Cruickshank); song, a twitter, *"speedz-sweet, speedz-sweet"* (Dickey), much repeated; "when alarmed . . . *te-a-rr* [reminiscent] of the Common Tern" (Stoner).
**Eggs**
3–7; white; 0.7 x 0.5 in. (1.8 x 1.3 cm). Nest is in colonies, in hole excavated in bank along river, shoreline, right-of-way, or gravel pit.
**Range**
Breeds from Labr. and n. Man. s. to Va. and Tex.; winters in S. America.

## ROUGH-WINGED SWALLOW
*Stelgidopteryx ruficollis*                                          32:5

**Description**
Size, 5–5¾ in. (12.7–14.6 cm). Uniform • *grayish-brown above* and dusky on throat, white below.
**Similarities**
Smaller Bank Swallow has white throat, distinct breastband. Young Tree has white throat.
**Habitat**
Open areas, creeks, ponds, rivers.
**Habits**
Flight more direct, less erratic than Bank's; wingbeats deeper and slower, more glides and sails; more solitary, less colonial than Bank.
**Voice**
Twitter similar to, but lower-pitched than, Bank's; call, "a *trit, trit-trit,* or *tri-ri-ri-rit*" (Saunders); a "rasping squeak . . . *quiz-z-z-zeep, quiz-z-z-zeep*" (Dickey); unmusical *"burp-burp"* (Cruickshank).
**Eggs**
4–8; white; 0.7 x 0.5 in. (1.8 x 1.3 cm). Nest is singly or in small groups, in holes in banks, caves, or crevices, usually near water.
**Range**
Breeds throughout most of U.S. from N.B. w. to Mich. and s. to Fla. and S. America; winters from Mexico southward.

## BARN SWALLOW
*Hirundo rustica*                                      **32:2**

**Description**
Size, 7 in. (17.8 cm). Adult: iridescent steel-blue above, chestnut
forehead and throat, thin necklace, white on inner tail feathers,
buffy below, • *long, forked tail*. Immature: paler, with buffy-brown
throat; lacks tail streamers.

**Habitat**
Meadows, marshes, ponds, coast, often near water or buildings.

**Habits**
Flight strong, swift, graceful; drives through air, wings close to and
parallel with body at end of stroke; gregarious; large flocks migrate
along coast.

**Voice**
Various cheerful twitterings, *"sweeter-sweet, sweeter-sweet"*
(Cruickshank); a *kittik, kittik* or *kvik, kvik,* repeated.

**Eggs**
3–6; white, spotted with red-brown; 0.8 x 0.6 in. (2.0 x 1.5 cm).
Nest is of mud, often inside barns, boathouses, or other structures,
on rafter or against wall, usually in small colonies.

**Range:**
Breeds from Maritime Provinces s. to S.C. and w. to Ark.; winters
in S. America.

## CLIFF SWALLOW
*Petrochelidon pyrrhonota*                             **32:7**

**Description**
Size, 5–6 in. (12.7–15.2 cm). Adult: creamy forehead, chestnut
face, gray collar, • *buffy rump;* rest of upperparts blue, underparts
white, • *tail square*. Immature: more brownish, rump paler.

**Similarities**
Barn has long forked tail.

**Habitat**
Meadows, marshes, open area.

**Habits**
Colonial, often quite local.

**Voice**
Squeaky, creaking, grinding but pleasant series of chirrupings,
huskier than Barn; a nasal *y-e-a-a*.

**Eggs**
4–6; white, thickly marked with reddish-brown, 0.7 x 0.6 in. (1.9 x
1.7 cm). Nest is of mud, gourd-shaped, on outside of buildings
under eaves, cliffs, bridges.

**Range**
Breeds from N.S. w. to Ont. and s. to Va., Mo., and C. America;
winters in S. America.

## PURPLE MARTIN
*Progne subis*                                         **32:1**

**Description**
Size, 7½–8½ in. (18.4–21.6 cm). The largest swallow. Male: • *all-
purple* (looks black at a distance). Female and immature: grayish
throat and breast, whitish underparts, some gray on dark upperparts.

**Similarities**
Immature Tree Swallow is smaller and with white below.

**Habitat**
Seashore, meadows, about houses, even common in some midwestern
towns.

### Habits
Wings triangular in flight; rapid wingbeats alternate with sails, often in a great arc; flight vaguely suggests Starling, but wings much longer; often heard at night; prior to migration collects in big flocks in groves, where their united voices sound like escaping steam.

### Voice
Rich, liquid, loud chirruping, a somewhat guttural *too-too* and *too-too-too-weadle;* call, "a harsh *zhupe, zhupe*" (Stone); alarm, a *kerp*.

### Eggs
3–5; white; 0.9 x 0.7 in. (2.3 x 1.8 cm). Nesting is colonial, originally in holes in trees; then also in hollow gourds put up by the Indians; now largely in artificial birdhouses.

### Range
Breeds from N.S. and Man. s. to Fla., the Gulf Coast, and the W. Indies; winters in S. America.

# JAYS, MAGPIES, AND CROWS
Family Corvidae

These medium to large birds, collectively called corvids, have a stout, pointed bill; rounded wings; and a rounded or wedge-shaped tail. Crows and ravens are black, jays usually colorful in blue and green, magpies black and white with long tails. These birds are omnivorous, eating insects, berries, nuts, and seeds.

## GRAY JAY
*Perisoreus canadensis*                                              **35:9**

### Description
Size, 10–13 in. (25.4–33.0 cm). Adult: white forehead, throat, and collar; • *black cap* on back of head; • *rest of plumage gray,* darker above. When fluffed up, suggests a giant chickadee. Immature: dark slate, head blackish.

### Similarities
Northern Shrike has black wings and tail.

### Habitat
Coniferous forest and clearings.

### Habits
Tame, inquisitive, bold; takes food about camps, robs traps, stores food. In flight, seems to float in air; glides more than most jays, wingbeats deeper; sails from top of one tree to bottom of another, then hops up branches in a spiral around tree, and repeats; on ground, hops.

### Voice
"A loud, hawklike whistle, very like that of the Red-shouldered" (Brewster); a querulous *quee-ah, kuoo* or *whah* (Knight); other varied notes, some suggesting Blue Jay or Red Squirrel; a Catbird-like song in spring; a raucous rattle; somewhat of a mimic.

### Eggs
3–5; gray, wreathed with brown markings at large end; 1.1 x 0.8 in. (2.8 x 2.0 cm), laid while snow is still on ground. Nest is of twigs lined with moss; large, 4–30 ft. (1.2–9.1 m) up, on branch near trunk of conifer.

### Former name
Canada Jay.

### Range
Resident from Labr. s. to New England and N.Y. and w. along n. U.S.; some populations move south in winter as far as Pa. and Great Plains.

# BLUE JAY
*Cyanocitta cristata* **35:6**

**Description**
Size, 12 in. (30.5 cm). • *Crested;* tail long, wedge-shaped; • *violet-blue above, azure on wings and tail;* white face, throat, wing bars, and outer tail tips; pale below and under tail.

**Habitat**
Woods, edges, suburbs.

**Habits**
Active, aggressive, drives other birds away from feeder; its loud cries warn of intruders, but it is furtive about its own nest; flight steady, slow, on level plane, and with primaries spread; group flies across open space one at a time; collects in flocks in migration; stores food by burying acorns, etc.; hops, does not walk.

**Voice**
Noisy, varied; a shriek, *jay, jay;* a scream, *teearr,* like Red-shouldered Hawk; a creaky *tea-cup, tea-cup;* a *chillak, chillak;* also notes like a toy trumpet; song, a low mixture of whistles and sweet notes, often imitating various small birds.

**Eggs**
3–6; variable, often olive-blotched with brown; 1.1 x 0.9 in. (2.8 x 2.3 cm). Nest is of twigs, ragged, bulky, usually 8–20 ft. (2.4–6.1 m) up, near trunk.

**Range**
Mainly resident e. of Rockies from s. Canada s. to Gulf of Mexico.

Note: The **STELLER'S JAY,** *Cyanocitta stelleri,* size, 12 in. (30.5 cm), occurs along the western fringe of the eastern range, from Colorado to the Black Hills, and casually farther east. It is a blue-black, crested jay lacking the Blue Jay's head patterning and is white on the wings, tail, and underparts. In Florida a form of the **SCRUB JAY,** *Aphelocoma coerulescens* **(35:8),** size, 11½ in. (29.2 cm), is found, far isolated from the mainly western range of this species. It is blue and gray, lacks a crest, and has a blue-bordered white throat. This jay is social, often nesting in small groups.

# BLACK-BILLED MAGPIE
*Pica pica* **35:11**

**Description**
Size, 17½–22 in. (44.5–55.9 cm); tail 9½–12 in. (24.1–30.5 cm). A large • *black-and-white bird.* Wings short, rounded; • *tail glossy green, long.*

**Habitat**
Foothills, roadsides, streamsides, thickets, fields, pastures, ranches.

**Habits**
Often found in small groups; conspicuous from car or train as one approaches Rockies; suspicious but inquisitive; spends much time on ground, twitches tail constantly in its jerky walk, hops when in a hurry; flight short, wavering, jaylike rather than crowlike.

**Voice**
Noisy; a harsh *ca-ca-ca,* higher than Common Crow; "a nasal querulous *maag?* or *aag-aag?*" (Peterson); other chattering and whistling notes.

**Eggs**
6–9; grayish, blotched with brown; 1.3 x 0.9 in. (3.3 x 2.3 cm). Nest is sometimes in colonies; huge round mass of sticks with hole in one side, lined with grass, often in willow thickets.

**Range**
Resident chiefly in W., but e. to Alta. and Man. and s. to n. Ariz.; in winter, casual to Iowa.

## COMMON RAVEN
*Corvus corax*                                         *Fig. 13*

**Description**
Size, 21½–27 in. (54.6–68.6 cm). Largest species in Order
Passeriformes. Larger and with longer, thicker neck than Common
Crow; all-black, with heavier bill; throat "bearded." In flight,
wings angled at shoulder, outspread primaries, horizontal wings,
• *wedge-shaped tail* (longer than Crow's).
**Habitat**
Seashore, mountains.
**Habits**
Aggressive but wary; flight with heavy slow wingbeats, often
hawklike aerial revolutions in courtship; hovers, drops shellfish
from height to break it open; a scavenger; sometimes mobbed by
crows; seems to pair for life.
**Voice**
Various croaks, such as *c-r-r-r-u-u-k*, and others (young sometimes
*caw* like crow); also a high-pitched *tok-tok-tok*.
**Eggs**
4–7; greenish, spotted with brown; 1.8 x 1.4 in. (4.6 x 3.6 cm).
Nest is very large, of sticks, near top of conifer or on ledge.
**Range**
Resident throughout much of Canada and in w. U.S. from n.
Canada and Greenland s. to n. New England, w. to Daks.

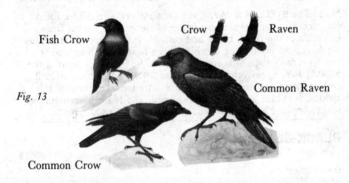

Fish Crow

Crow

Raven

*Fig. 13*

Common Raven

Common Crow

## COMMON CROW
*Corvus brachyrhynchos*                                 *Fig. 13*

**Description**
Size, 17–21 in. (43.2–53.3 cm). All-black, tail gently rounded.
**Similarities**
Tail of Common Raven is longer; see also Fish Crow.
**Habitat**
Fields, woods, coasts, parks.
**Habits**
Wary, intelligent; hops and walks; frequents parkways feeding on
animal life killed by cars; often seen flying overhead with steady,
deep wingbeat; soars with wings in shallow V; congregates in great
roosts in winter.
**Voice**
*Caw, caw* and variations thereof. Young: *car, car* (somewhat like
adult Fish Crow but louder).

**Eggs**
3–6; greenish, blotched with brown; 1.7 x 1.2 in. (4.3 x 3.0 cm).
Nest is bulky, of sticks, 10–50 ft. (3.0–15.2 m) up in a tree.
**Range**
Breeds from Newfoundland, s. to Fla. and Gulf Coast and w.
across Canada and U.S.; winters from n. U.S. s. to Mexico and
southward.

## FISH CROW
*Corvus ossifragus*                                              *Fig. 13*

**Description**
Size, 17 in. (43.2 cm). All-black; tail gently rounded.
**Similarities**
Common Crow is virtually indistinguishable except by voice and
direct size comparison, although Common is slimmer and wings
are more pointed at tips and broader at base.
**Habitat**
Seashores, river valleys, occasionally vicinity of inland waters.
**Habits**
Feeds on beach, hovers over food, engages in aerial revolutions,
soars more and flies with more alternate sailing and flapping than
Common Crow; somewhat more gregarious at all seasons.
**Voice**
Nasal *car, car* much like squawk call of young Common Crow,
actually "closer to a Black-crowned Night Heron's *quowk*"
(Pough); "hoarse *craaa-craaa* and a staccato *ca-a*" (Cruickshank).
**Eggs**
3–5; greenish, blotched with brown; 1.5 x 1.1 in. (3.8 x 2.8 cm).
Nest is of sticks, often in small colonies, in swamps or coastal
pines.
**Range**
Resident along Atlantic and Gulf coasts, from Mass., s. to Fla. and
w. to Tex.; also found inland along large rivers from Va. to Tenn.
and Ark.

Note: Two other crowlike birds of western North America reach
the western fringe of our region. The **PINYON JAY**, *Gymnorhinus
cyanocephalus*, size, 10 in. (25.4 cm), is a dull bluish, short-tailed
bird with a thinner bill than other jays. It is a flocking species with
crowlike habits, sporadically reaching the western Plains. The
**CLARK'S NUTCRACKER**, *Nucifraga columbiana* **(35:10)**, size, 12 in.
(30.5 cm), of western mountains resembles the Pinyon Jay in
shape, but is pale gray with black and white wings and tail. It
wanders eastward in some numbers after breeding.

# CHICKADEES AND TITMICE
Family Paridae

These are small, plump, big-headed birds with short, straight bills;
beady black eyes; rounded wings; and soft, fluffy plumage,
predominantly gray, black, and white in color. They are very active
and often hang head down from a branch as they feed. They are
confiding and inquisitive, and can be attracted by squeaking or by
imitations of their notes. Their natural foods are insects, insect eggs
or larvae, seeds and berries, but they readily frequent feeding trays.

## BLACK-CAPPED CHICKADEE
*Parus atricapillus*                                        **34:10**

**Description**
Size, 4¾–5¾ in. (12.1–14.6 cm). • *Black cap and bib*; gray above
with white edges on wing feathers; white below, pale chestnut
wash on flanks; • *white cheeks* noticeable at a distance.

**Similarities**
Narrow line of white feather edges on folded wing and somewhat
larger size distinguish this species from the similar Carolina
Chickadee when together, but their voices provide the best
identifying feature. Eastern Phoebe's *fee-bee* note is hoarse, not
whistled, as in this species. Spring male Blackpoll Warbler does
not have black bib.

**Habitat**
Woods, edges, gardens, towns.

**Habits**
Tame, may be taught to eat from hand; somewhat migratory; tends
to wander after breeding season in mixed flocks with Downy
Woodpeckers, nuthatches, kinglets.

**Voice**
Call, a *chick-a-dee-dee-dee;* song, a sweet whistled *fee-bee* or
*SPRING's-come,* or *FEE-be-be,* the first note higher; hisses if
disturbed on nest.

**Eggs**
4–9; white, finely spotted with reddish-brown; 0.6 x 0.5 in. (1.5 x
1.3 cm).

**Range**
Breeds from Newfoundland and cen. Man. s. to s. N.J. and Mo.,
mts. to Tenn.; winters from breeding range s. to Md. and Tex.

## CAROLINA CHICKADEE
*Parus carolinensis*                                        **34:7**

**Description**
Size, 4½ in. (15.7 cm).

**Similarities**
Resembles Black-capped Chickadee but shows • *little or no white in
wings;* best distinguished by voice.

**Habitat**
Coastal-plain swamps; woods, especially pinewoods; residential
areas.

**Habits**
Similar to Black-capped Chickadee; roams in mixed species flocks
when not breeding.

**Voice**
*Chick-a-dee-dee-dee,* higher pitched than Black-capped and faster;
song is a short, whispered sibilant *su* preceding each syllable of *fee-
bee* song, thus *sufee-subee;* "the pumphandle strain . . . spee-deedle-
dee-deedle-dee"* (Dickey).

**Eggs**
Similar to Black-capped's.

**Remarks**
There is virtually no overlap in breeding season, but Black-cappeds
move into the adjacent range of the Carolinas in winter.

**Range**
Resident from cen. N.J. and Mo. s. to Fla., Tex., and Okla.

## BOREAL CHICKADEE
*Parus hudsonicus*                                    **34:6**

**Description**
Size, 5–5½ in. (12.7–14.0 cm). Brownish above, • *brown cap,*
whitish below, bib black, flanks chestnut.
**Similarities**
Black bib is smaller, chestnut flanks brighter than in Black-capped.
**Habitat**
Northern evergreen forest.
**Habits**
Similar to Black-capped Chickadee's, but not as lively and works
closer to tree trunk.
**Voice**
Somewhat like a Black-capped Chickadee, but notes slower,
wheezier, and blurred; *chick-a-day-day;* song, a warble.
**Eggs**
5–9; white, finely spottted with reddish-brown; 0.5 x 0.4 in. (1.4 x
1.1 cm).
**Former name**
Brown-capped Chickadee.
**Range**
Resident in much of Canada, from Labr. w. to Man. and s. to n.
New England; some wander s. to Md., Ohio, and Ill. in winter.

## TUFTED TITMOUSE
*Parus bicolor*                                    **34:4**

**Description**
Size, 5–6 in. (12.7–15.2 cm). A small • *gray bird with a crest.* Gray
above, cheeks and underparts whitish, chestnut wash on flanks.
**Similarities**
Cedar Waxwing, also crested, is larger, brownish, and with bright
markings on wings and tail.
**Habitat**
Swamps, deciduous woodlands, residential areas.
**Habits**
Somewhat like Black-capped Chickadee; visits feeders; after
breeding season roams in mixed flocks with chickadees, Downy
Woodpeckers, nuthatches; not migratory.
**Voice**
Loud whistled *peter, peter, peter,* often confused at a distance with
Carolina Wren or Northern Oriole; a chickadeelike *dee-dee-dee;*
and whisperings; a scolding *ya-ya-ya.*
**Eggs**
4–7; white, thickly dotted with reddish-brown; 0.7 x 0.5 in. (1.8 x
1.3 cm).
**Range**
Resident in much of e. U.S. from Maine and Wis. s. to Fla., Gulf
Coast, and Tex., s. to Mexico.

# NUTHATCHES
Family Sittidae

Nuthatches are small, stout birds with thin, sharp bills; short,
square tails; and strong legs and feet. They clamber about tree
trunks and branches, often head down. They nest in cavities, which
they frequently excavate themselves in the soft wood of dead
stumps, stubs, or branches, from 4–120 feet (1.2–36.6 m) up. Their
food consists of insects, seeds, nuts, berries, and fruit.

## WHITE-BREASTED NUTHATCH
*Sitta carolinensis* **34:2**

**Description**
Size, 5–6 in. (12.7–15.2 cm). Black cap and nape, beady black eye.
White completely encircling eyes; blue-gray back, white
underparts, chestnut undertail coverts, white markings on outer tail
feathers. Female often has grayer crown.

**Similarities**
Red-breasted has black eye stripe.

**Habitat**
Large trees in deciduous woods and about habitations.

**Habits**
Tame, comes to feeders, likes suet; stores food in crevices; except in
breeding season, often travels in mixed flocks; flight slightly
undulating.

**Voice**
Nasal *yank-yank-yank;* a short, high, conversational *hit-hit;* song, a
low-pitched *"to what what what what"* (Thoreau).

**Eggs**
4–8; white spotted with reddish-brown; 0.6–0.8 x 0.5–0.6 in (1.5–
2.0 x 1.3–1.5 cm).

**Range**
Resident from N.S. and Ont. s. to Fla., Gulf Coast, and Mexico.

## RED-BREASTED NUTHATCH
*Sitta canadensis* **34:5**

**Description**
Size, 4½ in. (11.4 cm). Both a black and a white face stripe. Male:
black cap, white line over eye, • *black line through eye,* dark blue-
gray upperparts, reddish below. Female: paler.

**Similarities**
White-breasted is larger, has no eye stripe.

**Habitat**
Northern evergreen forest, also deciduous woods. In winter, when
food supplies fail in north, given to irregular southward incursions,
and it then appears in all sorts of places.

**Habits**
More active than White-breasted, forages to end of branches,
captures insects in air; tame, frequents feeders, stores food; flight
undulating; smears entrance to nesting hole with pitch; except in
breeding season, often travels in small mixed species flocks.

**Voice**
High-pitched nasal *ink-ink-ink,* like a little tin horn; much higher
than White-breasted's *yank;* a *hit-hit* similar to White-breasted;
song somewhat similar to that of White-breasted but faster, higher,
reedier.

**Eggs**
4–8; white, spotted with reddish-brown; 0.6–0.8 x 0.5–0.6 in. (1.5–
2.0 x 1.3–1.5 cm).

**Range**
Breeds from Newfoundland and Man. s. to N.J., N.C., and w.
across U.S.; winters from breeding range s. to Gulf Coast and
Mexico.

# BROWN-HEADED NUTHATCH
*Sitta pusilla* **34:3**

**Description**
Size, 4½ in. (11.4 cm). A southern pinewoods species lacking any nasal calls. • *Brown cap coming down to eye,* blue-gray back, underparts and spot on nape white.

**Similarities**
Red- and White-breasteds are larger.

**Habitat**
Southern pinewoods, especially loblolly; also mixed woods, cypress swamps.

**Habits**
Forages to end of twigs; flight gently undulating; gregarious and sociable, often traveling with chickadees, titmice, and woodpeckers.

**Voice**
Fast, high *yit, yit, yit;* a chickadeelike *dee-dee-dee.*

**Eggs**
4–8; white, spotted with reddish-brown, 0.6–0.8 x 0.5–0.6 in. (1.5–2.0 x 1.3–1.5 cm).

**Range**
Resident from Del., Mo., and e. Tex. s. to Fla. and Gulf Coast.

Note: The western mountain pine forest **PYGMY NUTHATCH,** *Sitta pygmaea,* size, 4 in. (11 cm), is possibly only a subspecies of the Brown-headed Nuthatch. It reaches the western fringe of the eastern region in the Black Hills. It is somewhat grayer-capped but otherwise closely resembles the Brown-headed—their ranges do not meet.

# CREEPERS
Family Certhiidae

Creepers are small, stiff-tailed, slender birds. They creep up and around tree trunks, probing for bark insects with slightly curved bills.

# BROWN CREEPER
*Certhia familiaris* **34:1**

**Description**
Size, 5½ in. (14.0 cm). Creeps up tree trunks. Streaked brownish above, white below; pale band across relatively long wing; • *bill thin, curved.*

**Habitat**
Woodlands, swamps, shade trees.

**Habits**
Using long tail as prop, it creeps up trunk of tree often in a spiral, or along underside of horizontal branch, feeding as it goes; then drops to base of neighboring tree and repeats; solitary, but often travels with groups of chickadees, nuthatches, woodpeckers, kinglets; by means of its protective coloration, it often escapes detection; comes to feeders for suet, chopped peanuts, peanut butter.

**Voice**
Long, thin, high *tseee,* longer than Golden-crowned Kinglet's and not quickly repeated in series; song, *seee-see-see, swee-swee.*

**Eggs**
4–9; white, with reddish-brown spots; 0.6 x 0.5 in. (1.5 x 1.3 cm). Nest is under loose-hanging slab of bark low on a tree trunk.

### Range
Breeds from Newfoundland and Ont. s. to C. America; winters from n. U.S., s. to Gulf Coast and Fla.

Note: The **DIPPER**, *Cinclus mexicanus*, size, 8 in. (20.3 cm), is the only member of its family, the Cinclidae, found in North America. It is an unmistakably plump, short-tailed, large gray wrenlike bird, and is a resident along mountain streams in which it feeds. It is found in the Black Hills, and from the Rocky Mountains west.

# WRENS
Family Troglodytidae

Wrens are plump, vivacious brown birds, generally smaller than sparrows, that often carry the tail cocked over the back. They have thin, curved bills and short, rounded wings. The sexes look alike. They spend most of their time on or near the ground, feeding on insects and spiders.

## HOUSE WREN
*Troglodytes aedon*                                    **34:14**

### Description
Size, 4½–5¼ in. (11.4–13.3 cm). Gray-brown above, grayish below, brownish flanks, faint barring on wings and tail, longish tail, • *no eye stripe.*
### Similarities
Winter Wren is smaller, has shorter tail, and dark barring below; see also Bewick's Wren.
### Habitat
Woods, thickets, gardens, farming country. Common.
### Habits
Aggressive, scolds intruders; frequents bird bath and feeder; will nest in wren house; male builds extra dummy nests.
### Voice
Song, a long gushing, bubbling melody; alarm note, a grating chatter like a Common Yellowthroat's.
### Eggs
5–12; pinkish, heavily dotted with reddish-brown; 0.6 x 0.5 in. (1.5 x 1.3 cm). Nest is of twigs, etc., in virtually any cavity or bird box.
### Range
Breeds from Maine and Ont. s. to N.C.; winters from S.C. s. to Gulf Coast and Fla.

## WINTER WREN
*Troglodytes troglodytes*                              **34:11**

### Description
Size, 4 in. (10.2 cm). Dark brown above and on belly, breast lighter; • *belly barred with black;* • *tail stubby;* faint eye stripe.
### Similarities
See House Wren.
### Habitat
In summer, northern or mountain forest; in migration, thickets, brush piles.
### Habits
Secretive, mouselike; bobs up and down and cocks tail; feeds along stone walls, woodpiles, and banks of streams; keeps near cover.
### Voice
Song is one of the most beautiful in North America, a long, wild,

ringing, tinkling melody, first part often ending in a trill, second part ending in a trill an octave higher; call, a *tik, tik* like chip of Song Sparrow; also a *chirr*.

**Eggs**
4–10; white, dotted with red-brown and purple; 0.7 x 0.5 in. (1.8 x 1.3 cm). Nest is of sticks and moss in root mass of fallen tree or other tangle near ground.

**Remarks**
This is the only species of wren in the Old World and the Wren of English literature.

**Range**
Breeds from Newfoundland and s. Man. s. to Conn. and Mich.; also in Ga. mountains; winters from s. parts of breeding range, s. to Gulf Coast.

## BEWICK'S WREN
*Thryomanes bewickii*                                    **34:13**

**Description**
Size, 5½ in. (14.0 cm). Slim; • *fanlike tail longer than wings, with white corners;* brown above, whitish below.

**Similarities**
House Wren lacks white eye stripe, white underparts, and white tail corners. Rock Wren is grayer, with a streaked breast.

**Habitat**
Gardens, farms, woodpiles, thickets, woodlands.

**Habits**
Less aggressive than House Wren, movements more deliberate than House Wren or Carolina, comes to feeders, often fans and flirts tail.

**Voice**
Song is loud, beautiful; often confused with that of Song Sparrow, but longer, more varied, and ends on a trill; alarm, a chatter somewhat like Carolina's; a low *pit*.

**Eggs**
4–7; white or pinkish, heavily dotted with red-brown; 0.7 x 0.5 in. (1.8 x 1.3 cm). Nest is of twigs, etc., in cavities virtually anywhere; also in boxes.

**Range**
Resident throughout much of s. U.S. from cen. Pa. and Nebr. s. to Gulf of Mexico and Mexico, excluding Atlantic Coast.

## CAROLINA WREN
*Thryothorus ludovicianus*                                **34:17**

**Description**
Size, 6 in. (15.2 cm). Largest eastern wren. • *White eye stripe,* buffy belly, and no white in the tail; • *reddish-brown above,* buffy below; chin white; undertail coverts barred; narrow black barring on wings and tail; bill rather long and curved.

**Similarities**
Smaller Bewick's Wren is similar but is whitish below and has white edges of the tail.

**Habitat**
Woods, swamps, thickets, brushy slopes, gardens, orchards.

**Habits**
Teeters body, jerks tail, often escapes by running mouselike through cover; inquisitive but quick to hide; likes brush piles; does not migrate, cold winters kill off many in northern portions of their range; comes to feeders.

**Voice**
Song is a clear whistled *tea-kettle, tea-kettle, tea-kettle,* or *sweet-william, sweet-william, sweet-william* with many variations; mimics other birds; sings throughout year. Alarm is a buzzy chatter; many other notes.

**Eggs**
4–8; creamy white, heavily spotted with reddish-brown; 0.8 x 0.6 in. (2.0 x 1.5 cm). Nest is of vegetation in any cavity or nesting box, up to 40 ft. (12.2 m) high.

**Range**
Resident throughout se. U.S. from N.Y., Mich., and Wis. s. to Fla. and Gulf Coast.

## LONG-BILLED MARSH WREN
*Cistothorus palustris*                                                34:9

**Description**
Size, 4¾–5½ in. (12.1–14.0 cm). Dark brown above, with a blackish, unstreaked crown; white line over eye; whitish below, including undertail coverts; buffy flanks; • *blackish back with white lines.*

**Similarities**
Bewick's has longer white-cornered tail and no white lines on back. Short-billed has shorter bill, streaked crown, buffy breastband, and brownish undertail coverts.

**Habitat**
Freshwater cattail or salt marshes.

**Habits**
Sings from stalks in marsh, often rises 6–10 ft. (1.8–3.0 m) in air and flutters down singing, climbs nimbly about reeds; males often polygamous and given to building extra dummy nests.

**Voice**
Short, rich, bubbling, somewhat guttural song; alarm, a chatter, often suggesting a Red-winged Blackbird; sometimes sings at night.

**Eggs**
3–7; pale to dark brown, often with darker dots; 0.6 x 0.5 in. (1.5 x 1.3 cm); 2 broods. Nest is a ball of grasses with a side entrance, attached to reeds in marsh.

**Range**
Breeds from N.B. and Man. s. to Fla., the Gulf Coast, and Mexico; winters from N.J. along Atlantic and Gulf coasts to Mexico.

## SHORT-BILLED MARSH WREN
*Cistothorus platensis*                                                34:8

**Description**
Size, 4–4½ in. (10.2–11.4 cm). Tiny. Streaked crown and back, bill and tail short, brownish above, buff band across breast, throat and belly white, undertail coverts and flanks brownish.

**Similarities**
Long-billed has eye stripe.

**Habitat**
Wet meadows and grassy marshes, fresh or salt; unevenly distributed, common in some areas, absent in others.

**Habits**
Secretive, stays near ground, sings from weed stalks; if flushed, flight is short, bird soon drops into marsh; male builds extra dummy nests.

**Voice**
Short series of clinking notes, somewhat like knocking pebbles together; *chap, chap, chapperm-chapperm-chapper,* running down scale and increasing in tempo.

**Eggs**
4–10; white; 0.6 x 0.5 in. (1.5 x 1.3 cm). Nest is round ball of grass with side entrance, attached to growing grass stalks.

**Range**
Breeds from N.B. and Man. s. to Del., Mo., and Kans. and s. to S. America; winters from N.J. and Ill. s. to Fla. and Mexico.

Note: The **CANYON WREN,** *Catherpes mexicanus,* size, 5½–5¾ in. (14.0–14.6 cm), can be seen in rocky canyons and ravines in the western fringe of the eastern region. It is a rusty-brown wren with a clear, white throat and upper breast.

## ROCK WREN
*Salpinctes obsoletus*                                              **34:16**

**Description**
Size, 5–6¼ in. (12.7–15.9 cm). Upperparts gray-brown; white line over eye; • *underparts white with dusky streaked breast;* tail rounded, and with black band just before buffy tip.

**Similarities**
Bewick's Wren lacks breast streaks.

**Habitat**
Open rocky places, including deserts, and mountains above timberline.

**Habits**
Bobs up and down on rock, stays close to ground. In flight, quick, jerky.

**Voice**
Song sometimes suggests a Brown Thrasher; a "harsh, grating . . . *kerEE kerEE kerEE . . . chair chair chair chair, deedle deedle deedle deedle, tur tur tur tur, kerEE kerEE kerEE trrrrrrrrrr"* (Nice); call, *tick-ear,* often given with a bob; also a loud, purring trill.

**Eggs**
5–8; white, lightly dotted with red-brown; 0.7 x 0.5 in. (1.8 x 1.3 cm). Nest is of grass, moss; cup-shaped, often in rocky cranny.

**Range**
Breeds chiefly in W., but e. to N. Dak., southward; winters from Calif. and Tex. southward.

# MOCKINGBIRDS AND THRASHERS
Family Mimidae

Members of this family are slim, medium-sized birds with short, rounded wings and long, rounded tails, which they pump when excited. They feed on wild berries, insects, and fruit on the ground, in thickets, or in the open.

## NORTHERN MOCKINGBIRD
*Mimus polyglottos*                                                **35:2**

**Description**
Size, 9–11 in. (22.9–27.9 cm). Slender; gray above, whitish below; no black on face; outer tail feathers with much white; • *white wing patches* conspicuous in flight; long tail.

**Similarities**
Shrikes are stouter, with hooked bills, shorter tails; black on face, wings, and tail; and less white in wings.
**Habitat**
Gardens, farms, roadsides, towns.
**Habits**
Active, aggressive; runs, hops, and feeds on ground with tail raised; but sings from fence post, TV antenna, top of building, or roadside wire; frequently raises wings, flies with rapid wingbeats, long tail noticeable; frequents feeders (likes raisins, nutmeats, suet, crumbs); chases other birds.
**Voice**
Song is varied, mellifluous; notes often repeated 3 or more times. This species is one of the world's great songsters and an accomplished mimic; often sings in flight and by moonlight; call, a harsh *chat*.
**Eggs**
3–6; greenish-blue, spotted with red-brown; 1.0 x 0.8 in. (2.5 x 2.0 cm). Nest is bulky, of twigs, in shrub or thicket, 2–20 ft. (0.6 x 6.1 m) up.
**Range**
Resident from N.S. throughout much of U.S. s. to W. Indies and Mexico.

## GRAY CATBIRD
*Dumetella carolinensis* 35:3

**Description**
Size, 8–9½ in. (20.3–24.1 cm). • *Slate-gray,* with a black cap. • *Undertail coverts chestnut.*
**Habitat**
Thickets, wet or dry; hedges.
**Habits**
Friendly, inquisitive; comes to feeders; often nests near houses; usually sings from concealed perch; common.
**Voice**
Song is squeaky, simple version of Northern Mockingbird's or Brown Thrasher's, but always with some catlike mews and phrases not repeated; somewhat of a mimic; often sings at night, one of earliest songs heard at dawn; note, a catlike mew.
**Eggs**
3–6; greenish-blue; 1.0 x 0.7 in. (2.5 x 1.8 cm); 2 broods. Nest is of twigs, in thickets, 2–8 ft. (0.6–2.4 m) up.
**Range**
Breeds from N.S. and Man. s. to Fla., Tex.; winters from Gulf Coast to C. America.

## BROWN THRASHER
*Toxostoma rufum* 35:1

**Description**
Size, 10½–12 in. (26.7–30.5 cm). • *Reddish-brown above;* whitish below with • *brown streaks on breast;* eye yellow; bill long, curved; 2 white wing bars.
**Similarities**
Wood Thrush is smaller; is less reddish and more brown; has shorter tail, and has straight bill, brown eye, and no wing bars.
**Habitat**
Dry thickets.

**Habits**
Aggressive; will not brook intruders at nest; sings from lofty perch; scratches for food on ground in thicket; walks, runs, or hops, likes dust baths and water baths; less common about habitations than Gray Catbird.

**Voice**
Long, somewhat like Northern Mockingbird's and Gray Catbird's, but few imitations and each phrase repeated twice, as *"drop it, drop it, cover it, cover it, I'll pull it up, I'll pull it up"* (Thoreau); call, a distinctive kissing sound.

**Eggs**
3–6; whitish, heavily dotted with red-brown; 1.1 x 0.8 in. (2.8 x 2.0 cm); often 2 broods. Nest is bulky, of twigs, on or near ground in thicket.

**Range**
Breeds throughout much of e. U.S. from n. New England, Ont., and Man. s. to Fla. and Gulf Coast; winters in s. parts of breeding range.

Note: The sagebrush-dwelling **SAGE THRASHER**, *Oreoscoptes montanus* **(35:4)**, size, 9 in. (22.9 cm), occurs from southern Wyoming and eastern Colorado west, with an isolated population in southern Saskatchewan (rarely wanders eastward). It is a gray, open-country bird with white wing bars and a streaked breast.

# THRUSHES, BLUEBIRDS, AND ALLIES
Family Turdidae

Most thrushes are woodland birds that forage on the ground. The young all have spotted breasts. This family is known for its songsters. They feed on insects, spiders, worms, grubs, wild fruits, berries, and seeds.

## AMERICAN ROBIN
*Turdus migratorius*                                                **36:2**

**Description**
Size, 9–11 in. (22.9–27.9 cm). Slaty back, • *reddish breast,* bill yellow, head black, belly and tips of outer tail feathers white; female paler than male; juvenal with black spots on buffy breast.

**Habitat**
Lawns, gardens, suburbs, fields, swamps, clearings.

**Habits**
Adapts well to civilization, hops or runs on lawn, cocks head, nests near houses, likes to bathe; flocks spend winter in woods and swamps, sometimes gather in huge roosts; one of first species to sing at dawn. Flight is direct, with deep, fairly rapid wingbeats, straight back, and white rear flanks noticeable.

**Voice**
Song is a carol, *cheerily cheer-up, cheerily cheer-up;* notes include a querulous *cuk, cuk;* a *ssssp;* a nervous *bup, bup;* a rattling *chi, il, il, il, il;* and others.

**Eggs**
3–4; Robin's-egg blue; 1.1 x 0.8 in. (2.8 x 2.0 cm); 2 broods. Nest is cup-shaped, of grass and mud, in crotch or branch of tree, or in man-made nook, usually 5–20 ft. (1.5–6.2 m) up.

**Range**
Breeds throughout most of Canada and U.S., from Newfoundland and Man. s. to S.C., Ark., and C. America; occasional breeder along Gulf Coast; winters from Newfoundland, s. Ont., s. to Gulf Coast and southward.

# THRUSHES, BLUEBIRDS, AND ALLIES

Note: The **VARIED THRUSH (36:1)**, size, 9–10 in. *Zoothera naevia* (22.9–25.4 cm), is a western bird that sporadically wanders into the eastern region during migration. It resembles an American Robin, but both sexes of the Varied Thrush have a buffy or rusty eye stripe and wing bars; the male has a black band across its orange-brown breast.

## WOOD THRUSH
*Hylocichla mustelina*                                                    36:11

**Description**
Size, 8 in. (20.3 cm). Plump; bright brown above, becoming
• *reddish on head, neck, and upper back;* breast creamy with large, round, dark spots.
**Similarities**
Brown Thrasher is longer tailed and yellow eyed.
**Habitat**
Shade trees, moist woods.
**Habits**
Originally a woodland bird, now common in suburbs and seen on lawn like American Robin; aggressive, less furtive than other brown thrushes.
**Voice**
Song is an unhurried, liquid, flutelike melody, *"eeohlay-ayolee-ahleelee-ayleahlolah-iloliee"* (Saunders); alarm note, *quirt-quirt;* a rattling *putt, putt, putt.*
**Eggs**
3–4; Robin's-egg blue; 1.1 x 0.7 in. (2.8 x 1.8 cm). Nest is of mud, grass, and paper, 3–12 ft. (0.9–3.7 m) up in tree or shrub.
**Range**
Breeds throughout much of e. U.S. from N.S. and Ont. s. to Fla. and Gulf of Mexico; winters from Mexico s. to C. America, but some winter from s. Fla. and Tex.

## HERMIT THRUSH
*Catharus guttatus*                                                      36:5

**Description**
Size, 6½–7¾ in. (16.5–19.7 cm). Russet-brown above; • *rump and tail rufous;* hint of rufous in primaries; below whitish with small dark wedge-shaped spots on breast.
**Similarities**
Fox Sparrow has thick bill, entire upperparts reddish-brown.
**Habitat**
Damp, mixed woods; in winter, edges, thickets.
**Habits**
Raises and slowly lowers tail on alighting; usually seen near ground, but sings from treetop; hardy, a few linger in north in winter; the only brown thrush to winter in U.S.; visits feeders.
**Voice**
Song is a clear, flutelike introductory note, then various 5–12 note phrases, a pause, then another introductory note on a different pitch, and so on; notes are sweet, clear, bell-like; also a soft *chuck,* a catlike *pay,* a scolding *tuk-tuk.*
**Eggs**
3–5; greenish-blue; 0.9 x 0.7 in. (2.3 x 1.8 cm). Nest is of moss, twigs, grasses, on or near ground.
**Range**
Breeds from Labr. and Sask. s. to N.Y. and Mich.; also in mountains to Va.; winters from s. New England s. to Fla. and C. America.

## SWAINSON'S THRUSH
*Catharus ustulatus*                                          36:9

### Description
Size, 6½–7¾ in. (16.5–19.7 cm). Above, uniform olive-brown;
breast light buff with small dark spots; belly whitish; • *buffy eye-ring* and cheek contrasting with rest of head.

### Similarities
Gray-checked Thrush has gray cheeks, approximately same color
as rest of head, breast whitish; it is sometimes hard to distinguish
these 2 species in fall.

### Habitat
Breeds in spruce and fir of Canadian zone; in migration, woods,
shrubbery, even suburban lawns.

### Habits
If flushed, flies to low branch and remains motionless, then flits on
farther; sometimes pumps tail, but not habitually as Hermit does;
sings from treetop.

### Voice
Song is simpler than Wood or Hermit Thrush; seems to ascend in
spirals; *"whao-whayo-whiyo-wheya-wheeya"* (Saunders). Notes, a
*whit;* a catlike *twee-ur;* in migration at night overhead, a short
*heap.*

### Eggs
3–5; greenish-blue, spotted with reddish-brown; 0.9 x 0.7 in. (2.3 x
1.8 cm). Nest is of grass, moss, twigs, usually in spruce or fir, 4–15
ft. (1.2–4.6 m) up.

### Range
Breeds in Canada from Newfoundland and Man. s. to n. New
England and Mich.; also in mountains to W. Va.; winters in S.
America.

## GRAY-CHEEKED THRUSH
*Catharus minimus*                                          36:10

### Description
Size, 7 in. (17.8 cm). Above, olive-brown; below, whitish with
small dark spots on breast; eye-ring absent or faint; • *cheeks gray,*
approximately as dark as rest of head.

### Similarities
See Swainson's Thrush.

### Habitat
Willows or alders in tundra; Hudsonian zone of northern
coniferous forest; in migration, woods and shrubbery.

### Habits
Similar to Swainson's, but shyer.

### Voice
Least impressive of eastern thrushes; fainter than Swainson's and
without upward spiral; a subdued, faraway-sounding *"wee-oh,
chee-chee-wee-oh, wee-oh"* (Gillespie); also, *"chook-chook, wee-o,
wee-o, wee-i-ti-t-ter-ee"* (Wallace); also a similar flight song. Call
is a melancholy *pheu,* or penetrating *queep,* often heard from
migrants overhead at night.

### Eggs
3–5; greenish-blue, spotted with red-brown; 0.9 x 0.7 in. (2.3 x 1.8
cm). Nest is of grass, twigs, on low limb or bush.

### Range
Breeds in n. Canada from Newfoundland and n. Ont. s. to N.Y.;
winters in W. Indies, C. and S. America.

## VEERY

*Catharus fuscescens*                                          **36:8**

**Description**
Size, 7 in. (17.8 cm). • *Above bright cinnamon-brown;* below light buff with • *inconspicuous spots on throat;* throat appears white at a distance.

**Similarities**
Swainson's is same size, but has broad, buffy eye-ring and browner, less rufous upperparts. Hermit has reddish tail. Gray-cheeked has more gray-brown, no buff on chest.

**Habitat**
Undergrowth in swamps, wet woodlands, ravines, and dry hillsides.

**Habits**
Keeps close to ground, even when singing; progresses by jumping, stands upright.

**Voice**
Song is a series of hollow *whree-u*'s descending the scale, or *"ta-weel'ah, ta-weel'ah, twil'uh, twil'ah"* (Ridgeway); dawn chorus in June on breeding grounds is hauntingly beautiful; call, a distinctive *view.*

**Eggs**
3–5; greenish-blue; 0.9 x 0.7 in. (2.3 x 1.8 cm). Nest is of grass and twigs on ground.

**Range**
Breeds from Newfoundland and s. Man. s. to N.J. and Ind.; also some in mountains to Ga.; winters in S. America.

## EASTERN BLUEBIRD

*Sialia sialis*                                                **36:6**

**Description**
Size, 6½–7½ in. (16.5–17.8 cm). Spring male: • *bright blue above, breast reddish,* belly white. Spring female: grayish, with bright blue on wings and tail. Juvenal: spotted above and below. Fall adult: duller.

**Habitat**
Orchards, wood edges, roadsides, farmlands.

**Habits**
Often seen on roadside wire, looks hunched when perched; catches some insects in air, drops from perch to pounce on others.

**Voice**
Song is a soft mellow warble, *purity, purity;* call, *oola* or *aloola.*

**Eggs**
3–7; light blue; 0.8 x 0.6 in. (2.0 x 1.5 cm). Nest is hole in stump or tree, often abandoned woodpecker's hole 3–30 ft. (0.9–9.1 m) up, or in bird box 3–6 ft. (0.9–1.8 m) up, with entrance hole 1½ in. (3.8 cm) in dia.

**Range**
Breeds locally throughout e. U.S. from s. Canada s. to Gulf of Mexico and w. to Rockies; winters from s. New England and Mo. s. to Mexico.

## MOUNTAIN BLUEBIRD

*Sialia currucoides*                                           **36:7**

**Description**
Size, 6½–7¾ in. (16.5–19.7 cm). Male: paler below, belly white. Female: brownish gray, gray below, with blue on wings and tail. Both duller in winter.

**Similarities**
Eastern Bluebird is not as slim; Mountain sits more erect. Juvenal has much less spotting than Eastern.
**Habitat**
Open woods in mountains; ranches, open country.
**Habits**
Seizes insects in air, or hovers and then drops on them on ground.
**Voice**
Usually silent; near dawn "a sweet, clear *trually tru-al-ly* like that of the Eastern species, and a mellow warble" (Wheelock); note, a low *turr*.
**Eggs**
4–6; pale blue; 0.8 x 0.6 in. (2.0 x 1.5 cm). Nest is of grass, in hole in stump or tree, or beneath eaves of building; also in nesting box.
**Range**
Breeds chiefly in W. but e. to s. Man. and s. to w. S.Dak. and Nebr.; winters from Nebr. s. to Tex. and Mexico.

Note: The **COMMON WHEATEAR,** *Oenanthe oenanthe,* size, 6 in. (15.2 cm), is a small gray and white bird with black wings. It breeds in the Arctic and winters in Africa, but is seen in migration occasionally in Canada. The **TOWNSEND'S SOLITAIRE,** *Myadestes townsendi,* size, 9 in. (22.9 cm), is found in mountain canyons and woods from the Black Hills west. It is an erect-perching, grayish thrush with a white eye-ring that forages for insects by flycatching. It somewhat resembles the Northern Mockingbird, but has white edges on its outer tail feathers, a shorter bill, and buff rather than large white wing patches.

# GNATCATCHERS, KINGLETS, AND OLD-WORLD WARBLERS
Family Sylviidae

This is a family of tiny, active woodland birds with small bills adapted to a diet of insects. Gnatcatchers are distinguishable by their comparatively long tails, kinglets by their brightly colored crowns.

## BLUE-GRAY GNATCATCHER
*Polioptila caerulea*                                        **34:12**

**Description**
Size, 4–5 in. (10.2–12.7 cm). Slender; blue-gray above, eye-ring and underparts white; male has black forecrown; • *long, floppy tail* conspicuous in flight; • *tail white-bordered.*
**Similarities**
Resembles a tiny Northern Mockingbird.
**Habitat**
Thickets, swamps, tall trees.
**Habits**
Active, graceful, tail always in motion; often hovers in front of a twig, frequently catches insects in air.
**Voice**
Call is a shrill, high-pitched, scolding *spee, spee, spee;* song, "rarely heard, a low, pleasant, exquisite, warbling ditty . . . *zee-u, zee-u, ksee-ksee-ksee-ksee-ksee-ksee-ksu*" (Simmons).

**Eggs**
4–5; bluish-white, spotted with red-brown and slate; 0.6 x 0.4 in. (1.5 x 1.1 cm). Nest is tiny felted cup in desert shrub.

**Range**
Breeds from N.H., s. Ont., and cen. Minn. s. to Bahamas, Gulf Coast, and Ark.; winters from N.C. and the Gulf Coast southward.

## GOLDEN-CROWNED KINGLET
*Regulus satrapa*                                                    **41:15**

**Description**
Size, 3½–4 in. (8.9–10.2 cm). Smaller than any wood warbler. Black-bordered orange or yellow crown. Bill tiny; tail short, notched; olive-gray above, underparts, 2 wing bars, and line over eye white; • *crown yellow* in female, and with • *orange center* in male (sometimes concealed).

**Similarities**
Ruby-crowned has broken white eye-ring, no white line over eye, male crown patch scarlet.

**Habitat**
Coniferous and mixed woods; deciduous trees, bushes, as well as evergreens in migration and winter.

**Habits**
Tame, confiding; very active, clambers about or flutters at ends of twigs, often flips out wing tips over back; catches insects in air as well as on twigs; in winter, frequently travels in mixed species flocks.

**Voice**
Call is a high, thin *see-see-see*, shorter than note of Brown Creeper and rapidly repeated; song, similar but longer *"zee, zee, zee, zee, zee, why do you shilly-shally?"* (Stanwood).

**Eggs**
5–10; creamy, marked with brown and lavender; 0.5 x 0.4 in. (1.3 x 1.0 cm). Nest is of moss, lichens; round, suspended from twig of conifer, 4–50 ft. (1.2–15.2 m) up.

**Range**
Breeds from Newfoundland and Man. s. to Mass. and Mich.; also into mountains of N.C.; winters from s. parts of breeding range to Fla., Gulf Coast, and C. America.

## RUBY-CROWNED KINGLET
*Regulus calendula*                                                  **41:13**

**Description**
Size, 3¾–4⅓ in. (9.5–11.4 cm). A very small, green, short-tailed bird with a • *white eye-ring*. Bill tiny; tail notched; olive-gray above, whitish below; 2 white wing bars, rear one with black border; male has concealed patch of ruby on crown, visible when bird erects it in display; eye-ring gives black beady eye a staring look.

**Similarities**
Any kinglet without a visible crown patch is this species; *Empidonax* flycatchers have white eye-rings, but they sit still, do not flit about.

**Habitat**
Breeds in northern bogs and forests; in migration, shrubbery, orchards, second growth, thickets.

**Habits**
Tame, active, restless, flips wing tips out over back; often associates with warblers and nuthatches in migration.
**Voice**
Ringing song, loud for so small a bird, seems to go up and down, *"see-see-see-you-you-you-just-look-at-me just-look-at-me"* (Cruickshank); also a harsh, wrenlike note.
**Eggs**
5–11; creamy white, usually finely dotted with red-brown; 0.5 x 0.4 in. (1.4 x 1.1 cm). Nest is of moss and lichens; suspended, or on limb of conifer, 2–50 ft. (0.6–15.2 m) up.
**Range**
Breeds from Newfoundland and Man. s. to n. New England and Ont.; winters from s. New England s. to Fla., Gulf Coast, and C. America.

# PIPITS
Family Motacillidae

Pipits are slender, larklike open country birds with thin bills and long hind claws. They are brownish above, and have streaked breasts and white outer feathers on their tails, which they frequently pump. They eat insects, grubs, small mollusks, and crustaceans.

## WATER PIPIT
*Anthus spinoletta*                                              **48:5**
**Description**
Size, 6–7 in. (15.2–17.8 cm). Unstreaked dark back and dark legs. Spring: gray-brown above, • *buffy below* with streaks on upper breast, white outer tail feathers, bill slender. Fall: darker above, buffier below, breast more heavily streaked.
**Similarities**
Vesper Sparrow has short, thick bill and hops, does not walk or wag tail. Horned Lark has pipitlike flight, but is heavier, has white belly, different pattern, and narrower white edges on black tail.
**Habitat**
Breeds on mountains above timberline, tundra; plowed fields, dunes, flats.
**Habits**
Terrestrial, restless, nods head like dove as it walks, usually seen in flocks; perches on trees, fences, wires; wags tail more than Sprague's.
**Voice**
Note, heard in migration, "a sharp *tsip-tsip, tsip-it* . . . [the lark's is lower . . . more rolling . . . *sleek-slik-seezik*]" (Cogswell); song on breeding grounds simple, pleasant, often with trills; also has a flight song.
**Eggs**
4–7; whitish, heavily marked with chocolate; 0.8 x 0.6 in. (2.0 x 1.5 cm). Nest is of grass on ground in shelter of rock or hummock.
**Range**
Breeds in Arctic regions of Greenland, Newfoundland, and Hudson Bay, s. to n. Maine; winters from s. Pa., w. Va., s. to Fla., Gulf Coast, and C. America.

## SPRAGUE'S PIPIT
*Anthus spragueii* 48:4

**Description**
Size, 6½–7 in. (16.5–17.8 cm). Paler than Water; head, neck, and
• *back streaked with black and buff;* white below, washed with buff,
finely streaked; more white in tail than Water Pipit; bill lighter;
legs yellowish.
**Similarities**
Vesper Sparrow is heavier, has short, thick bill; hops, does not
walk. Horned Lark is heavier, has less white in tail in flight.
Water Pipit is darker, more grayish, has unstreaked back, more
streaks on breast, pumps tail more. Longspurs have thicker bills.
**Habitat**
Prairie, plains.
**Habits**
Male spends much time in air ascending in circles; often occurs
singly or in pairs; flight more bouncy than Water Pipit's.
**Voice**
Famous flight song delivered at average height of 300 ft. (91 m),
"As it flies around, its flight rises and falls. Each time it rises the
bird sings; when it falls, he is silent . . . . The song is clear, sweet,
and musical" (Saunders); call, single notes, harsher than paired
notes of Water Pipit.
**Eggs**
4–5; grayish-white, thickly blotched with purplish-brown; 0.8 x 0.6
in. (2.0 x 1.5 cm). Nest is of grass, on ground.
**Range**
Breeds from cen. Man. s. to Minn. and Mont.; winters from Miss.
and Tex. s. to Mexico.

# WAXWINGS
Family Bombycillidae

Waxwings are slim, crested brown birds, with sleek plumage; black
chins and foreheads; red, waxlike tips to their secondaries; and a
yellow band across the end of their squared tails. They eat berries,
especially cedar and mountain ash, and seeds and insects.

## CEDAR WAXWING
*Bombycilla cedrorum* 36:4

**Description**
Size, 6½–8 in. (16.5–20.3 cm). Above, reddish-brown; flanks
yellow, undertail coverts white; • *crested.* Juvenal: duller, grayer
with blurry streaking below.
**Similarities**
Throat black more restricted toward chin than in Bohemian;
redder brown above.
**Habitat**
Open woods, edges, orchards.
**Habits**
Social; occurs in small flocks here today, gone tomorrow; flock
alights on tree in compact body, birds remain upright, motionless,
before feeding; sometimes catches insects in air, occasionally gets
drunk on overripe berries; wing shape similar to Starling's; a late
nester.
**Voice**
Thin, high hiss; once learned it is the best means of detecting their
presence.

**Eggs**
3–6; often pale-bluish, spotted with blackish; 0.9 x 0.7 in. (2.3 x 1.8 cm). Nest is bulky cup of twigs, grass, moss on horizontal branch 4–40 ft. (1.2–12.2 m) up.
**Range**
Breeds from Cape Breton Is. and cen. Man. s. to Ga. and Ark.; winters from New England s. to Fla., Gulf Coast, and C. America.

## BOHEMIAN WAXWING
*Bombycilla garrulus*                                          36:3

**Description**
Size, 7½–8¾ in. (19.1–22.2 cm). Throat black; flanks gray;
• *undertail coverts chestnut; yellow markings* on tail and wings, white streak in wings.
**Similarities**
Grayer above than Cedar Waxwing.
**Habitat**
Breeds in northwestern coniferous forests, flocks visit northern United States occasionally in winter.
**Habits**
Gregarious at all seasons, flight undulating in tight flocks; catches insects in air; may come to feeder for raisins.
**Voice**
"Alarm note *tzee-tzee*" (Forbush); no song.
**Eggs**
3–6; grayish, spotted with dark brown; 0.9 x 0.7 in. (2.3 x 1.8 cm). Nest of twigs, grass, in conifer, 8–50 ft. (2.4–15.2 m) up.
**Range**
Breeds chiefly in nw. N. America, but e. to ne. Man. and s. to s. Alta.; winters s. and e. of breeding range, to s. Ont.

# SHRIKES
Family Laniidae

American shrikes are gray, black, and white predators with heavy heads, stout hooked beaks, black masks, and slim tails. They are birds of the open country, where there are scattered trees. They prey upon insects, small mammals, and birds, and impale their prey on thorns.

## NORTHERN SHRIKE
*Lanius excubitor*                                             35:7

**Description**
Size, 9–10¾ in. (22.9–27.3 cm). Less extensive than Loggerhead.
• *Black mask;* feathers just over the bill are whitish; bill light below at base except in spring and summer. Adults: gray above, rump light, wings and tail black with some white on wings and outer tail feathers; underparts whitish, faint barring on breast in fall and winter. Immature: browner, plainly barred below, bill brown; this plumage worn through first winter.
**Similarities**
Loggerhead Shrike is darker above, whiter below, with black feathers over bill.
**Habitat**
Breeds in openings in northern forest; in winter, open country, swamps, orchards.

#### Habits
Bold, aggressive; perches on exposed observation post or wire like a
Kestrel and watches for prey; sometimes hovers, occasionally
pumps tail; flight slightly undulating; usually catches birds on the
wing; this species more apt to be seen in northern part of our area
and in winter; range may barely overlap with next only in winter.
#### Voice
Call is "a harsh shrieking *jo-ree*" (Knight); song prolonged, a
Robin-like or Catbird-like carol with some harsh notes; also
mimicking notes of other birds; both sexes sing.
#### Eggs
4–7; whitish, with darker markings; 1.1 x 0.8 in. (2.8 x 2.0 cm).
Nest is of twigs and leaves in thorny bush or on limb of tree, 5–20
ft. (1.5–6.1 m) up.
#### Range
Breeds in n. Canada from Labr. w. to n. Man.; winters from s.
parts of breeding range s. to Va. and Tex.

## LOGGERHEAD SHRIKE
*Lanius ludovicianus*                                     **35:5**
#### Description
Size, 8–10 in. (20.3–25.4 cm). • *Black mask* is complete; feathers
just over the bill are black. Dark above, unbarred below; bill all-
black; immature faintly barred on pale gray underparts.
#### Similarities
Slimmer Northern Mockingbird lacks face mask; Northern Shrike
is larger.
#### Habitat
Farming country with scattered trees.
#### Habits
Similar to Northern Shrike; much more apt to be seen in southern
part of eastern area; range may barely overlap with the Northern
only in winter.
#### Voice
Harsh *chack-chack; tsirp-see, tsirp-see;* song somewhat resembles
Gray Catbird's or Brown Thrasher's, but lower, harsher.
#### Eggs
4–8; white to greenish-gray, variously marked; 1.0 x 0.7 in. (2.5 x
1.8 cm). Nest is of rootlets, twigs, in bush or tree 5–20 ft. (1.5–6.1
m) up.
#### Range
Breeds locally throughout most of U.S. from Maritime Provinces
and s. Man. s. to s. Fla., Gulf Coast, and Mexico; winters from
Va. southward.

# STARLINGS
Family Sturnidae

This is a large and varied Old World family of gregarious, sharp-
billed, short-tailed birds, many looking like blackbirds. Only one
introduced species occurs in East. They are omnivores, eating
insects, worms, grubs, seeds, fruits.

## STARLING
*Sturnus vulgaris*                                       **42:3**
#### Description
Size, 7½–8½ in. (19.1–21.6 cm). • *Short-tailed,* chubby, black bird.
Spring adult: iridescent black, white dots on back; strong, pointed

• *yellow bill*; squared tail. Fall adult: similar, but liberally covered with white spots, some brown spots; black bill. Juvenal: grayish, blurry whitish streaking on throat and belly, black bill.

**Similarities**
Young have longer bills, shorter tails than female cowbirds.

**Habitat**
Cities, suburbs, farmlands, edges, lawns, pastures; even deserts.

**Habits**
Gregarious, persistent, wary, aggressive; comes to feeders; walks or waddles energetically and erratically on ground; collects in great flocks in cities, suburbs, and marshes; dense flocks in air perform intricate revolutions in concert; looks triangular or spindle-shaped in flight with short, pointed, fast-beating wings, alternately flapping and sailing; often flocks with Red-winged Blackbirds and other American blackbirds; usually feeds on ground.

**Voice**
Varied, including whistles, squeaking, squealing, harsh and grating notes; mimics many other species.

**Eggs**
4–7; pale blue; 1.2 x 0.9 in. (3.0 x 2.3 cm); 2 broods. Nest is of sticks, grass, in hole in tree or cactus, or other natural or man-made cavity, 10–25 ft. (3.0–7.6 m) up.

**Range**
Resident throughout much of Canada and most of U.S., from Que., s. to Gulf Coast and Mexico; introduced from Europe in 1890, now well established.

# VIREOS
Family Vireonidae

Vireos are sparrow-sized arboreal birds, olive-gray above, white or yellowish below, with thick, slightly hooked bills. They are larger than warblers and less active. Most are quite deliberate in their movements as they creep about the outer branches gleaning insects; hence, they are more often heard than seen. Vireos all build dainty cuplike nests suspended between forked twigs; the male helps in incubation and sings on the nest. Their diet consists of insects.

## VIREO COMPARISON CHART

| Species | Wing Bars | Breast | Eye Markings |
|---|---|---|---|
| Red-eyed | no | white | black and white line |
| Warbling | no | white | light line |
| Philadelphia | no | yellowish | light line |
| Bell's | faint | white | white spectacles |
| Solitary | yes | white | white spectacles |
| White-eyed | yes | white | yellow spectacles |
| Yellow-throated | yes | yellow | yellow spectacles |

## WHITE-EYED VIREO
*Vireo griseus*        **40:18**

**Description**
Size, 6 in. (15.2 cm). Only vireo with 2 white wing bars, • *white breast*, and • *yellow spectacles*. Greenish-olive below, eyes white, flanks pale yellow; immature has dark eyes.

**Similarities**
Bell's is smaller, has faint wing bars, dark eyes, and faint white spectacles.

**Habitat**
Thickets, often near water.

**Habits**
Aggressive, restless, noisy, inquisitive, fearless; but also shy, hard to see, retiring; pumps tail; often victimized by cowbird.

**Voice**
Emphatic *"Chick! ticha WHEEyo chick!"* (Saunders); and many variations of these, some ventriloquial; also notes mimicking other birds; a harsh mew, loud whistle, and a short *tik.*

**Eggs**
4; white, faintly dotted with purple and brown; 0.8 x 0.6 in. (2.0 x 1.5 cm). Nest is in shrub, 2–8 ft. (0.6–2.4 m) up.

**Range**
Breeds throughout much of se. U.S. from cen. New England and Nebr. s. to s. Fla. and Mexico; winters from Gulf States s. to W. Indies and C. America.

## BELL'S VIREO
*Vireo bellii*                                                      **40:15**

**Description**
Size, 4½–5 in. (11.4–12.7 cm). Nondescript. Olive above, rump brighter, white below, faint white eye-ring, 1 or 2 pale wing bars, pale yellowish on sides.

**Similarities**
See White-eyed Vireo. Warbling lacks any eye-ring and wing bars. Ruby-crowned Kinglet has shorter bill, distinct eye-ring.

**Habitat**
Bottomland shrubbery, mesquite thickets, hedgerows.

**Habits**
Similar to White-eye; active, keeps out of sight, but inquisitive, fearless about nest.

**Voice**
Song is *"whillowhee, whillowhee, WHEE"* (Nice); this phrase is repeated, first rising as if asking a question, then falling as if answering it; also a hoarse scolding note; call, not as harsh as White-eyed's.

**Eggs**
3–5; white, sparingly spotted with brown; 0.7 x 0.5 in. (1.9 x 1.4 cm). Nest is low in briar patch.

**Range**
Breeds from Ind. and S.Dak. s. to Tex. and C. America; winters in C. America.

## YELLOW-THROATED VIREO
*Vireo flavifrons*                                                 **40:16**

**Description**
Size, 6 in. (15.2 cm). Greenish above, grayer on rump and tail; • *yellow throat* and spectacles, white belly.

**Similarities**
Pine Warbler is more slender, has no yellow spectacles, and frequents pines.

**Habitat**
Shade trees, orchards.

**Habits**
Sluggish; feeds slowly, deliberately, climbing about upper branches.

**Voice**
Song is faintly reminiscent of American Robin, includes as its main

phrase a distinctive nasal *three-eight;* much deeper and less continuous than Red-eyed's; also a scolding note.

**Eggs**
3–4; white, spotted with purplish-brown; 0.9 x 0.6 in. (2.3 x 1.5 cm). Nest is in tree, 3–50 ft. (0.9–15.2 m) up.

**Range**
Breeds throughout much of e. U.S. from s. New England and Minn. s. to Fla. and Gulf states; winters from Mexico to S. America.

## SOLITARY VIREO
*Vireo solitarius*                                           **40:13**

**Description**
Size, 5–6 in. (12.7–15.2 cm). Olive-green back contrasts sharply with • *blue-gray head;* • *white throat, spectacles,* wing bars, and underparts; sides yellowish; western races lack bluish cast on head.

**Habitat**
Coniferous and mixed open forests and edges.

**Habits**
Tame, sedate, an early migrant; wanders about when singing, unlike Red-eyed, which moves little; commonly victimized by cowbird.

**Voice**
Song is in character between that of Red-eyed and Yellow-throated, sweeter than Red-eyed's, with longer pauses between phrases and including a rising *tu-wee-tu;* and a scolding note.

**Eggs**
3–5; white, dotted with red-brown and umber; 0.8 x 0.6 in. (2.0 x 1.5 cm). Nest is in tree, 5–12 ft. (1.5–3.7 m) up.

**Former name**
Blue-headed Vireo.

**Range**
Breeds from Newfoundland and Man. s. to Conn. and Mich.; also into mountains of Ga.; winters from N.C. and Gulf Coast s. to C. America.

## RED-EYED VIREO
*Vireo olivaceus*                                           **40:14**

**Description**
Size, 5½–6½ in. (14.0–16.5 cm). A "whiskerless" vireo with • *black and white eye stripe.* Upperparts olive, grayer on crown; white below; • *no wing bars;* red eye inconspicuous at a distance.

**Similarities**
Warbling is smaller, grayer above and lacks black eye stripe. Philadelphia is yellowish below, dark eyed. Tennessee Warbler has thinner bill, lacks conspicuous black eye stripe.

**Habitat**
Virtually any woodland or shade trees.

**Habits**
Movements deliberate, but less sluggish than Yellow-throated and Solitary; sings through heat of day; often parasitized by cowbird; very common and widespread.

**Voice**
Song is a continuing repetition of little phrases, remotely Robin-like, *now you see me, now you don't, here I am, there I go;* or *"Hello, hello, Are you there? Can you hear me? This is Vireo. Yes, yes, Vireo, Vireo"* (Taylor). Alarm, a descending, catlike *kway.*

**Eggs**
3–4; white, sparingly spotted with blackish; 0.9 x 0.6 in. (2.3 x 1.5 cm). Nest is in sapling or tree, 4–50 ft. (1.2–15.2 m) up.
**Range**
Breeds throughout much of e. U.S. from Gulf of St. Lawrence and Ont. s. to Fla., Gulf Coast, and Tex.; winters in S. America.

Note: The **BLACK-WHISKERED VIREO,** *Vireo altiloquus* **(40:17)**, size, 5 in. (12.7 cm), has extended its range from southern to northern Florida recently. It very closely resembles the Red-eyed, but it has a fine, black whisker mark; its song is like that of Red-eyed, but phrases paired.

## PHILADELPHIA VIREO
*Vireo philadelphicus*                                    **40:12**

**Description**
Size, 4½–5½ in. (11.4–12.7 cm). • *Yellowish breast* and • *no wing bars*. Olive above, light white line over eye, dark spot between bill and eye, chin and belly whitish.
**Similarities**
See Red-eyed Vireo. Warbling Vireo is whitish below, but sometimes has yellowish wash on sides; also, Warbling has light, not dark, spot between bill and eye. Tennessee and Orange-crowned Warblers sometimes similar in color, but more slender, with thin bill and warblerlike build and actions. Female Black-throated Blue Warbler has small, white square on wing.
**Habitat**
Edges of woods, low trees, and shrubs, often near water.
**Habits**
Like Warbling Vireo; tame, when feeding, rather active for a vireo, sometimes flutters in front of leaves, hangs upside down in twig, or catches insects in air; uncommon, but frequently overlooked, in migration in East.
**Voice**
Similar to Red-eyed's, but a trifle higher, less often repeated, and containing frequently a characteristic abrupt, rising 2-syllable note.
**Eggs**
4; white, sparingly spotted with umber; 0.8 x 0.5 in. (2.0 x 1.3 cm). Nest is in shrub or tree, 9–40 ft. (2.7–12.2 m) up.
**Range**
Breeds in s. Canada, s. to n. U.S.; winters in C. America.

## WARBLING VIREO
*Vireo gilvus*                                            **40:10**

**Description**
Size, 4½–5½ in. (11.4–12.7 cm). Olive-gray above, white spot between bill and eye, sometimes a yellowish wash on sides, whitish breast, • *no wing bars*.
**Similarities**
See Red-eyed and Philadelphia Vireos.
**Habitat**
Shade trees and large forest-edge trees, often in upper branches.
**Habits**
Sings much, but not as consistently as Red-eyed; often parasitized by cowbird.
**Voice**
Pleasant, continuing warble, somewhat reminiscent of Purple Finch, but softer, more languid; call, a snarl, *myee*.

**Eggs**
3–5; white, often spotted with red-brown; 0.7 x 0.5 in. (1.9 x 1.4
cm). Nest is in tree, 20–70 ft. (6.1–21.3 m) up.
**Range**
Breeds throughout much of U.S., from N.E. and Man. s. to N.C.,
La., and Mexico; winters in C. America.

# WOOD WARBLERS
Family Parulidae

Wood Warblers are smaller and more slender than sparrows, with
thin, straight, unhooked bills and wings usually longer than the
tail. Most are active and restless, seeking insects in diverse habitats.
They do not warble, but have distinctive songs. All wood warblers,
except the Yellow-rumped and Palm, are normally transients or
summer visitors to the eastern area, although an occasional Yellow-
breasted Chat or Common Yellowthroat may linger into the
winter. Warblers are sometimes confused with vireos, which
average larger and have heavier bills and sluggish movements; or
with kinglets, which are smaller, shorter-tailed, and rounder and
given to flirting their wings.

Many warblers in fall have a plumage quite different from that of
spring; immatures may also differ. Illustrated and described are the
principal plumages. For greatest help in identification, check both
text and illustration. Occasionally, a bird in molt fails to match
either and may offer real problems in identification.

## BLACK-AND-WHITE WARBLER
*Mniotilta varia*                                              **38:11**

**Description**
Size, 4½–5½ in. (11.4–12.7 cm). • *Black and white,* with a striped
crown. Male: • *striped black and white;* 2 white wing bars. Female
and immature: paler, more white below, lacking black throat.
**Similarities**
See Blackpoll Warbler.
**Habitat**
Woodlands, preferably deciduous, moist or dry.
**Habits**
Creeps along trunks and branches like a nuthatch, head up or
down; sometimes catches insects in air.
**Voice**
Thin, wiry, repeated *zee, zee, zee,* or *we-see, we-see, we-see,*
spiraling upward, other similar songs; call, a short *pip,* a louder
*jink.*
**Eggs**
4–5; creamy white, well spotted with dark brown; 0.7 x 0.5 in. (1.8
x 1.3 cm). Nest is of grasses and rootlets, on ground at foot of tree,
stump, or rock.
**Range**
Breeds throughout most of e. U.S. from s. Canada s. to s. U.S., e.
of Rockies; winters from s. Gulf states s. to S. America.

## PROTHONOTARY WARBLER
*Protonotaria citrea*                                         **39:9**

**Description**
Size, 5½ in. (14.0 cm). Bill long (for a warbler) and slightly
curved. Male: olive back; • *blue-gray wings,* rump, and tail, latter

with white on outer feathers; no wing bars. Unmarked • *golden head and breast.* Female and immature: duller.

**Similarities**
Smaller Yellow Warbler is lighter above, tail corners yellow.

**Habitat**
Swamps, moist bottomlands, edges of watercourses.

**Habits**
Active, tame, low-ranging; creeps nuthatchlike about trunks; flight swift, direct, like a waterthrush.

**Voice**
Song is a loud, clear *tweet-tweet-tweet* of 7–12 notes, all on same key, reminiscent of Solitary Sandpiper; also a low, sweet flight song, *"che-wee—che-wee chee-chee, chee-chee-che-wee—che-wee"* (Walkinshaw), given while hovering; call, a soft chip, alarm note sharp, like Louisiana Waterthrush.

**Eggs**
4–6; creamy white, often liberally marked with brown and purple; 0.7 x 0.5 in. (1.9 x 1.4 cm). Nest is in woodpecker hole, cavity, or bird box, 1–30 ft. (0.3-9.1 m) up.

**Range**
Breeds throughout much of e. U.S., from N.Y. and Minn. s. to Fla. and Gulf Coast; winters from Mexico to S. America.

## SWAINSON'S WARBLER
*Limnothlypis swainsonii*                                    **37:10**

**Description**
Size, 6 in. (15.2 cm). Bill rather long. Adult: underparts and • *line over eye whitish;* brownish-olive, unstreaked above; red-brown crown; no wing bars. Immature: olive above, yellowish eye stripe and underparts, red-brown wings.

**Similarities**
Worm-eating has black stripes on crown. Waterthrushes are streaked below. Palm Warbler has red-brown cap, but has streaks on sides, some yellow below, and wags tail.

**Habitat**
Coastal swamps; rhododendron-clad Appalachian hillsides, bottomlands, and swamps, up to 3000 ft. (914 m).

**Habits**
Low-ranging; walks like an Ovenbird on ground, sings from perch 1–20 ft. (0.3–6.1 m) up; elusive, more often heard than seen.

**Voice**
Song is loud, ringing, suggesting Louisiana Waterthrush but shorter, *"whee, whee, whee, WHIP-poor-WILL"* (Brooks); call, a clear *churp.*

**Eggs**
3; bluish-white; 0.8 x 0.6 in. (2.0 x 1.5 cm). Nest is bulky, of leaves, pine needles, in cane or bush, 2–10 ft. (0.6–3.0 m) up.

**Range**
Breeds in se. U.S. from Md. and Ill. s. to Fla. and Gulf Coast; winters in Caribbean, southward.

## WORM-EATING WARBLER
*Helmitheros vermivorus*                                     **37:9**

**Description**
Size, 5 in. (12.7 cm). • *Head buff,* with 4 bold • *black stripes;* rest of upperparts brownish-olive; underparts buffy white; no wing bars; sexes similar.

**Similarities**
Lacks golden crown and spotted breast of Ovenbird.
**Habitat**
Dry, wooded hillsides; damp woods; ravines.
**Habits**
Low-ranging, shy; bobs head as it walks on ground with tail high,
often sings from stub or sapling, sometimes clambers about trunk
of tree like Black-and-White.
**Voice**
Song is dry trill somewhat like a Chipping Sparrow's but faster,
buzzier; also a goldfinchlike flight song; call, a sharp *pip*, like a
Black-and-White's; alarm, a *chip-chip-chip*.
**Eggs**
4–5; white, spotted with brown and lavender; 0.7 x 0.5 in. (1.8 x
1.3 cm). Nest is of leaves, on steep hillside in dense underbrush.
**Range:**
In se. U.S. from Mass. and Iowa s. to Fla. and Gulf Coast; winters
in W. Indies and from Mexico to C. America.

## GOLDEN-WINGED WARBLER
*Vermivora chrysoptera* 39:7

**Description**
Size, 4½–5 in. (11.4–12.7 cm). Spring male: • *black bib* and ear
patch; • *gold on crown and wings;* rest of upperparts blue-gray;
underparts white. Spring female: like male, but bib and ear patch
dull gray. Fall adults and immature: greener, duller; black or gray
areas obscured.
**Habitat**
Bogs, wet woodland edges; also wet or dry hillside thickets.
**Habits**
Tame; feeds from low shrubbery to tall treetops, often hanging
upside down from end of twig.
**Voice**
Song is *see, buzz-buzz-buzz,* sometimes 3 or 5 notes; similar song
of Blue-winged normally has 2 notes (but these 2 species
occasionally interchange songs); also a *"th-th-th-th-th-th-three"*
(Tyler); alarm note, a sharp *chip*.
**Eggs**
3–5; white, speckled with brown; 0.7 x 0.5 in. (1.8 x 1.3 cm). Nest
is bulky, of leaves, bark strips, in tussock of grass, goldenrod, bush.
**Range**
Breeds from N.H. and Minn., s. to N.J. and Iowa; also to
mountains of Ga.; winters from Mexico. s. to S. America.

Note: The Golden-winged and Blue-winged Warblers overlap and
hybridize frequently in a very complex fashion. The hybrids, two
of the more typical forms, "Lawrence's" and "Brewster's" **(39:8,
39:10),** are variable, mixing colors of the parental forms, except
that they either have the face and throat patch of the Golden-
winged, or lack it completely as the Blue-winged. Blue-winged-like
hybrids with a face and throat patch, usually found where Blue-
wingeds are commoner, are known as **LAWRENCE'S WARBLER,**
originally described as a distinct species, *Vermivora lawrencei.* The
more common hybrids lacking the face and throat patch but
tending more or less toward the Golden-winged in other features
are called **BREWSTER'S WARBLER,** also originally described as a
species, *Vermivora leucobronchialis.* The hybrid males may sing
like Blue-winged or Golden-winged, or sing both songs. Hybrids
have habits like the two parental forms, and may be expected
wherever Golden-wingeds or Blue-wingeds occur.

## BLUE-WINGED WARBLER
*Vermivora pinus*                                    **39:12**

**Description**
Size, 4½–5 in. (11.4–12.7 cm). Spring male: bright yellow head
and underparts, olive-green back, blue-gray wings with 2 usually
• *white wing bars*, gray tail with some white, white undertail
coverts, • *black line through eye*. Spring female, fall adults, and
immature: duller, crown more olive, wing bars white to yellowish.

**Habitat, Eggs**
Same as Golden-winged.

**Habits**
Forages near ground, sings from exposed branch, 12–30 ft. (3.7–
9.7 m) up, motions somewhat deliberate.

**Voice**
Song, a buzzy *seee-buzzzz,* second note lower (similar song of
Golden-winged normally has 4 notes); also a "longer song, *wee-
chi-chi-chi-chi, chur, chee-chur"* (Chapman).

**Range**
Breeds from cen. New England and Nebr. s. along Atlantic Coast
states to S.C. and inland to s. Ill.; winters from Mexico to C.
America.

## BACHMAN'S WARBLER
*Vermivora bachmanii*                                **39:15**

**Description**
Size, 4¼ in. (10.8 cm). Male: • *black throat and crown patch;* gray
nape, wings and tail; back olive, tips of outer tail feathers white;
eye-ring, breast and belly yellow; no wing bars. Female: similar,
but black replaced by gray on crown, yellow on throat; eye-ring
and bend of wing yellow. Immature: male, less black on throat;
female, less black, and back grayer.

**Similarities**
In the larger male Hooded a complete black hood joins black
crown and throat. Female Nashville has more gray on head and
white eye-ring. Female Wilson's, Yellow-throated and Hooded lack
yellow eye-ring of female Bachman's and have olive, not gray,
crown; also female Hooded is larger and has white tail corners.

**Habitat**
Heavily wooded swamps.

**Habits**
Very active; forages somewhat like a Parula from 2 to 40 ft. (0.6–
12.2 m) up; sometimes hangs upside down like a chickadee; often
flies hundreds of yards between feeding spots; sings from tops of
tall trees, seen there also in migration; elusive, very hard to find.
Very rare.

**Voice**
Song is *"zee zee zee zee zee see zee zee chew"* (Borror), *zee's*
somewhat similar to those of Golden-winged; "wiry and insect-like,
strongly suggesting a 'listless' Parula Warbler . . . ventriloquial."
*"Zrr, zrr, zrr, zrr, zrr, zrr* [sometimes ending in a] *chwit"*
(Barnes).

**Eggs**
3–4; white; 0.6 x 0.5 in. (1.5 x 1.3 cm). Nest is of grass and leaves,
in bush or cane, 2–5 ft. (0.6–12.2 m) up.

**Range**
Breeds locally in se. U.S., in S.C., Ala., Ark., and Mo.; winters in
Cuba and Isle of Pines.

## ORANGE-CROWNED WARBLER
*Vermivora celata*                          **37:7, 40:9**

**Description**
Size, 4½–5½ in. (11.4–14.0 cm). Spring adult: perhaps the most
nondescript spring warbler; olive-green above, yellowish-green
below; pale eye stripe; • *faint stripes on flank;* hidden patch of
orange in crown; yellow undertail coverts; no wing bars. Fall adult
and immature: similar but duskier, often quite grayish, and hardly
any paler below; crown patch absent or not visible.

**Similarities**
Nashville is yellow below, unstreaked. See fall Tennessee.
Philadelphia Vireo is larger, sluggish, has thicker bill, light stripe
over eye, no side streaks, and is an early fall migrant.

**Habitat**
Open woodlands with heavy brush, especially on slopes or near
water.

**Habits**
Forages low. Rare in East in spring, less uncommon in late fall
(October).

**Voice**
Song is somewhat like a Chipping Sparrow's, a series of 18–22 *si*
notes, dropping slightly in middle, rising and fading toward end;
note, a distinctive *chip.*

**Eggs**
4–6; white, dotted with reddish-brown; 0.7 x 0.5 in. (1.8 x 1.3 cm).
Nest is of grass, rootlets, on ground in shrubbery on brushy (or
other) hillside.

**Range**
Breeds from Labr. and Man. s. to Tex.; winters from S.C. and
Ark. s. to C. America.

## NASHVILLE WARBLER
*Vermivora ruficapilla*                      **39:11, 40:5**

**Description**
Size, 4–5 in. (10.2–12.7 cm). Spring male: olive back, chestnut
crown patch (often concealed); gray head, undertail coverts whitish,
• *yellow underparts, white eye-ring;* no wing bars, no white in tail;
female duller. Fall adults and immature: duller; yellow below but
with color of head and back blending; immature quite brownish
above.

**Similarities**
Connecticut, Mourning, and MacGillivray's are somewhat similar
in color but are larger and throats not yellow.

**Habitat**
Breeds in northern bogs and second growth, up to 15 ft. (4.6 m)
high; in migration, edges of woods, orchards, shade trees.

**Habits**
Active, forages at all levels; relatively early spring and late fall
migrant.

**Voice**
Song is *"see it see it see it see it see it, ti-ti-ti"* (Gunn); note, a *chip.*

**Eggs**
3–5; white, dotted with red-brown; 0.6 x 0.5 in. (1.5 x 1.3 cm).
Nest is of grass, leaves, concealed in or under tussock on ground.

**Range**
Breeds from N.S. and Man. s. to Conn. and Minn.; winters from
Fla. and Mexico, s. to C. America.

## TENNESSEE WARBLER
*Vermivora peregrina*                                                37:8, 40:11

**Description**
Size, 4½–5 in. (11.4–12.7 cm). Spring adult male has gray crown,
• *white line over eye,* and dark line through it, rest of upperparts
greenish, white below; female has crown more olive, underparts
somewhat yellowish. Fall adult and immature: bright greenish
above, unstreaked yellowish below, yellowish line over eye (grayer
above, whiter below in male), pale wing bars, undertail coverts
white (sometimes with a trace of yellow).

**Similarities**
Fall Nashville has white eye-ring, yellow undertail coverts. Fall
Orange-crowned is darker, dingier, with faint streaks on sides,
yellow undertail coverts, no line over eye; occurs in late (not early)
fall. Vireos are larger, have heavier bill, and are sluggish. Red-
eyed has more conspicuous eye lines.

**Habitat**
Breeds in openings in northern coniferous forest; in migration,
woodlands, edges, shade trees, brush.

**Habits**
Restless; prefers upper branches in spring, any height in fall; on
breeding grounds sings from outer branch of top of tree 5–30 ft.
(1.5–9.1 m) up.

**Voice**
Song is loud, in 3 parts, last somewhat like a Chipping Sparrow's;
*tenne-tenne-tenne-tenne, chip-chip, ssee-ssee-ssee-ssee-ssee;* or
"*tsip-pit-tsip-PIT, tsip-PIT, tsip-PIT, tsip-pit-tsee, tsee, tsee, tsee,*
more rapid and higher pitched toward end" (Harrison);
reminiscent of Nashville's song but latter's is only 2-parted.

**Eggs**
4–7; creamy white, with red-brown spots; 0.6 x 0.5 in. (1.5 x 1.3
cm). Nest is cup-shaped, of vegetable fibers and grass, close to
ground in conifer.

**Range**
Breeds from Labr. and Man. s. to Maine and Wis.; winters from
Mexico to S. America.

## NORTHERN PARULA
*Parula americana*                                                37:5, 41:12

**Description**
Size, 4½ in. (11.4 cm). Smallest warbler; 2 white wing bars.
Spring male: • *Yellow throat,* breast, and eye-ring; greenish-yellow
patch on back, • *blue above with a chestnut breastband.* Spring
female: blue above, yellow below, breastband narrower or missing.
All plumages are duller in fall.

**Habitat**
Breeds in swamps, wood edges, orchards, and parks where trees
bear *Usnea* lichen (gray moss) or Spanish moss; in migration, any
trees or shrubs.

**Habits**
Tame; sometimes hangs head down at end of limb, forages more
often at medium and upper levels, often sings from treetops.

**Voice**
Song is a rising, sizzling trill, *zzzzzzzz-ZIP;* or a *bz-bz-bz-bz-
zzzzzz-ZIP;* call, a *chip.*

**Eggs**
4–5; white, wreathed at large end with brown spots; 0.7 x 0.5 in.
(1.8 x 1.3 cm). Nest is gourd-shaped, of *Usnea* or Spanish moss,
near end of branch, 5–30 ft. (1.5–9.1 m) up.

### Range
Breeds throughout much of e. U.S. from se. Canada s. to Gulf of Mexico; winters from s. Fla. s. to W. Indies and C. America.

## YELLOW WARBLER
*Dendroica petechia*                                    **39:6, 40:1**

### Description
Size, 4½–5¼ in. (11.4–13.3 cm). • *All-yellow.* Male: darker yellow above, lighter below, chestnut streaks along sides, • *yellow wing bars*, slightly duller in fall. Female and immature: similar but darker above, paler below, and fewer streaks on sides. Alaskan race in fall is duller, grayer above.

### Similarities
Fall immature Yellow is almost as dusky as Orange-crowned, but has yellow tail corners. American Goldfinch has black wings and tail. Female Wilson's has all-dark tail.

### Habitat
Swamps, shrubbery, willows or alders near water; orchards, gardens.

### Habits
Tame, a low forager; an early fall migrant; frequently victimized by Brown-headed Cowbird, but often builds new floor in nest over egg of Cowbird so as to incubate only its own eggs.

### Voice
Song is *sweet, sweet, sweet, I'm so sweet,* with a goldfinchlike ending; call, a loud *chip.*

### Eggs
3–5; greenish-white, marked with brown and purple; 0.7 x 0.5 in. (1.8 x 1.3 cm). Nest is cup-shaped, of silver-gray plant fibers in fork of low bush or sapling.

### Range
Breeds from Newfoundland and n. Que. s. to S.C. and Tex. and s. to S. America; winters from Mexico to S. America.

## MAGNOLIA WARBLER
*Dendroica magnolia*                                    **38:2, 41:10**

### Description
Size, 5 in. (12.7 cm). From below, • *tail looks white with black band at end.* Spring male: gray crown and wings; yellow rump and underparts; black ear patch, back, and streaks on breast and sides; 2 white wing bars. Spring female and fall adults: similar but duller, back grayer, side streaks less pronounced. Immature: even duller; olive above, side streaks faint but has white eye-ring and tail pattern of adult.

### Similarities
Fall Nashville lacks yellow rump and white tail band.

### Habitat
Breeds principally in coniferous forest; in migration, woods, edges, gardens.

### Habits
Active, a low forager; often droops wings, spreads tail.

### Voice
Song is short, variable; W. W. G. Gunn records 8 different varieties on his warbler song record; one common form he renders as *chew-chew-chew-chew wit-see,* another as *wit wit witty swit;* call, a sharp *chip.*

**Eggs**
4–5; white, spotted with brown and lavender; 0.7 x 0.5 in. (1.8 x 1.3 cm). Nest is cup-shaped, of twigs, grass, on limb of conifer, 3–35 ft. (0.9–10.7m) up.

**Range**
Breeds from cen. Canada s. to ne. U.S. and into mountains of Va.; winters in W. Indies and C. America.

## CAPE MAY WARBLER
*Dendroica tigrina*                                    38:3, 41:7

**Description**
Size, 5–5½ in. (12.7–14.0 cm). Spring male: olive with black streaks above, • *chestnut cheek patch,* yellow with black streaks below, side of neck and rump yellow, wing patch and tail corners white. Spring female and fall adults: much duller; cheek patch gray, light eye stripe, • *dull yellowish patch behind ear,* 2 narrow white wing bars, streaks fainter, underparts white. Immature: similar to female, back not streaked, wing bars buffy, rump duller yellow, sparsely streaked below with whitish belly.

**Similarities**
Pine has fainter side streaks, lacks yellow rump. Fall Yellow-rumped is browner above with a streaked back, yellow undertail coverts, and virtually no wing bar.

**Habitat**
Breeds in northern coniferous forests; in migration seeks edges, shrubbery.

**Habits**
Sings from treetops, often at extreme pinnacle of spruce or fir; catches some insects in air; found in migration at all levels.

**Voice**
Song is a thin, high *see-see-see-see-see;* also, less commonly, a series of 6–8 *to-be*'s preceding a *see-see-see;* call, a faint *chip.*

**Eggs**
4–9; greenish-white, dotted with browns and lilac; 0.7 x 0.5 in. (1.8–1.3 cm). Nest is large for warbler; of sphagnum moss, in end clump of vegetation, near top of spruce or fir, 30–60 ft. (9.1–18.3 m) up.

**Range**
Breeds from Que. and Man. s. to N.S., Maine, N.Y., Mich., and N.Dak.; winters from W. Indies s. to C. America.

## BLACK-THROATED BLUE WARBLER
*Dendroica caerulescens*                               37:4, 40:7

**Description**
Size, 5 in. (12.7 cm). Small • *white wing spot* in all adult plumages. Male: • *throat and sides black,* contrasting sharply with • *white underparts;* duller, but essentially similar in fall. Female: very different; brownish above, buffy below with dark cheek and light line over eye; immature fall females lack white wing spot.

**Similarities**
Fall Tennessee is much greener above, yellower below; has yellowish line over eye; lacks dark cheek and wing spot.

**Habitat**
Breeds in deciduous or mixed woodlands with underbrush, especially on slopes; in migration, shrubs, trees, gardens, orchards.

**Habits**
Tame, movements deliberate; while perched often partly opens its wings; forages at all levels, sometimes catches insects in air.

**Voice**
Song is a hoarse, buzzy, deliberate *bee-tee-bee* (BTB, its initials), or *zero, zero, zeee,* last note in each case rising.
**Eggs**
4–5; whitish, marked with browns and lavender; 0.7 x 0.5 in. (1.8 x 1.3 cm). Nest is cup-shaped, of bark shreds, rootlets in bushes.
**Range**
Breeds from se. Canada to ne. U.S. and into mountains of n. Ga.; winters s. of U.S.

## YELLOW-RUMPED WARBLER
*Dendroica coronata*                    38:1, 41:8

**Description**
Size, 5–6 in. (12.7–15.2 cm). Spring male: rump, sides, crown, and in West, throat, bright yellow; blue-gray above, white below; heavily streaked with black above; black cheek patch, breastband, and heavy streaking on sides; 2 white wing bars east, or white wing patch in west; white tail corners. Spring female, immature and fall adults: gray replaced by brown, less yellow on sides, markings below fainter.
**Similarities**
Magnolia and spring Cape May have yellow rumps, but are yellow below.
**Habitat**
Breeds in northern coniferous forest; in migration, almost anywhere, in trees, shrubs, even on ground; common in winter in coastal bayberry bushes.
**Habits**
Visits feeders, catches insects in air; flight zigzaggy, jerky; an early migrant, winters in southern part of our area.
**Voice**
Song is a thin, colorless jumble, weaker in spring, a stronger rising or falling trill in summer; call, a highly characteristic *chip.*
**Eggs**
3–5; white, spotted brown and purple; 0.7 x 0.5 in. (1.8 x 1.3 cm). Nest is cup-shaped, rather bulky, of bark shreds, twigs, in conifer, 3–40 ft. (9.1–12.2 m) up.
**Remarks**
Formerly Myrtle and Audubon Warblers, now recognized as single species.
**Range**
Breeds from cen. Que. and n. Man. s. to Maine, Mass., n. N.Y., and Mich.; winters from s. parts of breeding range s. to C. America.

## BLACK-THROATED GREEN WARBLER
*Dendroica virens*                    38:6, 41:4

**Description**
Size, 4½–5 in. (11.4–12.7 cm). Contrasting • *yellow cheeks* in all plumages; olive above; 2 white wing bars. Spring male: • *black throat* and streaks on sides, white below and on tail. Spring female and fall male: similar, but duller and no black throat, back unstreaked.
**Similarities**
Fall Blackburnian has a streaked back, darker cheek patch, yellowish breast.
**Habitat**
Breeds in northern coniferous or mixed forest, or in cypress swamp in south; in migration, prefers conifers.

**Habits**
Forages at all levels, but usually well up; migrates across Gulf of Mexico.

**Voice**
Song is a dreamy *zee-zee-zee-zoo-zee;* or *please, please, pass the cheese,* second and third notes from end lower.

**Eggs**
4–5; white, marked with various shades of brown; 0.6 x 0.5 in. (1.5 x 1.3 cm). Nest is cup-shaped, of twigs, birchbark, bound with spider webs, on limb of conifer, 15–70 ft. (4.6–21.3 m) up.

**Range**
Breeds from Newfoundland and Ont. s. to N.J., Ohio, and Minn., and into mountains in Ga.; winters from Fla. and Tex. s. to S. America.

Note: The **BLACK-THROATED GRAY WARBLER,** *Dendroica nigrescens* **(38:9)**, size, 4½–5 in. (11.4–12.7 cm), regularly wanders eastward somewhat from its western breeding range. It is black, white, and gray; the head and throat pattern of the male closely resembles that of the Black-and-white Warbler, but its back is gray with fine black streaks, not black-and-white streaked.

## CERULEAN WARBLER
*Dendroica cerulea*                                              37:6

**Description**
Size, 5 in. (12.7 cm). Male: sky-blue above, white below; black streaks on back and sides; • *blue-black breastband* (incomplete in fall); dusky cheek patch; 2 wing bars and spots on tip of tail white. Female: greenish above with bluish tinge especially on crown; yellowish wash below; pale streak over eye, dark line through it; no black streaks on back or sides; no breastband. Immature: similar, but duller, more olive-green above.

**Similarities**
Female resembles female Black-throated Blue, but has 2 wing bars, not single white wing spot. Immature resembles fall Blackpoll, but is brighter green above and with more distinct eye stripe. Immature Northern Parula is deeper yellow below, no line over eye.

**Habitat**
Large trees in relatively open situations, as along river valleys.

**Habits**
Keeps to treetops, sometimes catches insects in air, or creeps along branches like Black-and-White; migrates across Gulf of Mexico.

**Voice**
"The songs of the Parula and Cerulean are very similar but . . . the pattern is reversed. . . . the Parula's song is *bzz, bzz, bzz, trill,* while that of the Cerulean is *trill, trill, trill, bzz.* The Cerulean's song can be expressed by the phrase *Just a little sneeze!"* (Murray); note, a sharp *chip*.

**Eggs**
3–4; white, spotted with red-brown; 0.7 x 0.5 in. (1.8 x 1.3 cm). Nest is shallow, cup-shaped, of grasses, bound with spider webs, decorated with lichens, suspended in fork of branch, 20–90 ft. (6.1–27.4 m) up.

**Range**
Breeds from N.Y. and Great Lakes s. to se. U.S. and cen. states; winters in S. America.

## BLACKBURNIAN WARBLER
*Dendroica fusca*                                        **38:5, 4l:6**

**Description**
Size, 4½–5½ in. (11.4–14.0 cm). Spring male: • *flaming orange throat* and breast with black streaks on sides, orange on crown and sides of head, back black with white streaks, large white wing patch, white tail corners. Spring female and fall adults: similar but duller and paler, 2 white wing bars. Immature: even duller; similar pattern in brownish and dull yellow, 2 white wing bars.

**Similarities**
Immature resembles immature Black-throated Green, but has dark ear patch, light stripes on back, and yellow throat.

**Habitat**
Breeds in northern and mountain forest, especially in conifers.

**Habits**
Active, forages well up in trees, sings from treetops; but found at all levels in migration.

**Voice**
Song is a high, thin, wiry *tsee-tsee-tsee-tsee-tsee-zi-zi-zi-zi;* call, a faint *tseeck*.

**Eggs**
4–5; blue-green to whitish, spotted with lilac and brown; 0.7 x 0.5 in. (1.8 x 1.3 cm). Nest is of twigs in conifers near end of limb, 10–80 ft. (3.0–24.4 m) up.

**Range**
Breeds from se. Canada s. to ne. U.S., and s. in mountains to Ga.; winters in S. America.

## YELLOW-THROATED WARBLER
*Dendroica dominica*                                        **38:7**

**Description**
Size, 5½ in. (14.0 cm). Bill rather long. Spring adult: • *Yellow bib;* unstreaked • *gray back;* white wing bars; forehead, cheeks, and stripes along sides black; line over eye, underparts, and tail corners white. Fall adult: similar, but washed with brownish above. Immature: still browner above, throat paler yellow.

**Similarities**
Female Blackburnian has yellow stripes on head.

**Habitat**
Bottomlands with Spanish moss; pines.

**Habits**
Creeps along branches like Black-and-White, forages in middle and upper levels, movements somewhat deliberate; likes to bathe.

**Voice**
Song is quite suggestive of Indigo Bunting, especially in opening phrases, *"tee-ew, tew tew tew tew—tew wi* or [western race] *see-wee, see-wee, see-wee, swee swee swee swee . . .* [or] *sweetie sweetie, sweetie"* (Sprunt).

**Eggs**
3–4; whitish, with dark dots and spots; 0.7 x 0.5 in. (1.9 x 1.4 cm). Nest is of twigs, bark strips, on branch of pine, oak, or sycamore, 10–120 ft. (3.0–36.6 m) up.

**Range**
Breeds from s. N.J. and Ill. s. to cen. Fla., Gulf states, and Tex.; winters from Gulf states s. to C. America.

## CHESTNUT-SIDED WARBLER
*Dendroica pensylvanica*                                   **37:3, 41:14**

**Description**
Size, 4½–5¼ in. (11.4–13.3 cm). Spring male: • *yellow crown* and
black mark below eye, back olive with black stripes; • *sides
chestnut*; 2 yellow wing bars; cheeks, underparts, and tail corners
white. Spring female and fall adults: similar but duller, with less
chestnut and black. Immature: bright lemon-green above, white
below; white eye-ring; gray cheeks; buffy wing bars.

**Similarities**
Spring Bay-breasted has chestnut crown and throat.

**Habitat**
Second growth, shrubbery, dry thickety hillsides.

**Habits**
Tame; forages low where breeding, but at all levels in migration;
often victimized by Brown-headed Cowbird; may migrate across
Gulf of Mexico.

**Voice**
Song resembles Yellow's but with emphatic final double note,
*pleased, pleased, pleased t'MEETcha* or *I want to see Miss
BEECHer;* many variations around this theme; also a quite
different late-season song; call is a soft *tsip* or louder *chip*.

**Eggs**
Usually 4; white, spotted chiefly at larger end with brown and
lavender; 0.6 x 0.5 in. (1.7 x 1.3 cm). Nest is loose, of bark strips
and plant fibers, in low bush.

**Range**
Breeds from s. Canada s. to s.-cen. U.S. and into Appalachians;
winters from C. to S. America.

## BAY-BREASTED WARBLER
*Dendroica castanea*                                        **37:2, 41:1**

**Description**
Size, 5½ in. (14.0 cm). Spring male: • *chestnut (or bay) on back of
head,* throat, and sides; black face; dark crown; • *buffy neck patch;*
above gray with black stripes; wing bars, underparts, and outer tail
tips white. Spring female: similar but much less chestnut, face
grayish. Fall adults and immature: olive-green above, buffy below,
dusky cheek patch, black streaks on back, 2 white wing bars, legs
normally black (but on occasion light), often a trace of chestnut on
flanks, undertail coverts buffy.

**Similarities**
The similar fall Blackpoll has a greenish-yellow tinge below,
usually appears streaked; and has light legs, white undertail
coverts, stronger streaking above, and no chestnut. Fall Pine is
unstreaked above.

**Habitat**
Breeds in bogs and openings in spruce forests; relatively
uncommon.

**Habits**
In breeding season, forages at all levels; movements somewhat
deliberate; in migration, keeps lower, often occurs in woods,
particularly pine; a late spring and early fall migrant (August,
September), before most Blackpolls.

**Voice**
Song is a high, thin *"seetzy-seetzy-seetzy-seetzy-see"* (Gunn);
reminiscent of Blackburnian, Cape May, and Golden-crowned
Kinglet; call, a fine *tsip*.

**Eggs**
4–7; often bluish-green, spotted with brown; 0.7 x 0.5 in. (1.8 x 1.4 cm). Nest is of twigs and moss on branch of conifer, 3–20 ft. (0.9–6.1 m) up.

**Range**
Breeds in se. Canada and ne. U.S.; winters in C. and S. America; migrates along Atlantic Coast.

## BLACKPOLL WARBLER
*Dendroica striata*                                          **38:10, 41:2**

**Description**
Size, 5½ in. (14.0 cm). Black with white wing bars in all plumages. Spring male: olive above, white below, black stripes on back and sides, cheek and throat white, • *black cap.* Spring female: cap olive streaked with black, streaks less conspicuous, below white, cheeks gray. Fall adults and immatures: olive-green above, white below with tinge of greenish-yellow, black streaks on back and faintly on sides, legs pale, undertail coverts white.

**Similarities**
See Bay-breasted Warbler for difficult distinction between these 2 species in fall; use caution. Black-and-White Warbler has a black-and-white striped crown.

**Habitat**
Breeds in northern and mountain coniferous forest, especially by bogs and in areas of stunted trees.

**Habits**
Movements rather deliberate, a good flycatcher; forages and sings chiefly at middle and upper levels in migration; a late-spring night migrant, many hit lighthouses, beacons, and TV towers.

**Voice**
Song is easy to learn and remember, a high, penetrating *zee-zee-Zee-Zee-Zee-Zee-ZEE-ZEE-Zee-Zee-zee-zee,* of even pitch, becoming louder in middle, fading away at end; call, a sharp *chip.*

**Eggs**
3–5; whitish, boldly spotted with red-brown; 0.7 x 0.5 in. (1.8 x 1.3 cm). Nest is of twigs, moss, low in spruce.

**Range**
Breeds from n. Canada s. to s. Canada and ne. U.S.; winters in S. America; migrates through W. Indies.

## PINE WARBLER
*Dendroica pinus*                                            **38:12, 41:3**

**Description**
Size, 5½ in. (14.0 cm). All plumages have white wing bars and tail corners, dark legs. Spring male: unstreaked yellow-green above, • *yellow breast* with obscure streaking, gray wings and tail, yellow line over eye, white undertail coverts. Spring female and fall adults: much duller, less yellow on breast. Immature female: duller, more brownish above; whitish below, often with little trace of yellow.

**Similarities**
Similar Yellow-throated Vireo has yellow spectacles and is more deliberate in its movements. Palm Warbler has red-brown cap. Orange-crowned lacks wing bars. In fall, similar Bay-breasted and Blackpoll have streaked backs; Bay-breasted has buffy undertail coverts; Blackpoll has yellowish legs.

**Habitat**
Pines, but any trees or shrubs in migration.

**Habits**
Rather tame; often creeps along branches like Brown Creeper; occasionally feeds on ground; catches insects in air; an early spring and late fall migrant, often seen with Palm Warblers.

**Voice**
Song is a Chipping Sparrow-like trill but slower, softer, with intervals between notes longer than in Chipping, much longer than in Worm-eating; call, a *chirp*.

**Eggs**
4; whitish, boldly marked with lilac and purple; 0.7 x 0.5 in. (1.8 x 1.3 cm). Nest is of bark strips, pine needles, on pine limb, 10–50 ft. (3.0–15.2 m) up.

**Range**
Breeds from s. Canada s. to Gulf of Mexico; winters from s. parts of breeding range s. to W. Indies.


## KIRTLAND'S WARBLER
*Dendroica kirtlandii*                                          38:8, 41:5

**Description**
Size, 5¾ in. (14.6 cm). Very local and declining. Rarely seen except where breeding. Spring male: blue-gray above, yellow below; black streaks on back, breast, and sides; incomplete white eye-ring; blackish mask; 2 obscure white wing bars; dark legs. Spring female: duller and without mask. Fall adults: head, back, and sides brownish; all markings duller.

**Similarities**
Canada has unmarked back, black necklace, light legs.

**Habitat**
Breeds in open stands of jack pines, 3–18 ft. (0.9–5.5 m) high. Extremely rare.

**Habits**
Tame; jerks tail, often forages on ground by flying or hopping along, sometimes catches insects on wing.

**Voice**
Song is a loud, clear *"chip chip chip-chip-chip-teoo-teoo-weet-weet"* (Borror); call, a *tsip*.

**Eggs**
4; white, finely spotted with brown; 0.7 x 0.5 in. (1.8 x 1.4 cm). Brown-headed Cowbirds severely parasitize the nests of this warbler. Nest is of bark and vegetable fiber, on ground, in groves of jack pines; usually in loose colonies.

**Range**
Breeds exclusively in n.-cen. Mich.; winters in Bahamas.


## PRAIRIE WARBLER
*Dendroica discolor*                                          38:14, 41:11

**Description**
Size, 5 in. (12.7 cm). Spring male: olive-green above, with patch of • *chestnut streaks on back*; 2 yellowish wing bars; white tail corners and undertail coverts; all-yellow below with bold • *black streaks down sides*. Spring female, fall adults, and immature: similar pattern but duller, grayer above; chestnut spots concealed or missing; wing bars and side streaks inconspicuous or absent.

**Similarities**
Fall Pine is larger, has more conspicuous wing bars.

**Habitat**
Second-growth low pine and scrub oak.

**Habits**
Forages low, wags tail (but not as much as Palm), catches insects in air, also hovers before leaf, sometimes sings from top of medium-sized tree.
**Voice**
Song is a distinctive, easy-to-learn *zee-zee-zee-zee-zee-zee*, rising in pitch throughout; call, a soft *chip*.
**Eggs**
3–5; white, spotted with brown; 0.6 x 0.5 in. (1.5 x 1.3 cm). Nest is of bark, plant stems, in fork near top of low shrub.
**Range**
Breeds from cen. New England and S.Dak. s. to Fla., Gulf Coast, and Okla.; winters from s. Fla. s. to C. America.

## PALM WARBLER
*Dendroica palmarum*                                    **38:13, 41:9**

**Description**
Size, 4½–5½ in. (11.4–14.0 cm). Spring adult: olive-brown above with chestnut cap; rump greenish-yellow; line over eye; throat, upper breast, and coverts yellow; belly yellow to white; gray cheek patch; light chestnut streaks on breast; white tail corners. Fall adult and immature: browner above, concealing chestnut cap; line over eye, throat, and breast white; obscure streaks below. Belly always white in western race; chestnut brighter and belly always yellow in eastern race.
**Similarities**
Young Cape May has white undertail coverts.
**Habitat**
Breeds in northern bogs and muskeg; seen everywhere in migration, especially lawns, fields, swamps, and in undergrowth of open woods.
**Habits**
Pumps tail, has gliding walk, hops, runs, spends much time on ground or in low bushes; often seen with Yellow-rumped or Pine Warblers, kinglets and chickadees, or sparrows in field; hardy, a few winter north to New England.
**Voice**
Song is "a short rapid series of thin, lisping notes. . . . Very similar to a Chipping Sparrow; *thi, thi, thi, thi, thi*" (Griscom); call, a *chip*.
**Eggs**
4–5; creamy white, spotted with brown and lilac; 0.7 x 0.5 in. (1.8 x 1.3 cm). Nest is of grass, in moss at foot of bush or tree.
**Range**
Breeds in cen. Canada s. to n. U.S.; winters from s. U.S. s. to C. America.

## OVENBIRD
*Seiurus aurocapillus*                                    **37:11**

**Description**
Size, 5½–6½ in. (14.0–16.5 cm). Olive above, with a black-bordered • *rufous-buff crown;* eye-ring and underparts white; breast and sides heavily streaked with black; legs pinkish; no wing bars.
**Similarities**
Waterthrushes lack rusty crown.
**Habitat**
On or near ground in open deciduous or mixed woodlands; in migration, shrubbery in gardens, parks.

**Habits**
Inquisitive; suggests a small thrush, feeds and walks daintily on forest floor, sings from branch.
**Voice**
Song is a loud, ringing crescendo often described as *tea-cher, tea-cher,* plus a varied, melodious, and gushing evening flight song sometimes also given at night; alarm note, a sharp hard *chik.*
**Eggs**
3–6; creamy white, finely spotted with red-brown and lilac; 0.8 x 0.6 in. (2.0 x 1.5 cm). Nest is arched, and with side entrance; bulky, of grass and leaves, concealed, on forest floor.
**Range**
Breeds from cen. Canada s. to n. Gulf states; winters from Gulf of Mexico s. to S. America.

# NORTHERN WATERTHRUSH
*Seiurus noveboracensis*                                            **37:12**

**Description**
Size, 5½–6½ in. (14.0–16.5 cm). Dark olive-brown above, greenish-yellow below; heavy, • *dark streaks on breast and sides;* fine spots on throat; legs pinkish, no wing bars.
**Similarities**
Louisiana lacks spots on throat, is whiter below, has white eye line, but, in fall, western subspecies of Northern also has white eye line.
**Habitat**
Breeds in northern bogs and swamps and places with quiet water; in migration, found along streams, in shrubbery anywhere.
**Habits**
Feeds on ground, sings from branch; constantly teeters like Spotted Sandpiper as it walks or runs, often along edge of pond or watercourse; often sings at middle level, birds chase each other through woods in fast zigzag flights; an early fall migrant.
**Voice**
Song is a loud, carrying *"sweet sweet sweet swee-wee-wee-chew-chew-chew-chew"* (Gunn), last *chew* notes lower, diagnostic; or *"hurry, hurry, hurry, pretty, pretty, pretty"* (Sutton); call, a sharp distinctive *clink.*
**Eggs**
4–5; pinkish-white, spotted and scrawled with red-brown and lavender; 0.8 x 0.6 in. (2.0 x 1.5 cm). Nest is of moss, in moss, at base of stump or tree, near water.
**Range**
Breeds from n. Canada s. to n. U.S.; winters from Mexico s. to S. America.

# LOUISIANA WATERTHRUSH
*Seiurus motacilla*                                                 **37:13**

**Description**
Size, 5¾–6¾ in. (11.4–17.1 cm). Similar to Northern, but averages larger, whiter below, often with buffy flanks; • *white eye line;* • *unmarked throat* diagnostic (at close range).
**Similarities**
See Northern Waterthrush.
**Habitat**
Breeds in ravines, gorges, near flowing streams; frequents swamps and other wet areas in southern portion of range and on migration.
**Habits**
Similar to Northern Waterthrush, sometimes sings in flight.

**Voice**
Song is wild, ringing; 3 loud introductory notes, then a sweet, jumbled, descending warble; more penetrating than song of Northern; also an evening flight song; call, like Northern's.

**Eggs**
4–6; white, marked with brown and lilac; 0.8 x 0.6 in. (2.0 x 1.5 cm). Nest is bulky, of leaves, grass, hidden in cavity in stream bank.

**Range**
Breeds from cen. New England and Minn. s. to Ga. and Tex.; winters from Mexico s. to S. America.

## KENTUCKY WARBLER
*Oporornis formosus*                                    **39:1**

**Description**
Size, 5½ in. (14.0 cm). Olive-green above, black forehead, • *yellow spectacles,* pale legs; male has more • *black in mustache* and on forehead than female or immature.

**Similarities**
Common Yellowthroat is smaller, has white belly, and no yellow spectacles.

**Habitat**
Swampy, wet woods.

**Habits**
Walks, occasionally teeters like waterthrush.

**Voice**
Song is a loud, ringing *turtle, turtle, turtle* somewhat like a Carolina Wren's; alarm note, a metallic *chip.*

**Eggs**
4–5; white, spotted with red-brown and lilac, chiefly at larger end; 0.8 x 0.6 in. (2.0 x 1.5 cm). Nest is bulky, of leaves, grass, well-hidden on ground in woods.

**Range**
Breeds from N.J., Pa., Ind., and Iowa s. to se. U.S.; winters from Mexico s. to S. America.

## CONNECTICUT WARBLER
*Oporornis agilis*                                  **39:14, 40:4**

**Description**
Size, 5¼–6 in. (13.3–15.2 cm). Olive above, yellow below; yellow undertail coverts are ⅔ the length of tail; • *white eye-ring complete.* Male: • *gray hood.* Female: duller. Immature: even duller; crown, throat, upper breast, and flanks brownish; eye-ring slightly buffy.

**Similarities**
Nashville has yellow throat. Male Mourning has black on breast. Fall Mourning has incomplete eye-ring.

**Habitat**
Forest bogs, brush; in migration, thickets, overgrown fields.

**Habits**
If flushed, flits to low branch and sits motionless.

**Voice**
Song is a loud ringing *"chuckety chuckety chuckety chuck"* (Allen); call, a metallic *pink.*

**Eggs**
3–5; creamy white, with a few dark spots wreathed around larger end; 0.8 x 0.6 in. (2.0 x 1.5 cm). Nest is of grasses, in moss, on ground in tamarack swamps.

**Range**
Breeds from s.-cen. Canada s. to Mich., Wis., and n. Minn.; winters in S. America.

## MOURNING WARBLER
*Oporornis philadelphia*                                   **39:13, 40:6**

**Description**
Size, 5–5¾ in. (12.7–14.6 cm). Olive above, yellow below; • *gray hood, black apron;* undertail coverts less than half length of tail; legs pinkish; no eye-ring. Male: gray hood with black apron, no eye-ring. Female and immature: similar to larger female and immature Connecticut, but usually no eye-ring in spring and eye-ring broken in fall.

**Similarities**
Spring male Connecticut has gray, not black, apron.

**Habitat**
Wet woods, thickets.

**Habits**
Very late migrant.

**Voice**
Song is a somewhat low *"wee surree surree surree surree* with a falling inflection" (Allen); also an emphatic *"cheese cheese cheese, cheese sweet"* (Gunn); call, "a strong almost bi-syllabic *tchip,* a bit sharper than a Yellowthroat's . . . quite distinctive" (Peterson).

**Eggs**
Indistinguishable from Kentucky's. Nest is of plant stalks, bark shreds, on ground in tussock in woods.

**Range**
Breeds from N.Dak. and Alta. s. to s. Canada and n. U.S.; winters from C. to S. America.

Note: The **MacGILLIVRAY'S WARBLER,** *Oporornis tolmiei,* size, 5½ in. (14.0 cm), replaces the Mourning Warbler in the Black Hills, the Cypress Hills, and Rocky Mountains. The two are indistinguishable in fall, but in spring the MacGillivray's has a white eye-ring, broken in front and at the rear. These two very similar forms interbreed in Alberta, and may belong to the same species.

## COMMON YELLOWTHROAT
*Geothlypis trichas*                                        **39:2, 40:8**

**Description**
Size, 4½–5¾ in. (11.4–14.6 cm). Olive upperparts, yellow throat, white belly, yellow undertail coverts, no white wing bars or tail corners. Male: distinctive black mask. Female: similar but with whitish eye-ring and no black mask. Immature: similar but throat often buffy and sides brownish (males may show trace of black mask). Fall females and immatures among the most nondescript of warblers.

**Similarities**
Nashville has gray crown and all-yellow underparts. Connecticut and Mourning have dark, not yellow, throat.

**Habitat**
Damp undergrowth, marshes, thickets.

**Habits**
Active, inquisitive, wrenlike; skulks near ground and in underbrush.

**Voice**
Song is a distinctive, easy to learn, but quite variable *WITCH-i-ty, WITCH-i-ty, WITCH,* or *which IS it, which IS it, which IS it?* also a flight song; alarm note, a chatter like House Wren; call, a *check.*

**Eggs**
Usually 4; creamy white, marked with red-brown and lilac; 0.7 x
0.5 in. (1.8 x 1.3 cm). Nest is large, cup-shaped, of grass, leaves,
underbrush in marsh.
**Range**
Breeds from Newfoundland and Ont. s. to Fla., Gulf Coast, and
Mexico; winters from Gulf states southward.

## YELLOW-BREASTED CHAT
*Icteria virens*                                                    39:3

**Description**
Size, 6½–7½ in. (16.5–19.1 cm). The largest and most
unwarblerlike warbler; bill stout, tail long. Olive above; bright-
• *yellow throat and breast;* • *white spectacles,* mustache, belly, and
undertail coverts.
**Similarities**
Yellow-throated Vireo is smaller, has 2 white wing bars, frequents
upper level of trees.
**Habitat**
Thickets, especially dry hillsides, briar patches.
**Habits**
Rises in air from thicket and sings with wings flopping and feet
and tail dangling; secretive and hard to see, but is attracted by
hand-kissing; actions and behavior seem clownish, eccentric.
**Voice**
Song is a highly variable medley of calls, clucks, mews, whistles,
and gurgles, often with long pauses, remotely suggesting a
Northern Mockingbird; a long bugle note; often sings at night.
**Eggs**
3–5; white, marked with brown and lilac; 0.9 x 0.7 in. (2.3 x 1.8
cm). Nest is large, cup-shaped, of grass, leaves; well hidden in
thicket 1–5 ft. (0.3-1.5 m) up.
**Range**
Breeds locally from Mass. and Ont., s. to Fla., Gulf Coast, and
Mexico; winters from Mexico, s. to C. America.

## HOODED WARBLER
*Wilsonia citrina*                                              39:5, 40:3

**Description**
Size, 5½ in. (14.0 cm). Olive above, • *yellow face* patch about beady
dark eye, forehead, and underparts, including undertail coverts;
white in outer tail feathers; • *no wing bars.* Male has black hood
and throat; female and immature with less or no black on hood,
ear region dusky.
**Similarities**
Very similar female Wilson's has no white in tail. Female Yellow
has yellow tail corners, less conspicuous eye. Immature Canada has
dark forehead and white undertail coverts; see also Bachman's
Warbler.
**Habitat**
Lower story of wet woods.
**Habits**
Often opens and shuts tail, revealing white spots on outer tail
feathers; flycatches.
**Voice**
Song is a loud, sweet ringing *"Monte Monte video"* (Sprunt); or
*"wit a wit TEOO* or *toway toway toway TEE-TOO"* (Saunders);
call, a *chick.*

**Eggs**
3–5; white, spotted and blotched with chestnut and lilac; 0.7 x 0.5 in. (1.8 x 1.3 cm). Nest is cup-shaped, of leaves, bark shreds, in bush, 1–5 ft. (0.3–1.5 m) up, in upland or marsh.
**Range**
Breeds from n. U.S. s. to Gulf Coast, winters from Mexico s. to C. America.

## WILSON'S WARBLER
*Wilsonia pusilla*                                                39:4, 40:2

**Description**
Size, 4¼–5 in. (10.8–12.7 cm). Male: olive-green above, bright yellow below; • *black cap; no wing bars* or white in tail. Female and immature: yellow line over beady black eye; olive about ears; may or may not show trace of black cap.
**Similarities**
See Hooded, Yellow.
**Habitat**
Low trees and shrubs, underbrush.
**Habits**
Catches insects, twitches tail, flips wings.
**Voice**
Song is a quick, dry chatter, last half faster, lower, *chee-chee-chee-chee-chee, chi-chi-chi-chi-ch-ch;* call, a *chip.*
**Eggs**
3–6; white, with brown spots, wreathed around large end; 0.6 x 0.5 in. (1.5 x 1.3 cm). Nest is of grass, at base of tree in shrubbery.
**Range**
Breeds from Newfoundland and n. Man. s. to n. New England and Minn.; winters from Mexico s. to S. America.

## CANADA WARBLER
*Wilsonia canadensis*                                              38:4

**Description**
Size, 5–5¾ in. (12.7–14.6 cm). Slate-blue above, bright yellow below; white undertail coverts; no white on wings or tail. Male: black forehead, trim • *black necklace of spots, yellow spectacles.* Female and immature: duller with only faint necklace and spectacles.
**Habitat**
Shrubbery, understory of woods; but at all levels in migration.
**Habits**
Active, flycatches.
**Voice**
Song is a short jumble, often containing a *witchity* note like yellowthroat's introduced by a *chip,* same as call note.
**Eggs**
3–5; white; speckled with red-brown, mainly around large end; 0.7 x 0.5 in. (1.8 x 1.3 cm). Nest is of grass, leaves, well hidden on ground in woods.
**Range**
Breeds from s. Canada s. to n. U.S., e. of Rocky Mountains, and into mountains of n. Ga.; winters in S. America.

## AMERICAN REDSTART
*Setophaga ruticilla*                                    37:1

**Description**
Size, 4½–5¾ in. (11.3–14.6 cm). In any plumage, an interrupted
broad orange or yellow band across mid-tail. Adult male: black
with • *salmon-orange patches on shoulder,* wings, and tail; white
belly and undertail coverts. Adult female: salmon-orange replaced
by yellow, head gray, upperparts olive, underparts white. First-
year male: like female, but orange where female is yellow.

**Habitat**
Deciduous woodlands, swamps, shrubbery, clearings.

**Habits**
Most restless warbler, frequently spreads patterned tail, droops
wings; good flycatcher.

**Voice**
Song is a thin, incisive series of single notes with emphasized note
near end, *zee-zee-zee-zee-Zee zee-oo (zee-oo* descending); or of
double notes, *teetza-teetza-teetza-teetza, teetza;* sometimes
alternates these songs; alarm note, a sharp *chik.*

**Eggs**
3–5; whitish, marked with lilac and brown, wreathed at larger end;
0.6 x 0.5 in. (1.7 x 1.3 cm). Nest is cup-shaped, of bark, shreds,
leaf stalks, in fork of bush or tree, 3–30 ft. (0.9–9.1 m) up.

**Range**
Breeds from Newfoundland and cen. Man. s. to Ga. and se. Okla.;
winters from Mexico s. to S. America.

# WEAVER FINCHES
Family Ploceidae

This is a widely distributed, sparrowlike family of the Old World,
two species of which were introduced during the nineteenth century
into North America. They feed on grain, fruit, seeds, garbage, and
insects.

## HOUSE SPARROW
*Passer domesticus*                                    45:2

**Description**
Size, 6 in. (15.2 cm). Bill stout; tail short, slightly notched. Male:
white cheek and wing bars; chestnut nape; red-brown and gray
above, grayish white below; • *black throat* mixed with gray in
winter and in immature. Female: streaked, buffy; gray-and-brown
above, pale brownish-gray below; • *buffy line over eye.* Albinos of
all degrees are not uncommon, giving rise to reports from beginners
of unidentifiable birds.

**Habitat**
Cities, suburbs, farms, ranches, and other haunts of man. Common.

**Habits**
Tame, gregarious, aggressive, hardy, prolific; hops, does not walk;
frequents feeders, likes dust baths; does not "kick" feet in foraging
like many native sparrows; gathers in large roosts except when
breeding.

**Voice**
Calls, a *chissik, chissik;* a chirp; alarm note, *tell, tell;* and a noisy
chatter.

**Eggs**
5–6; gray-white marked with brown and gray; 0.9 x 0.6 in. (2.3 x
1.5 cm). Nest is bulky, of straw, debris, in ivy, tree, or cranny in
habitations.

**Range**

Resident throughout much of Canada and most of U.S., from Newfoundland and n. Man. s. to Mexico.

Note: Unlike the related House Sparrow, the **EUROPEAN TREE SPARROW**, *Passer montanus (Fig. 14)*, size, 6 in. (15.2 cm), has not spread from the vicinity of St. Louis, where it was introduced in 1870. Found only about that city in Missouri and immediately adjacent Illinois, it can be told from the House Sparrow by its black throat in all plumages, its chestnut crown, and its black ear spot.

*Fig. 14*                    European Tree Sparrow

# MEADOWLARKS, BLACKBIRDS, AND ORIOLES
Family Icteridae

This family is quite varied in character; but its members all have more or less conical, pointed bills and rounded, not notched, tails. They feed on seeds and insects.

## BOBOLINK
*Dolichonyx oryzivorus*                    **48:1**

**Description**

Size, 6–8 in. (15.2–20.3 cm). Bill rather short, tips of tail feathers look worn. Spring male: head and • *underparts black;* nape buffy; large • *white areas on wings, back,* and rump. Females and fall male: rich buff, darker above; crown black with buff stripe down middle; buff and brown stripes on back.

**Similarities**

Female Dickcissel has slightly notched tail and rusty shoulders. Female Red-winged Blackbird has longer bill and heavy streaks on breast. Stripes on Bobolink's back suggest giant Grasshopper Sparrow.

**Habitat**

Breeds in wet grassy fields, meadows, and along river valleys; in migration, marshes, rice fields.

**Habits**

Sings as it flies low or hovers over meadow; perches on grass stalk, shrub, tree, or fence post; gathers in flocks in migration, flies high.

**Voice**

Song, a rich, bubbling medley *"bob-o-link, bob-o-link, spink, spank, spink"* (William Cullen Bryant) given in flight; call, a metallic *pink,* often heard overhead in migration.

**Eggs**

4–7; very variable, gray to red-brown, spotted and blotched with browns and purples; 0.8 x 0.6 in. (2.0 x 1.5 cm). Nest is of grass, weed stems, in tall grass in meadows, hidden.

**Range**

Breeds from Newfoundland and Man. s. to Md., Ohio, and Kans.; winters in S. America.

# EASTERN MEADOWLARK
*Sturnella magna*                                                    48:2

**Description**
Size, 8½–11 in. (21.6–27.9 cm). Plump; bill rather long; tail short, wide, rounded. Black V on yellow breast; wide black stripes on crown, white cheeks, brown upperparts, streaked sides; • *white outer tail feathers* conspicuous in flight.

**Similarities**
Dickcissel is much smaller, slimmer, and has short bill. See Western Meadowlark.

**Habitat**
Prairies, hayfields, pastures, coastal marshes.

**Habits**
Flicks tail as it walks on ground; sings from tree or fence post, early spring to late fall. Flight quaillike (rapid wingbeats, then a glide), usually low over a field, frequently sailing with wings outstretched and pointed slightly down; silhouette in flight looks like a Starling. Gathers in small flocks in winter, migrates by night; males are polygamous.

**Voice**
Song is a sweet, whistled *"ah-tick-seel-yah"* (Thoreau); sometimes rendered as *spring-o'the-YE-ar;* call, an emphatic *dzhert;* alarm note, a rapid, guttural chatter (often as it flies away).

**Eggs**
3–7; white, speckled with brown and lavender; 1.0 x 0.8 in. (2.5 x 2.0 cm). Nest is arched saucer of grasses and weeds, under tuft of grass in field.

**Range**
Breeds from N.S. and se. Ont. s. to Fla., Gulf Coast and Mexico; winters from New England s. to C. America.

# WESTERN MEADOWLARK
*Sturnella neglecta*

**Description**
Size, 8½–11 in. (21.6–27.9 cm).

**Similarities**
Closely resembles Eastern Meadowlark; a little paler, yellow of chin carried on to mustache area near base of bill, but cannot be safely identified in field except by voice.

**Habitat**
Prefers drier grasslands than Eastern.

**Habits**
Same as Eastern.

**Voice**
Song is varied, different from Eastern's; louder, warbled, variously rendered as *"Hip! Hip! Hurrah! boys; three cheers!"* (Dawson and Bowles); or *"tung-tung-tung-ah, tillah'-tillah'-tung"* (Ridgeway); call, a metallic *tuk,* sharper, harsher than Eastern's; alarm note, a chatter.

**Eggs**
Same as Eastern's.

**Range**
Breeds from Man. s. to La. and cen. Tex., spreading eastward; winters from Nebr. and Utah southward.

## YELLOW-HEADED BLACKBIRD
*Xanthocephalus xanthocephalus*                                    **42:1**

**Description**
Size, 8½–11 in. (21.6–27.9 cm). Male: black, base of bill black,
spot on wing white, • *yellow head* and neck. Female and immature:
dark brown; yellowish cheek and line over eye; yellow throat and
breast; female smaller than male, lacks white in wing.

**Similarities**
Grackles have longer tails.

**Habitat**
Fields, marshes, farmyards.

**Habits**
Gregarious; flight slow, deliberate, undulating like the Red-winged;
flocks long, loose (not wide like Red-winged); often flocks with
Red-wingeds, grackles, cowbirds; in fall, gathers in great roosts;
follows plow, climbs about reeds.

**Voice**
*"Oka WEE wee"* (Bent), *oka* guttural, *WEE wee*, loud whistles;
an unmusical *"klick-kluck-klee-klo-klu-klel-kriz-kri-zzzzzzz-
zeeeeee"* (Taverner), delivered with much effort; alarm, a vehement
*"klookoloy, klookoloy, klook ooooo"* (Dawson); call, a low *kek*.

**Eggs**
3–7; white to greenish, heavily marked with red-brown and gray,
and with some hairlines; 1.0 x 0.7 in. (2.5 x 1.8 cm). Nest is in
colony; of blades of sedge and grass attached to cattails, reeds; 1–3
ft. (.3–.9 m) up, in marshes.

**Range**
Breeds chiefly in W., but e. to cen. Wis. and s. to Ark.; winters
from Tex. southward.

## ORCHARD ORIOLE
*Icterus spurius*                                                  **43:3**

**Description**
Size, 6–7 in. (15.2–17.8 cm). Slender. Adult male: has 1 white
wing bar; black with • *chestnut breast*, belly, rump. Female: olive-
green above, 2 white wing bars, rump and underparts yellow.
Immature male: as female, but with black face and chin. Later
intermediate plumages show scattered chestnut.

**Similarities**
Female Northern is brownish-olive or grayer above, dull orange
below; some have black throats.

**Habitat**
Orchards, tall hedgerows, and farmlands with scattered trees.

**Habits**
Lively; pumps tail, may hang head downward; migrates by day;
males precede females; sometimes nests in a loose colony.

**Voice**
Song is a lively, pleasant warble, suggestive of Purple Finch or
Blue Grosbeak, quite variable, often ending in a tanagerlike
downward-slurred note, *"Look here, what cheer, what cheer, whip
yo, what cheer, wee yo"* (Townsend); not as loud, but longer than
Northern's; call, a *chak* like Red-winged's, a longer rattle.

**Eggs**
4–6; bluish-white, marked with browns and purples; 0.8 x 0.6 in.
(2.0 x 1.5 cm). Nest is of grass, 3 in. (7.6 cm) deep, hanging from
small branch usually of fruit tree.

**Range**
Breeds from cen. New England and s., Minn. s. to Gulf Coast, and
w. to Nebr. and Tex.; winters from Mexico s. to S. America.

## NORTHERN ORIOLE
*Icterus galbula*                                    **43:1**

**Description**

Size, 7–8 in. (17.8–20.3 cm). Adult male: bright • *orange; black head and throat;* patterned black and yellow tail; white wing bar; colors conspicuous in flight. Female: variable; brownish-olive or grayish above, brighter on tail; dull orange or yellowish rump; underparts yellowish-orange to whitish; 2 white bars on blackish wings; sometimes some black on head. Immature male (to second summer, may breed): as female but duskier above and shows black on throat.

**Similarities**

Female Scarlet Tanager has no wing bars. Female and immature Northerns always have some orange, lacking in female and young Orchards.

**Habitat**

Shade trees and edges.

**Habits**

Sings from treetop or upper branch; migrates by day, flying high.

**Voice**

Song is loud, robust whistles; highly varied, usually a series of 1- and 2-note phrases; sometimes sings on wing; call note, a whistled *pe-ter*; alarm, a chatter.

**Eggs**

4–6; grayish-white, streaked and scrawled with browns and black; 0.9 x 0.6 in. (2.3 x 1.5 cm). Nest is of grasses and string, gourd-shaped, carefully woven, hanging from branch.

**Remarks**

Former (eastern) Baltimore and (western) Bullock's Orioles are now regarded as 1 species.

**Range**

Breeds from N.S. and Sask. s. to Ga., La., and Mexico; winters from Mexico southward.

Note: In the central Great Plains, from Saskatchewan to Oklahoma, there is a zone of hybridization in which variable hybrids of the eastern (Baltimore) and western (Bullock's) subspecies are abundant. The free interbreeding and the existence of the hybrid population indicates that these two birds belong to the same species. The hybrids show diverse mixtures of the patterns of the two forms, typical individuals of which are illustrated on Plate 43. From the western plains westward, orioles of this species are of the Bullock's type.

In the Miami, Florida, area is found the **SPOT-BREASTED ORIOLE,** *Icterus pectoralis* **(43:2)**, size, 8 in. (20.3 cm), introduced from Central America. Both sexes are orange with an all-black tail, black and white wings, a black bib and area in front of the eye, and a sprinkling of black spots on the sides of the breast; immature much duller.

## RED-WINGED BLACKBIRD
*Agelaius phoeniceus*                                **42:2**

**Description**

Size, 7–9 in. (17.8–24.1 cm). Adult male: black with • *red-and-yellow shoulder patch.* Female: like a large, dark sparrow, with a sharp, pointed bill; heavily streaked below, light stripe over eye. Immature: like a dark female with a reddish shoulder patch. When perched, especially in winter, red on shoulder sometimes concealed by the yellow.

191

# MEADOWLARKS, BLACKBIRDS, ORIOLES

**Similarities**
See Bobolink.
**Habitat**
Fields, marshes, edges.
**Habits**
Noisy, gregarious; sings from reeds, tree, or fence post, spreading wings and tail; comes to feeders; on the ground, walks, runs, or hops. Flight undulating, a gentle rise on beating wings, a short sharp drop on a glide; flocks advance on a wide front and wheel in unison; often occurs in mixed flocks with cowbirds, grackles, and Starlings; in spring, males arrive before females.
**Voice**
Song is a "pleasing *conk-er-EEE* or *oolong TEA*" (Collins); call, a loud *chak*; also a *tee-urr*; and others.
**Eggs**
3–5; pale, bluish-green variously marked with brown, purple, and black; 1.0 x 0.7 in. (2.3 x 1.8 cm). Nest is bulky, bowl-shaped, of grasses attached to reeds in marsh or in bush or grass.
**Range**
Breeds from Newfoundland and cen. Man. s. to Fla., Gulf Coast, and Mexico; winters from Pa. s. to C. America.

## RUSTY BLACKBIRD
*Euphagus carolinus*                                     **42:5**

**Description**
Size, 8½–9¾ in. (21.6–24.8 cm). Eye yellow, tail of medium length. Spring male: black head with dull green sheen. Spring female: gray. Fall and winter adults: barred with rusty on head and body, black wings and tail. Immature: similar but even rustier.
**Similarities**
Brewer's female has dark eye, head of male has purple sheen. Grackles have longer tails.
**Habitat**
Swamps, marshes, woods.
**Habits**
Gregarious, sometimes in mixed flocks; on ground, walks and runs, nodding head; flirts tail, often gives voice with bodily contortions.
**Voice**
Song is an even, rhythmic *"tolalee-eek-tolalee-eek*; [also a rapid] *kawicklee kawicklee"* (Saunders); flock in song utters a creaky chorus; call, a *kik* and a rattling, a *chuck* like call of wood frog.
**Eggs**
3–6; light bluish-green, blotched with chocolate and gray; 1.0 x 0.8 in. (2.5 x 2.0 cm). Nest is bulky; of leaves, grass, in alder or willow, 1–2 ft. (0.3–0.6 m) above water in swamp.
**Range**
Breeds from n. Canada s. to ne. U.S.; winters from cen. New England and s. Man. s. to Fla., Gulf Coast, and Tex.

## BREWER'S BLACKBIRD
*Euphagus cyanocephalus*                                 **42:6**

**Description**
Size, 8–10 in. (20.3–25.4 cm). Tail of medium length. Male: black with an • *iridescent purplish head*; body with greenish gloss, but looks black at distance; eye yellow. Female: gray, paler over eye and on throat, eye dark. Immature: male may show slight grayish edgings.

**Similarities**
Rusty Blackbird's plumage is rusty in fall and winter. Grackles are iridescent, have long tails. Starling's tail is short. Female Brown-headed Cowbird is smaller, lighter, and has short, conical bill.
**Habitat**
Fields, ranches, farmyards.
**Habits**
Wanders in flocks over countryside; gathers in roosts at night; follows plow, nods head as it walks.
**Voice**
Laidlaw Williams recognized a scolding *tschup*, a hoarse, whistled *squee*, a clear, whistled *tee-uuu*; also a *kit-tit-tit-tit*; and, when elevating tail, a *chug-chug-chug*, *tucker-tucker-tucker*, or *tit-tit-tit*.
**Eggs**
4–6; grayish, often heavily blotched with brown; 1.0 x 0.7 in. (2.5 x 1.9 cm). Nest is in small colonies; bulky; of twigs, bark, and mud, in bush or tree, 1–30 ft. (0.3–9.1 m) up, often in colonies.
**Range**
Breeds chiefly in w. U.S., but e. to Great Lakes and s. to Ind. and Tex. winters from Wis. s. to Mexico.

## BOAT-TAILED GRACKLE
*Quiscalus major*                                                    42:8
**Description**
Size, male, 16 in. (40.6 cm), female, 12 in. (30 cm), tail, 7 in. (17.8 cm). Male: • *Long, wide tail;* tan, gray, or pale yellow eyes. Female: paler and smaller.
**Similarities**
Much larger than Common Grackle.
**Habitat**
Coastal salt marshes and adjacent towns.
**Habits**
Singing males have wing-flutter display as they sing.
**Voice**
Male has 3-part song with harsh *tireet* notes, rapid series of ascending notes, then another *tireet* series; warning note in colonies, a high-pitched *kle-teet* or just *teet;* no whistled call of males to other males (Selander and Giller).
**Eggs**
3–4; pale blue, with umber and vinaceous clouding at smaller end and with fine black and brown blotches and marks, most prominent at larger end; 1.3 x 0.9 in. (3.3 x 2.3 cm). Nest is invariably in salt marsh, of marsh grasses, reeds, etc., with some paper, rags, feathers, and mud or dung wrapped in it.
**Range**
Resident along Atlantic and Gulf coasts from N.J., s. to La.; also inland in Fla.

Note: The **GREAT-TAILED GRACKLE**, *Quiscalus mexicanus*, male, size, 17 in. (43.2 cm), is a sibling (i.e., virtually identical) species that occurs from Mexico to southwest Louisiana. Although very similar, males differ in being larger, darker, more uniform-colored; with more glossy purple and less blue-green; and with longer, wider tail than Boat-tailed. However, the main field difference in the area of overlap is its bright yellow eyes. Females also have bright yellow eyes and are darker, especially below, than Boat-tailed. Less strictly coastal than Boat-tailed, it inhabits marshes, ponds, lakes, irrigated desert, farms, and towns. The male's song is a 4-part crackling, hissing, ending usually in 2 (1–5) long, piercing *cha-we* notes, the end phrase often being given alone; the warning note in the colonies is a loud, low *clack*. Its nest is usually in trees or man-made structures, similar in structure to Boat-tailed's, but of sticks with papers, moss, rags, feathers, mud or dung, when in trees. The eggs have a darker blue tone and with greater clouding of vinaceous and umber and more numerous black and brown blotches and lines at smaller end.

## COMMON GRACKLE
*Quiscalus quiscula*                                                    **42:7**

**Description**
Size, 11–13½ in. (27.9–34.3 cm). Eye yellow, • *tail wedge-shaped.*
Male: black with iridescent hues above of purple, blue, green, and bronze. Female: smaller, duller, iridescent only on front of body. Immature: dull brown, with a brown eye.

**Similarities**
Rusty Blackbird and Brewer's Blackbird have shorter tail. Various races have either a purplish or bronze back, with many intermediates. See Boat-tailed.

**Habitat**
Lawns, shade trees.

**Habits**
Walks; often gives voice with bodily contortions. Flight is direct, steady (not rising and falling).

**Voice**
Medley of harsh squeaking and guttural noises, some suggesting a rusty hinge, but not altogether unmusical; call, a *chack,* louder, lower than Red-winged's.

**Eggs**
4–6; greenish-white to reddish-brown, variously marked with brown; 1.2 x 0.9 in. (3.0 x 2.3 cm). Nest is bulky; of twigs, grass, and mud; on branch or in fork of conifer; 5–40 ft. (1.5–12.2 m) up.

**Former name**
Purple Grackle, Bronzed Grackle.

**Range**
Breeds from Newfoundland and cen. Man. s. to Fla., the Gulf states, and Tex.; winters from n. U.S. s. to Mexico border.

## BROWN-HEADED COWBIRD
*Molothrus ater*                                                        **42:4**

**Description**
Size, 6–8 in. (15.2–20.3 cm). Eye dark; bill short, almost sparrowlike. Male: black with iridescent reflections of purple and green on upper back, • *head and neck coffee-brown.* Female and immature: gray; rather streaked below in juvenal plumage.

**Similarities**
Starling has longer bill, shorter tail. In mixed flocks can be recognized by smaller size; female Rusty and Brewer's are larger and with longer, thinner bills.

**Habitat**
Fields, open woods, edges.

**Habits**
Gregarious, often flocks with Starlings, Red-wingeds, grackles; walks with tail up; associates with cattle; spreads wings and tail as it squeaks; flight like Red-winged's.

**Voice**
Male, a characteristic squeak, like the swinging of an unoiled gate; also a liquid *glub, glub, glee;* female, a rattling chatter; call, a *chuk.*

**Eggs**
4–5; white, speckled with brown; 0.9 x 0.7 in. (2.3 x 1.8 cm). No
nest; lays eggs in other birds' nests—only parasitic species in east.
**Range**
Breeds from Newfoundland and cen. Man. s. to Va., La., and
Mexico; winters from Md. southward.

# TANAGERS
Family Thraupidae

Tanagers are arboreal birds of tropical origin. They have notched
tails and are somewhat like finches in appearance, but have longer,
often notched bills. The males are brilliantly plumaged, the females
modestly so. Tanagers are deliberate, even sluggish in their
movements. They feed on insects, fruits, and berries.

## WESTERN TANAGER
*Piranga ludoviciana*                                           **43:5**

**Description**
Size, 6¼–7½ in. (15.9–19.1 cm). Adult male: yellow with • *red
head* and black back. Black also on wings and tail, 2 pale yellow
wing bars. Yellow rump, face largely yellow in fall. Female and
immature plumages: as those of Scarlet Tanager, olive above,
yellow below, but with 2 pale yellow wing bars.
**Similarities**
Female orioles have sharp, pointed bills and lighter cheeks.
**Habitat**
Upper canopy of open pine or mixed forests and wood edges.
**Habits**
Often catches insects on wing; flight straight, wingbeats fairly
rapid; tame, visits feeders.
**Voice**
Song is similar to Scarlet Tanager's; call, *pit-ik* or *pit-itik.*
**Eggs**
3–5; similar to Scarlet's. Nest is saucer-shaped, of bark shreds and
grass, on low branch.
**Range**
Breeds chiefly in w. U.S., but e. to Nebr. and w.-cen. Tex.;
winters from Mexico southward; occasional birds farther e. in
migration.

## SCARLET TANAGER
*Piranga olivacea*                                              **43:4**

**Description**
Size, 7 in. (17.8 cm). Spring male: in spring, scarlet with • *black
wings* and tail; in fall, has scarlet replaced by olive-green above,
yellow-green below; wings blackish; summer birds in molt show
patches of red and green. Female: like fall male, but wings dark
olive-gray (not black); undertail coverts canary yellow.
**Similarities**
Male Summer Tanager has red wings. Male Cardinal has crest
and red wings. Female orioles and Western Tanager have wing
bars.
**Habitat**
Canopy of deciduous woodlands, especially where oaks are present.
**Habits**
In migration, sometimes seen low or on ground.
**Voice**
Song is like a hoarse Robin's; call, a burry *KIP-purr.*

**Eggs**
3–5; pale greenish-blue, spotted with brown; 1.0 x 0.7 in. (2.5 x 1.8 cm). Nest is cup-shaped, thin; of bark shreds, twigs; near end of branch, 10–30 ft. (3.0 –9.1 m) up.

**Range**
Breeds from se. Canada s. to e.-cen. U.S.; winters in S. America.

## SUMMER TANAGER
*Piranga rubra*                                                    43:6

**Description**
Size, 7½ in. (19.1 cm). Adult male: entirely • *rose-red* with • *no black*. Female: above olive-green washed with yellow; below yellow with orange tinge, overall more orange-yellow, less yellow-green than female Scarlet. Immature male: as female, or with varying patches of rosy-red (but lacks black wings of similarly changing Scarlet Tanager).

**Similarities**
Female Scarlet is darker, greener above, more yellow below, and has dark gray wings. Orioles have wing bars.

**Habitat**
Shade trees; deciduous, pine, and mixed woodlands.

**Habits**
Catches insects in air.

**Voice**
Song is a sweet, long carol suggesting both a Robin and a Rose-breasted Grosbeak; call, a *chicky-chucky-tuck*.

**Eggs**
3–4; bluish-green, spotted with purplish-brown; 0.9 x 0.7 in. (2.3 x 1.8 cm). Nest is cup-shaped, thin; of bark shreds and grass; near end of limb, 5–30 ft. (1.5–9.1 m) up.

**Range**
Breeds from Del. and Wis. s. to Gulf Coast and Mexico; winters from Mexico southward.

# GROSBEAKS, FINCHES, SPARROWS, LONGSPURS, AND BUNTINGS
Family Fringillidae

Members of this large family have strong, short, conical bills adapted for eating seeds, a major item in their diet, which often includes berries, fruits, buds, and insects. Grosbeaks, finches, and buntings tend to be arboreal and brightly colored; sparrows, terrestrial and brownish. The term "winter finches" includes the Pine and Evening Grosbeaks, Pine Siskin, crossbills, and redpolls. Most of these northern birds are seen only in the colder months. Most sparrows are streaked brown above, pale and often streaked below. They are mainly ground-dwelling, but frequently sing from an elevated perch. Longspurs are sparrowlike birds of the open country. They walk and run rather than hop.

## CARDINAL
*Cardinalis cardinalis*                                            44:1

**Description**
Size, 7½–9 in. (19.1–22.9 cm). Bill red-orange. Male: crested; • *red* with a black face and throat. Female: reddish hue confined to crest, wings, and tail; back brownish-gray; face blackish; underparts buffy.

**Similarities**
Male Scarlet Tanager has black wings and tail. Male Summer Tanager has no crest.
**Habitat**
Thickets, gardens.
**Habits**
Tame, frequents feeders (likes sunflower seeds); sings from high perch.
**Voice**
Song is a loud, clear whistle, *wheat-wheat-wheat, what-cheer, what-cheer, what-cheer*; call, a sharp *tik*.
**Eggs**
3–4; whitish, blotched with brown and lilac; 1.1 x 0.8 in. (2.8 x 2.0 cm). Nest is ragged; of twigs, bark shreds; in low bush.
**Range**
Resident from N.S. and S.Dak. s. to Fla., Gulf Coast and s. Tex., s. to Mexico.

## ROSE-BREASTED GROSBEAK
*Pheucticus ludovicianus*                              43:12
**Description**
Size, 7–8½ in. (17.8–21.6 cm). Spring male: heavy whitish bill, black head and upperparts, white on wings and lower back conspicuous in flight, • *rose-red triangle on breast*. Fall male: black of head and back replaced by streaked dark brown, rose and white of underparts spotted with dusky. Female: like a large brown sparrow; white line over eye, 2 white wing bars, heavy buff bill. Immature male: like female, but with some rose on breast.
**Similarities**
Female Purple Finch is much smaller, with smaller bill.
**Habitat**
Wood edges, open woods, shade trees, thickets.
**Habits**
Sluggish; male sings fron high perch (also on nest), feeds from near ground to treetops.
**Voice**
Song is a beautiful rolling warble, like a Robin that has had singing lessons! Call, a distinctive sharp *ink*.
**Eggs**
3–5; bluish-green, blotched with brown and lilac; 1.0 x 0.7 in. (2.5 x 1.8 cm). Nest is saucer-shaped; of grass, twigs; in bush or tree, 5–20 ft. (1.5–6.1 m) up.
**Range**
Breeds from S. Canada s. to cen. U.S., and in mountains to n. Ga.; winters from Mexico s. to S. America.

## BLACK-HEADED GROSBEAK
*Pheucticus melanocephalus*                              43:11
**Description**
Size, 6½–7¾ in. (16.5–19.7 cm). A Western species. Male: black head; orange-brown underparts, nape, and rump; • *white wing bars,* conspicuous in flight. Female: streaked brown above; buffy brown below, especially across breast, virtually unstreaked; white line over eye and under cheek; pale heavy bill. Both sexes: yellowish belly, yellow wing linings.
**Similarities**
Female Rose-breasted is whiter with more streaked breast.

**Habitat**
Deciduous and mixed woods, thickets, edges.
**Habits**
Comes to feeders and picnic tables.
**Voice**
Song similar to Rose-breasted's; call, a sharp *tik,* somewhat like Rose-breasted's.
**Eggs**
Same as Rose-breasted's.
**Range**
Breeds chiefly in W., but e. to S.Dak. and cen. Kans. where hybrids occur with Rose-breasted; winters from La. s. to C. America. Rarely reaches e. Coast in fall.

## BLUE GROSBEAK
*Guiraca caerulea*                                                    **43:7**

**Description**
Size, 6–7½ in. (15.2–19.1 cm). Bill heavy. Male: Purplish-blue (looks black in poor light) with • *2 rufous wing bars.* Female and immature: brown above, often with some bluish; buffy brown below; bill tan-colored; • *2 buffy wing bars;* immature male in spring is like female, but with considerable blue.
**Similarities**
Indigo Bunting is smaller, small-billed, lacks wing bars, and male is all blue.
**Habitat**
Scattered shrubs in dry fields, thickets near water, farms.
**Habits**
Sluggish, shy; sings from bush top.
**Voice**
Song is a finchlike warble suggesting Indigo Bunting and also Orchard Oriole; call, a loud *chuk.*
**Eggs**
3–5; light blue; 0.9 x 0.6 in. (2.3 x 1.7 cm). Nest is of grass, rootlets, snakeskin; in shrub or on low branch.
**Range**
Breeds from N.J. and Mo. s. to Fla., Gulf Coast, and Mexico; winters in C. America.

## DICKCISSEL
*Spiza americana*                                                    **48:6**

**Description**
Size, 6–7 in. (15.2–17.8 cm). Spring male: black crescent on yellow breast; streaked brown above; head gray; bill bluish; throat white; yellow line over eye; dark, thin mustache streak; chestnut on shoulder; belly whitish. Fall male: often lacks black bib. Female and immature: paler, no black V, breast with some streaks; suggestion of male head pattern; chestnut shoulder, yellow wash on breast.
**Similarities**
Female fall Bobolink, male are buffier, have head stripes, and lack chestnut on shoulder. Female House Sparrow lacks chestnut shoulder, blackish mustache streak, yellow on breast.
**Habitat**
Fields, especially alfalfa; prairies.
**Habits**
Breeds in loose colonies; sings from wire, bush, or stalk, persistently and into late summer; gathers in flocks in migration.

**Voice**
Song is *dick, dick, dick-cissel,* much repeated, with varying numbers of *dick*'s and *cissel*'s.
**Eggs**
3–5; greenish-blue; 0.8 x 0.6 in. (2.0 x 1.5 cm). Nest is of grass; on ground by tussock of grass.
**Range**
Breeds from n.-cen. U.S. s. to Gulf states; winters from Mexico s. to S. America.

## INDIGO BUNTING
*Passerina cyanea*                                                    43:9

**Description**
Size, 5½–5¾ in. (14.0–14.6 cm). Bill small. Spring male: • *rich blue;* looks black in poor light; in fall, strong mixture brown on back and head, and whitish below (molting birds blue and brown). Females and immature: sparrowlike; warm brown above, paler below, faintly streaked; often some blue in tail and wings in adult female. First-year males: often have white belly, otherwise blue.
**Similarities**
Blue Grosbeak is larger, bill heavier, male with buffy brown wing bars.
**Habitat**
Dense brush in bottomlands, clearings, edges, roadsides.
**Habits**
Sings from exposed perch on wire or near top of tree, persistently and through heat of day.
**Voice**
Loud song, notes paired, seems to "change gears" in middle; a "high, strident whistle, *sweea sweea sit sit seet seet sayo*" (Saunders); alarm note, a sharp *chit.*
**Eggs**
3–4; pale blue; 0.8 x 0.6 in. (2.0 x 1.5 cm); often 2 broods. Nest is cup-shaped; of grass, weeds; in bush.
**Range**
Breeds from New England s. to Gulf states and w. to Utah and Tex.; winters from Mexico s. to C. America.

## PAINTED BUNTING
*Passerina ciris*                                                    43:8

**Description**
Size, 5¼ in. (13.3 cm). No wing bars. Adult male: • *violet head* and nape, • *yellow-green back, red rump* and underparts. Female: dull green, paler below than above. Immature male: in spring, like female but with some blue on head.
**Habitat**
Dry thickets, hedgerows, abandoned fields, streamsides, gardens.
**Habits**
Shy; sings from elevated perch.
**Voice**
Song suggests that of Indigo, but weaker; call, a sharp *chip.*
**Eggs**
3–5; white, marked with red-brown; 0.7 x 0.5 in. (1.9 x 1.4 cm). Nest is cup-shaped; of leaves, grass; in bush or tree in river flood plains
**Range**
Breeds from N.C. and Mo. s. to se. states and w. to Okla.; winters from Gulf states s. to Cuba, and from Mexico to C. America.

## LAZULI BUNTING
*Passerina amoena*                                              43:10

**Description**
Size, 5–5½ in. (12.7–14.0 cm). A western species. Adult male:
• *azure-blue* head and rump, • *orange-brown breastband,* white
belly and wing bars. Immature male: often has blue on head only.
Female: head and back unstreaked gray-brown, some blue on
wings, rump, and tail, light wing bars, underparts pale with buffy
wash on breast, no streaking.
**Similarities**
Female Blue Grosbeak is larger, paler, has a much bigger bill, and
brownish, not light, wing bars.
**Habitat, Habits, Voice, Eggs**
Like those of Indigo Bunting.
**Remarks**
Observers in the plains should be aware that Indigo and Lazuli
hybridize; mixed traits of the two may be seen in hybrids on
breeding grounds or migration in eastern Plains states.
**Range**
Breeds chiefly in W., but e. to e. S.Dak. and e. Nebr.; winters in
Mexico.

## EVENING GROSBEAK
*Coccothraustes vespertina*                                     44:3

**Description**
Size, 7–8½ in. (17.8–21.6 cm). Chunky body, stout yellowish-white
bill, short notched tail. Male: golden-yellow body, forehead; dusky
brown head; black wings and tail; large white wing patches.
Female: much of brown and yellow replaced by gray; dingier white
on wings, white tail.
**Similarities**
Female Pine Grosbeak longer, bill is stubby, black; no white in
tail.
**Habitat**
Breeds in northern coniferous forests; in winter, found among trees
or by edges.
**Habits**
Gregarious; wanders widely in winter; tame, frequents feeders
(likes sunflower and hemp seeds). Flight is swift, undulating;
attracted by salt.
**Voice**
Song is a sweet warble; also a metallic cry; a call like an oversized
House Sparrow's and a chattering.
**Eggs**
3–4; bluish-green, marked with gray and brown; 0.9 x 0.6 in. (2.3
x 1.7 cm). Nest is saucer-shaped; of twigs, grass; in top of conifer,
15–20 ft. (4.6–6.1 m) up.
**Range**
Breeds from N.S. and s. Man. s. to n. New England and Minn.;
winters from breeding range s. to S.C. and Tex.

## HOUSE FINCH
*Carpodacus mexicanus*                                          44:7

**Description**
Size, 5½ in. (14.0 cm). Male: more orange-red than similar Purple
Finch and streaked below. Female and immature: like small, dingy,
less-patterned Purple Finch.

**Similarities**
See Purple Finch.

**Habitat**
Cities, towns, open country.

**Habits**
Gregarious; comes to feeders.

**Voice**
Song is a continuous, variable warble, higher, longer than that of
Purple Finch; notes various, some like House Sparrow's; also a
chatter.

**Eggs**
4–5; pale blue, with some black spots; 0.7 x 0.5 in. (1.8 x 1.3 cm).
Nest is of grass, paper, rags; in vines, conifers, often near houses.

**Remarks**
A number of House Finches were apparently surreptitiously
released on western Long Island in 1940. The first bird in the
wild, presumably one of those released, was reported on April 11,
1941, and the first nest reported in 1943. They spread rapidly, and
by 1970 were established from Connecticut and Long Island,
through southern New York, New Jersey, to parts of Delaware,
Maryland, and Virginia. They may be partly replacing the Purple
Finch, whose numbers have diminished during this period.

**Range**
Resident originally in W., but introduced to e. U.S. and now
established from Maine s. to S.C.

## PURPLE FINCH
*Carpodacus purpureus*                                    **44:5**

**Description**
Size, 5½–6¼ in. (14.0–15.9 cm). Stout bill, notched tail; Male:
• *raspberry-colored,* no streaks on side or belly. Female and
immature: sparrowlike; heavily streaked all over; gray-brown
above, white below; pale line over eye; dark stripe below dark
cheek.

**Similarities**
Pine Grosbeak is much larger. Male House Finch has dark streaks
on sides and belly and much less red overall. Female House Finch
is less distinctly marked; face unpatterned; ventral streaks paler,
narrower, and less contrasting, hence appearing dingy.

**Habitat**
Woods, especially conifers; wood edges, shade trees.

**Habits**
Sings from treetops; wanders in fall, frequents feeders (likes seeds
of sunflower, hemp, millet).

**Voice**
Song is a mellifluous warble like Warbling Vireo, but louder; call,
a *tik, tik;* in flight, a sharp *pink.*

**Eggs**
4–5; blue-green, with dark spots chiefly around larger end; 0.8 x
0.6 in. (2.0 x 1.5 cm). Nest is cup-shaped, frail; of dark shreds,
rootlets; in conifer.

**Remarks**
The Purple Finch is easily confused with the House Finch. In
addition to the above characters, it is larger, chunkier, and less trim
than the House Finch.

**Range**
Breeds from Newfoundland and Que. s. to N.J. and Minn.;
winters from N.S. s. to Fla., Gulf Coast, and Tex.

## PINE GROSBEAK
*Pinicola enucleator*                                          **44:2**

**Description**
Size, 8–10 in. (20.3–25.4 cm). Big and chunky. Tail long, notched; bill, wings, and tail black; 2 white wing bars. Adult male: rose-red and gray. Female: rosy-red replaced by olive-yellow and gray. Immature male: like female but often with some red or orange-brown.

**Similarities**
White-winged Crossbill is smaller, has crossed bill. Purple Finch lacks white wing bars.

**Habitat**
Coniferous woods, edges, shrubbery.

**Habits**
Tame; in winter occurs in flocks, varying widely in numbers from year to year; flight undulating; likes to bathe in soft snow.

**Voice**
Song is warble and whistle reminiscent of Robin and of Purple Finch; a whistled *yew, yew, yew* somewhat like the Greater Yellowlegs's; flight call, *pee-ah.*

**Eggs**
3–5; greenish, spotted with purple; 0.9 x 0.6 in. (2.3 x 1.5 cm). Nest is of twigs, rootlets, grass; in conifer.

**Range**
Breeds from Newfoundland and n. Que. s. to n. New England and Man.; winters from s. parts of breeding range s. to Pa. and Kans.

## GRAY-CROWNED ROSY FINCH
*Leucosticte tephrocotis*                                      **48:13**

**Description**
Size, 5¾–6¾ in. (14.6–17.1 cm). Tail notched; bill and forehead black; crown gray; foreparts deep chestnut brown; wings, belly, and rump pink; bill yellow in winter; females paler than males.

**Similarities**
Sparrows and finches with which females might be confused lack dark, unmarked breast.

**Habitat**
In summer, in mountains above timberline; in winter, open mountain areas, foothills, plains.

**Habits**
Active; ground feeder, flycatches at times; wanders in flocks after breeding.

**Voice**
Song is a high-pitched series of *chips;* flight note, a *"chee-chee-chi-chi-chi"* (Hoffmann).

**Eggs**
4–5; white; 0.9 x 0.6 in. (2.5 x 1.5 cm). Nest is cup-shaped; of grasses; in rocky crevice above timberline.

**Range**
Breeds chiefly in nw. North America; winters e. sporadically to w. fringe of Great Plains.

## COMMON REDPOLL
*Carduelis flammea*                                            **44:8**

**Description**
Size, 5¼ in. (13.3 cm). Tail notched. Winter male: • *red cap on forecrown;* streaked gray brown above, paler on rump; 2 white

wing bars, whitish below; • *chin blackish;* breast pink. Female and immature: similar, but no pink on breast. Browner in spring plumage.

**Similarities**
Male Purple Finch has entire head reddish. Pine Siskin is more heavily streaked. See Hoary Redpoll.

**Habitat**
Tundra, openings in northern forest; fields, swamps.

**Habits**
Gregarious, active; clings to weed stems; sings from perch or in air; flocks twitter as they feed; frequents feeders (likes rolled oats and seeds of hemp, millet, and sunflowers). Flight is goldfinchlike; flocks wheel in unison.

**Voice**
Call, a dry *ch-ch,* repeated; song, same alternated with a trill.

**Eggs**
3–6; pale blue, spottted with brown and lavender; 0.7 x 0.5 in. (1.8 x 1.4 cm). Nest is of grass, plant down; in low bush or tree.

**Range**
Breeds from n. Que. s. to Newfoundland and n. Man.; winters from s. parts of breeding range s. to N.C. and Okla.

## AMERICAN GOLDFINCH
*Carduelis tristis*                                              **44:10**

**Description**
Size, 4½–5½ in. (11.4–14.0 cm). Notched tail, no streaks. Summer male: • *yellow,* with • *black wings, tail, forehead;* wings and tail with white markings; bill pink. Female and immature: yellow replaced by a brownish olive-yellow, no black forehead; female somewhat browner, duller in winter. Winter male: like female, but with yellow shoulder.

**Similarities**
Pine Siskin and redpolls are streaked. Warblers have thin bills. Yellow Warbler is yellow all over.

**Habitat**
Open country, weedy fields, edges, gardens.

**Habits**
Lively, gregarious; characteristic roller-coaster flight, utters flight note on top of rise; nests late, often into August; in winter often flocks with redpolls, siskins, crossbills.

**Voice**
Song is long, pleasing, somewhat canarylike; also a medley of trills and other notes, often including a *swee;* or a *dee-ar;* flight note, *per-CHIK-o-ree.*

**Eggs**
3–7; bluish-white; 0.6 x 0.5 in. (1.7 x 1.3 cm). Nest is cup-shaped; of grass, bark shreds, thistledown; in fork of bush or sapling.

**Range**
Breeds from Newfoundland and s. Man. s. to Ga., Ark., cen. Okla.; winters from n. U.S. s. to Gulf Coast and Mexico.

Note: The southwestern **LESSER GOLDFINCH,** *Carduelis psaltria* **(44:11),** size, 4 in. (10.2 cm), is found in open areas in Colorado to eastern Texas. Males are black above, yellow below, and retain a black cap even in immature plumage. Females are like the American Goldfinch, but the rump is colored like back, not paler as in American.

# HOARY REDPOLL
*Carduelis hornemanni*

**Description**
Size, 5 in. (12.7 cm). Virtually identical to Common Redpoll;
distinguished by white, unstreaked rump under optimum conditions
only.
**Habitat, Habits, Voice, Eggs**
Same as Common Redpoll.
**Range**
Breeds in Arctic; winters from Arctic s. to n. U.S.

# PINE SISKIN
*Carduelis pinus*                                    44:9

**Description**
Size, 4½–5¼ in. (11.4–13.3 cm). A • *heavily streaked* finch with a
flash of • *yellow in wings and tail;* tail notched, wings with 2 white
bars and yellow notch at base of primaries; base of outer tail
yellow.
**Similarities**
American Goldfinch is never streaked. Redpolls are paler, streaks
lighter. Female Purple and House finches are larger, have larger
bills, and lack yellow in tail.
**Habitat**
Breeds in conifers; at other times found almost anywhere.
**Habits**
Active, gregarious; often occurs in flocks with goldfinches,
crossbills, redpolls; flight goldfinchlike, frequently recognized by
*shree* note as it flies overhead; frequents feeders (likes millet).
**Voice**
Song is like American Goldfinch's, but lower, longer, rougher;
calls, a rasping, rising *shre-e-e-e-e;* a soft *it-it-it;* a louder *klee-
up;* "2 or 3 very rapid, twangy syllables that suggest striking a
tightly stretched wire" (Harrison).
**Eggs**
3–6; pale green-blue, spotted with brown; 0.7 x 0.5 in. (1.8 x 1.3
cm). Nest is saucer-shaped; of grass and twigs; in conifer, 8–30 ft.
(2.4–9.1 m) up.
**Range**
Breeds from cen. Que. and s. Man. s. to N.S. and Nebr., and in
mountains to N.C.; winters from s. parts of breeding range, s. to
Fla., Tex., and Mexico.

# RED CROSSBILL
*Loxia curvirostra*                                  44:4

**Description**
Size, 5¼–6½ in. (13.3–16.5 cm). Crossed mandibles; bill at a
distance looks slender; tail short, notched. Adult male: • *brick-red,*
brightest on rump; wings and tail dusky. Female: olive-gray, rump
and breast olive-yellow, wings and tail dusky. Immature male:
intermediate in color between female and adult male.
**Similarities**
Crossed bill, lack of wing bars distinguish it from all other reddish
birds.
**Habitat**
Conifers, especially pine and spruce.
**Habits**
Tame, actions parrotlike; extracts seeds from evergreen cones with
its crossed mandibles; sings from treetop. Flight is undulating;

travels in small flocks, occurrence irregular both in summer and winter; nests from January to July.

**Voice**
Song is *"too-tee too-tee, too-tee, tee, tee"* (Hoffmann); also a trill; call, a *pip*, or *pip-pip*.

**Eggs**
4–5; pale greenish, spotted with brown and lavender; 0.8 x 0.6 in. (2.0 x 1.5 cm). Nest is of evergreen twigs and moss; usually in conifer, 5–20 ft. (1.5–6.1 m) up.

**Range**
Breeds from Newfoundland and Man. s. to N.C. and Wis.; winters from breeding range s. to Gulf Coast.

## WHITE-WINGED CROSSBILL
*Loxia leucoptera*                                                    44:6

**Description**
Size, 6–6¾ in. (15.2–17.1 cm). Crossed mandibles, • *2 white wing bars* conspicuous in flight. Adult male: • *rose-red*, brightest on rump; wings and tail black with bold, white wing bars. Female: olive-gray, yellowish on rump; more streaked than Red Crossbill; wings as male. Immature male: intermediate in color between female and adult male.

**Similarities**
Pine Grosbeak is much larger.

**Habitat**
Conifers, especially spruce.

**Habits**
Travels in small flocks, occurrence irregular winter or summer.

**Voice**
Song is loud, long, varied, finchlike, sometimes given in air; calls, a clear *cheap*, a dry *chif-chif.*

**Eggs**
2–4; pale bluish-green, marked with brown and lavender; 0.8 x 0.6 in. (2.0 x 1.5 cm). Nest is of twigs, birch bark, moss; in tip of conifer branch.

**Range**
Breeds from n. Que. and cen. Man. s. to Newfoundland and s. Ont.; winters in breeding range and s. to New England and N.Y.; occasional to N.C.

## RUFOUS-SIDED TOWHEE
*Pipilo erythrophthalmus*                                             45:6

**Description**
Size, 7–8½ in. (17.8–21.6 cm). Tail long, rounded; eyes red, white in southeast Atlantic Coast race. Male: black upperparts, hood, and breast; rufous flanks; white in wing and on belly; white tips of outer tail feathers conspicuous in flight. Female: black replaced by brown. Juvenal (summer): like a large, slender sparrow with white tail corners, 2 buffy wing bars, streaked below. Western "spotted" form (found in Great Plains river valleys) has much white spotting on back and wings.

**Habitat**
Dry woods, especially second growth; edges; thickets.

**Habits**
Forages on ground, scratching noisily among dead leaves; sings from bush or low branch; opens and shuts tail.

**Voice**
Song is *drink your tea-e-e-e, see towhee*, or sometimes just *teeeee;* call, a *che-WINK!*

**Eggs**
3–6; white, finely dotted with red-brown; 1.0 x 0.8 in. (2.5 x 2.0 cm). Nest is of leaves, bark shreds; well hidden on ground in woods.

**Range**
Breeds from Maine and s. Man. s. to Fla., La., and C. America; winters from Md. and Nebr. southward.

Note: The **GREEN-TAILED TOWHEE**, *Pipilo chlorurus* **(45:5)**, size, 7 in. (17.8 cm), occurs from southern Wyoming and central Colorado west. It is a bird of brushy mountain slopes, rusty-capped, olive above, gray below and on face, and with a distinct white throat. It winters to central Texas and occasional strays are recorded eastward.

# LARK BUNTING
*Calamospiza melanocorys*                                    48:3

**Description**
Size, 6 in. (15.2 cm). Spring male: black with large • *white wing patch* conspicuous in flight. Female: streaked brown above, white below; brown on cheeks; white wing patch smaller than in male. Fall male: similar to female, but chin, wings, and tail black.

**Similarities**
Female chunkier than most sparrows. Male Bobolink has white on base of wing and back, and yellow hindneck.

**Habitat**
Grasslands.

**Habits**
Sings from elevated perch or from air; gregarious when not nesting; flocks wheel in unison.

**Voice**
Song is a series of warbled trills; flock often sings in chorus; flight call, a sweet *whoo-ee.*

**Eggs**
4–6; pale blue; 0.8 x 0.6 in. (2.2 x 1.7 cm). Nest is of grass, plant down; in tussock of grass.

**Range**
Breeds from Minn. and Man. s. to Tex.; winters from s. Tex. and Ariz. s. to Mexico.

# GRASSHOPPER SPARROW
*Ammodramus savannarum*                                   46:11

**Description**
Size, 4½–5¼ in. (11.4–13.3 cm). Head large, flat; short tail bristly tipped. Adult: buff stripe through dark crown; yellow spot before eye and on bend of wing; back heavily streaked, throat and • *breast unstreaked buff.* Immature: breast buffier than adult.

**Similarities**
Savannah has notched tail, streaked breast.

**Habitat**
Meadows, dry fields, prairies.

**Habits**
Runs mouselike through grass, sings from top of weed stalk, often quite persistently. Distribution uneven. In flight, close to ground, fluttering, wrenlike.

**Voice**
Song is an insectlike buzz, *tit-zeeeeeeeeee;* also a long, low, varied song later in season, like elfin music; call, a *tlik.*

**Eggs**
3–5; white, sparingly spotted with brown; 0.7 x 0.5 in. (1.8 x 1.4 cm). Nest is cup of grass in grass.

**Range**
Breeds from N.H. and Man. s. to Fla., W. Indies, and Mexico; winters from N.C. and Tex. southward.

## BAIRD'S SPARROW
*Ammodramus bairdii* 46:7

**Description**
Size, 5–5½ in. (12.7–14.0 cm). Appearance somewhat like commoner Grasshopper Sparrow, but with necklace of sharp black streaks on buffy breast; head yellow-brown streaked with black and prominent ochre crown stripe; tail notched.

**Similarities**
Savannah has narrower, lighter crown stripe and more extensive breast streakings.

**Habitat**
Dry long-grass prairies.

**Habits**
Sings from weed stalk.

**Voice**
Song is somewhat like Savannah's, but more musical and with more body—several *chip's*, then a trill.

**Eggs**
3–5; whitish, marked with red-brown and black; 0.8 x 0.6 in. (2.0 x 1.5 cm).

**Range**
Breeds from Man. s. to Minn. and Mont.; winters in Tex. and Mexico.

## HENSLOW'S SPARROW
*Ammodramus henslowii* 46:6

**Description**
Size, 5 in. (12.7 cm). Large bill and head; short, spiky tail; pale stripe on buffy-olive crown, no yellow before eye; hindneck greenish; back and wings tinged with chestnut, streaked with white; upper breast pale buff with fine black streaks.

**Similarities**
Adult Grasshopper Sparrow has unstreaked, buffy breast and less prominent white streaks on back.

**Habitat**
Fields and meadows, often near water. Distribution irregular; only locally common.

**Habits**
Scurries mouselike through grass; flies low and jerkily, twisting tail; sings from top of weed stalk or fence post, often quite persistently.

**Voice**
Song is a short *s'LICK*, often heard at night.

**Eggs**
Similar to Grasshopper's.

**Range**
Breeds from ne. U.S. s. to e.-cen. states; winters from Ga. and Gulf states, southward.

## SAVANNAH SPARROW
*Passerculus sandwichensis*                                    46:5

**Description**
Size, 4½–5¾ in. (11.4–14.6 cm). • *Yellow in front of eye;* light stripe through crown; breast streaked with black, and with dark central spot; underparts whiter than most sparrows; legs pinkish; • *tail short,* notched. The Sable Island, Nova Scotia race, called the "Ipswich Sparrow" and recognized by some as a full species, is recognizably larger, and paler above; it winters on Atlantic Coast beaches.

**Similarities**
Song Sparrow has larger breast spot; wider and browner streaks; longer, rounded tail, lacks yellow on face. Vesper has white outer tail feathers. Other similar sparrows have rounded tails.

**Habitat**
Fields, fresh or salt meadows, beaches, dunes.

**Habits**
Hops, rarely walks; runs through cover with head low; if flushed, makes short zigzag, undulating flight revealing notched tail, then drops back into meadow; sings from low perch or wire; feigns injury to lure intruder from nest.

**Voice**
Song is a buzzy *tsip-tsip-tsip, seeee-saaay* or *tsip-tsip-tsip, saaay-seee;* call, a *thlip.*

**Eggs**
4–6; pale greenish-white spotted with red-brown and purple-brown; 0.8 x 0.6 in. (2.0 x 1.5 cm).

**Range**
Breeds from Labr. s. to N.J. and Mo.; winters from Mass. along coast and from cen. Ohio s. to Gulf Coast.

## SEASIDE SPARROW
*Ammospiza maritima*                                         46:10

**Description**
Size, 6 in. (15.2 cm). Dark olive-gray above, faintly streaked with gray below; large head and bill; • *yellow mark before eye;* black jawline separates white mustache from white throat. Distinctive races (formerly considered species) occur locally in parts of coastal eastern and southern Florida.

**Similarities**
Female Dickcissel has heavier, conical bill. See Sharp-tailed.

**Habitat**
Salt marshes.

**Habits**
Skulks mouselike in grass; if flushed, soon drops again into marsh; wades like a sandpiper; sometimes sings in flight, rising in air; inquisitive, can be attracted by hand-kissing.

**Voice**
Song is a "buzzy *tip-tip-ZEE-reeeeeeeee* with a *z* sound in gradually fading trill" (Saunders); *ZEE* note distinguishes it from Sharp-tailed; call, a harsh, Red-winged-like *check.*

**Eggs**
4–6; grayish-white, coarsely spotted with red-brown; 0.8 x 0.6 in. (2.0 x 1.5 cm).

**Remarks**
Dusky Seaside Sparrow and Cape Sable Seaside Sparrow now considered part of same species.

**Range**
Breeds along Atlantic and Gulf coasts from s. New England s. to Fla. and Tex.; winters in s. parts of breeding range.

## LE CONTE'S SPARROW
*Ammospiza leconteii*                                    **46:9**

**Description**
Size, 4½–5½ in. (11.4–14.0 cm). Tail short, spiky-tipped; crown
stripe white; hindneck and nape with purple-chestnut (in young,
buffy and unstreaked); • *eye line buffy-ocher,* throat and breast
yellow-brown, sides streaked.

**Similarities**
Henslow's breast much less buffy. Henslow's and Grasshopper
have buffy crown stripe. Grasshopper lacks streaked sides. Sharp-
tailed lacks white crown stripe (but young Sharp-taileds have
bright, buffy breasts).

**Habitat**
Prairie marshes, fields with matted cover, wet meadows.

**Habits**
Secretive, prefers running to flight; if flushed, flies jerkily for short
distance, then drops back into cover; sings from top of weed or
bush.

**Voice**
Song is thin, high, grasshopperlike, with a *chip* at end.

**Eggs**
4–5; whitish, heavily spotted with red-brown; 0.7 x 0.5 in. (1.8 x
1.3 cm). Nest is grassy cup in grass.

**Range**
Breeds from s.-cen. Canada s. to n.-cen. U.S.; winters in cen. and
s. states.

## SHARP-TAILED SPARROW
*Ammospiza caudacuta*                                    **46:8**

**Description**
Size, 5–6 in. (12.7–15.2 cm). • *Gray ear patch bordered with buff;*
short, spiky-tipped tail. Adult: dark cap (no obvious central stripe),
gray nape; light streaks on dark back; more or less streaked on
buffy breast. Young: much buffier and with dark streaking above
and below. Subspecies (interior and coastal races) are variable in
brightness of facial lines, amount of breast streaking, and overall
dark or light coloration.

**Similarities**
Young Bobolink is similar but larger than young Sharp-tailed.
When sparrows are flushed in salt marsh, a Sharp-tailed looks
brownish above with a bristly rounded tail; a Savannah looks
brownish above with a notched tail; a Seaside looks dark grayish
above.

**Habitat**
Salt- or freshwater marshes.

**Habits**
Scurries mouselike through grass; if flushed, soon drops again into
marsh, tail down; has flight song in spring.

**Voice**
Song is a "wheezy *tsup tsup shreeeeeeeee* with a *sh* sound in
gradually fading trill" (Saunders).

**Eggs**
4–5; grayish-white, finely dotted with brown; 0.7 x 0.6 in. (1.9 x
1.5 cm). Nest is loosely woven grass cup in marsh.

**Range**
Breeds locally from cen. Canada to Minn. and along Atlantic Coast
from lower St. Lawrence R., s. to N.C.; winters along Atlantic and
Gulf coasts, from S.C. s. to Fla. and w. to Tex.

## VESPER SPARROW
*Pooecetes gramineus*
46:2

**Description**
Size, 5–6½ in. (12.7–16.5 cm). • *White outer tail feathers,* conspicuous in flight; streaked grayish-brown above, whitish below, with dark streaks ending sharply on lower breast; no breast spot; • *bend of wing chestnut; pale eye-ring* and rather distinct ear patch.

**Similarities**
Song Sparrow is browner above, has dark breast spot, lacks white outer tail feathers. Water Pipit is more slender; walks, does not hop. Lark Sparrow has white tail corners.

**Habitat**
Upland fields, pastures, fencerows.

**Habits**
Hops, does not walk; likes dust baths; sings more toward evening and at dawn; has courtship flight song; seen in small flocks in migration.

**Voice**
Song is *"ah ah ay ay tetetetetetatatatata toto tu"* (Saunders). The introductory notes are long and low (those of Song Sparrow are short and high); call, a short *tsi.*

**Eggs**
4–6; grayish-white dotted with red-brown; 0.8 x 0.6 in. (2.0 x 1.5 cm). Nest is grass cup in grass.

**Range**
Breeds from N.S. and Ont. s. to N.C. and Tex.; winters from New England s. to Gulf Coast and Mexico.

## RUFOUS-CROWNED SPARROW
*Aimophila ruficeps*
47:8

**Description**
Size, 5–6 in. (12.7–15.2 cm). Adult: reddish-brown and gray streaks on back (appears dark); crown solid rufous; eye line gray; black whiskerlike streak bordering sides of throat; breast unstreaked grayish-white; tail rounded. Juvenal: crown brown, streaked; thin, dark streaks on breast.

**Similarities**
Chipping Sparrow has white line over eye, notched tail, black stripes on back. Swamp Sparrow is more rufous and with black stripes above, grayer below.

**Habitat**
Dry, grassy hillsides with low scrub, open pine-oak woods.

**Habits**
Sometimes occurs in small, loose colonies; skulks mouselike on ground, sings from low perch, looks hunched when sitting.

**Voice**
Song is several *mew*'s, then a gurgling warble (first part rising, last part falling), somewhat like House Wren's; call, a musical *deer, deer* or a nasal *chur, chur, chur.*

**Eggs**
3–5; bluish-white; 0.8 x 0.6 in. (2.0 x 1.5 cm). Nest is grassy cup on ground.

**Range**
Resident in sw. U.S., but e. rarely to w.-cen. Okla. and Tex., s. to Mexico.

# BACHMAN'S SPARROW
*Aimophila aestivalis*                                    47:4

**Description**
Size, 6 in. (15.2 cm). Brownish above, streaked with some gray and black; dull white below, unstreaked, with • *tan wash across breast;* pale bill.

**Similarities**
Field Sparrow has pink, not horn-colored, bill. Grasshopper has much shorter tail.

**Habitat**
Open pine woods, old brushy fields, cut-over areas, old orchards.

**Habits**
Shy, secretive, but sings from low branch or top of bush, February to August.

**Voice**
Song is clear, tuneful, of 5–12 phrases of different pitches and with pauses between, each phrase consisting of a long note and a trill; somewhat suggests pattern of Hermit Thrush; alarm note, a snakelike hiss.

**Eggs**
4; white; 0.7 x 0.6 in. (1.9 x 1.5 cm).

**Former name**
Pinewoods Sparrow.

**Range**
Breeds locally in se. U.S. from Pa., Ohio, and Ill. s. to cen. Fla., Gulf Coast, and Tex.; winters along Atlantic Coast from N.C., s. to Fla. and along Gulf Coast to s. Tex.

Note: **CASSIN'S SPARROW,** *Aimophila cassinii,* size, 5¼–5¾ in. (13.3–14.6 cm), breeds in the southwestern area from west-central Kansas and central Colorado, south and west. A bird of arid grasslands, it is plain-breasted; streaked, gray-brown above, with a streaked crown and white eye stripe. The **BLACK-THROATED SPARROW,** *Amphispiza bilineata* **(45:1),** size, 5 in. (12.7 cm), is a characteristic desert bird of the Southwest that reaches arid southeastern Colorado and the southern Texas coast. It is distinguished by its striking face pattern of stripes plus a black throat.

# DARK-EYED JUNCO
*Junco hyemalis*                                    45:4

**Description**
Size, 5½–6½ in. (14.0–16.5 cm). Solid gray, brown-gray, or blackish head, pinkish-white bill. Females, and especially immatures, tend to be duller and browner. Juvenals are streaked below. Dark tail with white outer feathers, white belly. Variously gray above and on breast and sides (eastern "slate-colored" form); gray with 2 white wing bars ("white-winged" Black Hills form); pinkish or brown sides with red-brown back, and in males gray (n. Rocky Mt. "pink-sided" form) to black (western "Oregon" form) hood.

**Habitat**
Diverse conifers, mixed, deciduous woods, edges, brushy fields, suburban gardens.

**Habits**
Tame; winters in flocks, often with other sparrows; sings from tree.

**Voice**
Song is a Chipping Sparrow-like trill, but more musical; also a warbled song; call, a *click*.

**Eggs**
3–6; grayish-white, thickly spotted with lilac and brown; 0.8 x 0.6 in. (2.0 x 1.5 cm).

**Former names**
White-winged Junco, Oregon Junco, Slate-colored Junco, Gray-headed Junco.

**Remarks**
Diverse forms of this variable species hybridize freely wherever they are in contact, much as in the Common Flicker. Intermediate patterns of color among hybrids are common. Although typical individuals of the different forms can be distinguished by the observer, the birds appear to disregard these differences in mating; hence, we consider them as one species. Individuals of western races wander east in winter, even to the Atlantic Coast. This is one of the commonest wintering species and a common visitor at bird feeders.

**Range**
Breeds from Newfoundland and Man. s. to Ga. and Mexico; winters from s. parts of breeding range s to Gulf Coast and Mexico.

## LARK SPARROW
*Chondestes grammacus*                                              **45:3**

**Description**
Size, 5½–6¾ in. (14.0–17.1 cm). Adult: • *chestnut crown stripes* and ear patch; white eye line and mustache; black line on either side of chin; dark spot in middle of unmarked white breast; tail black, • *fan-shaped,* and with • *white sides and corners.* Juvenal: fine, dark streaks replace dark breast spot.

**Similarities**
Vesper has white outer tail feathers.

**Habitat**
Grassy areas near trees; farms, ranches.

**Habits**
When sitting, raises crown at intervals; sings from elevated perch, or hovering in air.

**Voice**
Song is "somewhat like that of Indigo Bunting, but louder, clearer, much finer" (Forbush).

**Eggs**
3–5; white, spotted and scrawled with black; 0.8 x 0.6 in. (2.0 x 1.5 cm). Nest is grassy cup on ground or in low bush.

**Range**
Breeds from s. Ont. and n. Minn. s. to s. S.C., cen. Ala., and Tex.; winters from Gulf Coast s. to C. America.

## TREE SPARROW
*Spizella arborea*                                                 **47:6**

**Description**
Size, 5½–6½ in. (14.0–16.5 cm). Rufous cap; • *black spot on plain breast;* bill dark above, light below; tail slightly notched; streaked brownish above, breast whitish; gray on sides of face; 2 white wing bars; a little white on tail.

**Similarities**
Lark Sparrow has a striped crown, much white in tail.

**Habitat**
Low trees, brush, weedy fields, gardens.

**Habits**
Gregarious in winter; visits feeders, often sings in concert.
**Voice**
Song is a sweet *"eee eee tay tititee tay"* (Saunders); call, a *tee-lo* and a *tseet*.
**Eggs**
4–5; light green, dotted with light brown; 0.7 x 0.5 in. (1.9 x 1.4 cm).
**Range**
Breeds from n. Que. and n. Man. s. to Newfoundland, cen. Que., and cen. Man.; winters from n. U.S. s. to N.C. and Ark.

## CHIPPING SPARROW
*Spizella passerina*                                          46:13

**Description**
Size, 5–5¾ in. (12.7–14.6 cm). Trim and slim; tail notched. Spring adult: streaked brown above; breast grayish-white, unmarked; chestnut cap; • *black line through eye,* white line over eye; 2 thin white wing bars. Immature: brown cap with pale central stripe and black streaks; buffy eye line; brown ear patch (not bordered with black), gray rump.
**Similarities**
See Clay-colored.
**Habitat**
Pinewoods, wood edges, lawns, gardens, country roadsides.
**Habits**
Tame, feeds on ground, sings from wire or high branch; often victimized by Brown-headed Cowbird.
**Voice**
Song, a simple unmusical trill, all on one pitch, suggesting songs of Worm-eating and Pine Warblers, junco, Swamp Sparrow, and others, but is the most mechanical; call, a *tsip*.
**Eggs**
3–5; greenish-blue with wreath of blackish dots about larger end; 0.7 x 0.5 in. (1.8 x 1.3 cm). Nest is hair-lined cup in tree or bush.
**Range**
Breeds from Newfoundland and Man. s. to Fla. and C. America; winters from Md. and Tenn., southward.

## CLAY-COLORED SPARROW
*Spizella pallida*                                            46:12

**Description**
Size, 5½ in. (14.0 cm). Adult: head mainly white with black molar streak, • *whitish center crown stripe* separating brown areas, black-bordered • *brown ear patch,* white stripe over eye; nape unmarked, pale gray; back gray-brown striped with black; 2 fine, white wing bars; entire underparts white. Immature: much browner.
**Similarities**
Immature Chipping has brown cheek patch, more bordered with black, and buffier rump. Larger Lark Sparrow has somewhat similar head pattern, but with dark spot on breast.
**Habitat**
Fields, open brush and underbrush, often near water.
**Habits**
Feeds low or on ground; in breeding season sings persistently from bush top.
**Voice**
Song is 3–4 slow insectlike buzzes; call, a *chip*.

**Eggs**
3–5; light green-blue, with brown spots about larger end; 0.6 x 0.5
in. (1.7 x 1.3 cm). Nest like Chipping Sparrow's.
**Range**
Breeds from n.-cen. Canada and Ill. s. to Nebr. and Mont.; winters
from s. Tex., southward; occasional e. to w. N.Y., Conn.

Note: The westernmost fringe of our area in summer is visited by
the **BREWER'S SPARROW,** *Spizella breweri,* size, 5 in. (12.7 cm),
which resembles the Clay-colored, but has a finely streaked rather
than broadly striped crown.

## FIELD SPARROW
*Spizella pusilla*                                                    **47:5**

**Description**
Size, 5¼–6 in. (13.3–15.2 cm). • *Pink bill;* tail slightly notched;
reddish ear patch; dull gray border to narrow white eye-ring; 2
white wing bars; underparts unmarked grayish-white, with brown
wash on breast and sides.
**Similarities**
Tree Sparrow has dark spot on breast, stronger eye stripe,
mustache stripe, grayer face and nape. Chipping Sparrow has
white line over eye.
**Habitat**
Brushy hillsides, edges of fields, pastures, near bushes, trees.
**Habits**
Sings from low perch, sometimes by moonlight; gregarious in
winter.
**Voice**
Song is a sweet, musical *twee-twee-twee, te-te-te-te-te-te,* first notes
slow, trill becoming faster, and sometimes rising or dropping in
pitch; call, a *tsip.*
**Eggs**
4–5; grayish-white dotted with red-brown and lilac; 0.7 x 0.5 in.
(1.8 x 1.3 cm).
**Range**
Breeds from n. New England, s. Wis., and N.Dak. s. to Ga.,
Miss., La., and cen. Tex.; winters from Mass., Ohio and Kans. s.
to Fla. and Gulf Coast.

## HARRIS' SPARROW
*Zonotrichia querula*                                                **47:1**

**Description**
Size, 7½ in. (19.1 cm). Spring adult: • *black cap,* face, throat and
upper breast; bill pinkish. Brownish above, whitish belly, sides
streaked; cheeks gray; 2 white wing bars. Fall adult: black crown,
partly veiled with gray. Immature: spotted brown crown; buffy
face, flanks and undertail coverts; white throat; necklace of brown
streaks.
**Similarities**
Spring male Lapland Longspur is smaller, has chestnut nape.
**Habitat**
Breeds in dwarf timber near tundra; in migration and winter,
thickets.
**Habits**
Similar to White-crowned and White-throated.
**Voice**
Song is 1–5 clear whistles, usually on same pitch, then a pause,
followed by several notes on another pitch; alarm note, a loud *winck.*

**Eggs**
3–5; greenish-white, spotted and blotched with brown; 0.8 x 0.6 in.
(2.4 x 1.7 cm).
**Range**
Breeds from n. Canada s. to n. Man.; winters from Iowa s. to Tex.

## WHITE-CROWNED SPARROW
*Zonotrichia leucophrys*                                47:3

**Description**
Size, 5½–7 in. (14.0–17.8 cm). Black-and-white striped crown,
pearly-gray breast, • *pinkish bill.* Adult: streaked brownish above, 2
white wing bars. Immature: similar, but crown striped with red-
brown and buff, underparts washed with brown.
**Similarities**
White-throated is darker billed, adult has whiter throat sharply set
off by black lines at sides, yellow spot before eye; immature much
dingier on breast.
**Habitat**
In summer, shrubby mountainsides, dwarf willows; in migration,
brush, edges, thickets, roadsides.
**Habits**
Has dignified bearing, can erect crown; forages on ground; sings
from elevated perch, sometimes at night; often found with other
species of sparrows.
**Voice**
Song is a clear, whistled *aaaa-ee-aay;* a husky, descending *see-say-
so;* call, a *chink.*
**Eggs**
3–5; greenish-white, heavily spotted with red-brown; 0.8 x 0.6 in.
(2.2 x 1.7 cm).
**Range**
Breeds from Labr. and Newfoundland w. across n. Canada and s.
in far West; winters from N.J. s. to Ga., Gulf Coast, and Mexico.

## WHITE-THROATED SPARROW
*Zonotrichia albicollis*                                47:2

**Description**
Size, 6–7 in. (15.2–17.8 cm). • *Black-and-white crown,* dark bill.
Adult: black-and-white striped crown with yellow spot before eye,
back brownish streaked with black, 2 white wing bars, cheeks and
underparts gray; white throat sharply set off by black malar line.
Immature: similar, but usually with some blurry streaking below,
crown stripes brown and buff, white on throat duller. There is
considerable variation with sharply marked, brighter, and duller,
less white-faced phases.
**Similarities**
White-crowned may have grayish-white throat, but less distinct,
not bordered by dark line as in White-throated. Adult Swamp has
rusty crown; immature Swamp is smaller, very faint wing bars.
**Habitat**
Various woodlands, slash piles, brushy pastures, thickets.
**Habits**
Active; common; visits feeders, forages noisily on ground using feet
to kick aside debris; sings from ground or low perch, sometimes at
night and on spring migration; gathers in small flocks in fall.

**Voice**
Song is a set of clear whistles, often written as *Old Sam PEAbody, PEAbody, PEAbody;* notes, a distinctive, longish *ssst;* also a *chip.*
**Eggs**
4–5; grayish-white, dotted with red-brown; 0.8 x 0.6 in. (2.0 x 1.5 cm).
**Range**
Breeds from Newfoundland and cen. Que., s. to Pa., Wis., and N.Dak.; winters from s. New England and Kans., s. to Gulf Coast and Mexico.

## FOX SPARROW
*Passerella iliaca*                                             **46:1**

**Description**
Size, 6¼–7¼ in. (15.9–18.4 cm). Large, stocky; • *streaked red-brown and gray above,* rusty brightest on wings, rump, and tail; white below, heavily streaked with red-brown, forming a more or less conspicuous spot on the breast.
**Similarities**
Hermit Thrush has thinner, longer bill, no streaks on back.
**Habitat**
Open woods, edges, thickets.
**Habits**
Scratches noisily on ground with both feet at once, sings from elevated perch, visits feeders; gregarious except in breeding season.
**Voice**
Song is "loud, beautiful whistled melody, *hear hear I sing-sweet sweeter most-sweetly*" (Cruickshank); alarm note, a *smack;* call, a *sssp.*
**Eggs**
4–5; pale green, heavily spotted with rusty brown; 0.9 x 0.7 in. (2.3 x 1.8 cm). Nest is grass cup, feather lined, in bush or on ground.
**Range**
Breeds from n. Que. and n. Man. s. to N.B. and n. U.S.; winters from cen. states s. to Fla. and Gulf Coast.

## LINCOLN'S SPARROW
*Melospiza lincolnii*                                          **46:3**

**Description**
Size, 5–6 in. (12.7–15.2 cm). Appearance like a slender, elegant Song Sparrow; streaked grayish-brown above, white below; • *fine dark streaks largely confined to buffy band across breast,* seldom merge into dark spot; brown crown with light-gray central stripe; narrow white eye-ring.
**Similarities**
Song Sparrow is heavier, browner above, with broader, more extensive streaks below forming a spot on the breast; juvenal Song is finely streaked below but has no eye-ring or buffy wash on breast. Swamp is browner above, grayer below with rusty cap; immature Swamp has less distinct breast streaks than Lincoln's. Baird's is shorter-tailed, has buffy crown stripe, no buffy on breast. Savannah has shorter, notched tail; yellow before eye; no buff on breast.
**Habitat**
Breeds near bogs and water in northern forest; in migration, wet areas; dense thickets, edges; stone walls, bushy fences.

**Habits**
Secretive; does not sing in migration; skulks or runs mouselike along ground and under cover; hard to see, but responds to squeaking.
**Voice**
Song "suggests the bubbling, guttural notes of the House Wren, combined with the sweet, rippling music of the Purple Finch" (Dwight); call, a low *tsip*.
**Eggs**
4–5; white, heavily spotted with brown; 0.8 x 0.6 in. (2.0 x 1.5 cm).
**Range**
Breeds from Labr. and Newfoundland w. to n. Que. and s. to New England and n. Wis.; winters from Gulf Coast s. to C. America.

## SWAMP SPARROW
*Melospiza georgiana*                                                        47:7

**Description**
Size, 5–5¾ in. (12.7–14.6 cm). • *Rusty crown;* •*gray-bordered, white throat,* tail rounded; eye stripe, grayish-white line over eye; mustache stripe; gray wash on breast; rusty wings with no wing bars. Spring adult dark rufous above, very faint bars; rusty crown; cheeks, nape, and underparts gray; white throat clearly outlined with gray. Fall adult: often buffy on sides and underparts, crown streaked with black, light center stripe. Immature: crown brown with light center stripe; belly, breast, sides buffy; faint streaks on breast.
**Similarities**
See Lincoln's Sparrow. Song Sparrow has longer, more rounded tail; is lighter above; and has streaked breast. Chipping is more slender, has notched tail, and black-and-white eye stripes. Tree and Field have white wing bars.
**Habitat**
Swamps, freshwater marshes; in migration, also weedy fields.
**Habits**
Sings from reed or bush in marsh.
**Voice**
Song is a 1-pitch trill, slower, louder, sweeter than Chipping Sparrow's; note, a *chip,* similar to White-throated's.
**Eggs**
4–5; variable, often bluish-white, heavily spotted with brown; 0.8 x 0.6 in. (2.0 x 1.5 cm).
**Range**
Breeds from e.-cen. Canada s. to e.-cen. U.S.; winters from N.Y. and Nebr. s. to Fla. and Gulf Coast.

## SONG SPARROW
*Melospiza melodia*                                                          46:4

**Description**
Size, 5–6 in. (12.7–15.2 cm). One of the commonest sparrows. Brown streaked below with streaks converging into spot on breast. Adult: tail long, rounded; variable in tone; light line across top of brown crown and over each eye; blackish mustache stripe; white below; occasionally without central black spot. Juvenal: streaked buffy band across breast; hard to distinguish from Lincoln's or young Swamp Sparrow.
**Similarities**
Savannah has shorter, notched tail.

**Habitat**
Thickets, shrubbery, especially in damp or wet areas.

**Habits**
Visits feeder, likes to bathe; pumps tail up and down in flight, often hunches back, lowers tail before landing; forages on ground, sings from elevated perch.

**Voice**
Song is variable, but easy to learn: *"Maids! maids! maids! hang up your teakettle-ettle-ettle"* (Thoreau); call, a *tsak*.

**Eggs**
3–7; variable, often whitish, spotted with red-brown; 0.8 x 0.6 in. (2.2 x 1.5 cm); 2 broods. Nest is grassy cup on ground or in low bush.

**Range**
Breeds from Newfoundland and n. Man., s. to S.C. and N.Dak.; winters from s. Canada, s. to Gulf Coast and Mexico.

## McCOWN'S LONGSPUR
*Calcarius mccownii*                    **48:12**

**Description**
Size, 6 in. (15.2 cm). Black inverted T on rather short, white tail. Spring male: black crown, whisker, and small breast patch; white eye line, throat, and belly; • *gray collar;* streaked brown-gray above, chestnut shoulder patch. Fall male: more tawny above, black largely replaced by gray. Female and immature: streaked brownish and buffy, similar to female and young Chestnut-collared, but tail pattern different.

**Similarities**
Spring male Chestnut-collared has chestnut collar. Female Smith's is buffier below, has white on outer rectrices only. Lapland has more distinct breast streaks, white outer rectrices only. Horned Lark has thin bill, black sideburns.

**Habitat**
Short-grass plains.

**Habits**
Flight undulating; after breeding season, gregarious; often found with Horned Larks and Chestnut-collared Longspurs.

**Voice**
Pleasant warble, with twitters, uttered hovering in air with wings seemingly straight up; call, a *chirrup-chirrup*.

**Eggs**
3–4; pale greenish, dotted with brown; 0.8 x 0.6 in. (2.0 x 1.5 cm).

**Range**
Breeds from se. Man. s. to N.Dak. winters from Nebr. s. to Mexico.

## LAPLAND LONGSPUR
*Calcarius lapponicus*                    **48:9**

**Description**
Size, 6–7 in. (15.2–17.8 cm). Legs dark; white outer tail feathers in all plumages. Spring male: crown, face, throat, sides, and tail black; collar chestnut; upperparts brownish, streaked; wing bars, stripe from eye down sides, underparts, and outer tail feathers white. Fall male and female: black replaced by dark smudge on sides of lower throat and ear; legs black. In flight, wings look dark.

**Similarities**
Other longspurs have more white on tail, different tail patterns, except Smith's, which is buffy, not white, below. Snow Bunting is much lighter, has light, not dark, wings. Horned Lark has thin

bill, yellow and black throat markings, longer tail, less undulating flight.

**Habitat**
Tundra, plains, fields, beaches.

**Habits**
Gregarious, except in breeding season; often clings to weed stalks; hard to detect when it remains motionless on a field; flies like Snow Bunting, often seen with it.

**Voice**
Song is short *"tee-tooree, tee-tooree, teereeoo"* (Snyder), uttered on wing; in winter, a hoarse *churrr;* and a rattle followed by a whistled *dicky-dick-do.*

**Eggs**
6; variable, often greenish-gray, heavily blotched with red-brown; 0.8 x 0.6 in. (2.2 x 1.5 cm).

**Range**
Breeds from Greenland and Canadian Arctic s. to n. Que.; winters from n. U.S. s. to N.Y. and n.-cen. states.

## SMITH'S LONGSPUR
*Calcarius pictus*                                                48:11

**Description**
Size, 6½ in. (16.5 cm). Underparts buffy, bill thinner and more pointed than in other longspurs, legs paler; white outer tail feathers. Spring male: crown and cheek triangle black; unmistakable pattern of black-and-white face, buff nape, and underparts. Female, fall male, and immature: similar, but head streaked like rest of upperparts; underparts paler, but still completely buffy; throat and breast often somewhat streaked; fall male often shows small, white shoulder patch of spring plumage.

**Similarities**
See McCown's Longspur.

**Habitat**
Tundra, plains.

**Habits**
Gregarious in fall and winter, often occurs with other longspurs and Horned Lark; sings from ground.

**Voice**
Sharp clicking notes when flushed; song longer, but includes similar notes.

**Eggs**
4–6; whitish, spotted and blotched with purple-brown; 0.8 x 0.6 in. (2.0 x 1.7 cm).

**Range**
Breeds from n. Ont. and n. Man. s. to Hudson Bay; winters from Nebr. s. to Tex.

## CHESTNUT-COLLARED LONGSPUR
*Calcarius ornatus*                                               48:10

**Description**
Size, 6 in. (15.2 cm). Black triangle on white tail. Spring male: • *black cap* and underparts; white line over eye; chestnut collar; streaked brown above, whitish cheek and throat. Fall male: black and chestnut replaced by brown. Female and immature: streaked brown above, buffy below, with faint streaks on sides and breast.

**Similarities**
McCown's lacks breast streaks.

**Habitat**
Plains, prairies.
**Habits**
Similar to Lapland and McCown's; gregarious, often found with other longspurs; flight undulating, showing much white on tail.
**Voice**
Song is short, high, weak, and twittery, uttered on wing; remotely suggests Western Meadowlark; flight call, a twitter.
**Eggs**
4; pale greenish, dotted with brown; 0.8 x 0.6 in. (2.0 x 1.5 cm).
**Range**
Breeds from Man. s. to Minn.; winters from Iowa s. to Tex. and Mexico.

## SNOW BUNTING
*Plectrophenax nivalis*                                    **48:8**

**Description**
Size, 6–7¼ in. (15.2–18.4 cm). Summer male: • *white;* with back, bend of wing, wing tips, and middle tail feathers black. Summer female: similar, but black replaced by dusky. Winter male: white; cap, ear patch, and spots on rump rusty; bend of wing, wing tips, midtail, and spots on back black; winter female rustier; Immature: the rustiest, least white of all. In flight (in winter) looks white below including tail, dark above; wings always flash some white.
**Similarities**
In flight, Water Pipit and Horned Lark show dark wings and black tails.
**Habitat**
Mountain slopes, tundra; fields, salt marshes, flats, beaches in winter.
**Habits**
Walks, runs, hops; gregarious except in breeding season; compact flocks wheel in unison, and settle on field like snowflakes; often seen with Lapland Longspurs; in north, frequents camps, villages; called "House Sparrow of the Arctic."
**Voice**
Song is *"turee-turee-turee-turiwee"* (Snyder); notes include a high whistled *teer;* also a purring note.
**Eggs**
4–7; variable, whitish spotted with brown; 0.8 x 0.6 in. (2.0 x 1.5 cm).
**Range**
Breeds in Arctic, from Greenland w. across Canada and s. to n. Que.; winters from n. Canada s. to Pa., Ind., and Kans.

# Mammals

**Consulting Editor**
Sydney Anderson
Curator of Mammalogy
American Museum of Natural History

**Illustrations**
Plates 49–52, 54, 56–58, 60, 61 Nina L. Williams
Plates 53, 55, 59, 63 John Hamberger
Plate 62 Jennifer Emry-Perrott
Text Illustrations by Nina L. Williams and Jennifer Emry-Perrott

# Mammals
## Class Mammalia

Mammals are warm-blooded, air-breathing, milk-producing vertebrates with hair. They give their young a period of parental care, and all but the platypus and echidna, which lay eggs, bring forth their young alive. Most have several different kinds of teeth.

*Warm-blooded* means the animal is capable of maintaining a relatively constant internal body temperature irrespective of the outside air or water temperature. The normal internal temperature for most active mammals lies in the range of from 90° to 104° Fahrenheit. In hibernating individuals, however, the temperature is reduced and may be only a few degrees above freezing. All mammals have hair, although with some mammals the hair may be in the form of bristles. Even whales, at least embryonic whales, have some bristles. In contrast, the reptiles, from which mammals descended, are cold-blooded vertebrates with scales or plates instead of hair, and they do not produce milk or give their young parental care. Reptilian teeth are less differentiated, as well.

Most mammals also have four feet and a tail. In the seals, whales, and manatees, the forefeet are transformed into flippers; and in the whales and manatees, the hind limbs have totally disappeared externally. In a few species the tail is greatly reduced or missing, as in bears and human beings.

Early mammals had five digits on each foot. This number was gradually reduced in many evolutionary lines; in the horse only one digit, bearing the hoof, remains. Human beings retain the original five, which of course, has been of crucial importance in their evolution.

### Evolution
Mammals arose from therapsid reptiles in the late Triassic period and early Jurassic. In addition to the differences already mentioned, mammals represent an advance over reptiles in various other ways. Being warm-blooded means that, unlike reptiles or amphibians, mammals can occupy cold climates and deal with freezing winters by continuing activity or hibernation. Having differentiated teeth, they can chew their food and prepare it for quick digestion, increasing its availability as a source of energy. This makes mammals more active than reptiles, which, after swallowing their prey unchewed, may be sluggish for a week or two as the meal is slowly digested.

Increased activity means greater ease in catching prey and escaping from predators, both desirable in the struggle for existence. This activity is aided also by the development of a diaphragm, which promotes deep breathing, hence more rapid aeration of the blood and quicker metabolism. A reptile's rib-controlled breathing can never provide oxygen as quickly as the bellows action of the mammalian diaphragm. Activity is also furthered by a four-chambered heart, which separates arterial blood from venous blood, helping in the more rapid and efficient utilization of oxygen. The red blood cells of mammals are nonnucleated, thus improving oxygen-carrying capacity.

Evolution has given rise to three quite different groups of living mammals. It is presumed that the subclass Prototheria arose in the Jurassic period. The present representatives of this subclass are the primitive egg-laying monotremes, the platypus and echidna, which are confined to Australia and New Guinea.

In the succeeding period, the Cretaceous, two new groups arose.
The less advanced were the Metatheria, the marsupials, or pouched
mammals, which bring forth living young, but in an embryonic
condition. In most species, these are kept and nourished in an
external pouch for a protracted period. The more advanced were
the Eutheria, the placental mammals, which retain the young for a
longer period within the body of the mother and which lack any
outer pouch. At the end of the Cretaceous and in the Paleocene,
great development took place among the placentals. This group
then proved superior to the marsupials, which have survived to the
present almost exclusively in Australia and South America, continents
long isolated from the world of the more aggressive placentals.

### Adaptations

Mammals have dominated the globe for 70 million years. In that
time they have become adapted to every major type of environment.
From a central, walking terrestrial stock, various running, or
cursorial, forms progressively lost digits as they increased their speed,
until they evolved into the fleet antelope with two hoofs and the horse
with one. From the central stock some forms, such as the pocket
gophers and moles, developed adaptations for a life underground.

Others, such as the squirrels and monkeys, took to the trees, where
they now lead semiarboreal or arboreal lives. Flying squirrels
suggest an attempt at the conquest of the air by rodents, a triumph
achieved by bats in the Paleocene or earliest Eocene. Notable as
well were the adaptations required for successful reinvasion by
mammals of the sea, the ancestral home of all life. Seals and
walruses have not yet freed themselves from dependence on the
land, on which they still must bring forth their young; but the
whales and sirenians live entirely in the water, to which they are
almost as well adapted as fish, save for their dependence on the air
above for oxygen. Along with their adaptations to all environments,
mammals have successfully adapted to all altitudes, latitudes, and
climates, and have grown to various sizes.

It is in intelligence, however, that mammals are supreme. Compared
with most other mammal genera, *Homo* is a newcomer, perhaps
two or three million years old, compared, for example, with the
eight-million-year-old lineage of his companion, *Felis,* the house cat.

### Conservation

With mammals, as with almost all the wildlife covered in this
book, the greatest enemy today is people. Once shooting and
trapping were the main dangers; they may still be for the bighorn
and grizzly, the mountain lion and fisher. More important now,
however, is loss of habitat through the expansion of civilization.
This loss continues and the adverse pressure increases, particularly
on many of the larger mammals. A great deal still needs to be
known about the habits and ecologic relationships of most
mammals, including even some of the more common species.

### Field Study

Field study of mammals is quite different from that of birds. A few
common species, such as rabbits, squirrels, skunks, woodchucks,
and some mice and rats, are frequently encountered. But most are
not usually seen. Many are nocturnal; most are secretive; the larger
ones are quite wary; sea mammals are usually seen by accident.

Therefore mammal study is usually part of a general field trip on
which some mammals and the tracks, scats, and signs of others
may be seen. A flashlight will surprise some mammals at night.
Taking a "stand" (that is, sitting quietly at a likely place in the
woods) will often afford a view of some of the shyer species.

### Habitat
The best places to see mammals on land include woods, wood edges, stone walls, the unkempt corners of suburbs and countryside (small mammals), the edges of lakes and streams (moose, beaver, otter, muskrat), deep woods and swamps (bear, bobcat), caves and old mines (bats), and old houses and outbuildings (mice, squirrels).

Mammals in general may be seen throughout the year; many, however, appear to be most active in spring and fall. The hibernators, of course, are hidden away in winter, although if their roosts are known, they may be seen readily.

### Teeth and Food
The primitive placental mammalian tooth count was forty-four, consisting, in each jaw, of three incisors, one canine, four premolars, and three molars on either side. In this chapter the dental formula will be written as 3143/3143. The digits of the first number refer to the number of incisors, canines, premolars, and molars, respectively, on each side of the upper jaw; the second number, to those on each side of the lower. Incisors are for cutting, canines for tearing, premolars and molars for grinding or shearing. Mammals with different diets require different kinds of teeth, and their tooth structure has developed along with their diet. Omnivorous human beings have the formula 2123/2123. The rodents, which gnaw but do not tear, have lost their canines; armadillos, which neither gnaw nor tear, have lost both incisors and canines; many mice have no premolars; some whales have no teeth at all. The food of an animal determines its habits; its teeth reflect its food. Thus, given in the chapter are the dental formulae, as a matter of interest, and as an aid in the identification of skulls that the naturalist sometimes encounters in the field, in caves, or in the pellet of a raptor.

### Reproduction
Placental mammals are so named from the placenta through which the young are nourished inside the body of the mother. This provides a safe first home in which young mammals can develop for a longer time and hence reach a more advanced state of development. This characteristic has enabled these mammals to concentrate on producing a few or only one young, instead of the dozens or hundreds of young or eggs that reptiles require in order to assure the perpetuation of their kind.

Despite the prolonged internal development, newborn mammal young are often quite helpless. Many require parental, or at least maternal, care for a period after birth. This affords the parent an opportunity to teach the young, and the species thus acquires a more varied behavioral pattern than the stereotyped responses characteristic of the reptiles. The increased emphasis on learned responses provides greater opportunity for the use and development of intelligence.

### Taxonomy
The names and organization of the Class Mammalia in this chapter follow the most recent checklist of J. Knox Jones, Jr., Dilfold C. Carter, and Hugh H. Genoways of Texas Tech University, entitled *Revised Checklist of North American Mammals North of Mexico, 1979.*

### Range and Scope
There are about 4000 mammal species in the world, representing 122 families; of these, there are in North America, north of Mexico, about 410 species in 40 families, of which 183 nondomesticated species occur in the region covered in this chapter.

The area of coverage includes that region of North America east of the 100th meridian and north of the 25th parallel. This encompasses the United States and Canada from the Atlantic Ocean west to the Dakotas, Kansas, Nebraska, Oklahoma, and east Texas; and from northern Canada to southern Florida. Distributions apply to the species as a whole only within the areas of geographic coverage of the chapter.

Species may vary widely in color over different parts of their range, or vary as a result of age, sex, or season. Those illustrated are of a typical color, and the text describes the range of variation. Longevities given are mostly for captive species; maximum age in the wild is usually much less.

### Nomenclature

The scientific names used in the chapter are from Hall and Kelson's *The Mammals of North America,* Ronald Press, 1959, as revised in scientific literature through 1976. Common English names follow *Revised Checklist of North American Mammals North of Mexico,* 1979, by J. Knox Jones, Jr., D. C. Carter, and H. H. Genoways, of the Museum of Texas Tech University, Lubbock, Texas.

## USEFUL REFERENCES

Anderson, S., and Jones, J. K., Jr., eds. 1967. *Recent Mammals of the World.* New York: Ronald Press. (453 pp., 70 figs.)

Blair, W. F., et al. 1968. *Vertebrates of the United States.* New York: McGraw-Hill. (616 pp., illustrated.)

Burt, W. H. 1972. *Mammals of the Great Lakes Region.* Ann Arbor: Univ. of Michigan. (246 pp., 54 figs., many maps.)

Burt, W. H., and Grossenheider, R. P. 1976. *A Field Guide to the Mammals.* 3rd ed. Boston: Houghton Mifflin. (319 pp., 24 col. pl. and other illustrations.)

Cahalane, V. H. 1947. *Mammals of North America.* New York: Macmillan. (682 pp., illustrated.)

Hamilton, W. J., Jr. 1943. *The Mammals of Eastern United States.* Ithaca, N.Y.: Comstock. (Cornell Univ.) (432 pp., illustrated.)

Jackson, H. H. 1961. *Mammals of Wisconsin.* Madison: Univ. Wisconsin Press. (504 pp., 81 maps, illustrated.)

Lowery, G. H. 1974. *The Mammals of Louisiana and Its Adjacent Waters.* Baton Rouge: Louisiana State Univ. Press. (565 pp., illustrated.)

Murie, O. J. 1954. *A Field Guide to Animal Tracks.* Boston: Houghton Mifflin. (374 pp., illustrated.)

Palmer, E. L. 1957. *Fieldbook of Mammals.* New York: Dutton. (321 pp., illustrated.)

Palmer, R. S. 1954. *The Mammal Guide.* New York: Doubleday. (384 pp., 30 col. pl. and other illustrations.)

Peterson, R. L. 1966. *The Mammals of Eastern Canada.* Toronto: Oxford Univ. Press. (465 pp., 233 figs., 107 maps.)

Schwartz, C. W., and Schwartz, E. R., 1972. *The Wild Mammals of Missouri.* Columbia: Univ. Missouri Press and Missouri Conserv. Comm. (341 pp., illustrated.)

# Marsupials
Order Marsupialia

## NEW WORLD OPOSSUMS
Family Didelphidae

### VIRGINIA OPOSSUM
*Didelphis virginiana*

**51:6**
*Fig. 15*

**Description**
Size: head and body, 15–20 in. (38.1–50.8 cm); tail, 9–13 in.
(22.9–33 cm); weight, 6–12 lb. (0.27–0.54 kg); 50 teeth, 5134/
4134; mammae, usually 13. Fur coarse, usually gray above, but
varies from white to near-black, only slightly paler below; face
white, ears and inner part of tail black. Snout long, legs short; 5
toes on each foot; first toe on hind foot opposable (thumblike) and
lacks claw; tail is long, scaly, prehensile.
**Habitat**
Variable, but prefers woodlands.
**Habits**
Nocturnal; climbs well; feigns death or "plays possum" if
frightened.
**Reproduction**
2 litters per year; nest located in a burrow, hollow log, hollow tree;
gestation 12½ days; 8–18 (occasionally more) young that weigh
$\frac{3}{100}$–$\frac{7}{100}$ oz. (1–2 g), born at an early stage of development; young
travel immediately to pouch, or *marsupium,* where they attach
firmly to a teat and remain for 55–70 days; pouch young usually
number 6–9.
**Remarks**
The Virginia Opossum, which is the only marsupial that occurs
north of Mexico, has extended its range northward in both
prehistoric and recent times. The family Didelphidae dates back to
Cretaceous times, making it one of the oldest of living mammalian
families.
**Range**
Throughout e. U.S. and southernmost Ont.
**Other name**
Common Opossum.

Virginia Opossum

*Fig. 15*

Nine-banded Armadillo, p. 241

# Insectivores

Order Insectivora

The insectivores of the eastern United States are small terrestrial or fossorial mammals with small, sharp teeth; long snouts; and flat, claw-bearing feet. Of all placental mammals living today, these are regarded as the most primitive, and therefore closest to the ancestral mammalian stock.

## SHREWS

Family Soricidae

At first glance a shrew looks like something halfway between a mouse and a mole. It is small, mouse-colored, and has a long, pointed nose. Its inconspicuous eyes and ears distinguish it from a mouse, and it has five toes on each foot. Most mice have only four on the front foot. Shrews lack the strong digging forelimbs of moles, and are swift runners above ground. They are small, three and one-half to six inches (8.9–15.2 cm) long, usually grayish or brownish, generally lighter beneath; the sexes look alike. Females have six mammae, usually two abdominal, four inguinal. The teeth, brown-tipped, number thirty-two, formula 3133/1113, except in the Least and Desert Shrews.

Favorite habitats include moist forest floors, swamps, marshes, bogs, rocks, tundra, and mountains, but a few species inhabit dry areas. Shrews make small, round holes in leaf litter about rocks or logs and use the surface runways of mice and voles. They do not hibernate. In disposition they are aggressive, irascible, and nervous.

The diet includes insects, earthworms, grubs, snails, other invertebrates, and mice, as well as berries and soft vegetation. They are voracious feeders and many apparently eat their own weight in food every twenty-four hours just to sustain life. Shrews breed early in the year and many species have more than one litter; young are brought forth in a little round nest of shredded vegetation concealed in leaves, rocks, logs, or a burrow. Their gestation averages eighteen to twenty-two days, and litters range from four to ten. The young are often independent within three weeks. Maximum age rarely exceeds two years. Many utter tiny, high-pitched squeaks. Young shrews have a pencil of hairs at the tip of the tail. These wear off with age, and the tail tip of an old shrew is nearly naked.

It is difficult to distinguish the individual species of shrews.

## LONG-TAILED SHREWS

Genus *Sorex*

Shrews of the Genus *Sorex* are widespread in North America, occurring discontinuously from the Bering Strait to Guatemala. These shrews are also widely distributed in northern Eurasia.

### MASKED SHREW

*Sorex cinereus*                                                    **52:14**

#### Description
Size: head and body, 2½ in. (6.4 cm); tail, 1¼ in. (3.2 cm); weight, $\frac{1}{10}$–$\frac{1}{5}$ oz. (2.8–5.6 g). Grayish-brown above, paler below; tail dark above, buffy below. Pelage in winter darker above than in

summer, paler below. Five unicuspids on each side of upper jaw, third and fourth slightly smaller than first and second.

**Habitat**
Moist areas from salt marshes to alpine meadows; may often be found in drier areas.

**Habits**
Can eat 3 or 4 times its own weight in 24 hours.

**Range**
Most widely distributed in the n. part of the E., s. to Ill. and the Appalachian Mts.

**Other name**
Cinereus or Common Shrew.

## SOUTHEASTERN SHREW
*Sorex longirostris*                                          *Fig. 16*

**Description**
Size: head and body, 2½ in. (6.4 cm); tail, 1½ in. (3.8 cm); weight, ⅛ oz. (3.5 g). Reddish-brown, paler below; tail relatively short, slightly bicolored.

**Habitat**
Cool, open fields and woods, grassy edges of ponds and bogs.

**Range**
In s. states, from Ill., southward, and from Coast, w. to Mo.

Long-tailed Shrew, p. 231

Southeastern Shrew

## WATER SHREW
*Sorex palustris*                                            *Fig. 17*

**Description**
Size: head and body, 3⅓–3⅖ in. (8.5–8.7 cm); tail, 2½–3 in. (6.4–7.6 cm); weight, ⅓–½ oz. (9.5–14.2 g). Largest long-tailed shrew, with stiff hairs on the sides of its large hind feet. Glossy black or blackish-gray above; gray, grayish-brown, or silver below; tail bicolored. Middle toes partly webbed.

Pygmy Shrew, p. 231

Smoky Shrew, p. 230

*Fig. 17*

Water Shrew

**Habitat**
Bogs, edges of streams, ponds, and lakes in northern and
mountainous areas.
**Habits**
Swims, dives, and can run on surface of water.
**Range**
Throughout e. Canada, s. through Appalachians and w. to e.
Mont. and Wyo.
**Other name**
Northern Water Shrew.

## SMOKY SHREW
*Sorex fumeus*                                              *Fig. 17*

**Description**
Size: head and body, 2½–3 in. (6.4–7.6 cm); tail, 1¾–2 in. (4.4–5.1
cm); weight ⅕–⅖ oz. (5.7–11.3 g); all 6 mammae inguinal. Dull
brown, grayish in winter, with bicolored tail and pale feet.
**Habitat**
Damp, shady forests, especially in deep moss, leaf litter, and
rotting logs.
**Habits**
Feeds primarily on earthworms and insects; seems to experience
more fluctuations in numbers than do most shrews.
**Range**
Throughout e. Canada, s. through Appalachians and w. into cen.
Midwest.

## ARCTIC SHREW
*Sorex arcticus*                                              **52:13**

**Description**
Size: head and body, 3 in. (7.6 cm); tail, 1¼–1⅔ in. (3.2–4.1 cm);
weight, ¼ oz. (7.1 g). Brightly tricolored; back dark brown, sides
light brown, underparts pale; brighter in winter.
**Habitat**
Spruce and tamarack swamps in southern parts of range and
tundra farther north.
**Range**
Throughout ne. U.S. and Canada.

## GASPÉ SHREW
*Sorex gaspensis*

**Description**
Size: head and body, 2 in. (5.1 cm); tail, 2 in. (5.1 cm). Similar to
Long-tailed, but smaller.
**Habitat**
Moist streamsides.
**Range**
Shickshock Mts. of Gaspé Peninsula, Que., and nearby Mt.
Carleton, N.B.

## LONG-TAILED SHREW
*Sorex dispar*                                                    *Fig. 16*

**Description**
Size: head and body, 2¾ in. (7 cm); tail, 2–2½ in. (5.1–6.4 cm);
weight, ⅕ oz. (5.7 g). Tail proportionally long for the genus. In
summer, dark gray above, only slightly paler below; in winter,
slate-gray above and below.
**Habitat**
Damp, rocky situations in coniferous forests.
**Range**
Throughout Appalachians, from s. Que. to e. Tenn.
**Other name**
Rock Shrew.

# OTHER SHREWS
Genera *Microsorex, Blarina, Cryptotis,* and *Notiosorex*

## PYGMY SHREW
*Microsorex hoyi*                                                 *Fig. 17*

**Description**
Size: head and body, 2–2½ in. (5.1–6.4 cm); tail, 1–1⅖ in. (2.5–
3.6 cm); weight, $\frac{1}{12}$ oz. (2.4 g); 32 teeth, 3133/1113. Smallest
shrew in the East. Gray or grayish-brown above, paler below; 5
unicuspids, but only 3 are easily visible in lateral view.
**Habitat**
Wet and dry woods and adjacent grass clearings.
**Remarks**
Regarded as America's smallest living mammal, 6 of these tiny
shrews do not equal ½ ounce (14.2 g).
**Range**
Throughout most of e. Canada, southward into e. U.S., to Pa.
and Ohio, w. to Wis. and e. N. and S. Dakota.

## SHORT-TAILED SHREW
*Blarina brevicauda*                                             **52:11**

**Description**
Size: head and body, 3–4 in. (7.6–10.2 cm); tail, ¾–1 in. (1.9–2.5
cm); weight, ⅖–1 oz. (11.3 –28.4 g). Slate-gray; eyes barely
apparent; ears small and hidden in pelage; short tail; stocky build;
muzzle not as pointed as long-tailed shrews; skull ridged, other
shrews have smoother skulls. Teeth dark chestnut; first and second
unicuspids larger than third and fourth.
**Habitat**
Forests and moist grassy areas having copious vegetation and litter.
**Habits**
Makes flattened runways, 1 x ¾ in. (2.5 x 1.9 cm), through leaf
mold.
**Remarks**
One of the few poisonous mammals; the saliva in its bite can
poison and paralyze its prey, and it is painful but not otherwise
dangerous to humans.
**Range**
Throughout entire ne. U.S. and adjacent areas of s. Canada.

## SOUTHERN SHORT-TAILED SHREW
*Blarina carolinensis*

**Description**
Size: head and body, 2½–3½ in. (6.4 –8.9 cm); tail, ½–¾ in. (1.3–1.9 cm); weight, ¾ oz. (21.3 g). Like Short-tailed except smaller. The short-tailed shrews of the Dismal Swamp, Virginia, may be of a separate species.
**Habitat and Habits**
Like those of Short-tailed.
**Range**
S. of the range of Short-tailed, throughout se. U.S.

## LEAST SHREW
*Cryptotis parva*                                    **52:12**

**Description**
Size: head and body, 2⅕–2½ in. (5.6–6.4 cm); tail, ½–¾ in. (1.3–1.9 cm); weight, ⅐–⅛ oz. (4.1–5.7 g); 30 teeth, 3123/1113, dark-tipped. Small, short-tailed. Ashy below; young are dark slaty-gray.
**Habitat**
Brushy and grassy situations, dried marshes, damp woods.
**Habits**
In captivity, when ample food is available, are docile and nest in groups; also known in wild; have been observed to hunt in packs and jointly attack animals many times their own size.
**Range**
Throughout cen. and s. parts of e. U.S., n. from N.Y. and S.Dak.

## DESERT SHREW
*Notiosorex crawfordi*

**Description**
Size: head and body, 2½–2¾ in. (6.4–7 cm); tail, 1–1½ in. (2.5–3.8 cm), weight, ⅛–¼ oz. (4.7–7.1 g); 28 teeth, 3113/2013. Pale ashy. Three unicuspids: only shrew having less than 30 teeth.
**Habitat**
Desert areas and short grass plains.
**Habits**
Poorly known; often found in association with wood-rat dens or apiaries.
**Range**
In e. Okla., e. Tex., and w. Ark.

# MOLES
Family Talpidae

Moles are small, burrowing animals with dark, soft, velvety fur that rubs both ways. They weigh one and one-half to four ounces (43–115 g) and have small eyes and external ears. Their outwardly turned forefeet make them strong, powerful diggers. They can utter a high-pitched squeak. Moles occupy all humid habitats from sea level to above timberline and are active at all hours and seasons; they do not hibernate. The tunnels of the Eastern and Hairy-tailed Moles make ridges on top of the ground.

The mole diet is largely earthworms and other invertebrates, but may include some plant material. The nest is a six-inch (15.2 cm) agglomeration of dry vegetation in a sheltered or underground earth mound. The female brings forth her two to six naked young

from April to June. Moles check wasteful runoff by churning up
the forest or meadow floor, thus helping rain soak into the ground.
The family dates back to the Eocene.

## HAIRY-TAILED MOLE
*Parascalops breweri*                                            **52:15**

**Description**
Size: head and body, 4¼ in. (10.8 cm); tail, 1-1½ in. (2.5-3.8 cm);
44 teeth, 3143/3143. Fur blackish, sometimes yellowish on belly or
snout. Nose naked, pointed; forefeet equally broad as long.
Distinguished from other moles in East by its hairy tail.
**Habitat**
Well-drained soil with wooded cover, lawns, golf courses, to 3000
ft. (915 m).
**Habits**
Ridges less conspicuous than those of Eastern Mole; also makes
low mounds; reputed to be capable of eating 3 times its own weight
in earthworms in 24 hours.
**Range**
New England, throughout Appalachians, and adjacent parts of
Canada.

## EASTERN MOLE
*Scalopus aquaticus*                                            **52:16**

**Description**
Size: head and body, 4¼-6½ in. (10.8-16.5 cm); tail, 1-1½ in.
(2.5-3.8 cm); 36 teeth, 3133/2033. Most widespread and
commonest eastern mole. Color ranges from dark gray-brown to
pale whitish. Nose naked; nostrils open upward; forefeet broader
than long, and much larger than hind feet; tail naked.
**Habitat**
Well-drained soil in open areas.
**Habits**
Makes circular mounds; Pocket gophers' mounds are crescent-
shaped. Most active in spring and fall and after summer rains.
**Range**
Entire e. U.S., from coast to 100th meridian.

## STAR-NOSED MOLE
*Condylura cristata*                                            **52:17**

**Description**
Size: head and body, 4¼-5 in. (10.8-12.7 cm); tail, 3-3⅓ in. (7.6-
8.5 cm); 44 teeth, 3143/3143. A "star nose," or ring of 22 fleshy
pink tentacles at end of nose which move as the animal searches for
food. Fur dark brown or black. Eyes visible; forefeet as long as
broad. Tail scaly with hairs thickest in middle, narrower near
body; thicker in winter.
**Habitat**
Moist ground near water.
**Habits**
Makes mounds in muck 1 ft. (0.3 m) wide, can swim and dive.
**Range**
In ne. U.S. and se. Canada.

# Bats

Order Chiroptera

Bats are the only mammals that fly. They do this by means of thin membranes (double-layered extensions of the body skin) that connect all fingers with the arm, sides, legs, and tail. The middle finger usually is the longest; the thumb is free. The sexes generally look alike, but young frequently are darker in color.

Bats have certain peculiar structures. The interfemoral membrane joins leg and tail. The calcar is a cartilage that extends from the foot along the outer edge of this membrane and acts as a brace. If some membrane extends beyond the calcar, the calcar is said to be keeled. The tragus is a leaflike formation inside the ear.

Bats utter high-pitched squeaks, sometimes fifty or more per second, which are inaudible to humans. Aided in part by the tragus, they hear the echoes of these sounds as they bounce back off insects and nearby obstacles; thus, bats can find their prey in the dark and avoid hitting obstructions. Bats are crepuscular (twilight-flying) and nocturnal. They roost by day, hanging upside down by their feet in caves, mine shafts, rocky crevices, hollow trees, thick foliage, or buildings, or behind loose bark or shutters. Some species are solitary; others are colonial. In winter bats either hibernate, often in colonies or caves, or, like birds, migrate south. Some species both migrate and hibernate.

The diet of bats in eastern North America is exclusively insects, which they capture on the wing, either in their mouths or in the "dragnet" of their tail membrane when cupped and used as a scoop.

In the females of some hibernating species the sperm or embryo can lie dormant for several months; thus an early fall mating may give rise to a late spring young.

Females often assemble in "maternity roosts" away from the males before giving birth to the young, usually one or two, which are born from May to July. Young babies often cling to their mother when she flies; later they hang by themselves at their roost until they are able to fly, which is usually in five to six weeks.

Bats have few natural enemies. When not hibernating, bats seem unable to endure long fasts; protracted cold and windy, rainy weather, which keeps insects from flying, may result in some mortality. Bats are complementary to birds in providing a natural check on insects. Birds feed on day-flying insects, bats on nightfliers.

Bats should be encouraged, and not feared, because they are valuable insect eaters.

# COMMON OR PLAIN-NOSED BATS

Family Vespertilionidae

In this family, the tail protrudes only slightly beyond the edge of the interfemoral membrane. There are two mammae, except in the Red, Yellow, and Hoary Bats, which have four. In most, the interfemoral membrane is not furred or only slightly so. The number of teeth ranges from thirty to thirty-eight. In winter, vespertilionids either hibernate or migrate southward. The family dates back to the middle Eocene. In the Genus *Myotis* the thirty-eight teeth are all 2133/3133. There is usually one offspring born in late spring or early summer, and it weans in about six weeks.

## LITTLE BROWN MYOTIS
*Myotis lucifugus*                                              **49:5**

**Description**
Size: forearm, 1½ in. (3.8 cm). Small. Dark brownish above, buffy-gray below; calcar unkeeled.

**Habitat**
Hangs in caves, hollow trees, buldings, under loose bark or shingles, and behind shutters.

**Habits**
Colonial; flits on rounded wings over fields, forests, wood edges or water; flight is erratic, feeble; sometimes flies by day in forest clearings; hibernates.

**Range**
Throughout most of E., from tree limit s. to cen. Ga.

**Other name**
Little Brown Bat.

## SOUTHEASTERN MYOTIS
*Myotis austroriparius*

**Description**
Size: forearm, 1½ in. (3.8 cm). Small, fur thick. Medium orange-brown to buff-brown above, dull buff below.

**Habitat**
Principally caves and old mines.

**Range**
In se. U.S., s. from Ind. and Ill. to Gulf Coast.

## GRAY MYOTIS
*Myotis grisescens*

**Description**
Size: forearm, 1¾ in. (4.4 cm). Medium-sized, fur velvety. Color is russet smoky-brown. Hairs not banded from base to tip, same color throughout; wing membrane attached on ankle, not to base of toes, as in other small eastern bats.

**Habitat**
Roosts in caves, sometimes sewers or buildings.

**Habits**
Colonial.

**Range**
In se. U.S., from Ky., w. to Mo. and Kans., s. through Ala.

## CAVE MYOTIS
*Myotis velifer*

**Description**
Size: forearm, 1¾ in. (4.4 cm). Medium-sized. Dull brown above, paler below; base of hairs dark. Hair has "woolly" appearance; wing membrane starts at base of toes.

**Habitat**
Roosts in caves and, in summer, buildings.

**Habits**
Colonial.

**Range**
In e. range, se. Kans.

## KEEN'S MYOTIS
*Myotis keenii*                                                  *Fig. 18*

**Description**
Size: forearm, 1½ in. (3.8 cm). Small. Fur dark yellowish-brown.
Ears, if laid forward, extend slightly beyond end of nose.
**Habitat**
Wooded regions.
**Habits**
Hangs in caves, trees, buildings; flight direct, strong.
**Range**
Throughout e. U.S., from s. Canada to n. Ala.

*Fig. 18*

Keen's Myotis

## INDIANA MYOTIS
*Myotis sodalis*

**Description**
Size: forearm, 1½ in. (3.8 cm). Small, fur glossy. Hairs 3–banded
from base to tip, giving "purple" hue to pelage.
**Similarities**
Little Brown is similar but calcar not keeled and feet are larger.
**Habitat**
Limestone caves and nearby areas; rare.
**Range**
In n.-cen. U.S. from New England s. through w. N.C. to se. Ala.
and w. to s. Mich. and s. Wis., s. to Mo.
**Other name**
Social Myotis.

## SMALL-FOOTED MYOTIS
*Myotis leibii*                                                    **49:3**

**Description**
Size: forearm, 1½ in. (3.8 cm). Smallest myotis; fur long, silky.
Color is golden brown in contrast to blackish ears and membranes;
blackish mask across face.
**Habitat**
Caves, especially in hemlock forests in mountains or hilly country,
particularly in rocky areas.
**Range**
From w. Maine and w. N.C., westward; also in extreme s.
Canada.
**Other name**
Sometimes called *Myotis subulatus*.

## SILVER-HAIRED BAT
*Lasionycteris noctivagans*                                    **49:6**

**Description**
Size: forearm, 1⅔ in. (4.2 cm); 36 teeth, 2123/3133. Medium-
sized. Brownish to blackish front frosted with silvery-white;
females somewhat paler than males. Ears, membrane black; tragus
blunt; basal half of tail membrane furred.
**Similarities**
Hoary Bat is larger and has buffy throat.
**Habitat**
Lakes, streams in wooded regions, up to evergreens in mountains.
**Habits**
Hangs in hollow trees, under loose bark, amid thick leaves, in
buildings. Flight distinctive, fluttery with short sideways turns,
sallies, and glides. In summer, males solitary, females gregarious;
in winter, many presumably migrate south, while some hibernate
in buildings.
**Range**
Throughout most of e. U.S., from Hudson Bay, s. to Ga. and La.

## EASTERN PIPISTRELLE
*Pipistrellus subflavus*                                    **49:8**

**Description**
Size: forearm, 1⅛ in. (3 cm); 34 teeth, 2123/3123. Smallest
eastern bat. Yellowish-brown to pale brownish, somewhat grizzled;
ears and forearms pinkish-brown contrasting with blackish wing
membranes. Fur tricolored; darkish brown at base, then lighter
yellowish-brown with tips darker; can be seen by blowing on fur to
separate the hairs. Thumb large, ⅛ length of forearm; tragus
blunt.
**Habitat**
Arid lowlands to evergreen mountain forests; woods, canyons, cliffs,
rocky hillsides; waterways; buildings.
**Habits**
Comes out early, often before sundown; flight erratic, weak,
mothlike, often undulating. Hibernates in caves, never in clusters.
**Reproduction**
Females congregate as do those of other bat species to bring forth
single young in late spring; mating takes place in fall and
occasionally in winter.
**Range**
Throughout much of ne. U.S., from s. Canada to n. Va. and w. to
Nebr. and Kans.

## BIG BROWN BAT
*Eptesicus fuscus*                                    **49:2**

**Description**
Size: forearm, 2 in. (5.1 cm); wingspread, 12 in. (30.5 cm); 32
teeth, 2113/3123. Largest brown bat. Uniformly dark brown above
in the East to pale brown in the West, paler below; ears,
membranes black. Tragus blunt.
**Habitat**
Varied, cities to wilderness, sea level to mountains; caves, mines,
crevices, buildings.
**Habits**
Flight strong with sudden, frequent changes in direction; often seen
about street lights; enters hibernation late, emerges early; may fly

on mild, sunny middays in winter. Commonest large bat about buildings, in which it often hibernates.

**Voice**
Like escaping steam; also a click.

**Reproduction**
Mates in fall; single young (East) or twins (West) born in late May or June.

**Age**
To 9 years.

**Range**
S. Canada, s. throughout U.S. to S.A.

## RED BAT
*Lasiurus borealis*                                        **49:10**

**Description**
Size: forearm, 1⅔ in. (4.1 cm); 32 teeth, 1123/3123; mammae, 4. Medium-sized; only reddish bat. Male is light reddish-orange; female much duller and with frosted appearance. Fur soft, fluffy; ears short, tragus blunt; wings narrow; tail membrane fully furred above.

**Habitat**
Forested regions.

**Habits**
Roosts in vegetation, often low; flies early, often high and descending in spirals. Migrates April–May, September–late November, sometimes flying in numbers like birds. Common; migrates in winter.

**Reproduction**
Mates in flight in August; 1–4 young, born late May–mid-June.

**Range**
Throughout s. Canada, s. through e. U.S. to Gulf Coast.

## SEMINOLE BAT
*Lasiurus seminolus*                                       **49:11**

**Description**
Size: forearm, 1½ in. (3.8 cm); 32 teeth, 1123/3123. Resembles Red Bat, but color darker, rich mahogany-brown, lightly frosted with grayish-white.

**Habitat**
Forested regions.

**Habits**
Like Red Bat.

**Reproduction**
Mates in flight in August; 1–4 young, born late May–mid-June.

**Range**
Along Atlantic Coast states, from s. N.Y. to cen. Fla. and Gulf Coast; w. to cen. Pa., n. Ga., and e. Tex.

## HOARY BAT
*Lasiurus cinereus*                                        **49:4**

**Description**
Size: forearm, 2+ in. (5.1+ cm); wingspread, 14 in. (35.6 cm) or more; 32 teeth, 1123/3123; mammae, 4. Largest eastern bat, recognized by large size, swift flight, and narrow, pointed wings. Fur frosted, general appearance grayish-white; hairs are dark brown or blackish at base, followed by band of golden yellow, band

of dark brown or blackish subterminally, and a white tip. Tragus
blunt; tail membrane fully furred above.
**Similarities**
Silver-haired is colored similarly but lacks buffy throat.
**Habitat**
Forested regions.
**Habits**
Hangs alone in trees, bushes; emerges late in evening; frequents
meadows, watercourses; flight direct, purposeful; seldom seen.
**Reproduction**
1–2 young (usually twins) born in late May or June, weigh nearly
⅙ ounce (5.0 g) each at birth; carried by female in flight until old
enough to be left hanging in tree when she forages.
**Range**
Throughout s. Canada and e. U.S. in warm months; migrates
southward in winter.

## NORTHERN YELLOW BAT
*Lasiurus intermedius*                                    **49:9**

**Description**
Size: forearm, 1⅞ in. (4.8 cm); 30 teeth, 1113/3123; mammae, 4.
Yellowish to yellowish-brown; tail membrane fully furred.
**Similarities**
Red and Hoary are similar but Yellow is larger than former,
somewhat smaller than latter, and differs in color and absence of
small upper premolar.
**Habitat**
Forested regions.
**Habits**
Much the same as other members of genus.
**Reproduction**
1–2 young (usually twins) born in late May or June.
**Range**
Principally in se. U.S. and Mexico, but extends from N.J.,
southward through coastal states.

## EVENING BAT
*Nycticeius humeralis*                                    **49:1**

**Description**
Size: forearm, 1½ in. (3.8 cm); 32 teeth, 1123/3123. Medium-
sized. Fur sparse; dark brown above, washed with buff below; ears
and membranes black. Ears thick; tragus curved, blunt.
**Similarities**
Little Brown Myotis is confusingly similar, but Evening is slightly
larger and has fewer teeth.
**Habitat**
Buildings and tree hollows.
**Habits**
Hangs in hollow trees, under bark, in buildings; flight steady,
straight, slow; often flies high to travel, low to feed; migratory east
of 100th meridian.
**Reproduction**
Usually 2 young, born May–July.
**Range**
Throughout s. and cen. parts of e. U.S., from s. Canada s. to Fla.
and w. to Iowa and Nebr.

## TOWNSEND'S BIG-EARED BAT

*Plecotus townsendii* **49:7**

**Description**
Size: forearm, 1¾ in. (4.4 cm); 36 teeth, 2123/3133. Medium-sized. Fur long, woolly. Warm brown to cinnamon above, pinkish-buff to pale brownish below; 2 conspicuous lumps on nose. Tragus pointed; ears extremely long, joined at base.
**Habitat**
Limestone caves, forested regions.
**Habits**
Hangs, with long ears spiraled alongside neck like a ram's horn, in hollow trees, mine shafts, buildings, and semidark areas of dry caves; emerges after dark; flight agile, swift; can hover; wary. Hibernates.
**Reproduction**
Single young born in spring and early summer; able to fly at 3 weeks of age.
**Range**
From w. Va. and W.Va., westward in band through Ky., Ind., s. Ill., to Mo. and s. Kans.; also in Black Hills of S.Dak. and adjacent areas.
**Other name**
Lump-nosed Bat.

## RAFINESQUE'S BIG-EARED BAT

*Plecotus rafinesquii* *Fig. 19*

**Description**
Resembles Townsend's Big-eared Bat but tips of ventral hairs whitish instead of brownish.
**Habits**
Much the same as for Townsend's.
**Range**
In se. U.S., from cen. Va. and W. Va. s. to Fla. and w. to Ind. and Mo.

*Fig. 19*

Rafinesque's Big-eared Bat

# FREE-TAILED BATS

Family Molossidae

Bats in this family have tails that extend beyond the interfemoral membrane, hair that is short, ears that are blunt, and tragi that are very small. The family dates back to the late Oligocene.

## BRAZILIAN FREE-TAILED BAT

*Tadarida brasiliensis* *Fig. 20*

**Description**
Size: forearm, 1⅘ in. (4.6 cm); 32 teeth, 1123/3123, upper incisors converge at tips. Medium sized. Fur is dark brown, short, not glossy. Upper lip grooved; ears not united at base and, when laid forward, do not extend beyond end of nose; wings long, narrow.

**Habitat**
Caves.
**Habits**
Hangs in caves, crevices, buildings; highly colonial. Flies over open areas; take-off from cave mouth is straight. Flight rapid with abrupt changes; perhaps the most rapid flier among bats. Migrates southward from range.
**Range**
In se. U.S., w. along Gulf Coast and to e. Kans. and se. Nebr., with some n. to Ohio.

Fig. 20

Brazilian Free-tailed Bat

# BIG FREE-TAILED BAT
*Tadarida macrotis*

**Description**
Size: forearm, 2½ in. (6.4 cm); 30 teeth 1123/2123. Large; similar to Brazilian Free-tailed, but larger, base of hairs whitish; ears united at base, and, if laid forward, extend well beyond end of nose. Upper incisors do not converge at tips.
**Habitat**
Caves, crevices in cliffs, buildings.
**Habits**
Nocturnal, colonial.
**Range**
In e. range, Iowa and e. Kans.

# Edentates
Order Edentata

# ARMADILLOS
Family Dasypodidae

## NINE-BANDED ARMADILLO                          **51:8**
*Dasypus novemcinctus*                            Fig. 15

**Description**
Size: head and body, 15–17 in. (38.1–43.2 cm); tail, to 16 in. (40.6 cm); weight (adults), to 17 lb. (7.7 kg); 28–36 teeth, 007-9/007-9. Teeth degenerate, premolars indistinguishable from molars. Upper parts tan or yellowish. Covered with horny armor, with 9 flexible bands in the center. Snout long; feet strong, well-clawed, with 4 toes in front, 5 behind; upperparts and ears nearly naked.

**Habitat**
Brush, open woodlands.
**Habits**
Most active by night; a burrower.
**Voice**
Grunt.
**Food**
Insects, other small invertebrates, some vegetable matter.
**Reproduction**
Young, usually 4, identical quadruplets, born March–April in a burrow.
**Range**
From Kans., s. to Tex., extending its range in some se. states where it has been introduced.

# Hares, Rabbits, and Pikas
Order Lagomorpha

# HARES AND RABBITS
Family Leporidae

Hares and rabbits are distinguished by their soft fur, long legs and ears, and short tails, usually white below. The larger species are called hares or jackrabbits, the smaller ones rabbits or cottontails. The classical distinction between hares and rabbits is that young of the former are precocial, whereas those of the latter are altricial. The fur is generally brownish or grayish above, sometimes white in the north or in winter, and paler or white below. They have twenty-eight teeth, formula 2033/1023. One pair of upper incisors is directly behind the other.

These mammals occupy all habitats from desert to moist forest, from sea level to beyond tree line, and from the northern limit of land southward to the Gulf. They do not hibernate. Hares generally like open country; rabbits prefer shrubby cover. Both run rapidly and jump well with their powerful hind legs. They rest in their "forms" or protected hiding places, read the wind with their noses, and thump on the ground with their hind feet to warn others of danger. All are timid, inoffensive animals that rely for safety on hiding or flight.

Their diet is vegetarian and includes a wide variety of plant food, from grass and garden greens to buds, twigs, bark, and leaves of trees. By breeding early and often they withstand the heavy tolls levied by fox and owl, gun and auto. Many species have periodic fluctuations in population numbers. The family dates back to the Eocene and has endured well.

# COTTONTAILS
Genus *Sylvilagus*

Cottontails are smaller and have shorter ears and hind legs than hares; they usually seek safety by hiding. After a gestation period of about twenty-eight days, usually four or five naked, blind, and helpless young are born in a fur-lined nest on the ground; they are weaned in less than three weeks. Females often breed after nine months of age and produce several litters a year.

## EASTERN COTTONTAIL
*Sylvilagus floridanus*                                       **50:12, 51:1**

**Description**
Size: head and body, 14–17 in. (35.6–43.2 cm); ear, to 3 in. (7.6 cm). Grayish-brown, underside of tail cotton-white; rusty nape. Ears short.
**Habitat**
From swampy wood to upland thickets and farmlands.
**Habits**
Most active at dawn and dusk; timid, restricts movements to area of only a few acres.
**Range**
Entire e. U.S., except n. New England.

## NEW ENGLAND COTTONTAIL
*Sylvilagus transitionalis*                                   **50:14, 51:1**

**Description**
Size: head and body, 15–17 in. (38.1–43.2 cm); ear, to 2½ in. (6.4 cm). Small. Summer: reddish-brown. Winter: reddish-gray, with only a pale rusty nape patch. Ear tips and patch between ears black, feet whitish.
**Similarities**
Eastern is larger, with longer ears. Snowshoe Hare has longer ears.
**Habitat**
Primarily forests.
**Habits**
More nocturnal than Eastern; prefers thicker cover.
**Range**
New England and through Appalachians to ne. Ala.

## SWAMP RABBIT
*Sylvilagus aquaticus*                                        **50:15**

**Description**
Size: head and body, 14–17 in. (35.6–43.2 cm); ear, to 4 in. (10.2 cm). Fur short, sleek. Rich brownish-gray, with small nape patch; hind feet rusty above. Tail slender; toes large, splayed, and with sharp nails.
**Similarities**
Eastern Cottontail is smaller; has longer, less sleek fur, with more pronounced rusty nape patch, whitish hind feet, and thicker tail.
**Habitat**
Moist bottomland.
**Habits**
Often walks rather than hops, stays very close to cover, swims well.
**Range**
In s.-cen. states, n. Ga., w. to e. Tex. and Gulf of Mexico; also from Ind. and Ill., w. to s. Kans.

## MARSH RABBIT
*Sylvilagus palustris*                                        *Fig. 21*

**Description**
Size: head and body, 14–16 in. (35.6–40.6 cm); ear, to 3 in. (7.6 cm). Dark brown. Small hind feet, ears, and tail. Differs from all other cottontails in the East by having brown or grayish underside of the tail, rather than white.

**Habitat**
Swamps and marshes, possibly not occurring above 500 ft. (152.4 m).
**Habits**
Nocturnal, secretive; often walks instead of hopping while feeding; uses own trails through dense marshy vegetation; swims readily.
**Range**
From the Dismal Swamp of Va., s. through Fla.

*Fig. 21*

Marsh Rabbit

# HARES AND JACKRABBITS
Genus *Lepus*

Hares and jackrabbits are larger and have longer ears and hind legs than do cottontails; they usually seek safety in flight. The female makes no nest. After a gestation period of thirty to forty-three days, three to eight young are born. They are furred, with eyes open, and are able to move about within a few minutes. There are usually two or more litters a year.

# SNOWSHOE HARE
*Lepus americanus*                                   **50:16, 51:3**
**Description**
Size: head and body, 13–18 in. (33–45.7 cm); ear, to 4 in. (10.2 cm). Smaller than jackrabbits, larger than cottontails. Summer: brown above, white below; tail dark above; feet brownish, not whitish as in cottontails. Winter: only tips of hairs white, bases dark; ear tips dark; tail all white. Large, long-furred feet permit travel on heavy snow, whence its name.
**Habitat**
Northern and mountainous swamps, forests and brush; laurel and rhododendron thickets.
**Habits**
Spends day in cover, feeds at night in open; uses own trails.
**Age**
5 years in wild, 8 in captivity.
**Remarks**
This is a classic North American example of a species with widely fluctuating numbers. During 1 peak year in Ontario a density of 3400 individuals to a square mile was reached. At the 10-year low, however, there may be only 1 or 2 per square mile.
**Range**
Entire Canada, s. through U.S. to New England, n. Mich., and n. Minn.; also s. throughout Appalachians.
**Other name**
Varying Hare.

# ARCTIC HARE
*Lepus arcticus*                                           *Fig. 22*

**Description**
Size: head and body, 17–24 in. (43.2–61 cm); ear, to 4½ in. (11.4 cm). Large. Winter: white; also white all year in Greenland, Ellesmere Island, and northern Baffin Island, hairs white to base. Winter: gray or brown farther south. Ear tips black, tail always all-white.

**Habitat**
Tundra and rocky slopes from tree limit north.

**Habits**
Usually in groups; often unwary; sometimes stands on hind feet only and hops on them. Migratory, moving south in winter and returning to north in summer.

**Range**
Arctic regions, especially around Hudson Bay and Baffin Is.; also in Greenland.

*Fig. 22*

winter        summer

Arctic Hare

European Hare, p. 246

# WHITE-TAILED JACKRABBIT
*Lepus townsendii*

**51:2**
*Fig. 23*

**Description**
Size: head and body, 18–22 in. (45.7–55.9 cm); ear, to 6 in. (15.2 cm). Large. Summer: brownish-gray. Winter: paler in south, almost white in north. Tail all-white below and above or with dusky or buffy middorsal stripe that does not extend onto back. Ear tips black.

**Similarities**
Snowshoe Hare is smaller, with shorter ears, and has preference for brush and thickets. Black-tailed Jackrabbit is smaller and with only partially white tail.

**Habitat**
Open country and exposed mountain slopes.

**Habits**
Most active at dawn and dusk, but may be seen infrequently during the day and may be active at any time of night.

**Range**
In n. parts of Kans., Iowa, and Minn., westward.

*Fig. 23*

summer        winter

White-tailed Jackrabbit

## BLACK-TAILED JACKRABBIT
*Lepus californicus*                                      **50:13, 51:2**

**Description**
Size: head and body, 17–21 in. (43.2–53.3 cm); ear, to 7 in. (17.8 cm). Large. Grayish-brown above, pure white below; ear tips, rump, and top of tail black.
**Habitat**
Open country.
**Habits**
Relies on speed and rapid changes of direction to elude predators.
**Range**
In e. range, from w. Mo. and e. Tex., w. to s. S.Dak.

## EUROPEAN HARE
*Lepus capensis*                                              *Fig. 22*

**Description**
Size: head and body, 25–27 in. (63.5–68.6 cm); ear, to 5 in. (12.7 cm). Large. Brownish-gray above, pure white below; tail black above, white below.
**Habitat**
Prefers areas where open fields and pastures are bordered by hedgerows and woodlots.
**Habits**
Does not make burrow or den but hides in dense vegetation by day.
**Age**
To 12 years in wild.
**Range**
In Great Lakes region, from s. Canada, s. to New England and e. Pa.; introduced from Europe.
**Other name**
Cape Hare.

# Rodents
Order Rodentia

Rodents are small- to medium-sized gnawing mammals distinguished by possessing two, and only two, incisors in each jaw, which continue to grow throughout life. They lack canine teeth and instead have a conspicuous space between the incisors and the grinding cheek teeth. Most have four toes on each front foot, five behind. Members of the only other order of gnawers, the Lagomorpha, which includes the hares and rabbits, have two pairs of incisors in the upper jaw, one small pair just behind the big front pair. Considered as a whole, rodents are one of the most successful orders of mammals. They are widely and abundantly distributed from the tropics to the tundra, from the sea beach to the alpine meadow.

# SQUIRRELS
Family Sciuridae

Members of this family include woodchucks; prairie dogs; ground, tree, and flying squirrels; and chipmunks. All have hairy, sometimes bushy, tails. The tooth formula is 1023/1013, except for the chipmunks and the Fox and Red Squirrels, in which the formula is 1013/1013. Incisors are yellow except for the Woodchuck, in which they are white.

These mammals live either on the ground or in trees, and they range throughout the East, north to the Arctic Ocean and up to the tops of high mountains. In winter, to the north and in high altitudes the ground squirrels hibernate. All but the flying squirrels are diurnal. All nest in the ground in burrows or under logs or rocks except the tree-nesting flying squirrels and tree squirrels. Most members of the family are capable of vocalization, such as a high-pitched whistle, barking, or chattering. The family dates back to the Miocene.

## EASTERN CHIPMUNK
*Tamias striatus*                                    **50:11, 54:8**

**Description**
Size: head and body, 5–6 in. (12.7–15.2 cm); tail, to 4 in. (10.2 cm). Small. Reddish-brown above, white below; stripes on back and sides end at reddish rump; stripes extend onto face. Has internal cheek pouches.

**Similarities**
Thirteen-lined Ground Squirrel has no face stripes.

**Habitat**
Hardwood forests and edges, rocks, stone walls, outbuildings.

**Habits**
Unsocial; a ground dweller that burrows and climbs, often sits upright on stumps when frightened or, in a hurry, runs with tail straight up or at raised angle; hibernates in North.

**Food**
Nuts, berries, seeds, insects, and small animals; stores food.

**Reproduction**
Usually 3–5 young, born in an underground chamber; usually 2 litters yearly.

**Age**
To 8 years.

**Range**
Throughout most of E. from s. Canada, s. to cen. Miss. and w. to e. Daks., Nebr., and Kans.

## LEAST CHIPMUNK
*Eutamias minimus*                                        **50:10**

**Description**
Size: head and body, 3½–4½ in. (8.9–11.4 cm); tail, to 4½ in. (11.4 cm). Yellowish, reddish, or grayish with stripes on face, back, and sides; stripes on back and sides extend to base of tail. Has internal cheek pouches.

**Habitat**
Rocks and openings in evergreen or mixed woodland; brush, edges of woods, lakes; outbuildings, sawmills.

**Habits, Food, Reproduction**
Similar to Eastern Chipmunk.

**Age**
To 8 years.

**Range**
In Canada, from s. of Hudson Bay to n. Mich. and Wis.; w. to w. Daks. and w. Nebr.

## WOODCHUCK
*Marmota monax*                    **50:6, 54:2**

**Description**
Size: head and body, 16–20 in. (40.6–50.8 cm); tail, to 7 in. (17.8 cm). Large. Above dark brown to yellowish-brown; grizzled; paler, sometimes rusty below. No white between eyes; sides of neck same color as back; feet black to dark brown. Heavy-bodied, ears small, legs and tail short.

**Habitat**
Dry woods and adjacent open spaces; brushy ravines, rocky slopes; fields, mowed parkway borders.

**Habits**
Most active in early morning or late afternoon; hibernates in winter, not always too deeply; can climb and swim; frequently hunted; main den entrance frequently has fresh dirt about it, especially in spring.

**Food**
Clover, alfalfa, other perennial plants.

**Reproduction**
Grass-lined nest in side tunnel; young, usually 4, born naked, blind, and weighing about 1 ounce.

**Age**
To 5 years.

**Remarks**
This species has been made famous by folklore relating to Ground Hog Day, February 2 each year.

**Range**
In most of e. U.S., from Hudson Bay to n. Ala. and w. to edge of prairies.

# GROUND SQUIRRELS
Genus *Spermophilus*

Ground squirrels have unstriped faces and internal cheek pouches. They live near burrows, which they dig; in winter they hibernate. Their food includes seeds, plants, and insects. The young, as many as thirteen in some species, are born blind and naked, but they become independent within a month. By churning up the soil, ground squirrels check run-off, but they also damage crops and pastures.

## RICHARDSON'S GROUND SQUIRREL
*Spermophilus richardsonii*                    *Fig. 24*

**Description**
Size: head and body, 8–9½ in. (20.3–24.1 cm); tail, to 4½ in. (11.4 cm). Large. Dull; buffy gray, paler below; no spots or stripes. Tail light brown, bordered with buff or white.

**Similarities**
Franklin's is larger, darker, and has longer tail.

*Fig. 24*

Richardson's Ground Squirrel        Plains Pocket Gopher, p. 253

**Habitat**
Meadows, prairies, sagebrush.
**Habits**
Colonial; adults stay underground from July to late winter; young go below in September.
**Reproduction**
Usually 1 litter per year, normally consisting of 6–7 young that are born in mid-May.
**Range**
From Minn., w. to 100th meridian in Daks. and w. Nebr.
**Other name**
Picket Pin.

## THIRTEEN-LINED GROUND SQUIRREL
*Spermophilus tridecemlineatus*                              **50:8**

**Description**
Size: head and body, 4½–6½ in. (11.4–16.5 cm); tail, to 5¼ in. (13.3 cm). Only striped ground squirrel. Back with a series of alternating dark (brownish or blackish) and pale longitudinal stripes; a row of squared whitish spots in each dark dorsal stripe; stripes on sides less well defined than those on back.
**Habitat**
Prairies, pastures, brushy edges of woods, golf courses, roadsides.
**Habits**
Most active during warmer hours of the day; often seen scurrying across highways; runs with tail straight out behind. Burrow is a small, round hole without telltale dirt, hidden in vegetation.
**Reproduction**
May have 2 litters per year, the first in late April or May; 4–12 young per litter.
**Range**
From cen. Man., s. to Ohio and Mo., w. to 100th meridian.

## SPOTTED GROUND SQUIRREL
*Spermophilus spilosoma*                                     **50:9**

**Description**
Size: head and body, 5–6 in. (12.7–15.2 cm); tail, to 3½ in. (8.9 cm). Small; only spotted ground squirrel. Pale reddish-brown or grayish-brown with squarish light spots on back; white below. Tail not bushy.
**Habitat**
Semi-arid plains; prefers sandy soils.
**Reproduction**
2 litters per year, 3–8 young per litter.
**Range**
Plains of w.-cen. Nebr. and Kans.; adjacent areas in S.Dak., Wyo., Colo.

## FRANKLIN'S GROUND SQUIRREL
*Spermophilus franklinii*

**Description**
Size: head and body, 9–10 in. (22.9–25.4 cm); tail, to 6 in. (15.2 cm). Gray with brownish wash, only slightly lighter below; no stripes or spots.
**Similarities**
Richardson's is smaller, has shorter tail. Gray Squirrel has longer, bushier tail, longer ears, and sides not buffy.

**Habitat**
Tall-grass prairies, edges of fields, open woodland, pastures;
frequently in tall and dense vegetation.
**Habits**
Most active on sunny days; can climb trees; forms small colonies.
**Reproduction**
1 litter of 4–10 young, born in May or June.
**Range**
From Man., s. to w. Ind. and w. to w. Nebr.

## ROCK SQUIRREL
*Spermophilus variegatus*                                   **50:5**

**Description**
Size: head and body, 10–11 in. (25.4–27.9 cm); tail, to 10 in. (25.4
cm). Large. Varying shades of gray washed with brownish, giving
mottled effect; light gray or buffy below; sometimes almost black on
head, back, or entire upper parts. Tail bushy and about as long as
head and body.
**Habitat**
Rocky areas.
**Habits**
Often sits on top of boulder; climbs bushes, trees.
**Reproduction**
About 6 young, born in spring and summer.
**Age**
To 9 years.
**Range**
In e. range, only se. Colo.

# PRAIRIE DOGS
Genus *Cynomys*

## BLACK-TAILED PRAIRIE DOG
*Cynomys ludovicianus*                                      **50:7**

**Description**
Size: head and body, 11–13 in. (27.9–33 cm); tail, to 4 in. (10.2
cm). Yellowish above, whitish below; outer third of tail black.
Stout; short-legged; ears small. Has internal cheek pouches.
Females slightly smaller than males; young paler than adults.
**Habitat**
Short-grass prairie of Great Plains.
**Habits**
Highly colonial; sits erect by entrance to burrow, a 1–2-ft. (0.3–0.6
m)-high mound; hibernates in cold weather.
**Reproduction**
Usually 5 young, naked, blind, born in spring in grass-lined nest
inside tunnel of burrow.
**Age**
To 10 years.
**Range**
In w.-cen. parts of Daks., Nebr., and Kans.

# TREE SQUIRRELS
Genera *Sciurus, Tamiasciurus,* and *Glaucomys*

Tree squirrels have no internal cheek pouches. They are primarily arboreal, but are also seen on the ground. They do not hibernate. Their food, some of which they store, includes nuts, berries, fruits, seeds, buds, twigs, bark, eggs, fungi, and insects. They usually nest in tree cavities or leaf nests. Gestation takes forty to forty-five days. The one to seven young are born naked and blind; there are one to two litters a year.

## GRAY SQUIRREL
*Sciurus carolinensis*                                    **50:2, 54:11**

**Description**
Size: head and body, 8–10 in. (20.3–25.4 cm); tail, 8–10 in. (20.3–25.4 cm). Fur gray above, often washed with fulvous, whitish below; light eye-ring; in winter, no fulvous, back of ears white; long, bushy tail of same general tone as body; also has an all-black phase. More graceful than the larger Fox Squirrel; head not squarish.

**Habitat**
Usually found in heavily forested areas, especially bottomlands; also in suburbia and city parks.

**Habits**
Most active early morning and late afternoon; highly arboreal; more secretive than Fox Squirrel.

**Age**
To 15 years.

**Range**
Along s. border of Canada and throughout e. U.S. w. to e. parts of Daks., Nebr., and Kans.

## FOX SQUIRREL
*Sciurus niger*                                                **50:4**

**Description**
Size: head and body, 10–15 in. (25.4–38.1 cm); tail, 9–14 in. (22.9–35.6 cm). Color varies: in North, generally yellowish-rust above, pale yellow to orange below, tail with rusty or buffy border; in South, grayish or grizzled; dark head with white nose and ears; in mid-Atlantic states, an all-steel-gray phase occurs. Intermediate variations also exist; a black phase also known. Heavy-bodied, squarish face.

**Habitat**
Groves of oak, longleaf pine; borders of cypress swamps, thickets; in western part of range restricted to trees along streams, in woodlots, and in urban areas.

**Habits**
Walks or climbs without speed or grace of Gray; tolerates more open situations than Gray; active throughout day.

**Age**
To 10 years.

**Range**
Throughout entire e. U.S., except New England, N.Y., and n. parts of Wis., Minn., N.Dak.

# RED SQUIRREL
*Tamiasciurus hudsonicus*                                    **50:1**

**Description**
Size: head and body, 7–8 in. (17.8–20.3 cm); tail, 4–6 in. (10.2–
15.2 cm). Reddish, sometimes yellowish; white below; tawny
border to tail; black line along sides; in winter, paler above and
with ear tufts.
**Similarities**
Gray and Fox are larger.
**Habitat**
Evergreen forests, less common in hardwoods.
**Habits**
Leaves piles of pine-cone cuttings on rocks, logs; sometimes
migrates.
**Age**
To 9 years.
**Remarks**
Most commonly seen mammal of the Great North Woods.
**Range**
From tree limit n., s. throughout Appalachian region, and w. to
Ind. and Iowa.

# NORTHERN FLYING SQUIRREL
*Glaucomys sabrinus*                                         *Fig. 25*

**Description**
Size: head and body, 5½–6 in. (14–15.2 cm); tail, to 5½ in. (14
cm). Similar to Southern Flying Squirrel, but is slightly larger and
heavier. Belly hairs lead-colored at base, white only at tips.
**Habitat**
Mixed woods and conifers.
**Habits**
As Southern.
**Range**
From tree limit, s. to Appalachian region and s. Wis. and N.Dak.

*Fig. 25*

Northern Flying Squirrel

# SOUTHERN FLYING SQUIRREL
*Glaucomys volans*                                           **50:3**

**Description**
Size: head and body, 6 in. (15.2 cm); tail, 4 in. (10.2 cm). Small.
Fur thick, soft, glossy. Olive-brown above, white below; belly hairs
white to base. Eyes large, tail flattened.
**Habitat**
Woodlands with tall trees.
**Habits**
Nocturnal; gregarious; sleeps by day in hole in tree (or attic); can
glide 125 ft. (38.1 m) given a high enough takeoff point. Lands
with an audible thump, as on a camp roof.
**Age**
To 13 years.

**Remarks**
Flying squirrels have a fold of loose, furred skin along their sides from wrist to ankle. When extended this enables them to glide (not fly) from tree trunk to tree trunk.
**Range**
Throughout most of e. U.S., w. to e. Kans. and e. Nebr.

# POCKET GOPHERS
Family Geomyidae

Chunky, toothy, big-headed, these brownish, rat-sized burrowers are noted for their underslung jaws and external, fur-lined cheek pouches, which they use for carrying food to storage. There are twenty teeth, formula 1013/1013. The long-clawed, sturdy forefeet and the front teeth are adapted for fast and protracted digging. Eyes and ears are small; the mouth closes with the upper incisors outside. The tail is sensitive to touch, nearly naked, and shorter than the body.

Even where gophers are common and their mounds abundant, they themselves are rarely seen. When they are seen, it will probably be the front end only as the animal hurriedly peers from its burrow. Most of their lives are spent underground. Their mound is fan-shaped, the mole's is round. The hole beneath leads off at an angle, whereas the mole's smaller hole leads straight down.

Gophers eat roots, bulbs, and other vegetation. Extensive surface tunnels are used in foraging; a smaller, deeper system contains the nest, food-storage chambers, and toilet chamber. These mammals are unsocial. If they meet—except to mate—they fight. Gestation is presumed to be about twenty-eight days; the litter size is usually one to five, with one to three litters a year. Species are sometimes hard to identify when ranges overlap: if the front teeth have two distinct grooves, it is a Plains Pocket Gopher or Southeastern Pocket Gopher; if no grooves (or very faint), it is a Northern Pocket Gopher. The family dates back to the Oligocene.

## NORTHERN POCKET GOPHER
*Thomomys talpoides*

**Description**
Size: head and body, 5-6½ in. (12.7–16.5 cm); tail, to 3 in. (7.6 cm); mammae, usually 10. Males larger. Brownish, lightly washed with black or gray; nose black or brown; patches on back of ears black; 1 faint groove on front of each incisor; claws relatively short.
**Habitat**
Uplands and mountains with thin soil.
**Range**
From s. Canada, southward and from e. Daks., westward.

## PLAINS POCKET GOPHER
*Geomys bursarius*                                        *Fig. 24*

**Description**
Size: head and body, 5½-9 in. (14–22.9 cm); tail, to 4½ in. (11.4 cm); mammae, 6. Tawny in West to nearly black in Illinois; albinos and white-spotted individuals are known. Claws long; 2 grooves on incisors.

**Habitat**
Deep soil, frequently sandy, in treeless regions, hay and alfalfa
fields, pastures and edges of highways, and railroad rights-of-way.
**Range**
Throughout plains from cen. Ind. and Ill., w. to Daks., and from
Minn. and N.Dak., s. to Kans.

## SOUTHEASTERN POCKET GOPHER
*Geomys pinetis*                                              Fig. 26

**Description**
Similar to Plains Pocket Gopher. Three other closely related
species with very small ranges occur in the same general area and
can be distinguished only by specialists.
**Habitat**
Dry sandy soils.
**Range**
In s. Ga., n. and cen. Fl., and se. Ala.

*Fig. 26*                                    Southeastern Pocket Gopher

# POCKET MICE AND KANGAROO RATS
Family Heteromyidae

This family is found only west of the Mississippi. It consists of
small-eared rodents with hind limbs much larger than forelimbs.
These small, long-tailed, nocturnal, burrowing rodents have
external fur-lined cheek pouches which they use to carry food to
storage. They have twenty teeth, 1013/1013. Their habitat is arid
or semiarid plains and prairies. They make underground tunnel
systems with sleeping, nesting, and food-storage chambers and
inconspicuous surface holes that are plugged up during the day.

Their principal foods are seeds and greens, which they store in
underground chambers for future use. They are inactive in cold
weather. Populations are subject to periodic fluctuations. The one
to eight, usually four, young are born in spring or summer; there
are one or more litters a year. The family dates back to the
Oligocene.

# POCKET MICE
Genus *Perognathus*

Pocket mice are the smallest family members, with moderately long
and untufted tails. Their faces are unmarked; a buffy lateral stripe
separates the darker back from the white belly. These mice inhabit
sandy soil; they are poor jumpers. Their voice is a thin, high
squeak. The soles of the hind feet are naked.

## OLIVE-BACKED POCKET MOUSE
*Perognathus fasciatus*

**Description**
Size: head and body, 3 in. (7.6 cm); tail, 2½ in. (6.4 cm). Fur soft, silky. Olivaceous above; yellow spots behind ears; yellowish lateral line.
**Range**
From Canadian prairies, s. through Daks.

## PLAINS POCKET MOUSE
*Perognathus flavescens*

**Description**
Size: head and body, to 3¾ in. (9.5 cm); tail, to 2⅝ in. (6.6 cm). Back pale buff, variously washed with blackish; spot behind ear and lateral line clear buff; belly white.
**Range**
In plains, from Minn. and e. N.Dak., s. to nw. Mo. and Kans.

## HISPID POCKET MOUSE
*Perognathus hispidus*                                   **52:8**

**Description**
Size: head and body, 4½–5 in. (11.4–12.7 cm); tail, to 4½ in. (11.4 cm). Largest pocket mouse in East. Hair coarse. Back ochreous, mixed with blackish; sides only faintly paler than back. Lateral line distinct; uncrested tail equal to, or shorter than, head and body.
**Range**
From cen. N.Dak., S.Dak., Neb., and Kans., westward and southward through Tex.

# KANGAROO RATS
Genus *Dipodomys*

Kangaroo rats all have long tails tufted at the tip, very long hind legs, and distinct facial markings. They prefer arid or semiarid country and easily worked soil. The belly is white and the soles of the hind feet are moderately haired.

## ORD'S KANGAROO RAT
*Dipodomys ordii*                                   **52:5, 54:3**

**Description**
Size: head and body, 4–4½ in. (10.2–11.4 cm); tail, to 6 in. (15.2 cm). Fur silky; eyes large. Pale to bright orange-brown above, white below. Face with white crescent through eye and under ear; lower incisors rounded, not flat across front. White lateral stripe crosses thigh and extends out along side of long, tufted tail. Dark tail stripes wider than white ones. Hind legs long, strong; 5 toes on each hind foot; soles have short, stiff fur. Young darker than adults.
**Habitat**
Arid areas, sandy, soft, or hard soils.
**Habits**
Nocturnal; runs by leaping with its hind feet. Seldom seen above-ground in very cold (or very hot) weather.
**Range**
Extreme s.-cen. N.Dak. southward and westward.

# BEAVER
Family Castoridae

The beaver is the largest rodent in the United States of America and the only land mammal with a broad, flat tail. The body is thickset and compact, the legs short, ears small, hind feet large, toes webbed.

## BEAVER                                            51:5
*Castor canadensis*                                  *Fig. 27*

**Description**
Size: head and body, 25–30 in. (63.4–76.2 cm); tail, to 10 in. (25.4 cm); weight, to 60 lb. (27.2 kg); 20 teeth 1013/1013. Fur dense, brown, waterproof; head massive; incisors large, chestnut-colored; grinding teeth high-crowned; hind feet webbed, tail scaly.
**Habitat**
Streams and lakes.
**Habits**
Builds relatively watertight dam of sticks and mud across a stream, a cone-shaped house in a pond; trees gnawed a foot from the ground are beaver signs. Most active at dawn, dusk, and by night.
**Voice**
Various sounds made only within the lodge; outside, slaps water with its tail as a sign of warning.
**Food**
Bark of aspen, alder, birch, maple, willow, and other vegetation.
**Reproduction**
Young, called kits, usually 2–4, are born from April-July in the lodge.
**Range**
Throughout entire e. U.S., absent from the Gulf Coastal Plain.

*Fig. 27*

Porcupine, p. 269

Beaver

# NEW WORLD RATS AND MICE
Family Cricetidae

These small rodents have sixteen teeth, 1003/1003. The rear foot has five toes. Mice and rats generally have large eyes and ears, long tails, and four toes on the front foot. Voles and lemmings have small eyes and ears, short tails, and four or five toes on the front foot. All groups have one, sometimes two or more, litters a year of three to four young, usually in a nest of vegetation on the ground; none hibernate. The family dates back to the Oligocene.

## MARSH RICE RAT
*Oryzomys palustris*                                     **53:13**

**Description**
Size: head and body, 5 in. (12.7 cm); tail, to 7 in. (17.8 cm). Fur short, coarse. Above grayish-brown, underparts and feet whitish. Tail long, scaly, paler beneath. Two rows of tubercles on molars.
**Similarities**
Norway and Black Rats have 3 rows of tubercles on molars; tails are longer and not lighter below. Wood rats are larger and have smoother fur.
**Habitat**
Saltwater and freshwater marshes and meadows, and vicinity; upland clearings.
**Habits**
Nocturnal, social; swims, dives, and makes runways in salt meadows.
**Food**
Plants, seeds, crustaceans, snails.
**Range**
In se. U.S., from cen. N.J., s. to Fla. and from Va., w. to s. Ill. and s. Mo.

# HARVEST MICE
Genus *Reithrodontomys*

These are brownish mice with a longitudinal groove on each upper incisor, conspicuous ears, no external cheek pouches, and medium-sized, thinly haired tails. The young are darker than the adults. Preferred habitats are overgrown fields, marshes, and woods edges. They are nocturnal and good climbers. Their voice is a high ventriloquistic bugling. The diet consists of plant cuttings, seeds, and insects.

## PLAINS HARVEST MOUSE
*Reithrodontomys montanus*                               **52:10**

**Description**
Size: head and body, 2–3 in. (5.1–7.6 cm); tail to 2⅝ in. (6.6 cm). Fur pale gray above with a faint tawny cast, middle of back often darker; pale gray to whitish below; narrow blackish dorsal stripe on tail.
**Similarities**
Tail of Western usually more than 2½ in. (6.4 cm); that of Fulvous more than 3⅓ in. (8.5 cm).
**Habitat**
Principally an upland species, especially where grasses are relatively short and sparse.
**Range**
From e. Nebr. and e. Kans., westward.

## EASTERN HARVEST MOUSE
*Reithrodontomys humulis*                                **53:6**

**Description**
Size: head and body, 3 in. (7.6 cm); tail, to 2½ in. (6.4 cm). Fur rich brown, grayish below, often with pinkish hues; tail bicolored, feet light.
**Habitat**
Brier patches, roadside ditches, pasture, bogs.
**Range**
From Md., s. to Fla. and w. to s. Ohio and e. Tex.

## WESTERN HARVEST MOUSE
*Reithrodontomys megalotis*                    *Fig. 28*

**Description**
Size: head and body, 3 in. (7.6 cm); tail, 2⅓–3⅛ in. (5.8–8.1 cm).
Color varies geographically from pale gray to warm brown above,
often with darker stripe down back, less pronounced than in
Plains; buff to white below. Dorsal stripe on tail broad, blackish.
**Similarities**
Plains has thin-stripe tail, usually under 2½ in. (6.4 cm). Fulvous
has bright tawny sides and tail more than 3⅓ in. (8.5 cm).
**Habitat**
Grassy areas, also dense grasses, arid regions in riparian situations.
**Range**
From Canada, s. to Minn., w. Wis., Mo. and westward.

*Fig. 28*                    Western Harvest Mouse

---

## FULVOUS HARVEST MOUSE
*Reithrodontomys fulvescens*                    **53:5**

**Description**
Size: head and body, 3⅛ in. (8.1 cm); tail, 3⅓–4 in. (8.5–10.2 cm).
Largest harvest mouse in the East. Above grayish-brown, sides
bright orange-buff, below whitish, tail darker above.
**Habitat**
Grassy and brushy lowlands along streams.
**Range**
From se. Kans. and s. Mo., s. to the Gulf.

---

# WHITE-FOOTED MICE
Genus *Peromyscus*

These common mice have no grooves on their upper incisors. They
are usually brownish above, sometimes grayish, with white feet and
underparts. The base of the hairs is slaty. Tails are long and hairy.
Those mice that live in woodlands are dark; those that live in the
open are pale. The young are grayish, difficult to tell apart. Most
are terrestrial; all are nocturnal. Food consists of seeds, berries,
fruit, insects, carcasses.

---

## DEER MOUSE
*Peromyscus maniculatus*                    **52:3, 54:4**

**Description**
Size: head and body, 2⅘–4 in. (7.1–10.2 cm); tail, 2–5 in. (5.1–
12.7 cm). Color varies geographically from gray to brown to pale
yellowish and in between, sometimes with a dark, dorsal stripe; tail
sharply bicolored. Tufts in front of ears often whitish. Hind foot
relatively short in comparison with that of White-footed Mouse.
**Habitat**
Woods, prairies, dry uplands, sand beaches, farmland.

**Habits**
Ground-dwelling, but can climb.
**Range**
Areas e. of 100th meridian except coastal areas from R.I. southward and most of the s. coastal states.

## FLORIDA MOUSE
*Peromyscus floridanus*                                        53:3
**Description**
Size: head and body, 4½–5 in. (11.4–12.7 cm); tail, about 3½ in. (8.9 cm). Relatively large. Color brownish-gray.
**Habitat**
Sandy, upland ridges.
**Habits**
Often uses burrows of other animals.
**Range**
Most of peninsular Fla.

## OLDFIELD MOUSE
*Peromyscus polionotus*                                        53:7
**Description:**
Size: head and body, about 3½ in. (8.9 cm); tail, about 1½–2½ in. (3.8-6.4 cm). Small, often pale.
**Habitat**
Sand beaches and old fields.
**Habits**
Nocturnal; makes burrows in reasonably dry areas.
**Range**
In se. states, from S.C., s. to parts of Fla. and w. to Ala.

## COTTON MOUSE
*Peromyscus gossypinus*
**Description**
Size: head and body, 3½–4½ in. (8.9–11.4 cm); tail, 2¾–3½ in. (7-8.9 cm). Dark grayish-brown to tawny above.
**Similarities**
White-footed is smaller, with brighter whitish underparts.
**Habitat**
Lowland woods, swamps; invades camps, houses.
**Range**
From Dismal Swamp of Va., s. along coast, and w. to Ky., Ill., s. Mo.

## TEXAS MOUSE
*Peromyscus attwateri*                                        53:4
**Description**
Size: head and body, 3⅗–4⅛ in. (9.1-10.1 cm); tail, 3⅗–4⅖ in. (9.1-11.2 cm). Brown or grayish-brown above, tawny sides. Well-haired tail dark above, light below; tail usually longer than head and body.
**Habitat**
Brushy areas and rocky situations.
**Habits**
Nests under rocks, in crevices, or in piles of brush or sticks.
**Range**
In s. Mo. and se. Kans.

## WHITE-FOOTED MOUSE

*Peromyscus leucopus*                                    **52:1**

**Description**
Size: head and body, 3½–4⅛ in. (8.9–10.7 cm); tail, 2–4 in. (5.1–10.2 cm). Reddish-brown above, white below; occasionally brown or gray; often with darker dorsal stripe; upper lip dusky. Tail frequently not sharply bicolored, shorter than head and body, less hairy than in Deer Mouse.
**Similarities**
Cotton is darker, heavier, without white beneath.
**Habitat**
Mixed woods, hardwoods; thickets, streamsides.
**Age**
To 5 years.
**Range**
Throughout e. U.S. except the coastal plain from S.C. to La.

# OTHER MICE

Genera *Ochrotomys, Onychomys, Baiomys,* and *Sigmodon*

## GOLDEN MOUSE

*Ochrotomys nuttalli*                                    **53:9**

**Description**
Size: head and body, 3⅖–3⅘ in. (8.6–9.7 cm); tail, 3–3½ in. (7.6–8.9 cm). Thick, soft fur. Golden above, white below.
**Habitat**
Evergreen forests, hardwoods, thickets, swamps.
**Habits**
Largely arboreal; builds grapefruit-sized nests of Spanish moss, grasses, and other vegetation, up to 10 ft. (3 m) aboveground in shrubs, vines, or trees.
**Range**
In se. U.S., along coast from Va. to cen. Fla., and w. to Ky., s. Ill.

## NORTHERN GRASSHOPPER MOUSE

*Onychomys leucogaster*                             **52:7, 54:5**

**Description**
Size: head and body, 4–5 in. (10.2–12.7 cm); tail, to 2½ in. (6.4 cm). Two phases, with varying shades of gray or pinkish-cinnamon above, white below, giving bicolored pattern. Fur short, dense; body stocky.
**Habitat**
Prairies.
**Habits**
Nocturnal.
**Range**
From s. Man., s. to w. Iowa and e. Kans. and westward.

## NORTHERN PYGMY MOUSE

*Baiomys taylori*                                       **53:1**

**Description**
Size: head and body, 2–2½ in. (5.1–6.4 cm); tail, about 1½ in. (3.8 cm). A tiny gray mouse. Length of hind foot ⅝ in. (1.5 cm). Color

grayish, like a House Mouse or young *Peromyscus,* but has tail paler beneath than House Mouse and more delicate feet than young *Peromyscus.*

**Habitat**
Grassy areas.

**Habits**
Uses small runways in the grass, eats seeds.

**Range**
Mexican-U.S. border areas, well into s. Tex.

## HISPID COTTON RAT
*Sigmodon hispidus*                                    53:10

**Description**
Size: head and body, 5–8 in. (12.7–20.3 cm); tail, 3–5 in. (7.6–12.7 cm). Fur long, coarse. Grizzled brown and black above, whitish beneath; darker in the East, paler in the West; feet gray. Thinly haired tail dark above, light below, and shorter than head and body. Hind foot longer than in Prairie Vole or Southern Bog Lemming; molars somewhat rounded, not angular.

**Habitat**
Moist grassy areas, meadows, ditches; also dry places in areas with ample ground cover.

**Habits**
Active at all hours; strews little piles of cut grasses along its runways.

**Range**
Along Atlantic and Gulf Coasts, and w. to Tenn., Nebr., and Kans.

# WOODRATS
Genus *Neotoma*

These rats have conspicuous ears and eyes, soft fur, white feet, hairy tails, and flat molars. In comparison, Old World rats have smaller ears, dusky feet, scaly tails, and cusped molars. When alarmed, woodrats thump with their hind feet. Scats are frequently deposited in large piles. Their food is seeds, fruits, leaves, berries, cactus pulp, grass, and insects. These nocturnal rats collect unusual objects, such as cans, silver, belt buckles, and so forth at their nest sites and sometimes replace objects they take with other items. Other names include pack rat and trade rat.

## EASTERN WOODRAT
*Neotoma floridana*                                    53:11

**Description**
Size: head and body, 8–9 in. (20.3–22.9 cm); tail, 6–8 in. (15.2–20.3 cm). Above brown; underparts, throat, and chest white; midline of abdomen pure white; sides of belly white, with hairs gray at the roots; tail shorter than head and body, hairy but not bushy, dark above, light below. Norway and Black Rats have longer, scaly tails and their underparts are never white.

**Habitat**
Caves, cliffs, rocky outcrops, hedgerows; nests on ground in West, on ground or in trees in East.

**Range**
From N.Y., s. to Va. and w. to Nebr. and Kans.; also s. to Gulf Coast.

## SOUTHERN PLAINS WOODRAT
*Neotoma micropus*                    **53:12**

**Description**
Size: head and body, 7½–8½ in. (19–21.6 cm); tail, 5½–6½ in.
(14–16.5 cm). Above all-gray; below whitish; throat, chest, and feet
white; tail darker above than below.
**Habitat**
Dry plains, usually in areas where cactus abounds.
**Range**
In s. Kans.

# VOLES
Genera *Clethrionomys* and *Phenacomys*

## SOUTHERN RED-BACKED VOLE
*Clethrionomys gapperi*                **52:9, 55:6**

**Description**
Size: head and body, 3⅔–4⅔ in. (9.3–11.8 cm); tail, to 2 in. (5.1
cm). Three phases: red, gray, and black. Red has reddish back and
gray sides. Gray lacks reddish back and is hard to distinguish
externally from other voles. All phases are silvery below. Tail thin;
black above, gray below, lacking bristles except at tip.
**Habitat**
Northern forests, clearings, and swamps.
**Habits**
Terrestrial; a good climber; active at all hours.
**Voice**
High, musical squeak.
**Food**
Plants, seeds, nuts, berries, lichens.
**Range**
Throughout Canada and s. to Appalachians, S.Dak., and Iowa.
**Other name**
Boreal Red-backed Mouse.

## HEATHER VOLE
*Phenacomys intermedius*                **55:2**

**Description**
Size: head and body, 3½–4¾ in. (8.9–12.1 cm); tail, to 1⅔ in. (4.2
cm). Nose and (sometimes) face yellowish, tail bicolored.
**Similarities**
Rock Vole has longer tail. Red-backed and Meadow and Prairie
Voles lack yellow nose.
**Habitat**
Grassy areas in spruce forests; by streams.
**Habits**
Terrestrial; rare.
**Food**
Twigs and needles of conifers.
**Range**
Most of Canada, and s. to n. Minn.

# MEADOW MICE AND OTHER VOLES
Genus *Microtus*

These voles are best known by the popular name of "meadow mice." They have long, grayish-brown fur; short ears and tails; and small eyes. They have ungrooved teeth, whereas the bog lemmings have grooved teeth. Their tails usually are more than an inch long, and are not brightly colored. They live on the ground usually in grassy terrain, where they make runways and leave cut grass stems and deposit brown droppings. They are active at all hours and can swim and dive. In the winter they make round holes to the surface through the snow. Except for the Woodland Vole, their voice is a high-pitched squeak. They eat grass, roots, bark, and seeds.

## MEADOW VOLE
*Microtus pennsylvanicus*                    **52:4, 54:7, 55:4**

**Description**
Size: head and body, 3½–5 in. (8.9–12.7 cm); tail, 1⅖–2⅖ in. (3.6–6.6 cm). Commonest and most widespread vole. Grizzled fur varies from gray in the West to a rich brown in the East. Underparts vary from silvery to buff to dark gray; tail bicolored. Ears almost hidden in fur.

**Habits**
Population of 15 to an acre may increase to 250 in 2 years.

**Range**
Throughout Canada and n. U.S., from Hudson Bay, s. to s. Va. and n. Kans.

Note: The **BEACH VOLE,** *Microtus breweri* (**55:1**), is similar to the Meadow Vole, but larger, longer-tailed, and paler. It is the only vole found on Muskeget Island, Massachusetts, and is the sole mammal species peculiar to that state.

## ROCK VOLE
*Microtus chrotorrhinus*                    **55:5**

**Description**
Size: head and body, 4¾ in. (12.1 cm); tail, 2 in. (5.1 cm). Fur grayish-brown; nose yellow to reddish-orange, but color not always conspicuous; grayish below.

**Habitat**
Lives in small colonies in rocky places; rather uncommon.

**Range**
From coastal Labr., n. Minn. and s. to N.C.

**Other name**
Yellow-nosed Vole.

## PRAIRIE VOLE
*Microtus ochrogaster*                    **55:7**

**Description**
Size: head and body, 3½–5 in. (8.9–12.7 cm); tail, 1⅗ in. (4.1 cm); mammae, 6. Common vole of the prairies. Fur grayish to blackish-brown with yellowish grizzle, paler to the Northwest, darker to the South and East; tawny below; yellowish-rusty at base of short tail.

**Similarities**
Meadow usually has a longer tail, darker back, and silver-gray belly. Woodland has long, lax fur.

**Habitat**
Prairies, fencerows, fields.
**Habits**
Active at all hours, burrows where cover is scant.
**Range**
From W.Va., w. to 100th meridian and from s. Alta. to Kans.

## WOODLAND VOLE
*Microtus pinetorum*                                               **55:3**

**Description**
Size: head and body, 3–4 in. (7.6–10.2 cm); tail, ⅔–1 in. (1.7–2.5 cm); mammae, 4. Fur soft, thick; lacks guard hairs, has no set, can be rubbed forward or backward without ruffling. Above auburn, with buffy sides in Northeast; gray or buff below; ears and tail short. Not grizzled like the longer-tailed Meadow or Prairie Voles.
**Habitat**
Deciduous forests, orchards, gardens, fields.
**Habits**
Fossorial; tunnels through leaf mold, pushing up the mold into little ridges. Frequently causes damage to orchards by girdling trees beneath snow in winter.
**Voice**
*Chitter*'s and *chirr*'s; alarm is a thrushlike *cheer, cheer.*
**Range**
From Maine to n. Fla. and from Wis. to e. Kans.
**Other name**
Pine Vole

# WATER RATS
Genera *Neofiber* and *Ondatra*

## ROUND-TAILED MUSKRAT
*Neofiber alleni*                                               *Fig. 29*

**Description**
Size: head and body, 7–8 in. (17.8–20.3 cm); tail, 4–6 in. (10.2–15.2 cm). Legs and ears short; fur dense. Color is a rich brown.
**Habitat**
Shallow, grassy marshes.
**Habits**
Builds small, dome-shaped houses of grass about 7–25 in. (17.8–63.5 cm) across, also feeding platforms of packed grass fragments.

*Fig. 29*

Nutria, p. 270

Muskrat

Round-tailed Muskrat

**Food**
Plants.
**Range**
Most of Fla., except the w. peninsula and Okefenokee vicinity in southern Ga.
**Other name**
Florida Water Rat.

# MUSKRAT
*Ondatra zibethicus*

51:4
Fig. 29

**Description**
Size: head and body, 10–14 in. (25.4–35.6 cm); tail, to 11 in. (27.9 cm); mammae, 8–10. Above thick, dark brownish or blackish fur with coarser guard hairs; silvery below. Tail black, naked, laterally compressed; hind feet partly webbed.
**Habitat**
Fresh water and saltwater marshes; lakes, ponds, watercourses.
**Habits**
Aquatic, active at any hour; often seen swimming with head appearing wedge-shaped, while Beaver's head appears squarish. Builds a cone-shaped lodge, 5 ft. (1.5 m) in diameter at base, up to 3 ft. (0.9 m) above the water, of mud and sticks in marsh; has an underwater entrance; burrows in banks, has feeding platform on a mat of cattails.
**Food**
Stems of cattails, grasses; mussels.
**Voice**
Moans, squeals, chatters from within lodge.
**Reproduction**
5–7 born in lodge.
**Range**
Throughout the East; absent from Fla. and adjacent areas.

# LEMMINGS
Genera *Lemmus* and *Synaptomys*

These lemmings have curved upper incisors with shallow grooves near the outer edges and very short tails. Their long, grizzled, grayish-brown fur almost hides their ears. The thumb has a large digging claw. They are colonial and active day or night.

# BROWN LEMMING
*Lemmus sibiricus*

55:9

**Description**
Size: head and body, 4½–5½ in. (11.4–14 cm); tail, to 1+ in. (2.5+ cm). Thick, long hair. Upper incisors not grooved. Grayish head, reddish body, brown rump; cream to medium brown below, no dark dorsal stripe.
**Habitat**
Tundra and adjacent forests.
**Habits**
Leaves piles of dung pellets in runways; active at all hours.
**Range**
N.W.T. and n. Man.

## SOUTHERN BOG LEMMING
*Synaptomys cooperi*                                            **55:8**

**Description**
Size: head and body, 3⅖-4⅖ in. (8.6-11.2 cm); tail, ⅜-⅞ in. (0.95-2.2 cm). Grizzled brown above, underparts grayish and lighter.
**Habitat**
Bogs, swamps, and meadows with thick ground vegetation.
**Range**
From s. Canada throughout ne. U.S. to Tenn. and sw. Kans.

## NORTHERN BOG LEMMING
*Synaptomys borealis*                                           **52:6**

**Description**
Size: head and body, to 4¾ in. (12.1 cm); tail, to 1 in. (2.5 cm). Pale to dark brown above, lighter below. Long thick claws on front toes in winter.
**Similarities**
Southern has shorter tail.
**Habitat**
Locally distributed in open or wooded, moist or dry areas.
**Range**
N.W.T. and Alta., s. to Maine, N.H., and n. Minn.

# COLLARED LEMMINGS
Genus *Dicrostonyx*

The members of this circumpolar genus are the only rodents that turn white in winter. Another distinctive feature of the group is the bulbous central extensions of the third and fourth claws of the forefeet in winter, probably an adaptation for grooming the long winter pelage and digging in snow; these structures are lost in summer. Collared lemmings are not known to occur south of the treeless parts of the Arctic. One to eight (usually three to four) young are born in each of two litters annually (June and July). The upper incisors are not grooved and the ears are entirely concealed in fur. The voice has been reported as a squeal or chuckling note.

## COLLARED LEMMING
*Dicrostonyx torquatus*                                        **55:10**

**Description**
Size: head and body, 4-5½ in. (10.2-14 cm); tail, to ⅘ in. (2 cm). Summer: above brownish-black with some buff; tawny color across throat; creamy-buff below.
**Similarities**
Brown Lemming is less stout, lacks dark dorsal stripe, and stays brown in winter.
**Habitat**
Dry gravelly or sandy tundra.
**Range**
Greenland, N.W.T., and n. Man.
**Other name**
Greenland Collared Lemming.

## LABRADOR COLLARED LEMMING
*Dicrostonyx hudsonius*

**Description**
Size: head and body, 5–6 in. (12.7–15.2 cm); tail, ⅔–1 in. (1.7–2.5 cm). Summer: buffy-gray above, dark stripe down middle of back; flanks, throat, and underparts buffy or rusty.
**Habitat**
Open tundra and rocky hillsides.
**Range**
Labr. and n. Que. (Ungava Bay).

# OLD WORLD RATS AND MICE
Family Muridae

Old World rats and mice are grayish-brown to black above, usually grayish below. They have even-colored, nearly naked, long scaly tails; sixteen teeth, 1003/1003; and molars with three longitudinal rows of tubercles. They frequent buildings, dumps, ships, or fields; are active at all hours; swim; and do not hibernate. Their varied diet includes grain, groceries, garbage, and meat.

These animals are our unwanted companions, having entered the Western Hemisphere on ships from the Old World. For survival they rely on fecundity. Their nests are of anything soft in almost any kind of hole. Gestation takes from nineteen to twenty-three days; there are five to nine young, and several litters a year.

## BLACK RAT
*Rattus rattus*                                              *Fig. 30*

**Description**
Size: head and body, 7 in. (17.8 cm); tail, 9 in. (22.9 cm). Black to tawny above, slate-colored to white below; tail naked, scaly, usually longer than head and body.
**Similarities**
Norway Rat has less slender body and shorter ears and tail.
**Habitat**
Includes ships, trees.
**Habits**
Climbs; commonest in the South.
**Remarks**
Fleas that infest this rat are the carriers of the plague, or "black death."
**Range**
S. and e. temperate climates, especially large coastal cities.

*Fig. 30*

Norway Rat, p. 268                    Black Rat

## NORWAY RAT
*Rattus norvegicus*

**54:10**
*Fig. 30*

**Description**
Size: head and body, 7–10 in. (17.8–25.4 cm); tail, to 8 in. (20.3 cm). Fur coarse, ears short. Underparts usually grayish but sometimes white; tail naked and scaly, usually shorter than head and body.
**Similarities**
Woodrats have hairy tails and white underparts and feet. Black Rat has a longer tail.
**Habitat**
Cities and farmyards.
**Habits**
Destructive; yearly damage runs into millions.
**Voice**
A squeal.
**Range**
All areas inhabited by humans from Hudson Bay, southward.

## HOUSE MOUSE
*Mus musculus*

**53:2**

**Description**
Size: head and body, 3⅖ in. (8.6 cm); tail, 2¾–3⅖ in. (7–9.7 cm). Dull gray to grayish-brown above, underparts grayish or buffy; fur short; ears nearly naked; notch on posterior tip of upper incisors best seen in side view. Notch separates the House Mouse from all the native mice in the East.
**Habitat**
In or near buildings.
**Habits**
Prolific; can breed when 40 days old.
**Voice**
A squeak.
**Remarks**
The white strain is kept as a pet and for experimentation.
**Range**
Throughout most of e. U.S. in and around habitations of humans; feral populations known in many places as well.

# JUMPING MICE
Family Zapodidae

These are small, delicate mice with very long tails and hind legs by which they can leap more than six feet. They differ from pocket mice and kangaroo rats by the absence of external cheek pouches; their tails are scantily haired and are never tufted at the tip. They are yellow- or orange-brown above with darker, somewhat bristly hairs in a band along the back; the underparts are whitish. The tail is dark above and white underneath. Each upper incisor has a groove down the front surface.

Swamps, bogs, meadows, and woodlands are their preferred habitat. They do not occur in arid regions. Jumping mice hibernate from October or November to April or May, depending on the latitude, in winter nests located in burrows at least a foot below the surface in well-drained ground. Insects, seeds, berries, and other fleshy fruits make up the diet. Usually two litters of four to six young are produced each summer. The summer nests, rounded in shape, are made of grasses and leaves, and are lined with finer materials; they are placed on the ground at the end of short burrows, under logs and other debris, in tufts of grass or in clumps of shrubs.

Jumping mice are secretive, nocturnal, and uncommon; they apparently make no runways. There is no mound of earth at the burrow entrance, but the entrance may be plugged with earth.

## MEADOW JUMPING MOUSE
*Zapus hudsonius* **52:2**

**Description**
Size: head and body, 3–3½ in. (7.6–8.9 cm); tail, 4–5¾ in. (10.2–14.6 cm); teeth, 1013/1003. Sides grayish-yellow, dorsal band dark; tail long, bicolored, lacks white tip.
**Habitat**
Mesic areas, especially grassy lowlands along streams or around lakes.
**Habits**
Swims readily.
**Age**
To 2 years.
**Range**
Throughout most of e. N.A. with the exception of extreme n. parts and s. of N.C. and Mo.

## WOODLAND JUMPING MOUSE
*Napaeozapus insignis* **53:8**

**Description**
Size: head and body, 3–4 in. (7.6–10.2 cm); tail, 4½–6 in. (11.4–15.2 cm); teeth, 1003/1003. Sides bright orange, dorsal band dark; tail with white tip.
**Habitat**
Moist, cool, deep forests and forest clearings, frequently along lake and stream beds.
**Habits**
Dives into water when frightened; climbs vines and bushes for berries.
**Range**
From New England and Canada s. through Pa. and Appalachians.

# PORCUPINE
Family Erethizontidae

The spiny-quilled porcupines are large, blackish rodents about the size of a small dog. Most of the body, especially the rump and tail, is thickly set with long, sharp, needle-tipped spines.

## PORCUPINE
*Erethizon dorsatum* **51:7**
*Fig. 27*

**Description**
Size: head and body, 18–22 in. (45.7–55.9 cm); tail, 7–9 in. (17.8–22.9 cm); weight, to 30 lb. (13.6 kg); 20 teeth, 1013/1013.
Underfur short, soft, and covered by longer, coarse guard hairs, amid which grow quills. Underfur blackish, guard hairs and quills often light-tipped.

**Habitat**
Forests, preferably with conifers or poplars.
**Habits**
Usually nocturnal, but sometimes seen by day shuffling over forest
floor or hunched into a large ball in a tree. If bothered, the animal
erects its spines, arches its back, and tucks its head between its
front legs; keeping its back toward the tormentor, it strikes out
with its tail, heavily armed with spines; the spines are not thrown
or "shot."
**Voice**
Snort and bark; groaning or crying sounds.
**Food**
Twigs, leaves, and buds in summer; inner bark of conifers and
hardwoods in winter.
**Age**
To 10 years.
**Reproduction**
1 young; well-developed, can climb trees and eat leaves when 2
days old. Nest is in a rocky den, burrow, or hollow log.
**Range**
Throughout N., except for treeless regions; most abundant where
conifers abound; s. to Pa., Mich., Wis.

# NUTRIAS
Family Capromyidae

## NUTRIA
*Myocastor coypus*                                              *Fig. 29*

**Description**
Size: head and body, 22–25 in. (55.9–63.5 cm); tail, 12–17 in.
(30.5–43.2 cm); weight, 15–18 lb. (6.8–8.2 kg). Underfur
brownish, fine, covered by longer coat of coarse overhair.
Somewhat resembles a Muskrat, but is larger and has a long,
scantily haired, rounded tail.
**Habitat**
Swamps, marshes, and rivers.
**Habits**
Semiaquatic, swims well; makes shallow burrows in banks with an
enlarged nesting chamber at the rear.
**Food**
Aquatic vegetation.
**Reproduction**
2–8 young.
**Range**
Watercourses in s. U.S.; introduced.
**Other name**
Coypu.

# Cetaceans
Orders Odontoceti and Mysticeti

Whales and their allies, collectively known as cetaceans and comprising two orders, Odontoceti (toothed) and Mysticeti (baleen), are the most completely aquatic mammals. They are distinguished by nostrils (blowholes) set high on the head; fins instead of forelimbs; a laterally flattened tail bearing horizontal flukes; the absence of external hind limbs; a thick, subcutaneous layer of blubber; and a body nearly devoid of hair. The largest animal that ever lived, the Blue, or Sulphur-bottomed, Whale, reaches a length of more than 100 feet (30.5 m) and a weight in excess of 100 tons (90.7 t).

The larger species can be located by their characteristic "blow." Cetaceans frequently occur together in groups; groups of whales are called gams or pods; groups of porpoises or dolphins are called schools. The latter are often seen leaping through the waves in coastal areas and playing about or following ships.

The habitat of many whales is the open sea; smaller species are sometimes also seen in bays or near shore. They concentrate where their food is most abundant, which may be the plankton-rich edge of the polar ice or more temperate latitudes rich in squid and fish. Certain species have special calving grounds, often in shallow tropical waters where they repair in the breeding season. Some migrate regularly, following their sources of nourishment.

Some cetaceans indulge in extensive courtship before mating. The newborn must immediately learn to breathe, hold its breath under water, suckle without drowning, and swim to keep up with the mother. Cetaceans are quite intelligent, show parental affection, and can be taught to perform. Many, perhaps all, species employ sonarlike echolocation for navigation and finding food.

The toothed whales, as their name implies, are equipped with teeth. Additionally, they have a single blowhole, an asymmetrical skull, and nasal bones that form no part of the roof of the narial passage; the skull is highly telescoped. The suborder ranges in geologic time from the late Eocene. It contains five living families and more than thirty living genera.

In contrast, the baleen whales lack teeth, being equipped instead with dense sheets of baleen (whalebone) that grow downward from the upper jaws and form an effective sieve by which microorganisms are strained from the sea. The narial openings, or blowholes, are paired. The suborder dates back certainly to the middle Oligocene.

Mystery still surrounds the immediate ancestors of the whales and the steps these ancestors took in abandoning the land, for scientists are unanimous in assigning them a terrestrial origin. Probably they split from primitive eutherian stock in the Cretaceous period. Ancestral cetaceans may have been inhabitants of freshwater lakes and streams. Perhaps adaptation to the deep by the proto-cetaceans was relatively rapid. Upon the disappearance of the marine reptiles of the Mesozoic, there must have been a vast and attractive food supply there and a relative scarcity of enemies. But little is known, and it makes a fascinating field of inquiry for the paleontologist.

Whales today are threatened with extinction. Within the last 100 years whalers have virtually extirpated them, in commercial quantities, from the Northern Hemisphere, and in the past few years have greatly reduced populations in the Southern Hemisphere.

# BEAKED WHALES
Family Ziphiidae

Beaked whales have tapering heads and "beaks," and a dorsal fin placed behind the midpoint of the back. They are medium-sized to moderately large cetaceans and are characteristically brownish or grayish but sometimes dark above and paler on the sides and below. The functional teeth are reduced in number; the remaining are vestigial. Little is known of the biology of beaked whales. Evidently preferring the cold waters of open oceans, they are seldom seen near shore. Their food probably consists of squid and other cephalopods, but some fish also are eaten. The family dates back to the early Miocene.

# RARE BEAKED WHALES
Genus *Mesoplodon*

This is a genus of rare whales that are encountered usually only when occasionally one is cast ashore. Adults range in overall length from fifteen to twenty-two feet and are generally blackish in color. Little is known of the natural history of these whales since they are of no commercial value. Females are thought to bear a single calf, as is the rule in most cetaceans. Members have only two teeth and two grooves on the throat that meet in front. The size and positioning of the teeth are useful in distinguishing between the four species which occur in eastern waters, but are not a sure means of identifying every individual of each species.

## NORTH ATLANTIC BEAKED WHALE
*Mesoplodon bidens*                                        *Fig. 31*

**Range**
Principally North Sea and adjacent areas in the eastern Atlantic; occasional records along Atlantic Coast of N.A., from Newfoundland to Mass.

*Fig. 31*

North Atlantic    Tropical    Gervais'    True's

## TROPICAL BEAKED WHALE
*Mesoplodon densirostris*                                  *Fig. 31*

**Range**
Along Atlantic Coast, from N.S., s. to Carolinas.

## GERVAIS' BEAKED WHALE
*Mesoplodon europaeus*                                     *Fig. 31*

**Range**
Along Atlantic Coast, from Long Is., N.Y., southward.

## TRUE'S BEAKED WHALE                                     **56:6**
*Mesoplodon mirus*                                         *Fig. 31*

**Range**
North Atlantic Ocean, from N.S., occasionally to Fla.

# OTHER BEAKED WHALES
Genera *Ziphius* and *Hyperoodon*

## GOOSE-BEAKED WHALE
*Ziphius cavirostris*                                    **56:7**

**Description**
Size: length, to 28 ft. (8.5 m). Body thick; beak short, forehead
fairly prominent; 2 teeth, rounded or conical, at tip of lower jaw.
Color varies from gray to black, sometimes with white on head or
back; sides brownish or spotted; belly whitish; pattern occasionally
reversed. No other beaked whale has a projecting lower jaw and a
distinct ridge from the dorsal fin to the tail.
**Habits**
Travels in gams of 30 or more, but often solitary; some are taken
commercially.
**Range**
Along Atlantic Coast, from R.I., s. to Fla.

## NORTH ATLANTIC BOTTLE-NOSED WHALE
*Hyperoodon ampullatus*                                 **56:3**

**Description**
Size: length, 20–30 ft. (6.1–9.1 m); 1 large pair of teeth at front of
lower jaw. Snout sharp, short. Only beaked whale with a bulging
forehead. Color varies from black to yellowish, becoming lighter
with age; head and belly whitish. Males have high forehead with a
white patch and a white dorsal fin.
**Habits**
Highly sociable, travels in gams; migratory, often stranded;
individuals can remain submerged for 30–45 minutes.
**Range**
Arctic and Atlantic oceans, s. at least to R.I.

# GIANT AND PYGMY SPERM WHALES
Families Physeteridae and Kogiidae

These families contain three species: the sperm whale, and two
pygmy sperm whales. The latter, regarded in the past as
representing a distinct family, Kogiidae, are now considered a
subfamily of Physeteridae. Physeterids have a fossil history dating
back to the early Miocene.

## SPERM WHALE
*Physeter macrocephalus*                                **56:5**

**Description**
Size: length, to 60 ft. (18.3 m); weight, to over 110,000 lb. (50,000
kg). Males much larger. Head enormous, squared, containing
huge, barrel-shaped spermaceti organ filled with high-grade
translucent oil. Upper jaw toothless, lower jaw slender and
containing about 24 functional teeth. Single narial passage on left
and directed forward, old right narial passage modified as sound-
producing organ. Dorsal fin obtuse or rounded. Uniform slate-
gray.
**Habits**
Migratory, gregarious, polygamous; males may move to edge of
pack ice in summer but females generally remain in tropical or
temperate waters.

**Food**
Principally squid.
**Reproduction**
Sexual maturity reached in about 8 years; females bear single calf every 4 or 5 years after gestation of about 15½ months; young nurse up to 2 years.
**Range**
All oceans, except polar ice fields; most common between 50°N and 50°S.

## PYGMY SPERM WHALE
*Kogia breviceps*                                                   **56:4**

**Description**
Size: length, to 13 ft. (4 m); weight, to 881 lb. (400 kg). Head blunt, rounded; lower jaw narrow and with 8–11 teeth on each side. Dorsal fin small, curved, and posterior to center of back. Blowhole as in Sperm Whale. Blackish above, light below.
**Habits**
Probably solitary, rarely seen, occasionally stranded.
**Food**
Squid, octopi.
**Reproduction**
Sexual maturity reached in about 2 years; mating occurs in late summer and females bring forth a single calf in spring.
**Range**
Worldwide in tropical and temperate waters; north along Atlantic Coast to N. S.

Note: Little is known of the biology of Pygmy Sperm Whales. A second species, the **DWARF SPERM WHALE**, *Kogia simus,* also may occur in coastal areas in the southern part of the United States. *K. simus* is smaller than *K. breviceps:* length, to 9 feet (2.7 m), weight, to 595 pounds (270 kg). It has a high dorsal fin, and usually 12 to 16 teeth (as opposed to 8 to 11) in each lower jaw.

# WHITE WHALE AND NARWHAL
Family Monodontidae

These are small whales without a dorsal fin and with a single blowhole. The family contains but two species, each of which occurs only in Arctic and north temperate waters. These are gregarious animals that move southward in winter beyond the ice cap or congregate in areas kept free of ice by tides, currents, or winds. The family is known only from the Pleistocene.

## WHITE WHALE
*Delphinapterus leucas*                                             **56:1**

**Description**
Size: length, to 18 ft. (5.5 m); weight, to 4000 lb. (1814.4 kg). Both jaws with 8–10 teeth on each side. The only white whale. Young are dark gray, then mottled, then yellowish or bluish before becoming white as adults.
**Habits**
Travels in gams; migrates, often ascends rivers.
**Food**
Squid, fish, crustaceans.
**Reproduction**
Young born April–June; nurse 8 months, mature in 2 to 3 years.

**Remarks**
This is one of the few cetaceans that enter rivers. Schools of them live in the lower St. Lawrence, where the alert boat passenger may occasionally glimpse them from the bow. White Whales are hunted by Eskimos for their meat, oil, and hides.
**Range**
Arctic and subarctic seas of Atlantic and Pacific, rarely s. to Mass.
**Other name**
Beluga Whale.

# NARWHAL
*Monodon monoceros*                                                 56:2

**Description**
Size: length, to 18 ft. (5.5 m); weight, to 3000 lb. (1360.8 kg). Snout blunt, eye small; ridge along mid-back. Above mottled gray, below white; young are gray, may be confused with young White Whale. Male has single, forward-projecting, spiraled tusk, which is the left, rarely the right, canine tooth, up to 9 ft. (2.7 m) long. Females occasionally have small tusks, but there are no other teeth in either sex.
**Habits**
May dive to 1200 ft. (365.8 m); use of tusk is unknown.
**Voice**
Females may roar in calling young.
**Food**
Shrimp, cuttlefish, fish.
**Range**
Arctic seas; virtually limited to n. parts of Hudson Bay and coast of Labr.

# PORPOISES AND DOLPHINS
Family Delphinidae

Members of this family are small whales, usually with teeth in both jaws. They have a single blowhole far back from the snout, and the dorsal fin is near the middle of the back. They are generally dark above and paler below. Adults range in size from less than six feet in the Atlantic Harbor Porpoise to more than thirty feet in the Killer Whale. The family dates back to the Miocene.

# STRIPED PORPOISE
*Stenella caeruleoalba*

**Description**
Size: length, to 8 ft. (2.4 m); teeth, 44–50 in each half of each jaw. Dark above, white below; black bands from eye to base of flipper and from eye to vent.
**Range**
Widely distributed in temperate and tropical waters the world around.

# ATLANTIC SPOTTED DOLPHIN
*Stenella plagiodon*                                               Fig. 32

**Description**
Size: length 5–7 ft. (1.5–2.1 m). The only really spotted dolphin, although other species may show various light marks. Small and

numerous white spots on back, paler below. Young are not spotted.
Moderately long snout separated from forehead, as in Bottle-nosed
Dolphin.
**Range**
Coastal waters, from N.C., southward; fairly common in Gulf of
Mexico.

*Fig. 32*

Atlantic Spotted Dolphin

## ROUGH-TOOTHED PORPOISE
*Steno bredanensis*

**Description**
Size: length, to 8 ft. (2.4 m); 20–27 rugose teeth in each half jaw.
Beak not distinctly set off from forehead. Purplish-black above,
yellowish or whitish spots on sides, whitish to purplish-pink below.
**Range**
Tropical and warm temperate seas; s. from coastal Va.

## COMMON DOLPHIN
*Delphinus delphis*                                                    **57:6**

**Description**
Size: length, to 8½ ft. (2.6 m); weight, to about 180 lb. (81.6 kg);
teeth small, conical, approximately 40–50 in each half of both jaws.
Only delphinid that is black above, white below, and with
yellowish flanks. Jaws slender; eye-rings white and connected by 2
white lines. Flippers black; flanks with yellowish, grayish, or
whitish bands.
**Habits**
Gregarious; frequently travels in schools of up to 100 individuals;
follows ships, makes graceful leaps out of water; normally
submerges for 1–3 minutes.
**Age**
To about 15 years.
**Reproduction**
Single calf born in spring after gestation of 9–10 months.
**Range**
Atlantic Coast, from N.S., southward.
**Other name**
Saddle-backed Dolphin.

## BOTTLE-NOSED DOLPHIN
*Tursiops truncatus*                                                   **57:2**

**Description**
Size: length, to 12 ft. (3.7 m); weight, to 400 lb. (181.4 kg); teeth
large, smooth, 19–26 in each half of both jaws. The most common
dolphin. Transverse groove between beak and rounded forehead;
lower jaw longer than upper; dorsal fin prominent; belly paler than
back. Color, grayish; short snout, and "bottle nosed."
**Habits**
Migratory, travels in schools, leaps in air at surface.

**Food**
Fish.
**Reproduction**
Single calf, 3–4 ft. (0.9–1.2 m) long and weighing 25 lb. (11.3 kg); born every other year after gestation of about 12 months.
**Range**
Atlantic Coast from N.B. to Mexico.

## WHITE-BEAKED DOLPHIN
*Lagenorhynchus albirostris*                                    57:1

**Description**
Size: length, to 10 ft. (3 m); weight, to 600 lb. (272.2 kg); teeth, 22–25 in each half of both jaws. Only dolphin with white beak. Above black, pale stripe along side, belly whitish.
**Habits**
Travels in schools, sometimes of 1000 or more individuals.
**Range**
North Atlantic, rarely s. to Gulf of Mexico.

## ATLANTIC WHITE-SIDED DOLPHIN
*Lagenorhynchus acutus*                                        57:3

**Description**
Size: length, to 9 ft. (2.7 m); weight, to 500 lb. (226.8 kg); teeth, 30–34 in each half of both jaws. Snout short, blunt; dorsal fin prominent. Back black, belly white, bright whitish stripe on sides.
**Habits**
Gregarious.
**Food**
Fish and squid.
**Range**
North Atlantic, s. to Mass.; most common off Newfoundland.

## GRAMPUS
*Grampus griseus*                                              57:7

**Description**
Size: length, to 13 ft. (4 m); functional teeth, 2–7 in each half of lower jaw only. Head bulging, yellowish; no beak. Blunt, but not projecting, nose and slender flippers. Males bluish-white above, females brownish; flippers mottled grayish to blackish, belly grayish-white.
**Habits**
Solitary or in small schools.
**Food**
Fish, cuttlefish.
**Reproduction**
A single calf, about 5 ft. (1.5 m) long, born in winter.
**Range**
All temperate and tropical seas from Mass., southward.
**Other name**
Risso's Dolphin.

## KILLER WHALE
*Orcinus orca*                                            **57:5**

**Description**
Size: length, 15–30 ft. (4.6–9.1 m); weight of males, to about
10,000 lb. (4535.9 kg), of females, to about 5000 lb. (2268 kg);
teeth long, narrow, 10–14 in each half of both jaws. Body stout,
rounded; nose blunt; eye large; belly and spot behind eye white;
recognized by its tall dorsal fin, glossy black back, and white rear
flank.

**Habits**
Travels in schools of up to 40 or more, a fast swimmer.

**Reproduction**
Length of a single calf at birth 7–9 ft. (2.1–2.7 m).

**Range**
Along Atlantic Coast.

## COMMON PILOT WHALE
*Globicephala melaena*                                    **57:8**

**Description**
Size: length, to 28 ft. (8.5 m); weight, to about 6000 lb. (2721.6
kg); teeth, 10 in each half of both jaws. Only dolphin that is almost
uniformly black and has a bulging forehead. Body cylindrical;
dorsal fin large, curved, set forward of midback; flippers ⅕ length
of body. Belly with a light stripe that enlarges to a heart-shaped
area on throat.

**Habits**
Spouts to 5 ft. (1.5 m); migratory, travels in large schools, which
are occasionally stranded.

**Food**
Squid and some fish.

**Reproduction**
Breeds in spring; single (usually) calf born after gestation of 16
months; nurses up to 2 years.

**Range**
Coastal waters from Greenland to Va.

## SHORT-FINNED PILOT WHALE
*Globicephala macrorhynchus*

**Description**
Size: length, 15–20 ft. (4.6–6.1 m); teeth 7–9 in each half of both
jaws. Similar to larger Common Blackfish, but flippers only ⅙
length of body.

**Habits, Food, Reproduction**
Same as Common Pilot Whale

**Range**
Principally tropical, from N.J. along Atlantic Coast.

## HARBOR PORPOISE
*Phocoena phocoena*                                       **57:4**

**Description**
Size: length, to 6 ft. (1.8 m) weight, to about 50 lb. (22. 7 kg);
teeth, 22–27 in each half both jaws, have spadelike crowns. Only
small dolphin with a triangular dorsal fin. Body small, thick;
mouth round, eye small. Back black, sides pinkish, belly white;
dark line from corner of mouth to flipper.

**Habits**
Spout short, vertical; makes puffing sound; active before squalls; travels in schools; frequents shorelines, harbors.
**Food**
Fish, squid.
**Range**
In North Atlantic from Greenland to N.J.

## FALSE KILLER WHALE
*Pseudorca crassidens*                                    *Fig. 33*

**Description**
Size: length, 13–18 ft. (4–5.5 m); teeth, relatively large, 8–10 on each side above and below. Black, slender, small, without bulging forehead. Dorsal fin relatively small, projecting slightly backward and slightly ahead of the middle of the body; snout bluntly rounded, longer than lower jaw, without beak or bulging forehead.
**Food**
Fish and squid.
**Remarks**
Mostly known from mass strandings.
**Range**
Atlantic Coast from N.C., southward.

False Killer Whale

*Fig. 33*

# FINBACK WHALES
Family Balaenopteridae

Members of this family are commonly called fin whales or rorquals. They generally are characterized by short, broad plates of whalebone, or baleen, hanging in rows from each side of the upper jaw, and they possess no teeth as adults. They have a dorsal fin, many furrows in the throat, and a double row of blowholes. These whales have been the mainstay of the recent whaling industry and, after being ruthlessly exploited in northern waters, are now being systematically hunted in the Southern Hemisphere. The family dates back to the early Miocene.

## FIN WHALE
*Balaenoptera physalus*                                    *Fig. 34*

**Description**
Size: length, to over 80 ft. (24.4 m); weight, to 70 tons (63.5 t); baleen, to 3 ft. (0.9 m), purple and white, sometimes grayish; 70–80 throat furrows. Up to about 400 plates of baleen on either side of upper jaw. Head flat; dorsal fin small, with slightly concave rear edge, just in front of flukes. Above gray, below white.
**Habits**
Spout 15–20 ft. (4.6–6.1 m) high, inclined forward, narrow at start, then elliptical, made with loud whistling sound; fastest of all whales; sometimes comes near ships. Gregarious, usually occurring

in pods of 10–15 but there are reports of many hundreds flocking together.
**Food**
Plankton, small crustaceans.
**Reproduction**
Gestation period 1 year; a single calf, occasionally 2, weighing up to 4000 lb. (1814.4 kg) and 20–22 ft. (6.1-6.7 m) long at birth.
**Range**
All oceans from Greenland to the Caribbean.

## SEI WHALE
*Balaenoptera borealis*

**Description**
Size: length, to 60 ft. (18.3 m); weight, rarely exceeds 20 tons (18.1 t); baleen, to 3 ft. (0.9 m), blackish; 40–60 throat furrows. Somewhat similar to Fin Whale, but body smaller, dorsal fin relatively larger and set farther forward. Back bluish-black, sides gray, belly white.
**Habits**
Spout to 14 ft. (4.3 m), vertical; migratory; uncommon.
**Food**
Small marine invertebrates.
**Reproduction**
Single calf born about November after gestation of 1 year; weaned after 5 months when 26–30 ft. (7.9–9.1 m) long.
**Range**
All oceans; from coast of Labr., southward.

## MINKE WHALE
*Balaenoptera acutorostrata*                                  *Fig. 34*

**Description**
Size: length, to 33 ft. (10.1 m); weight, to about 3 tons (2.7 t); baleen, to 8 in. (20.3 cm), whitish; 50–70 throat furrows. The only cetacean with a broad white band across the flipper. Snout triangular from above; dorsal fin far back, tip curved. Varying shades of dark above, white below.
**Habits**
Spout faint, vertical; migratory, travels singly or in small groups; frequents coastal waters.
**Food**
Mostly small school fish.
**Reproduction**
Single calf.
**Range**
Widely distributed in all oceans; from Baffin Bay, s. to Fla.
**Other name**
Little Piked Whale.

*Fig. 34*

Fin Whale, p. 279

Minke Whale

# BLUE WHALE
*Balaenoptera musculus*

**Description**
Size: length, to 106 ft. (32.3 m); weight, to 140 tons (127 t);
baleen, to 40 in. (101.6 cm), bluish-black; 80–100 throat furrows.
Identified by its U-shaped snout (from above) and great size.
Dorsal fin small, set far back. Above slaty- to bluish-gray; below
yellowish or whitish.

**Habits**
Spout to 20 ft. (6.1 m), vertical; travels singly or in gams; migrates
to warm water to breed; at other times frequents water near pack
ice.

**Food**
Mainly small, shrimplike invertebrates (krill).

**Reproduction**
Probably monogamous, mating under water from May–August;
gestation period probably 10–11 months; calves (25 ft. [7.6 m] long
and 2–4 tons [1.8–3.6 t] at birth) are weaned at 5–7 months when
about 50 ft. (15.2 m) long.

**Remarks**
This is the largest animal that ever lived on land or sea. Weighing
a ton or more a foot, a 106-ft. (32.3-m) specimen would be twice
or thrice the presumed weight of *Brontosaurus*, the largest
dinosaur.

**Range**
All oceans; occurs along entire Atlantic Coast.

# HUMP-BACKED WHALE
*Megaptera novaeangliae*                                        *Fig. 35*

**Description**
Size: length, to 50 ft. (15.2 m); weight, 25–45 tons (22.7–40.8 t).
Baleen varies in length; shorter at front of mouth, less than 3 ft.
(0.9 m) long; all-black, with black or olive-black bristles. No other
finback has a humped back and scalloped flippers and flukes.
Flipper ⅓ length of body; dorsal fin small.

**Habits**
Spout to 20 ft. (6.1 m) in an expanding column; travels singly or in
small schools; can leap clear of the water.

**Food**
Small crustaceans.

**Reproduction**
Mating takes place in late winter or spring; gestation period about
10 months; calves 9–15 ft. (2.7–4.6 m) long at birth, weigh up to
4000 lb. (1814.4 kg).

**Range**
All oceans; occurs along entire Atlantic Coast.

Black Right Whale, p. 282

*Fig. 35*

Hump-backed Whale

# RIGHT WHALES
Family Balaenidae

These are the only whales with flexible plates of whalebone, or baleen, smooth throats, double blowholes, and no dorsal fin. When diving deeply (sounding), the flukes are thrown clear of the water. The family dates back to the early Miocene.

## BLACK RIGHT WHALE
*Balaena glacialis*                                              *Fig. 35*

**Description**
Size: length, to 60 ft. (18.3 m), weight, to 60 tons (54.4 t); baleen, to 9 ft. (2.7 m), black. Only baleen whale with high arched mouth and head about ⅕ the total length of the body. Body stout, blackish, sometimes with patches of white below.

**Habits**
Spout 10–15 ft. (3–4.6 m) high, in 2 columns which diverge to form a V; slow-moving, easy to capture; rare.

**Food**
Plankton and other small invertebrates.

**Reproduction**
Little known; gestation thought to last about 10 months; 1 calf 10–20 ft. (3–6.1 m) long at birth and nurses for about 1 year.

**Range**
Temperate waters of both hemispheres, from at least Gulf of St. Lawrence southward; said once to have occurred as far n. as Greenland.

## BOWHEAD WHALE
*Balaena mysticetus*

**Description**
Size: length, to 65 ft. (19.8 m); weight, to over 50 tons (45.4 t); baleen, to 14 ft. (4.3 m), black. Only whale whose head is more than ⅓ of body length. Body very stout; lower jaw bowed gently upward, not arched, as in Right Whale. Above dark grayish-brown, belly may be spotted with white.

**Habits**
Spout V-shaped; does not migrate; not gregarious, normal submersion 20–30 minutes.

**Food**
Small crustaceans.

**Reproduction**
Little known; 1 calf, 10–18 ft. (3–5.5 m) long, born in late winter or early spring; bluish-gray in color, later blackish.

**Range**
Polar seas, s. to St. Lawrence R.; now rare.

# Carnivores
Order Carnivora

Terrestrial carnivores are distinguished by their large canine teeth (although the Virginia Opossum, a marsupial, also has large canines) and their strong jaws, legs, and claws. All have five toes on the front foot and four or five on the hind foot. Their sight and sense of smell are often keen, and many are fleet of foot. The smallest, the Least Weasel, weighs less than two ounces; the largest, the Grizzly Bear (a western species), weighs up to 850 pounds (385.6 kg). Females are smaller than males, sometimes by as much as one-third. Carnivores occur in all latitudes and altitudes and from low, hot deserts to cool, damp woods. As a recognizable order, carnivores go back to the early Paleocene. By the Oligocene most of the principal modern families had arisen.

A few carnivores, such as the bears, become inactive during the winter, but most are active at all seasons. Many have voices which, like the cry of the wolf, can be heard from afar. All are flesh- or fish-eaters, but a number vary their diet with berries, fruits, and nuts. Females breed once or twice a year, bringing forth a litter of blind and usually furred young (called cubs, pups, or kits) in a secluded den, hollow log, or burrow. These usually stay with the mother into the summer or fall or, in the case of some species, the next year. Some mustelids have a prolonged gestation period (up to one year), but most embryo growth takes place in final few weeks.

# COYOTES, WOLVES, AND FOXES
Family Canidae

All members of this family are doglike. They have four toes on each hind foot. The dental formula of all eastern species is 3142/3143. The young are born in a hollow log, den, or burrow, and the gestation period is seven to ten weeks. The family dates back to the Eocene. The mammae always occur in pairs.

## COYOTE
*Canis latrans*                                    **59:1, 60:5, 60:8**

**Description**
Size: head and body, to 54 in. (137.2 cm); tail, to 16 in. (40.6 cm); weight, to 50 lb. (22.7 kg). Buffy-gray with rusty legs, feet, and ears; underparts whitish. Nose and ears pointed, nose pad narrow; legs long; long, bushy tail with black tip.
**Similarities**
Gray Wolf is larger, grayer, and has broader nose pad.
**Habitat**
Brush country, grasslands, farmlands; prefers open spaces.
**Habits**
Largely nocturnal, but frequently active by day; carries tail drooping when on run.
**Voice**
High-pitched bark and howl, often in evening chorus.
**Food**
Small mammals and ground-nesting birds; in summer, also fruits and berries.
**Reproduction**
Usually 6–7 pups.
**Range**
In much of U.S., from New England westward, in se. states.

## RED WOLF
*Canis rufus*                                                60:3

**Description**
Size: head and body, to 49 in. (124.5 cm); tail, to 17 in. (43.2 cm);
weight, to 80 lb. (36.3 kg). Similar to Gray Wolf, but smaller,
with color varying from grayish-brown to reddish and black. Light-
colored individuals can be distinguished from Coyote by their
broader nose patch (over 1 in. [2.5 cm] wide).
**Habitat**
Plains, forests.
**Habits**
Runs with tail horizontal.
**Voice**
Like Gray Wolf.
**Food**
Small or weak deer, small mammals.
**Reproduction**
Usually 4–6.
**Range**
Formerly most se. states, now reduced to remnants in Tex. and La.

## GRAY WOLF
*Canis lupus*

60:3, 60:9
*Fig. 36*

**Description**
Size: head and body, to 43–48 in. (109.2–121.9 cm); tail, to 19 in.
(48.2 cm); weight, to 170 lb. (77.1 kg). Usually gray but varies
from white to gray to black; pure white on Arctic islands and
Greenland; all above colors in Alaska. Fur long.
**Similarities**
Coyote has less rounded ears and narrower nose pad; is less doglike
in appearance.
**Habitat**
Wilderness only; tundra, plains, forests.
**Habits**
Travels singly or in pairs or packs; often hunts cooperatively;
carries tail level or upraised when on the run.
**Voice**
Loud howl; various barks and growls.
**Food**
Young, sick, and aged deer; mountain sheep, bison, and musk oxen;
small mammals; also berries and fruits.
**Reproduction**
First breeds at 2 years of age; 1–13 pups born in spring (usually
April), after gestation of 60–63 days.
**Range**
Formerly occurred throughout eastern U.S., now restricted to
wilderness areas in extreme n. U.S. and Canada.
**Other name**
Timber Wolf.

*Fig. 36*

Gray Wolf

## ARCTIC FOX
*Alopex lagopus*

60:4
Fig. 37

**Description**
Size: head and body, to 31 in. (78.7 cm); tail, to 15 in. (38.1 cm);
weight, 7–15 lb. (3.2–6.8 kg). Two color phases: blue phase in
summer is brownish or gray above with yellowish-white sides and
belly; in winter, slate-blue, sometimes with brown on head and
feet, no white tip to tail. White phase in summer is same as blue
phase; in winter, all-white. Feet heavily furred; ears short and
rounded.
**Habitat**
Tundra and Arctic coast.
**Habits**
Unsuspicious; burrows in snow for a temporary den; solitary or
travels in pairs, but never in packs.
**Voice**
High-pitched yapping bark.
**Food**
In winter, dead seals, walruses, whales; in summer, rodents, birds,
eggs, berries.
**Reproduction**
Reaches sexual maturity in 10 months; mates in March to April;
litter of 6–7 kits born following gestation of 52–57 days.
**Range**
N.W.T. and adjacent areas in Canada; coastal areas of Greenland.

Fig. 37
Arctic Fox

summer                winter

## SWIFT FOX
*Vulpes velox*

59:3, 60:4

**Description**
Size: head and body, to 31 in. (78.7 cm); tail, to 12 in. (30.5 cm);
weight, 4–6 lb. (1.8–2.7 kg). Gray to pale yellowish-brown. Body
slender, ears large, tip of tail black; blackish spot on each side of
snout.
**Habitat**
Plains, dry foothills, deserts.
**Habits**
Nocturnal, shy.
**Voice**
Weak bark.
**Food**
Mice and other small mammals, insects, some fruit.
**Reproduction**
Mates in December and January; gestation about 50 days; litter
size usually 4–7.
**Remarks**
Once nearly extirpated from much of its range, now seems to be
increasing substantially in number.
**Range**
Principally Great Plains; reaches Prairie Provinces of Canada.

## RED FOX
*Vulpes vulpes*            **59:2, 60:4, 60:7**

**Description**
Size: head and body, to 42 in. (106.7 cm); tail, to 16 in. (40.6 cm);
weight, 8–15 lb. (3.6–6.8 kg). Two color phases: red phase is
reddish above, white below, feet black; black phase is all-black
except white tip on tail. Cross phase is reddish-brown, with heavy
black markings on shoulder. Ears large, triangular; tail large,
bushy, always having a white tip. Somewhat resembles small collie
or Coyote, but both lack white-tipped tail.

**Habitat**
Extremely adaptable, including the ability to thrive in proximity to
humans; probably prefers sparsely inhabited areas combining
forests and open country.

**Habits**
Primarily nocturnal but occasionally active by day; males assist in
rearing young.

**Voice**
Male, "a short yelp, ending in a *yurr,* as if gargling"; female, "a
yapping scream" (Palmer).

**Food**
Rabbits, rodents, snakes, berries, and fruit.

**Reproduction**
Reaches sexual maturity in about 1 year; mates in January or
February; gestation period is about 52 days; single litter each year
containing 4–5 kits.

**Range**
Throughout most of e. U.S. except the Gulf Coastal plain.

## GRAY FOX
*Urocyon cinereoargenteus*            **59:5, 60:4**

**Description**
Size: head and body, to 44 in. (111.8 cm); tail, to 16 in. (40.6 cm);
weight, 7–13 lb. (3.2–5.9 kg). Above salt-and-pepper gray; sides,
legs, and feet rusty; underparts gray; tail black-tipped and bushy,
with a prominent blackish area down middorsal line.

**Habitat**
Brush, wooded lowlands; swamps.

**Habits**
Largely nocturnal, solitary; can climb trees.

**Voice**
Harsh, rarely heard.

**Food**
Rabbits, rodents, reptiles, berries, and fruits.

**Reproduction**
Most individuals breed in January or February in their first year
of life; the 1 annual litter (3–5 kits) born after 63-day gestation.

**Range**
From Canadian border southward, w. to S.Dak. and w. Nebr.

# BEARS
Family Ursidae

Bears are the largest living carnivores. They have five toes on each
foot, and their short tails are almost hidden in long fur. The tooth
formula is 3142/3143 and the gestation period is seven to nine
months. Bears walk on their entire foot, like human beings. The
family dates back to the Miocene.

# BLACK BEAR
*Ursus americanus*

60:1, 61:3

**Description**
Size: length, to 6 ft. (1.8 m); height at shoulder, to 3 ft. (0.9 m); weight, usually to 350 lb. (158.8 kg). The smallest and most common bear. Black phase most common in East, cinnamon (all-brown) phase most common in West; face always brown, breast often has patch of white. Snout long, face profile straight; front claws short, rounded; no hump on shoulders.

**Habitat**
Forests, swamps, mountains.

**Habits**
Solitary, quarrelsome; inactive in winter but does not hibernate; marks home range by clawing boundary trees.

**Voice**
Various whines, grunts, and huffs.

**Food**
Omnivorous; flesh, fish, carrion, fruits, and berries.

**Reproduction**
2 cubs, born late January or February; weight 8 oz. (226.8 g) or less; breeds every other year.

**Range**
Formerly occurred throughout entire E. except in Greenland and extreme n. Canada; no longer found in heavily populated areas.

# POLAR BEAR
*Ursus maritimus*

*Fig. 38*

**Description**
Size: length, to 8 ft. (2.4 m); height at shoulder, to 4 ft. (1.2 m); weight of males, usually to 800 lb. (362.9 kg). Coat white, often with a yellow tinge; young whiter than adults. Eye, nose pad, and foot pad black. Head small, neck long.

**Habitat**
Arctic coast and ice floes.

**Habits**
Roams widely, digs winter dens in snow, sand, or dirt; a powerful fighter.

**Voice**
A roar.

**Food**
Seals, young walruses, stranded whales.

**Reproduction**
Cubs (usually 2) born in November or December in winter den; females probably breed every other year, may have prolonged gestation.

**Range**
Arctic Coast of Canada and Greenland.

*Fig. 38*

Polar Bear

# WALRUS
Family Odobenidae

Walruses are pinnipeds, aquatic carnivorous mammals that come to shore to breed and sometimes to rest. Their limbs have evolved into flippers, with which they can haul themselves out on the beach or ice floe; otherwise they are poorly adapted for locomotion on land because the limbs are enclosed within the body to the level of the elbow and knee; the digits of the paddle-shaped hand and foot usually retain claws. They have short tails. The males are usually much larger than females.

Walruses evolved directly from terrestrial carnivores, and are first reported from the Miocene. Being still tied to the land to breed, they are less adapted to marine life than the whales, whose ancestors may have entered salt water 30 million years earlier.

## WALRUS
*Odobenus rosmarus*                                          **62:2**

**Description**
Size: length, to 12 ft. (3.7 m); weight, to 3000 lb. (1360.8 kg); 18 teeth, 1130/0130. Only marine mammal with 2 large white tusks, smaller in cows than bulls. Body tan to gray, almost hairless; hide thick, in folds; no external ears; hind limbs bend forward when not in water.
**Habitat**
Ice floes, usually over shallow water; seldom goes ashore.
**Habits**
Gregarious, feeds standing on head in water, may inflate neck and float on water when asleep.
**Voice**
Bellow like voice of St. Bernard dog; elephantlike trumpeting.
**Food**
Mollusks.
**Reproduction**
Gray calf, born in spring, lacks tusks; travels on cow's neck.
**Range**
Arctic coasts, s. in Atlantic rarely as far as Maine.

# RACCOONS AND ALLIES
Family Procyonidae

With the exception of the Lesser Panda, all recent procyonids are restricted to the New World. The dental formula is 3142/3142 for the species in the East; the cheek teeth are less shearing than those of canids but not so flattened as the teeth of bears. Procyonids walk on the entire flattened surface of the foot, have five toes on each foot, and nonretractile or semiretractile claws. The family is first known from the Oligocene.

## COATI
*Nasua narica*                                              *Fig. 39*

**Description**
Size: head and body, 20–25 in. (50.8–63.5 cm); tail, 20–25 in. (50.8–63.5 cm); weight, to about 20 lb. (9.1 kg). Long-nosed, long-tailed, with indistinctly ringed tail tapering toward end. White spots near eyes; general color grizzled brown.

**Habitat**
Shrubby hillsides and ravines.
**Habits**
Tough nose pad and movable snout used in rooting out food; active day or night; forages in groups of various sizes; tail often carried aloft.
**Voice**
Loud grunts when alarmed.
**Food**
Invertebrates, lizards, small rodents.
**Reproduction**
Gestates 72 days; 2–7 young; weaned at 4 months, mature at 2 yrs.
**Range**
Chiefly Mexican, but into s. Tex. and border areas farther w.

*Fig. 39*

Coati

# RACCOON
*Procyon lotor*                                          **54:6, 61:7**

**Description**
Size: head and body, to 33 in. (83.8 cm); tail, to 12 in. (30.5 cm); weight, 6–35 lb. (2.7–15.4 kg). Fur is gray, brown, and black, giving a grizzled appearance; face with black mask, tail ringed. Body stout, fur long; snout and ears pointed.
**Habitat**
Woods, swamps.
**Habits**
Nocturnal, sometimes washes food; a good fighter, climber, and swimmer.
**Voice**
Varied; includes barks, growls, a throaty cry, a whine, and an owllike quaver.
**Food**
Fish, crayfish, birds, eggs, corn, vegetables, fruit.
**Reproduction**
Females mature sexually in first year, males in second year; litter of 3–4 young weighing less than ½ lb. (226.8 g) born in den in ground or hollow tree in April or May after 63-day gestation period; some females have second litter in August, especially if first is lost; weaned at 7–16 weeks.
**Range**
Throughout e. U.S. and adjacent Canada.

## RINGTAIL CAT
*Bassariscus astutus*                                         *Fig. 40*

**Description**
Size: head and body, to 17 in. (43.2 cm); tail, to 19 in. (48.2 cm); weight, 30–45 oz. (870–1300 g). Buff above, overcast with blackish or brownish; white or buffy-white below; eye ringed with black or brown.
**Habitat**
In northern part of range, rocky outcrops, frequently along streams.
**Habits**
Agile, good climber, nocturnal.
**Food**
Omnivorous; eats small mammals, birds, eggs, berries, and other fruit.
**Voice**
"Spits" or "growls" defensively.
**Reproduction**
3–4 young born in den in rocks or a hollow tree in May or June; mature at one year, weaned at 4–8 weeks.
**Range**
In s. Kans.

*Fig. 40*

Ringtail Cat

# WEASELS, SKUNKS, AND ALLIES
Family Mustelidae

These mammals are often predominantly brown, but vary considerably in color and size. They usually have long bodies, short legs, rounded ears, and large, paired scent glands. The family dates back to the Oligocene.

## MARTEN
*Martes americana*                                         **58:1, 59:7**

**Description**
Size: head and body, to 17 in. (43.2 cm); tail, to 9 in. (22.9 cm); 38 teeth, 3141/3142. Above yellowish-brown to dark brown (in Labrador almost black); head, belly paler; has a pale buff patch on the throat and chest; legs, tail, and ears dark. Head small, feet partly furred.
**Similarities**
Mink is darker, and with shorter tail and small white patch on chin. Fisher is darker and larger.
**Habitat**
Coniferous forests.
**Habits**
Largely arboreal, but also hunts on the ground and occasionally in water; does not hibernate.
**Voice**
High-pitched screams and squeals.

**Food**
Variable, including squirrels, mice, birds, eggs, and fish.
**Reproduction**
3–5 young, born in leaf-lined nest in hollow tree, in March or
April, after protracted gestation of 220–265 days.
**Range**
Formerly throughout boreal forest of Canada and n. U.S.; now
rare or extirpated over much of s. part of former range.
**Other name**
Sable.

# FISHER
*Martes pennanti*                                    58:6, 59:9

**Description**
Size: head and body, 25 in. (63.5 cm); tail, to 15 in. (38.1 cm); 38
teeth, 3141/3142. Blackish-brown, head and shoulders grizzled;
legs and tip of tail black; tail bushy, tapering.
**Similarities**
Marten is smaller, lighter, and has light patch on chest.
**Habitat**
Dense mixed forests.
**Habits**
Mainly nocturnal; somewhat arboreal, solitary.
**Voice**
Scream and hiss.
**Food**
Squirrels, mice, raccoons, rabbits, some vegetable matter, carrion.
**Reproduction**
1–5 (usually 3) young, born in late March or April; females mate
shortly after birth of young, delayed implantation accounting for
the prolonged period of gestation.
**Range**
Same as Marten.

# WEASELS
Genus *Mustela*

Weasels have small heads, long necks, and a dental formula of
3131/3132. They are terrestrial, largely nocturnal, and ferocious
for their size. They are carnivorous, feeding chiefly on small
animals, both warm- and cold-blooded.

# ERMINE
*Mustela erminea*                                    59:6

**Description**
Size: head and body, to 9 in. (22.9 cm); tail, to 4 in. (10.2 cm). In
summer brown above; feet and underparts white. In winter, white
(in the East); end of tail always black. Spring and fall molts show
transition between brown and white upperparts.
**Habitat**
Field borders, open woodlands, brushy and rocky places.
**Habits**
Primarily terrestrial, but good swimmer and climber; sometimes
hunts in pairs.
**Voice**
Varied squeals, barks, hisses, and chatters.

**Reproduction**
Usually 4–7 young, born in fur-lined nest on ground; weight at birth, $\frac{1}{14}$ oz. (2 g).
**Range**
Throughout Canada, coastal Greenland, s. into U.S. to Md. and w. to Iowa.
**Other name**
Short-tailed Weasel.

## LEAST WEASEL
*Mustela nivalis*                                          58:9, 59:4

**Description**
Size: head and body, 6½ in. (16.5 cm); tail, to 1½ in. (3.8 cm). In summer, brown above, white below. In winter, white in northern part of range, some individuals brown above in southern areas. Tip of tail not black, but may have few black hairs.
**Habitat**
Open woods, brush piles, grassy areas.
**Habits**
Home range about 2 acres; can swim.
**Voice**
A weak bark or shriek.
**Reproduction**
3–10 (usually 5) young, evidently born in any season, in underground nest lined with mouse fur or grass.
**Range**
Hudson Bay, southward, excepting New England states; s. to N.C. and w. to n. Kans.

## LONG-TAILED WEASEL
*Mustela frenata*                                          58:8, 61:4

**Description**
Size: head and body, 10½ in. (26.7 cm); tail, to 6 in. (15.2 cm). In summer, brown above, yellowish-white below; tail black-tipped. In winter, in northern part of range, white, with black-tipped tail; in southern part of range, same as summer.
**Similarities**
Summer Ermine has white, not brown, feet; winter Ermine is similar but smaller. Mink is dark below.
**Habitat**
Variable, including farmlands, prairies, woodlands, swamps.
**Habits**
Primarily nocturnal, but often seen by day, usually solitary; climbs well.
**Voice**
Hisses, screams, purrs.
**Reproduction**
4–9 young, born in nest in stump, rock pile, or borrow.
**Range**
Throughout e. U.S., from N.B., Que. and s. parts of Prairie Provinces of Canada to s. U.S.

## BLACK-FOOTED FERRET
*Mustela nigripes*                                    **59:8**

**Description**
Size: head and body, to 18 in. (45.7 cm); tail, to 6 in. (15.2 cm).
Only eastern weasel with a black mask and black feet. Above,
buffy-yellow; underparts paler, tip of tail black.
**Habitat**
Plains, prairies coinciding with range of prairie dogs, on which it
feeds.
**Habits**
More active at night than by day; wary.
**Voice**
Hiss and chatter.
**Reproduction**
Mates in March and April; gestation 42–45 days; litter of 1–5
(usually 3–4).
**Remarks**
Formerly abundant, now extremely rare.
**Range**
Canada, s. to e. Nebr. and cen. Kans.

## MINK
*Mustela vison*                                    **58:7, 61:5**

**Description**
Size: head and body, to 17 in. (43.2 cm); tail, to 9 in. (22.9 cm).
Only uniformly dark brown weasellike animal in east with white
patch on its chin. No seasonal color change.
**Habitat**
Near water; streams, marshes; in winter, woods.
**Habits**
Nocturnal, solitary, wary; spends much time in water; emits strong
odor when cornered; stores food in den.
**Voice**
Various hisses, screams, barks, purrs.
**Reproduction**
Single litter of 5–8 young, born in nest in rocks or other natural
cavity, often in bank of stream, river, or lake.
**Range**
Throughout most of e. U.S.

# OTHER CARNIVORES
Genera *Gulo, Taxidea, Spilogale, Mephitis,* and *Lutra*

## WOLVERINE
*Gulo gulo*                                    **58:2, 59:10**

**Description**
Size: head and body, to 32 in. (81.3 cm); tail, to 9 in. (22.9 cm);
weight, to 40 lb. (18.1 kg); 38 teeth, 3141/3142. Color varies from
yellowish-brown to almost black, paler on head; has a broad light
stripe on each side. Fur long, ears small, back arched, feet large,
tail bushy.
**Habitat**
Wilderness, chiefly brushlands, forests, and mountains.
**Habits**
Active at all times of day and year; mainly terrestrial, but can
climb.

**Voice**
Snarl and growl.
**Food**
Any animal it can kill; also carrion.
**Reproduction**
Usually 2–3 young, born in March or April after a long gestation period involving delayed implantation.
**Range**
Formerly in Canada and n. U.S., s. to Pa., Ind., and Nebr., but now limited to n. wilderness areas; may be increasing in abundance in some areas.

## BADGER
*Taxidea taxus*                                       **54:1, 61:8**

**Description**
Size: head and body, to 22 in. (55.9 cm); tail, to 12 in. (30.5 cm); 34 teeth, 3131/3132. No other mammal has white stripe from its nose back over the top of its head. Above grayish-yellow, giving grizzled appearance; underparts paler; head black and white; feet black; tail short, yellowish. Fur long; body flattened, stout; front claws very long.
**Habitat**
Dry, open prairie and farmlands.
**Habits**
Nocturnal, but also abroad by day; a strong digger, lives in an underground den.
**Voice**
Grunts and growls, but usually silent.
**Food**
Ground squirrels, other small mammals, birds, eggs, reptiles.
**Reproduction**
1–5 (usually 2) young, born in April or May in grass-lined, subterranean nest.
**Range**
Throughout prairies, from w. N.Y. and Alta., southward; absent in e. Canada and se. part of the region.

## EASTERN SPOTTED SKUNK
*Spilogale putorius*                                  **58:3, 61:2**

**Description**
Size: head and body, 13½ in. (34.3 cm); tail, to 9 in. (22.9 cm); 34 teeth, 3131/3132. Stripes on neck, back, and sides; spots on head; and tip of tail all variable, including white.
**Habitat**
Brushy areas, edges of woods, rocky outcrops.
**Habits**
Nocturnal; terrestrial, but also a good swimmer, climber, and burrower; very playful.
**Food**
Rodents, birds, eggs, insects, fruits.
**Reproduction**
2–6 (usually 4) young, born in a dry, vegetation-lined den in early spring.
**Range**
In s. states, from n. Minn., Ind., and s. Pa., southward.

## STRIPED SKUNK
*Mephitis mephitis*                                    **58:4, 61:1**

**Description**
Size: head and body, to 18 in. (45.7 cm); tail, to 10 in. (25.4 cm);
34 teeth, 3131/3132. Narrow white stripe up forehead; tail bushy,
may have white tip; scent glands well developed; can often be
identified by its characteristic pungent odor.

**Habitat**
Woods, plains, meadows, suburbs.

**Habits**
Hunts at night; protects itself by using scent glands if molested.

**Voice**
Low *churr*'s and growls.

**Food**
Rats, mice, chipmunks, insects, fruits, berries.

**Reproduction**
4–7 (usually 5) young, born in vegetation-lined den or sometimes
under a building.

**Range**
Throughout e. U.S. and adjacent Canada.

Note: The **HOG-NOSED SKUNK,** *Conepatus mesoleucus,* has been
reported east from extreme southwestern Texas. It resembles the
Striped Skunk in size, but differs in having a single, broad, white
stripe on the back and a tail that is often entirely white.

## RIVER OTTER
*Lutra canadensis*                                    **58:5, 61:6**

**Description**
Size: head and body, to 30 in. (76.2 cm); tail, to 19 in. (48.2 cm);
weight, to 30 lb. (13.6 kg); 37 teeth, 3141/3132. Above brown to
grayish; chin and throat grayish-white; underparts with silvery
sheen. Snout broad, rounded; tail round, stout, tapering; hind feet
webbed.

**Habitat**
Watercourses and edges of bodies of water.

**Habits**
Makes earthen slides into the water, on which it slides apparently
in sport; active at all times of day and year.

**Voice**
Varied, including grunts, chatters, chuckles.

**Food**
Crayfish, frogs, and fish.

**Reproduction**
1–4 (usually 4) tiny young, born in vegetation-lined den near water
in March or April; animals mature sexually in 2 years, but males
frequently do not breed successfully until 5–6 years old.

**Range**
Formerly from n. Que. and n. Mackenzie, southward; extirpated
over much of s. part of former range.

# HAIR SEALS
Family Phocidae

The hind limbs of these mammals extend behind the tail and cannot be rotated forward. They have short, thick necks; claws on all their digits; and no external ears. Their habitat is coastal and offshore marine waters and ice floes. They feed on fish, squid, crustaceans, and mollusks. The family dates back to the Miocene.

## RINGED SEAL
*Phoca hispida*                                                     **62:1**

**Description**
Size: length, to 6½ ft. (2 m); weight, to at least 200 lb. (90.7 kg); teeth, as in Harbor Seal. Color variable; above usually grayish-brown with ringed spots and streaks, often continuous along back; below, white or yellow, sometimes spotted; area around eyes black; ring marks on sides white to yellowish-brown.

**Similarities**
Harbor has more pronounced pattern and lacks streaks and rings around spots.

**Habits**
Inhabits open, marine waters and pack ice; congregates in winter in sheltered bays and inlets.

**Reproduction**
Pup is white, born February-April in cavity in snow.

**Remarks**
Of all mammals, this species regularly ranges the farthest north.

**Range**
Arctic Ocean, s. to Labrador.

## HARP SEAL
*Phoca groenlandica*                                               **62:4**

**Description**
Size: length, to 6 ft. (1.8 m); weight, to about 400 lb. (181.4 kg); teeth, as in Harbor Seal. No other light-colored seal has a dark saddle across its shoulders and along its sides, and sometimes dark spots on its feet and neck. Adult male has upperparts yellowish to grayish, head and saddle brownish-black, underparts white. Adult female has upperparts brownish to dull white, flippers and saddle brownish-black; face lighter than in male, underparts white to pale brownish. Bedlamer (2–3 years) is above gray or brown, spotted; head and flippers darker. Rusty (under 1 year) is pale gray above with light spots; lighter below. Pup is whitish; eyes and muzzle dark.

**Habits**
Gregarious, migrates south in winter; on way north in April adults and bedlamer gather on ice to fast and molt.

**Reproduction**
Usually 1 pup, born in February or March on ice.

**Voice**
Males emit a long "hoh" and also roar.

**Remarks**
This seal migrates up to 1000 miles (1609.3 km) each way each year from Greenland to Newfoundland. There, on the pack ice offshore and in the Gulf of St. Lawrence, the young are born in March, and thousands are taken each year for pelts by the sealers. For sustained yield this controversial harvest may need to be curtailed. The Canadian population has been estimated to total more than 3 million.

**Range**
Arctic, s. to Gulf of St. Lawrence, occasionally farther.

## HARBOR SEAL
*Phoca vitulina*                                                    62:5

**Description**
Size: length, to about 5 ft. (1.5 m); weight, to 255 lb. (115.7 kg);
34 teeth, 3141/2141; mammae, 2. Yellowish with brown spots, or
brownish with yellow spots. Face doglike.

**Similarities**
Ringed has streaks as well as spots; its spots are often ringed; and
its pattern is generally softer. Gray is larger, grayish, and with
squarer, heavier snout.

**Habits**
Hunts alone along bottom of shallow water, frequently in quiet
harbors and bays; congregates in loosely organized colonies when
on land.

**Voice**
Unimpressive grunts and barks.

**Reproduction**
Pup is usually grayish above, white below, born in spring or early
summer.

**Remarks**
Only seal found regularly south of Nova Scotia, and the only
spotted seal south of Labrador.

**Range**
Arctic and Atlantic coasts, s. to N.C.

## BEARDED SEAL
*Erignathus barbatus*                                              62:6

**Description**
Size: length, to 12 ft. (3.7 m); weight, to 1000 lb. (453.6 kg); teeth,
as in Harbor Seal; mammae, 4. Only eastern seal with thick tufts
of bristles on each side of its muzzle. Above gray to yellow and
brown; below silvery to yellowish; spots absent or obscure. Third
finger on forelimb longest.

**Habitat**
Solitary, sluggish, nonmigratory; feeds on bottom of ocean shallows.

**Reproduction**
Pup to almost 5 ft. (1.5 m), gray, born April–May on ice floes.

**Range**
Arctic, s. occasionally to Newfoundland.

## GRAY SEAL
*Halichoerus grypus*                                              62:3

**Description**
Size: length, to 12 ft. (3.6 m); weight, to 800 lb. (362.9 kg); teeth,
as in Harbor Seal. Adults: above light gray to almost black, often
with obscure spots; lighter below. Yearlings: yellowish with gray
blotches on sides. Body stout; snout long, squarish; upper teeth
hooked; skin has odor of tar.

**Similarities**
Harbor is smaller and not as heavy, with less square head. Hooded
has hood.

**Habits**
Gregarious, but rare in w. Atlantic; more deliberate in movements
than Harbor Seal; slapping of water if alarmed can be heard from
afar. Frequents rough waters with strong currents near rocks, cliffs
or reefs.

**Voice**
"Gloomy . . . between a moo and a howl" (Palmer).

**Reproduction**
1 pup (rarely twins), woolly-white, born on a ledge, sand bar, or ice, in January–February; molts into yellowish coat after 6 weeks.
**Range**
Greenland to Labr.

## HOODED SEAL
*Cystophora cristata*                                               **62:7**

**Description**
Size: length, to 10 ft. (3 m); weight, to 850 lb. (385.6 kg); 30 teeth, 2141/1141. The commonest seal in the North Atlantic. Only seal with an inflatable sac or hood on top of its snout. Adults: blue-gray to slaty-black above, lighter below; muzzle black; sides with white spots on dark blotches. Female paler, hood smaller. Young: called "graybacks," bluish above, flippers blackish-white below; sometimes confused with Harbor or Gray Seal.
**Habits**
Gregarious, migratory; bull, when excited, inflates hood and puffs out mucous membrane into crimson bladders; can keep farther out of water than any other species of seal.
**Voice**
Growl.
**Reproduction**
1 pup, born on ice in February or March, weighs 25–30 lb. (11.3–13.6 kg), nurses for 2–3 weeks.
**Range**
Greenland to N.S.

# CATS
Family Felidae

Members of this family, which includes the domestic cat, have short faces and small, somewhat rounded ears. Five toes are present on each front foot, four on each hind foot; claws are retractile. The dental formula is 3131/3121. The family dates back to the Pliocene.

## MOUNTAIN LION                                          **60:11**
*Felis concolor*                                               *Fig. 41*

**Description**
Size: head and body, 42–54 in. (106.7–137.2 cm); tail, to 36 in. (91.4 cm); height at shoulder, to 31 in. (78.7 cm); weight, to 260 lb. (117.9 kg). Females, ⅓ smaller. The largest North American cat, and only one that is unspotted when adult and that has a long tail. Above yellowish-brown; underparts whitish. Fur short, head small and round, whiskers prominent.
**Habitat**
Remote mountains, plains, woodlands, and swamps.
**Habits**
Active at all hours, ranges widely for food, climbs trees when chased.
**Voice**
Prolonged scream.
**Food**
Mainly deer, also smaller animals.
**Reproduction**
1–6 (usually 2–3) buffy, spotted kits, with ringed tails, are born in

a cave or den in dense woods; litters usually born in late spring, but summer births not uncommon; gestation period approximately 70 days.

**Range**
Formerly throughout U.S., as well as adjacent regions of Canada, from N.B., Ont., and Alta., southward; presently exists only in remote areas of former range.

**Other names**
Cougar, Puma, Panther.

Fig. 41

Mountain Lion

# LYNX
*Felis lynx*                                       Fig. 42

**Description**
Size: head and body, to 40 in. (101.6 cm); tail, to 4 in. (10.2 cm). Color grayish-buff, lightly spotted. Ear tufts prominent; feet large. Somewhat similar to Bobcat, but the Lynx's tail is black only on the tip.

**Habitat**
Forests, thickets, swamps, tundras.

**Habits**
Furred feet give indistinct footprints, larger than those of the Bobcat; climbs and swims well, usually hunts alone.

**Voice**
Catlike, seldom heard.

**Food**
Small mammals, mainly Snowshoe Hare.

**Reproduction**
1–4 (usually 3) kits, which are more strongly marked than adults, born in well-hidden den in some natural cavity in early spring.

**Range**
Formerly throughout most of Canada and n. U.S., s. to Ind. and Nebr.; extirpated over much of s. part of former range.

**Other name**
Canada Lynx.

Fig. 42

Bobcat                                       Lynx

# BOBCAT                                       60:10
*Felis rufus*                                       Fig. 42

**Description**
Size: head and body, to 30 in. (76.2 cm); tail, to 5 in. (12.7 cm). Above pale brown to reddish-brown with black streaks and spots; below whitish with dark spots; ear tufts small. Tail black only above, white at extreme tip.

**Habitat**
Forests, swamps, deserts, mountains.
**Habits**
Nocturnal; swims, climbs well.
**Voice**
Catlike.
**Food**
Hares, rabbits, and other small mammals; ground birds.
**Reproduction**
Breeds in first year; usually 4 (1–6) kits, born in March or April, in den in ledge, log, or underbrush; usually 1 but occasionally 2 litters per year.
**Range**
Throughout e. U.S. and also in s. parts of adjacent provinces of Canada.

# Dugong and Manatee
Order Sirenia

## MANATEES
Family Trichechidae

These are aquatic animals not related to whales, considered to date back to the Pleistocene. There are three species known, but only the Manatee is found in eastern U.S. waters.

### MANATEE
*Trichechus manatus*                                          *Fig. 43*

**Description**
Size: head and body, 10–13 ft. (3–4 m). Body bulky, head broad; lips thick and bristled. No hind limbs; front limbs are flippers and tail is broadly rounded and horizontally flattened. Skin wrinkled; hairs inconspicuous. Color is gray to brown.
**Habitat**
Both fresh and salt water, especially shallow waters near coast.
**Habits**
Relatively slow-moving, hence sometimes injured by motor boats.
**Food**
Aquatic vegetation.
**Range**
Warm waters of Fla. and s. states.
**Other name**
Sea Cow.

*Fig. 43*

Manatee

# Even-Toed Ungulates
Order Artiodactyla

Mammals of this order usually have either two or four toes on each foot. Each toe ends in a hoof. This group is virtually worldwide in distribution. Many artiodactyls have been domesticated.

## OLD WORLD SWINE
Family Suidae

### WILD BOAR
*Sus scrofa*                                                    *Fig. 44*

**Description**
Size: length, to 5 ft. (1.5 m); height at shoulder, to 3 ft. (0.9 m); 44 teeth, 3143/3143 (usually not all teeth are present, however); mammae, normally 12. A wild pig. Tusks to 1 foot, upcurved. Color pale gray to blackish. Legs long.
**Habitat**
Mountains, forests.
**Habits**
Good swimmer and fierce fighter; will breed with domestic swine.
**Voice**
Various grunts.
**Food**
Roots, tubers.
**Reproduction**
3–12 piglets, dark-spotted and striped lengthwise; active very soon after birth.
**Range**
Introduced from Europe into parts of N.H., N.C., and elsewhere as game.

*Fig. 44*

Wild Boar

# DEER
Family Cervidae

These are the only hoofed animals that have true antlers, which they shed every year. Females are appreciably smaller than males and lack antlers, except in the Caribou. Deer hear well and have an excellent sense of smell. The family dates back to the Oligocene.

## ELK
*Cervus elaphus*                                          63:2

**Description**
Size: length of bull, to 9½ ft. (2.9 m); height at shoulder, to 5 ft. (1.5 m); weight, to 1100 lb. (499 kg); 34 teeth, 0133/3133. Antlers unpalmated, branched. In summer, light brown, head and limbs darker, rump buffy, hair short, mane slight. In winter, grayish-brown, head and limbs dark, rump buffy, mane longer and darker. Calf primarily brown with light spots till early fall, rump buffy.
**Habitat**
Semiopen woodlands.
**Habits**
Summers in mountains, winters in valleys; alert, curious.
**Voice**
Far-carrying bugling; a loud bark of alarm.
**Food**
Grass, leaves, twigs.
**Reproduction**
Usually 1 calf, born May–June.
**Remarks**
The elk was extirpated in the East long ago. It has since been reintroduced from the West. The male is the second largest deer, moose is larger.
**Range**
Once common over much of the temperate part of E.; now occurs only in semicaptivity in small, reintroduced populations.
**Other name**
Wapiti.

## MULE DEER
*Odocoileus hemionus*                                54:9, 63:5

**Description**
Size: length, to 6½ ft. (2 m); height at shoulder, to 3½ ft. (1.1 m); weight, to 400 lb. (181.4 kg); 32 teeth, 0033/3133; mammae, 4. Brownish in summer, brown-gray in winter; belly, throat patch, and rump patch white. Ears large, tail ropelike and with blackish tip, scent glands on legs large. Principal tines of antlers branch equally. Fawns have white spots.
**Similarities**
Antlers of White-tailed are prongs from common base.
**Habitat**
Open forests, brushy areas, rock uplands.
**Habits**
Has a jumping gait; carries tail down at all times; more gregarious than White-tailed; summers in mountains or hilly areas, winters in valleys.
**Voice**
A *baaaa*.
**Food**
Grass, forbs, moss, leaves, twigs.

**Reproduction**
Usually 2 (1–3) fawns, born late spring, bred in November–
December; lie hidden to 6–8 weeks.
**Range**
From Canada, southward, and from Minn. and Iowa, w. to 100th
meridian.

# WHITE-TAILED DEER
*Odocoileus virginianus*                                    **54:9, 63:6**
**Description**
Size: length, to 6 ft. (1.8 m); height at shoulder, to 3¾ ft. (1.1 m);
weight, to 400 lb. (181.4 kg); teeth, same as Mule Deer. Only
eastern deer with tail white below and the same color as the back
above. Adults tawny above in summer, blue-gray in winter; white
below. Antlers have erect, unbranched tines, arising from a main
base. Fawn, white spots on reddish coat for 3½ months.
**Similarities**
Mule Deer has the first tines branched; color duller in summer.
**Habitat**
Low mixed woodlands, forest edges, second growth.
**Habits**
Secretive, alert; in flight, raises tail to show white flag of
underside; in winter of heavy snow, congregates and keeps packed-
down "yards" at feeding grounds.
**Voice**
Whistling snort when startled.
**Food**
Grass, leaves, twigs.
**Reproduction**
Usually 2 (1–4) fawns, born in hideaway in brush, in May or June.
**Range**
Throughout E., w. to 100th meridian and s. from edge of tundra.

# MOOSE
*Alces alces*                                              **63:1**
**Description**
Size: length, to 10 ft. (3 m); height at shoulder, to 7½ ft. (2.3 m);
weight, to 1400 lb. (635 kg); 32 teeth, 0033/3133; mammae, 4.
This huge deer is distinguished by its overhanging snout and, in
the male, by the huge broad antlers. Coat blackish or brownish,
legs lighter. Antlers palmate and with small prongs at border,
spread to 6 feet; shed annually. Bull has mane and also bell of skin
and hair hanging from throat. Cow lacks antlers, mane, and bell;
smaller than males. Calf is dull reddish-brown.
**Habitat**
Wilderness forests near shallow lakes.
**Habits**
Feeds in shallow water at dawn, dusk; a good swimmer; hunters
call bull on moose horn by imitating voice of cow.
**Voice**
Bull, a rising *moo*; cow, softer, more like domestic cow.
**Food**
Water plants, leaves, and twigs.
**Reproduction**
Usually 2 (1–3) calves, born mid-May–early June.
**Range**
Formerly edge of tundra, s. to Pa., Minn., and N.Dak.; now rare
or extirpated over parts of former range.

## CARIBOU
*Rangifer tarandus* **63:4**

**Description**
Size: length, to 8 ft. (2.4 m); height at shoulder, to 4 ft. (1.2 m); weight, to about 600 lb. (272.2 kg); 34 teeth, 0133/4033. Unique member of the deer family in that both males and females possess antlers. Only deer in which part of the antlers project toward the nose. Color varies from almost pure white to blackish-brown; white band above hooves.

**Habitat**
Tundra, muskegs, coniferous woodlands; mountains to above timberline.

**Habits**
Migratory, travels in herds; when in motion, feet make clicking sound; gait, loping and running to bounding.

**Voice**
Various snorts and grunts; when in rut, bucks roar.

**Food**
Lichens, reindeer moss, grasses, browse of willows and birches.

**Reproduction**
Usually 1 calf, born in May or June, unspotted, weight 10 to 15 lb. (4.5–6.8 kg).

**Range**
From Greenland, s. to n. border of U.S.; now extirpated in some parts of former range, particularly in s.

# PRONGHORNS
Family Antilocapridae

## PRONGHORN
*Antilocapra americana* **63:3**

**Description**
Size: length, to 4½ ft. (1.4 m); height at shoulder, to 3½ ft. (1.1 m); weight, to 140 lb. (63.5 kg); 32 teeth, 0033/3133; mammae, 4. Only deerlike animal with white rump and 2 white throat bands. Above tan, grayish in winter; below white. Both sexes horned, horn sheath (composed of fused hairs) shed each year. Horn sheath branched about two-thirds of the distance from base to tip; bony horn core unbranched, laterally flattened. The horn core is never shed. Hoofs pointed, lacking dewclaws. Buck with face and patch on side of neck black; horns longer than ears. Doe with black mask and patch almost absent, horns seldom longer than ears. Kid (to 3 months) with coat gray, wavy.

**Habitat**
Plains, prairies, sagebrush flats, deserts.

**Habits**
Active day and night; travels in bands; erects white hairs of rump patch when disturbed; said to run up to 50 mph (80 kph).

**Food**
Browse plants, sagebrush, weeds, grasses.

**Reproduction**
1–2 kids, born in June.

**Range**
From s. Canada, s. to high plains of w. Kans., w. Nebr., and the Daks.; formerly found as far east as Iowa and Minn.

**Other name**
Antelope.

# WILD CATTLE
Family Bovidae

Members of this family have, in both sexes, true, unbranched horns that are not shed. The tooth formula for the wild cattle is 0033/3133. The family dates back to the Miocene.

## BISON
*Bison bison*

54:12
Fig. 45

**Description**
Size: length, to 11½ ft. (3.5 m); height at shoulder, to 6 ft. (1.8 m); weight, to 2000 lb. (907.2 kg). Head huge, neck short, high shoulder hump. Hair on shoulders and front legs shaggy, forming a beard on chin, tail tufted. Bull has brownish-black on head, neck, tail, and shoulder; remainder of body brown to light brown. Cow smaller, more evenly colored, less bearded, smaller hump, horns more slender and curved. Calf reddish-yellow, legs and belly lighter. Adult in winter, yellow-brown.

**Habitat**
Grasslands and open woodlands.

**Habits**
Highly gregarious, formerly migratory; eyesight fairly poor, hearing and sense of smell keen; likes to wallow in mud or dust.

**Voice**
A bellow.

**Food**
Grass.

**Reproduction**
Usually 1 calf, born in May; hump not noticeable until fall; stays up to 3 years with cow.

**Range**
Formerly throughout e. U.S., except for Atlantic Coast, New England, and ne. Canada; extirpated over most of former range and at present restricted to small herds on parks and game preserves, notably Wind Cave National Park and Custer State Park in the Black Hills of S.Dak.

**Other name**
Buffalo.

Bison

*Fig. 45*

## MUSKOX
*Ovibos moschatus*

Fig. 46

**Description**
Size: length, to 6 ft. (1.8 m); height at shoulder, to 5½ ft. (1.7 m); weight, to 900 lb. (408.2 kg). Deep brown or blackish, nose and patch behind shoulders pale, long fur hangs down almost to its feet. Horns flat, broad, close to skull, and with curved tips pointing

forward; shoulders with slight hump; legs and neck short; tail very
short. Cow frequently has paler face and more slender horns.

**Habitat**
Arctic and subarctic coast, tundra, and foothills.

**Habits**
Gregarious; habit of forming circle with calves inside, if attacked,
has made it an easy prey to the rifle; species now dangerously
reduced in numbers.

**Voice**
A bellow; also various snorts.

**Food**
Sedges, grasses, leaves, twigs.

**Reproduction**
Usually 1 calf, born April–June on open tundra.

**Range**
From Greenland and n. Canada, s. to 60° latitude.

*Fig. 46*

Muskox

# Reptiles

**Consulting Editor**
Patricia Gardner Haneline
Andrew Mellon Fellow
Department of Biological Sciences, University of Pittsburgh

**Illustrations**
Plates 64–69 John Cameron Yrizarry
Text Illustrations John Cameron Yrizarry and
Jennifer Emry-Perrott

# Reptiles
Class Reptilia

A reptile is a cold-blooded vertebrate that lays amniotic eggs covered with a hard shell on land, or that bears living young on land or in the sea. But it is not dependent, as are amphibians, on the return to the water to breed. Fertilization is internal, there is no free larval stage, and the young, as well as the adult, has a dry skin bearing scales or horny plates. The term cold-blooded, in the strict sense, is a popular misnomer. It means not that the blood necessarily is "cold," but that the body temperature varies with that of the surrounding atmosphere, and is not maintained internally at a constant level, as it is in birds and mammals. Ectothermic in a better term. A controversy is currently raging over the physiology of dinosaurs, and whether they were warm-blooded like birds and mammals. If so, they were probably more active than has been popularly portrayed. Eastern American reptiles certainly cannot survive freezing temperatures and must hibernate in holes or burrows below the frost line in climates with cold winters. Reptiles are more abundant both as to species and individuals in the tropics and subtropics than in the temperate zones.

Of all land vertebrates, reptiles have enjoyed the longest sway. During the entire Mesozoic era, the age of dinosaurs, they ruled the land and sea and air. Brontosaurus in the swamp, Tyrannosaurus on land, Pteranodon in the air, Ichthyosaurus and Plesiosaurus in the sea were the largest and most powerful creatures earth had ever seen. Compared with this seemingly endless era of 130 million years, the subsequent reign of the birds and mammals has been short-lived, and that of man is but as minutes out of a twenty-four-hour day.

The group of reptiles known as dinosaurs has now long been extinct. But other groups still live, namely the orders of turtles, crocodilians, lizards and snakes, and rhynchocephalians. Of this last, only one species, the Tuatara, survives, a primitive lizardlike form virtually unchanged since the Jurassic. It is now confined to a few small islands off New Zealand.

### Size
The sizes given for identification are maximums. When an unusually large individual has been found, that record is listed separately.

### Range and Scope
In this chapter are described forty-two species of turtles, one crocodilian, eighteen lizards, and sixty snakes. These all occur within the range, which covers the United States from the Canadian border (50th parallel) to the south of Florida (26th parallel); and from the 100th meridian east to the Atlantic Coast.

### Nomenclature
The common and scientific names in this chapter follow *Standard Common and Current Scientific Names for North American Amphibians and Reptiles.*

## USEFUL REFERENCES

Anderson, P. 1965. *The Reptiles of Missouri.* Columbia: Univ. of Missouri Press.

Barbour, R. W. 1971. *Amphibians and Reptiles of Kentucky.* Lexington: University Press of Kentucky.

Bellairs, A. 1969. *The Life of Reptiles.* 2 vols. London: Weidenfeld and Nicolson.

Cagle, F. R. 1968. *Vertebrates of the United States.* 2d ed. pp. 213–68. New York: McGraw-Hill.

Carr, A. 1952. *Handbook of Turtles.* Ithaca, N.Y.: Comstock.

———. 1963. *The Reptiles.* New York: Time, Inc.

———. 1956. *The Windward Road.* New York: Knopf.

Collins, J. T. 1974. *Amphibians and Reptiles in Kansas.* Lawrence: University of Kansas Mus. of Nat. Hist.

Conant, R. 1975. *A Field Guide to Reptiles and Amphibians of Eastern and Central North America.* 2d ed. Boston: Houghton Mifflin.

Ernst, C. H., and Barbour, R. W. 1973. *Turtles of the United States.* Lexington: University Press of Kentucky.

Goin, C. J., and Goin, O. B. 1978. *Introduction to Herpetology.* 3d ed. San Francisco: Freeman.

Kauffeld, C. 1957. *Snakes and Snake Hunting.* New York: Hanover House.

———. 1969. *Snakes: The Keeper and the Kept.* New York: Doubleday.

Lazell, J. D., Jr. 1976. *This Broken Archipelago: Cape Cod and the Islands, Amphibians and Reptiles.* New York: Quadrangle.

Leviton, A. E. 1972. *Reptiles and Amphibians of North America.* New York: Doubleday.

Minton, S. A., and Minton, M. R. 1973. *Giant Reptiles.* New York: Scribner's.

Mount, R. H. 1975. *The Reptiles and Amphibians of Alabama.* Auburn: Auburn Univ. Agricultural Experiment Station.

Oliver, J. A. 1955. *Natural History of North American Amphibians and Reptiles.* Princeton, N.J.: Van Nostrand.

Pope, C. H. 1955. *The Reptile World.* New York: Knopf.

———. 1937. *Snakes Alive and How They Live.* New York: Viking.

———. 1946. *Turtles of the United States and Canada.* New York: Knopf.

Porter, K. R. 1972. *Herpetology.* Philadelphia: W. B. Saunders.

Pritchard, P. C. H. 1967. *Living Turtles of the World.* New York: T.F.H. Publications.

Schmidt, K. P. 1953. *A Check List of North American Amphibians and Reptiles.* 6th ed. Chicago: American Soc. of Ichthyologists and Herpetologists.

Schmidt, K. P., and Davis, D. D. 1944. *Field Book of Snakes.* New York: G. P. Putnam's Sons.

Schmidt, K. P., and Inger, R. F. 1957. *Living Reptiles of the World.* Garden City, N.Y.: Doubleday.

Smith, H. M. 1946. *Handbook of Lizards.* Ithaca, N.Y.: Comstock.

SSAR Committee on Common and Scientific Names. 1978. *Standard Common and Current Scientific Names for North American Amphibians and Reptiles.* Available from D. H. Taylor, Dept. of Zoology, Miami University, Oxford, Ohio.

Stebbins, R. C. 1954. *Amphibians and Reptiles of Western North America.* New York: McGraw-Hill.

Wright, A. H., and Wright, A. A. 1957. *Handbook of Snakes.* 2 vols. Ithaca, N.Y.: Comstock.

## GLOSSARY

*Anterior*  Toward the front (of the body).

*Carapace*  Upper shell of a turtle or tortoise, including bony plates and horny shields.

*Caudal scale*  Straplike scale extending across ventral surface of tail.

*Cloaca*  Area into which internal wastes discharge.

*Dewlap*  Skin fold hanging from neck region.

*Diurnal*  Daytime.

*Dorsolateral*  Pertaining to the upper sides of the animal.

*Exfoliation*  Scaling off in flakes.

*Frontals*  Bony membranes that form the forehead.

*Interspace*  Area of merging of two dorsal color patterns in lizards and snakes.

*Lateral*  Of or pertaining to the side of the body.

*Maxillary bone*  Bone on each side of head, forming side border of upper jaw and bearing most of the upper teeth.

*Occipitals*  Bony membranes that form the posterior part of the skull.

*Parietal bones*  A pair of membrane bones in the roof of the skull between the frontals and occipitals.

*Plastron*  Underpart of the shell of a turtle or tortoise.

*Posterior*  Toward the rear (of the body).

*Postocular*  Behind the eye.

*Postorbital*  Behind the eye.

*Reticulate*  Having the form or appearance of a net.

*Riparian*  Relating to the bank of a river, lake, or pond.

*Rugose*  Rough, wrinkled.

*Shield*  In turtles, any one of the horny plates that cover the shell.

*Temporal horns*  In horned lizards, horns toward the sides of the crown.

*Tubercle*  Any of various small knoblike prominences.

*Vent*  Opening on the surface of the cloaca.

*Venter*  Belly.

*Vertebral stripe*  Stripe down the midline of the back.

*Vertical pupil*  Eye in which pupil is elliptical; long axis is vertical.

# Crocodilians
Order Crocodilia

Crocodilians are large, lizardlike reptiles with an elongate skull, four sturdy limbs, and a long, powerful tail. They flourish in fresh- and saltwater habitats in tropical and subtropical regions, and alligators are found in certain temperate climates. Crocodilians are familiar to us through animal parks, museums, and books. In the South, the American Alligator was once struggling for its survival, but fortunately, enforcement of conservation laws during this last century has enabled it to regain healthy populations. To encounter one of these gigantic beasts in the field is a memorable experience.

**Features of Crocodilians**

Crocodilians are usually divided into three groups. The crocodiles occur in Africa, Asia, Australia, widely in the Pacific, and in the Americas. The alligators are found in North America and in the Yangtze drainage of China (if they are not already extinct), and their close relatives, the caimans, are widely distributed in Central and South America. The strangest species of all is the Gavial, or Gharial, which inhabits rivers in India and feeds largely on fish by lashing its long and extremely thin jaws back and forth in the water.

Beneath the thick, horny scales are bony plates called osteoderms on neck, back, and tail and, in some species, on the belly. Crocodiles are unusual among reptiles in that they have a four-chambered heart to prevent mixing of oxygenated and deoxygenated blood and a good secondary palate that allows them to eat and breathe simultaneously. Their teeth are implanted in sockets in the jaw and are generally similar in shape. Crocodilians are carnivorous, devouring prey whole or tearing it to pieces with their strong jaws. Their nostrils are located at the tip of the long snout, and the eyes are located at the top of the skull, enabling them to expose as little of the body as possible while cruising in the water looking for food. Crocodilians are also commonly seen basking or building nests on river banks and coastal shores.

**Identification**

The American Alligator is discussed below because it is widely distributed in the South. Problems of identification will occur only in Florida: The American Crocodile breeds in the Florida Keys and may travel up the coast, and the Spectacled Caiman has been introduced through the pet trade. In the American Crocodile the snout tapers to a point, and the fourth lower tooth is greatly enlarged and visible from the side when jaws are closed. In the American Alligator the snout is broad and is rounded in front, and the fourth tooth, while enlarged, fits into a pit in the upper jaw and is not exposed at the side when jaws are closed. The Spectacled Caiman is small and has a bony ridge between the eyes.

**Reproduction**

Crocodilians are unusually active in the care of their young. The mating and nesting behavior of the American Alligator is well known and will be described below, but all species exhibit similar behaviors. During courtship the male strokes the female with his forelimbs and head and finally embraces her and bends his hindquarters under her body. The female builds a nest on land of soil and vegetation. First she beats down an area about four feet by six feet (1.2 m by 1.8 m) in size and then gathers twigs, leaves, and other plants, which are piled and compacted to a height of three feet (0.9 m). In a depression that she makes in the top of the mound the eggs are laid, and then they are covered over. The female remains near the nest and periodically moistens it. When, with the aid of an egg caruncle on the snout of the embryo, the young hatch, they usually make noises in concert and this alerts the female to come and open the nest. Some species have been seen carrying young in the mouth to water. In any case, the female may stay with them for as long as a year, sheltering them in the kind of burrow that all adult alligators make near river banks, consisting of a hole four to six feet (1.2–1.8 m) deep connected to an underground passage some fifty feet (15.2 m) in length and about ten feet (3.0 m) in width. Predation is heaviest on the eggs and young, common predators in the United States being the peccary and bears, skunks, and raccoons. When an alligator has reached adult size it has few enemies other than man.

# ALLIGATORS
Family Alligatoridae

## AMERICAN ALLIGATOR
*Alligator mississippiensis*          *Fig. 47*

**Description**
Length, to 19 ft. 2 in. (5.8 m); young 9 in. (22.9 cm) at hatching. Large, lizardlike, distinguished by its size and broad, flattened snout. Ground color brown with 8 rows of enlarged shields running down back. Limbs stout, well-developed webs between toes. Tail massive and dangerous. Young have yellow and black crossbands.

**Habitat**
Usually near water: ponds, swamps, rivers.

**Habits**
Often seen submerged in water amidst aquatic vegetation. Hibernates in the North in burrows during cold weather.

**Voice**
Adults bellow frequently and young make high-pitched noises when in danger.

**Food**
Young—insects, small vertebrates such as frogs or fish; adults— fish, turtles, snakes, and small mammals, wild and domestic. Large adult can take deer or cattle.

**Reproduction**
Lays 2–88 eggs; average 42.

**Range**
Both coastal and inland waters; in Coastal Plain from se. N.C. s. through Fla., and w. to coastal Tex.; n. to s. Ark. Introduced in the Rio Grande of Tex.

*Fig. 47*                        American Alligator

# Turtles
Order Testudines

Turtles are extraordinary animals, distinguished for their ancient ancestry, conservatism, and tenacity. Today man is their number one enemy, and due to his predation and destruction of habitats, many species in recent years have suffered a reduction in numbers. There are in the world today only about 255 species, divided into eleven families. Of these, 42 species are covered here from the eastern United States. Turtles are hardly as diverse as lizards or snakes, but they are widespread in warm and temperate climates and immediately recognizable.

## Features of Turtles

A turtle is simply a reptile with a shell. The shell may be rigid and made of a bony layer overlaid by large horny scales called scutes; or it may be pliable and covered with a leathery skin. Vertebral column and ribs between neck and sacrum, in all turtles except the Leatherback, are fused with the bone outside of the hip and shoulder girdles, a situation unique among vertebrates. The clavicle is incorporated into the plastron. Partly because of this bizarre arrangement, it has been said that no scientist could make an accurate restoration from fossil remains alone. It is known from embryological studies that the turtle starts growing normally, but during development the girdles migrate inward, and muscles that normally occur above the vertebral column degenerate. The bones of the shell consist of a middorsal row and variable rows to the side and around the edge above, and both paired and unpaired series below, with a bridge often connecting them at the sides. The bones of this dorsal carapace and ventral plastron fuse into a solid shell in most turtles. Sometimes a hinge is present from side to side on the plastron to effect complete closure of the extremities into the protective shell. The seams between scutes are not arranged in line with the bone sutures below, thereby adding strength to the shell construction. In sea turtles the bone has been reduced to lighten the mass.

A turtle has four well-developed limbs and the shape of the feet reflects the environment in which it lives. Terrestrial tortoises have stout elephantine limbs with claws and no webbing. Aquatic fresh-water species often have webbed feet and long claws, and pelagic sea turtles have limbs modified into true flippers with elongated and bound digits. The forelimbs are always larger than hindlimbs, and act as paddles, while the hindlimbs are used in steering. Another characteristic of all living turtles, which proves quite effective in procuring food and repelling predators, is the presence of a horny beak covering the jaw bones. The solidly built head has some features unusual in reptiles. The palate is fused to the braincase, and the skull is cut away from behind to accommodate the large muscles necessary to move the jaws.

Turtles, of course, are air breathers and have lungs. The mechanics of inhalation and exhalation have been solved by the retention of abdominal muscles within the immovable shell. Respiration is aided in aquatic forms by the mouth and cloaca, which possess respiratory surfaces similar to that of a gill, capable of absorbing oxygen from water passed over them. This permits a turtle to stay underwater a long time without surfacing for air, although it also uses anaerobic metabolism.

Turtles do not hear well, but they do smell and see well, and they are sensitive to touch on both shell and body. They are the most intelligent of the reptiles. Most eastern American turtles are silent, although the Leatherback, if so inclined, can utter a bellow audible for a quarter mile; some other species emit hisses, barks, or grunts.

### Identification
Terms used in describing the turtle shell are illustrated below. A turtle that has retreated into its shell might emerge again when placed in a container of water. Note coloration of shell and skin, retractability of the head, number and configuration of the scutes, keels and knobs on the carapace, and size. Sizes given in the text approach the maximum, although average lengths may be smaller. When an individual far exceeds the maximum length of most specimens, both that figure and the record length are given. For measurement of length, take the straight line distance along the midline of the carapace.

### Habitat
Turtles like the edges of bodies of water, such as lakes, ponds, rivers, and swamps. They also like to bask on offshore logs or rocks. Terrestrial turtles are commonly found in open woods, at the edges of fields, and in abandoned pastures and gardens. Turtles are most active in spring, when they move about looking for mates or nesting sites after emerging from their winter hibernation. They are also likely to be seen after rains. Summer heat will drive them into retirement, or in some cases into aestivation. In early fall, turtles are again on the move and more readily seen than in midsummer.

Successful strategies for finding turtles include using field glasses to observe turtles as they bask in ponds, seining lakes, attaching bucket traps to basking sites, and diving under water to catch them by hand. They may be revealed by using a bright flashlight while wading through shallow water at night. Baited traps can also be used. An experienced companion can often point out well-known haunts and habits in your area, and the curator of your local zoo or natural history museum can give you an idea where and when to find them.

### Measurements
The sizes given are for maximum lengths of *carapace*.

*Fig. 48*
Parts of a Typical Turtle

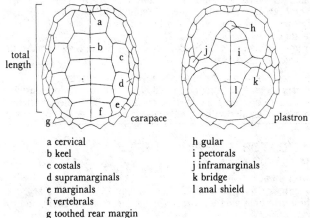

carapace                    plastron

a cervical            h gular
b keel               i pectorals
c costals            j inframarginals
d supramarginals     k bridge
e marginals          l anal shield
f vertebrals
g toothed rear margin

# SNAPPING TURTLES

Family Chelydridae

Members of this family have a small, cross-shaped plastron, a greatly enlarged, nonretractile head, powerful jaws with a hooked beak, paired barbels, and a very long tail with erect bony scales. The family is confined to the Americas from Canada to Ecuador, and is represented in the East by two species. Authorities disagree about whether a third species exists in Florida.

## SNAPPING TURTLE

*Chelydra serpentina*                                                64:1

### Description
Length, to 12 in. (30.5 cm); record size 18½ in. long and 86 lbs. (47.0 cm and 39.0 kg). Shell brown to olive above, lighter below. Large head covered with skin, not plates; eyes can be seen from directly above (not so in Alligator Snapping); neck and legs warty; tail has upperside saw-toothed. Carapace has 3 ridges of knobby plates, usually flatter in older specimens, rear margin toothed, and often algae-covered; plastron small and cross-shaped. Young have sharper keels on carapace than adults.

### Similarities
Alligator Snapping Turtle is larger, keels on carapace are higher; has extra row of scutes on sides of carapace.

### Habitat
A common species, found in any aquatic situation, preferably with mud and abundant vegetation.

### Habits
Vicious predator, actively seeks prey at night, but during day may lie on bottom at a depth where head can reach air for breathing. Basking uncommon. When encountered on land will lunge at humans.

### Food
Plants, earthworms, crayfish, clams, fish, frogs, salamanders, snakes, turtles, birds, mammals, and carrion.

### Reproduction
Lays 11–83 round, hard eggs, average 25, May–September (usually in June) in hole about 6 in. (15.2 cm) deep in well-drained soil some distance from shore; incubation 8–18 weeks; young may overwinter in nest.

### Remarks
The most widely distributed and consumed North American turtle. When carried, it should be held by hindlimbs with plastron facing the carrier and well away from human legs.

### Range
From Canada and e. U.S. to Tex. and w. to the Rockies. Also in Mexico and S. America.

Note: A subspecies, the **FLORIDA SNAPPING TURTLE,** *Chelydra serpentina osceola,* is found in peninsular Florida and is considered by some to be a separate species. It has granular scales and long, pointed tubercules on back of head.

## ALLIGATOR SNAPPING TURTLE

*Macroclemys temmincki*                                             64:2

### Description
Length, to 26 in. (66.0 cm); record weight 219 lbs. (99.3 kg). Similar to Snapping Turtle, but much larger and heavier, and eyes

cannot be seen from directly above. Check for extra row of 3–5 supramarginal scutes at each side above marginals. Light gray plastron small; brown carapace has 3 keels, persistent in old individuals; rear margin deeply and evenly toothed. Head covered with plates; nose pointed; beak with pronounced hook; fleshy pink extension, or "lure," on tongue. Tail very long, about length of plastron. Young rougher than adults.

**Similarities**
Snapping Turtle has keels lower, head covered with skin.

**Habitat**
Deep water of rivers, oxbows, sloughs, canals, swamps, sluggish streams.

**Habits**
Forages for food at night but by day "fishes" by lying motionless on bottom and wiggling lure with mouth open.

**Food**
Omnivorous, including fish, shellfish, carrion.

**Reproduction**
Lays 15–50 eggs, April–July in hole 4–7 in. (10.2–17.8 cm) deep, closer to water than Snapping Turtle.

**Remarks**
Largest freshwater turtle in the world; consumed in the South. Can give dangerous bite.

**Range**
Se. U.S. from sw. Ga. and n. Fla. w. to e. Tex., n. through se. Kans., and up Miss. R. to sw. Ind. and se. Iowa.

# MUD AND MUSK TURTLES
Family Kinosternidae

Distributed widely in the New World, members of this family have odoriferous glands at either side of bridge region, have ten or eleven scutes on the plastron, and the pectoral scutes are not in contact with the marginals. Their jaws have smooth edges and they have claws on their limbs. Two genera are present in the United States, and they can be separated as follows: mud turtles, Genus *Kinosternon,* have a triangular pectoral scute and two transverse hinges across the large plastron; musk turtles, Genus *Sternotherus,* have squarish pectoral scutes and a single, indistinct hinge across the reduced plastron.

# EASTERN MUD TURTLE
*Kinosternon subrubrum* 65:4

**Description**
Length, to 4 in. (10.2 cm). Two distinct transverse hinges on plastron, and marginal scutes all approximately same height. Carapace unstriped, yellow to black, below yellow to orange or reddish. Head skin variably marked with 2 light stripes behind eye, sometimes broken into spots. Carapace smooth, unkeeled or with slight median keel, depressed; plastron wider, hinged at both ends. Male has nail at tip of a thicker tail and horny scales on inner side of thigh and below knee. Young have ridges on plates above and a large, yellow to orange spot on each marginal plate.

**Habitat**
Meadow brooks, ponds, ditches; also brackish and salt water.

**Habits**
Can exude offensive musk (but less obnoxious than that of Stinkpot); often forages on land.

**Food**
Omnivorous; aquatic insects, algae, crustaceans, carrion.
**Reproduction**
Lays 2–5 eggs under logs or in hole in ground, June–July.
**Range**
Sw. Conn., s. throughout se. U.S., excluding Appalachian region; n. into s. Ill. and w. Ind. and w. to s.-cen. Okla. and e. Tex., with a locality in w.-cen. Mo.

Note: The **STRIPED MUD TURTLE**, *Kinosternon bauri,* is distinguished by light stripes on the head and carapace. It occurs in extreme southern Georgia, south through Florida.

## YELLOW MUD TURTLE
*Kinosternon flavescens*                                          65:2
**Description**
Length, to 6 in. (15.2 cm). The 9th and 10th marginal scales are much higher than the 8th. Head, neck, and throat show considerable bright yellow. Carapace brown to olive with no markings; plastron yellowish-brown. Carapace smooth, unkeeled, low, broad. Male has horn-tipped tail and 2 patches of horny scales on inner surface of hindleg. In young, 9th and 10th marginals lower than 7th and 8th, but 9th has pointed top.
**Habitat**
Pools, ponds, marshes, canals, cattle tanks; prefers a mud bottom.
**Habits**
Basks; may forage on land, particularly in rainy season; migrates overland if pools dry up; hibernates in burrows during winter, aestivates under soil during drought. Can emit a strong odor.
**Reproduction**
Lays 2–4 eggs, July–August.
**Range**
S. Nebr. and e. Colo., s. to n. Mexico, and w. to Ariz. Disjunct populations in extr. se. Kans. and adj. Mo., and in w. Ill. and adj. states; also in cen. Ill.

## STINKPOT
*Sternotherus odoratus*                                          65:3
**Description**
Length, to 4½ in. (11.4 cm). Lacks distinct hinge below and has 2 conspicuous light stripes on each side of dark head and neck, 1 above and 1 below eye. Chin and throat have little projections of skin (barbels). Above dark brown or black, below lighter, yellow to brown; light stripes on sides of head may be indistinct. Carapace high, unkeeled in adults, often algae-covered; scutes do not overlap. Male has blunt nail on end of tail and horny scales on inner surface of hindleg. Young have a higher, keeled carapace.
**Habitat**
Any permanent water.
**Habits**
Pugnacious, thoroughly aquatic but may climb up trees above water; stalks prey on bottom at night.
**Food**
Aquatic plants, carrion, mollusks, insects, and crayfish.
**Reproduction**
Lays 1–9 hard, pebbly eggs, May–August under logs, brush piles, at base of a tree, or in shallow hole dug in loose soil.

**Remarks**

Stinkpot is the common name for this species, although Common Musk Turtle is also used. Name derives from 2 pairs of glands at bases of limbs just below carapace, which emit a powerful, disagreeable, musky odor.

**Range**

S. Ont. and most of e. U.S. from s. Maine w. into s. Wis., e. Kans., and e. Tex.

Note: In the **RAZOR-BACKED MUSK TURTLE,** *Sternotherus carinatus,* a keel is always present, and the upper part of the carapace forms a very pointed top. It occurs in the southern part of the Mississippi Valley, to southeastern Oklahoma and eastern Texas.

## LOGGERHEAD MUSK TURTLE
*Sternotherus minor*

**Description**

Length, to 4½ in. (11.4 cm). Above brown, usually with irregular darker markings; below yellow and unmarked. Skin gray-brown with dark brown or black markings, head with dark stripes or spots. Carapace smooth, high, with some indication of median keel throughout life and sometimes 2 lateral keels; scutes are overlapping (imbricate). First vertebral scute does not touch second marginal; 10th and 11th marginal scutes raised; 1 scute on throat. Barbels on chin only. Male has spine on tail and horny scales on legs.

**Similarities**

Stinkpot has barbels on neck.

**Habitat**

Lakes, cypress swamps, and streams; found near submerged logs and debris.

**Habits**

Aquatic; bites when disturbed and should be held by rear of shell.

**Range**

Extr. w. Va., cen. Tenn. s. through n. Fla. and w. Miss.

Note: A subspecies, *Sternotherus minor minor,* lacks stripes, but has dark spots over a light ground on the neck and head. Males have a greatly enlarged head; young have 3 distinct keels, which are lost with age. It occurs in central and southern Georgia through northern Florida and southeastern Alabama.

Note: The **STRIPE-NECKED MUSK TURTLE,** *Sternotherus minor peltifer,* has a neck with dark stripes over a light ground color. There is a single middorsal keel in adults. It occurs in extreme southwestern Virginia, perhaps in adjacent Kentucky, through eastern Tennessee, northwestern Georgia, and south and west through Alabama to extreme eastern Louisiana.

# SOFTSHELL TURTLES
Family Trionychidae

Members of this peculiar family, so different from all other freshwater turtles, split off from the main evolutionary line sometime in the Cretaceous, perhaps 100 million years ago. They are all aquatic. They have a flat, leathery shell with cartilaginous, and therefore pliable, edges. Softshells can breathe (in part at least) through the leathery skin of this shell and through highly vascularized folds in the pharynx (neck region) by absorbing oxygen from the surrounding water as it is flushed in and out of the nostrils or mouth. The nostrils are at the end of a long, tubelike extension of the snout. Males are smaller than females. Specimens should be handled with care by the rear of the carapace and away from the body, because the long neck allows a great range for powerful jaws, and the clawed, webbed feet are quite dangerous. Softshells are distributed in Africa, North America, Asia, and through many localities in the Pacific to New Guinea.

## SPINY SOFTSHELL
*Trionyx spiniferus*                                    **64:3**

**Description**
Length, to 18 in. (45.7 cm), but males to only 9 in. (22.9 cm). Has spines or warts on front margin of leathery shell and ridges projecting from the septum separating nostrils. Carapace gray, low, flat, almost round, and with spotted dark marginal line, but no ridge. Prominent markings in juveniles of dark rings or dots (or white dots, depending on area of origin) fade into lichenlike markings in adult females. Two light lines along each side of head through eye and back from jaw. Feet strongly patterned. Carapace has gritty or sandy surface.

**Habitat**
A freshwater species, in large rivers, ponds, lakes, small streams.

**Habits**
Lies in ambush for prey under mud or sand at bottom of quiet, shallow water; extends tubular nostrils to surface periodically to breathe, but can submerge a long time; a great basker, sometimes in groups, on shore or floating at surface.

**Food**
Plants, insects, crayfish, mollusks, earthworms, frogs, some carrion.

**Reproduction**
Lays 4-32 round eggs, June-July in flask-shaped hole up to 10 in. (25.4 cm) deep near water; hatchlings August-October.

**Remarks**
Vicious to handle; can give nasty bite and scratches with its sharp claws.

**Range**
Most of cen. U.S. n. to cen. Minn. and cen. Wyo., and in e. U.S. to w. N.Y. and to Atlantic Coast in Ga.; s. to Fla. panhandle and w. along Gulf Coast. Isolated populations in the East in ne. N.Y. and adj. N.H. and Canada, and in N.J. Sporadic distribution in the West and n. Mexico.

## SMOOTH SOFTSHELL
*Trionyx muticus*                                    **64:4**

**Description**
Length, to 14 in. (35.6 cm), but males average half that size. Has leathery shell with smooth front margin above. Carapace nearly circular, olive to brown; plastron clear and light; light line behind eye. Old females develop a lichenlike mottling, while males and juveniles have spots and streaks. Nostrils circular, no ridge projects from septum. Feet without markings.

**Habitat**
A freshwater species, in permanent water: large, slow rivers; lakes; clear, sandy creeks.

**Habits**
Similar to those of Spiny Softshell, but more aquatic; less of a basker, slightly less pugnacious.

**Food**
Crayfish, insect larvae and other invertebrates, fish, frogs, tadpoles, and some vegetation.

**Reproduction**
Lays 4–33 round eggs, spring or late summer, in holes excavated in sandbars or banks; incubation about 70 days.

**Range**
W. Pa. and e. Nebr. s. to Gulf Coast in w. Fla. and e. Tex. (farthest westward record in e. N.Mex.); and following large rivers n. into Minn. and s. N.Dak.

Note: The large **FLORIDA SOFTSHELL**, *Trionyx ferox,* is the heaviest and bulkiest of all North American turtles, but has the most limited range. It is dark brown with large, vague, dark spots and a bumpy carapace. It occurs in the Coastal Plain and all of Florida except the Keys.

# Freshwater and Box Turtles
Subfamily Emydinae

This subfamily comprises the largest number of turtle species. The toes are webbed, or partly so, and the species are generally aquatic or semiaquatic, although the box turtles have been secondarily adapted to a life on land. The plastron contains twelve scutes, with the pectoral scute connected to the marginals. They are widely distributed in the Northern Hemisphere and absent from Australia and southern parts of Africa.

## WOOD TURTLE
*Clemmys insculpta*                                          *Fig. 49*

**Description**
Length, to 9 in. (22.9 cm). Has strongly keeled carapace, and orange on underside of legs and throat. Above dark, sometimes with black and yellow markings. Carapace low, broad, keeled; scutes on back sculptured with concentric grooves into pyramidal blocks, rear edges flaring. Plastron rigid, scutes yellow, each with black blotch at outer rear section. Legs and neck dark, sometimes with red dots. Beak notched. Male has thicker tail, longer claws, and larger scutes on forelimbs than female.

**Habitat**
Likes to swim but may be terrestrial and may feed some distance from water; found in bogs and marshes.

**Habits**
Sometimes hibernates in Muskrat den.

**Food**
Algae, fish, tadpoles, insect larvae, mollusks, earthworms; in captivity, berries, meat.

**Reproduction**
Lays 4–12 eggs in June.

*Fig. 49*

Wood Turtle

**Remarks**
Unusually intelligent; makes a good pet and will eat from the hand.
**Range**
Ne. U.S. and extr. se. Canada, s. through most of N.Y. to n. Va. and W.Va. and w. around Great Lakes in north of range to e. Iowa and Minn.

## SPOTTED TURTLE
*Clemmys guttata*                                                    **65:1**

**Description**
Length, to 5 in. (12.7 cm). Small low carapace has black ground color with various numbers of conspicuous yellow spots; plastron light with black blotches usually on outer part of plates. Head, neck may show yellow markings. Male has brown eye and chin, concave plastron. Female has orange eye, yellow chin, and flat or convex plastron.
**Habitat**
Bogs, brooks, pastures.
**Habits**
Emerges from hibernation early in spring, stays out March-June; during day basks and forages actively, nesting females may be out of water at night. Often wanders across roads and is killed by careless (or malicious) motorists.
**Reproduction**
Lays 1-5 eggs in June.
**Range**
S. Maine, w. through most of lower Mich. and into e. Ill., but s. of Pa. Restricted to Atlantic Coastal Plain through cen. Fla.

## BOG TURTLE
*Clemmys muhlenbergi*                                              *Fig. 50*

**Description**
Length, to 4½ in. (11.4 cm). Has big, bright orange ear patch and low keel. Ground color above brown to black, plastron lighter, both usually with some red or yellow markings, but not dotted. Skin often mottled with orange or reddish. Carapace long, domed, straight-sided, wider at rear; scutes somewhat sculptured; plastron long and broad. Male has narrower rear notch on plastron than female.
**Habitat**
Sphagnum bogs, wet meadows, small streams; often seen on land. Man's destruction of bog habitats has made this species rare.
**Habits**
Aestivates as well as hibernates; likes to burrow.

Bog Turtle

*Fig. 50*

Blanding's Turtle, p. 331

**Food**
Berries, insects, but apparently omnivorous; feeds in captivity on chopped meat, earthworms, greens, fruit.
**Reproduction**
Lays 3–5 eggs, June–July.
**Range**
Very restricted and patchy in e. U.S., including cen. N.Y., R.I., w. Mass., N.J., parts of Pa., and the meeting of Va., Tenn., and N.C.

# EASTERN BOX TURTLE
*Terrapene carolina*                                          *Fig. 51*

**Description**
Length, to 6 in. (15.2 cm); record 8½ in. (21.6 cm). Slightly keeled, oblong, boxlike carapace, highest in the middle, plastron dark or mottled, but never striped. Shell brown to olive and yellowish (see below for description of subspecies variations). Head, neck, limbs irregularly marked with yellow. No bridge; wide plastron with single hinge; can completely enclose head, limbs, and tail; tail short. Male has bright red eyes, female yellow-brown. Young without hinged plastron.
**Habitat**
Open woodlands, wood edges, fields, stream banks, moist areas.
**Habits**
In hot, dry weather, burrows in mud or seeks shade; active in morning and after rain.
**Food**
Young mostly carnivorous (earthworms, snails, insect larvae); adults mostly herbivorous (berries, fungi, fruit); in captivity fresh meat, canned dog food, fruits, greens.
**Reproduction**
Usually mates soon after emergence from hibernation, but nests throughout summer; lays 2–8 eggs, May–July, in flask-shaped hole in loose soil 3 in. deep; incubation 69–136 days. Young may emerge and then hibernate, or overwinter in nest.
**Remarks**
Gentle, makes a good pet. Many are crushed by cars while crossing highways. This is the familiar turtle of woodlands, prairies, and plains. The plastron has a single hinge that allows complete closure of front and rear parts of the shell.
**Range**
S. New England, w. to lower Mich., s. Ill., and cen. Kans., s. along Atlantic and Gulf coasts to cen. Tex.

Note: Where the following 4 subspecies meet, intergradation occurs and precise identification may be impractical.

*Fig. 51*

Eastern Box Turtle

Western Box Turtle, p. 324

*Terrapene carolina carolina,* is found in northern part
of range west to the Mississippi River, and south to
northern Florida. The shell is highest at the rear and is decorated
with yellow spots. The **GULF COAST BOX TURTLE,** *Terrapene
carolina major,* is restricted to the Coast from Louisiana to
northern Florida. The carapace is dark on this, the largest,
subspecies. The **FLORIDA BOX TURTLE,** *Terrapene carolina bauri,*
occurs in peninsular Florida and the Keys. There are radiating
yellow lines on the blue-black carapace; plastron is unmarked
yellow. The hindfoot usually has 3 toes. The **THREE-TOED BOX
TURTLE,** *Terrapene carolina triunguis,* occurs from Missouri and
Kansas south, east to extreme southwestern Georgia, and west to
Texas. The hindfoot usually has 3 toes (but sometimes 4). Head is
unicolor brown or olive with bright spots; male sometimes has a
red head.

## WESTERN BOX TURTLE
*Terrapene ornata*                                    *Fig. 51*

### Description
Length, to 5⅞ in. (14.9 cm). Carapace flat-topped, unkeeled.
Plastron hinged, with striking pattern of light yellow lines on
darker background; carapace has radiating yellow lines over dark
to reddish-brown; 4 toes on hindfoot. Male has red iris and
enlarged first toe on hindfoot. Young almost round; flatter and
darker than adults.

### Habitat
Short-grass plains, prairies, savannas; sandy, semiarid regions;
woodlands, swamps.

### Habits
Burrows, emerges in morning, at dusk, and during rains; many
killed crossing highways.

### Food
Insects, earthworms, some carrion; berries, melons, and other
plants. Captives also take mealworms, fresh meat.

### Reproduction
Lays 2–8 eggs, usually in June.

### Remarks
This species has an interesting courtship pattern. The male chases
the female (if a turtle may be said to "chase"); upon reaching her,
he raises himself on his hind legs and hurls the front of his
plastron at the rear of her carapace, emitting from each nostril a
stream of fluid, which he sprays on her back. After half an hour of
such wooing, the female yields.

### Range
Cen. U.S. from w. Ind., s. Wis., and extr. se. Wyo. s. and w. to
Gulf Coast in La. and w. to se. Ariz., into adj. n. Mexico.

## DIAMONDBACK TERRAPIN
*Malaclemys terrapin*                                 *Fig. 52*

### Description
Length, to 9 in. (22.9 cm). Handsome polygonal scutes with
concentric ridges on carapace and spotted skin distinguish this
species. Carapace low, broad, slightly wider to rear; plastron
oblong. Head and forelimbs white with dark spots, lips white.
Female 2–3 in. (5.1–7.6 cm) longer than male, head heavier and
blunter, shell deeper.

### Habitat
Brackish water along coast, tidal shores, salt marshes.

**Food**
Crustaceans, mollusks, insects, carrion.

**Reproduction**
Lays 4–18 eggs, averaging 10 or less, with average clutch numbers decreasing from north to south; laid in holes in sand to depth of 6 in. (15.2 cm), April–July; often laid on a barrier beach.

**Remarks**
This species has been divided into several subspecies that differ in external characters, but which form a continuum of races of smaller size with smoother vertebral scutes in the North, to those of larger size with rougher, keeled vertebrals on the Gulf Coast. Once heavily consumed in fine restaurants, these turtles have only recently recovered from overcollecting.

**Range**
Coastal and adj. waters from Cape Cod through most of Gulf Coast.

*Fig. 52*

Diamondback Terrapin

# MAP TURTLE
*Graptemys geographica*

**65:10**
*Fig. 53*

**Description**
Length, to 10¾ in. (27.3 cm). Carapace low, flat, low keel lacks distinct knobs; rear margin slightly toothed; netlike patterns of yellowish lines and circles over dark green color resemble streams on a map (best seen when moist); marginals below yellow with round olive marks. Plastron smooth, usually plain solid yellow or cream color, but males and juveniles may have dark line along seams of scutes. Head wide, large, jaws strong; green or yellow spot of variable shape behind eye. Young have wider, stronger keels on carapace.

**Habitat**
Rivers and large bodies of water, also less abundantly in ponds and marshes.

**Habits**
Timid, hard to catch; basks on rocks, logs, and (which is unusual) on sand or grass beaches. Active into October.

**Food**
Mollusks (eaten especially by females), crayfish, insect larvae, and carrion.

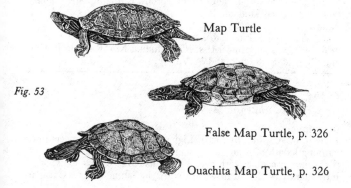

Map Turtle

*Fig. 53*

False Map Turtle, p. 326

Ouachita Map Turtle, p. 326

**Reproduction**
Lays 10–20 eggs, June–July in flask-shaped hole; hatchlings
emerge August–September, or May–June from overwintering
clutches.
**Remarks**
Does not thrive in captivity.
**Range**
Nw. Vt. and n. N.Y., w. through s. Mich., Wis., and e. Minn.; s.
through Miss. Basin to ne. Okla., La., and cen. Ala.

Note: A number of species of map turtles occur in the southeastern
United States, not all of which can be covered here. Those with
narrow, sharp-edged jaws include the **RINGED MAP TURTLE,**
*Graptemys oculifera;* the **YELLOW-BLOTCHED MAP TURTLE,**
*Graptemys flavimaculata;* and **BLACK-KNOBBED MAP TURTLE,**
*Graptemys nigrinoda,* all of which are found in much the same
area, from Louisiana to western Florida, as are the **ALABAMA MAP
TURTLE,** *Graptemys pulchra,* and the **BARBOUR'S MAP TURTLE,**
*Graptemys barbouri.* The last two species have large jaw surfaces
suitable for coping with crustaceans and shellfish, much as in the
Map Turtle, above. The **FALSE** and **OUACHITA MAP TURTLES,**
described below, have narrow jaws that are adapted for a more
vegetarian diet and for insects. Two additional species are found
only in central Texas—**CAGLE'S MAP TURTLE,** *Graptemys caglei,*
and **TEXAS MAP TURTLE,** *Graptemys versa.*

---

## FALSE MAP TURTLE
*Graptemys pseudogeographica*

**65:9**
*Fig. 53*

**Description**
Length, to 11 in. (27.9 cm). Recognized by variable orange-yellow
crescent behind eye and alternating green and yellow lines on
throat that reach eye. Carapace with low keel without prominent
knobs; rear margin saw-toothed. Color olive with 1 or more dark
blotches encircled with yellow or orange toward rear of each plate,
often faded in older specimens. Marginals from below marked with
concentric blotches, and plastron with concentric yellow and dark
green circles, which may fade to mottling in adults.
**Habitat**
Very aquatic; likes rivers and lakes, usually in slow-moving water
with dense vegetation.
**Habits**
Wary; often basks in large groups with other map turtle species;
strong swimmers.
**Food**
Young are carnivorous; adults omnivorous, eating aquatic plants,
mollusks, insects, and carrion.
**Range**
From Wis. and Minn. s. to La. and e. Tex. in large river
drainages emptying into Miss. R., including Mo. R. w. to N.Dak.,
and Ohio R. e. to Ohio and W.Va.

---

## OUACHITA MAP TURTLE
*Graptemys ouachitensis*

*Fig. 53*

**Description**
Length, to 10¼ in. (26.0 cm). Light blotch behind eye meets lines
running length of head above; sometimes broken with 1–9 stripes
that come in contact with eye. Throat has four yellow spots, or,
infrequently, yellow transverse stripes (in Sabine River). Carapace

and plastron similar in color to False Map Turtle, with dark spots on rear of each scute encircled by latticework of yellow and orange. Color variable, may be faded in older specimens, or even dark and almost patternless. As in False Map Turtle, carapace has high keel with prominent black knobs at center of vertebral scutes.

**Habitat**
Rivers only, in sections with swift currents and aquatic vegetation.

**Habits**
Basks; females feed on surface while males feed below. Spends night on bottom.

**Food**
More herbivorous than other map turtles, but also takes insects, carrion, and a few mollusks.

**Reproduction**
Lays 8–15 eggs, May–July.

**Range**
Wis. and Minn., s. through La. in Miss. Basin, e. to Ind., and w. to Okla. A distinctive population in Sabine R. of La. and Tex.

# PAINTED TURTLE
*Chrysemys picta*                                                    65:6

**Description**
Length, to 7 in. (17.8 cm); record 9⅞ in. (25.1 cm). Has smooth-edged oval shell; a notch, bordered by cusps, in upper jaw; and red and yellow marks over limbs, neck, and carapace. Carapace smooth, low, gently arched; margin rounded and often with light yellow or red markings. Marginal scutes always contain red. Plastron has 6 pairs of yellow or buff plates; plastron sometimes marked with central dark blotch. Neck and limbs have red streaks. Male's shell lower and foreclaws longer than female's.

**Habitat**
A common species, found on quiet lake shores, in ponds, slow streams, marshes, ditches, or even brackish tidal water; to 6000 ft. (1828.8 m).

**Habits**
Timid, but easily tamed; often basks in groups or floats with head sticking up through water amid lilies or duckweed; sometimes migrates short distances across land.

**Food**
Adults omnivorous—plants, insects, mollusks, crustaceans, mosquito larvae, carrion; juveniles carnivorous; feed under water or by skimming surface with mouth open.

**Reproduction**
In courtship male swims backward in front of female tickling her face with his claws. Lays 2–20 elliptical, soft eggs, May–mid-July, 2–5 in. deep within 10 yards (9.1 m) of water; incubation 8–12 weeks or until following spring; female often lays 2 clutches.

**Range**
Most of e. and n. U.S. and in extr. s. Canada, s. in the East to La.; but absent from s. Coastal Plain and from most of Southwest including Tex.

Note: Four eastern subspecies can be distinguished in the field.
**EASTERN PAINTED TURTLE,** *Chrysemys picta picta,* with plastron unmarked, has seams across carapace continuous in a straight line and with wide, light edges. It occurs in Nova Scotia, southern Maine, and southeastern Quebec, south through northeastern Alabama (but not in Coastal Plain). Intergrades with Midland subspecies (see below).

# FRESHWATER AND BOX TURTLES

**SOUTHERN PAINTED TURTLE,** *Chrysemys picta dorsalis,* with plastron unmarked, has carapace seams staggered, and with wide, light edges; a wide red or yellow stripe down middle of back. It occurs in Mississippi Valley from southern Illinois south, east to central Alabama, and west to extreme southeastern Oklahoma.

**MIDLAND PAINTED TURTLE,** *Chrysemys picta marginata,* with plastron narrow, has dusky figure down center, carapace seams staggered and with dark edges; sometimes with narrow, pale stripe down back. It occurs from southern Maine through southern Ontario, west to eastern Wisconsin, and south through Tennessee. Intergrades with Western and Eastern subspecies over broad areas.

**WESTERN PAINTED TURTLE,** *Chrysemys picta belli,* has plastron with large branching pattern; carapace seams are staggered with dark edges, but light lines decorate centers of scutes. It occurs in sw. Ont., upper Mich. and Ill., and w. through much of n. Central Plains and Pacific Northwest. Intergrades with Midland subspecies (see above).

## COOTER
*Pseudemys floridana*                                                65:7

**Description**
Length, to 15⅞ in. (40.3 cm). Carapace dark with network of light markings, but no "C" on 2nd costal. Plastron unspotted, yellow to orange or greenish; marginals below may have eye-shaped markings. Head and neck have parallel yellow lines that are variously broken and interconnected. Front of broad-domed, oval carapace is higher, rear margin toothed. Furrows present over shell. Lower jaw flattened.
**Similarities**
River Cooter has figure "C" on 2nd costal scute.
**Habitat**
Permanent bodies of water; large lakes, ponds, swamps, springs, or rivers.
**Habits**
Basks, often in groups, on logs, rocks; slides off quickly when approached.
**Food**
Young carnivorous, adults vegetarian.
**Reproduction**
Lays 12–29 eggs in flask-shaped hole in loose, open soil.
**Range**
Coastal Plain from e. Va. s. and w. to cen. Okla., n. through s. Miss. Valley to s. Ill.

## SLIDER
*Pseudemys scripta*                                                65:8

**Description**
Length, to 8 in. (20.3 cm). Carapace dark with yellowish and dusky stripes running crosswise on costal and marginal scutes, lengthwise on first two vertebrals. Marginals yellow below with dark spots between sutures. Plastron yellow with variable dark markings. Carapace smooth, weakly keeled; often algae-covered; margin slightly toothed. Beak notched, but not cusped; edges of both jaws smooth, lower jaw rounded. Male smaller, foreclaws larger. Old males often blackish, also lose red behind ear when present. Young have carapace nearly round and with vivid markings.

**Habitat**
Quiet water in a variety of situations.
**Habits**
Aquatic and wary, wanders through aquatic vegetation; basks 2- or 3-deep on logs, or floats with head just above water; disappears into bottom mud if frightened.
**Food**
Young are carnivorous, adults omnivorous.
**Reproduction**
Lays 2–22 soft eggs, April–July in hole 4–8 in. (10.2–20.3 cm) deep near water; incubation 8–14 weeks or more; some young overwinter in the nest.
**Range**
Most of se. U.S., se. Va., s. through n. Fla., w. through most of Ill. and into N.Mex., with isolated localities in Mich., Ohio, W.Va., s. Fla., and into Mexico.

Note: The **RED-EARED SLIDER,** *Pseudemys scripta elegans,* the commonest of the Slider's 3 subspecies in the East, is the only turtle with a wide red band behind eye. The **YELLOW-BELLIED SLIDER,** *Pseudemys scripta scripta,* and **CUMBERLAND SLIDER,** *Pseudemys scripta troosti,* subspecies, with a yellow patch and line respectively, occur in the southeastern United States.

# RIVER COOTER
*Pseudemys concinna*

**Description**
Length, to 12 in. (30.5 cm); record 16⅜ in. (41.6 cm). Carapace variously marked but always with light "C" whose arms touch rear margin of 2nd costal. Plastron spotted variously, always has more spotting on bridge and marginals than Cooter. Head lines not broken and interconnected. Carapace broad and somewhat flattened; rugose, furrowed. Lower jaw flattened.
**Similarities**
Cooter's shell high-domed, has less spotting on bridge and marginals.
**Habitat**
Rivers, streams, springs, less frequently ditches; in fairly quiet water with vegetation.
**Habits**
Shy; often seen basking.
**Reproduction**
Lays 18 soft eggs, May–July.
**Range**
S. U.S., from se. Va. through w. Fla., w. to se. N.Mex. and n. Mexico, and n. in Miss. Valley to s. Ill. and Ind.

# RED-BELLIED TURTLE
*Pseudemys rubriventris*                               **65:5**

**Description**
Length, to 15¾ in. (40.0 cm). Has much red on carapace and plastron. Reddish band through costals and marginals above and below, but some individuals dark overall; plastron with pinkish or reddish. Carapace highest in middle, widest at rear, often furrowed lengthwise, rear margin virtually smooth. Similar to Cooters and Slider, but shell more oblong, and beak markedly notched. Jaw edges saw-toothed, lower jaw flattened. Foreclaws of male straight, longer than those of female. Young are rounder and brighter-colored.

**Habitat**
Lakes, ponds, streams (clear, muddy, or brackish).
**Habits**
Throughout most of its range, the only basking turtle. Wary, basks in places difficult to reach and dives into deep water when disturbed.
**Food**
Omnivorous.
**Reproduction**
Lays 12–35 eggs in flask-shaped nest 4 in. (10.2 cm) deep, often in cornfields.
**Remarks**
Young make good pets but should be released, as with all other animals, if not thriving. (Be sure to release your pet exactly where you found it.)
**Range**
Two populations—in the North limited to Plymouth Co. and Naushon Is. in e. Mass.; in the South, continuous from s. N.J. s. to N.C. and w. through Md. to extr. e. W.Va.

Note: Two species, the **FLORIDA RED-BELLIED TURTLE,** *Pseudemys nelsoni,* of the Florida peninsula, and the **ALABAMA RED-BELLIED TURTLE,** *Pseudemys alabamensis,* of southwestern Alabama, are similar to the Red-bellied Turtle, but have much more limited distributions.

## CHICKEN TURTLE
*Deirochelys reticularia*                                    *Fig. 54*

**Description**
Length, to 6 in. (15.2 cm); record 10 in. (25.4 cm). Has very long, striped neck (almost as long as carapace), and vertically striped rump. Carapace ground color dark with yellow lines and often a yellow border on margin. Plastron hingeless, yellow to orange, sometimes with pattern along seams. Dark skin has yellowish or greenish lines.
**Habitat**
Lowland cypress swamps, lakes, ponds, or ditches and other temporary water.
**Habits**
Semiterrestrial, often found wandering overland.
**Food**
Plants, tadpoles, crayfish; adults eat more vegetation.
**Reproduction**
Lays 5–8 oblong, leathery eggs.
**Range**
Coastal Plain from se. Va. to e. Tex. and up Miss. Valley to se. Mo., but absent from Miss. R. itself.

*Fig. 54*

Gopher Tortoise

Chicken Turtle

## BLANDING'S TURTLE
*Emydoidea blandingi*                                    *Fig. 50*

**Description**
Length, to 9 in. (22.9 cm). Only turtle with combination of a
dotted carapace, hinged plastron, and bright yellow patch on lower
jaw and under neck. Dots may blend together. Plastron yellow,
with large scutes, each with black blotches at outer rear corners;
closure incomplete, hind end notched. Carapace long, narrow,
elevated. Beak notched; feet webbed. In young, carapace is dark
and unmarked, plastron black with yellow margin.
**Habitat**
Ponds, backwaters, small streams, moist land.
**Habits**
Timid; if surprised basking, will plunge to bottom and remain
there for hours; a good swimmer.
**Food**
Omnivorous; includes crustaceans, insects.
**Reproduction**
Lays 6–11 eggs, June–July; hatchlings appear September.
**Remarks**
Makes a good pet.
**Range**
Spotty and discontinuous distribution, in scattered localities along
Atlantic Coast from N.S. to s. N.Y., and more or less continuously
from extr. sw. Que. and s. Ont. around cen. Great Lakes area w.
to cen. Nebr. and possibly s. to s. Ind.

# Tortoises
Subfamily Testudininae

This group is closely related to the Emydinae. Found on most
continents except Asia and Australia, they have elephantine
hindlimbs and usually high, domed shells. They are strictly
terrestrial and are commonly sold as pets.

## GOPHER TORTOISE
*Gopherus polyphemus*                                    *Fig. 54*

**Description**
Length, to 10 in. (25.4 cm); record 14½ in. (36.8 cm). Recognize
this tortoise by the broad, domed (but somewhat flattened)
carapace, stoutly elephantine hindlimbs, and hingeless plastron.
Carapace brown to black, scutes with concentric growth rings;
plastron yellow. All feet unwebbed. Forelimbs shovellike, used for
burrowing, covered with hard scales that remain exposed when
tortoise retreats into shell. Unable to close up tight because hinge
in plastron lacking; it joins elbows in front of head and seals front
opening with forearms.
**Habitat**
Sandy areas often far from water; strictly terrestrial.
**Habits**
An accomplished burrower, making tunnels up to 40 ft. (12.2 m)
long. Male defends burrows against other males, but females may
pass into them freely.
**Food**
Vegetarian; grazes on grasses; fruits and berries; may accept meat
in captivity.

**Reproduction**
Lays 4–7 eggs in sandy hole dug far from burrow, infrequently at burrow entrance.
**Remarks**
This "gopher" usually shares its burrow with all manner of uninvited but tolerated guests: rattlesnakes, indigo snakes, frogs, opossums, burrowing owls, and insects. *Do not reach into a burrow with your hand* under any circumstances.
**Range**
Coastal Plain from s. S.C. through suitable habitat in Fla. w. to e. La.

# SEA TURTLES
Family Cheloniidae

In these turtles the shell has evolved to be lighter in weight, the short, heavy neck cannot be completely drawn back into the shell, and the forelimbs are paddlelike. They are completely marine, the females returning to land to lay their eggs in holes they excavate on sandy beaches above the high-water mark. They are found in all tropical and warm temperate oceans.

## GREEN TURTLE
*Chelonia mydas*

**64:6**
*Fig. 55*

**Description**
Length, to 48 in. (121.9 cm); record 55 in. (139.7 cm), and weight 650 lbs. (294.8 kg). The only sea turtle with 4 large plates (costals) on each side of back, *and* 1 pair of plates (prefrontals) behind nostrils (between eyes). Ground color mottled brown or gray. Carapace broad, low, and unkeeled; plates do not overlap. In males forelimb usually has 1 claw only, shell usually longer and tapering, and tail very long (8 in. or 20.3 cm). Young have white edges on carapace and limbs.
**Similarities**
Hawksbill's plates overlap.
**Habitat**
Pelagic, migrating long distances in open ocean; feeds on vegetation in shallows near mouths of rivers.
**Habits**
Sometimes basks on surface; normal speed in water to 1.4 mph. (2.3 km/hr.), but clumsy on the beaches where females lay eggs.
**Food**
Adult mainly herbivorous; young more carnivorous.
**Reproduction**
Lays 10–200 (average 107) eggs in hole about 18 in. (45.7 cm) deep.
**Range**
Warm tropical and subtropical seas, occasionally n. to Mass.

*Fig. 55*          Green Turtle

Loggerhead

## HAWKSBILL
*Eretmochelys imbricata*                                      **64:5**

**Description**
Length, average of 24 in. (61 cm), and weight 50 lbs. (22.7 kg); to
length 36 in. (91.4 cm), and weight 160 lbs. (72.6 kg). The only
sea turtle with overlapping plates (except when old), 2 pairs of
prefrontals on snout, and 4 pairs of big costals. Plates variegated in
color. Carapace shield-shaped, keeled, with toothed margin.
Forelimb has 2 claws. Male has longer claws than female.
**Habitat**
Reefs and shallow coastal waters.
**Food**
Omnivorous; captives thrive on fish and meat.
**Reproduction**
Lays 53–206 eggs, normally in June–July.
**Remarks**
Plates formerly used to make tortoiseshell ornaments.
**Range**
Tropical warm seas, found only occasionally off ne. U.S. coast.

## ATLANTIC RIDLEY
*Lepidochelys kempi*                                         **64:8**

**Description**
Length, to 29½ in. (74.9 cm), and weight to 110 lbs. (49.9 kg).
Carapace heart-shaped or round, gray. Inframarginal plates 4 or
more on each side and costals 5. Large head has 2 pairs of
prefrontals.
**Similarities**
Loggerhead has 3 inframarginals, Green Turtle and Hawksbill
have 4.
**Habitat**
Shallow waters of tropical shoreline; nests in small numbers on
southern Texas coast.
**Habits**
Irascible, violent when caught, cannot be kept on its back as can
other sea turtles.
**Food**
Crabs and other animal food, details not well known.
**Reproduction**
Only one major nesting area exists, on northeast coast of Mexico,
concentrated on a small stretch of beach. At peak of breeding
season up to 40,000 Ridleys may gather off this beach.
**Remarks**
A separate species inhabits Pacific Ocean.
**Range**
Gulf of Mexico and up into coastline of n. U.S. states.

## LOGGERHEAD
*Caretta caretta*                                            **64:7**
                                                             *Fig. 55*

**Description**
Length, to 45 in. (114.3 cm); record length 48 in. (121.9 cm), and
weight 500 lbs. (226.8 kg). Color reddish-brown. Has 3
inframarginal plates and 5 or more costals. Head very large with 2
pairs of prefrontals; forelimb has 2 claws. Male has tapering shell
and long, thick tail. Young have 3 keels above, 2 below.
**Similarities**
Ridley has 4 or more inframarginals; Green Turtle and Hawksbill
have 4 costals.

**Habitat**
Coastal bays; brackish streams; creeks and, of course, the high seas.
Most widespread and common of the sea turtles.

**Habits**
Sometimes vicious if molested; an oceanic wanderer; basks at
surface of water and infrequently on beaches.

**Food**
More carnivorous than other sea turtles, consuming crabs, conchs,
fish, clams, oysters, sponges, jellyfish, turtlegrass.

**Reproduction**
Lays 60–300 round, soft, leathery eggs, April–August on our
southern coasts, and more recently reported on a New Jersey beach
(formerly commonly nested to Virginia). Incubation 1–3 months;
hatchlings emerge from nest en masse, usually at night, and head
straight for the water.

**Remarks**
This hardy and tenacious species is the only sea turtle that still
nests regularly on the South Atlantic coast. It is almost as pelagic
as the Leatherback, and there have been reports of size up to 7 ft.
(2.1 m) and 1000 lbs. (453.5 kg).

**Range**
Atlantic Ocean n. to N.S.

# LEATHER-BACKED TURTLES
Family Dermochelyidae

The Leatherback is the sole member of this family.

## LEATHERBACK
*Dermochelys coriacea*                                    64:9

**Description**
Length, to 72 in. (182.9 cm), and weight to 1600 lbs. (725.6 kg).
The largest living turtle and only sea turtle with a leathery shell
lacking large plates. Smooth black carapace has 7 long ridges;
plastron mainly white, has 5 long ridges; limbs lack claws.

**Habitat**
Pelagic, coming on land only to breed.

**Habits**
Strong, fights with both jaws and flippers. Able to live in summer
in northern waters and warm its body to 80°F (26.7°C), even in
45°F (7.2°C) water, by a countercurrent system of blood vessels to
and from extremities. (This idea is disputed by some herpetologists,
who maintain that Leatherbacks in northern waters are strays that
will eventually die.)

**Voice**
Utters loud sounds when agitated.

**Food**
Jellyfish and other marine life.

**Reproduction**
Lays 50–170 round, soft eggs, June–July in hole 1–3 ft. (0.3–0.9
m) deep excavated at night by female in sand above high-water
mark; incubation 8–10 weeks.

**Remarks**
The leathery shell of this species makes its classification uncertain.
Therefore, it may deserve a family of its own.

**Range**
Warm seas, straying to Newfoundland.

# Lizards and Snakes
Order Squamata

## Lizards
Suborder Sauria

### Features of Lizards

A lizard is a scaly reptile that usually has visible eardrums, movable eyelids, a nonexpandable jaw, and four legs, each with five clawed toes. Most of them also have several rows of scales on the underside. But in certain lizards some of these characteristics may be missing. For example, glass lizards do not have legs and superficially they resemble a snake—but snakes do not have ears, eyelids, or legs, and they have an enormously expandable jaw and usually just one row of broad belly scales.

Do not confuse lizards with salamanders. Salamanders are amphibians and have a moist, scaleless skin and unclawed toes.

The tail is often fragile and breaks off when a lizard is caught; the discarded fragment wriggles for several minutes. When a tail is lost in this way, a new, shorter one grows in its place, but the vertebral column is replaced by cartilage, and the scale and color patterns may be noticeably unlike the originals.

Many lizards have excellent eyesight, especially adapted for detecting moving objects, and a good sense of smell. Despite the general presence of external ears, many seem oblivious to sound, and lizards have little, if any, voice. Teeth that are generally all alike are present on jaw edges and sometimes on the palate. They are commonly used to grab and hold their prey rather than for chewing. No poisonous lizards occur in the eastern United States, but in fact only two poisonous lizards are known, one being the Gila Monster of the American Southwest. Lizards mainly eat insects, spiders, snails, worms, and other lizards, but are also known to be herbivorous. Most eastern lizards are terrestrial and diurnal.

*Fig. 56*

Parts of a Typical Lizard

335

# LIZARDS

### Identification
Lizards in the eastern United States are a small group. With the exception of some of the skinks, the individual species are relatively easy to distinguish. First, determine which species occur in your state. Where several species are found, they may fall into certain groupings. If a species has spiny scales, it is either one of the fence or horned lizards (the only lizard with "horns"). If a species has granular scales, it is an anole. If a species has smooth, polished scales, it is probably a skink. The only skink with diagonal scale rows on its sides is the Great Plains Skink. Only three lizards with legs have the tail more than twice the length of the body, namely the Collared Lizard, the Green Anole, and the Six-lined Racerunner (which is slender with a pointed snout). The Collared Lizard is our only species that habitually runs on its hind legs.

Nonspiny lizards with legs can also be divided by preferred habitat:

| *Moist, Humid, Wooded Places* | *Dry, Sandy, Rocky, Grassy Places* |
|---|---|
| Ground Skink | Collared Lizard |
| Five-lined Skink | Lesser Earless Lizard |
| Broad-headed Skink | Six-lined Racerunner |
| Southeastern Five-lined Skink | Great Plains Skink |
| Coal Skink | Many-lined Skink |
| Green Anole | Prairie Skink |

### Habitat
Almost everyone has seen some lizards. The anoles; the collared, horned, and fence lizards; and the racerunners are relatively conspicuous. Others, for instance the skinks and glass lizards, are wary or secretive and are seldom seen unless searched for.

Look for lizards basking or running along stone walls and fences, on stumps or logs or in brush piles, amid dead leaves, around old piles of sawdust, in deserted buildings and clearings, on limestone hills and rocky outcrops, in canyons, and on flat sandy or rocky soil with sparse vegetation. Look for some on tree trunks, for others in moist situations near water. Study lizards through binoculars, adjusted for close focus. You can approach many species to within a few feet; from there, good binoculars can bring them into beautiful scale-counting view.

Most lizards hibernate in the colder months. They are apt to be particularly active soon after emergence in the spring. Lizards can stand a good deal of heat. They prefer sunny days, particularly mornings. Cool or cloudy days keep them under cover and many retreat to their hiding places in late afternoon even while the sun is still up.

### Measurements
The sizes given are the overall maximum lengths from *snout to end of tail*. The second measurement gives the snout to vent length because in many cases lizards lose their tails.

### Range
Lizards are often seen amid ruins and in deserts. They are also found in moist and dry woods, and in flat or rocky uplands. Lizards are particularly abundant in the southern United States. In Florida, in addition to long-established endemic forms, recently introduced lizard species have arrived from the West Indies, particularly colonizing the area around Miami; and in Texas the desert faunas from the southwestern United States and Mexico swell the number of species. Included in this section are the most common and abundant species.

# IGUANAS
Family Iguanidae

This is the dominant family of New World lizards. Its members are varied in form, habits, and habitat, but in all the teeth lie in grooves on the inner surface of the jaws. All have a scaly body and four well-developed limbs. Some reach a length of five feet (1.5 m) and are ornamented with crests and dewlaps. Some forms have taken to water; however, those of the eastern United States are somewhat less exotic. Iguanids have a characteristic habit of bobbing up and down, which signals territoriality or willingness to mate. The family ranges from Canada to Argentina, with many species in the West Indies as well as in Madagascar and the Fiji Islands of the South Pacific.

## GREEN ANOLE
*Anolis carolinensis*                                                      *Fig. 57*

**Description**
Length, to 8 in. (20.3 cm); snout-vent length to 3 in. (7.6 cm). Recognize these "chameleons" by the male's pink or red throat fan, or dewlap, and the expanded pads near tips of digits. Color green to brown, and true to its misnomer (chameleons occur in the Old World), can effect rapid color change. Throat fan is longitudinal flap of skin extended by cartilaginous rod during mating or territorial displays. Small granular scales over body. Toe pads, or expanded lamellae, allow anole to gain purchase on arboreal perch.
**Habitat**
Various situations, including wooded areas and cypress swamps, but also close to human dwellings; a climber, found in trees and large bushes or on fence posts, but juveniles found on ground.
**Habits**
Acrobatic and visible, have complex social displays; sleep on vegetation, often high in trees.
**Food**
Insects and other invertebrates.
**Reproduction**
Lays a single egg, buries it in soft earth, every 10–14 days, May–January, but egg rarely discovered.
**Remarks**
The only native representative of a very interesting and diverse genus found mainly in the West Indies and Central and South America, where about 300 species are known. Through recent human activities, other anole species have been introduced into very limited localities in the United States; they are variously distributed in peninsular Florida, mainly near Miami.
**Range**
N.C. throughout Fla. and w. to se. Okla. and e. Tex. Disjunct locality in extr. s. Tex.

*Fig 57*

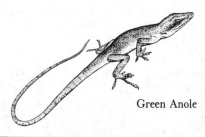

Green Anole

## COLLARED LIZARD
*Crotaphytus collaris*         **66:2**

**Description**
Length, to 14 in. (35.6 cm); snout-vent length 4½ in. (11.4 cm).
Stout-bodied lizard with a long, thin, tapering tail; large head; and
thin neck with 2 black collars. Much variation in color; males often
green or yellow with yellowish to red dots, throat often orange,
inside throat black. Blue spots on back, black collars broken in
front. Females often gray, with red spots. Both sexes brightest for
short time only during breeding season. Dark bands on tail and big
hindlegs. Young duller, crossbands of cream and gray more
conspicuous.

**Habitat**
Arid and semiarid regions, limestone-topped hills and bluffs,
prairie rocks and canyons, unshaded hillsides.

**Habits**
Ground-dwelling, wary, agile, a good jumper and bipedal runner,
but pugnacious when cornered.

**Food**
Grasshoppers, spiders, moths, beetles; other lizards and small
snakes; flowers.

**Reproduction**
Lays 2–21 (usually fewer than 12) eggs, May–July in sand or
tunnels to depth of 4 in. (10.2 cm).

**Remarks**
The sight of one of these highly colored lizards running on its
hind legs, using its long tail for a balancer, suggests a tiny dinosaur.

**Range**
S. Mo. and nw. Ark., w. to Utah and Ariz., and s. through n.-cen.
Mexico.

## EASTERN FENCE LIZARD
*Sceloporus undulatus*         **66:9**

**Description**
Length, to 7¼ in. (18.4 cm); snout-vent length to 3¼ in. (8.3 cm).
Above dark gray or brown with curved, barlike pattern across back
and tail, or longitudinal dark stripes. Stout body, scales keeled.
Male shows blue under sides and on throat; female more boldly
marked above.

**Similarities**
Collared Lizard larger and has collars; Lesser Earless Lizard
stouter, shorter-tailed, and paler.

**Habitat**
The most common iguanid lizard, found in dry, piny, and
deciduous woods; clearings, brushlands, grasslands, sandy areas; on
fences, old houses, brush heaps, fallen tree trunks or rocks.

**Habits**
Climbs; a fast runner and artful dodger about tree trunks; most
active on sunny days; often seen basking on fence posts; defends
territories and has complex social system.

**Food**
Insects, spiders.

**Reproduction**
Lays 3–13 eggs, May–July to depth of 4 in. (10.2 cm) in sandy
soil or rotten logs or under bark; incubation about 8 weeks.

**Range**
Most of cen. and s. U.S. from s. N.Y., n. Mo., and s. S.Dak.,
s. and w. to Utah and Ariz., and into n.-cen. Mexico. Absent from
s. half of peninsular Fla., s. La., and extr. s. Tex.

Note: The **FLORIDA SCRUB LIZARD**, *Sceloporus woodi,* somewhat resembles the Eastern Fence Lizard, but its lateral stripe is brown rather than black, and the markings, though similar, are more clearly defined. This species occurs in 4 distinct and local areas in the Florida peninsula: on the east-central and southwestern coasts, and in 2 interior areas, 1 in the north and the other in the center.

## LESSER EARLESS LIZARD
*Holbrookia maculata*                                                    **66:10**

**Description**
Length, to 5 in. (12.7 cm); snout-vent length to $2\frac{7}{16}$ in. (6.2 cm). Tail is about equal in length to body; no ear opening. Above pale, back has gray stripe bordered with dark spots; 2 short, black bars low on each side; below plain whitish. Body short and plump but somewhat flattened, scales tiny, dorsal scales smaller than ventrals. Fleshy fold between chest and throat (gular fold).
**Habitat**
Plains, dry rocky or sandy areas with sparse vegetation; chalk beds.
**Habits**
Particularly active on hot days; makes short dashes; inquisitive; often seen in the shade of vegetation.
**Food**
Grasshoppers, true bugs, spiders, beetles, grubs.
**Reproduction**
Lays 2–8 eggs, July–early August; hatchlings August–September.
**Range**
Central Plains and sw. U.S. from s. S.Dak. and extr. se. Wyo. s. through n. Tex. and w. through much of Ariz. In cen. and w. mainland Mexico. Questionable record from Mo.

## TEXAS HORNED LIZARD
*Phrynosoma cornutum*                                                    *Fig. 58*

**Description**
Length, to 4 in. (10.2 cm); record 7⅛ in. (18.2 cm); snout-vent length to 5⅛ in. (13.0 cm). Only lizard with horns projecting from back of head. Above tan to brown with light-bordered, black blotches to either side and light streak down middle of back; below pale with black spots. Spiny, short-tailed, squat.
**Habitat**
Dry, flat land with sparse vegetation.
**Habits**
Most active during heat of day, though in hottest summer months active morning and evening; found alongside ant trails in the open; reluctant to move when approached by a predator, but will eventually waddle away in a comical gait.
**Food**
Ants; also other insects, and spiders.
**Reproduction**
Lays 14–37 (average 26) eggs, May–July.

Short-horned Lizard, p. 340

*Fig. 58*

Texas Horned Lizard

**Remarks**
These horned lizards have no problems of predation during periods
of exposure because of their concealing coloration and spiny,
widened bodies, which make them hard to spot and to swallow.
However, they are not free from predation by some birds, such as
shrikes and hawks, and by large snakes.
**Range**
S.-cen. U.S. from se. Colo., Kans., and extr. sw. Mo. s. through
Tex. and nw. La., and w. in e. and s. N.Mex. to se. Ariz. through
n. Mexico. Introduced in Fla. and elsewhere in the South.

## SHORT-HORNED LIZARD
*Phrynosoma douglassi*                                    *Fig. 58*

**Description**
Length, to 4½ in. (11.4 cm); snout-vent length to 2½ in. (6.4 cm).
Has ridge on back of head that lacks large spines, and short tail,
banded above. Spiny, squat.
**Similarities**
Texas Horned Lizard has prominent horns and more distinct
blotches and line on back.
**Habitat**
Semiarid, short-grass plains; hardpan, light soil, sandy and rocky
terrains.
**Food**
Ants; also grasshoppers, beetles, insect larvae.
**Reproduction**
Bears 5–31 (average 16) living young in August.
**Range**
Only marginally in eastern U.S.; s. Alta., Mont., and se. N.Dak.,
s. to Ariz. and N.Mex. and into n.-cen. Mexico; disjunct
populations in w. Tex., in Pacific Northwest.

# GLASS LIZARDS
Family Anguidae

Members of this family in North America have a groove of
granular scales along each side, presumably to allow for expansion
of the armored body after big meals or during egg development. In
the eastern United States only the legless glass lizards occur,
whereas in the West the alligator lizards, with four well-developed
limbs, are found. This family comprises residents of the New
World and Eurasia, but not of the tropics of Asia or Africa.

## EASTERN GLASS LIZARD
*Ophisaurus ventralis*                                    *Fig. 59*

**Description**
Length, to 42⅝ in. (103.8 cm); snout-vent length to 12 in. The
largest legless lizard, superficially resembling a snake, but with
movable eyelids and ventral scales same size as dorsal scales.
Broad, dark band on side and white at edges of scales. Lateral fold
to either side below. Above brown to olive with no middorsal stripe
in adults and with white bars on neck. Below pale, no markings
below lateral fold.
**Similarities**
Slender Glass Lizard has narrow dark lines on side, white in
middle of scales.

**Habitat**
A burrower; usually in moist, grassy areas in pine flatwoods and in hummocks; in places with loose soil, cavities under roots or stones, or in burrows.

**Habits**
Basks in mild sun, comes out after rain. Moves by undulating from side to side stiffly.

**Food**
Grasshoppers; other insects, spiders, snails, earthworms.

**Reproduction**
Lays 4–17 white eggs with flexible shell, June–July; female broods eggs under cover. (See details of incubation habits under Slender Glass Lizard, below.)

**Remarks**
If this or following species is attacked or captured, its tail, which is ⅔ of total length, comes off, wriggles, and breaks into pieces, thus diverting attention and enabling owner to escape. Body regrows new, shorter tail.

**Range**
Coastal Plain from N.C. through Fla. and w. to e. La. Disjunct populations in e.-cen. Mo. and extr. se. Okla.

Eastern Glass Lizard

*Fig. 59*

Slender Glass Lizard

# SLENDER GLASS LIZARD
*Ophisaurus attenuatus*                     *Fig. 59*

**Description**
Length, to 42 in. (106.7 cm); snout-vent length to 11⅜ in. (38.9 cm). Has narrow dark lines along each lower side, white marks in centers of scales (may form a series of stripes). Usually with stripe along middle of back, and stripes below lateral fold on lower sides of body.

**Similarities**
Eastern Glass Lizard has broad band on side, white at edges of scales.

**Habitat**
Found in drier situations than Eastern Glass Lizard—oak and pine woodland, fields and grassy areas, brush piles.

**Reproduction**
Lays 6–17 eggs. Brooding female of this species and Eastern Glass Lizard curls body around eggs and guards them against small predators during incubation. From time to time she basks in sun and returns to warm eggs with her newly warmed body; pays no attention to hatchlings.

**Range**
Se. U.S. from se. Va. through most of Fla. and w. to cen. Tex.; a broad northward extension up the Miss. Valley to w. Ind. in east of range, s. Wis. in north, and se. Nebr. to w.

Note: The **ISLAND GLASS LIZARD,** *Ophisaurus compressus,* is distinguished from the Eastern and Slender Glass Lizards by the lack of dark stripes *below* the lateral groove. It does, however, have one solid dark stripe on each side of its body. It occurs in coastal South Carolina and Georgia, and in all of peninsular Florida except for the extreme southern tip.

# SKINKS
Family Scincidae

The skinks make up a very large, cosmopolitan family. Eastern species are ground lizards with smooth, flat, polished-looking scales and short, but fast, legs. They can easily slip through a hand when picked up. They feed diurnally on insects and spiders, hibernate in winter, and lay eggs that the female often broods.

## GROUND SKINK
*Scincella lateralis*                                                      66:8

**Description**
Length, to 5⅛ in. (14.0 cm); snout-vent length to 1¹⁵⁄₁₆ in. (4.9 cm). The smallest eastern North American lizard, with tiny limbs; look for transparent "window" in eyelid. Back golden-brown, sides of body from snout to base of tail dark brown, below light, often with yellow. Scales smooth. Tail breaks off readily if seized.

**Habitat**
Has wide ecological tolerance; found in dry woods, pine or deciduous; in clearings, under leaves, bark, fallen tree trunks; also moist places near streams.

**Habits**
Very secretive, crawls quietly amid leaves or runs with snakelike sideways movements; may jump into shallow water.

**Food**
Small insects, earthworms and other invertebrates.

**Reproduction**
Lays 1–7 hard eggs, June–August in rotten logs, stumps, humus.

**Range**
Se. U.S. from s. N.J. w. through s. Ohio, Ill., and Mo. into e. Kans., and s. through most of Fla. to s. Tex., with a record in n. Mexico. Absent from most of Appalachian Region.

## FIVE-LINED SKINK
*Eumeces fasciatus*                                       66:3

*Figs. 60, 61*

**Description**
Length, to 8¹⁄₁₆ in. (20.5 cm); snout-vent length to 3⅜ in. (8.6 cm). Any lizard seen in New England or Michigan will be this species. This and following two species, however, are difficult (sometimes impossible) to tell apart in the field. In the hand, look for 7 upper labial scales with 4th below the eye, and 2 enlarged postlabials. Middle row of scales underneath original tail widened; 5 light lines down black back (but old male loses these lines, body turns olive-brown, head and throat turn reddish); 26–30 rows of scales around middle of body, rows at side parallel to rows on back. Head of breeding male wider, redder than that of female. Young have yellow head stripes and blue tails that turn brown with age.

**Habitat**
Dry to damp woods, old sawdust and brush piles, dead leaves, rotten logs, stumps; on shores of Great Lakes in decayed driftwood.

**Habits**
Very agile; basks on logs and stumps, changing position frequently; seldom climbs trees (two following species do); bites if captured.

**Food**
Insects, earthworms, spiders, snails.

**Reproduction**
Distinguishes its mate by behavior rather than appearance. Courting male rushes around with open mouth toward a

conspecific. If the lizard attacks back, it is a male; if it runs away or stays still, male grasps female's back in his mouth and mates with her. Lays 4–15 eggs, May–July in rotten logs, stumps, or loose soil; brooded by female for 4–7 weeks.

**Range**

Isolated and old records in Mass.; from se. N.Y. and adj. cen. Vt. s. to n. Fla., w. through s. and cen. Pa., s. Ont., around Lake Mich. to cen. Wis.; s. Ill. to extr. se. Nebr., e. Kans. to e. Tex. Isolated populations reach to N. Dak., Minn., and Iowa.

Fig. 60     Broad-headed Skink     Five-lined Skink

regenerated tail

Fig. 61     Broad-headed Skink     Southeastern
and Five-lined Skink     Five-lined Skink

## BROAD-HEADED SKINK     66:4
*Eumeces laticeps*     *Figs. 60, 61*

**Description**

Length, to 12¾ in. (32.4 cm); snout-vent length to 5⅝ in. (14.3 cm). Large, with 8 upper labials, the 5th below the eye, widened ventral scales underneath original tail. Similar to Five-lined and old Southeastern Five-lined Skinks, but body much thicker; males have broader yellowish or reddish heads than males of other two species. One or 2 small postlabials or none at all. Thirty to 32 scale rows around middle of body. Young sometimes have 7 instead of 5 stripes.

**Habitat**

Arboreal, but also found in habitat similar to that of Five-lined Skink.

**Reproduction**

Lays 13–16 eggs, June–July.

**Range**

Extr. se. Pa. through s. W.Va. and s. Ohio, w. to se. Kans., and s. to Atlantic and Gulf coasts, but only to cen. Fla.

## SOUTHEASTERN FIVE-LINED SKINK
*Eumeces inexpectatus*     *Fig. 61*

**Description**

Length, to 8½ in. (21.6 cm); snout-vent length to 3⅓ in. (8.9 cm). Striped, scales underneath original tail all same size (middle row not widened as in Five-lined and Broad-headed species). Hard to distinguish in field from two preceding species if of this size, but light stripes on head are orange-red; sublateral light line may be present. In juveniles, blue color of tail extends beyond hip.

**Habitat**

Much as for two preceding species, but often in drier habitats.

**Habits**

More arboreal than Five-lined, less so than Broad-headed.

**Reproduction**

Lays 11 eggs.

**Range**

Coastal Plain from s. Md. and cen. Va. to Miss. R. in cen. Miss., n. through cen. Tenn. into s. Ohio.

## GREAT PLAINS SKINK
*Eumeces obsoletus*                                                          66:1

**Description**
Length, to 13¾ in. (34.9 cm); snout-vent length to 5⅝ in. (14.3 cm). Adults light gray to brown above with black-bordered scales; below unmarked cream. Young black with blue tail and rows of gold dots on top of head, white spots on labial scales.

**Habitat**
Flat limestone rocks on grassy hillsides, also in grasslands and woods; under rocks, bark, boards, logs, but often digs burrows; most common in eastern Kansas.

**Habits**
Vicious; secretive.

**Food**
Insects, spiders, smaller lizards.

**Reproduction**
Lays 7–17 eggs, June–July in hollows under rocks. Female broods until hatching in August.

**Range**
Extr. e. Mo. through s. Nebr. to extr. se. Wyo. and s. into cen. and w. Tex., s. and e. N.Mex., and se. Ariz., to ne. Mexico. Isolated records in ne. Mo.

## MANY-LINED SKINK
*Eumeces multivirgatus*                                                      66:5

**Description**
Length, to 7⅝ in. (19.4 cm); snout-vent length to 2⅞ in. (7.3 cm). Slender, has many light lines running length of body, but check for light line on 3rd scale row from midline. Wide, light middorsal stripe from head to tail; chin area, tail light, underparts gray, 24–26 rows of scales around middle of body. Tail 1½ times body length. Young have blue or lavender tail and only 5 light stripes, but range does not overlap that of other five-lined skinks.

**Habitat**
Short-grass prairies; under boards in urban vacant lots, under cow chips in prairie-dog towns.

**Habits**
Lays 3–5 eggs, otherwise little is known.

**Range**
Sw. S. Dak. and se. Wyo., s. through w. Nebr., and across Colo. to se. Utah, ne. Ariz., and much of N.Mex. Also found in disjunct populations in w. Tex.

## COAL SKINK
*Eumeces anthracinus*                                                        66:7

**Description**
Length, to 7 in. (17.8 cm); snout-vent length to 2¾ in. (7.0 cm). Broad stripe on back bordered by white stripe and below it a wide dark brown lateral stripe 2½–4 scales wide, bordered below by another white stripe. Four light stripes above. Olive-brown band on back covers 4 or more scale rows; bordering white stripe extends onto tail but not onto head. Below bluish or gray. Scale rows 24 around middle of body. One postmental scale on chin. Young often all black with reddish spots on head; tail blue-violet.

**Habitat**
Under stones, brush piles, leaves, logs, or on wooded hillsides near water; also near swamps.

**Food**
Small insects.
**Reproduction**
Lays 8–9 eggs, usually late June; brooded by female 4–5 weeks.
**Range**
Spotty, unevenly distributed from w. N.Y. s. to w. Fla. and w. to
cen. Kans. and ne. Tex., but absent entirely in Ill., Ind., and Tenn.
Discontinuity of range indicates that this species once occupied a
larger area.

## PRAIRIE SKINK
*Eumeces septentrionalis*                                             **66:6**

**Description**
Length, to 8⅛ in. (20.6 cm); snout-vent length to 3¼ in. (8.3 cm).
Somewhat similar to Coal Skink but lateral dark stripe covers at
most 2 scale rows, and 2 postmental scales present. Rows of scales
28 around middle of body, 2 white stripes on each side, 4 dark
stripes on back, the 2 middle ones lighter than 2 outer ones. Light
lines on 4th or 5th scale row from midline.
**Similarities**
In the hand, Many-lined Skink shows light lines on 3rd scale row
from midline. Coal Skink has lateral dark line covering 2½–4 scale
rows.
**Habitat**
Flat rocks on grassy hillsides and in oak woodlands; sandy or
gravelly areas.
**Habits**
Wary, fast; usually found under some cover.
**Food**
Small insects, snails, spiders.
**Reproduction**
Lays 5–13 eggs, May–June under boards, logs, bark, stone, or in
loose, moist soil; incubation about 45 days.
**Range**
Extr. se. N.Dak., Minn., and w. Wis., s. through e. Tex. Scattered
localities in n. Ill., e. Wis., and s. Man.

# WHIPTAILS
Family Teiidae

This largely South American family contains varied forms, ranging
from tiny, elongate, limbless creatures to large lizards called tegus
(Genus *Tupinambis*), three feet (0.9 m) in length, which prey on
small birds and mammals. North American species are active and
small. Parthenogenesis, or asexual reproduction in an all-female
species, is common in some of the western forms and has made
their classification complex.

## SIX-LINED RACERUNNER
*Cnemidophorus sexlineatus*                                          **66:11**

**Description**
Length, to 10½ in. (26.7 cm); snout-vent length to 3⅜ in. (8.6 cm).
The only North American 4-limbed lizard with a very long tail,
small scales above, and large rectangular plates below. Above dark
with 3 light lines along each side of back; below white (with green
or blue in adult male). Graceful; head, neck, and body almost same
width; pointed snout. Tail twice or more length of body and with
rings of large scales around it.

**Habitat**
Various terrains up to 2000 ft. (609.6 m) but prefers dry regions, flat or hilly; sandy or rocky soil; short grass, thin woods, dusty roadsides.

**Habits**
Reaches amazing speeds; stops and starts suddenly; digs its own burrows with forelegs; home range about ¼ acre (0.1 hectare); lives in colonies.

**Food**
Insects, spiders, snails.

**Reproduction**
Lays 1–6 eggs, June–August under rocks or logs or in holes in sandy soil a few inches deep. Incubation about 8 weeks.

**Range**
S. Md., e. Va., and e. Ind., s. throughout most of se. U.S.; follows Miss. R. into Wis. and Minn., and w. to s. S.Dak., se. Wyo., and e. N.Mex.

# Snakes
Suborder Serpentes

### Features of Snakes
A snake, or serpent, is a limbless reptile with expandable jaws; slender, curved teeth; no ear openings or movable eyelids; and a single row of large belly scales. The flickering of its forked tongue picks up dust particles that give the snake sensations of both taste and smell. Its sense of smell is excellent. So, in general, is its sight. A snake "hears" by means of vibrations that its body picks up, not through the air, but through the ground. Snakes cast their skins, usually several times a year, by crawling out of them head first. These shed skins, or sheds, are often seen in suitable snake habitats.

Snakes in the eastern United States fit well the general description given in the introduction; however, certain groups, such as the boas in the West, have flaps at each side of the vent, a condition that is in contrast to the limblessness of most of the others. However, in no case do snakes have vestiges of forelimbs or a pectoral girdle. In snakes, one lung is always reduced, and the remaining lung extends far down the length of the body. Their long, flexible form is well adapted for the undulating waves that move along the body

*Fig. 62*
Parts of a Typical Snake

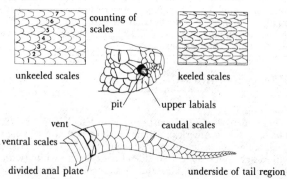

counting of scales

unkeeled scales

keeled scales

pit    upper labials

vent    caudal scales

ventral scales

divided anal plate    underside of tail region

and propel the snake forward whenever resistance is met in the substrate. Without rocks, stones, furrows on logs, or even soil, the snake would not be able to move.

The distensible lower jaws spread apart to accommodate oversized prey, and the skin between the scales is very elastic. The teeth are not used to chew food, but make very effective grappling hooks to hold prey, usually still struggling, in the mouth. Although certain teeth may be specialized for the injection of venom, many species (in fact the majority) kill their prey by constricting it with folds of the body, or by engulfing it while it is still alive.

### Identification
Individual snakes of the same species often vary widely in color in different parts of their range, or even in the same locality. Thus pattern—lines, blotches, or spots; shape of body—long, short, slender, or stout; and shape of head are important for identification, often more so than color. Indeed, colors may vary from light and patterned in young, to dark in old individuals. Unlike most other groups of animals, subspecies of snakes are often easily recognizable and may differ quite widely in appearance within the same species. Whenever this is the case in this chapter, the subspecies are discussed.

### Habitat
Favorite snake habitats are the edges of watercourses and bodies of water, on stone walls, rocky ledges, and causeways (especially the marsh side), in tangles of downed timber, on stumps and logs, and at the edges of bogs. Many serpents are nocturnal, and they are often attracted by highways, which retain warmth after the sun goes down. However, serpents like warmth only to a point; they cannot long tolerate hot sun. Therefore, the hours of sunshine early in the season, and early in the day later on, are the best times to find snakes on the surface. They tend to disappear in daytime in the hottest months, when their activity is restricted to the hours of darkness. In the fall, just prior to hibernation, they become more common again by day.

### Habits
Snakes have no voice, though some are able to make a hissing sound, and certain pit vipers can buzz with the rattles on the ends of their tails. Snakes live exclusively on animal food: frogs, toads, salamanders, rabbits, rats, mice, shrews, birds, and worms (there are no vegetarian serpents). They digest slowly, and after feeding they will often be sluggish for several days; meals may exceed 50 percent of their own body weight. Snakes hibernate in the colder months. In spring, they come out of hibernation and become active, starting once more to feed after the winter's fast, and to mate.

### Measurements
Length refers to the *straight line measurement* from the tip of the snout to the vent. Tail length is not included.

# BLIND SNAKES
Family Leptotyphlopidae

The pinkish, semitransparent members of this family look much like earthworms. Their tiny degenerate eyes, only a pair of black dots at one end of the body, are not used, and their scales are the same size all around the body, not enlarged on the belly as in other North American snakes.

## TEXAS BLIND SNAKE
*Leptotyphlops dulcis*                                      **67:1**

**Description**
Length, to 10¾ in. (27.3 cm). Very slender, pale, resembles an
earthworm more than a vertebrate, has reduced eyes. Scales on
underside not enlarged; body elongate, tail short. Brown above and
white or pink below.

**Habitat**
In sandy areas and prairies.

**Habits**
A burrower, most often found under stones or logs, but sometimes
abroad in early evening.

**Food**
Termites, ants, insect larvae.

**Reproduction**
Lays 2–7 eggs.

**Range**
S.-cen. U.S. from extr. sw. Kans. and cen. Okla. s. through Tex.
and w. into se. Ariz. and Mexico.

# COMMON HARMLESS SNAKES
Family Colubridae

This one family embraces all the North American nonpoisonous
snakes except for the Texas Blind Snake (see above). All members
have eyes and broad belly scales; most have a spine on the end of
the tail. None has pits behind the nostrils, as do the pit vipers, nor
the permanently erect front upper fangs of the coral snakes. Some
species have grooved rear fangs that conduct venom to prey seized
in the mouth, but in the East these species are harmless to humans.

# WATER AND CRAYFISH SNAKES
Genera *Nerodia, Regina,* and *Clonophis*

These snakes have stout bodies and keeled scales. They may have
wide heads, often quite large, very distinct from the neck. All swim
well, have a cold-blooded diet, and usually do not kill their prey by
constriction. They are often vicious when captured. All North
American species bear living young, but closely related European
and Asiatic members of this group lay eggs. The name *Natrix* has
now been restricted to the Eurasian forms and American species
are classified in different genera.

## GREEN WATER SNAKE
*Nerodia cyclopion*                                        **67:8**

**Description**
Length, to 60 in. (153.4 cm). Has ring of small scales around eye.
Appearance often plain, but above dark green or olive-brown with
many narrow black bands across back; alternating black blotches
on sides, most pronounced to rear; below yellowish to brownish,
especially anteriorly, and clear or spotted. Old individuals are
uniform dull olive above. Head long, upper lip swollen; eyes placed
high and forward.

**Similarities**
Resembles the Diamondback Water Snake, but lacks diamond
pattern.

**Habitat**
Small lakes, marshes, swamps.

**Habits**
Vicious; climbs branches; will lie with head flattened on ground, giving head triangular appearance.
**Food**
Fish, amphibians.
**Reproduction**
Bears 7–19 living young, July–September.
**Range**
Coastally from s. S.C. to se. Tex., continuously throughout Fla. and up Miss. Valley to extr. w. Ky., s. Ill., and se. Mo.

## PLAIN-BELLIED WATER SNAKE
*Nerodia erythrogaster*                                    **67:6**

**Description**
Length, to 48 in. (121.4 cm); record 62 in. (157.5 cm). Unmarked yellow or reddish belly, above is usually patternless. Head slightly wider than neck. Scale rows 21–25, lower labials 10. Most adults gray, brown, or black above; top of head dark. Young have brownish blotches.
**Similarities**
Northern Water Snake has smaller eyes.
**Habitat**
Near water, usually ponds, swamps, or lakes.
**Habits**
Hard to catch, shy; much more wary than Northern Water Snake; active at night.
**Food**
Fish, amphibians, crayfish.
**Reproduction**
Bears 5–27 living young, August–October.
**Range**
Mostly se. U.S. from s. Delmarva Peninsula along Atlantic and Gulf Coastal Plains (excluding peninsular Fla.) to Tex. N. in cen. U.S. to Kans. and s. Mich.

Note: Four subspecies in the eastern U.S. are the following:

**RED-BELLIED WATER SNAKE,** *Nerodia erythrogaster erythrogaster,* is uniform rusty brown above, occasionally with faint banding; below unmarked vermilion. It occurs in the Coastal Plain in southern half of Delmarva Peninsula and continuously from southern Delaware to southeastern Alabama and northern Florida.

**YELLOW-BELLIED WATER SNAKE,** *Nerodia erythrogaster flavigaster,* is plain gray above, plain yellow below. It is found in north-central Georgia and extreme western Florida to eastern Texas and extreme southeastern Oklahoma, and north in Mississippi Valley to western Illinois and southeastern Iowa.

**COPPER-BELLIED WATER SNAKE,** *Nerodia erythrogaster neglecta,* is darker than Red-bellied Water Snake, red to orange below. There are scattered populations from southern Michigan and central Ohio to southeastern Illinois, western Kentucky, and northwestern Tennessee, and to southeastern Iowa.

**BLOTCHED WATER SNAKE,** *Nerodia erythrogaster transversa,* has dark dorsal saddles and dark spots on sides; adults essentially retain a juvenile pattern; plain yellow below. It ranges west of Mississippi Valley from western Missouri and northwestern Arkansas south and west through central Texas and into northern Mexico and eastern New Mexico.

## NORTHERN WATER SNAKE
*Nerodia sipedon* **68:8**

**Description**
Length, to 53 in. (134.6 cm). Variable ground color gray to brown with dark crossbands anteriorly (usually about 30) that are wider than interspaces; alternating dorsal and side blotches toward rear of body. Below white or yellow with red to gray half-moons and dark dots that are more widely spaced at front of body and continue onto tail. Pattern varies in subspecies and becomes obscure in old individuals. Heavy-bodied; short-tailed; head distinct from neck. Lacks stripe from eye to corner of mouth. Very young are brightly marked, patterned jet-black on gray.

**Habitat**
Near water, especially streams, marshes, rivers, and lakes.

**Habits**
Diurnal to nocturnal depending on temperature, but active at temperatures lower than those tolerated by other snakes.

**Food**
Fish and anuran amphibians.

**Reproduction**
Bears 8–48 living young, with record litters of up to 99, June–September.

**Range**
Most of e. U.S. and se. Canada, but absent from n. New England and Atlantic Coastal Plain; in Gulf Coastal Plain from extr. w. Fla. to Miss. R.; nw. only to se. Minn. and extr. s. S.Dak. and to Colo. in Platte Valley.

Note: Four subspecies occur in the East:

A subspecies, *Nerodia sipedon sipedon,* as described above. It is found in North Carolina, Tennessee, Missouri, and northeastern Oklahoma northwards except in Mississippi River and Ohio River valleys in southern Illinois and Indiana; throughout most of northeastern U.S. and to southern Quebec.

**LAKE ERIE WATER SNAKE,** *Nerodia sipedon insularum (Fig. 63),* has no pattern, is plain grayish above, plain whitish below, or with faint evidence of pattern. It occurs in Pt. Pelée and Put-in-Bay archipelago in Lake Erie (between southern Ontario and northern Ohio).

**CAROLINA SALT MARSH SNAKE,** *Nerodia sipedon williamengelsi,* has dark overall coloration with black half-moons on belly. Found in salt-marsh areas, on outer banks of North Carolina and in nearby Pamlico and Core sounds.

**MIDLAND WATER SNAKE,** *Nerodia sipedon pleuralis (Fig. 63),* has brown to reddish ground above with narrower dark bands, overall light appearance. Below cream with reddish-brown belly spots in 2 rows. It occurs in southern Illinois and Indiana, west to eastern Oklahoma and northern Arkansas, and east of Mississippi River in lowland areas to northern Georgia and northern South Carolina.

*Fig. 63*

Lake Erie
Water Snake

Midland
Water Snake

# SOUTHERN WATER SNAKE
*Nerodia fasciata*

**Description**
Length, to 42 in. (106.7 cm); record 62½ in. (158.8 cm). Broadly banded throughout length of body and has dark stripe from eye to corner of mouth. Belly spots squarish in most subspecies, but check below.

**Similarities**
Cottonmouth has no red spots below; Salt Marsh Snakes (see below) often lack bands and have longitudinal stripes.

**Habitat**
Permanently aquatic situations, ponds, marshes, swamps, lakes, also in temporary water, more often near shallow, still water.

**Habits**
Nocturnal, activity often correlated with large choruses of frogs. Like the Cottonmouth, defends itself when cornered, but bite is not venomous.

**Food**
Frogs, small fish, salamanders, tadpoles. (Salt Marsh Snakes feed on fish and crabs.)

**Reproduction**
Bears 9–57 living young, born June–August.

**Range**
Coastal Plain from N.C. to e. Tex. and up Miss. R. valley to extr. s. Ill.

Note: Four subspecies are found in the East:

**BANDED WATER SNAKE,** *Nerodia fasciata fasciata,* has dark crossbands over a tan to reddish-brown ground color, with crossbands rarely broken. Below yellow with reddish-brown blotches alternating irregularly from side to side, each blotch limited to 1 ventral. It is found in the Coastal Plain, except peninsular Florida, from North Carolina to Mississippi.

**FLORIDA WATER SNAKE,** *Nerodia fasciata pictiventris,* is similar to above, with gray, brown, or red crossbands, wormlike belly markings, and dull blotches between crossbands at sides. It occurs in all of eastern and peninsular Florida and into adjacent southeastern Georgia.

**BROAD-BANDED WATER SNAKE,** *Nerodia fasciata confluens,* has dark crossbands very wide above, almost running together, outlined with orange (or red) and white. Below yellowish-white with large, squarish, dark spots covering 3 or more scales. It occurs in the Mississippi Valley from extreme southern Illinois south; eastward only narrowly, and westward to southeastern Oklahoma and eastern Texas.

**GULF COAST SALT MARSH SNAKE,** *Nerodia fasciata clarki,* is dark gray to brown with 2 prominent longitudinal yellow stripes on each side and a duller middorsal stripe. Below blackish with a midventral light stripe. Attains a length of only 36¾ in. (93.3 cm), occurs in salt marsh habitats, and may intergrade with the freshwater species in areas of intermediate salinity. It is found on the Gulf Coast from central Florida to southern Texas.

## DIAMONDBACK WATER SNAKE
*Nerodia rhombifera*                                                   **67:7**

**Description**
Length, to 48 in. (121.9 cm); record 63 in. (160.0 cm). No other
water snake has diamond pattern above. Above yellow to olive with
pattern of interconnecting diamonds of dark brown lines enclosing
areas of ground color; dark bands at sides extend from bottom of
diamonds to edge of belly. Below yellowish, ventral scales edged
with crescent-shaped dark spots, especially near tail. Body large,
heavy in adults.

**Habitat**
Near water; in open and under cover on lake shores and marshes;
on branches overhanging water; by cattle tanks.

**Habits**
Aquatic, nocturnal; a good climber and vicious fighter. Emerges
early from hibernation.

**Food**
Fish, frogs, turtles.

**Reproduction**
Bears 14–62 living young, August–November.

**Range**
Sw. Ind., s. Ill., extr. se. Iowa, Mo., and s. and e. Kans., s. to Gulf
Coast, and into e. Mexico. Absent from Ozark Plateau and
Ouachita Mts.

## BROWN WATER SNAKE
*Nerodia taxispilota*                                                  **67:9**

**Description**
Length, to 69 in. (175.3 cm). Largest water snake, distinguished by
rectangular black-bordered blotches on rusty or brownish-green
upperparts. Dark blotches on sides alternate with rectangles above,
but pattern may be irregular posteriorly. Below pied, white or buff
with numerous dark markings. Young have paler ground color
with black bands. Body very stout; head long, narrow; tail tapers
rather abruptly.

**Similarities**
Cottonmouth either plain dark or brown or with wide, dark
crossbands on paler ground.

**Habitat**
Near water; rivers, creeks, swamps; on branches overhanging
water; sometimes frequents brackish water.

**Habits**
Arboreal as well as aquatic.

**Food**
Frogs, fish.

**Reproduction**
Bears 14–58 living young, June–November, but principally in August.

**Range**
Coastal Plain from se. Va. through Fla to se. Ala.

## GRAHAM'S CRAYFISH SNAKE
*Regina grahami*                                                       **67:3**

**Description**
Length, to 28 in. (71.1 cm); record 47 in. (119.4 cm). Distinguished
by broad yellow band on first 3 scale rows on each side, bordered
below with thin, black, zigzag stripe. Above brown with paler dorsal
stripe; belly yellowish, sometimes with median dark dots or darkish
stripe. Head small, narrower than body, but wider than neck.

**Habitat**
Edges of creeks in prairie country; sloughs, ponds, ditches; often under logs, boards, or rocks, or in crayfish holes; occasionally seen on branches overhanging water.
**Habits**
Generally nocturnal; emerges early from hibernation, dens up late.
**Food**
Primarily crayfish; also fish, amphibians.
**Reproduction**
Bears 9–39 living young, July–September.
**Remarks**
Most snakes sleep in shelter of some protective nook, but this species and other water snakes often sleep on a branch over water.
**Range**
Ill. to se. Nebr., southward w. of Miss. R. to La. and e. Tex., and e. into n. Miss. Absent from Ozark Plateau and Ouachita Mts.

## GLOSSY CRAYFISH SNAKE
*Regina rigida*                                                   67:5

**Description**
Length, to 31⅜ in. (79.7 cm). Dark, shiny, with traces of black lines down back and 2 rows of prominent dark spots on yellow belly. Top of head dark, upper lip yellow; underside of tail unmarked or may have irregular mid-stripe. No light stripes on sides. Scales have polished look. Body small, stout, head small.
**Habitat**
Edges of lakes, ponds, streams, swamps, marshes; often in mud.
**Habits**
Very aquatic; timid; a mud burrower.
**Food**
Hard-shelled crayfish, which it immobilizes by body constriction.
**Range**
Coastal Plain from N.C. through se. Okla. and e. Tex., but only in n. half of Fla. Disjunct population in New Kent Co., Va.

## QUEEN SNAKE
*Regina septemvittata*                                           67:4

**Description**
Length, to 24 in. (61 cm); record 36¼ in. (92.1 cm). Relatively slender, with 4 dark stripes on yellow belly and very small head. Above dark olive-brown with yellow band on side; upper lips and nose yellow; 3 faint narrow black stripes on mid-back. Belly pattern often distinct only anteriorly. Head barely distinct from neck; tail slender, tapering to a point.
**Habitat**
Swift, rocky, and shallow streams; edges of lakes, ponds, ditches, canals, sloughs; old quarries; often under rocks, branches, or debris, overhanging water.
**Habits**
Spends much time in water, tends to congregate before hibernating.
**Food**
Crayfish that have just molted.
**Reproduction**
Bears 5–23 living young, July–September.
**Range**
S. Great Lakes region from w. N.Y. to se. Wis., s. to Gulf Coast from panhandle of Fla. to Miss., e. to Del. and se. Pa. in mid-Atlantic region. Disjunct part of range in w. Mo. and Ark.

## KIRTLAND'S WATER SNAKE
*Clonophis kirtlandi*                                                    *Fig. 64*

**Description**
Length, to 24 in. (61 cm). Above with 4 rows of blotches down
body, mid-belly red with 2 distinct rows of black spots. Top of
head black; lips, chin, and throat yellowish. Upright dark spots on
sides alternate with blotches on back; background fairly dark, thus
giving generally dark appearance. Head small, not distinct from
neck; tail slender.

**Habitat**
Prairie country near streams and ponds; under cover in marshes,
damp woods, ravines, pastures.

**Habits**
Probably nocturnal; when alarmed flattens itself and becomes
immobile.

**Food**
Minnows, amphibians, earthworms.

**Reproduction**
Bears 4–13 living young, August–September.

**Range**
S. Great Lakes area from se. Wis. and s. Mich. s. to w. Ky., e. to
w. Pa., and w. through cen. Ill. into Mo.

*Fig. 64*

Fox Snake, p. 376

Butler's Garter
Snake, p. 358

Kirtland's
Water Snake

Note: The **STRIPED CRAYFISH SNAKE,** *Regina alleni,* is similar in
coloring to the garter and ribbon snakes, but can be distinguished
by the fact that it has unkeeled scales and a divided anal plate. It
occurs in extreme southern Georgia and the entire Florida
peninsula.

# SWAMP SNAKES
Genus *Seminatrix*

## BLACK SWAMP SNAKE
*Seminatrix pygaea*                                                    *Fig. 65*

**Description**
Length, to 18½ in. (47 cm). Shiny black above and reddish below
with regular prominent black bars to sides of ventrals only. Scale
rows 17; anal divided; scales smooth. Above smooth, with narrow
light line through middle of lower scale rows.

**Habitat**
Swamps, ponds, lakes with emergent vegetation, especially where
water hyacinths are abundant.

**Food**
Invertebrates, tadpoles, small frogs, salamanders, small fish.

**Reproduction**
Bears 3–11 living young, August–October.

**Range**
Coastal Plain from N.C. through all Fla. except for w. tip of
panhandle.

*Fig. 65*                Black Swamp Snake

# PINE WOODS AND EARTH SNAKES
Genera *Rhadinaea* and *Virginia*

## PINE WOODS SNAKE
*Rhadinaea flavilata*                                      *Fig. 66*

**Description**
Length, to 15⅞ in. (40.3 cm). Upper lip yellow and dark line from
snout through eye to mouth. Scales smooth; scale rows 17 at
midbody; upper labials 7; anal divided. Above tan to yellowish or
red, sometimes with obscure and irregular longitudinal stripes.
Below yellow.
**Habitat**
Damp pine flatwoods, under rotting logs or stones.
**Habits**
Secretive.
**Food**
Small amphibians and reptiles.
**Reproduction**
Lays 2–4 eggs, May–August; female may produce 2 clutches per
egg-laying period.
**Remarks**
The Pine Woods Snake is rear-fanged and mildly venomous, but
poses no danger to humans. It is the sole member of the Genus
*Rhadinaea* in the U.S. Over 20 related species occur southward
through central South America.
**Range**
Coastal Plain marginally in states from N.C. to Miss. R. and
throughout Fla. except s. tip of peninsula.

*Fig. 66*                Pine Woods Snake

## ROUGH EARTH SNAKE
*Virginia striatula*                                      **68:4**

**Description**
Length, to 12¾ in. (32.4 cm). Head small, thick, sharply pointed,
distinct from neck. Brown or reddish, with strongly keeled scales
and yellow or pink belly. In younger individuals a light band
across back of head. Upper labials 5; anal divided.
**Similarities**
Worm Snake has smooth scales, pink belly, but is darker above,
and head is not distinct from neck. The Red-bellied Snake has
dark head.
**Habitat**
Rocks, hillsides, woods, rubbish piles; near cities; wooded
bottomlands, under bark, logs, or trash.
**Habits**
A burrower, secretive; stays under cover, but sometimes found on
roads.
**Food**
Earthworms, snails, insects.
**Reproduction**
Bears 2–8 living young in August.
**Range**
Se. U.S., along Coastal Plain from s. Va. to cen. Tex., but absent
in peninsular Fla. and s. La.; n. to s. Tenn., cen. Mo., se. Okla.,
and se. Kans.

## SMOOTH EARTH SNAKE
*Virginia valeriae*                                                68:5

**Description**
Length, to 13¼ in. (33.7 cm). Small, brown or gray, with cream-
colored belly and smooth or weakly keeled scales. Back may be
spotted with lines of tiny black dots, a faint line through each scale.
Head wider than neck, but small, flat, narrow; snout pointed.
Upper labials, 6; 2 small scales behind eye; anal divided.
**Habitat**
Fields, pastures, dry or moist woodlands; vacant lots, abandoned
buildings, under tarpaper, rubbish, leaves, boards.
**Habits**
A burrower; secretive; stays under cover.
**Food**
Earthworms, snails, insects.
**Reproduction**
Bears 4–12 living young in August.
**Range**
Much of e. U.S. from N.J. and Pa. s. through s. Fla., and w. in
lower elevations to s. Iowa, cen. Okla., and cen. Tex. Absent from
Gulf Coast w. of Miss. R.

# BROWN AND RED-BELLIED SNAKES
Genus *Storeria*

## BROWN SNAKE
*Storeria dekayi*                                                  68:2

**Description**
Length, to 13 in. (33.0 cm); record 20¾ in. (52.7 cm). Two lines of
small, dark spots down brown back, and dark brown blotches on
each side of head behind eye. Scales keeled, 17 scale rows on body;
anal divided. Above gray to brown with faint, light middorsal
stripe. Young are darker and have light rings around neck;
distinguish from young Ringneck Snake by keeled scales.
**Habitat**
Moist and dry situations, under stones, boards, leaves, trash heaps,
roofing paper; rurally and in urban vacant lots.
**Habits**
Nocturnal and secretive.
**Food**
Earthworms, snails, slugs, insects, spiders, treefrogs.
**Reproduction**
Bears 3–27 living young, July–September.
**Remarks**
Formerly known as Dekay's Snake.
**Range**
Extr. s. Canada and e. U.S. from s. Maine through Fla., w. to cen.
Minn. and extr. ne. S.Dak., se. Nebr., and cen. Tex.; also into e.
Mexico.

## RED-BELLIED SNAKE
*Storeria occipitomaculata*                                       68:3

**Description**
Length, to 16 in. (40.6 cm). Plain reddish belly, and 3 light spots
on nape. Fifteen scale rows; scales keeled; anal divided. Ground
color gray, black, brown, to red-orange; head dark; faint middorsal

stripe. Belly edged with slate-gray. Sometimes small dark spots form lines down back.

**Habitat**
Moist woodlands, especially those with rocks; under stones, logs, boards, leaves, bark; in abandoned dwellings.

**Food**
Insects, earthworms, slugs.

**Reproduction**
Bears 1–13 living young, June–September.

**Range**
E. U.S. and s. Canada from N.S. to extr. se. Sask., e. N.Dak., and s. through extr. e. Tex., with patchy distribution in Ind. and Mo. Absent from extr. n. Maine and s. Fla. A disjunct population in S. Dak. and Wyo.

# LINED SNAKES
Genus *Tropidoclonion*

## LINED SNAKE
*Tropidoclonion lineatum*                                    **67:2**

**Description**
Length, to 21 in. (53.3 cm). Greenish-white underparts with 2 rows of dark triangular dots running length of body. Scales keeled. Above brownish with 3 yellowish stripes; 2 rows of dots flank middorsal stripe. Light lateral stripe on 2nd and lower part of 3rd scale row. Head small, with 3 yellowish stripes; 2 rows of dots flank middorsal stripe. Head small, pointed, not distinct from neck; tail short.

**Habitat**
Near water; in old fields, rocky places, under boards and litter; urban vacant lots.

**Habits**
Secretive.

**Food**
Earthworms, insects.

**Reproduction**
Bears 2–12 living young in August.

**Range**
Continuously from se. S.Dak. and w. Iowa s. through e. Nebr. to cen. Tex. Spotty distribution in east of range to Ill. and Mo. and in west to Colo. and N.Mex.

# GARTER SNAKES
Genus *Thamnophis*

Garter snakes are characterized by three narrow, light stripes on a darker ground color above, but coloration is variable. Typically, their heads are distinct from their necks and the scales are keeled. Scale rows are fewer than twenty-seven, and a single anal plate is found in most.

## SHORT-HEADED GARTER SNAKE
*Thamnophis brachystoma*                                    *Fig. 67*

**Description**
Length, to 22 in. (55.9 cm). Only garter snake with 17 rows of dorsal scales throughout length of body; yellow or orange lateral

stripes on 2nd and 3rd scale rows toward front. Scales keeled; anal single. Body stout, tapers forward to long, thin neck not distinct from head; head very small. Above brown or black, light stripe on back may or may not be conspicuous. Black blotches between lateral stripes reduced to thin bars or absent.

**Habitat**
Meadows and pastures near water; among stones near marshes; not in undisturbed areas of forest.

**Food**
Earthworms, plus frogs, fish, toads, mice, insects.

**Reproduction**
Numbers of living young and season of birth not yet known.

**Range**
In nw. Pa. from Meadville area n. to Lake Erie and eastward, barely into s. N.Y.; along Allegheny R. drainage.

Western Terrestrial Garter Snake, p. 360

*Fig. 67*

Common Garter Snake

Short-headed Garter Snake, p. 357

# BUTLER'S GARTER SNAKE
*Thamnophis butleri*                               *Fig. 64*

**Description**
Length, to 27¼ in. (69.2 cm). Small head, not distinct from neck. Above, 3 yellow stripes on black or brown ground, lateral stripes on upper 2nd, all of 3rd, and lower 4th scale rows; dark blotches sometimes present between stripes; top of head black. Below greenish-white with some black spots on edges. Nineteen rows of dorsal scales at neck, 17 just before anus. Tail short, stout, tapers to point.

**Habitat**
Near water, swamp edges, stream banks; in dry grass.

**Habits**
Mild-tempered.

**Food**
Earthworms, leeches, amphibians, insects.

**Reproduction**
Bears 4–16 living young, June–September.

**Range**
Great Lakes area in extr. s. Ont., e. Mich., ne. Ind., and w. and cen. Ohio. A disjunct population in se. Wis.

# CHECKERED GARTER SNAKE
*Thamnophis marcianus*                             **67:12**

**Description**
Length, 24 in. (61 cm); record 42½ in. (108 cm). Yellow crescent on each side of head. Pale; above straw-colored with indistinct yellowish back and side stripes and dark, square blotches between stripes. Lateral stripe on 3rd scale row anteriorly, on 2nd (or 2nd and 3rd) to rear. Black vertical stripes on labial scales. Below whitish, often with small dark spots on ends of ventrals. Scales keeled, anal single.

**Habitat**
Near water, stream edges; river bottoms in desert regions.
**Food**
Mice, frogs, toads, fish, lizards, earthworms.
**Reproduction**
Bears 6–18 living young, July–August.
**Range**
Sw. U.S. from sw. Kans., s. through cen. and w. Tex., and w. to
extr. se. Calif. Also through most of cen. Mexico and to the south.

## PLAINS GARTER SNAKE
*Thamnophis radix*                                       **67:11**

**Description**
Length, to 28 in. (71.1 cm); record 40 in. (101.6 cm). Three
distinct bright stripes with 2 rows of squarish black spots between
them; lateral stripes on 3rd and 4th scale rows in front, on 2nd and
3rd rows to rear. Above light olive to dark brown or black, side
stripes may be lighter than yellow-orange stripe on back; upper lip
yellow, barred with black. Below white to greenish with black
spots on ends of ventrals. Scales keeled; anal single, body stout;
head broad, distinct; tail ¼, but no more, of total length.
**Similarities**
Checkered Garter Snake has lateral stripe on 3rd scale row only
anteriorly.
**Habitat**
Wet prairies, roadside ditches, shores of bodies of water and
watercourses.
**Habits**
Aggressive; like most garter snakes, attempts to bite upon capture.
**Food**
Frogs, fish, toads, mice, earthworms, insects.
**Reproduction**
Bears 5–92 (generally not more than 40) living young, late July–
September.
**Range**
N. Plains states w. of Great Lakes from Wis., Ind., and Ill., w.
through Kans. to ne. N.Mex.; in north of range through w. Minn.
and s. Man. to se. Alta.; also in cen. Ohio and extr. nw. Ark.

## COMMON GARTER SNAKE
*Thamnophis sirtalis*                          **67:10, 68:7**
                                                  *Fig. 67*

**Description**
Length, to 26 in. (66 cm); record to 48¾ in. (123.8 cm). Variably
colored, with light lateral stripes not involving 4th scale row, but
on 2nd and 3rd scale rows anteriorly. Above green to orange-
brown or black with 3 yellowish or greenish stripes, sometimes
broken into alternating square black spots. Below greenish-white or
yellow, infrequently with black spots on tips of ventrals.
**Similarities**
Ribbon snake more slender, has longer tail, lacks alternating spots
between stripes.
**Habitat**
Widely distributed in various habitats, including fields, meadows,
marshes, roadsides, gardens; often near water.
**Habits**
First snake to emerge in spring, last to disappear in fall.
**Food**
Earthworms, amphibians, mice, young birds, insects, spiders, fish.

**Reproduction**
Bears 3–96 living young, June–September.

**Remarks**
The northernmost-ranging reptile in North America, this snake
reaches about 60° N latitude in the Northwest Territories. It has
an interesting aggregative behavior during spring in northern parts.
In Manitoba, up to 10,000 Red-sided Garter Snakes may gather to
overwinter in limestone sinks. Immediately after emergence in the
spring they may form balls of up to 200 intertwined individuals to
conserve heat in response to low temperatures. Makes a good pet.

**Range**
Throughout e. U.S. and s. Canada, although distribution patchy in
w. Tex., Okla., and Kans.; also widely distributed in the
Northwest and in Calif.

Note: The major subspecies in the East are:

**EASTERN GARTER SNAKE,** *Thamnophis sirtalis sirtalis,* has no red
on sides and has 2 rows of dark spots distinct on upperside. It
occurs in Mississippi Valley eastward.

**RED-SIDED GARTER SNAKE,** *Thamnophis sirtalis parietalis,*
generally has red skin in between scales on sides (best seen when
stretched), and 2 rows of dark spots at each side coalesce. It ranges
west of Mississippi Valley.

## WESTERN TERRESTRIAL GARTER SNAKE    67:14
*Thamnophis elegans*                                    Fig. 67

**Description**
Length, to 35½ in. (90.2 cm). A discontinuous yellow back stripe
and dull yellow stripe on each side on 2nd and 3rd scale rows.
Above brownish; may have rows of dark squarish spots; yellow
spot in front of eye, black band across neck.

**Habitat**
Near or in water, flowing or still, but also in terrestrial places.

**Food**
Small vertebrates, including frogs and fish; earthworms, insects.

**Reproduction**
Bears 8–19 living young, June–September.

**Remarks**
Eastern subspecies is called Wandering Garter Snake, *T. e.
vagrans.*

**Range**
Widely distributed through the West, including w. S.Dak. and
Nebr.; also in extr. w. Okla.

## EASTERN RIBBON SNAKE
*Thamnophis sauritus*                                    67:13

**Description**
Length, to 25 in. (64.3 cm); record 40 in. (101.6 cm). Lateral
stripe on 3rd and 4th scale rows only, and dark lower lateral stripe
that usually runs onto sides of ventral scales. Ground color brown
to black with 3 prominent stripes along body, and occasionally
vague dark spots between stripes. Spots to rear on top of head
sometimes present, but dull and never fused. Relatively large, long-
tailed, the tail more than ¼ total length; scale rows 19 over most of
body.

**Habitat**
Marshes, lake shores, stream edges, meadows.

**Habits**
Semiaquatic, quick, able to climb to low vegetation.
**Food**
Amphibians, fish.
**Reproduction**
Bears 3–20 living young, July–August.
**Remarks**
Will bite; secretes fluid from anal glands when caught.
**Range**
E. of Miss. R. from s. Maine, s. Ont., and lower Mich., s. throughout Fla. to Miss. R., from Ky. to La. Disjunct populations in s. N.S. and ne. Wis.

# WESTERN RIBBON SNAKE
*Thamnophis proximus*                                    *Fig. 68*

**Description**
Length, to 30 in. (76.2 cm); record 48½ in. (123.2 cm). Similar to Eastern Ribbon Snake, but with 2 bright spots, usually fused, at rear of top of head, and only a narrow lower lateral dark stripe. Ground color of brown to black or olive with 3 stripes running along body. Middorsal stripe yellow, orange, or red. Lateral stripe on 3rd and 4th scale rows. Ventral scales lack dark pigment. Large, long tail; scale rows 19; scales keeled; anal single.
**Habitat**
Near most permanent water—streams, swamps, lakes, and rivers.
**Habits**
Diurnal; may try to bite on capture.
**Food**
Frogs, toads, salamanders, fish.
**Reproduction**
Bears 4–27 living young, July–September.
**Range**
S. Wis. and Ind., s. through Miss. Valley and w. to e. Nebr., se. Colo., and e. N.Mex.; also into e. Mexico.

 *Fig. 68*   Western Ribbon Snake

# SMALL WOODLAND SNAKES
Genera *Diadophis* and *Carphophis*

# RINGNECK SNAKE
*Diadophis punctatus*                                    **68:1**

**Description**
Length, usually to 18 in. (45.7 cm). Unicolor above with yellow or orange ring around neck (though ring sometimes broken or even absent). Above dark gray to black; below yellow to orange or reddish, sometimes with one or more rows of black dots. Scales smooth, plain; anal divided; body slender, head not distinct from neck, top of head flat. Hatchlings blackish with distinct yellow around neck.
**Habitat**
Under bark or logs in woods, near water, rock-covered hillsides, forest paths, field edges.

**Habits**

Nocturnal; secretes foul-smelling defensive substance; in some populations, coiling of tail and exposing of red underside when disturbed is characteristic.

**Food**

Subspecies in the East feed on earthworms, insects, salamanders, small frogs and toads; larger western subspecies feed on lizards and snakes, and may be mildly venomous.

**Reproduction**

Lays 1–10 eggs, June–August; hatchlings August–September.

**Range**

Most of e. U.S. and in extr. se. Canada w. to se. S. Dak., se. Colo., through much of Southwest. Also n. on Pacific Coast and s. into Mexico.

Note: There are 5 subspecies in the East:

**SOUTHERN RINGNECK SNAKE**, *Diadophis punctatus punctatus (Fig. 69)*, has yellow neck ring broken, series of half-moon-shaped black dots down center of yellow or orange belly. It ranges in Coastal Plain from New Jersey through Florida and Upper Keys to Alabama.

**KEY RINGNECK SNAKE**, *Diadophis punctatus acricus,* has no neck ring. It occurs on Florida Keys only.

**PRAIRIE RINGNECK SNAKE**, *Diadophis punctatus arnyi (Fig. 69)*, has neck ring normally unbroken, black dots scattered or in pairs over yellow belly. It occurs in northern Mississippi Valley from extreme southeastern Minnesota to Missouri, west through eastern Nebraska, eastern New Mexico, and northern Texas.

Northern Ringneck Snake

Southern Ringneck Snake

*Fig. 69*

Mississippi Ringneck Snake

Prairie Ringneck Snake

**NORTHERN RINGNECK SNAKE**, *Diadophis punctatus edwardsi (Fig. 69)*, has neck ring 1–3 scales wide, only occasionally broken; belly plain yellow or orange, sometimes with faint black dots down middle. It occurs in Appalachians and Piedmont from northeastern Alabama and northern Georgia, north through southern Canada, and west to southern Illinois through Wisconsin around most of Great Lakes, to northeastern Minnesota.

**MISSISSIPPI RINGNECK SNAKE**, *Diadophis punctatus stictogenys (Fig. 69)*, like *D. p. edwardsi,* above, but belly strongly spotted. Occurs in lower Mississippi Valley from western Alabama to eastern Texas and north to extreme southern Illinois.

## WORM SNAKE
*Carphophis amoenus*

**68:6**

**Description**

Length, to 14¾ in. (37.5 cm). Small, slender, dark above and pink below without any markings or pattern. Scales smooth, opalescent. Above brown to black; pale belly color extends to 3rd scale row at each side. Thirteen scale rows; 5 upper labials; head not distinct from neck, snout pointed, eye very small; tail short, pointed. Young darker than adults and have red belly.

**Habitat**

Under logs, stones, slabs, leaves, in wooded areas, abandoned fields, pastures.

**Habits**

Semiburrowing; has limited home range.

**Food**

Earthworms primarily, but also slugs, insects, snails.

**Reproduction**

Lays 1–12 eggs, late June–July; hatchlings August–September.

**Remarks**

Its common name refers not only to dietary specialization but also appropriately describes appearance.

**Range**

E.-cen. U.S. from w. Mass. s. coastally to Ga. and w. below Great Lakes area to se. Nebr. in the North and extr. ne. Tex. and n. La. Absent in most of w. Miss. Valley, but present on Gulf Coast from Miss. R. to Ala.

# MUD AND RAINBOW SNAKES

Genus *Farancia*

## MUD SNAKE

*Farancia abacura*

68:16

**Description**

Length, to 81 in. (205.7 cm). Large, stout, shiny black above, black and coral-red below, the red forming bars or triangles on lower sides. Scales smooth, glossy. Body rounder than in rat snakes; head long, barely distinct from neck; spinelike scale at end of tail. Young may show signs of red crossbands on back.

**Similarities**

Rainbow Snake has stripes.

**Habitat**

Marshes, swamps, ditches.

**Habits**

A mud burrower; nocturnal; can jab with the spine of its tail, but spine is not poisonous as many who fear snakes would believe.

**Food**

Mainly amphiumas and sirens, but also frogs, fish.

**Reproduction**

Lays 4–104 eggs, July–September; female remains coiled around eggs during incubation; hatchlings, September–October.

**Remarks**

Like many other brilliantly colored animals, this snake has a characteristic warning posture that displays the red belly. In actuality a harmless snake, it is sometimes confused with poisonous snakes or with a fictitious "hoop snake," which rolls like a wheel while holding its tail with its mouth.

**Range**

Coastal Plain from s. Va. to e. Tex. and up Miss. Valley to s. Ill.

## RAINBOW SNAKE

*Farancia erytrogramma*

68:15

**Description**

Length, to 66 in. (167.6 cm). Three red stripes above and 3 rows of black spots on red belly. Ground color above black; scales

smooth, glossy; body heavy, head not distinct from neck, spine at end of short tail.
**Habitat**
Mud or sand; sandy fields near marshes, springs.
**Habits**
A burrower; also aquatic; rarely surfaces; able to jab with spine on its tail.
**Food**
Eels; also frogs, fish, amphiumas, sirens, worms.
**Reproduction**
Lays 20–52 eggs, July–August in hole 4–6 in. (10.2–15.4 cm) deep in sandy field; hatchlings in September.
**Range**
Coastal Plain of North America from s. Md. to Miss. R. In s. Fla. present only near Lake Okeechobee.

# HOGNOSE SNAKES
Genus *Heterodon*

Hognose snakes have an upturned, keeled snout. When threatened, a hognose snake puffs itself out, flattens the head, hisses, and strikes but never bites. If these menacing tactics fail, it feigns death by rolling over on its back and letting its tongue loll out. Thus fooled, potential predators relax their attention, and the snake escapes. However, the hognose snake betrays itself by its behavior when turned upright; it promptly repeats its performance and rolls belly-up.

## WESTERN HOGNOSE SNAKE
*Heterodon nasicus*          *Fig. 70*

**Description**
Length, to 35¼ in. (89.5 cm). Blotched with dark olive or brown above and underside of tail as black as belly. Yellowish ground above, blotches alternating with those on sides. Body heavy, scales keeled; head broad, barely distinct from neck; tail very short.
**Similarities**
Eastern Hognose darker, uniformly colored, has larger eye, snout less sharply upturned.
**Habitat**
Dry, sandy areas; prairies.
**Habits**
Diurnal. See also genus description, above.

Eastern Hognose Snake

*Fig. 70*

Western Hognose Snake

Southern Hognose Snake

**Food**
Toads, frogs, lizards, snakes, rodents, shrews, small birds.
**Reproduction**
Lays 5–24 eggs in August.
**Range**
Central Plains from s.-cen. Canada to Mexico, with easternmost localities in Ill., and a patchy distribution in Minn., Iowa, Mo., and ne. Tex.; w. to cen. Mont., cen. Colo., and se. Ariz.

## EASTERN HOGNOSE SNAKE                          68:9
*Heterodon platyrhinos*                          *Fig. 70*

**Description**
Length, to 45½ in. (115.6 cm). Underside of tail lighter than belly. Above color very variable, sometimes uniform gray, brown, black, or green; if blotched, patterns may be gray, black, brown, or even reddish on a lighter ground. Dark line across top of head, dark line from eye to corner of mouth. Below whitish or yellowish with black markings. Body heavy, scales keeled, head flat on top, eyes large.
**Similarities**
Western Hognose has snout more upturned.
**Habitat**
Dry, sandy areas, open woods, uplands, hillsides, fields.
**Habits**
See genus description, above.
**Food**
Toads, frogs, insects, lizards.
**Reproduction**
Lays 4–61 eggs, June–July in earth, rotten logs, etc.; female often coils about eggs while they incubate, 6–9 weeks.
**Range**
S. Ont. and e. U.S. from se. N.H. through most of Fla. and w. through s. N.Y., lower Mich., and most of Wis. to cen. Minn., e. and s. Iowa, and ne. Nebr.; also se. S.Dak. and Kans. s. to Gulf Coast and e. Tex.

## SOUTHERN HOGNOSE SNAKE
*Heterodon simus*                                *Fig. 70*

**Description**
Length, to 24 in. (61 cm). Generally light in color with uniformly gray belly, the same color continuing under tail. Belly either uniformly colored or speckled with gray-brown. Dorsal ground gray, brown, yellowish with some red; dark blotches with smaller blotches laterally.
**Similarities**
Eastern Hognose is darker, larger, has snout scale less sharply upturned.
**Habitat**
Sandy lowland areas, usually in the open.
**Habits**
See genus description, above.
**Food**
Toads.
**Reproduction**
Lays 6–10 eggs, about which little is known.
**Range**
Coastal Plain from cen. N.C. to s. Miss., but absent s. of Lake Okeechobee in Fla.

# SCARLET SNAKE
Genus *Cemophora*

### SCARLET SNAKE
*Cemophora coccinea*                                        **69:8**

**Description**
Length, to 20 in. (50.8 cm); record 32¼ in. (81.9 cm). Pale,
unmarked belly and wide blotches of red on back and sides, with
narrow black edges separated by narrow yellow or white in the
formula black-yellow-black-red-black. Snout and top of head red
with black band across eyes. Body slender and round, scales
smooth; head not distinct from neck; snout pointed; eye small. In
young, yellow is replaced by white.

**Similarities**
Pattern of banded Scarlet Kingsnake. See under Coral Snake.
Blotches of Scarlet Snake do not go onto ventral scales.

**Habitat**
Underground in earth, sandy soil, or muck; on surface under slabs,
stones, logs.

**Habits**
Secretive; a burrower.

**Food**
Young mice, lizards, small snakes, snake eggs, insect larvae.

**Reproduction**
Lays 3–8 eggs in June.

**Range**
Se. U.S.; cen. N.J. and Delmarva Pen.; s. Md., s. along Coastal
Plain to extr. e. Tex.; northern limits in Va., Ky., extr. s. Ind., Ill.,
extr. s. Mo., and Okla. Disjunct populations at periphery of range,
especially s. Tex., cen. Mo., Ky., and Va.

# KINGSNAKES AND MILK SNAKES
Genus *Lampropeltis*

This large group of snakes varies widely in pattern, color, and size.
Their heads are typically small and barely distinct from neck, the
bodies are cylindrical, the scales smooth, and they have a single
anal scale. The various species are often strikingly colored in
patterns of rings or bands. All are powerful constrictors, and some
species are notable in their immunity to the venom of rattlesnakes.
All lay eggs.

### COMMON KINGSNAKE
*Lampropeltis getulus*                                      **69:4**

**Description**
Length, varies according to subspecies. A highly variable species:
ground color from black or dark blue to yellow, with speckled,
banded, or striped pattern. Top of head usually dark, lips white
with dark edges. Body stout.

**Habitat**
Woods, hayfields, pastures, meadows, roadsides; near water;
around farm buildings; rocky places.

**Habits**
Nocturnal and diurnal, above and below ground; uses holes made
by other animals.

**Food**

Snakes, reptiles' and birds' eggs, young birds, rodents, lizards, amphibians.

**Reproduction**

Lays 5–17 eggs, June–July; hatchlings August–September.

**Remarks**

Immune to venom of native poisonous snakes.

**Range**

From s. N.J. in the East and Nebr. in the West, s. throughout e. U.S.; into the Southwest and Mexico. Subspecies also in West.

Note: There are 5 subspecies in the East. Color pattern and ranges are as follows:

**EASTERN KINGSNAKE,** *Lampropeltis getulus getulus,* length, to 82 in. (208.3 cm), has white chain pattern above on a black background; below blotched with black and white. It occurs in eastern coastal and Appalachian U.S. from southern New Jersey through North Carolina south to central Florida and coastal Alabama. Intergrades with Florida Kingsnake (see below).

**FLORIDA KINGSNAKE,** *Lampropeltis getulus floridana,* length, to 66 in. (167.6 cm), is generally pale with vague bands. It is found in peninsular Florida, intergrading with Eastern Kingsnake in central Florida, and separate populations on the eastern Georgia and Florida borders and in south-central area of Panhandle.

**BLACK KINGSNAKE,** *Lampropeltis getulus niger (Fig. 71),* length, to 58 in. (147.3 cm), is black with traces of yellow or white spots, or bands, above; throat white. It occurs in southern Ohio, western West Virginia, and southeastern Illinois, south to northeastern Mississippi, northern Alabama, and northwestern Georgia.

*Fig. 71*

Black Kingsnake

**SPECKLED KINGSNAKE,** *Lampropeltis getulus holbrooki* **(69:7),** length, to 72 in. (182.9 cm), has yellow center in each dark scale, giving a salt-and-pepper appearance. It occurs in the Mississippi Valley from central Illinois and western Alabama, west to Nebraska and eastern Texas. Intergrades with Desert Kingsnake (see below).

**DESERT KINGSNAKE,** *Lampropeltis getulus splendida,* length, to 60 in. (152.4 cm), has dark saddles above, and light-centered scales with occasional dark spots on each side. Black below. It intergrades with Speckled Kingsnake over large area from southern Nebraska to central Texas, but is found farther west than Speckled Kingsnake into southeastern Arizona and northern Mexico.

# MILK SNAKE
*Lampropeltis triangulum*

**69:3**
*Fig. 72*

**Description**

Length, varies according to subspecies. A highly variable species; usually has dark-bordered red, brown, or gray blotches, sometimes forming rings around body over a white, gray, or yellowish ground. Body slender.

# KINGSNAKES AND MILK SNAKES

**Habitat**
Fields, open woods, river bluffs, rocky hillsides, prairies, pine-barren bogs, sandy soil; under cover or in open.
**Habits**
Secretive, but may be encountered during day in early spring.
**Food**
Rodents, snakes, lizards, frogs, eggs and young birds, fish, earthworms.
**Reproduction**
Lays 5–16 parchmentlike, adhesive eggs, June–July; hatchlings August–October.
**Range**
Most of e. U.S. from s. Maine w. into Mont. and s. to Atlantic and Gulf Coasts, into Mexico. Subspecies in w. U.S. and w. coast of Mexico.

*Fig. 72*

Milk Snake, p. 367          Red Milk Snake

Note: There are 7 subspecies in the area covered by this chapter, 3 east of the Mississippi, along with Eastern and Red Milk Snakes, and 4 others to the west.

**EASTERN MILK SNAKE,** *Lampropeltis triangulum triangulum,* length, to 52 in. (132.1 cm), has reddish-brown saddles with black borders on a milky to gray background, black and white checkered belly. There is a light triangle or Y on back of head and neck. The young are more brilliantly colored. It occurs in northeastern U.S., southern Ontario, and Quebec, from southern New England south through Appalachians and west to southeastern Minnesota and Northern Illinois. Intergrades with Scarlet Kingsnake (see below).

**SCARLET KINGSNAKE,** *Lampropeltis triangulum elapsoides* **(69:6),** length, to 27 in. (68.6 cm), broad red bands bordered with black, on a yellow or white ground; red and black encircle body. Found in southeastern U.S., intergrading with Eastern Milk Snake from southern New Jersey to northern North Carolina, and in central Kentucky and Tennessee; along Coastal Plain from North Carolina to Mississippi River, north to Kentucky. (See Coral Snake.)

**RED MILK SNAKE,** *Lampropeltis triangulum syspila (Figs. 72, 73),* length, to 42 in. (106.7 cm), resembles Eastern Milk Snake, but saddles are red; black spots on edge of belly; no triangle. It is found in central Mississippi Valley from southern Indiana west to extreme southeastern South Dakota and south to northwestern Mississippi and northeastern Oklahoma.

*Fig. 73*

Red Milk Snake

**CENTRAL PLAINS MILK SNAKE,** *Lampropeltis triangulum gentilis,* length, to 36 in. (91.4 cm), has red bands bordered by black which encircle body, over a white background. Black head with lighter snout. It occurs in southern Nebraska, western and central Kansas, western and central Oklahoma, and northern Texas into north Colorado.

**LOUISIANA MILK SNAKE,** *Lampropeltis triangulum amaura,* length, to 31 in. (78.7 cm), has broad red crossbands on body; black head, but snout normally light; black over much of belly. It ranges west of Mississippi in southwestern Arkansas, southeastern Oklahoma, Louisiana, and eastern Texas.

**PALE MILK SNAKE,** *Lampropeltis triangulum multistrata,* is similar to above, but with orange snout; body rings and center of belly white. Found in western South Dakota and western and central Nebraska into extreme northeastern Colorado and north-central Montana.

**MEXICAN MILK SNAKE,** *Lampropeltis triangulum annulata,* length, to 39 in. (99.1 cm), has wide red rings, a black snout and belly. Black rings usually wide. It occurs in central Texas southward into Mexico, with many subspecies extending into Central and South America.

The **SHORT-TAILED SNAKE,** *Stilosoma extenuatum,* may be confused with the Milk Snakes because of its similar habits and habitat. However, its coloration is distinct, with alternating black and orange blotches, and it is particularly remarkable for being extremely slender. It occurs only in northern and central peninsular Florida.

# PRAIRIE KINGSNAKE
*Lampropeltis calligaster*                                             **69:5**

**Description**
Length, to 52⅛ in. (132.4 cm). Smooth scales, very narrow transverse markings on back and dark blotches on yellow belly. Above dark blotches, black-edged and somewhat dumbbell-shaped; two rows of alternating round blotches on flank; pattern becomes obscure in older individuals; may appear as 4 indistinct stripes or be all brownish. Dark band across front of head and from eye to angle of mouth, arrowhead mark on back of head. Also an all-black phase. Head blunt, anal plate single.

**Similarities**
Glossy Snake is unmarked below.

**Habitat**
Prairie country, in pastures, fields, roadsides, hay or grain shocks, open woods.

**Food**
Rodents, frogs, toads, lizards, small snakes.

**Reproduction**
Lays 6–13 adhesive eggs, June–August in holes in fields; hatchlings from August on.

**Range**
Se. U.S. from s.-cen. Md. s. to n. Fla. and w. to se. Nebr., Kans. and e. Tex. *Lampropeltis calligaster calligaster* is found in the Mississippi Valley west from central Kentucky and western Indiana through central Illinois, southern Iowa, and Kansas, south to western Louisiana and eastern Texas.

Note: The **MOLE KINGSNAKE,** *Lampropeltis calligaster rhombomaculata,* a burrower, is found east of the Mississippi Valley from southern Maryland through eastern and central Virginia and into central Tennessee, south to northern Florida and central Mississippi and Louisiana, to the Mississippi River.

# GREEN SNAKES
Genus *Opheodrys*

## ROUGH GREEN SNAKE
*Opheodrys aestivus*                                                68:10

**Description**
Length, to 45⅝ in. (115.9 cm). Unicolor bright green dorsum,
keeled scales; underparts yellow to green. Body slender, head
distinct, but narrow, tail long.
**Similarities**
Smooth Green Snake has stouter body, shorter tail.
**Habitat**
Bushes, small trees in streamside and roadside locations where it is
well camouflaged.
**Habits**
Arboreal, diurnal, even-tempered.
**Food**
Insects, spiders, snails, salamanders.
**Reproduction**
Lays 3–12 long, smooth, hard, adhesive eggs, July–August;
hatchlings August–September.
**Range**
S. N.J., w. to e. Kans., s. throughout e. U.S. except Appalachians
of W.Va. to cen. Tex. and into e. Mexico.

## SMOOTH GREEN SNAKE
*Opheodrys vernalis*                                              *Fig. 74*

**Description**
Length, to 26 in. (66 cm). Green above with lighter green or
yellowish belly, scales smooth.
**Similarities**
Rough Green Snake has slender body, longer tail, keeled scales.
**Habitat**
Grassy fields, meadows, low bushes, bogs.
**Habits**
Less arboreal than Rough Green Snake; secretive; gentle.
**Food**
Much as for Rough Green Snake.
**Reproduction**
Lays 2–8 thin, cylindrical, blunt eggs, June–September (mainly
July–August); hatchlings August–early October.
**Range**
Primarily ne. U.S. and se. Canada from N.S. w. to extr. se. Sask.,
s. to s. N.C., the Appalachians in Tenn., n. Mo., Kans., and Nebr.
Also in disjunct populations in n. and se. Tex., western states, and
Mexico.

*Fig. 74*                                    Smooth Green Snake

# RACERS AND COACHWHIP
Genera *Coluber* and *Masticophis*

## RACER
*Coluber constrictor*

68:11

**Description**
Length, to 73 in. (185.4 cm). Large snake, uniformly bluish or black above and below (except in Yellow-bellied and Buttermilk subspecies) with smooth, satiny, lustrous scales. Throat and chin white. Body long, slender; tail very long; scales smooth, anal plate divided, scale rows 15 just in front of anus. Young pale gray above with dorsal saddles—large brownish blotches on back and black spots on sides.

**Habitat**
Varied, often near water, but includes brush, stone walls, rocky outcrops, where it may hibernate; trash piles, roadsides, swamps, cultivated fields.

**Habits**
Agile; will fight when cornered and can give a vicious, but not dangerous, bite; may be found in bushes and trees.

**Food**
Reptiles, birds, mammals, amphibians, insects.

**Reproduction**
Lays 1–28 eggs, June–July in soil or rotten wood; hatchlings late July–September.

**Range**
Most of e. U.S. except in extr. North, w. to Tex. and into Pacific Northwest. Also along e. coast of Mexico.

Note: Four subspecies of the East are:

**BLACK RACERS,** *Coluber constrictor constrictor* and *C. c. priapus,* are gun-metal black above and below. They occur in southern New England, south through most of Florida, and west through Ohio Valley and south-central Mississippi Valley, southeastern Missouri, extreme southeastern Oklahoma, and extreme northeastern Texas; in Coastal Plain west only to Mississippi River.

**BLUE RACER,** *Coluber constrictor foxi,* is bluish above, lighter below. Found in southern Ontario and west and central Ohio, west through southern Wisconsin and Illinois to southeastern Minnesota and eastern Iowa. A disjunct population occurs in southern upper Michigan.

**EASTERN YELLOW-BELLIED RACER,** *Coluber constrictor flaviventris* **(68:12),** is bluish to brownish above and yellowish below. It occurs in Iowa, Missouri, and northwestern Arkansas, west through the western part of the Dakotas to Montana, south to northern and eastern Texas, and into the Central Plains. Also coastally in western Louisiana. Subspecies in the West and in Mexico.

**BUTTERMILK RACER,** *Coluber constrictor anthicus,* is a racer with contrasting light spotting variably sprinkled over body. It ranges west of Mississippi River in extreme southern Arkansas, Louisiana, and eastern Texas.

There are seven "black" snakes in the East:

**BLACK RACERS** are all black above and below except for white chin, have smooth scales with a gunmetal gloss, a slender body with long, narrow head.

371

**BLACK RAT SNAKE** is all black above and whitish below, has highly polished, keeled scales, and a stouter body, with a broader, flatter, more angular head.

**PINE SNAKE** has scales strongly keeled, is all black, but often with brown markings on lips and trace of a pattern. Its range is limited: from southern Alabama to extreme eastern Louisiana.

**BLACK KINGSNAKE** is all black but shows some yellow or white above.

**INDIGO SNAKE** has large glossy scales and patterned belly. Black phase of **EASTERN HOGNOSE** is all black above; it has an upturned nose. Black phase of **PRAIRIE KINGSNAKE** is all black above and below; it has no white chin or light marks on back.

## COACHWHIP
*Masticophis flagellum*          **68:17**

**Description**
Length, to 60 in. (152.4 cm); record 102½ in. (260.4 cm). A large, slender whipsnake, 2-toned, dark toward the head, light toward the rear, both above and below. Ground color reddish to gray or black. Tail markings resemble braided coachwhip; tail very long and thin. Scales smooth; 17 scale rows in front to about 12 near vent; anal scale divided; 8 labials above. Head narrow with large eyes. Populations from central Kansas south and west lack 2-toned appearance and are light brown or pinkish over entire body, somewhat paler below or with crossbands. Young have dark crossbands on anterior part of body.

**Habitat**
Warm, dry uplands; semiarid regions; pine flatwoods, pastures, fields, roadsides.

**Habits**
A fast mover; terrestrial, diurnal.

**Food**
Rodents, snakes, lizards, young birds, and turtles.

**Reproduction**
Lays 4–24 pebbly surfaced eggs, June–July in holes in fields.

**Remarks**
Can give a nasty bite when handled.

**Range**
Se. U.S. from s. N.C. through most of Fla. and w. to Tenn. and Miss. Hiatus in distribution along Miss. R. Western range from s. Mo. and w. La. to sw. Nebr., Wyo., and e. N.Mex. continuing into sw. U.S. and most of Mexico.

# INDIGO SNAKES
Genus *Drymarchon*

## INDIGO SNAKE
*Drymarchon corais*          *Fig. 75*

**Description**
Length, to 103½ in. (262.9 cm). Large-bodied with smooth, shiny scales blue to black dorsally; variable ventral coloration of alternating dark markings to either side of rear of scales, or with dark markings at edge of each. Sometimes reddish on chin, throat, and sides of head.

**Habitat**
Moist localities in pine woods, dry glades, tropical hummocks, flatwoods; also sandy hills, pine barrens, and oak ridges.
**Habits**
Slow-moving; may overwinter in tortoise burrows. Mild-tempered, but when cornered can give vicious appearance.
**Food**
Snakes, frogs, small mammals, birds, young turtles.
**Reproduction**
Lays 5–11 eggs, April–May.
**Range**
Disjunct population in s. Coastal Plain from se. Ga. through most of peninsular Fla. and Keys, and on Ala.–Fla. border; main population in s. Tex., s. through Mexico; Central and South America.

Indigo Snake

# PINE, BULL, AND GOPHER SNAKES
Genus *Pituophis*

## BULLSNAKE
*Pituophis melanoleucus*

**69:1**
*Fig. 76*

**Description**
Length, to 72 in. (182.9 cm); record 100 in. (254.0 cm). Variably colored ground, white to yellow or gray, with 30–60 dark blotches. Body stout, head small, snout pointed, tail ends in a spine. Scales keeled on back, smooth on sides, a single anal scale, and approximately 29 scale rows.
**Habitat**
Prairies, fields, roadsides; farmlands, farm buildings; dry, sandy pine woods; occasionally swamps, rhododendron slopes.
**Habits**
Diurnal; when cornered hisses and vibrates its tail.
**Food**
Small rodents, birds, and birds' eggs.
**Reproduction**
Lays 3–24 eggs, May–August; hatchlings August–September.
**Remarks**
Takes well to captivity; is highly beneficial to the farmer in pest control. The Gopher Snake of the West is a subspecies.
**Range**
As described for subspecies below and throughout w. U.S. and Mexico.

Bullsnake

Pine Snake, p. 374

Note: *Pituophis melanoleucus sayi* has dark yellow head with dark band across top, dark bar below eye; alternating lateral black, brown, or dark reddish blotches on a yellow ground. Venter is yellow with 2 rows of dark blotches. Central Plains continuously from s. Wis., e. Ill., and n. and w. Mo., w. through w. N.Dak. to s. Alta. and through cen. and w. Tex. into Mexico. Disjunct populations in nw. Minn., e. Ill., and Ind.

**PINE SNAKE,** *P. m. melanoleucus* (*Fig. 76*), and other subspecies—have fewer blotches on back and alternating small blotches on sides; venter is marble-white, but in some areas blotches above are obscure, or individuals totally dark. They occur in western Virginia and adjacent southern West Virginia to southern Kentucky, south and east to the Coastal Plain in extreme southern South Carolina through eastern Louisiana. Through all Florida except southern tip. Disjunct populations in New Jersey Pine Barrens and west-central Louisiana into adjacent eastern Texas.

# RAT SNAKES
Genus *Elaphe*

Rat snakes are distinguished by their flattened abdomens, an indication of climbing habits. They have stout bodies, broad and distinct heads, and boxlike snouts. All are powerful constrictors, vibrate the tail, and usually try to bite if caught. They show a diet preference for frogs and lizards when young, and rodents, birds, and eggs when older. All members of the genus lay eggs.

## CORN SNAKE
*Elaphe guttata*                                    68:14

**Description**
Length, to 48 in. (121.9 cm); record 72 in. (182.9 cm). Forty rich, reddish, saddle-shaped blotches, sometimes faded, above a reddish-brown or yellowish ground. Belly checkered dark and light with a pair of stripes under tail. Red stripe across forehead, red bar edged with black from eye through mouth; neck bands meet in middle of head, forming a spear pointing to snout; 2 rows of small blotches flank saddles. Body stout, head small, scales weakly keeled on at least 5 rows.
**Similarities**
Milk Snake has smaller head that is less distinct from body and weaker colors.
**Habitat**
Woods, wood edges, cornfields, outbuildings, roadsides, prairies, plains, mixed terrain, and even towns (in the South).
**Habits**
A ground snake, but can climb; nocturnal. Look for it on highways by night, under cover by day.
**Reproduction**
Lays 3–21 adhesive eggs, June–July.
**Remarks**
Makes a good pet.
**Range**
S. U.S. in N.J. Pine Barrens, s. Delmarva Pen., and s. Md. and Va., s. through Fla. Keys, and w. to se. Nebr. and into Colo., s. through Tex. into Mexico. Absent from most of Miss. R. and Ohio R. valleys n. of La.

Note: A subspecies, the **GREAT PLAINS RAT SNAKE**, *Elaphe guttata emoryi* (*Fig. 77*), record length to 60¼ in. (153 cm), is similar to the Corn Snake, but has no red; its blotches are dark on a gray background. (See Prairie Kingsnake.)

Gray Rat Snake

*Fig. 77*

Great Plains Rat Snake

# RAT SNAKE
*Elaphe obsoleta*

**Description**
Length, to 72 in. (182.9 cm); record 101 in. (256.5 cm). Highly variable, with upper surface blotched, striped, or all black. No spear on head, no regular stripes under tail. Body stout and flat ventrally in cross section; tail short, ending in a spine. Dark band from in front of eyes to mouth. Young often with H-shaped saddles on back over gray to brown ground.

**Habitat**
Trees, shrubs, ledges, hollow logs, fence rows, or stumps in a variety of situations, dry and moist.

**Habits**
A climber; emerges early from hibernation.

**Food**
As common to genus, plus small mammals in general.

**Reproduction**
Lays 5–44 eggs, June–July, in earth or logs; hatchlings late August–October.

**Range**
From extr. s. New England and into s. Ont., through the East and Midwest to sw. Wis., se. Nebr., e. Kans., and s. Tex.

Note: There are four major subspecies in the East:

**GRAY RAT SNAKE**, *Elaphe obsoleta spiloides* (*Fig. 77*), is gray above with brown saddles, which are often H-shaped, particularly on neck; below is white, peppered with gray forward, all gray to rear. Essentially it retains juvenile coloration. It occurs in Mississippi Valley to southern Illinois, and on Gulf Coast in Mississippi through Florida panhandle.

**YELLOW RAT SNAKE**, *Elaphe obsoleta quadrivittata*, has 4 dark stripes running length of the body, which has a yellow to light brown ground color. It occurs in southern North Carolina through most of peninsular Florida.

**TEXAS RAT SNAKE**, *Elaphe obsoleta lindheimeri*, is much like Gray Rat Snake above, but dark blotches, over the gray to yellow ground, are not as well defined. It occurs in all but the northern part of Louisiana to eastern Texas.

**BLACK RAT SNAKE**, *Elaphe obsoleta obsoleta* **(67:13)**, black above with much white below, and has highly polished, keeled scales.

Body is stouter, head broader, and body less uniformly black than Racer; has reddish skin between scales, white spots sometimes showing in edges of back scales. A northern subspecies, occurring in Mississippi Valley south only to southern Illinois, but south through Appalachians to northern Georgia, and in the west of range, to northern Louisiana and extreme northeastern Texas.

## FOX SNAKE
*Elaphe vulpina*                                                    *Fig. 64*

**Description**
Length, to 54 in. (137.2 cm); record 70½ in. (179.1 cm). Above light brown or straw with 28–51 large brown blotches on mid-back, flanked by 2 rows of smaller blotches. Head conspicuously chestnut, red, or brown, with neck bands not joining into a spear on head, but appearing to. Below yellow and black in keyboard design. Stout body, scales weakly keeled; tail stout, ending in a sharp spine. Young have more red.

**Habitat**
Prairies, wood edges, fields.

**Habits**
A ground snake, but can climb.

**Food**
Small mammals.

**Reproduction**
Lays 7–29 adhesive eggs, June–August in hollow stumps, logs, sawdust piles, manure heaps; hatchlings August–October.

**Remarks**
Makes a good pet.

**Range**
S. Ont. and adj. Mich. and n. Ohio; separate range from nw. Ind. around Lake Mich. into upper Mich. and w. through n. Mo. and s. Minn. to e. Nebr. and se. S.Dak.

# GLOSSY SNAKES
Genus *Arizona*

## GLOSSY SNAKE
*Arizona elegans*                                                    **69:2**

**Description**
Length, to 36 in. (91.4 cm); record 54⅝ in. (138.7 cm). Series of dark brown blotches on light brown ground above, and a white, unmarked belly. Ground color ranges from brown to yellow; 2 rows of small, dark lateral blotches, dark streak from eye to angle of jaw. Scales polished, smooth; anal single; snout scale projects backward. Head narrow, pointed.

**Habitat**
Sandy soil in woods, fields, plains.

**Habits**
Nocturnal.

**Food**
Rodents, moles, and probably lizards.

**Reproduction**
Lays 3–23 eggs.

**Range**
Extr. sw. Nebr., e. Colo., s. through w. Kans., w. Okla., and most of cen. and w. Tex., and in sw. U.S. and n. Mexico.

# GROUND SNAKES
Genus *Sonora*

## GROUND SNAKE
*Sonora episcopa*                                          *Fig. 78*

**Description**
Length, to 15⅛ in. (38.4 cm). Smooth, shiny scales, 15 scale rows
around body and anal scale divided. Head broad, flattened above;
barely distinct from body. Color and markings extremely
variable—may be gray, dark brown, or red above, sometimes
uniformly brownish. Dark saddles or bands variably present,
sometimes black rings encircle body, sometimes just a black collar
on neck. Below cream-white or greenish, sometimes uniformly
light.

**Habitat**
Prairies, dry places, buried in soil, or under stones or boards;
roadsides, timbered hillsides.

**Habits**
Nocturnal.

**Food**
Ants, spiders, centipedes, scorpions.

**Reproduction**
Lays 4–6 eggs in June; hatchlings in August.

**Range**
Sw. Mo. and nw. Ark., w. to se. Colo., and s. through Okla. and e.
N.Mex. to cen. and w. Tex. and into n. Mexico.

*Fig. 78*

Flat-headed Snake, p. 378        Ground Snake

# LONG-NOSED SNAKE
Genus *Rhinocheilus*

## LONG-NOSED SNAKE
*Rhinocheilus lecontei*                                    *Fig. 79*

**Description**
Length, to 41 in. (104.1 cm). Red, white, and black pattern of
blotches along body. Pattern of red, white, or yellow with 26–46
dark blotches, giving speckled or mottled effect. Sometimes black
above with red squares—or orange half-rings; reddish pattern may
be dull. Below yellow or white with some black at sides. Scales
smooth, snout moderately long, conical; plates under tail form
single row.

*Fig. 79*

Long-nosed Snake

**Habitat**
Sandy deserts, rocky slopes, arid brush.
**Habits**
A burrower and constrictor; nocturnal.
**Food**
Small rodents, lizards, young snakes.
**Reproduction**
Lays 4–9 eggs.
**Range**
Sw. and cen. Kans., w. Okla., s. into Mexico and into the West.

# BLACK-HEADED SNAKES
Genus *Tantilla*

These are small snakes with uniform body color, fifteen rows of smooth scales, and a small, dark head. They are secretive burrowers that range from the southern United States down into South America.

## SOUTHEASTERN CROWNED SNAKE
*Tantilla coronata*                                                    *Fig. 80*

**Description**
Length, to 13 in. (33 cm). Black head and collar, separated by a light band. Above brownish, below pearl- to purplish-gray. Eye small.
**Habitat**
Rocky or wooded hillsides under flat stones, logs, boards, bark.
**Food**
Insects, earthworms.
**Reproduction**
Lays 2–3 oval eggs in rotten wood of stump or log.
**Range**
E. U.S. e. of Miss. R. from s. Va., s. to cen. Ga., and w. in lowland areas to w. Ky. and extr. s. Ind. and along Gulf Coast from w. half of Fla. panhandle.

Southeastern Crowned Snake

*Fig. 80*    Plains Black-headed Snake

Flat-headed Snake

## FLAT-HEADED SNAKE
*Tantilla gracilis*                                              *Figs. 78, 80*

**Description**
Length, to 9⅝ in. (24.4 cm). Above shiny and brownish, gradually becoming only slightly darker on head; no collar. Below pinkish-orange, white toward head. Smooth scales; anal divided. Head flattened, barely wider than neck.
**Habitat**
High, flat grassy plains; rocky, sandy hillsides; under limestone rocks; by stream banks.
**Habits**
Probably nocturnal.

**Food**
Insects, spiders, centipedes, slugs.
**Reproduction**
Lays 1–4 eggs, mid-June–mid-July.
**Remarks**
Inoffensive when captured.
**Range**
Sw. Ill., s. Mo., and se. Kans., s. to n. La. and cen. and s. Tex.
Other records in ne. Mo., n. Tex., and Mexico.

## PLAINS BLACK-HEADED SNAKE
*Tantilla nigriceps*                                      *Fig. 80*

**Description**
Length, to 14¾ in. (37.5 cm). Convex or V-shaped black cap, with
apex pointing to rear. Above light brown or yellow; below white to
pinkish. Head flattened, barely distinct from neck; black area not
bordered by white band.
**Similarities**
Head color of Flat-headed Snake blends with, and is only slightly
darker than, body color.
**Habitat**
Under flat rocks on dry hillsides.
**Habits**
Nocturnal; secretive.
**Food**
Insects, spiders, earthworms.
**Range**
Sw. Nebr., cen. Kans., and w. Colo., s. through cen. Tex., w. to se.
Ariz., and into n. Mexico.

# NIGHT SNAKE
Genus *Hypsiglena*

## NIGHT SNAKE
*Hypsiglena torquata*                                      *Fig. 81*

**Description**
Length, to 20 in. (50.8 cm). Above with numerous dark spots,
lateral alternating with dorsal; dark blotch on either side of neck.
Dorsal ground color gray, pale brown, or yellowish; below light,
unmarked, scales smooth; head flat, pointed, distinct from neck; eye
pupil vertical.
**Habitat**
Extremely varied in arid and semiarid regions.
**Food**
Lizards, small snakes, frogs, earthworms.
**Reproduction**
Lays 3–12 eggs, April–July; hatchlings from June on.
**Remarks**
Venomous, fangs at rear of mouth, but not fatal to humans.
**Range**
E. of 100th meridian occurs only in se. Kans., w. Okla., and w.
and cen. Tex.; also w. to Colo. and into s. Mexico.

*Fig. 81*                        Night Snake

# CORAL SNAKES
Family Elapidae

These snakes are venomous. In the New World, this family is represented by only two genera, with most species restricted to Central and South America. They all have a conspicuous pattern of red, black, and white or yellow rings around the body. The upper front fangs are permanently erect and there are no pits behind the nostril.

## EASTERN CORAL SNAKE
*Micrurus fulvius*                                                        **69:10**

**Description**
Length, to 30 in. (76.2 cm); record 47½ in. (120.7 cm). Body rings in the formula black-yellow-red-yellow-black. Face black, wide band of yellow across middle of head. Yellow rings around body narrow, tail has no red rings. Scales smooth; head flat, blunt, not distinct from body. Young have similar pattern but paler colors.

**Habitat**
Underground, often in fields; in or near water.

**Habits**
Primarily diurnal.

**Food**
Small snakes, lizards, frogs.

**Reproduction**
Lays 2–4 soft eggs, May–July in loamy soil, humus, or logs; incubation 3 months.

**Range**
Se. N.C., w. to cen. Tex. in Coastal Plain, and on into e. Mexico; n. into cen. Ark.; absent in most of e. La. in Miss. Valley.

Note: Three snakes in the East have alternating body bands of red, yellow, and black. They may be distinguished as follows:

**EASTERN CORAL SNAKE** (above): Black-yellow-red-yellow-black (yellow narrow); face black; bands encircle body. The poisonous Coral Snake is the only one of these that wears a black mask on its face and has yellow rings touching red.

**SCARLET SNAKE**: Black-yellow-black-red-black (black and yellow narrow); bands do not encircle body; face red.

**SCARLET KINGSNAKE**: Black-yellow-black-red-black (black and yellow narrow); bands encircle body; face red.

# Pit Vipers
Subfamily Crotalinae

This subfamily belongs to the family Viperidae, whose other members are found exclusively in the Old World; the Crotalinae are found extensively in the New World and Asia. These venomous snakes have a heat-sensitive pit on either side of the head, behind, and much larger than, the nostrils. Each species has vertical pupils, a distinct, triangular head, and keeled scales. They all bear living young. Bites from these snakes may prove fatal to man; sometimes even severed heads can bite.

# COPPERHEADS AND COTTONMOUTHS
Genus *Agkistrodon*

## COPPERHEAD
*Agkistrodon contortrix*

**69:9**
*Fig. 83*

**Description**
Length, to 36 in. (91.4 cm); record 53 in. (134.6 cm). Distinguish this snake by its hourglass-shaped dark crossbands enclosing light lateral areas and coppery brown, red-brown, or pink head. Ground color above light brown or pink; below whitish with dark blotches; tongue red at base and with lighter tips. Young have bright yellow tail that is waved to attract food.

**Habitat**
Dry upland woods, rocky hillsides, ledges; mountains; moves to lowlands in summer.

**Habits**
Bites only if startled, though more aggressive in hot weather.

**Food**
Rodents, small birds, frogs, toads, lizards.

**Reproduction**
Bears 2–17 living young, August–October; 2–10 clutches per season.

**Range**
E. U.S. from sw. Mass. s. to extr. n. Fla., w. to extr. s. Iowa, e. Kans., and into cen. Tex., to Rio Grande.

Note: Four subspecies are:

**NORTHERN COPPERHEAD** (*Fig. 82*), *Agkistrodon contortrix mokasen,* has dark bands narrow on the back, wide on the sides, often with dark spots on the interspaces of ground color. It occurs in Massachusetts and Central Illinois, south to higher elevations in northeastern Mississippi and northern Georgia, and west to the Mississippi River.

**OSAGE COPPERHEAD,** *Agkistrodon contortrix phaeogaster,* is similar to northern subspecies, but the dark crossbands and immaculate lighter ground color contrast sharply; the belly is often dusky. It occurs west of the Mississippi River in extreme southeastern Iowa, northern and central Missouri into extreme southeastern Nebraska, eastern Kansas, and extreme northeastern Oklahoma.

**SOUTHERN COPPERHEAD,** *Agkistrodon contortrix contortrix* (*Fig. 82*), is similar, but has dark bands wider laterally, narrower dorsally; ground color is paler. It occurs in the lower Mississippi Valley to southern Illinois, and in the southern Coastal Plain from extreme southeastern Virginia to Texas, but is absent from most of Florida.

**BROAD-BANDED COPPERHEAD** (*Fig. 82*), *Agkistrodon contortrix laticinctus,* is brighter, has chestnut-brown crossbands very wide, broadest on sides. It occurs in the Great Plains from southern Kansas to central Texas.

Northern Copperhead

Southern Copperhead

*Fig. 82*

Broad-banded Copperhead

## COTTONMOUTH
*Agkistrodon piscivorus*                                   **69:11**

**Description**
Length, to 48 in. (121.9 cm); record 74 in. (188 cm). Largest pit
viper without a rattle. Has flattened head, eyes not visible from
above, and dark band from eye to angle of mouth bordered above
by pale line. Above brown with indistinct black bands that are
wider laterally. Below yellow, with or without dark markings. Old
adults may be entirely black on body. Inside of mouth white, but
this is not unique. Head chunky, tail short. Young brilliantly
marked and banded with dark on reddish-brown, very much
resembling close relatives, the copperheads.
**Similarities**
Deep, broad, spade-shaped head, eyes invisible from above, and
light lip marking distinguish Cottonmouth from water snakes.
**Habitat**
Near lakes, sloughs, streams, swamps, marshes, and other aquatic
situations.
**Habits**
Aquatic, usually nocturnal; gaping widely at foe is characteristic.
**Food**
Snakes, turtles, amphibians, fish, and mammals.
**Reproduction**
Bears 1–15 living young, August–October.
**Range**
Se. U.S. along Coastal Plain from se. Va. to cen. Tex., and n.
through w. Tenn., w. Ky., and e. Okla. to Mo. and extr. s. Ill.

# RATTLESNAKES
Genera *Sistrurus* and *Crotalus*

These snakes are all venomous, and all have a series of
interlocking, horny links on the tail that form a rattle. A new link
is added at each shedding of the skin, but due to wear and tear
rarely are more than ten segments found on an animal in the field.
They cannot be used to determine the animal's age. When
disturbed, a rattlesnake vibrates the tail, thus making the
characteristic buzzing noise for which they are famed. Members of
the Genus *Sistrurus* are small and have enlarged plates on the
head. Members of the Genus *Crotalus* have small scales on top of
the head between the eyes and have good-sized rattles.

## MASSASAUGA
*Sistrurus catenatus*                                      **69:14**

**Description**
Length, to 30 in. (76.2 cm); record 39½ in. (100.3 cm). Small,
brownish, with squarish dark blotches on back. (However, some
individuals are almost black with markings almost or entirely
blended into background.) Two or 3 series of small, dark blotches
on sides. Below dark and heavily blotched. Tail short, stout, and
ringed with dark brown and ground color. Rattle well developed,
medium-size.
**Habitat**
Damp lowlands; swamps, edges of streams, ponds, marshes,
meadows; fence rows and fields in summer.
**Habits**
Not usually aggressive.
**Remarks**
Venom deadly.

**Food**
Rodents, frogs, toads, birds, insects.
**Reproduction**
Bears 2–19 living young, late July–September.
**Range**
S. Ont. and w. N.Y., w. through lower Mich., n. Ind., and s. Wis.
to extr. se. Nebr. and e. and s. Kans., s. through much of Tex.,
and w. to se. Colo., N.Mex.; also in extr. n. Mexico.

## PIGMY RATTLESNAKE
*Sistrurus miliarius*                                  *Fig. 83*

**Description**
Length, to 22 in. (55.9 cm); record 31 in. (78.7 cm). Small, with 3
rows of vivid black blotches with reddish interspaces middorsally.
Ground color variable, gray to tan, even bright red-orange in some
North Carolina populations, and almost black in Florida. Black
stripe runs back from eye, middorsal blotches wider than long. Be-
low dark, blotched. Body stout, tail long, slender; rattle very small.
**Habitat**
Wet or moist locations; grass or woodlands; abandoned buildings,
debris.
**Habits**
Short-tempered, aggressive.
**Food**
Mice, lizards, frogs, small birds.
**Reproduction**
Bears 2–32 living young, August–September.
**Range**
In Coastal Plain from N.C., w. to e. Tex. and e. Okla., and n. to s.
Mo. and extr. sw. Ky.

*Fig. 83*

**Pit Vipers**

Pigmy Rattlesnake   Timber Rattlesnake   Copperhead , p.381

## TIMBER RATTLESNAKE
*Crotalus horridus*

**69:13**
*Fig. 83*

**Description**
Length, to 60 in. (152.4 cm); record 74 in. (188.0 cm). Has phases
varying from yellow or gray with irregular brown or black wavy
crossbands, to almost black in Northern populations, with bands
not discernible. (Phases are independent of sex, age, or season.)
Coastal Plain and Mississippi Valley populations with yellowish or
reddish dorsal band lengthwise on neck and body. Below yellowish,
with many minute black spots. Young are never black.
**Habitat**
Dry uplands, rocky ledges, hillsides, usually with southern
exposure; brambles, second growth, wastelands, swamp edges.
**Habits**
Shy, not aggressive. If sufficiently disturbed, may sound rattle as
warning.
**Food**
Small mammals, birds.

**Reproduction**
Bears 3–17 living young, August–October.
**Range**
S. New England, s. through Appalachians throughout most of e.
U.S. In s. N.J., but absent from cen. N.J. to Va., and absent
around Great Lakes except in N.Y. and scattered localities in Ohio
and Ont. Northward extension in west of range into se. Minn. and
Wis., and westward extension to extr. se. Nebr. through e. Tex.

## EASTERN DIAMONDBACK RATTLESNAKE
*Crotalus adamanteus*                                          *Fig. 84*

**Description**
Length, to 96 in. (243.8 cm). Yellow-edged, dark diamonds with
light centers along back, and vertical yellow stripe in front of each
nostril. Ground color olive to brown above; below yellow or white
with some brown pigment at sides. Dark stripe from eye back to
jaw. Large, heavy-bodied.
**Habitat**
Pine flatwoods and wooded hillsides, often resident in Gopher
Tortoise burrows. Dry areas generally.
**Habits**
May emerge on sunny days in winter to bask or search for food.
**Food**
Small mammals.
**Remarks**
Largest rattlesnake in the United States; extremely dangerous
when aroused to anger.
**Range**
Coastal Plain e. of Miss. R. from s. N.C. through Fla. Keys w. to
extr. e. La.

*Fig. 84*                                        Eastern
                                                 Diamondback
                                                 Rattlesnake

## WESTERN RATTLESNAKE
*Crotalus viridis*                                            **69:12**

**Description**
Length, to 57 in. (144.8 cm). Greenish-gray rattlesnake (but
sometimes yellow or red) with dark lozenge-shaped saddles down
middle of back. Saddles edged with white and flanked by 2 rows of
alternating smaller dark blotches; below yellow, mottled. Sides of
head have 2 white diagonal lines; tail has light and dark rings.
**Habitat**
Prairies, plains, badlands, open rocky ground.
**Habits**
If disturbed will usually sound rattle as warning.
**Food**
Rodents, small birds, frogs.
**Reproduction**
Bears 3–21 living young.
**Range**
Primarily Great Plains and the West, but occurs e. of 100th
meridian in sw. N.Dak. s. through w. Tex., with the farthest
eastward extension into Iowa near its border with S.Dak. and
Nebr.; also into Mexico.

# Amphibians

**Consulting Editor**
Patricia Gardner Haneline
Andrew Mellon Fellow
Department of Biological Sciences, The University of Pittsburgh

**Illustrations**
Plates 70–72, 73, 75 John Cameron Yrizarry
Plate 74 Jennifer Emry-Perrott
Text Illustrations John Cameron Yrizarry and
Jennifer Emry-Perrott

# Amphibians

Class Amphibia

Amphibians are cold-blooded vertebrates that have limbs and moist skin containing mucous glands. This is in contrast to the fins of fish and the dry, scaly skin characteristic of reptiles. The major difference between amphibians and reptiles, however, is in their reproductive habits. Amphibians generally lay small eggs surrounded by a jellylike capsule that swells when in contact with water; reptiles, on the other hand, possess an amniotic egg containing a large yolk, three extraembryonic membranes, and a leathery or hard shell. Amphibians are tied to moist places because of their skin and eggs, and furthermore usually go through a free-swimming aquatic larval stage after hatching. All but three of the eastern amphibian species possess four clawless limbs as adults. A distinguishing feature between orders of adult amphibians is the presence of a tail in salamanders and its absence in frogs and toads. However, because of their short bodies and their long legs adapted for jumping, or saltatory, locomotion, frogs will never be mistaken for anything else. Aquatic salamanders might be mistaken for eels at first glance, but you can tell the difference easily.

Amphibians were the first four-limbed land dwellers, or tetrapods. They arose in the Devonian period about 340 million years ago and were the dominant land animals for over 100 million years, reaching their peak of diversity in the Carboniferous. Amphibians evolved from the rhipidistian fishes of the order Crossopterygii, or lobe-finned fishes, whose sole surviving member is a rare coelacanth found only in deep water off the east coast of Africa. The first fossil amphibian known is *Ichthyostega* from the Upper Devonian of Greenland. Unlike the fins of their fish ancestors, amphibian limbs, fully articulated with pectoral and pelvic girdles, are used in supporting the body and in propulsive movements in walking or running. Early amphibians attained a large size, up to fifteen feet (4.6 meters) in the giants, and some had bony external armor covering parts of the body. The three surviving orders are in comparison small and secretive and are considered to be very closely related to one another, which is reflected in their grouping together in the Subclass Lissamphibia. Besides the two familiar orders of the frogs and toads, and the salamanders, there are the limbless, wormlike caecilians, some of which possess scales embedded in the skin. Caecilians occur only in the tropics.

### Distribution and Diversity

There are approximately 2500 species of amphibians worldwide. Amphibians occupy almost all land masses except the Arctic and Antarctic, but overall species diversity generally increases as one approaches the tropics. The factors ultimately limiting amphibian distribution are moisture and temperature.

### Habits

Amphibians breathe through skin, gills, or lungs, or a combination of them. Needed moisture is also absorbed through the skin. Like fishes and reptiles they are cold-blooded, or ectothermic; that is, their body temperature is maintained at high levels by external sources and is not internally controlled, as in birds and mammals.

### Reproduction

The typical pattern of reproduction is egg laying in or near water, although some species utilize moist areas on land to deposit eggs that hatch into miniature adults. The phenomenon of reproduction

387

around water often provides a herpetologist with the best opportunity to observe and collect amphibians, which are encountered either migrating to the breeding site or about the ponds, streams, and rivers themselves. Larval salamanders have external gills that disappear during metamorphosis and mouthparts similar to those of the adult. Larval frogs, called tadpoles, lack visible external gills and have a horny beak. Both types of larvae have a tail which is used in swimming, but in frogs this structure is later resorbed. Identification of larval forms, particularly in the early stages, is difficult and much too complex for this volume.

### Range and Scope

Discussed in this chapter are 107 amphibians, including fifty-nine species of salamanders and forty-eight species of frogs and toads. Coverage includes the North American land areas between the Canadian–United States border and southern Florida, and from the 100th meridian east to the Atlantic coastline.

### Nomenclature

The common and scientific names in this chapter follow *Standard Common and Current Scientific Names for North American Amphibians and Reptiles.*

## USEFUL REFERENCES

Amphibians and reptiles are generally considered together in books, and the reader is referred to the reptile references found on pages 309-311 in order to avoid duplication. Additional references are found below.

Altig, R. 1970. A Key to the Tadpoles of the Continental United States and Canada. *Herpetologica* 26: 180–270.

Bishop, S. C. 1943. *Handbook of Salamanders.* Ithaca, N.Y.: Comstock.

Blair, A. P. 1968. Amphibians. In Blair, Blair, Brodkorb, Cagle, and Moore, *Vertebrates of the United States.* New York: McGraw-Hill.

Cochran, D. M. 1961. *Living Amphibians of the World.* Garden City, N.Y.: Doubleday.

Logier, E. B. S. 1952. *Frogs, Toads and Salamanders of Eastern Canada.* Toronto: Clarke, Irwin.

Orton, G. L. 1952. Key to the Genera of Tadpoles in the United States and Canada. *American Midland Naturalist,* 47: 382–95.

SSAR Committee on Common and Scientific Names. 1978. *Standard Common and Current Scientific Names for North American Amphibians and Reptiles.* Available from D. H. Taylor, Dept. of Zoology, Miami University, Oxford, Ohio.

Wright, A. A., and Wright, A. H. 1949. *Handbook of Frogs and Toads of the United States and Canada.* 3rd ed. Ithaca, N.Y.: Comstock.

## GLOSSARY

*Adpressed* Front and back limbs extended toward each other, flat against sides; for measuring lengths.

*Anterior* Toward the front (of the body).

*Cloaca* Area into which internal wastes discharge.

*Costal groove* Vertical furrow on the flanks of salamanders, set off by the skin between successive grooves known as intercostal fold.

*Cranial crests* Raised area framing inner border of upper eyelids in toads.

*Dewlap* Skin fold hanging from neck region.

*Diurnal* Daytime.

*Dorsolateral* Pertaining to the uppersides of the animal.

*Frontals* Bony membranes that form the forehead.

*Gular fold* Fold of skin across posterior section of throat in salamanders.

*Intercostal fold* Skin fold bounded on each side by a costal groove.

*Larva* (*pl.* larvae) The earliest form that certain species take; unlike the parent (e.g., tadpole).

*Lateral* Of or pertaining to the side of the body.

*Maxillary bone* Bone on each side of head forming side border of upper jaw and bearing most of the upper teeth.

*Nasolabial groove* Groove from nostril to edge of upper lip.

*Neotenic* Retaining gills or other larval features throughout entire life.

*Nuptial pad* A patch of darkly pigmented, roughened skin that appears on certain digits during breeding season.

*Occipitals* Bony membranes that form the posterior part of the skull.

*Parotoid* One of two large wartlike glands found on the rear of the head in toads.

*Parietal bones* A pair of membrane bones in the roof of the skull between the frontals and occipitals.

*Posterior* Toward the rear (of the body).

*Postocular* Behind the eye.

*Postorbital* Behind the eye.

*Reticulate* Resembling the form or appearance of a net.

*Riparian* Relating to the bank of a river, lake, or pond.

*Rugose* Rough, wrinkled.

*Tubercle* Any of various small knoblike prominences.

*Tympanum* Thin round or oval external ear covering; prominent in frogs and many lizards.

*Vent* Opening on the surface of the cloaca.

*Venter* Belly.

*Vertebral stripe* Stripe down the midline of the back.

*Vertical pupil* Pupil that is elliptical; long axis is vertical.

*Vocal sac* Loose skin on the throat that can distend, forming a chamber which echoes the animal's vocalization.

*Vomerine teeth* Teeth anchored to the vomerine bones that form part of the forepart of the roof of the mouth.

# Salamanders
Order Caudata

Salamanders are seldom seen because of their generally small size, secretive habits, and slow movements. They are encountered only with active searching, and even then, with their concealing coloration, they may elude the naturalist.

Salamanders have a tail throughout life (in contrast to adult frogs), and have moist, relatively smooth skin, no external ear opening, and never more than four toes on their unclawed front feet. Lizards usually have five toes on their front feet, all their toes are clawed, and they have a dry, scaly skin. Many salamanders can regrow a lost limb or tail, and the latter is often sacrificed when grabbed by a predator.

## Identification

Features generally useful in identification are illustrated below. To start with, check for presence or absence of gills, nasolabial grooves from nostril to jaw (used in chemical sensing of the environment), costal grooves and their number (the grooves around the side of the body), and the gular, or throat, fold. Other important features include numbers of limbs and digits, and coloration, particularly if there are bands, stripes, or spots on the body; however, color and pattern are often variable. The species descriptions give common variations, but other listed diagnostic features should all be checked. The color often becomes darker and the pattern more obscure with age.

*Fig. 85*
Parts of a Typical Salamander

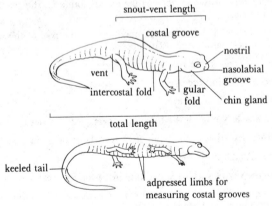

Measure the animal if possible, because size or body proportions may be important. The lengths given in the descriptions are the straight line *distances from the tip of the snout to the posterior edge of the vent*. Because the tail is often broken, the total length is not a reliable measurement. Amphibians grow continuously throughout life, although rapid growth occurs only in the first few years. Sizes, therefore, are the *maximum* size of any individual known, and in cases where the record size far exceeds that of most adults, both record and average are given.

The identification of salamanders provokes dissatisfaction even among herpetologists. One problem is the dearth of important distinguishing characteristics to measure and count, such as the useful scale patterns in reptiles. Many obvious differences in color or pattern are due to variations within one species, well illustrated in humans by variation in hair color. Other problems are the small size and secretiveness of salamanders, making it hard to observe all features without an examination in the hand.

Perhaps the worst problem in identification is one experienced by the salamanders themselves. At times they cannot tell who belongs to their own species when choosing a mate and they produce hybrid offspring derived from parents of two different, but usually closely related, species. Hybrids are often less likely to survive and reproduce themselves, but in certain areas they are successful.

## Habitat and Habits

Salamanders prefer a moist habitat and avoid direct sunlight. Some spend their entire lives in water; some spend most of their lives on land. Some are arboreal, others are cliff dwellers; some, usually blind, dwell in caves or in deep wells. Salamanders live from near sea level to an altitude of over 13,000 feet (3962.4 m). Where the winters are cold they hibernate, but they may be active underground. Unlike snakes and lizards, however, they prefer cool weather to warm, and some are active in icy water close to freezing.

Look for terrestrial salamanders in the spring and fall, particularly at or near the pools where they breed. After breeding they usually return to their hiding places under rocks, logs, bark, leaf litter, and debris left lying by human activities. Specific habitats of each species are listed in the descriptions, but the successful salamander hunter will overturn all cover, and then, of course, carefully replace it in its original position. Most species are nocturnal.

## Food

Salamanders are entirely carnivorous, and among other items will feed on aquatic or terrestrial insects, or other small invertebrates and their eggs and larvae, and on small vertebrates such as salamander larvae. In captivity they will eat earthworms, beef, or other meat, and insects when these are available.

## Reproduction

When male and female encounter each other during the mating season they usually go through an elaborate and species-characteristic courtship ritual accompanied by tactile and chemical sensory cues. You might observe this behavior in captive animals, given the proper sexes, the mating season, and no distractions from the observer. Fertilization is internal, except in the Hellbender, whose male fertilizes the eggs as they are released from the body of the female. In a few species the sperm is transferred to the female through the touching of vent to vent, but in most species after the courtship dance the male will deposit small spermatophores that contain sperm in a little sac on top of a jellylike pyramid. The female follows the male and picks up these sperm sacs with the sides of her vent and introduces them into her spermatheca, a specialized storage receptacle where sperm may be stored for several years. One clutch of eggs laid by the female can potentially be fertilized by sperm from two different males. It should be noted that mating and egg laying may not occur in the same season.

Eggs are deposited singly, in strings, or in masses either in the water or on land, where they are suspended from the roof of a moist retreat. The eggs may be brooded by the female. In due time, depending on the temperature, the eggs in aquatic situations hatch into larval salamanders with gills. Lower ambient temperatures will decrease the growth rate in these stages. The length of the larval period varies from a few days to over a year or two. In terrestrial species the larval period is passed within the egg, and upon hatching the salamander has only rudimentary gills, if any, which are shed shortly after birth. Another adaptation, neoteny, allows salamanders to spend their entire lives with gills and other larval features while coming to reproductive maturity. This last strategy is often found in an inhospitable or underground environment where the perils of transformation are too great.

# SIRENS
Family Sirenidae

Sirens are large aquatic salamanders with three pairs of external gills and forelimbs of moderate size. Hind limbs are entirely lacking.

## LESSER SIREN
*Siren intermedia*

**Description**
Length, to 27 in. (68.6 cm). Has no hind limbs and 3 pairs of external bluish gills. Body long, slender. Above dark brown, black, olive, or gray flecked with small black spots; below lighter and unspotted. Costal grooves 31–34 in area of overlap with Greater Siren, but 38 farther west.
**Similarities**
Greater Siren is generally larger.
**Habitat**
Shallow, warm, quiet ponds; ditches; cypress swamps and sometimes rivers and streams; under stones, logs, in matted vegetation.
**Habits**
Aestivates in cocoon when water dries up.
**Remarks**
May emit a distress cry when seized.
**Range**
Coastal Plain from se. N.C. to cen. Fla., w. to e. Tex. and into Mexico, and n. in Miss. Valley through cen. Ill. and sw. Mich.

## GREATER SIREN
*Siren lacertina*                                                **72:2**

**Description**
Length, to 30 in. (76.2 cm); record 38½ in. (97.8 cm). Distinguished by its large size, absence of hind legs, and 3 pairs of external greenish gills. Dorsum olive to light gray, lighter sides blotched with greenish-yellow. Venter paler with small flecks or mottling of greenish-yellow. Snout yellow or light brown. Costal grooves 36–39. Young darker.
**Habitat**
Shallow roadside ditches, under rocks in streams, weedy ponds and pools, muddy swamps.
**Habits**
Nocturnal; may be found during day under objects in water, or burrowed in mud, or in dense vegetation.
**Range**
Coastal Plain from e. Va. and adj. Md. through Fla. to s. Ala.

Note: The **DWARF SIREN**, *Pseudobranchus striatus*, has brown and yellow stripes on its body. It occurs from extreme southern South Carolina southward through Florida.

# AMPHIUMAS
Family Amphiumidae

Large eellike aquatic salamanders with small eyes, and four tiny limbs, without external gills and eyelids. Graceful in water but clumsy on land. Females lay eggs in strings under logs or rocks near water's edge or in water, then remain coiled around them through incubation period. CAUTION: Amphiumas will bite viciously when handled.

## THREE-TOED AMPHIUMA
*Amphiuma tridactylum*                                    *Fig. 86*

**Description**
Length, to 30 in. (76.2 cm); record 41¾ in. (106.1 cm). Three toes present on most of the tiny limbs. Has distinctly bicolored body, somewhat lighter below. Other distinguishing features are dark throat patch and oval outline of tail base in cross section; boundary between lighter ventral and dark dorsal coloration sharply defined.
**Habitat**
Lowland streams, pools, swamps, and cypress sloughs; also in water in nearby hilly regions.
**Reproduction**
Mates December to June; eggs found April to September in southern part of range, and August through midwinter in Tennessee.
**Remarks**
Mud Snake and sometimes Cottonmouth are major predators.
**Range**
Miss. Valley from extr. w. Ky. and se. Mo., s. through w. Ala., and w. to extr. se. Okla., and e. Tex.

*Fig. 86*

Two-toed Amphiuma     Three-toed Amphiuma

## TWO-TOED AMPHIUMA                                 **72:1**
*Amphiuma means*                                         *Fig. 86*

**Description**
Length, to 30 in. (76.2 cm); record 45¾ in. (116.2 cm). Has dark overall coloration and only 2 toes on the four tiny limbs. Body dark brown to slate with belly only slightly lighter than dorsum, color change gradual from side to below; no obvious throat patch. Tail keeled above and rounder in cross section at base.
**Similarities**
Three-toed Amphiuma is distinctly bicolored, with somewhat lighter venter, and has 3 toes on most of its tiny limbs.
**Habitat**
Lowland swamps, bayous, along sloughs, in drainage ditches, muddy pools, sluggish streams, and wet meadows.
**Habits**
Lives and feeds in crayfish burrows, often deep in the soil.
**Remarks**
Major predators are mud and rainbow snakes.
**Range**
Coastal Plain from se. Va., s. through Fla., and w. to extr. e. La.

# HELLBENDER AND MUDPUPPIES
Families Cryptobranchidae and Necturidae

All are aquatic forms with four short, stout legs and small eyes. The family Cryptobranchidae is represented in the United States by the Hellbender, which has a wrinkled skin and prominent lateral fold and lacks external gills. Its closest relatives live in Japan and China, where the Japanese Giant Salamander reaches over five feet (1.5 m) in length and is considered a great delicacy.

Members of the family Necturidae, the mudpuppies and waterdogs, have three pairs of dark red external gills. They are restricted to North America.

## HELLBENDER
*Cryptobranchus alleganiensis*                                    **72:4**

**Description**
Length, to 20 in. (50.8 cm); record 29⅛ in. (74.0 cm). Large, aquatic, with 1 pair of circular gill slits in the adult; has loose, wrinkled skin with prominent lateral fold. Legs short, stout; toes 4 on front- and 5 on hind-limbs. Head broad, flat; snout rounded. Color brownish, sometimes with darker blotches, especially in Ozark region.

**Habitat**
Rocky, fast-flowing waters of large streams and rivers under 2500 ft. (762.0 m) elevation.

**Habits**
Forages on bottom of water at night.

**Reproduction**
Up to 450 eggs laid in strings in saucer-shaped excavation below rock under water. Male guards nest, but cannibalism of eggs and larvae documented.

**Range**
Susquehanna R. drainage and w. N.Y. and Pa., s. to n. Ga., w. to s. Ill. along Ohio R. and Tenn. R. drainages, and to extr. ne. Miss. Disjunct populations in tributaries of Mo. R. and Black R. in Mo. and Ark.

## DWARF WATERDOG
*Necturus punctatus*

**Description**
Length, to 7 ⁷⁄₁₆ in. (18.9 cm). Small, slender. Color deep brown to gray, unspotted, or nearly so; throat white; belly bluish-white. Inconspicuous dark line passes through eye. Has external feathery red gills. Head and snout depressed; 4 toes on all 4 feet.

**Habitat**
Slower parts of streams near sandy or muddy banks and in ditches; hidden in bottom debris.

**Range**
Atlantic Coastal Plain from se. Va. to cen. Ga.

Note: The **ALABAMA WATERDOG**, *Necturus alabamensis,* occurs in west-central Georgia and the Florida panhandle to eastern and southwestern Mississippi. It is a variable reddish-brown and is distinguished by an unmarked central belly area.

## MUDPUPPY
*Necturus maculosus*                                72:3

**Description**
Length, to 17 in. (43.2 cm). Color dark olive or grayish-brown covered with blue-black spots. Has 4 short, stout limbs, each with 4 toes, and 3 pairs of external feathery red gills retained throughout life. Body and head flattened; tail compressed from side to side, keeled. Males in breeding season have swollen ridge around cloaca. Young have dark back bordered by narrow yellow line.

**Habitat**
Strictly aquatic; in clear or muddy streams, bayous, polluted canals, ditches, and reservoirs.

**Habits**
Active all year.

**Food**
Fish and fish eggs, also crayfish, insects, insect larvae, snails, and other amphibians.

**Reproduction**
Mates in fall; eggs laid singly, stuck to underside of an object in water.

**Range**
Most of ne. U.S. and se. Canada from s. Que. and nw. Vt., s. to n. Ga., and w. around most of Great Lakes region to se. Man., s. to e. Okla. and n. La. Introduced in New England.

# MOLE SALAMANDERS
Family Ambystomatidae

Large, stout-bodied salamanders distinguished by a lack of nasolabial grooves, presence of costal grooves, and lack of external gills in adults. Four sturdy limbs are capable of digging. Larvae have bushy external gills and a keel on the back and tail. Adults terrestrial except during the breeding season, and most frequently encountered migrating across roads on warm, rainy nights. Two species are terrestrial fall breeders and brooders; the remainder breed in temporary or semipermanent ponds that lack fish early in the spring. Males have swollen vents during the breeding season. Food includes worms and arthropods.

## SPOTTED SALAMANDER
*Ambystoma maculatum*                               72:9

**Description**
Length, to 9¾ in. (24.8 cm). Two rows of large, round, yellowish spots on black, bluish-black, slate, or brown back. Spots somewhat irregular and extending from behind eyes to tip of tail; rarely cream; first pair may be orangish. Ventral surface slate-gray, although sometimes lighter. Costal grooves usually 12 (11–13); tail at least ½ length of body.

**Habitat**
Hardwood and mixed deciduous forests.

**Habits**
Breeds in early spring; commonly found over its range throughout summer.

**Range**
Most of e. U.S. and se. Canada except absent from s. N.J., s. Delmarva Pen., most of Fla. and s. La. Western boundaries in Wis., forested parts of Ill., and se. Kans. s. to ne. Tex.

## MARBLED SALAMANDER
*Ambystoma opacum*                                   **72:10**

**Description**
Length, to 5 in. (12.7 cm). Shiny black, with irregular white
bands, giving it a marbled appearance. Body thick, short; legs
short, stout; tail about 40% of total length, tapered and somewhat
compressed. Light dorsal markings variable but usually bands
meeting laterally and enclosing black spots along back. In some
individuals crossbands are lacking or incomplete, or lateral
markings are interrupted. Light bands brighter in male than in
female. Below black.
**Habitat**
Under bark, logs, stones, or rubbish, along margins of ponds,
streams, and swamps during breeding season; at other times in dry
and shady areas away from water.
**Reproduction**
Breeds in fall; terrestrial breeding and brooding of eggs. Adults do
not become aquatic.
**Range**
Most of e. U.S. from se. N.H. to n. Fla., w. to s. Ill., se. Okla.,
and e. Tex. Disjunct populations on Lake Michigan border in se.
Wis. and Ind.-Mich. area.

## RINGED SALAMANDER
*Ambystoma annulatum*                                **72:12**

**Description**
Length, to 9¼ in. (23.5 cm). Slender and elongate. Color blackish-
brown, with 4–7 crossbands or rings of yellow. Rings on body may
be incomplete above, resembling spots; light line below at side;
rings never extend onto belly. Below slate-gray, with or without
white dots. Costal grooves 15; tips of adpressed limbs separated by
4 costal grooves.
**Habitat**
Under logs in mud; in tunnels beneath brush and leaves, in fields
near ponds, swamps, marshes.
**Reproduction**
Breeds following first heavy rain in the fall; lays eggs in loose
masses under water.
**Range**
On Ozark Plateau and in Ouachita Mts. in se. and cen. Mo., nw.
Ark., and e. Okla.

## TIGER SALAMANDER
*Ambystoma tigrinum*                                 **72:7**

**Description**
Length, to 8½ in. (21.6 cm), record 13 in. (33 cm). Large, dark,
with light, irregular dorsal blotches, bars, or reticulations above,
and yellowish with black marbling below. Body stout, head broad,
legs stout and long, eyes small. Tail laterally compressed.
Geographic variation in body pattern and color of spots; yellow
blotches sometimes tinged with olive or orange, may form bands on
lower sides and tail. Costal grooves 14 or fewer.
**Habitat**
A spring breeder often encountered at night or during rains
traveling to breeding sites of permanent shallow lakes, ponds,
ditches, or pools of rivers.

**Reproduction**
Eggs laid singly or in masses of up to 110.
**Remarks**
Neoteny, or the retention of some larval features into adult life, is common in many western populations where Tiger Salamanders retain their gills and aquatic life style.
**Range**
E. U.S., exclusive of New England and the Appalachian region, from s. N.Y. to cen. Fla., w. through Great Plains of Canada and U.S., and s. into e. Mexico. Rare in La. and adj. areas. (Also a disjunct population in Calif.)

## JEFFERSON SALAMANDER
*Ambystoma jeffersonianum*

**72:11**

**Description**
Length, to 8¼ in. (20.9 cm). Long-limbed. Color dorsally dark brown, brownish-gray, or lead-color with lighter belly; sometimes scattered blue flecks on limbs and lower sides; area around vent usually gray. Tips of adpressed limbs meet or overlap by 1–5 intercostal folds.
**Similarities**
Blue-spotted Salamander has shorter limbs and toes and more flecks. Female Jefferson Salamanders are similar to Silvery Salamander, and positive identification can be made only in the laboratory.
**Habitat**
Eastern deciduous forest.
**Remarks**
This is a bisexual diploid species (that is, both males and females and a normal set of paired chromosomes are present—in this case 2×14, or a total of 28).
**Range**
Se. Vt., s. to n. N.J., and through n. Va.; w. from s. of Lake Erie to s. Ind. and cen. Ky.

## SILVERY SALAMANDER
*Ambystoma platineum*

**Description**
Length, to record 7⅞ in. (20 cm). Coloration much like Jefferson Salamander's, but bluish flecks are scattered on dorsum, and may have larger spots on venter and dorsolateral surface. Area around vent usually gray.
**Habitat**
Eastern deciduous forest; in ponds or under logs, stones, or debris in moist areas.
**Remarks**
Although similar in appearance to Jefferson Salamander females, this is a triploid species (one extra set of chromosomes is present in the cells, making a total of 3×14, or 42), composed almost entirely of females.
**Range**
Within range of Jefferson Salamander in s. Mich. through s. Ohio, e. Ind. to extr. n. Ky.; also in n. N.J. and w. Mass.

## BLUE-SPOTTED SALAMANDER
*Ambystoma laterale*

**Description**
Length, to 5⅛ in. (13 cm). Black or grayish-black dorsum with numerous large bluish-white flecks covering body, especially along lower sides. Venter slightly lighter than dorsum. Area around vent is black. Limbs shorter than in Jefferson Salamander, which may be confused with it, but the adpressed limbs barely meet, if at all, or at most overlap by 1 intercostal fold.

**Similarities**
Shares same relationship with Tremblay's Salamander as does Jefferson Salamander with Silvery Salamander. Positive identification can be made only in the laboratory.

**Habitat**
Eastern deciduous forest.

**Remarks**
This is a bisexual, diploid species (see under Jefferson Salamander).

**Range**
Atlantic Coast from s. Que., s. to N.Y., w. through Canada to se. Man., and in U.S. in extr. nw. Ohio, n. Ind., and ne. Ill., nw. to ne. Minn. Disjunct localities in Labrador; Long Island, N.Y.; ne. N.J.; and e. Iowa.

## TREMBLAY'S SALAMANDER
*Ambystoma tremblayi*

**Description**
Length, to record of 6⅜ in. (16.2 cm). Dorsum dark gray to gray-black with lighter venter. Bluish-white markings on body most numerous along lower sides. Area around vent is black.

**Similarities**
Resembles female Blue-spotted Salamander. Light markings more diffuse than in Silvery Salamander.

**Habitat**
Eastern deciduous forest.

**Remarks**
Members of this triploid species (see under Silvery Salamander) are probably all female.

**Range**
Within range of Blue-spotted Salamander, known from scattered localities in N.S., Maine, Mass., s. Que., and from nw. Ohio and ne. Ind. n. through Mich. and ne. Wis.

## SMALL-MOUTHED SALAMANDER
*Ambystoma texanum*                                    **72:6**

**Description**
Length, to 7 in. (17.8 cm). Medium-sized, recognized by slender head and small mouth. Dorsum brownish-black or gray with lighter buff or grayish lichenlike blotches. Below paler, sometimes with light markings. Dark, unspeckled individuals may occur in northeastern part of range. Costal grooves 14 or 15; adpressed limbs separated by 2½ or 3 intercostal folds.

**Habitat**
Very diverse, from tall-grass prairie and farmed areas to woodland and dense hardwood forest; beneath logs or rocks, or buried in moist soil.

**Habits**
Breeds in early spring in temporary ponds or spring-fed streams and pools, but not in ponds with other mole salamanders.
**Range**
E. Ohio and se. Mich. to extr. se. Nebr., s. to Gulf from w. Ala. to e. Tex., but not in the Ozark Plateau or the Ouachita Mts.

## MOLE SALAMANDER
*Ambystoma talpoideum* **72:8**

**Description**
Length, to 4¹³/₁₆ in. (12.2 cm). Small, stout; very large, broad, flattened head. Dorsum of adults from dark gray to blue-black or brown, flecked with light gray or blue-gray markings. Below blue-gray with tiny white flecks.
**Habitat**
Woodlands at low elevation; adults live in burrows or under logs, debris, and leaf litter, often around a breeding pond.
**Range**
Atlantic Coastal Plain from cen. S.C. through n. Fla., w. to e. Tex., and n. in Miss. Valley to s. Ill. Disjunct populations in w. N.C., e. Tenn., and se. Okla.

Note: The **MABEE'S SALAMANDER**, *Ambystoma mabeei*, is distinguished by light flecks on the sides only. It occurs in the coastal Carolinas.

## FLATWOODS SALAMANDER
*Ambystoma cingulatum* *Fig. 87*

**Description**
Length, to 5¹/₁₆ in. (12.9 cm). Recognized by slender body, small head, and a black to brown dorsum with gray reticulations. Some individuals may lack reticulations. Ventral coloration dark with pattern of numerous light flecks or larger spots.
**Habitat**
Pine flatwoods. Adults often found beneath logs near cypress swamps.
**Habits**
One of two land-breeding mole salamanders.
**Reproduction**
Breeding takes place before mid-February, and when rains fill up depression where eggs are laid, larvae get head start on development.
**Range**
Atlantic Coastal Plain from S.C. to Okefenokee Swamp, Fla., to se. Miss.

*Fig. 87*

Flatwoods Salamander

# NEWTS
Family Salamandridae

In North America two groups of this widespread family occur, one over most of the East, the other confined to the Pacific Coast. Other members of the family live in eastern Asia and Japan, Europe, and North America. These brightly colored salamanders make attractive pets and in captivity feed on live insects, earthworms, and pieces of liver or other meat. Newts lack nasolabial grooves and costal grooves, and generally have thick, warty skins. Poisonous substances may be secreted from glands in the skin when these salamanders are handled.

## NEWT
*Notophthalmus viridescens*                                      **72:5**

**Description**
Length, to 5½ in. (14.0 cm). This salamander may be encountered in three stages. *Aquatic adult:* Skin smooth to finely granular, yellowish-green above with red spots, below yellow with many scattered black spots; tail strongly keeled. Vent of male enlarges during breeding season. *Immature terrestrial eft:* Length, to 3¼ in. (8.3 cm); skin warty and rough, orange-red above with red spots, yellow to orange below. *Larva*: With gills.

**Habitat**
Adult found in ponds, small lakes, marshes, swamps, quiet pools of streams, especially where vegetation is abundant. Eft frequents wooded areas in moist places.

**Habits**
Larva with gills is aquatic; after transformation it emerges onto land, where it remains as an eft for 2–3 years, then re-enters water and becomes permanently aquatic. Some newts omit the red eft stage.

**Reproduction**
Eggs laid singly on aquatic vegetation.

**Range**
Se. Canada and e. U.S., w. to n.-cen. Minn. and Lake Superior region in the North and to se. Kan. and e. Tex. in the South.

Note: The **STRIPED NEWT**, *Notophthalmus perstriatus*, has a continuous red dorsolateral stripe. It occurs in southern Georgia and northern Florida.

# LUNGLESS SALAMANDERS
Family Plethodontidae

These salamanders have costal grooves and a groove from nose to lip. They lack lungs, relying on skin and mouth for respiration. Those east of the 100th meridian usually have five toes on the hind feet (with two species having four), and four toes on the front feet. This is the most diverse family of salamanders, and except for two species confined to limestone areas in southern Europe, the family is restricted to the Americas. The eastern United States has the greatest generic diversity; however, the greatest number of species occurs in Central and South America. Reproduction in most species is terrestrial, and eggs are laid in secluded places on land. They develop directly into miniature adults without going through an aquatic larval stage in most cases, although the dusky salamanders and other groups have larvae. A salamander's larval stage is often subject to the heaviest predation, and in salamanders with terrestrial egg laying the average number of eggs laid per clutch or year is much smaller than in those with aquatic egg laying because fewer are lost.

# DUSKY AND SHOVEL-NOSED SALAMANDERS

Genera *Desmognathus* and *Leurognathus*

This group of salamanders is confined to the eastern United States. They all have a light line from eye to angle of the jaw, but the line in the Shovel-nosed Salamander is not well defined. The head points downward because the muscles above and behind it are quite enlarged. Unlike most other vertebrates, these salamanders have little mobility of the lower jaw, and the skull is raised to open the mouth. (These species may be rendered incapable of eating if their mouths are opened incorrectly or too roughly, so they should be treated with care when undergoing field identification.) The tongue is a pad attached in front. Most species are aquatic, but the smaller species have become increasingly terrestrial both in habitat and breeding sites. Food consists mainly of aquatic arthropods, but larval and small salamanders are also eaten.

## PIGMY SALAMANDER
*Desmognathus wrighti*                                                71:2

**Description**
Length, to 2 in. (5.1 cm). The smallest species of *Desmognathus*, with dark herringbone pattern on light dorsal band, which is light tan to reddish and with dark borders; belly flesh-colored. Tail round, slender, tapering.
**Habitat**
Spruce-fir forest at elevations of 2750 ft. (838.2 m) and up; under stones, bark, moss, and leaf litter.
**Habits**
Purely terrestrial; may overwinter in seepages and springs.
**Reproduction**
Eggs develop into miniature adults.
**Range**
High altitudes from Whitetop Mt., Va., s. along Tenn.–N.C. border.

Note: The **SEEPAGE SALAMANDER**, *Desmognathus aeneus*, occurs in the extreme south of the range of the Pigmy Salamander. It too is tiny, and can be distinguished by its kidney-shaped gland under the chin, smaller than that of the Pigmy Salamander.

## MOUNTAIN DUSKY SALAMANDER
*Desmognathus ochrophaeus*                                           71:2

**Description**
Length, to 4⅜ in. (11.1 cm). Small, with a straight-edged (or wavy in the South), light dorsal band from head onto tail and a round tail in cross section. Body slender, tail slim, tapering. *Light phase:* Color very variable—tan, gray, red, brown; within dorsal band there may be a row of dark spots. Infrequently, reddish patches on legs and/or cheeks. Sides blackish with tiny white spots (sometimes missing); below bluish-gray (darker than in Dusky Salamander). *Dark phase:* Uniformly black above, dark below.

**Habitat**
Moist woodland often near streams, under logs, rocks, and leaf litter, even under bark of trees. This species is not tied to streams as are most members of the genus.
**Habits**
Nocturnal.
**Range**
Mostly in higher elevations from cen. and s. N.Y., s. through n. and w. Pa. to Ga., w. to e. Ohio, e. Ky., and extr. e. Tenn. Disjunct population in ne. Ala.

## DUSKY SALAMANDER
*Desmognathus fuscus*                                               71:3

**Description**
Length, to 5⁹⁄₁₆ in. (14.1 cm). A tail keeled dorsally immediately behind vent and oval in cross section, and dorsum plain or blotched, along with bright and speckled lower sides, help distinguish this confusing species. Costal grooves 14; 2–5, usually fewer than 4, intercostal folds between adpressed limbs. Back tan to dark brown or green; plain, or commonly with 6–7 pairs of light blotches, often forming a stripe down back with somewhat wavy edges. Color of stripe variable. Below cream with brown blotches or spots. Edges of jaw in males not as sinuous as in Mountain Dusky Salamander.
**Habitat**
Shallow water under rocks in debris at stream edge, and in seepage areas, swamps.
**Habits**
Spends much time in water; active year-round but burrowed below frozen ground in gravel or limestone; mostly nocturnal; a good jumper when pursued.
**Food**
Terrestrial and aquatic insects, insect larvae, and worms.
**Remarks**
See Southern Dusky Salamander.
**Range**
New England and adj. Canada, s. through Appalachian region to Gulf Coast from w. Fla. to Miss. R. Absent from s. N.J. and Coastal Plain of se. Va. to cen. Fla. Populations w. of Miss. R. in ne. Ark., n.-cen. La., and adj. Ark.

## SEAL SALAMANDER
*Desmognathus monticola*                                            71:5

**Description**
Length, to 5⅞ in. (14.9 cm). Color and markings variable, light buff to grayish-brown above with many strong, dark-bordered pale blotches; belly pale to yellow and plain; lower sides mottled, often with a row of white spots from armpit to groin. Eyes large; limbs strong; tail short and compressed, keeled above.
**Habitat**
Under bark, logs, stones in or near mountain brooks or shallow muddy streams. Cool hemlock-shaded ravines on damp Appalachian mountainsides are favorite haunts.
**Range**
Sw. Pa., s. through the Appalachians to Ga. and Ala. Populations in s. Ga. and adj. w. Fla.

## BLACK MOUNTAIN SALAMANDER
*Desmognathus welteri*                               **70:12**

**Description**
Length, to 6¹¹⁄₁₆ in. (17 cm). Large, stout, similar to Seal
Salamander, but with mottled body, small, dark dorsal blotches,
and black on tips of toes. Black or brown blotches over back of
lighter brown. Change of coloration from back to belly gradual.
**Habitat**
Occurs only at higher elevations; found under stones in or near
brooks, springs, and roadside ditches in wooded areas.
**Range**
Black Mountain in Harlan Co., Ky., and nearby areas in w. Va.,
and e. Ky. Also in Warren Co. in cen. Ky.

## SOUTHERN DUSKY SALAMANDER
*Desmognathus auriculatus*                             *Fig. 88*

**Description**
Length, to 6⅜ in. (16.2 cm). Generally darker than Dusky
Salamander, with white or reddish spots prominent along sides in 1
or 2 rows, and below dark with white flecks; dorsally black with
traces of blotches, or patternless. Usually more than 4 intercostal
folds between adpressed limbs. Tail keeled above and compressed
from side to side.
**Habitat**
Around springs, swamps, cypress heads, muddy streams, and pools
in floodplains below Fall Line.
**Habits**
Often found in burrows.
**Remarks**
This species may hybridize with the Dusky Salamander where the
two occur together in the South, and animals with intermediate
characteristics may be found.
**Range**
Coastal Plain from extr. se. Va., s. to cen. Fla., and w. to se. Tex.

*Fig. 88*                        Southern Dusky Salamander

## BLACK-BELLIED SALAMANDER
*Desmognathus quadramaculatus*                     **70:1**

**Description**
Length, to 8¼ in. (21.0 cm). Large, with a dark or black belly.
Tail stout, less than ½ total length and sharply keeled dorsally.
Upperparts blackish with greenish-yellow blotches, sometimes with
rusty dorsal band; small light spots in 2 rows on sides from armpit
to groin.
**Habitat**
Along mountain streams above 2500 ft. (762.0 m), beneath stones
and logs in or by water. May be introduced at lower elevations by
fishermen releasing bait.
**Habits**
Highly aquatic, but in south of range may live on rock faces;
nocturnal.
**Range**
S. Appalachian Mts. from s.-cen. W. Va., s. to nw. S.C. and n.
Ga.

## SHOVEL-NOSED SALAMANDER
*Leurognathus marmoratus*                                    **70:6**

**Description**
Length, to 5¾ in. (14.6 cm). Resembles the Dusky Salamanders,
but can be distinguished in the hand by opening its mouth. It lacks
*round* nostrils, visible near front of the roof of mouth on each side.
White stripe from eye to angle of jaw is obscure. Color and pattern
variable; back usually dark brown to blackish with two rows of
light-colored blotches that are dull and blend into habitat. Snout
often darker than body. Two rows of spots along side of body;
below grayish with a light central area. Tail long and keeled.
**Habitat**
Found in mountain brooks and streams with sandy or gravelly
bottoms and rocks to hide under. Often occurs in same places
where the Black-bellied Salamander is found.
**Habits**
Strictly aquatic.
**Range**
Elevations over 1000 ft. (304.8 m) in s.-cen. Va., s. along the N.C.
and Tenn. borders into extr. nw. S.C. and ne. Ga.

# WOODLAND SALAMANDERS
Genus *Plethodon*

A high number of costal grooves, long and slender bodies, rounded
tails, and pronounced gular folds characterize the Woodland
Salamanders. These salamanders are very abundant in mostly
terrestrial places—in leaf mold and under rotting logs and rocks—
in the forests of the eastern United States, and other members of
the genus occur in the Pacific Northwest and in New Mexico. The
space needed for a salamander to crawl under an object is quite
small, and there are always other burrowing animals which help
make pathways. The eggs are usually suspended from above in logs
or from the stone, which arrangement seems to result from
gyrations of the female salamander as she lays them. Presumably,
getting them off the ground helps to protect them from the
elements and infections. Females may guard the clutch of eggs.
Embryos and newly hatched individuals are hard to tell apart.

## SOUTHERN RED-BACKED SALAMANDER
*Plethodon serratus*

**Description**
Length, to 4 in. (10.2 cm). Costal grooves 18–20. Edges of dorsal
stripe often serrated. Red often present on anterior belly and also
on back in the infrequent unstriped variant. Chin light. Check
Red-backed Salamander for description of the two color phases.
Rarely has individuals of the lead-backed color phase. Males have
large gland under chin.
**Similarities**
Zigzag Salamander has dorsal stripe with wavy edges, but fewer
costal grooves (normally 18).
**Habitat**
Under logs and stones in forests and also in fields; much as for
Red-backed Salamander, but often at higher elevations.
**Range**
Five disjunct localities—the s. Appalachians in extr. se. Tenn. and
sw. N.C.; the Piedmont of w. Ga. and adj. Ala.; se. Mo.; uplands
of se. Okla. and w.-cen. Ark.; and cen. La.

# Plates

**Note on the Bird Plates**

On each of the bird plates following (Plates 1 through 48) each species is designated by a different number. These guidelines will be of help in using the numbering system:

1. When a simple number (**1, 2, 3,** etc.) appears with an illustration, the bird depicted is either an adult male (designated ♂) or an adult female (designated ♀); in many instances, there are no immediately visible distinctions between males and females.

2. When the number is followed by a letter (**1a, 1b,** etc.) the bird depicted is a variant form. The major variants are immature or juvenal plumages, geographical races or subspecies, or color phases.

# PLATE 1
# LOONS AND GREBES

**1** Common Loon, **1a** summer, **1b** winter, p. 10. **2** Arctic Loon, **2a** summer,
**2b** winter, p. 10. **3** Red-throated Loon, **3a** summer, **3b** winter, p. 11.
**4** Red-necked Grebe, **4a** summer, **4b** winter, p. 12. **5** Western Grebe, p. 13.
**6** Horned Grebe, **6a** summer, **6b** winter, p. 12. **7** Eared Grebe, **7a** winter,
**7b** summer, p. 12. **8** Pied-billed Grebe, **8a** winter, **8b** summer,
**8c** immature, p. 13.

# PLATE 2
# PELAGIC BIRDS

**1** Parasitic Jaeger, **1a** light phase immature, **1b** light phase,
**1c** dark phase, p. 91. **2** Pomarine Jaeger, **2a** light phase, p. 91.
**3** Long-tailed Jaeger, **3a** light phase, p. 92. **4** Sooty Shearwater, p. 15.
**5** Greater Shearwater, p. 15. **6** Leach's Storm-Petrel, p. 16. **7** Northern
Fulmar, **7a** dark phase, **7b** light phase, p. 14. **8** Cory's Shearwater, p. 14.
**9** Wilson's Storm-Petrel, p. 16. **10** Manx Shearwater, p. 15.

## PLATE 3
## PELICANS, CORMORANTS, AND ALLIES

**1** Magnificent Frigatebird, **1a** immature, p. 20. **2** Anhinga,
**2a** breeding, p. 20. **3** Gannet, **3a** subadult, **3b** immature, p. 18.
**4** Brown Pelican, **4a** immature, p. 17. **5** Double-crested Cormorant,
**5a** immature, **5b** breeding, p. 19. **6** Great Cormorant, **6a** immature,
**6b** breeding, p. 18. **7** Neotropical Cormorant, **7a** breeding, p. 19.

# PLATE 4
# HERONS

1 Cattle Egret, **1a** immature, **1b** breeding, p. 24. **2** Great Egret, p. 21.
3 Louisiana Heron, p. 23. **4** Reddish Egret, **4a** dark phase,
**4b** white phase, p. 23. **5** Snowy Egret, **5a** immature, **5b** breeding, p. 22.
6 Little Blue Heron, **6a** immature, p. 22. **7** Yellow-crowned Night Heron,
**7a** immature, p. 24. **8** Black-crowned Night Heron, **8a** immature, p. 24.
9 Green Heron, **9a** immature, p. 21. **10** Least Bittern, p. 25.
11 American Bittern, p. 25.

# PLATE 5
## HERONS, IBISES, STORK, CRANES

**1** Great Blue Heron, **1a** breeding, **1b** "Great White" form, p. 21.
**2** Great Egret, **2a** breeding, p. 21. **3** Sandhill Crane, p. 65.
**4** Whooping Crane, p. 65. **5** White Ibis, **5a** immature, p. 27.
**6** Wood Stork, p. 26. **7** Limpkin, p. 66. **8** Glossy Ibis, **8a** breeding,
**8b** immature, p. 26. **9** Roseate Spoonbill, **9a** immature, p. 28.

**PLATE 6**
**GEESE AND SWANS**

**1** Snow Goose, **1a** "White" form, **1b** "White" form immature,
**1c** "Blue" form, **1d** "Blue" form immature, p. 31. **2** White-fronted Goose,
**2a** immature, p. 31. **3** Brant, p. 30. **4** Whistling Swan, **4a** immature, p. 29.
**5** Canada Goose, **5a** large race, **5b** small race, p. 30.

# PLATE 7
# POND DUCKS

1 Wood Duck, p. 36.  2 Fulvous Whistling-Duck, p. 31.
3 Eurasian Wigeon, p. 35.  4 American Wigeon, p. 35.  5 Common Teal, p. 34.
6 Northern Shoveler, p. 36.  7 Blue-winged Teal, p. 34.  8 Gadwall, p. 33.
9 Black Duck, p. 32.  10 Mallard, 10a "Mottled" race, p. 32.
11 Northern Pintail, p. 33.

## PLATE 8
## POND DUCKS IN FLIGHT—FEMALES

**1** Black Duck, p. 32.  **2** Mallard, p. 32.  **3** Gadwall, p. 33.
**4** Northern Pintail, p. 33.  **5** Common Teal, p. 34.
**6** Blue-winged Teal, p. 34.  **7** Northern Shoveler, p. 36.
**8** American Wigeon, p. 35.  **9** Wood Duck, p. 36.

# PLATE 9
## BAY AND SEA DUCKS IN FLIGHT—FEMALES

1 Redhead, p. 37. 2 Canvasback, p. 39. 3 Greater Scaup, p. 38.
4 Lesser Scaup, p. 38. 5 Common Goldeneye, p. 39. 6 Bufflehead, p. 40.
7 Hooded Merganser, p. 44. 8 Red-breasted Merganser, p. 45.
9 Oldsquaw, 9a winter, p. 41. 10 King Eider, p. 42.
11 White-winged Scoter, p. 43. 12 Surf Scoter, p. 43.

# PLATE 10
# BAY AND SEA DUCKS

**1** Barrow's Goldeneye, p. 40. **2** Bufflehead, p. 40. **3** Common
Goldeneye, p. 39. **4** Harlequin Duck, p. 41. **5** Oldsquaw, **5a** winter,
**5b** summer, p. 41. **6** Black Scoter, p. 44. **7** Surf Scoter, **7a** subadult, p. 43.
**8** White-winged Scoter, **8a** subadult, p. 43. **9** King Eider, **9a** immature, p. 42.

# PLATE 11
## POCHARDS, MERGANSERS, RUDDY DUCK

1 Redhead, p. 37.  2 Canvasback, p. 39.  3 Ring-necked Duck, p. 37.
4 Lesser Scaup, p. 38.  5 Greater Scaup, p. 38.  6 Hooded Merganser, p. 44.
7 Red-breasted Merganser, p. 45.  8 Common Merganser, p. 45.
9 Ruddy Duck, 9a summer, 9b winter, p. 46.

# PLATE 12
## VULTURES, EAGLES, CARACARA, OSPREY

1 Turkey Vulture, p. 47.  2 Black Vulture, p. 47.
3 Bald Eagle, 3a immature, p. 55.  4 Crested Caracara, p. 57.
5 Osprey, p. 56.  6 Golden Eagle, 6a immature, p. 55.

# PLATE 13
## ACCIPITERS, BUTEOS, HARRIER

**1** Northern Goshawk, **1a** immature, p. 49. **2** Cooper's Hawk, **2a** immature, p. 50. **3** Sharp-shinned Hawk, **3a** immature, p. 50. **4** Broad-winged Hawk, **4a** immature, p. 52. **5** Red-shouldered Hawk, **5a** immature, **5b** southern Florida forms, p. 52. **6** Northern Harrier, **6a** immature, p. 56. **7** Red-tailed Hawk, **7a** immature, p. 51. **8** Swainson's Hawk, **8a** immature, **8b** light phase, p. 53. **9** Rough-legged Hawk, **9a** light phase, p. 54.

## PLATE 14
## HAWKS IN FLIGHT

**1** Red-shouldered Hawk, p. 52. **2** Swainson's Hawk, **2a** dark phase, **2b** light phase, p. 53. **3** Broad-winged Hawk, p. 52. **4** Red-tailed Hawk, **4a** "Harlan's" race, dark phase, **4b** typical western race, **4c** eastern race, p. 51. **5** Short-tailed Hawk, **5a** light phase, **5b** dark phase, p. 53. **6** Rough-legged Hawk, **6a** light phase, **6b** dark phase, p. 54.

**PLATE 15
HAWKS IN FLIGHT**

**1** Crested Caracara, p. 57. **2** Osprey, p. 56. **3** Snail Kite, p. 48.
**4** Swallow-tailed Kite, p. 49. **5** Northern Harrier, **5a** immature, p. 56.

# PLATE 16
# FALCONS, KITES, GROUSE, QUAIL

**1** American Kestrel, p. 59. **2** Merlin, p. 58. **3** Mississippi Kite,
**3a** immature, p. 49. **4** White-tailed Kite, **4a** immature, p. 48.
**5** Gyrfalcon, **5a** gray phase immature, **5b** gray phase, p. 57.
**6** Peregrine Falcon, **6a** immature, p. 58. **7** Snail Kite, **7a** immature. p. 48.
**8** Spruce Grouse, p. 60. **9** Ruffed Grouse, **9a** gray phase,
**9b** tail of rufous phase, p. 60. **10** Bobwhite, p. 63. **11** Gray Partridge, p. 63.

# PLATE 17
## OPEN COUNTRY GROUSE AND PHEASANT

**1** Rock Ptarmigan, **1a** winter, **1b** summer, p. 61. **2** Willow Ptarmigan, **2a** summer, **2b** winter, **2c** spring, p. 61. **3** Greater Prairie Chicken, **3a** displaying, p. 62. **4** Sharp-tailed Grouse, **4a** displaying, p. 62. **5** Ring-necked Pheasant, p. 63.

# PLATE 18
# RAILS

1 Purple Gallinule, **1a** immature, p. 69. **2** American Coot, **2a** immature, **2b** chick, p. 70. **3** Common Gallinule, **3a** immature, **3b** chick, p. 69. **4** Clapper Rail, **4a** chick, p. 67. **5** King Rail, p. 66. **6** Virginia Rail, **6a** immature, p. 67. **7** Sora, **7a** immature, p. 68. **8** Black Rail, p. 68. **9** Yellow Rail, p. 68.

# PLATE 19
## SHOREBIRDS–FULL BIRDS IN SPRING PLUMAGES

**1** Piping Plover, **1a** fall, p. 72. **2** Semipalmated Plover, **2a** fall, p. 72.
**3** Killdeer, p. 73. **4** Snowy Plover, p. 73. **5** Wilson's Plover, p. 73.
**6** Black-bellied Plover, p. 74. **7** American Golden Plover, p. 74.
**8** Ruddy Turnstone, p. 76. **9** Upland Sandpiper, p. 77.
**10** Buff-breasted Sandpiper, p. 85. **11** Marbled Godwit, p. 87.
**12** Willet, p. 80. **13** Whimbrel, p. 76. **14** Long-billed Curlew, p. 77.
**15** Black-necked Stilt, p. 90. **16** Common Oystercatcher, p. 71.
**17** American Avocet, **17a** fall, p. 90.

## PLATE 20
## SHOREBIRDS—ALL BIRDS IN SPRING PLUMAGES

# PLATE 21
## SHOREBIRDS—MOST BIRDS IN FALL PLUMAGES

1 Semipalmated Sandpiper, p. 84. 2 Least Sandpiper, p. 82.
3 Western Sandpiper, p. 84. 4 Sanderling, p. 84. 5 Ruddy Turnstone, p. 76.
6 White-rumped Sandpiper, p. 81. 7 Purple Sandpiper, p. 80.
8 Curlew Sandpiper, p. 83. 9 Dunlin, p. 83. 10 Red Knot, p. 80.
11 American Golden Plover, p. 74. 12 Black-bellied Plover, p. 74.
13 Stilt Sandpiper, 13a spring, p. 85. 14 Short-billed Dowitcher, p. 86.
15 Hudsonian Godwit, p. 87. 16 Solitary Sandpiper, p. 78. 17 Willet, p. 80.
18 Spotted Sandpiper, 18a spring, p. 77. 19 Wilson's Phalarope, p. 88.
20 Lesser Yellowlegs, p. 79. 21 Greater Yellowlegs, p. 78.
22 Northern Phalarope, p. 89. 23 Red Phalarope, p. 88.

# PLATE 22
## FALL SHOREBIRDS IN FLIGHT

**1** White-rumped Sandpiper, p. 81. **2** Western Sandpiper, p. 84.
**3** Pectoral Sandpiper, p. 81. **4** Purple Sandpiper, p. 80. **5** Dunlin, p. 83.
**6** Spotted Sandpiper, p. 77. **7** Northern Phalarope, p. 89.
**8** Sanderling, p. 84. **9** Solitary Sandpiper, p. 78. **10** Red Phalarope, p. 88.
**11** Red Knot, p. 80. **12** Short-billed Dowitcher, p. 86. **13** Wilson's
Phalarope, p. 88. **14** Stilt Sandpiper, p. 85. **15** Lesser Yellowlegs, p. 79.

# PLATE 23
## SHOREBIRDS IN FLIGHT

1 Semipalmated Plover, **1a** spring, p. 72. **2** Piping Plover, **2a** spring, p. 72.
3 Killdeer, p. 73. **4** Buff-breasted Sandpiper, p. 85. **5** Upland
Sandpiper, p. 77. **6** American Golden Plover, **6a** fall, **6b** spring, p. 74.
7 Common Snipe, p. 75. **8** Black-bellied Plover, **8a** fall, **8b** spring, p. 74.
9 American Woodcock, p. 75. **10** Ruddy Turnstone, **10a** spring, p. 76.

PLATE 24
LARGER GULLS

1 Glaucous Gull, **1a** first winter immature, **1b** second winter immature, p. 9?

2 Great Black-backed Gull, **2a** first winter immature, **2b** winter, p. 96.

3 Iceland Gull, **3a** first winter immature, **3b** winter Kumlien's race,

**3c** nominate race, p. 94. 4 Herring Gull, **4a** winter,

**4b** first winter immature, **4c** second winter immature, p. 95.

5 Ring-billed Gull, **5a** winter, **5b** first winter immature, p. 95.

## PLATE 25
## SMALLER GULLS

1 Laughing Gull, **1a** winter, **1b** summer, **1c** first winter immature, p. 97.
2 Franklin's Gull, **2a** winter, **2b** summer, **2c** first winter immature, p. 97.
3 Bonaparte's Gull, **3a** summer, **3b** winter, **3c** first winter immature, p. 98.
4 Sabine's Gull, **4a** first winter immature, **4b** summer, p. 100.
5 Black-headed Gull, **5a** winter, p. 96. 6 Little Gull, **6a** winter,
**6b** first winter immature, p. 98. 7 Black-legged Kittiwake, **7a** winter,
**7b** first winter immature, p. 99.

## PLATE 26
## TERNS AND GULLS

1 Least Tern, **1a** fall, **1b** spring, p. 103. **2** Black Tern, **2a** fall, p. 105.
3 Common Tern, **3a** fall, **3b** spring, p. 102. **4** Forster's Tern, **4a** fall,
**4b** spring, p. 101. **5** Gull-billed Tern, **5a** fall, **5b** spring, p. 101.
6 Arctic Tern, **6a** spring, p. 102. **7** Roseate Tern, **7a** spring, p. 103.
8 Sandwich Tern, **8a** fall, p. 104. **9** Black-legged Kittiwake, **9a** spring, p. 99.
10 Ring-billed Gull, **10a** spring, p. 95. **11** Herring Gull, **11a** spring, p. 95.
12 Royal Tern, **12a** fall, p. 104. **13** Caspian Tern, **13a** fall, p. 105.

## PLATE 27
## TERNS—ADULTS IN SPRING PLUMAGES

**1** Royal Tern, p. 104. **2** Caspian Tern, p. 105. **3** Black Tern,
**3a** immature, p. 105. **4** Sandwich Tern, p. 104. **5** Gull-billed Tern, p. 101.
**6** Roseate Tern, p. 103. **7** Forster's Tern, **7a** immature, p. 101.
**8** Arctic Tern, p. 102. **9** Common Tern, **9a** immature, p. 102.
**10** Least Tern, **10a** immature, p. 103.

## PLATE 28
## ALCIDS

**1** Common Murre, **1a** summer, **1b** winter, p. 107. **2** Black Guillemot,
**2a** summer, **2b** immature, **2c** winter, p. 108. **3** Common Puffin,
**3a** summer, **3b** winter, p. 109. **4** Dovekie, **4a** winter, **4b** summer, p. 108.
**5** Razorbill, **5a** summer, **5b** immature, **5c** winter, p. 107.
**6** Thick-billed Murre, **6a** winter, p. 107.

## PLATE 29
## OWLS

**1** Burrowing Owl, p. 116.  **2** Boreal Owl, p. 119.  **3** Screech Owl,
**3a** rufous phase, **3b** gray phase, p. 115.  **4** Saw-whet Owl,
**4a** immature, p. 119.  **5** Hawk Owl, p. 116.  **6** Short-eared Owl, p. 118.
**7** Long-eared Owl, p. 118.  **8** Barn Owl, p. 114.  **9** Barred Owl, p. 117.
**10** Snowy Owl, p. 115.  **11** Great Horned Owl, p. 115.
**12** Great Gray Owl, p. 117.

# PLATE 30
## PIGEONS, SWIFTS, NIGHTJARS

**1** Rock Pigeon, p. 110. **2** White-crowned Pigeon, p. 110. **3** Mourning Dove, p. 110. **4** Common Ground Dove, p. 111. **5** Chimney Swift, p. 122. **6** Common Nighthawk, p. 122. **7** Chuck-will's-widow, **7a** tail, p. 120. **8** Poor-will, **8a** tail, p. 121. **9** Whip-poor-will, **9a** tails, p. 121.

# PLATE 31
# WOODPECKERS

**1** Pileated Woodpecker, p. 125. **2** Common Flicker, **2a** eastern
"Yellow-shafted" form, **2b** western "Red-shafted" form, p. 125.
**3** Downy Woodpecker, p. 128. **4** Hairy Woodpecker, p. 127. **5** Red-bellied
Woodpecker, **5a** immature, p. 126. **6** Red-cockaded Woodpecker, p. 128.
**7** Black-backed Woodpecker, p. 129. **8** Yellow-bellied Sapsucker,
**8a** immature, p. 127. **9** Three-toed Woodpecker, p. 129.
**10** Red-headed Woodpecker, **10a** immature, p. 126.

# PLATE 32
# SWALLOWS AND CUCKOOS

1 Purple Martin, p. 139. 2 Barn Swallow, p. 139. 3 Tree Swallow,
3a immature, p. 137. 4 Bank Swallow, p. 138. 5 Rough-winged
Swallow, p. 138. 6 Mangrove Cuckoo, p. 113. 7 Cliff Swallow, p. 139.
8 Yellow-billed Cuckoo, p. 112. 9 Black-billed Cuckoo, 9a immature, p. 113.

**PLATE 33
FLYCATCHERS**

1 Great Crested Flycatcher, p. 131. 2 Say's Phoebe, p. 132.
3 Eastern Phoebe, p. 132. 4 Scissor-tailed Flycatcher, p. 131.
5 Cassin's Kingbird, p. 131. 6 Western Kingbird, p. 130.
7 Eastern Kingbird, p. 130. 8 Gray Kingbird, p. 131.
9 Eastern Wood Pewee, p. 135. 10 Western Wood Pewee, p. 135.
11 Olive-sided Flycatcher, p. 136. 12 Alder/Willow Flycatcher, p. 133.
13 Least Flycatcher, p. 134. 14 Acadian Flycatcher, p. 133.
15 Yellow-bellied Flycatcher, p. 133.

# PLATE 34
# NUTHATCHES, TITMICE, WRENS AND OTHERS

**1** Brown Creeper, p. 147. **2** White-breasted Nuthatch, p. 146.
**3** Brown-headed Nuthatch, p. 147. **4** Tufted Titmouse, p. 145.
**5** Red-breasted Nuthatch, p. 146. **6** Boreal Chickadee, p. 145.
**7** Carolina Chickadee, p. 144. **8** Short-billed Marsh Wren, p. 150.
**9** Long-billed Marsh Wren, p. 150. **10** Black-capped Chickadee, p. 144.
**11** Winter Wren, p. 148. **12** Blue-gray Gnatcatcher, p. 157.
**13** Bewick's Wren, p. 149. **14** House Wren, p. 148. **15** Ruby-throated
Hummingbird, p. 123. **16** Rock Wren, p. 151. **17** Carolina Wren, p. 149.

# PLATE 35
## MIMIDS, CORVIDS, SHRIKES, KINGFISHERS

# PLATE 36
## THRUSHES AND WAXWINGS

**1** Varied Thrush, p. 154. **2** American Robin, **2a** juvenal, p. 153.
**3** Bohemian Waxwing, p. 161. **4** Cedar Waxwing, **4a** juvenal, p. 160.
**5** Hermit Thrush, p. 154. **6** Eastern Bluebird, **6a** juvenal, p. 156.
**7** Mountain Bluebird, **7a** juvenal, p. 156. **8** Veery, p. 156. **9** Swainson's
Thrush, p. 155. **10** Gray-cheeked Thrush, p. 155. **11** Wood Thrush, p. 154.

# PLATE 37
## WARBLERS—SPRING PLUMAGES

**1** American Redstart, p. 187. **2** Bay-breasted Warbler, p. 178.
**3** Chestnut-sided Warbler, p. 178. **4** Black-throated Blue Warbler, p. 174.
**5** Northern Parula, p. 172. **6** Cerulean Warbler, p. 176. **7** Orange-crowned
Warbler, p. 171. **8** Tennessee Warbler, p. 172. **9** Worm-eating
Warbler, p. 168. **10** Swainson's Warbler, p. 168. **11** Ovenbird, p. 181.
**12** Northern Waterthrush, p. 182. **13** Louisiana Waterthrush, p. 182.

# PLATE 38
# WARBLERS—SPRING PLUMAGES

1 Yellow-rumped Warbler, **1a** eastern "Myrtle" form, **1b** western "Audubon's" form, p. 175. **2** Magnolia Warbler, p. 173. **3** Cape May Warbler, p. 174. **4** Canada Warbler, p. 186. **5** Blackburnian Warbler, p. 177. **6** Black-throated Green Warbler, p. 175. **7** Yellow-throated Warbler, p. 177. **8** Kirtland's Warbler, p. 180. **9** Black-throated Gray Warbler, p. 176. **10** Blackpoll Warbler, p. 179. **11** Black-and-White Warbler, p. 167. **12** Pine Warbler, p. 179. **13** Palm Warbler, **13a** eastern race, **13b** western race, p. 181. **14** Prairie Warbler, p. 180.

## PLATE 39
## WARBLERS—SPRING PLUMAGES

**1** Kentucky Warbler, p. 183. **2** Common Yellowthroat, p. 184.
**3** Yellow-breasted Chat, p. 185. **4** Wilson's Warbler, p. 186.
**5** Hooded Warbler, p. 185. **6** Yellow Warbler, p. 173.
**7** Golden-winged Warbler, p. 169. **8** "Lawrence's" Hybrid, p. 169.
**9** Prothonotary Warbler, p. 167. **10** "Brewster's" Hybrid, p. 169.
**11** Nashville Warbler, p. 171. **12** Blue-winged Warbler, p. 170.
**13** Mourning Warbler, p. 184. **14** Connecticut Warbler, p. 183.
**15** Bachman's Warbler, p. 170.

# PLATE 40
# FALL WARBLERS AND VIREOS

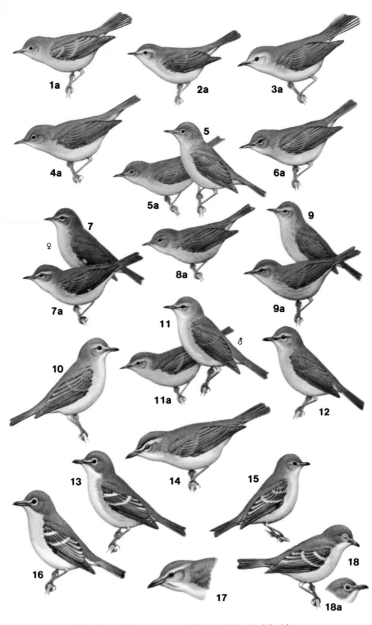

**1** Yellow Warbler, **1a** immature, p. 173. **2** Wilson's Warbler,
**2a** immature, p. 186. **3** Hooded Warbler, **3a** immature, p. 185.
**4** Connecticut Warbler, **4a** immature, p. 183. **5** Nashville Warbler,
**5a** immature, p. 171. **6** Mourning Warbler, **6a** immature p. 184.
**7** Black-throated Blue Warbler, **7a** immature, p. 175. **8** Common Yellowthroat
**8a** immature, p. 184. **9** Orange-crowned Warbler, **9a** immature, p. 171.
**10** Warbling Vireo, p. 166. **11** Tennessee Warbler, **11a** immature, p. 172.
**12** Philadelphia Vireo, p. 166. **13** Solitary Vireo, p. 165. **14** Red-eyed
Vireo, p. 165. **15** Bell's Vireo, p. 164. **16** Yellow-throated Vireo, p. 164.
**17** Black-whiskered Vireo, p. 166. **18** White-eyed Vireo, **18a** immature, p. 16

## PLATE 41
# FALL WARBLERS AND KINGLETS

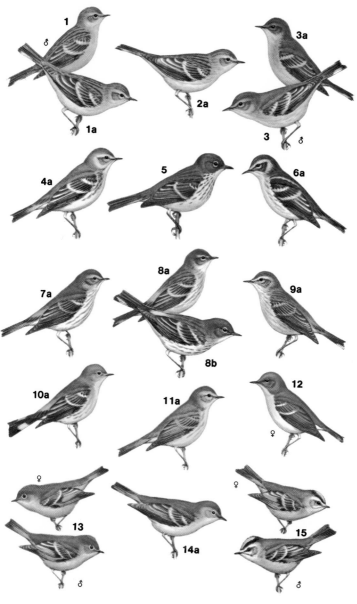

1 Bay-breasted Warbler, 1a immature, p. 178. 2 Blackpoll Warbler,
2a immature, p. 179. 3 Pine Warbler, 3a immature, p. 179. 4 Black-
throated Green Warbler, 4a immature, p. 175. 5 Kirtland's Warbler, p. 180.
6 Blackburnian Warbler, 6a immature, p. 177. 7 Cape May Warbler,
7a immature, p. 174. 8 Yellow-rumped Warbler, 8a immature eastern
"Myrtle" form, 8b immature western "Audubon's" form, p. 175. 9 Palm
Warbler, 9a immature, p. 181. 10 Magnolia Warbler, 10a immature, p. 173.
11 Prairie Warbler, 11a immature, p. 180. 12 Northern Parula, p. 172.
13 Ruby-crowned Kinglet, p. 158. 14 Chestnut-sided Warbler,
14a immature, p. 178. 15 Golden-crowned Kinglet, p. 158.

# PLATE 42
# BLACKBIRDS AND STARLING

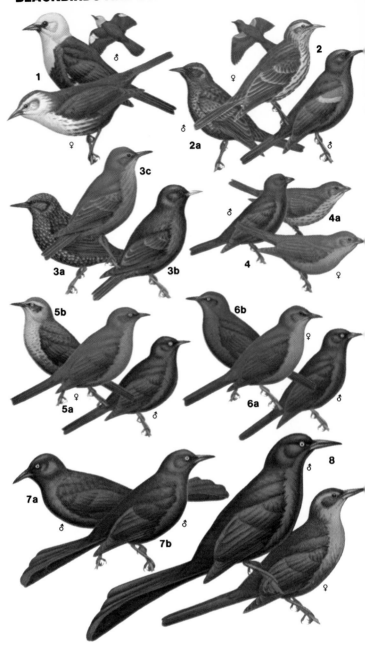

1 Yellow-headed Blackbird, p. 190. 2 Red-winged Blackbird,
2a immature, p. 191. 3 Starling, 3a fall, 3b spring, 3c immature, p. 162.
4 Brown-headed Cowbird, 4a juvenal, p. 194. 5 Rusty Blackbird, 5a spring,
5b fall immature, p. 192. 6 Brewer's Blackbird, 6a spring,
6b fall immature, p. 192. 7 Common Grackle, 7a "Purple" form,
7b "Bronzed" form, p. 194. 8 Boat-tailed Grackle, p. 193.

## PLATE 43 ORIOLES, TANAGERS, FINCHES—SPRING PLUMAGES

1 Northern Oriole, **1a** western "Bullock's" race, **1b** "Bullock's" immature,
**1c** eastern "Baltimore" race, **1d** "Baltimore" immature, p. 191.
**2** Spot-breasted Oriole, **2a** immature, p. 191. **3** Orchard Oriole,
**3a** immature, p. 190. **4** Scarlet Tanager, p. 195. **5** Western Tanager, p. 195.
**6** Summer Tanager, **6a** immature, p. 196. **7** Blue Grosbeak, p. 198.
**8** Painted Bunting, p. 199. **9** Indigo Bunting, p. 199.
**10** Lazuli Bunting, p. 200. **11** Black-headed Grosbeak, p. 197.
**12** Rose-breasted Grosbeak, **12a** immature, p. 197.

# PLATE 44
## FINCHES

1 Cardinal, p. 196. 2 Pine Grosbeak, 2a immature, p. 202.
3 Evening Grosbeak, p. 200. 4 Red Crossbill, 4a juvenal, p. 204.
5 Purple Finch, p. 201. 6 White-winged Crossbill, p. 205.
7 House Finch, p. 200. 8 Common Redpoll, 8a immature, p. 202.
9 Pine Siskin, p. 204. 10 American Goldfinch, 10a winter,
10b summer, p. 203. 11 Lesser Goldfinch, p. 203.

# PLATE 45
## SPARROWS, JUNCOS, TOWHEES

**1** Black-throated Sparrow, p. 211. **2** House Sparrow, **2a** summer,
**2b** immature, p. 187. **3** Lark Sparrow, p. 212. **4** Dark-eyed Junco,
**4a** eastern "Slate-colored" form, **4b** "Slate-colored" juvenal, **4c** Black Hills
"White-winged" form, **4d** no. Rocky Mt. "Pink-sided" form,
**4e** western "Oregon" form, p. 211. **5** Green-tailed Towhee, p. 206.
**6** Rufous-sided Towhee, **6a** western "Spotted" form, **6b** juvenal, p. 205.

# PLATE 46
# SPARROWS

1 Fox Sparrow, p. 216. 2 Vesper Sparrow, p. 210.
3 Lincoln's Sparrow, p. 216. 4 Song Sparrow, p. 217. 5 Savannah Sparrow,
5a "Ipswich" race, p. 208. 6 Henslow's Sparrow, p. 207.
7 Baird's Sparrow, p. 207. 8 Sharp-tailed Sparrow, 8a Interior race,
8b n.e. "Acadian" race, 8c Coastal race, p. 209. 9 LeConte's Sparrow, p. 209.
10 Seaside Sparrow, p. 208. 11 Grasshopper Sparrow, p. 206.
12 Clay-colored Sparrow, 12a immature, p. 213.
13 Chipping Sparrow, 13a spring, 13b immature, p. 213.

**PLATE 47**
**SPARROWS**

1 Harris' Sparrow, **1a** spring, **1b** immature, p. 214.
2 White-throated Sparrow, **2a** immature, p. 215. **3** White-crowned Sparrow,
**3a** immature, p. 215. **4** Bachman's Sparrow, p. 211. **5** Field Sparrow, p. 214.
6 Tree Sparrow, p. 212. **7** Swamp Sparrow, **7a** spring, **7b** immature, p. 217.
8 Rufous-crowned Sparrow, p. 210.

# PLATE 48 OPEN COUNTRY SONGBIRDS—
# MALES IN SPRING PLUMAGES

1 Bobolink, p. 188.  2 Eastern Meadowlark, p. 189.  3 Lark Bunting, p. 206.
4 Sprague's Pipit, p. 160.  5 Water Pipit, 5a fall, p. 159.  6 Dickcissel, p. 198.
7 Horned Lark, 7a "Yellow-faced" form, 7b "White-faced" form,
7c juvenal, p. 136.  8 Snow Bunting, 8a winter,  8b immature, p. 220.
9 Lapland Longspur,  9a fall, p. 218.  10 Chestnut-collared Longspur,
10a fall, p. 219.  11 Smith's Longspur,  11a fall, p. 219.
12 McCown's Longspur,  12a fall, p. 218.
13 Gray-crowned Rosy Finch,  13a immature, p. 202.

## PLATE 49
## BATS

**1** Evening Bat, p. 239. **2** Big Brown Bat, p. 237. **3** Small-footed
Myotis, p. 236. **4** Hoary Bat, p. 238. **5** Little Brown Myotis, p. 235.
**6** Silver-haired Bat, p. 237. **7** Townsend's Big-eared Bat, p. 240.
**8** Eastern Pipistrelle, p. 237. **9** Northern Yellow Bat, p. 239.
**10** Red Bat, p. 238. **11** Seminole Bat, p. 238.

# PLATE 50
# RABBITS AND SQUIRRELS

1 Red Squirrel, p. 252. 2 Gray Squirrel, 2a melanistic, p. 251.
3 Southern Flying Squirrel, p. 252. 4 Fox Squirrel, 4a eastern,
4b melanistic, 4c mid-Atlantic, p. 251. 5 Rock Squirrel, p. 250.
6 Woodchuck, p. 248. 7 Black-tailed Prairie Dog, p. 250.
8 Thirteen-lined Ground Squirrel, p. 249. 9 Spotted Ground Squirrel, p. 249.
10 Least Chipmunk, p. 247. 11 Eastern Chipmunk, p. 247.
12 Eastern Cottontail, p. 243. 13 Black-tailed Jackrabbit, p. 246.
14 New England Cottontail, p. 243. 15 Swamp Rabbit, p. 243.
16 Snowshoe Hare, 16a summer, 16b winter, p. 244.

# PLATE 51
## TRACKS—SMALLER MAMMALS

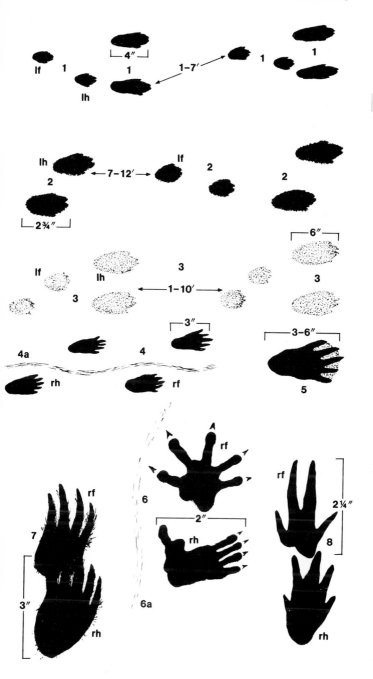

1 Cottontail, p. 243. 2 Jackrabbit, p. 245.
3 Snowshoe Hare, p. 244. 4 Muskrat, 4a tail mark, p. 265.
5 Beaver, hind foot covers 4 in. (10.2 cm) front to next track, p. 256.
6 Virginia Opossum, 6a tail mark, p. 227. 7 Porcupine, p. 269.
8 Armadillo, p. 241.

# PLATE 52
# MICE, SHREWS, MOLES, ALLIES

**1** White-footed Mouse, p. 260. **2** Meadow Jumping Mouse, p. 269.
**3** Deer Mouse, p. 258. **4** Meadow Vole, p. 263. **5** Ord's Kangaroo Rat, p. 255.
**6** Northern Bog Lemming, p. 266. **7** Northern Grasshopper Mouse, p. 260.
**8** Hispid Pocket Mouse, p. 255. **9** Southern Red-backed Vole, p. 262.
**10** Plains Harvest Mouse, p. 257. **11** Short-tailed Shrew, p. 231.
**12** Least Shrew, p. 232. **13** Arctic Shrew, p. 230. **14** Masked Shrew, p. 228.
**15** Hairy-tailed Mole, p. 233. **16** Eastern Mole, p. 233.
**17** Star-nosed Mole, p. 233.

## PLATE 53
## MICE AND RATS

1 Northern Pygmy Mouse, p. 260.  2 House Mouse, p. 268.
3 Florida Mouse, p. 259.  4 Texas Mouse, p. 259.  5 Fulvous Harvest
Mouse, p. 258.  6 Eastern Harvest Mouse, p. 257.  7 Oldfield Mouse, p. 259.
8 Woodland Jumping Mouse, p. 269.  9 Golden Mouse, p. 260.
10 Hispid Cotton Rat, p. 261.  11 Eastern Woodrat, p. 261.
12 Southern Plains Woodrat, p. 262.  13 Marsh Rice Rat, p. 257.

## PLATE 54
## TRACKS—SMALL AND HOOFED ANIMALS

1 Badger, p. 294.  2 Woodchuck, p. 248.  3 Kangaroo Rat, p. 255.
4 Deer Mouse, 4a tail mark, p. 258.  5 Grasshopper Mouse, running, p. 260.
6 Raccoon, p. 289.  7 Meadow Vole, bounding, p. 263.  8 Chipmunk, p. 247.
9 Deer, walking, p. 302.  10 Norway Rat, walking, p. 268.
11 Gray Squirrel, p. 251.  12 Bison, p. 305.

## PLATE 55
## VOLES AND LEMMINGS

**1** Beach Vole, p. 263. **2** Heather Vole, p. 262. **3** Woodland Vole, p. 264.
**4** Meadow Vole, p. 263. **5** Rock Vole, p. 263. **6** Southern Red-backed
Vole, p. 262. **7** Prairie Vole, p. 263. **8** Southern Bog Lemming, p. 266.
**9** Brown Lemming, p. 265. **10** Collared Lemming, **10a** summer,
**10b** winter, p. 266.

**PLATE 56**
**TOOTHED WHALES**

**1** White Whale, p. 274. **2** Narwhal, p. 275. **3** North Atlantic Bottle-nosed
Whale, p. 273. **4** Pygmy Sperm Whale, p. 274. **5** Sperm Whale, p. 273.
**6** True's Beaked Whale, p. 272. **7** Goose-beaked Whale, p. 273.

# PLATE 57
# DOLPHINS AND OTHER WHALES

**1** White-beaked Dolphin, p. 277. **2** Bottle-nosed Dolphin, p. 276.
**3** Atlantic White-sided Dolphin, p. 277. **4** Harbor Porpoise, p. 278.
**5** Killer Whale, p. 278. **6** Common Dolphin, p. 276.
**7** Grampus, p. 277. **8** Common Pilot Whale, p. 278.

# PLATE 58
# TRACKS—WEASELS AND ALLIES

1 Marten, p. 290.  2 Wolverine, p. 293.  3 Spotted Skunk, p. 294.
4 Striped Skunk, p. 295.  5 River Otter, p. 295.  6 Fisher, p. 291.
7 Mink, p. 293.  8 Long-tailed Weasel, p. 292.  9 Least Weasel, p. 292.

**PLATE 59**
**FOXES, WEASELS, ALLIES**

**1** Coyote, p. 283. **2** Red Fox, **2a** "cross" phase, p. 286. **3** Swift Fox, p. 285.
**4** Least Weasel, **4a** summer, **4b** winter, p. 292. **5** Gray Fox, p. 286.
**6** Ermine, **6a** summer, **6b** winter, p. 291. **7** Marten, p. 290.
**8** Black-footed Ferret, p. 293. **9** Fisher, p. 291. **10** Wolverine, p. 293.

# PLATE 60
## TRACKS—SMALL AND HOOFED ANIMALS

1 Badger, p. 375. 2 Woodchuck, p. 307. 3 Kangaroo Rat, p. 324.
4 Deer Mouse, 4a tail mark p. 331. 5 Grasshopper Mouse, running, p. 334.
6 Raccoon, p. 370. 7 Meadow Vole, bounding, p. 341. 8 Chipmunk, p. 303.
9 Deer, walking, p. 384. 10 Norway Rat, walking, p. 348.
11 Western Gray Squirrel, p. 315. 12 Bison, p. 387.

**PLATE 61
CARNIVORES**

**1** Striped Skunk, p. 295. **2** Eastern Spotted Skunk, p. 294.
**3** Black Bear, **3a** black phase, **3b** cinnamon phase, p. 287.
**4** Long-tailed Weasel, **4a** winter, **4b** intermediate, **4c** summer, p. 292.
**5** Mink, p. 293. **6** River Otter, p. 295. **7** Raccoon, p. 289. **8** Badger, p. 294.

# PLATE 62
## SEALS

**1** Ringed Seal, p. 296.  **2** Walrus, p. 288.  **3** Gray Seal, p. 297.
**4** Harp Seal, p. 296.  **5** Harbor Seal, p. 297.  **6** Bearded Seal, p. 297.
**7** Hooded Seal, p. 298.

# PLATE 63
## DEER AND PRONGHORN

**1** Moose, p. 303. **2** Elk, p. 302. **3** Pronghorn, p. 304. **4** Caribou, p. 304.
**5** Mule Deer, p. 302. **6** White-tailed Deer,
**6a** summer, **6b** winter, p. 303.

## PLATE 64
## SNAPPING TURTLES, SOFTSHELLS, SEA TURTLES

1 Snapping Turtle, p. 316.  2 Alligator Snapping Turtle, p. 316.
3 Spiny Softshell, p. 320.  4 Smooth Softshell, p. 320.
5 Hawksbill, p. 333.  6 Green Turtle, p. 332.  7 Loggerhead, p. 333.
8 Atlantic Ridley, p. 333.  9 Leatherback, p. 334.

**PLATE 65
OTHER TURTLES**

**1** Spotted Turtle, p. 322.  **2** Yellow Mud Turtle, p. 318.  **3** Stinkpot, p. 318.
**4** Eastern Mud Turtle, p. 317.  **5** Red-bellied Turtle, p. 329.
**6** Painted Turtle, p. 327.  **7** Cooter, p. 328.  **8** Slider, p. 328.
**9** False Map Turtle, p. 326.  **10** Map Turtle, p. 325.

## PLATE 66
## LIZARDS, SKINKS, WHIPTAILS

**1** Great Plains Skink, **1a** immature, p. 344. **2** Collared Lizard, p. 338.
**3** Five-lined Skink, **3a** immature, p. 342. **4** Broad-headed Skink, p. 343.
**5** Many-lined Skink, p. 344. **6** Prairie Skink, p. 345. **7** Coal Skink, p. 344.
**8** Ground Skink, p. 342. **9** Eastern Fence Lizard, p. 338.
**10** Lesser Earless Lizard, p. 339. **11** Six-lined Racerunner, p. 345.

## PLATE 67
## WATER, GARTER, OTHER SNAKES

1 Texas Blind Snake, p. 348. 2 Lined Snake, p. 357. 3 Graham's Crayfish
Snake, p. 352. 4 Queen Snake, p. 353. 5 Glossy Crayfish Snake, p. 353.
6 Plain-bellied Water Snake, p. 349. 7 Diamondback Water Snake, p. 352.
8 Green Water Snake, p. 348. 9 Brown Water Snake, p. 352.
10 Common Garter Snake, p. 359. 11 Plains Garter Snake, p. 359.
12 Checkered Garter Snake, p. 358. 13 Eastern Ribbon Snake, p. 360.
14 Western Terrestrial Garter Snake, p. 360.

# PLATE 68
# SNAKES

1 Ringneck Snake, p. 361. 2 Brown Snake, p. 356. 3 Red-bellied
Snake, p. 356. 4 Rough Earth Snake, p. 355. 5 Smooth Earth Snake, p. 356.
6 Worm Snake, p. 362. 7 Common Garter Snake, p. 359.
8 Northern Water Snake, p. 350. 9 Eastern Hognose Snake, p. 365.
10 Rough Green Snake, p. 370. 11 Racer, p. 371. 12 Eastern Yellow-bellied
Racer, p. 371. 13 Black Rat Snake, p. 375. 14 Corn Snake, p. 374.
15 Rainbow Snake, p. 363. 16 Mud Snake, p. 363. 17 Coachwhip, p. 372.

**PLATE 69
SNAKES**

**1** Bullsnake, **1a** Pine race, p. 373. **2** Glossy Snake, p. 376. **3** Milk
Snake, p. 367. **4** Common Kingsnake, p. 366. **5** Prairie Kingsnake, p. 369.
**6** Scarlet Kingsnake, p. 368. **7** Speckled Kingsnake, p. 367. **8** Scarlet
Snake, p. 366. **9** Copperhead, p. 381. **10** Eastern Coral Snake, p. 380.
**11** Cottonmouth, p. 382. **12** Western Rattlesnake, p. 384.
**13** Timber Rattlesnake, **13a** variant, p. 383. **14** Massasauga, p. 382.

## PLATE 70
## SALAMANDERS

1 Black-bellied Salamander, p. 403.  2 Pigmy Salamander, p. 401.
3 Spring Salamander, p. 412.  4 Grotto Salamander, p. 409.
5 Many-ribbed Salamander, p. 412.  6 Shovel-nosed Salamander, p. 404.
7 Many-lined Salamander, p. 409.  8 Dwarf Salamander, p. 411.
9 Ravine Salamander, p. 405.  10 Zigzag Salamander, p. 406.
11 Weller's Salamander, p. 405.  12 Black Mountain Salamander, p. 403.
13 Cheat Mountain Salamander, p. 407.  14 Wehrle's Salamander, p. 408.
15 Yonahlossee Salamander, p. 408.

**PLATE 71
SALAMANDERS**

**1** Red-backed Salamander, **1a** "Lead-backed" phase, p. 405. **2** Mountain
Dusky Salamander, p. 401. **3** Dusky Salamander, **3a** variant, p. 402
**4** Two-lined Salamander, p. 411. **5** Seal Salamander, p. 402. **6** Spring
Salamander, p. 412. **7** Slimy Salamander, p. 408. **8** Jordan's Woodland
Salamander, p. 407. **9** Green Salamander, p. 410. **10** Red Salamander,
**10a** old, p. 413. **11** Four-toed Salamander, p. 409. **12** Mud Salamander, p. 414.
**13** Cave Salamander, p. 410. **14** Long-tailed Salamander, p. 411.

# PLATE 72
# SALAMANDERS

1 Two-toed Amphiuma, p. 393. 2 Greater Siren, p. 392.
3 Mudpuppy, p. 395. 4 Hellbender, p. 394. 5 Newt, aquatic adult, 5a Eft,
terrestrial stage, p. 400. 6 Small-mouthed Salamander, p. 398.
7 Tiger Salamander, p. 396. 8 Mole Salamander, p. 399.
9 Spotted Salamander, p. 395. 10 Marbled Salamander, p. 396.
11 Jefferson Salamander, p. 397. 12 Ringed Salamander, p. 396.

PLATE 73
FROGS AND TOADS

**1** Plains Spadefoot, p. 418. **2** Eastern Spadefoot, p. 417.
**3** Southern Cricket Frog, p. 427. **4** Strecker's Chorus Frog, p. 430.
**5** Spotted Chorus Frog, p. 430. **6** Striped Chorus Frog, p. 428.
**7** Brimley's Chorus Frog, p. 429. **8** Mountain Chorus Frog, p. 429.
**9** Spring Peeper, p. 423. **10** Gray Treefrog, p. 425. **11** Pine Barrens
Treefrog, p. 424. **12** Barking Treefrog, p. 423. **13** Bird-voiced
Treefrog, p. 425. **14** Green Treefrog, p. 423. **15** Pine Woods Treefrog, p. 424.
**16** Squirrel Treefrog, p. 425.

## PLATE 74
## FROGS AND TOADS

**1** Western Spadefoot, p. 418. **2** Canadian Toad, p. 421. **3** Great Plains Narrow-mouthed Toad, p. 431. **4** Oak Toad, p. 420. **5** Southern Toad, p. 419. **6** Ornate Chorus Frog, p. 429. **7** Northern Leopard Frog, p. 435. **8** Southern Chorus Frog, p. 428. **9** Northern Cricket Frog, p. 426. **10** Plains Leopard Frog, p. 434. **11** Pig Frog, p. 433. **12** River Frog, p. 433.

**PLATE 75**
**FROGS AND TOADS**

**1** Pickerel Frog, p. 436. **2** Southern Leopard Frog, p. 435. **3** Bullfrog, p. 432.
**4** Green Frog, p. 432. **5** Wood Frog, p. 437. **6** Crayfish Frog, p. 434.
**7** Carpenter Frog, p. 434. **8** Woodhouse's Toad, p. 419. **9** American Toad, p. 419.
**10** Great Plains Toad, p. 420. **11** Red-spotted Toad, p. 422.
**12** Green Toad, p. 422. **13** Western Toad, p. 420.
**14** Eastern Narrow-mouthed Toad, p. 431. **15** Texas Toad, p. 421.

# PLATE 76
# SALTWATER FISHES

1 Northern Kingfish, p. 485.   2 Atlantic Silverside, p. 474.
3 Atlantic Mackerel, p. 494.   4 Spot, p. 485.   5 Redfish, p. 499.
6 Scup, p. 487.   7 Northern Puffer, p. 509.   8 Tautog, p. 489.
9 Haddock, p. 468.   10 Silver Hake, p. 470.   11 Pollack, p. 468.
12 Atlantic Cod, p. 467.   13 Menhaden, p. 463.   14 Tilefish, p. 477.
15 Atlantic Halibut, p. 505.

## PLATE 77
## OTHER SALTWATER FISHES

**1** White Marlin, p. 497. **2** Weakfish, p. 486. **3** Black Sea Bass, p. 476.
**4** Northern Sea Robin, p. 500. **5** Orange Filefish, p. 508.
**6** Atlantic Croaker, p. 484. **7** Striped Bass, p. 475. **8** Bluefish, p. 477.
**9** Dolphin, p. 482. **10** Swordfish, p. 498. **11** Atlantic Bonito, p. 496.
**12** Red Drum, p. 484. **13** Bluefin Tuna, p. 495. **14** Smooth Dogfish, p. 455.
**15** Summer Flounder, p. 503.

# PLATE 78
## SHARKS

**1** Bull Shark, p. 454. **2** Dusky Shark, p. 455. **3** Great Blue Shark, p. 454.
**4** Tiger Shark, p. 455. **5** Sandbar Shark, p. 455. **6** Smooth
Hammerhead, p. 456. **7** Spiny Dogfish, p. 456. **8** Thresher Shark, p. 454.
**9** White Shark, p. 453. **10** Porbeagle Shark, p. 452. **11** Shortfin Mako
Shark, p. 453. **12** Basking Shark, p. 453. **13** Sand Tiger, p. 451.
**14** Angel Shark, p. 457. **15** Greenland Shark, p. 457.

**PLATE 79**
**SKATES AND RAYS**

**1** Thorny Skate, p. 459. **2** Barn-door Skate, p. 458. **3** Little Skate, p. 458.
**4** Winter Skate, p. 459. **5** Clearnose Skate, p. 458. **6** Atlantic
Stingray, p. 460. **7** Atlantic Manta, p. 461. **8** Roughtail Stingray, p. 460.
**9** Smooth Butterfly Ray, p. 459. **10** Spotted Eagle Ray, p. 460. **11** Atlantic
Torpedo, p. 457. **12** Bullnose Ray, p. 460. **13** Cownose Ray, p. 461.

# PLATE 80 KILLIFISHES, PERCHLIKE FISHES, EELS, OTHERS

1 American Eel, 1a leptocephalus, p. 465. 2 Conger Eel, p. 466.
3 Mummichog, p. 473. 4 Striped Killifish, p. 473. 5 Cobia, p. 477.
6 Red Grouper, p. 476. 7 Lined Sea Horse, p. 475. 8 Sharksucker. p. 478.
9 Sheepshead Minnow, p. 473. 10 Oyster Toadfish, p. 470.
11 Halfbeak, p. 472. 12 Atlantic Needlefish, p. 472.

# PLATE 81
# JACKS

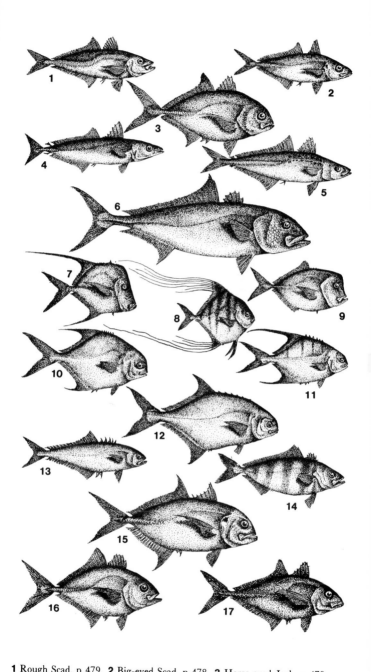

**1** Rough Scad, p. 479. **2** Big-eyed Scad, p. 478. **3** Horse-eyed Jack, p. 479.
**4** Mackerel Scad, p. 479. **5** Round Scad, p. 479. **6** Great Amberjack, p. 482.
**7** Lookdown, p. 480. **8** African Pompano, p. 480. **9** Atlantic
Moonfish, p. 480. **10** Permit, p. 481. **11** Palometa, p. 481. **12** Florida
Pompano, p. 481. **13** Leatherjacket, p. 481. **14** Banded Rudderfish, p. 482.
**15** Crevelle Jack, p. 480. **16** Yellow Jack, p. 479. **17** Blue Runner, p. 480.

# PLATE 82
# MACKERELS AND CROAKERS

1 Albacore, p. 496.  2 Skipjack Tuna, p. 495.  3 Spanish Mackerel, p. 493.
4 Cero, p. 493.  5 Chub Mackerel, p. 494.  6 Blue Marlin, p. 497.
7 King Mackerel, p. 494.  8 Little Tuna, p. 495.  9 Yellowfin Tuna, p. 496.
10 Red Goatfish, p. 488.  11 Banded Drum, p. 484.  12 Spotted
Sea Trout, p. 486.  13 Silver Perch, p. 484.  14 Black Drum, p. 486.

## PLATE 83
## PERCHLIKE FISHES

**1** Wrymouth, p. 492. **2** Atlantic Wolffish, p. 491. **3** Great
Barracuda, p. 489. **4** Striped Mullet, p. 490. **5** Northern Sennet, p. 489.
**6** White Mullet, p. 490. **7** Sheepshead, p. 487. **8** Pinfish, p. 487.
**9** Bermuda Chub, p. 488. **10** Cunner, p. 488.
**11** Atlantic Cutlassfish, p. 492. **12** American Sand Lance, p. 492.

## PLATE 84
## BOTTOM FISHES

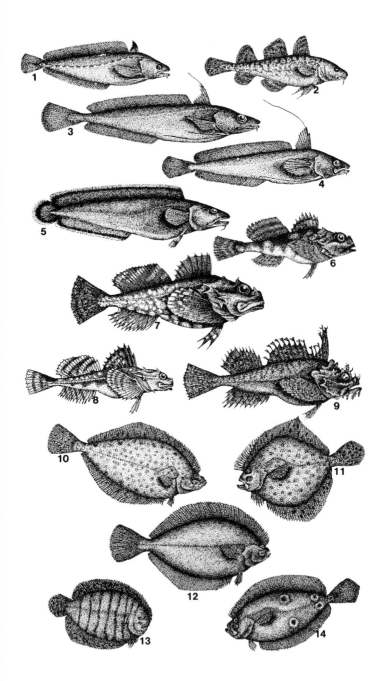

1 Spotted Hake, p. 469.  2 Atlantic Tomcod, p. 467.  3 White Hake, p. 469.
4 Red Hake, p. 469.  5 Cusk, p. 470.  6 Grubby, p. 501.  7 Shorthorn
Sculpin, p. 501.  8 Longhorn Sculpin, p. 501.  9 Sea Raven, p. 501.
10 Yellowtail Flounder, p. 506.  11 Windowpane, p. 503.  12 Winter
Flounder, p. 506.  13 Hogchoker, p. 507.  14 Fourspot Flounder, p. 503.

## PLATE 85
## HERRINGS AND SHADS

**1** Blueback Herring, p. 517. **2** Round Herring, p. 464.
**3** Atlantic Herring, p. 464. **4** Alewife, p. 518. **5** Spanish Sardine, p. 464.
**6** Skipjack Herring, p. 517. **7** American Shad, p. 518.
**8** Bay Anchovy, p. 465. **9** Gizzard Shad, p. 516. **10** Hickory Shad, p. 517.
**11** Striped Anchovy, p. 465. **12** Mooneye, p. 516.

## PLATE 86
## FRESHWATER FISHES

**1** Yellow Perch, p. 553. **2** Rainbow Darter, p. 555. **3** Shorthead
Redhorse, p. 531. **4** Northern Pike, p. 525. **5** Muskellunge, p. 526.
**6** Brown Trout, p. 519. **7** Brook Trout, p. 520. **8** Atlantic Salmon, p. 519.
**9** Lake Trout, p. 520. **10** Rainbow Trout, p. 520. **11** Walleye, p. 554.
**12** Freshwater Drum, p. 556. **13** White Bass, p. 547.
**14** Orange Spotted Sunfish, p. 552. **15** Lake Sturgeon, p. 513.

## PLATE 87
## FRESHWATER FISHES

**1** Chain Pickerel, p. 525. **2** White Sucker, p. 527. **3** Carp, p. 532.
**4** Bluegill, p. 551. **5** Pumpkinseed, p. 552. **6** Smallmouth Bass, p. 549.
**7** Rockbass, p. 548. **8** Yellow Bass, p. 547. **9** Black Crappie, p. 552.
**10** Southern Redbelly Dace, p. 534. **11** White Crappie, p. 553.
**12** Largemouth Bass, p. 549. **13** Black Bullhead, p. 540.
**14** Channel Catfish, p. 538. **15** Paddlefish, p. 513.

## PLATE 88
## STURGEONS AND GARS

1 Atlantic Sturgeon, p. 512. **2** Shortnose Sturgeon, p. 513.
3 Shovelnose Sturgeon, p. 512. **4** Pallid Sturgeon, p. 512.
5 Alligator Gar, p. 515. **6** Longnose Gar, p. 514, **7** Shortnose Gar, p. 515.
8 Spotted Gar, p. 514. **9** Bowfin, p. 514.

# PLATE 89
## HERRINGLIKE, CODLIKE, PERCHLIKE FISHES

**1** Burbot, p. 543. **2** Central Mudminnow, p. 524. **3** Troutperch, p. 542.
**4** Arctic Grayling, p. 523. **5** Pirate Perch, p. 543. **6** Rainbow
Smelt, p. 524. **7** Redfin Pickerel, p. 525. **8** Mottled Sculpin, p. 556.
**9** White Perch, p. 547. **10** Cutthroat Trout, p. 520. **11** Sauger, p. 553.
**12** Brook Silverside, p. 545. **13** Sockeye Salmon, **13a** breeding, p. 519.

## PLATE 90
## CHUBS AND SUCKERS

1 River Chub, p. 534. 2 Suckermouth Minnow, p. 536.
3 Spotted Sucker, p. 526. 4 Silver Chub, p. 535. 5 Creek Chub, p. 534.
6 Quillback, p. 529. 7 Fallfish, p. 533. 8 Lake Chubsucker, p. 527.
9 Black Buffalo, p. 529. 10 Mountain Sucker, p. 528.
11 Torrent Sucker, p. 531. 12 Northern Hog Sucker, p. 527.
13 Golden Redhorse, p. 532. 14 Goldfish, p. 533.

## PLATE 91
## DACE, SHINERS, MINNOWS

1 Silverjaw Minnow, p. 536. 2 Silvery Minnow, p. 536. 3 Bluntnose
Minnow, p. 537. 4 Satinfin Shiner, p. 538. 5 Spottail Shiner, p. 538.
6 Common Shiner, p. 538. 7 Brook Stickleback, p. 546. 8 Spotfin
Shiner, p. 538. 9 Redside Dace, p. 533. 10 Golden Shiner, p. 533.
11 Fathead Minnow, p. 536. 12 Emerald Shiner, p. 537.
13 Stoneroller, p. 537. 14 Blacknose Dace, p. 535. 15 Longnose Dace, p. 535.

# PLATE 92
# CATFISHES AND SUNFISHES

**1** Green Sunfish, p. 551. **2** Banded Sunfish, p. 548. **3** Warmouth, p. 551.
**4** Redbreast Sunfish, p. 550. **5** Blackbanded Sunfish, p. 548.
**6** Longear Sunfish, p. 550. **7** Stonecat, p. 541. **8** Margined
Madtom, p. 541. **9** Spotted Bass, p. 549. **10** White Catfish, p. 539.
**11** Tadpole Madtom, p. 541. **12** Blue Catfish, p. 539. **13** Yellow
Bullhead, p. 539. **14** Brown Bullhead, p. 540. **15** Flathead Catfish, p. 540.

## PLATE 93
## BLINDFISHES, KILLIFISHES, DARTERS

**1** Southern Cavefish, p. 542. **2** Northern Cavefish, p. 542.
**3** Starhead Topminnow, p. 544. **4** Blackstripe Topminnow, p. 545.
**5** Banded Killifish, p. 544. **6** Swampfish, p. 542. **7** Mosquitofish, p. 545.
**8** Johnny Darter, p. 555. **9** Blackside Darter, p. 554.
**10** Eastern Sand Darter, p. 554. **11** Greenside Darter, p. 555.
**12** Fantail Darter, p. 555. **13** Banded Darter, p. 555.
**14** Northern Swamp Darter, p. 555. **15** Logperch, p. 554.

# PLATE 94  LORICATES,
## KEYHOLE LIMPETS, LIMPETS, TOP SHELLS

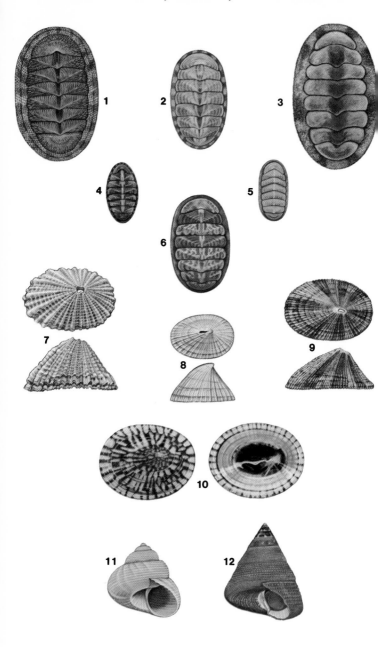

**PLATE 95**
**TURBANS, NERITES, PERIWINKLES, CHINK SHELLS**

**1** American Star Shell × 1, p. 570. **2** Long-spined Star Shell × 1, p. 569.
**3** Knobby Turban × 1, p. 569. **4** Bleeding Tooth × 1, p. 570.
**5** Zebra Nerite × 2, p. 571. **6** Tessellated Nerite × 1, p. 570.
**7** Variegated Nerite × 1, p. 570. **8** Rough Periwinkle × 2, p. 572.
**9** Smooth Periwinkle × 2, p. 571. **10** Salt Marsh Periwinkle × 1, p. 572.
**11** Knobby Periwinkle, p. 572. **12** Little Chink Shell × 2, p. 572.
**13** Common Periwinkle × 1, p. 571.

## PLATE 96 CERITHS, WENTLETRAPS, ALLIED FAMILIES

# PLATE 97
## BOAT SHELLS, MOON SHELLS, ALLIED FAMILIES

**1** Striate Cup-and-saucer × 1, p. 576. **2** Common Boat Shell × 1, p. 576.
**3** Little Boat Shell × 1, p. 577. **4** Flat Slipper Shell × 1, p. 576.
**5** Shark Eye × 1, p. 578. **6** Common Moon Shell × 1, p. 578.
**7** Miniature Natica × 2, p. 579. **8** Colorful Moon Shell × 1, p. 579.
**9** Coffee Bean Trivia × 1, p. 577. **10** Florida Fighting Conch × 1, p. 578.
**11** Common Baby's Ear × 1, p. 579. **12** Gray Cowry × 1, p. 580.
**13** Four-spotted Trivia × 2, p. 577. **14** Deer Cowry × 1, p. 579.

# PLATE 98
## TRITON, EGG, HELMET, FROG AND ROCK SHELLS

1 Flamingo Tongue × 1, p. 580. 2 Single-toothed Simnia × 1, p. 580.
3 Lace Murex × 1, p. 582. 4 Apple Murex × 1, p. 582.
5 Giant Eastern Murex × ½, p. 582. 6 Scotch Bonnet × 1, p 580.
7 White-mouthed Triton × 1, p. 581. 8 Hairy Triton × 1, p. 581.
9 Granular Frog Shell × 1, p. 581. 10 Florida Drill × 1, p. 583.
11 Oyster Drill × 1, p. 583. 12 Thick-lipped Drill × 1, p. 582.
13 Rock Purple × 1, p. 583. 14 Florida Rock Shell × 1, p. 583.

## PLATE 99
# WHELKS, DOG WHELKS, DOVE AND TULIP SHELLS

**1** Lightning Whelk × ⅓, p. 586. **2** Channeled Whelk × ⅓, p. 585.
**3** Knobbed Whelk × ⅓, p. 586. **4** Waved Whelk × ⅓, p. 584.
**5** Mud Snail × 1, p. 587. **6** Mottled Dove Shell × 1½, p. 584.
**7** New England Dog Whelk × 1, p. 587. **8** Greedy Dove Shell × 1½, p. 584.
**9** Mottled Dog Whelk × 1½, p. 586. **10** Brown-corded Neptune × ⅓, p. 585.
**11** Crown Conch × ⅓, p. 586. **12** Florida Banded Tulip × ⅓, p. 588.
**13** Tulip Shell × ⅓, p. 587. **14** Stimpson's Whelk × ⅓, p. 585.
**15** Florida Horse Conch × ¼, p. 588.

**PLATE 100**
**OLIVES AND ALLIED FAMILIES; BUBBLES, PYRAMS**

**1** Lettered Olive × ¾, p. 588. **2** Minute Dwarf Olive × 1½, p. 588.
**3** Rice Dwarf Olive × 1½, p. 589. **4** Atlantic Auger × 1, p. 591.
**5** Sallé's Auger × 1, p. 591. **6** Eastern Paper Bubble × 1½, p. 591.
**7** Common Nutmeg × 1, p. 589. **8** Beaded Miter × 1, p. 589. **9** Common
Atlantic Marginella × 1½, p. 590. **10** Interrupted Turbonille × 3, p. 592.
**11** Impressed Odostome × 3, p. 592. **12** Alphabet Cone × ¾, p. 590.
**13** Jasper Cone × 1, p. 590. **14** Junonia × 1, p. 589.

# PLATE 101
## TUSK SHELLS, AWNING, NUT AND ARK CLAMS

1 Comb Bittersweet × ¾, p. 597.  2 Atlantic Awning Clam × 1¼, p. 594.
3 Stimpson's Tusk × 1, p. 593.  4 Ivory Tusk × 1, p. 593.
5 File Yoldia × ¾, p. 595.  6 Pointed Nut Clam × 2, p. 595.
7 Atlantic Nut Clam × 2, p. 594.  8 Blood Ark × ½, p. 596.
9 Mossy Ark × ¾, 9a hinge, p. 595.  10 Transverse Ark × ¾, p. 596.
11 Heavy Ark × ¾, p. 596.  12 Turkey Wing × ¾, 12a hinge, p. 595.

# PLATE 102   MUSSELS,
## PEN SHELLS, TREE AND SPINY OYSTERS

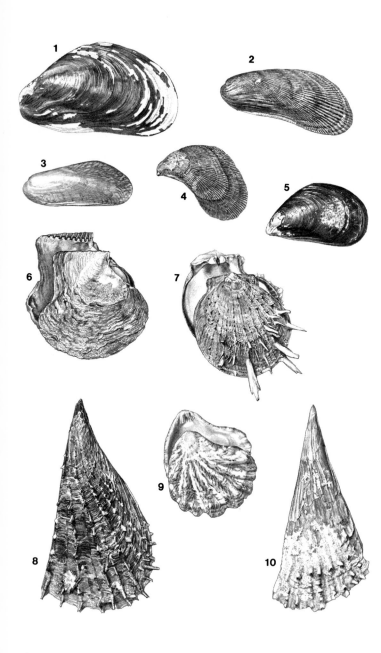

**1** Horse Mussel × ½, p. 598. **2** Ribbed Mussel × ½, p. 598.
**3** Paper Mussel × 1½, p. 597. **4** Hooked Mussel × ½, p. 597.
**5** Blue Mussel × ½, p. 597. **6** Flat Tree Oyster × ½, p. 599.
**7** Spiny Oyster × ½, p. 600. **8** Stiff Pen Shell × ⅓, p. 599.
**9** Kitten's Paw × 1, p. 601. **10** Flesh Pen Shell × 1, p. 598.

**PLATE 103**
## SCALLOPS, JINGLE SHELL, OYSTERS, LUCINES

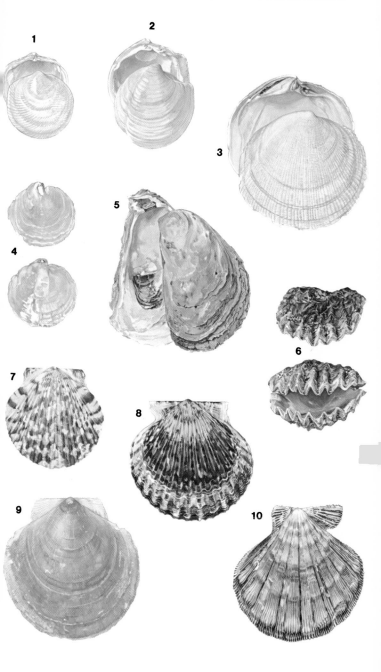

**1** Crosshatched Lucine × 1, p. 603. **2** Pennsylvania Lucine × ½, p. 603.
**3** Great White Lucine × ½, p. 603. **4** Common Jingle Shell × ½, p. 601.
**5** Virginia Oyster × ½, p. 602. **6** Coon Oyster × ½, p. 602.
**7** Calico Scallop × ½, p. 600. **8** Bay Scallop × ½, p. 600.
**9** Deep Sea Scallop × ⅓, p. 599. **10** Lion's Paw × ⅓, p. 600

## PLATE 104
## COCKLES AND OTHER CLAMS

**1** Boreal Astarte × ½, p. 604. **2** Broad-ribbed Cardita × ½, p. 602.
**3** Leafy Jewel Box × ½, p. 604. **4** Common Cockle × ½, p. 605.
**5** Great Heart Cockle × ½, p. 604. **6** Atlantic Surf Clam × ½, p. 611.
**7** Dwarf Surf Clam × 1, p. 611. **8** Ribbed Pod × ½, p. 607.
**9** Wedge Rangia × ½, p. 611. **10** Common Razor × ½, p. 607.

## PLATE 105
## TELLINS AND OTHER CLAMS

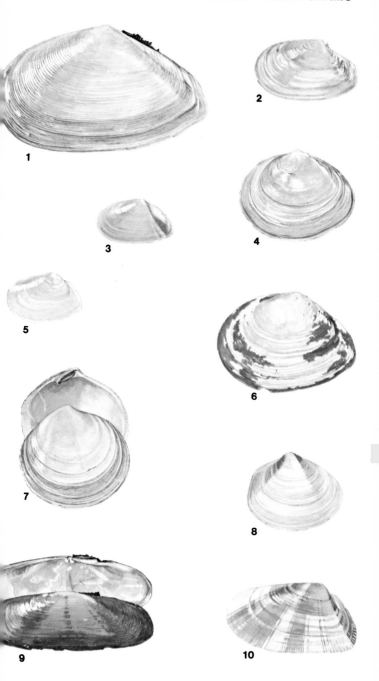

1 Lined Tellin × ¾, p. 607.  2 Tenta Macoma × 1½, p. 609.
3 Northern Tellin × 1½, p. 608.  4 Baltic Macoma × 1, p. 608.
5 DeKay's Tellin × 1½, p. 608.  6 Chalky Tellin × 1, p. 608.
7 White Semele × 1, p. 610.  8 False Coquina × 1½, p. 609.
9 Purplish Razor × 1, p. 610.  10 Southern Coquina × 1½, p. 609.

## PLATE 106
## VENUS CLAMS AND OTHER CLAMS

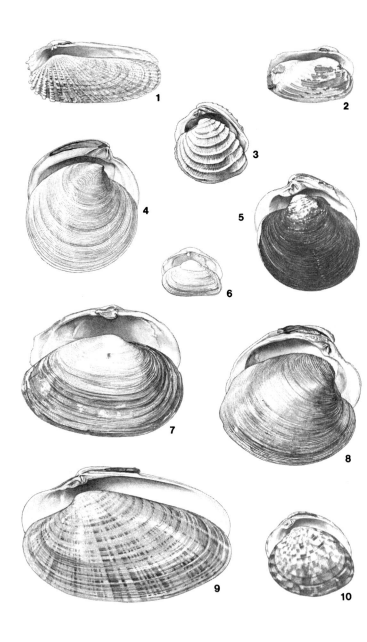

# PLATE 107
## PIDDOCKS AND OTHER CLAMS

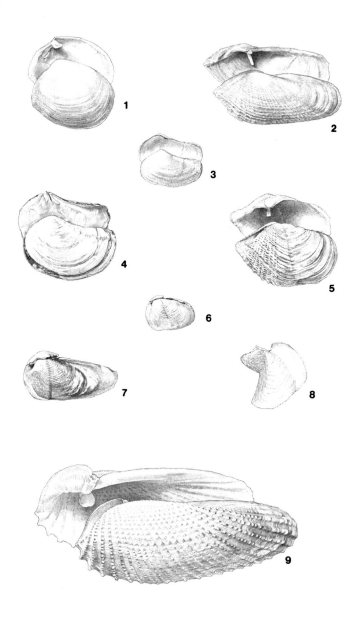

# PLATE 108
# CEPHALOPODS

**1** Atlantic Bob-tailed Squid × ⅜, p. 617. **2** Ram's Horn × ⅓, p. 617.
**3** Common Squid × 1, p. 618. **4** Sea Arrow × ¼, p. 618.
**5** Dwarf Squid × ¼, p. 618. **6** Paper Nautilus × ¼, p. 619.
**7** Atlantic Octopus × ¼, p. 619.

**PLATE 109**
**SEASHORE LIFE**

1 *Dactylometra quinquecirrha*, Sea Nettle, p. 630.
2 *Cyanea capillata* (Jellyfish), p. 631.
3 *Physalia physalis*, Portuguese Man-of-war, p. 630.
4 *Aurelia aurita*, Moon Jelly, p. 631. 5 *Tubularia crocea* (Hydroid), p. 628.
6 *Microciona prolifera*, Red Sponge, p. 627.
7 *Mellita quinquiesperforata*, Keyhole Urchin, p. 643. 8 *Arbacia punctulata*,
Purple Sea Urchin, p. 643. 9 *Henricia sanguinolenta* (Starfish), p. 640.
10 *Asterias forbesii* (Starfish), p. 641. 11 *Ophiothrix angulata*
(Brittle Star), p. 642. 12 *Ophioderma brevispinum*, Green Brittle Star, p. 642.
13 *Leptogorgia virgulata*, Whip Coral, 13a pod detail, p. 632.
14 *Astrangia danae* (Coral), p. 634. 15 *Orchestia platensis*, Beach Flea, p. 657.
16 *Pentamera pulcherrima* (Sea Cucumber), p. 645.

# PLATE 110
# SEASHORE LIFE

**1** *Pagurus pollicaris* (Hermit Crab), p. 661. **2** *Emerita talpoida*,
Mole Crab, p. 660. **3** *Ovalipes ocellatus*, Lady Crab, p. 663. **4** *Callinectes
sapidus*, Blue Crab, p. 663. **5** *Ocypode quadrata*, Ghost Crab, p. 664.
**6** *Uca pugilator*, Sand Fiddler Crab, p. 665. **7** *Cancer irroratus*,
Rock Crab, p. 663. **8** *Styela partita* (Sea Squirt), p. 667.
**9** *Lepas fascicularis*, Goose Barnacle, p. 653. **10** *Squilla empusa*,
Mantis Shrimp, p. 666. **11** *Crangon septemspinosa*, Sand Shrimp, p. 658.
**12** *Nereis virens*, Green Clam Worm, p. 647. **13** *Saccoglossus kowalevskii*
(Acorn Worm), p. 638. **14** *Diopatra cuprea* (Segmented Worm), p. 648.
**15** *Amphitrite ornata* (Segmented Worm), p. 649.

## RED-BACKED SALAMANDER
*Plethodon cinereus*                                    71:1

**Description**
Length, to 3⅝ in. (9.2 cm); record 5 in. (12.7 cm). Identified by straight-edged gray or red middorsal band and mottled belly with equal amounts of black and yellow or white pigment; no red below except occasionally between front legs. Body relatively stout; costal grooves 18–20. Legs short; intercostal folds between adpressed limbs 7–9. Two distinct color phases occur with variable frequency throughout range: *Red-backed* has dorsal band of some shade of red; sides brown, gray, or black. *Lead-backed* is uniformly dark gray or black above without a middorsal stripe, but with brassy red or white flecks. Males have a large gland under chin.

**Habitat**
A common species, found in and under rotted logs, in stumps, and among forest floor litter in relatively cool and mesic (moderately moist) coniferous and hardwood forests.

**Habits**
Active during above-freezing weather in Middle Atlantic States, but below ground at low elevations during midsummer heat.

**Range**
Ne. U.S. and se. Canada from N.S. s. through e. and n. N.C. and adj. border of Tenn., w. through Great Lakes to w.-cen. Minn., e. Ill., and extr. n. Ky.

## WELLER'S SALAMANDER
*Plethodon welleri*                                    70:11

**Description**
Length, to 3⅛ in. (7.9 cm). Small; has irregular dorsal band of silvery or golden blotches over black dorsal ground color. Sides sometimes with light spots. Throat grayish-white; belly dark slate with scattered small white spots. Costal grooves 16; 3–4 intercostal folds between adpressed limbs. Small gland under chin in males.

**Habitat**
Under logs, stones, or talus in forests, usually between 2300–5700 ft. (701.0–1737.4 m).

**Range**
Extr. nw. N.C. and extr. ne. Tenn., n. into s. Va.

## RAVINE SALAMANDER
*Plethodon richmondi*                                    70:9

**Description**
Length, to 5⅝ in. (14.3 cm). Distinguished by wide dorsal band and large number of costal grooves (20–22). Legs small, thin. Dorsally flecked with white and bronze to form a brown band, bordered by bluish-black; lower sides and cheeks spotted with white. Belly darker with fine, light mottling, as in Lead-backed Salamander. Often red pigment on cheeks, front legs, and anterior sides.

**Habitat**
Slopes of valleys and ravines.

**Range**
W. Pa., s. to nw. N.C. and ne. Tenn., and w. to se. Ind. In the southeast, range separated from Valley and Ridge Salamander by New R. in Va. and W.Va.

## ZIGZAG SALAMANDER
*Plethodon dorsalis*                                        70:10

**Description**
Length, to 4⅜ in. (11.1 cm). Similar to Red-backed Salamander, but red dorsal band has wavy black edges. Occasionally edges are straight, but then band on body is narrower than in Red-backed. Stripe sometimes absent, but flecks of red pigment in its place. Red often present on sides. Dorsally, brassy to silvery flecks to small white spots; ventrally mottled with equal amounts of white, red, and black pigment. Band down tail with straight edges. Slender body, head and legs. Normally 18 costal grooves; 6–7 intercostal folds between adpressed limbs. In males gland under chin small, round or oval.

**Habitat**
Often close to a moist retreat in a rock formation, escarpments, talus slopes, under logs and stones in woods. In and near mouths of caves, mostly under 1000 ft. (304.8 m) elevation.

**Habits**
Rarely found in summer at low elevations.

**Range**
Cen. Ind. and s. Ill., s. to nw. Ga. and n. Ala. Disjunct localities in s. Ohio, N.C., S.C., Ala., Miss., and the Ozark Plateau in ne. Ark. and adj. states.

## VALLEY AND RIDGE SALAMANDER
*Plethodon hoffmani*                                        *Fig. 89*

**Description**
Length, to 5⅜ in. (13.7 cm). Has unusually high number of costal grooves (20–21, usually 21); 11 intercostal folds between adpressed limbs. Gray mottled venter with red pigment confined to narrow area; light chin. Rarely has striped phase. Usually has brassy flecks dorsally. Feet slightly webbed.

**Habitat**
Wooded slopes.

**Range**
Valley and Ridge physiographic province of cen. Pa., s. through Md. and along Va.-W. Va. border to New R.

Valley and Ridge Salamander

*Fig. 89*

White-spotted Salamander

## JORDAN'S WOODLAND SALAMANDER
*Plethodon jordani*                                    **71:8**

**Description**
Length, to 5 in. (12.7 cm); record 7¼ in. (18.4 cm). Large, black, usually without dorsal spots. Dark brown to black ground color with lighter belly and chin. Some populations have red on cheeks and legs, lateral white or yellow spots, or brassy dorsal flecks; also red dorsal spots in young. Costal grooves 16, sometimes 15–17; 1–3 intercostal folds between adpressed limbs.

**Similarities**
Slimy Salamander has dorsal white spots, but variants of each species may look confusingly similar even to experts.

**Habitat**
Under rotten logs, bark, and leaf litter; sometimes under stones in heavily wooded areas.

**Remarks**
In southern part of range this species may hybridize with Slimy Salamander in areas of overlap.

**Range**
Normally above 3000 ft. (914.4 m) elevation in s. Appalachian Mts. from sw. Va. through w. N.C. and e. Tenn. to nw. S.C. and extr. ne. Ga.

## CHEAT MOUNTAIN SALAMANDER
*Plethodon nettingi*                                   **70:13**

**Description**
Length, to $4\frac{13}{16}$ in. (12.2 cm). Dark, flecked above with light, brassy red or yellow in a bandlike pattern. Below slate-gray to purplish-black with slight light spotting. Lighter under throat. Costal grooves 18–19, 7 intercostal folds between adpressed limbs.

**Similarities**
Weller's Salamander has 2–3 fewer costal grooves, shorter tail. Ravine Salamander has longer tail, 2–4 more costal grooves.

**Habitat**
Under rocks and logs in forests and in north-facing talus slopes, above about 3000 ft. (914.4 m).

**Range**
Three disjunct populations, each given subspecific status, at high elevations: e.-cen. W.Va.; the Peaks of Otter region of cen. Va.; and in Shenandoah National Park in nw. Va.

## WHITE-SPOTTED SALAMANDER
*Plethodon punctatus*                                  *Fig. 89*

**Description**
Length, to $6\frac{13}{16}$ in. (17.3 cm). Many large spots on the back; usually with 18, sometimes 17, costal grooves. Large dorsal spots like lateral yellow-white or yellow spots in coloration. Throat pale.

**Similarities**
Slimy Salamander has no yellow dorsal spots. Wehrle's Salamander has dorsal white spots smaller and less numerous and has red or brassy pigment on dorsum.

**Habitat**
High elevations, usually over 2800 ft. (853.4 m).

**Range**
Shenandoah and North Mts. along nw. Va.-W.Va. border.

## WEHRLE'S SALAMANDER
*Plethodon wehrlei*                                    **70:14**

**Description**
Length, to 6⁹⁄₁₆ in. (16.0 cm). Somewhat resembles Slimy
Salamander, but has light throat and light spots on lower sides
only. Above dark brown or black, sometimes with whitish specks or
paired red disklike spots. Lower sides bluish-brown with white
markings. Belly light gray or gray mottled. Throat light. Costal
grooves usually 17; 2–3 intercostal folds between adpressed limbs.
**Habitat**
Under rocks or stones in talus.
**Range**
Sw. N.Y., s. to sw. Va. and into N.C., w. to se. Ohio.

## SLIMY SALAMANDER
*Plethodon glutinosus*                                 **71:7**

**Description**
Length, to 8⅛ in. (20.6 cm). Shiny black above, dark slate below,
and usually with varying amounts of white, yellow, or brassy spots
on sides and back. Throat dark or light slate, somewhat mottled.
Spotting above may be on sides, head, back, tail, and limbs; or
individuals may lack spotting altogether. Costal grooves 16; 1–3
intercostal folds between adpressed limbs. Body cylindrical, legs long.
**Similarities**
Blue-spotted Salamander lacks nasolabial groove, has bluish-white
flecks, and rounded, rather than compressed, tail.
**Habitat**
Under large logs or stones in wooded areas, in crevices of rocks,
and in caves, under moist humus.
**Habits**
Nocturnal; burrows deeply in dry season; active throughout year in
the South during wet weather, but rarely seen in winter in the
North. Secretes a sticky substance from skin glands.
**Remarks**
See under Jordan's Woodland Salamander.
**Range**
S. N.Y. and extr. w. Conn., s. through N.J. and n. Md. to cen.
Fla., w. to Miss. R. in s. Ill. to La., and across Miss. R. in s. Mo.
to nw. Ark. and e. Okla. Disjunct localities in s. N.H. and adj.
Mass., n. Ind., cen. La., and cen. Tex.

## YONAHLOSSEE SALAMANDER
*Plethodon yonahlossee*                                **70:15**

**Description**
Length, to 7½ in. (19.1 cm). Dark gray, with bright chestnut,
sometimes in a band, dorsally from head to base of tail. Head
large; sides of head, body, and tail blotched with white. Venter
black with some white spots, especially on sides. Dark gray hind
legs, tail, and sometimes above lateral white spots. Costal grooves
16; 0–1 intercostal fold between toe-tips of adpressed limbs.
**Habitat**
Under logs or stones on wooded slopes between and in nearby
grassland.
**Range**
Blue Ridge Mts., in w. N.C., extr. ne. Tenn., and into s. Va. at
elevations from 2700–5700 ft. (823.0–1737.4 m) in the n., as low
as 1400 ft. (426.7 m) in se., part of range.

# FOUR-TOED, MANY-LINED, GROTTO, AND GREEN SALAMANDERS

Genera *Hemidactylium*, *Stereochilus*, *Typhlotriton*, and *Aneides*

## FOUR-TOED SALAMANDER
*Hemidactylium scutatum*                    **71:11**

**Description**
Length, to 4 in. (10.2 cm). Four toes on hind feet and a strong constriction at base of keeled tail. Head, body, and tail mottled reddish-brown, lower sides grayish; belly china-white with dark dots, especially near sides. Body short, costal grooves 13–14.

**Habitat**
Under logs, stones, leaf litter, or moss in wooded or open areas near sphagnum ponds, swamps, or bogs; larval life spent in pools or bogs.

**Habits**
Nocturnal, terrestrial.

**Remarks**
Constriction on tail marks a zone for tail breakage. In nature a predator may be fooled into eating the broken tail, which continues to wriggle, and in the process ignore the fleeing salamander.

**Range**
Continuous distribution from s. Maine through n. Ga. and Ala., w. through s. Ont. and Mich. to Wis. and w. Ky. to extr. ne. Miss. Spotty occurrence in N.S., n.-cen. Fla., La., and e.-cen. Mo.

## GROTTO SALAMANDER
*Typhlotriton spelaeus*                    **70:4**
                                            *Fig. 90*

**Description**
Length, to 5⅟₁₆ in. (13.5 cm). Adults, pale pinkish and blind. No gills; head flattened and wide; degenerate eyes appear as black spots. Body color grayish-lavender above, with dashed dorsal and lateral markings; below white. Costal grooves 16–19. Larvae darker, with functional eyes and external gills.

**Habitat**
In and near caves and underground limestone passages; larvae in cold open streams and springs.

**Range**
Sw. Mo. and nw. Ark. w. to extr. se. Kans. and ne. Okla.

Grotto Salamander                              larva

*Fig. 90*

Oklahoma Salamander, p. 412                     larva

## MANY-LINED SALAMANDER
*Stereochilus marginatus*
                                            **70:7**

**Description**
Length, to 4½ in. (11.4 cm). Recognized by narrow lines of yellow and brown along sides. Head small and pointed, with small sensory pores, remnants of lateral line system. Neck indistinct. Tail

409

flattened sideways with prominent dorsal keel. Dull yellow above, middorsal line dark; belly yellow with a few dark spots. On each side a dark line from the nostril runs onto tail with three lines below. Lines on tail have a meshlike appearance. Intercostal folds between adpressed limbs 8–9.

**Habitat**
Slow streams and pools in pine forests and in cypress and gum swamps; under dead leaves, rubbish, and logs in damp places and in sphagnum moss in the water.

**Range**
Se. Va. at least to s. Ga. in Coastal Plain.

## GREEN SALAMANDER
*Aneides aeneus*                                                    71:9

**Description**
Length, to 5½ in. (14.0 cm). This is the only truly green salamander. Expanded toe tips and the yellowish-green lichenlike blotches on the black back, tail, and limbs will also distinguish this beautiful salamander. Underparts pale. Depressed body; costal grooves 14–15. Limbs quite long and overlap by 1–3 intercostal folds. Tail long and round.

**Habitat**
Damp, but not wet, crevices in shaded rock outcrops and beneath rocks, bark, or logs in forests, up to 4400 ft. (1341.1 m) in N. Carolina. Rarely encountered.

**Reproduction**
Female broods eggs and young.

**Range**
Allegheny and Cumberland Mts. from sw. Pa. and extr. s. Ohio to n. Ala. and ne. Miss. Also in Blue Ridge Mts. in N.C., nw. S.C., and extr. ne. Ga.

# BROOK SALAMANDERS
Genus *Eurycea*

These salamanders are generally aquatic, but may be found near water in moist woodlands. They have slender bodies and moderate to long tails that are flattened sideways. Ground color generally yellow or orange. All have aquatic larvae.

## CAVE SALAMANDER
*Eurycea lucifuga*                                                  71:13

**Description**
Length, to 7⅛ in. (18.1 cm). Slender, reddish-orange, with long, whiplike tail (greater than ½ body length). Body cylindrical; legs long. Ground color varies from yellow to bright orange-red. Unspotted venter yellowish; black spots irregularly scattered over dorsum, sides, and tail, but forming no distinct pattern such as side bars on tail of Long-tailed Salamander. Eyes well developed. Costal grooves 14–15. Males have raised margins around cloacal opening at breeding time.

**Habitat**
In lighter regions of caves, limestone crevices around springs, under logs, stones, in damp places outside caves.

**Range**
Limestone areas from n. Va. to n. Ga., w. through s. Ind. and extr. ne. Miss.; in s. Ill. and w. to Ozark Plateau in Mo. and ne. Okla.

## TWO-LINED SALAMANDER
*Eurycea bislineata*                                              **71:4**

**Description**
Length, to 4¾ in. (12.1 cm). Light dorsal band bordered by two
black lines that continue onto tail as a row of spots; varies from
yellow to yellow-orange and has middle line of dark spots,
sometimes continuous. Above yellow to brownish or greenish; belly
always brilliant yellow. Well-developed eyes. Tail long,
compressed.
**Habitat**
In or near brooks and seepages, in moist soil, under stones, logs.
**Habits**
Aquatic, nocturnal; forages on land in rain.
**Food**
Insects and snails found at stream's edge.
**Range**
Se. Canada, excluding N.S., s. to n. Fla., and w. below Lake Erie
to e. Ill., e. of Miss. R., and s. to La.

## LONG-TAILED SALAMANDER
*Eurycea longicauda*                                             **71:14**

**Description**
Length, to 7¾ in. (19.7 cm). Yellowish ground color with small
dark spots over yellowish to orange dorsum and continuing onto
upper surfaces of limbs. Belly pale yellow to white, unspotted. Tail
very long, almost ½ again length of body, with dark vertical bars
on both sides. Eyes well developed. Costal grooves 14.
**Habitat**
In and under rotting logs, under stones near edges of streams, in
underground caves and crevices of rocks.
**Habits**
During breeding season males develop swollen snouts with long
projections called cirri.
**Range**
S. N.Y., s. through most of Ga. and Fla. panhandle, w. through s.
Ind. and La.; also n. in Ill. along Miss. R. and w. of river through
Ozark Plateau to s. Mo. and e. Okla.

Note: A subspecies, the well-named **DARK-SIDED SALAMANDER,**
*Eurycea longicauda melanopleura*, occurs in the Ozarks and north
in Illinois along the Mississippi River.

## DWARF SALAMANDER
*Eurycea quadridigitata*                                          **70:8**

**Description**
Length, to 3⁹⁄₁₆ in. (9.1 cm). Yellow dorsal color with dark side
stripe from snout to tip of tail, and 4 toes on hind foot distinguish
this small, slender species. Well-developed eyes; tail squarish in
cross section. Costal grooves 14–17.
**Habitat**
Under logs and debris in low, damp places, especially low pine
flatwoods.
**Range**
Atlantic Coastal Plain from se. quarter of N.C., s. to Lake
Okeechobee in Fla., and w. to e. Tex. Also in nw. S.C. and nw.
Ark. and in Mo.

## MANY-RIBBED SALAMANDER
*Eurycea multiplicata*                                    **70:5**

**Description**
Length, to 3 9/16 (9.1 cm). Some individuals are neotenic and may become larger; they are often pale gray stippled with darker pigment. Costal grooves 19–20. Body elongate. Broad brown median stripe often bordered by dark line on each side; occasionally a narrow dark vertebral line also present; lower sides have silvery flecks. Below pale gray to lemon-yellow.

**Habitat**
Under logs, gravel, stones, debris in streams, in springs, both in the open and in caves.

**Habits**
Mainly aquatic; neoteny associated with caves and cave streams.

**Range**
Extr. se. Kans. and in Ozark and Ouachita Mts. in sw. Mo., most of nw. Ark., and e. Okla.

## OKLAHOMA SALAMANDER
*Eurycea tynerensis*                                      *Fig. 90*

**Description**
Length, to 3⅛ in. (7.9 cm). Dorsum grayish with reticulate pattern of cream and black; pale, weakly pigmented body. Costal grooves 19–21.

**Habitat**
Loose gravel in small, cold, spring-fed streams.

**Habits**
Aquatic.

**Reproduction**
Eggs laid in fall or late spring.

**Remarks**
A neotenic species that retains gills throughout life.

**Range**
In drainage of Grand R. and Ill. R. in sw. Mo., nw. Ark., and adj. e. Okla.

# SPRING, RED, AND MUD SALAMANDERS
Genera *Gyrinophilus* and *Pseudotriton*

Members of this group of large salamanders are found mainly near or in water. The Red and Mud Salamanders are reddish with black spots. The Spring and West Virginia Cave Salamanders are duller and darker and have a light line from eye to nostril.

## SPRING SALAMANDER
*Gyrinophilus porphyriticus*                          **70:3, 71:6**

**Description**
Length, to 8⅝ in. (21.9 cm). Large, yellowish-brown to purplish-brown, with flesh-colored belly and a light line from eye to nostril bordered below by a darker one. Dorsal colors lack luster; markings variable, from dark mottlings to spots that may be in a row on the back and sides. Body stout and rounded. Head broad, flattened, snout rounded. Tail keeled above.

**Habitat**
Hill or mountain country in cool springs or streams, under logs, large stones, in moist areas away from streams; caves.

**Habits**
Nocturnal, will leave streams on rainy nights to forage nearby.
**Food**
Both terrestrial and aquatic insects; larvae, which reach up to 4 in. (10.2 cm) in length, will eat other salamander larvae, and immatures will eat Two-lined Salamander.
**Range**
Sw. Maine and s. Que., s. through Appalachians to Fall Line in S.C., Ga., and Ala., w. through N.Y. and Pa. to s. Ohio and cen. Tenn.

Note: The **TENNESSEE CAVE SALAMANDER**, *Gyrinophilus palleucus*, a neotenic cave dweller, is found only in caves of central and southeastern Tennessee, northern Alabama, and northwestern Georgia.

# WEST VIRGINIA CAVE SALAMANDER
*Gyrinophilus subterraneus*

**Description**
Length, to 4½ in. (11.4 cm). (Larvae attain size of adults.) Similar to Spring Salamander but restricted in range, with smaller eyes and 2–3 rows of pale spots along sides. Brownish-pink, darker dorsally, and with reticulated pattern over back and sides; light pink below with red gills. Head wider behind eyes; heavy-bodied. Transformed adults may lose pattern and have thinner head and body. Costal grooves 17, 7 between toe tips of adpressed limbs.
**Habitat**
Mud banks and stream in the one cave where they occur.
**Food**
Cave invertebrates.
**Range**
Known only from General Davis Cave, Greenbrier Co., W. Va.

# RED SALAMANDER
*Pseudotriton ruber*

**71:10**

**Description**
Length, to 7⅛ in. (18.1 cm). Red, with many irregular black spots; color darkens with age. Tail thick at base, keeled above. Newly transformed, it is bright coral red with black spots and plain pink belly. With age, it becomes purplish-brown above with dark spots fusing to give mottled appearance; belly becomes dull orange and develops black spots.
**Similarities**
Distinguished from somewhat larger Mud Salamander by generally redder color; brassy, gold, or silver-flecked iris; and irregular, more fused dark spots on back.
**Habitat**
In spring-fed brooks and in crevices and burrows in loose, moist soil nearby. Also, in more terrestrial situations, under logs, boards, and stones.
**Habits**
Less aquatic than Mud Salamander. As a self-protective device may mimic in coloration red eft, which is distasteful to predators.
**Range**
S. N.Y., s. to Ga. and Fla. panhandle, w. to Miss. R., and ne. to s. Ohio and Ind.

# MUD SALAMANDER
*Pseudotriton montanus*                                    71:12

**Description**
Length, to 6 in. (15.2 cm), record 8⅛ in. (20.6 cm). Similar to Red
Salamander, but brownish above. Iris dark; snout short, rounded.
Body stout; legs short, stout. Tail keeled above at tip. Above some
shade of brown or reddish with darker spots; sides and belly
salmon to flesh color, immaculate or flecked with round dark
markings. Old animals become darker with spots less prominent.

**Habitat**
In or near muddy springs, in wet lowlands under logs, bark.
Larvae aquatic. Adults usually beneath objects in water or along
stream banks.

**Habits**
These salamanders mimic distasteful red efts.

**Range**
Coastal Plain e. of Miss. R. from s. N.J. s. to n. Fla., and w. to
La.; in s. Ohio, w. W.Va. through e. and cen. Ky. and e. Tenn.
Also in disjunct localities in Pa., n. Ala., and cen. Miss.

# Frogs and Toads

Order Anura

Adult frogs and toads are distinguished from other amphibians by having short, squat bodies; four limbs, the hind limbs longer and more powerful than the forelimbs; and no tail. The use of the words "toad" and "frog" is often confusing. A true toad is a member of the Family Bufonidae, Genus *Bufo,* while a frog is any anuran; hence, a toad is also a frog. However, common names for several other groups incorporate the word "toad." Toads are recognizable by their warty skin, whereas most other frogs have relatively smooth skin. Toads also often have large parotoid glands at the shoulder.

The greatest diversity of amphibians occurs in the tropics. In the eastern United States there are seven families, fifteen genera, and sixty species, forty-seven of which are covered here. The total for all frogs is well over 2500 species divided into sixteen families, although authors will differ in their arrangements and numbers.

## Identification

Identification can be made with certainty for some forms only with the specimen in hand. Frogs and toads of the same species often vary widely in color depending on the environment, temperature, and humidity (the cooler and moister the environment, the darker the color). Note size or shape, the presence or absence of webbing or enlarged pads on the feet, exposed and hidden markings, or the presence of glands. Females, as a rule, are larger than males and may have a greatly distended belly full of eggs in the breeding season. Males often have dark throats and special clasping pads on their thumbs, fingers, forelegs, or chest. Sizes given reflect the *record adult sizes* and the length is measured from the *tip of the snout to the vent.* When record sizes are appreciably larger than the average maximum, such lengths are given separately.

*Fig. 91*

Parts of a Typical Frog and Toad

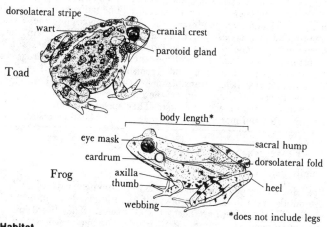

## Habitat

Look for frogs in wet areas, under cover or in water. They are easily caught by hand but a net may assist in water. Some anurans, especially toads, are found during the day in mountainous, forested, or prairie situations some distance from water.

## Vocalization

Vocalizations in frogs are usually limited to males, but in some species the female may utter a nonmating call, or both sexes may be silent. In addition to other visual, olfactory, and tactile cues, the mating call is a major means of mate and species recognition. Vocalizations may be divided into the following categories: mating calls, in males, often in a chorus heard over miles; release calls, given by a male or unreceptive female when clasped by a breeding male; a territorial call, which may not necessarily be associated with breeding activity; and various calls to indicate distress or give a warning. The cessation of calling may also be a signal.

## Food

Adults eat insects and other invertebrates, and bullfrogs are known to eat anything that moves and is of a suitable size, even the young bullfrogs. Tadpoles are usually vegetarian. Major changes in metamorphosis from tadpole to adult are in the shortening of the digestive tract and the disappearance of larval mouthparts.

## Reproduction

Eastern species generally reproduce by laying eggs in aquatic sites, although within the order Anura egg laying on land and live-bearing are known. The Greenhouse Frog, introduced in Florida, lays eggs that hatch into miniature adults under moist vegetation.

Frogs are somewhat less formal than salamanders when they encounter a potential mate, and the major device that prevents a mixing of species is the mating call. At the breeding grounds the males mount and clasp the females, holding this position of amplexus from a few minutes to over a day, at the end of which the male fertilizes the eggs as they are extruded from the female. Eggs are laid under water or on the surface, depending on the species. They may be laid singly, in long strings or cables, in globular masses, or in flat films. They may float or sink or be attached to vegetation. Individual clusters range from a few to hundreds; the total number laid by one female may go as high as 20,000 in the Bullfrog.

Eggs hatch into tadpoles in a short period, developmental rates often being dependent on the temperature. The tadpoles have horny beaks and often toothlike structures in rows around the mouth, and they breathe through gills covered by a flap of skin, the operculum. In general, most toad tadpoles are small and black and transform at very small sizes; at the other extreme, the true frogs have tadpoles that are larger and greenish, and which may live more than a year before transforming. At transformation they develop lungs, four limbs appear, the tail is resorbed, and structures related to feeding change. They spend less time in deeper water and more time on the edges of ponds and streams.

# SPADEFOOT TOADS
Family Pelobatidae

Spadefoots have generally smoother skin than other toads, vertical eye pupils, teeth in the upper jaw, and parotid glands small or absent. The hind feet have horny black "spades." They breed in temporary pools after rains, and their offspring may develop through egg and larval stages in as few as twelve days.

# EASTERN SPADEFOOT
*Scaphiopus holbrooki*

**73:2**
*Figs. 92, 93*

**Description**
Length, to 2⅞ in. (7.3 cm). The only spadefoot throughout the East. Has small, round parotoid glands. Above brownish, blackish, greenish, or yellowish, usually with 2 light stripes on the back forming hourglass pattern; other lines often continuing to the sides; sometimes with red spots. Below whitish.

**Habitat**
Deciduous or coastal pine forest, in sandy or light soils.

**Habits**
Digs burrows and is seen at surface only at night when breeding or foraging.

**Voice**
An explosive, harsh *ker-r-aw,* short in duration and repeated at intervals.

**Reproduction**
Breeds in temporary pools after heavy rains.

**Range**
Mass. and se. N.Y., throughout most of Fla., and w. to cen. Okla. and Tex., with large gap separating eastern and western populations in Miss. Flood Plain; also absent in higher areas in s. Appalachians.

*Fig. 92*

American
Toad, p. 419

Southern
Toad, p. 419

Woodhouse's
Toad, p. 419

Oak
Toad, p. 420

Great Plains
Toad, p. 420

Canadian
Toad, p. 421

Green
Toad, p. 422

Red-spotted
Toad, p. 422

Texas Toad,
p. 421

Plains Spadefoot,
p. 418

Eastern Spadefoot

Western
Spadefoot, p. 418

Eastern
Spadefoot

Plains
Spadefoot, p. 418

## PLAINS SPADEFOOT
*Scaphiopus bombifrons*

**73:1**
*Figs. 92, 93*

**Description**
Length, to 2½ in. (6.4 cm). Has protruding bump between eyes and a wedge-shaped black spade on each hindfoot. Lacks parotoid glands. Above light- to olive-gray with irregular dark spots; below whitish. Area between eyelids narrow. Male has dusky throat.

**Habitat**
Sandy soil of arid regions and prairies.

**Habits**
Burrows in sand and forages at night; seen mainly in breeding season after heavy rains.

**Voice**
Harsh squawking repeated about once or twice per sec.

**Reproduction**
Breeds in temporary pools.

**Range**
Great Plains as far e. as cen. Mo., from sw. Man. s. to cen. Okla. and e. Ariz. in the s. U.S. and into adj. Mexico. A disjunct population along Gulf of Mexico from extr. s. Tex. s.

Note: The **COUCH'S SPADEFOOT,** *Scaphiopus couchii,* which occurs in eastern Texas and southwestern Oklahoma, is distinguished by its sickle- rather than wedge-shaped spade.

## WESTERN SPADEFOOT
*Scaphiopus hammondi*

**74:1**
*Fig. 93*

**Description**
Length, to 2½ in. (6.4 cm). Distinguished by short, wedge-shaped black spade on hind foot and lack of boss between eyes. Dorsal color variable, tan or brown, sometimes with olive or green, and with small, light spots with dark edges. Usually several broad, light-colored bands running irregularly down back. Venter light with dark markings. Width of eyelid usually greater than distance between eyes.

**Habitat**
Lowland and foothill streams and temporary ponds in areas of loose soil.

**Voice**
Likened to purr of a cat, a *waa* lasting just under 1 sec., repeated about 80 times per min.

**Range**
Sw. U.S. from w. Okla. and cen. Tex., into sw. U.S., cen. mainland Mexico, and Baja Calif.

# TRUE TOADS
Family Bufonidae

This is an almost cosmopolitan family of thirteen genera, with only the largest, Genus *Bufo,* occurring in North America. They have warty skin, large parotoid glands, horizontal pupils, and often prominent and species-characteristic cranial crests. Teeth are lacking in the upper jaw. Eggs are laid in strings in the water. Identification of young toads is often complicated because cranial crests develop quite some time after metamorphosis, and because the newly transformed young are so small. Breeding males have a dark or discolored throat.

## AMERICAN TOAD
*Bufo americanus*

**75:9**
*Fig. 92*

**Description**
Length, to 4⅜ in. (11.1 cm). Has bony crests behind eye not touching parotoid gland or connected to it only by small spur; dark-spotted belly, and 1–2 warts in each large, dark dorsal spot. Color variable from gray to brown or reddish with dark spots on cream-colored belly. Large dark spots enclose up to 2 warts on the back; a light line frequently runs down back.

**Similarities**
Woodhouse's Toad also has light dorsal line, but spots on back enclose 3 or more warts.

**Habitat**
A common species, found in gardens, cultivated fields, and in wooded and open areas.

**Habits**
Active at dusk and into the night.

**Voice**
A high musical trill lasting 10–30 sec., about an octave lower than that of Southern Toad.

**Reproduction**
Breeds in shallow streams or ponds.

**Range**
E. Canada to e. U.S. from s. Labrador s. to n. Ga., w. to e. Man., and s. through n. La., e. Okla., and adj. w. Tex.

## SOUTHERN TOAD
*Bufo terrestris*

**74:5**
*Fig. 92*

**Description**
Length, to 3 in. (7.6 cm); record 4⁷⁄₁₆ in. (11.3 cm). Cranial crests form prominent knobs behind eyes to middle of head and do not touch parotoid gland. Belly unmarked or with a few spots near the arm. Coloration dorsally variable: gray, brown, or red, for example.

**Habitat**
Most common toad in the South, abundant in loose soil.

**Voice**
High-pitched trill, about an octave higher than and with a pulse rate twice as large as that of American Toad; call duration 8 sec. or less.

**Reproduction**
Breeds in late spring in small permanent ponds, woodland ponds, or flooded depressions.

**Range**
Coastal Plain from se. Va. through Fla. to La.

## WOODHOUSE'S TOAD
*Bufo woodhousei*

**75:8**
*Fig. 92*

**Description**
Length, 2–4 in. (5.1–10.2 cm). Distinguished by prominent bony crests that are connected with elongate parotoid glands behind eyes; by light belly with no or only one spot; and by conspicuous white stripe down back and dark dorsal spots, usually with 3 or more warts (may have fewer than 3 warts in populations from the Dakotas west). Above yellow-brown, greenish, grayish, or blackish. Below yellowish, occasionally with a breast spot. Parotoid glands smooth, narrow, 1½ times length of eyelid.

**Habitat**
Gardens, fields, woods, deserts; valleys in sandy, lowland areas.
**Habits**
Active day and night.
**Voice**
Suggests remotely the low *baa* of a sheep; 1–2½ sec. in length, repeated at 5–13-sec. intervals.
**Reproduction**
Breeds later in the year than American Toad, in shallow, quiet water.
**Range**
Much of e. and cen. U.S. from s. N.H. and se. N.Y. to n. S.C. and n. and w. Ga., w. to lower Mich., Ill., and the Great Plains; along Gulf Coastal Plain from extr. w. Fla. to s. Tex.; into much of w. U.S. and n.-cen. Mexico.

Note: The **WESTERN TOAD,** *Bufo boreas* (**75:13**), may occasionally wander east within the range of Woodhouse's Toad.

## OAK TOAD
*Bufo quercicus*

**74:4**
*Fig. 92*

**Description**
Length, to 1⁵⁄₁₆ in. (3.3 cm). Small, with prominent light line from tip of snout down middle of back. Brownish ground color with light vertebral line dissecting two blotches; white patch between legs ventrally that ranges from cream to orange in color. Dorsum evenly warty; cranial crests weak in back; parotoid glands round to oval and not parallel.
**Habitat**
Roadside ditches, ponds, pools in areas of sandy soil.
**Habits**
Generally adapted to digging and rarely encountered.
**Voice**
High-pitched *peeps,* usually heard in chorus; calls night and day in warm, rainy weather. Vocal sac sausage-shaped when inflated.
**Range**
Atlantic Coastal Plain from se. Va. through Fla., and w. to Miss. R. in La.

## GREAT PLAINS TOAD
*Bufo cognatus*

**75:10**
*Fig. 92*

**Description**
Length, to 4½ in. (11.4 cm). Has symmetrically placed rows of *large* dark spots on back, distinctly outlined with cream or yellow. Small, oval parotoid glands, cranial crests united to form knob between eyes, and paired, dark, distinctly outlined spots distinguish this species. Body large, broad; above brownish-yellow, greenish, or grayish, sometimes with narrow light stripe down back. Numerous small warts in each blotch. Below light, unmarked. Inner tubercle on foot long, blackened, and with free cutting edge; toes webbed.
**Habitat**
Desert and long- and short-grass prairies; farmlands; irrigation ditches.
**Habits**
Nocturnal but sometimes active by day.
**Voice**
Sustained, low-pitched trill in chorus, likened to sound of a riveting machine. Vocal sac sausage-shaped when inflated.

**Reproduction**
Up to 20,000 eggs laid in long strings, in early spring.
**Range**
Great Plains, extending e. in cen. Mo., from w. Minn. s. to cen.
Okla. and nw. Tex., n. from cen. Mont. into extr. s. Canada, and
through much of sw. U.S.; also into w. and cen. Mexico.

## CANADIAN TOAD                                      74:2
*Bufo hemiophrys*                                     Fig. 92

**Description**
Length, to 3¼ in. (8.3 cm). Large boss between eyes from rear of
eyelid to snout. Dorsally has brownish or greenish ovoid spots each
containing 2–5 warts on either side of a cream to yellow midline
stripe. Underparts buffy with small dark spots, a prominent one on
lower throat. Snout short.
**Habitat**
Streams and pond margins, usually near mud flats or sandy
beaches.
**Habits**
Breeds in May and calls into the summer.
**Voice**
A low trill similar to that of American Toad, but lower in pitch
and shorter, given about twice a min.
**Other name**
Dakota Toad.
**Range**
Nw. Minn., ne. S.Dak., and e. and n. N.Dak., w. to Mont., and n.
to the District of Mackenzie.

## TEXAS TOAD                                         75:15
*Bufo speciosus*                                      Fig. 92

**Description**
Length, to 3½ in. (8.9 cm). Two dark, sharp-edged "spades" on
foot, the inner one sickle-shaped. Dorsum greenish-gray, reddish,
or yellowish, with or without olive spots. Lacks light vertebral
stripe. Below yellowish or white and unspotted. Head ridges weak
or missing. Faint cranial crests. Parotoid glands small, oval, and
separated by 1½–2 times their own width. Male has buffy throat
with olive center.
**Habitat**
Arid, semiarid, or cultivated areas, mesquite woodlands, short-grass
plains.
**Voice**
Like call of Great Plains Toad, with a loud, shrill ½–1-sec. trill,
but higher pitched and with a faster trill rate. Vocal sac sausage-
shaped when inflated.
**Reproduction**
Breeds after rains in temporary or permanent pools.
**Range**
W. Okla. and adj. extr. s. Kan., s. through most of Tex., se.
N.Mex., and into n. Mexico.

Note: The **GULF COAST TOAD,** *Bufo valliceps,* is distinguished by
its broad, dark lateral stripe. This toad occurs along the Gulf Coast
east to Mississippi.

## GREEN TOAD
*Bufo debilis*

75:12<br>Fig. 92

**Description**
Length, to 2⅛ in. (5.4 cm). Small; has short legs and a greatly enlarged, kidney-shaped or elongate gland behind eye with indistinct cranial crests near eye. Above greenish or yellowish with many black spots and yellow warts. Below pale with pinkish or bluish cast, and may be spotted. Head narrow, body flat. Neck glands oblique on shoulders and almost as long as head.
**Habitat**
Short-grass prairie at higher elevations in dry areas.
**Habits**
Nocturnal, seldom seen except after rains.
**Voice**
Cricketlike; steady low trill, 3–7 sec. with interval of 5–9 sec.
**Reproduction**
Breeds in rain pools and ditches in spring and summer; eggs laid singly or in short strings.
**Range**
Sw. Kans. to se. Ariz., s. through cen. and w. Tex. to Gulf Coast and extr. ne. Mexico.

## RED-SPOTTED TOAD
*Bufo punctatus*

75:11<br>Fig. 92

**Description**
Length, to 3 in. (7.6 cm). Only eastern toad with *round* parotoid glands, and they are smaller than the eye. Has poorly defined cranial crests. Above gray, brown, or reddish with rusty warts, and without vertebral stripe. Underparts light, plain or with some small spots. Head and body flat; eyes widely spaced.
**Habitat**
Prairies and deserts, in rocky areas near water.
**Habits**
Nocturnal, active after heavy rainfall.
**Voice**
Clear, high-pitched, pleasing birdlike trill lasting 4–10 sec. with almost equal intervals between trills.
**Reproduction**
Eggs laid singly and adhere to objects in water.
**Range**
Sw. U.S. from w. Kans., w. Okla., and cen. Tex., w. to Calif., and s. through much of Mexico.

# TREEFROGS AND GRASS FROGS
Family Hylidae

North American members of this family are small, thin, and narrow-waisted. They have no neck glands or cranial crests. The tips of the digits are frequently enlarged into adhesive pads that enable these frogs to cling to smooth surfaces, even upside down. They are often found in bushes and trees. The species described all breed in the spring or summer, except the Spring Peeper, which breeds at the end of winter. In the South some species are active year-round.

## SPRING PEEPER
*Hyla crucifer*                                                    **73:9**

**Description**
Length, to 1⅜ in. (3.5 cm). The only small treefrog in the North.
Can be identified by the distinct, but often incomplete, dark X on
its back. Above variably greenish-gray to reddish-brown; dark band
between eyes and an irregular stripe from snout through eye and
tympanum; underparts light. Skin smooth, snout pointed. Toe disks
small but distinct. Male has brown throat.

**Habitat**
Marshes, swamps, pools, on low bushes and grasses in wooded
areas.

**Habits**
Nocturnal, but often calls during day; hibernates under leaves.

**Voice**
A musical series of ½-sec. *peeps* like a high, shrill whistle, rising
slightly at the end; often heard in a metallic chorus in early spring
in the North. Calls from ground at edge of, and from elevated
position in, tussocks or bushes.

**Reproduction**
Early breeder, peaks February–March.

**Range**
E. U.S. and se. Canada, but absent from s. half of peninsular Fla.;
w. to se. Man. in n. of range and e. Tex. in s. Introduced in Cuba.

## GREEN TREEFROG
*Hyla cinerea*                                                    **73:14**

**Description**
Length, to 2½ in. (6.4 cm). The only uniformly green treefrog,
with or without a light lateral stripe, often dark-bordered.
Individuals of the small Squirrel Treefrog may be confusingly
similar, however. Above bright green, often with tiny yellow dots;
below white or yellow. Skin smooth, head pointed.

**Habitat**
Males call from emergent vegetation in open, permanent bodies of
water; pools, edges of waterways.

**Voice**
Likened to a cowbell, a call of *quonk* or *quank,* repeated 30–60
times per min.

**Reproduction**
500–1000 eggs laid, adhering to floating vegetation.

**Range**
Atlantic Coastal Plain of se. U.S. from Delmarva Pen. s. through
Fla., and w. to cen. Tex.; also n. to s. Ill. and into se. Okla.;
disjunct populations in cen. Ky. and Tenn. Introduced in Puerto
Rico.

## BARKING TREEFROG
*Hyla gratiosa*                                                    **73:12**

**Description**
Length, to 2¾ in. (7.0 cm). The largest native treefrog. Has evenly
granular skin, a white stripe on upper jaw, and large disks on all
digits. Above dark olive, gray, or green, usually with conspicuous
round dark spots. Below creamy; groin and armpit yellow. Head
short, broad. Prominent fold on breast; hands large, toes webbed.

**Habitat**
Trees of hummocks and in pine barrens, in small permanent and
temporary bodies of water.

**Habits**
Arboreal, but also burrows.
**Voice**
Calls from water in a floating position, deep, low *boonk* or *moonk*
repeated 30–60 times per min.; also a "bark" call from position in
trees.
**Remarks**
Skin has strong-smelling secretion.
**Range**
Atlantic Coastal Plain from se. Va., s. through n. and cen. Fla.,
and w. to Miss. R. in La. Disjunct populations in s. N.J.
(introduced), Ky., Tenn., n. Ala., and nw. Ga.

## PINE BARRENS TREEFROG
*Hyla andersoni*                                                73:11

**Description**
Length, to 2 in. (5.1 cm). Has uniform green back with a light-
bordered plum-colored stripe on side and concealed bright orange
markings on legs. Below pale with yellow spots. Body stout, skin
smooth; tympanum about ⅓ size of eye. Toes partly webbed; toe
pads small. Males have dark throats.
**Similarities**
Green Treefrog is larger and more slender.
**Habitat**
Specialized habitat of acid hillside seepage bogs, shrub bogs, and
white cedar swamps; apparently needs acidic water conditions for
development of larvae.
**Habits**
Hardly to be found except during breeding season, May–July in
the North and April–September in Florida.
**Voice**
Calls from trees and bushes, a loud *quaak, quaak, quaak . . .*
lasting about 20 sec.
**Range**
Separate populations below Fall Line in s. N.J. Pine Barrens; in
se. N.C. and adj. S.C.; in e. Ga.; and in nw. Fla.

## PINE WOODS TREEFROG
*Hyla femoralis*                                                73:15

**Description**
Length, to 1¾ in. (4.4 cm). Conspicuous yellow spots on back of
thigh. Above reddish-brown, but sometimes greenish-gray, with
dark star on back; dark triangle between eyes. Below whitish,
throat sometimes dark. Back granular, belly rough. Head short,
snout rounded. Body slender, toe disks well developed.
**Habitat**
Pine woods.
**Habits**
Lives in tops of pine trees, but calls from bushes or small trees over
water from few inches above surface to head height.
**Voice**
Succession of harsh, high notes likened to "an amateur playing
with a telegraph key" (Conant).
**Reproduction**
Breeds in temporary pools or cypress ponds following heavy rains.
**Range**
Atlantic Coastal Plain from se. Va., s. through n. and cen. Fla.,
and w. to Miss. R. in La. Records from cen. Ala.

## BIRD-VOICED TREEFROG
*Hyla avivoca*                                                        73:13

**Description**
Length, to 2⅙ in. (5.5 cm). Resembles a small, slender Gray
Treefrog, but is more slender and has pale yellow or white
coloration in groin, and smoother skin. Above brownish, grayish, or
green with variable dark markings on the back; limbs barred.
Black-bordered light spot below eye, dark bar across head. Below
whitish, groin yellowish-green. Skin fairly smooth above, but
granular below.
**Habitat**
Flooded wooded swamps, river valleys, and lake shores.
**Voice**
Calls from vines or bushes at about waist to head height, in
repeated, piping, birdlike whistles.
**Range**
E. Miss. Valley from extr. s. Ill. and w. Ky., s. to Fla. panhandle
and La.; crosses Miss. R. in cen. La. Disjunct populations cen. Ga.
to adj. S.C. Also a record in Okla.

## GRAY TREEFROG
*Hyla versicolor* and *Hyla chrysoscelis*                            73:10

**Description**
Length, to 2⅜ in. (6 cm). Moderate-sized, with rough skin and a
distinct white spot below eye; usually a large dark blotch on back.
Above light brown to gray or green (can change color quickly), the
green phase usually lacking the large dark blotch on back; legs
usually banded; below light, rear of thigh and under hind parts
bright orange-yellow. Skin rough, warty; head short.
**Habitat**
Ubiquitous in permanent and temporary water; found on trees,
bushes, fences, vines in and at edges of woodlands.
**Habits**
Nocturnal, but may call on humid days; arboreal and acrobatic.
**Voice**
Calls in early evening from trees or bushes, but closer to water's
edge later in evening; a melodious, uniform trill, long in *versicolor*
and short in *chrysoscelis,* repeated.
**Reproduction**
Breeds for a short time in early spring.
**Range**
Most of e. U.S. and extr. se. Canada, from s. Maine s. to n. Fla.,
w. to s. Man. and e. Tex.

## SQUIRREL TREEFROG
*Hyla squirella*                                                     73:16

**Description**
Length, to 1⅝ in. (4.1 cm). Colors vary and can change rapidly. A
difficult species to identify. Above green, brownish, or pearly gray
with or without spots. Dark band between eyes, white line on
upper lip, light line from below eye to shoulder; underparts light
with no color on concealed surfaces of legs. Skin smooth above.
Head short; snout rounded.
**Habitat**
Breeds in temporary and permanent water in a variety of
situations: around buildings, gardens, fields, open woods.

**Habits**
Seen mainly during breeding season; calls by day during rains; occasionally transported outside its natural range while hidden on cultivated plants.
**Voice**
Calls from edge of shallow water or from emergent twigs or stems; harsh, but muffled, *aak, aak, aak* repeated once or twice per sec.
**Range**
Coastal Plain from se. Va., s. through Fla., w. along Gulf Coast to se. Tex. Disjunct populations in extr. se. Okla., e.-cen. Miss., and cen. Ala. Introduced in the Bahamas.

## LITTLE GRASS FROG
*Limnaoedus ocularis*                                    *Fig. 94*

**Description**
Length, to $^{11}\!/_{16}$ in. (1.8 cm). The smallest North American frog. Has dark stripe through eye and down side; triangle between eyes continuing as a line down back; also variable lines on each side. Color varies: brown, green, reddish.
**Habitat**
Moist areas around ponds.
**Habits**
Climbs only to low vegetation, breeds year-round in the South, and winter through summer in the North.
**Voice**
High-pitched tinkling, repeated.
**Range**
Coastal Plain from se. Va. through most of Fla. with western boundary extr. se. Ala. and the cen. panhandle.

*Fig. 94*

Little Grass Frog

# CRICKET AND CHORUS FROGS
Genera *Acris* and *Pseudacris*

Cricket frogs, Genus *Acris,* are not arboreal and lack expanded digit tips, but have long toes and considerable webbing for their semiaquatic habits at the water's edge. They have smooth skin below. Active both night and day as weather permits, they breed fairly late in the season.

Chorus frogs, Genus *Pseudacris,* are small with pointed snouts, reduced disks, and little webbing. They usually have a dark stripe from nostril through eye and a pale stripe above the upper lip. The skin is smooth above and granular below. Chorus frogs breed in winter or early spring.

## NORTHERN CRICKET FROG                          **74:9**
*Acris crepitans*                                        *Fig. 95*

**Description**
Length, ⅝–1⅜ in. (1.6–3.5 cm). Dark triangle between eyes. Uppermost black lengthwise stripe on rear of thigh irregular,

bordered above and below by white stripe. Body stout with rounded snout. Uniformly gray or brown, or with green, brown, yellow, or reddish markings bordering a stripe extending from triangle down middle of back; belly white. Alternating light and dark bars on upper lip and light stripe from eye to angle of jaw. Dorsum warty. Chin of males may be spotted and yellowish during breeding season.

### Similarities
Southern Cricket Frog has dorsum less warty, toes less webbed, thigh stripes with smooth edges.

### Habitat
Edges of small, shallow streams and ponds.

### Habits
Forages mostly on land; some individuals active well into fall.

### Voice
Long series of rapid, sharp notes *kick, kick, kick, kick,* like striking two stones together; calls from rim of pond in shallow water or on floating vegetation.

### Food
Terrestrial arthropods such as spiders and insects.

### Reproduction
Breeds April to July around lakes and ponds, marshes, small temporary bodies of water.

### Remarks
Enters water when frightened.

### Range
E. and cen. U.S. except New England and Appalachian region, and generally absent from Coastal Plain from N.C. s. through Fla. Northern boundaries in se. N.Y., lower Mich., s. Wis., and se. S.Dak., s. through extr. w. Fla. to w. Tex. and into extr. n. Mexico. Occurs primarily at low elevations.

Fig. 95

Southern Cricket Frog    Northern Cricket Frog

## SOUTHERN CRICKET FROG
*Acris gryllus*

73:3
Fig. 95

### Description
Length, to 1¼ in. (3.2 cm). Has clean-edged dark stripe (2 stripes in Florida) on rear of thigh. Above with variable markings; dark brown or black triangle between eyes. Body slender with pointed snout; 1st toe not fully webbed, disks on digits very small and indistinct.

### Similarities
Northern Cricket Frog has rough-edged thigh stripe, toes more webbed, dorsum wartier.

### Habitat
Grassy edges of small streams, ponds, and swamps.

### Voice
Short, explosive, rasping notes in a series, similar to the Northern Cricket Frog.

### Range
Coastal Plain from se. Va., s. through Fla., and w. to Miss. R. from extr. sw. Tenn. to La.

## STRIPED CHORUS FROG
*Pseudacris triseriata*                                   73:6

**Description**
Length, to 1½ in. (3.8 cm). Five dark stripes, sometimes not
continuous, down speckled or unicolor back and sides; frequently a
dark triangle between eyes with apex meeting the median dorsal
stripe. Dorsal ground color usually light grayish-brown to sooty-
gray (but sometimes olive to dark green) with darker stripes. A
dark gray or black stripe through eyes. Belly white, sometimes
lightly spotted. Dark band present on, at most, ½ dorsal surface of
lower leg. Body slender, snout round. Male has yellow vocal sac.
**Similarities**
Southern Chorus Frog also slender, but less robust.
**Habitat**
Temporary pools, flooded meadows, and roadside ditches in
wooded areas.
**Voice**
Calls only in spring, from grassy pond margins to more open areas,
may call day and night on cloudy days; a rising series of short
trills, *prreeep-preep . . . preep,* resembling the sound produced by
running thumbnail over teeth of a comb, somewhat resembles
Spring Peeper, but notes more trilled.
**Remarks**
Populations of this species may hybridize freely with Southern
Chorus Frog west of Alabama, and individuals with intermediate
characters will occur.
**Range**
Widely distributed in e. and cen. U.S., from n. N.J. s. to e. Tex.,
but absent from most of Coastal Plain. In n. and w. N.Y. and extr.
s. Que. and s. Ont. through Ohio, Ind., and w. to Northwest
Territories, Idaho, Utah, and Ariz.

## SOUTHERN CHORUS FROG
*Pseudacris nigrita*                                      74:8

**Description**
Length, to 1¼ in. (3.2 cm). White stripe on lip (but in peninsular
Florida black and white spots). Dorsal pattern of 5 longitudinal
black stripes, sometimes broken into spots; dark bands on lower leg
are broad with narrow bands of light in between. Head narrow,
snout pointed.
**Similarities**
Striped Chorus Frog has well-defined triangle between eyes and
stripes not as dark.
**Habitat**
Water in pine flatwoods and more open lowland areas; pools and
swamps; also lives in sandy soils.
**Voice**
Calls from grassy pond margins, sometimes concealed; call similar
to Chorus Frog but with a slower repetition rate.
**Remarks**
See under Chorus Frog.
**Range**
Coastal Plain from se. N.C., s. through Fla. and w. to se. Miss.

# ORNATE CHORUS FROG
*Pseudacris ornata*                                              74:6

**Description**
Length, to 1⅞₆ in. (3.7 cm). Broad dark stripe from tip of snout
through eye, interrupted at shoulder. Dark spot anterior to and
above groin, which is washed with yellow or orange. Usually with
2 wide, irregular stripes down back and a sharply defined stripe,
or often spots, on each side. Dark triangle between eyes. Ground
color varies from green to the usual reddish-brown. Heavy-bodied
with short legs.
**Habitat**
Ponds, roadside ditches, and flooded fields in well-drained areas.
**Voice**
Shrill, metallic *peep-peep* repeated up to 80 times per min., like
Spring Peeper, but without rise at the end. Male sits in clumps of
grass or on floating debris.
**Reproduction**
Very early breeder, December–March.
**Range**
Coastal Plain from se. N.C., s. to cen. Fla., and w. to extr. e. La.

# BRIMLEY'S CHORUS FROG
*Pseudacris brimleyi*                                            73:7

**Description**
Length, to 1¼ in. (3.2 cm). Distinguished by a dark line on side
from snout through eye to groin, by 3 broad, clean-edged stripes
down back and sides, and by dark spots prominent on chest. Above
yellowish- or reddish-brown with small warts. Middorsal stripe
brown; light line from upper jaw to shoulder; dark stripes
lengthwise on legs. Below yellow with some dark spots. Snout long,
round; no triangle between eyes. Disks on digits slightly dilated.
**Habitat**
Wooded pools, swamps, along rivers.
**Voice**
A short, rasping trill.
**Range**
Coastal Plain from se. Va to e. Ga.

# MOUNTAIN CHORUS FROG
*Pseudacris brachyphona*                                         73:8

**Description**
Length, to 1½ in. (3.8 cm). Dark triangle between eyes is not
white-edged behind. Usually has a pair of dorsolateral half-moons
on its back with ends curving outward to sides; sometimes absent
or replaced by dark flecks. Above gray or brown, with no
middorsal stripe or spots; dark bands down leg. Skin somewhat
rough.
**Habitat**
In hilly wooded, or partially wooded, areas; swampy areas along
small streams with abundant vegetation; and in seepage or roadside
ditches.
**Voice**
Calls day and night in rasping, fast notes; high-pitched, continuous
*wrrink* in chorus audible for ¼ mile (0.4 km).
**Range**
S. Appalachians from sw. Pa., s. to nw. Ga. and n. Ala. Disjunct
populations in e. W.Va. and extr. ne. Miss.

## STRECKER'S CHORUS FROG
*Pseudacris streckeri*           **73:4**

**Description**
Length, to 1⅞ in. (4.8 cm). Body stout, squat; coloration variable, but usually has a dark spot below eye and V-shaped mark between eyes. Above brownish to gray or green; dark stripe from snout to shoulder and then irregularly along body. Legs banded; below whitish to buffy. Head short, stout; tympanum small. Toes only slightly webbed and with small disks. Male has dark throat.

**Habitat**
Prairies, pastures, swampy areas, and flatwoods.

**Habits**
Usually nocturnal.

**Voice**
Shrill, clear, and with carrying power; sounds like a bell, sometimes likened to an ungreased wooden wheel. Sound in chorus resembles that of distant flock of geese.

**Reproduction**
Breeds in winter in the South and in early spring in southwestern Illinois and adjacent areas; in ditches, sloughs, and small ponds.

**Range**
E. and cen. Okla., s. to Gulf Coast, rarely into ne. Ark. and ne. La.; disjunct populations in cen. Ill.; and in vicinity of se. Mo. and adj. Ill. and Ark.

## SPOTTED CHORUS FROG
*Pseudacris clarki*           **73:5**

**Description**
Length, to 1¼ in. (3.2 cm). Dark-edged green spots irregularly placed, sometimes forming 3 broken stripes over lighter dorsal ground color of grayish or olive, and a light line on upper lip. Below white. Sometimes dark triangle, not bordered by white, between eyes. Dark band from nostril through eye and ear, widening down sides; legs barred above. Skin smooth dorsally, body slender. Snout projects beyond lower jaws. Males have dark throats.

**Habitat**
Open prairie grasslands and marshes at edge of woodlands.

**Habits**
Nocturnal, secretive, shy.

**Voice**
Males call from clumps of grass in shallow temporary water; repeated raucous and grating low-pitched notes at the rate of 2 notes per sec. with a rest between notes equal to note length.

**Range**
S.-cen. Kans., s. through cen. Tex. and extr. ne. Mexico. Record from Mont.

# NARROW-MOUTHED TOADS
Family Microhylidae

This family is widespread in the Southern Hemisphere, but in North America is limited to two genera and three species, and is primarily found in the South. Ant eaters, they have a small mouth, unwebbed toes, and a narrow, pointed snout; a fold is generally seen across the head behind the eyes.

## EASTERN NARROW-MOUTHED TOAD
*Gastrophryne carolinensis*                                    **75:14**

**Description**
Length, to 1½ in. (3.8 cm). Has narrow, pointed head, white belly
covered with dark mottling, and dorsum some shade of brown to
reddish-brown, sometimes with irregular dark markings. Broad
light stripes may be present dorsolaterally. Fold of skin across head
behind eyes; toes unwebbed. Males generally smaller, with vocal
pouch, blackish throat, and a gland on chest.

**Habits**
Active from early April to September or early October.

**Voice**
½-2½ sec. in duration, a nasal buzz, like bleating of a lamb, loud
in chorus; typically no *peep*.

**Food**
Mostly ants, also other insects.

**Reproduction**
Breeds in clear temporary bodies of water and around the shallows
of lakes, flood plains; lays up to 850 eggs in a mass on water's
surface. May hybridize with Great Plains Narrow-Mouthed Toad.

**Range**
Se. U.S. from s. Md. s. to Key West, Fla. w. through s. Ky., s. Ill.,
and s. Mo. and s. to e. Tex. Records in Iowa and s. Tex.
Introduced in the Bahamas.

## GREAT PLAINS NARROW-MOUTHED TOAD
*Gastrophryne olivacea*                                       **74:3**

**Description**
Length, to 1⅝ in. (4.1 cm). Has olive, gray, or tan dorsum,
generally plain, but may be marked with black dots or small
streaks. Belly unspotted, white. Throat, chest, and lower sides may
be marked. Fold of skin across head behind eyes; toes unwebbed.

**Similarities**
Eastern Narrow-mouthed Toad has brown to reddish-brown
dorsum with dark markings and belly with dark mottling.

**Habitat**
Dry rocky upland areas in open woods or woodland edges, also
river flood plains and cultivated fields.

**Habits**
Nocturnally active during humid or rainy weather; may be active
by day during rains. Hides during day under surface rocks or in
rodent burrows.

**Voice**
A *peep*, then a nasal buzz, like the bleating of a goat, 1–4 sec. in
duration.

**Reproduction**
Breeds May–August.

**Remarks**
May hybridize with Eastern Narrow-mouthed Toad.

**Range**
Extr. w. Mo. and se. Nebr. through e. and cen. Tex. to s. Ariz.
and in Mexico s. along both Atlantic and Pacific Coasts of
mainland.

# TRUE FROGS
Family Ranidae

Members of this family are generally widespread and successful, absent only from Antarctica. Typically they occupy semiaquatic situations. Although the family is diverse in the Old World, only the Genus *Rana* is found in North America. These frogs have long legs with broadly webbed hind feet and a large distinct eardrum or tympanum. They lack cranial crests, parotoid glands, and disks on the tips of their digits, and their skin is usually quite smooth. Usually a prominent dorsolateral fold is present down each side of the back. In some species the folds may be discontinuous, and the last half inch or so may be displaced toward the midline. Females lay a large number of eggs.

The Leopard Frogs have confounded scientists for decades. Because of their wide distribution and overall similarity, differences found in physiological tolerances, external appearance, and mating call among animals from various parts of the country were ascribed to variations within one species. The true story, and the story is by no means yet complete, is that several closely related species exist, and they may be found in the same breeding sites where their ranges overlap, with little or no hybridization. The species of the eastern United States, at least, are probably well-enough studied that new forms will not be found.

## GREEN FROG
*Rana clamitans*                                           **75:4**

**Description**
Length, to 4 in. (10.2 cm). Has dorsolateral folds not reaching groin. Above green to brownish, sometimes spotted with black; legs faintly barred; below white, occasionally with spots. Male has yellow throat and larger eardrums; female has white throat with spots.
**Similarities**
Resembles Bullfrog, but is smaller.
**Habitat**
Swamps, brooks, streams, ponds, and edges of lakes.
**Habits**
Solitary; hibernates in mud.
**Voice**
Low-pitched, like sound of banjo or double bass; throat swollen when calling.
**Range**
Se. Canada and e. U.S. except s. Fla. and cen. Ill., w. to extr. se. Man., cen. Minn., and e. Tex.

## BULLFROG
*Rana catesbeiana*                                         **75:3**

**Description**
Length, to 8 in. (20.3 cm). The largest North American frog. Lacks dorsolateral folds and has white belly, often mottled. Above green (especially forward), brown, or black with irregular dark spots; below whitish, often mottled; hind legs with dark crossbars. Skin fairly smooth; ridge running above, but stopping just behind, tympanum. Tympanum as large as eye in female, much larger in male.

**Habitat**
Permanent bodies of water: marshes, ponds, lakes, rivers.
**Habits**
Lives near deep water, hibernates in soft mud on bottom of ponds
and lakes; larvae live at least a year before metamorphosis.
**Voice**
Deep, hoarse, *jug o'rum, more rum* resounds throughout breeding
ponds.
**Food**
Invertebrates; also any vertebrates—fish, other frogs, snakes, baby
turtles, mice, ducklings—that it can catch and swallow, including
its own young.
**Range**
Extr. se. Canada throughout e. U.S. except s. Fla. and n. New
England, and w. to Wis., se. Wyo., and w. Tex. Also in numerous
localities throughout w. U.S. and s. along coastal Mexico, where it
has been introduced. Introduced in Cuba, Puerto Rico, Hispaniola,
and Jamaica.

## PIG FROG
*Rana grylio*                                                74:11

**Description**
Length, to 6⅜ in. (16.2 cm). Large frog with webbing almost to tip
of longest toe, no dorsolateral folds, and pointed snout. Above
brown to blackish-brown with black spotting; below heavily colored
with dark reticulations. Feet fully webbed; tympanum as large as,
or larger than, eye.
**Similarities**
Bullfrog has blunter snout and longest toe extends well beyond
webbing.
**Habitat**
Permanent open bodies of water with emergent vegetation.
**Voice**
Calls from water or from among masses of aquatic plants; loud,
resonant sound resembling grunting of a pig.
**Range**
S. Coastal Plain from s. S.C. through Fla. and w. to extr. se. Tex.
Introduced in the Bahamas.

## RIVER FROG
*Rana heckscheri*                                            74:12

**Description**
Length, to 5⁵⁄₁₆ in. (13.5 cm). Large greenish-black frog; venter
gray with light spots. Throat gray, white spots along lip. Skin
rough. No dorsolateral folds, or else poorly defined. Toes not fully
webbed; tympanum large.
**Similarities**
Pig Frog and Bullfrog have smoother skin and *light* venters with
dark markings.
**Habitat**
Swampy areas; shallows of streams and rivers.
**Voice**
Deep, variable, but likened to a snore and grunt.
**Range**
Atlantic Coastal Plain from se. S.C., s. through n. Fla., and w. to
se. Miss.

## CRAYFISH FROG
*Rana areolata*                                        75:6

**Description**
Length, to 4½ in. (11.4 cm). Recognize this secretive frog by 3 or 4
rows of round, dark spots with light gray edges overlying a
meshlike brownish or gray dorsal pattern, rough skin texture, and
continuous dorsolateral folds. Venter whitish and unspotted, legs
barred with brown and gray. In males paired vocal sacs over the
shoulder are conspicuous, bluish, like folds of skin when collapsed.

**Habitat**
Often in crayfish burrows, or "gopher" holes (made by tortoises),
by pools in lowland meadows and flood plains.

**Habits**
Nocturnal, subterranean; rarely seen except at breeding pools in
early spring (March and April).

**Voice**
Loud, deep snoring, deafening in chorus.

**Range**
Atlantic Coastal Plain from se. N.C. to cen. Fla., w. to extr. e. La.
Also from cen. Miss., n. to s. Ill. and sw. Ind., and w. through n.
Mo.; absent in Ozark region and e. La., but s. from se. Kans. to
Gulf Coast in w. La. and Tex.

## CARPENTER FROG
*Rana virgatipes*                                      75:7

**Description**
Length, to 2⅝ in. (6.7 cm). Lacks dorsolateral folds, but has 4
golden-brown stripes down the back contrasting with olive to black
ground color. Below yellowish or white with dark spots; back of
thighs with broad black stripes; throat yellow. Skin smooth; toes
not fully webbed. Male has 1 round vocal pouch on each side.
Young lighter overall than adults.

**Habitat**
Sphagnum bogs, cedar ponds; swamps and quiet streams; under
mats of vegetation.

**Habits**
Aquatic, seen at water's edge or on or under logs; also among
vegetation.

**Voice**
Similar to hammering of a nail (hence the name) repeated 3–6
times.

**Range**
Atlantic Coastal Plain; in Pine Barrens of cen. N.J. and s.
Delmarva Pen.; continuously s. from e. Va. to extr. ne. Fla.

## PLAINS LEOPARD FROG
*Rana blairi*                                          74:10

**Description**
Length, to 4⅜ in. (11.1 cm). Recognized by discontinuous light
dorsolateral folds—toward hind legs they break and are displaced
to midline. Has light spot in center of tympanum and only a few
dark spots on sides. Irregular dark spots dorsally on gray, tan, or
green ground color; hind limbs barred; belly white.Vocal sacs small
and paired above shoulder.

**Habitat**
Plains and prairies.

**Voice**
Call lasts less than 1 sec. at 60°F (15.6°C) and consists of 2–6 notes delivered at a rate of about 3 per sec.
**Reproduction**
Breeds in slow or standing water, even in temporary ponds, in early spring and into summer.
**Range**
W. Ind. to e. and s. Nebr., e. Colo., and s. through ne. N.Mex. and Okla. to cen. Tex.

# NORTHERN LEOPARD FROG
*Rana pipiens*                                                              74:7

**Description**
Length, to 4⅜ in. (11.1 cm). Above brown or green with rounded dark spots between dorsolateral folds; folds continuous, with other smaller folds between them. White stripe on lip. Males lack external vocal sacs, though internal pouches sometimes seen.
**Similarities**
Southern and Plains Leopard Frogs have fewer spots on lower sides.
**Habitat**
Permanent water; swamps, marshes, streams; also in meadows in summer.
**Voice**
Highly variable, but long, low-pitched trills lasting from 1–3 or more sec.; pulse rate about 20 per sec. at 60°F (15.6°C).
**Reproduction**
Breeds with warm temperatures of spring, about March or April in eastern U.S.
**Range**
S. Canada from Labrador to District of Mackenzie; and n. U.S. s. to cen. Conn., n. Pa., cen. Ky., s. Iowa, and s. Nebr. into N.Mex. and Ariz. Disjunct populations in Md., cen. Ill., and e.-cen. Okla. Occurs in most western states.

# SOUTHERN LEOPARD FROG
*Rana sphenocephala*                                                       75:2

**Description**
Length, to 3½ in. (8.9 cm); record 5 in. (12.7 cm). Has continuous dorsolateral folds, a pointed snout, and few spots on sides of body; also often light spots on tympanum. Above green or brown with dark brown spots; light line from shoulder to snout. Males have large external vocal sacs at each side above shoulder.
**Habitat**
In and around shallows of rivers, lakes, streams, marshes, or ponds.
**Voice**
At 60°F (15.6°C) the short trill lasts less than 1 sec., and consists of 4 or more pulses with a pulse rate of 13 per sec. or less.
**Range**
Se. N.Y., s. through Ga. and Fla., but absent from Appalachian region; n. through s. Ind. to s. and e. Mo. and se. Kans., w. in Plains and Gulf Coast to cen. Tex. Introduced in the Bahamas.

## PICKEREL FROG
*Rana palustris*                                              **75:1**

**Description**
Length, to 3⅞₆ in. (8.7 cm). Resembles a small brown Leopard
Frog, but has 2 regular rows of closely spaced, square to
rectangular, dark brown or black spots between the dorsolateral
folds, and underside of thighs is conspicuously bright yellow or
orange. Gray or tan above, spots outlined with light color. Dark
mark from eye to nostril and light line on lip. Hind limbs banded.
Vocal sac above shoulders on each side.

**Habitat**
Grassy meadows, streams, ponds in cool, clear water.

**Habits**
Active by day and night and much of the year; much time spent
out of, but near, water.

**Voice**
Long, low-pitched trills; at end of call frequency changes.

**Remarks**
Poisonous secretions of the skin affect other animals, even other
species of frogs kept in the same container.

**Range**
Se. Canada to S.C., w. to Wis., extr. se. Minn. in the North, and
e. Tex. in the South; absent from most of Coastal Plain and
prairies of Ill. and adj. states.

## MINK FROG
*Rana septentrionalis*                                        *Fig. 96*

**Description**
Length, to 3 in. (7.6 cm). In some ways a small edition of the
Green Frog, but with buff-bordered black markings. Above buff or
brown and green with mottled or spotted pattern; vermiculated
pattern on legs; below white or yellowish. Skin smooth, snout
rounded, body stout, eyes close together. Dorsolateral folds short or
absent. Feet fully webbed.

**Habitat**
Rivers, lakes, or peaty ponds; near water lilies. A common frog of
the northern forests.

**Habits**
Aquatic and solitary.

**Voice**
High, fast quacking with a metallic ring, *cuck-cuck-cuck-cuck.*

**Remarks**
Gives off strong minklike odor when annoyed.

**Range**
An exclusively northern distribution. E. Canada and extr. n. U.S.
from Maritime Provinces and New England s. to n. N.Y., w. above
Great Lakes to upper Mich., Wis., and ne. Minn., and to se. Man.

Mink Frog

*Fig. 96*

## WOOD FROG
*Rana sylvatica*                                              **75:5**

**Description**
Length, to 3¼ in. (8.3 cm). The only species with a wide black
mask extending from snout to tympanum, and a dark line from eye
to tip of snout. Above reddish-brown (either dark or light),
occasionally gray; middorsal white line may be present. Dark
markings on sides but rarely on back; below whitish. Dark mask
has narrow light line below it. Head pointed; prominent light
dorsolateral folds. Males with paired vocal sacs.

**Habitat**
Damp woods.

**Habits**
Terrestrial, hibernating in logs, stumps, under stones, rather than
under water.

**Voice**
A short, sharp, rasping clacking almost like clucking (not
quacking) sound of domestic ducks. Shorter than that of Leopard
Frogs, but not as loud or deep; calls only during breeding season.

**Reproduction**
Breeds in ponds very early in spring, even before ice has
completely melted; then returns to land.

**Range**
Most of Alaska and Canada, even n. of Arctic Circle and most of
ne. U.S. s. to Md.; in Appalachians of n. Ga. and ne. Tenn.
Isolated populations in Wyo. and Colo., in highlands of Ark. and
Mo., and in n. Idaho.

# Fishes

**Consulting Editor**
Franklin C. Daiber
Professor
College of Marine Studies
University of Delaware

**Illustrations**
Plates 76–93 Nina L. Williams
Text Illustrations by Jennifer Emry-Perrott and Nina L. Williams

# Fishes
Superclass Pisces*

There are in the world today more than 25,000 different species of fishes, more than there are of all other vertebrates put together. Included in this chapter are 164 species of saltwater and 141 freshwater fishes, or 305 species in all. These represent the food and game fishes that occur regularly in appreciable numbers and that can be readily identified in the field.

A fish is an aquatic, cold-blooded vertebrate with jaws and paired fins that breathes chiefly by means of gills. Most fishes also have scales. Fishes are grouped into two classes. The first, the Chondrichthyes, consists of sharks, skates, and rays that have a skeleton of cartilage. The second, the Osteichthyes, includes all the modern fishes that have a skeleton of bone.

Also included in this chapter are representatives of two more primitive classes of vertebrates, the lancelets (hagfishes) and the lampreys. These are also cold-blooded and aquatic, but they lack jaws, paired fins, scales, and plates; in the lancelets, lateral gill openings are also absent and the brain is rudimentary. Both are ancient groups of which only a few species, often parasitic among the lampreys, have survived to the present.

In degrees varying by species, fishes as a whole possess all five senses of sight, smell, taste, touch, and hearing, although they hear without external ears. Their mental powers are low. An air bladder serves in many species as a depth regulator and in some, notably the croakers, as a means of making a noise. In some species the males are the only sound makers, and some naturalists think the males use sound in the breeding season to attract the females, just as some birds use song. The subject of sound in fishes has been studied in recent years, yet more is to be learned.

Most fishes swim by flexing the body through lateral motions of the caudal peduncle and tail. Fishes with forked tails are the fastest swimmers. The dorsal and anal fins serve as a keel; the pectorals deflect the fish up or down. In a few species, however, the pectorals provide the driving power; and certain eellike forms progress by moving the whole body sideways in a series of undulations.

Fishes vary greatly in size, from the 2-inch (5.1-cm) length of the Naked Goby to the 45 feet (13.7 m) of the Basking Shark. Many continue to grow through much of their lives. Thus in any one species in any particular area, the larger fish are usually the older fish. In the data in this chapter, the maximum recorded size is given, whether caught by hook, gaff, or net. In some cases the all-tackle record is given as well. The average, of course, is substantially smaller.

Many fishes are strikingly beautiful in color, particularly the fishes of the coral reefs. There, seen through a glass-bottomed boat, finny gems of orange and blue, crimson, green, and violet flash by in kaleidoscopic profusion. In more northern waters, however, except in some darters and mummichogs in the breeding season, colors in general are not bright. Some fish go through spectacular color changes when dying; but colors usually fade quickly upon death. In certain species color may vary with the environment, but in most

---

* Also Classes Leptocardii and Agnatha.

*Fig. 97*
Parts of Typical Fishes

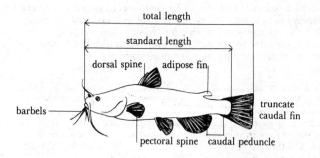

cases fish of the same kind tend to look much alike. There is some individual variation, but generally not as much as, for example, there is in snakes. However, many species go through a larval stage quite different in appearance from the adult.

## Habitat

Almost all groups of animals, as they evolve, tend to spread out by adaptive radiation into every available habitat. Fishes, being the oldest vertebrates, have had time to spread out widely. Some habitats, because they are richer in food or provide shelter, or for other reasons, naturally support more individuals and species of fish than do others. In the northern oceans, for example, on the fringe of the ice pack, waters rich in minerals support in abundance the food chains upon which many saltwater fishes depend.

The more varied the habitat, of course, the more species of fishes can live in it. Thus the coral reef boasts more varieties than the open sea. The tropics, in general, have many species but relatively few individuals of each; the boreal, or northern, seas count fewer species but many of these occur in stupendous numbers. In all waters, except the very deepest, whether the bottom be of sand, mud, shell, or rock, whether it be bare or covered with weed, there are always fishes.

In fresh waters as well, virtually all aquatic habitats are inhabited by some species of fish. Muddy river mouths appeal to catfish, clear streams to trout. Sunfish prefer calm, weedy pools, while salmon leap upstream through swift-flowing rapids and over waterfalls.

## Habits

The natural history of fishes is beyond the scope of this chapter; however, a few cases of special interest are mentioned. In a group as old as the fishes almost every imaginable way of life has been tried. At this time, one can see the culmination of all the experiments of the past that have succeeded.

Different habitats require different habits and different bodily adaptations. In the flatfishes that lie on their sides in the sand, one eye has migrated from the lower side to join the other on top of the head. The burrfish can puff itself up with air or water into a porcupinelike ball, highly unattractive to predators. The habits of other fishes are equally fascinating and almost infinitely varied.

Many species are migratory, moving north or inshore in spring and south or offshore in fall. Most have a favored range of temperature, pressure, and saltiness or muddiness of the water.

## Food

Many fishes are carnivorous. Most carnivorous fishes feed on other kinds of fishes or on marine invertebrates such as jellyfish, copepods, squid and other mollusks, and on plankton, the minute animal and plant life that floats in the top layers of the sea. Some fishes, however, are herbivorous and feed on seaweeds or on the microscopic one-celled diatoms. Other fishes are scavengers and help keep the sea and rivers clean by feeding on carrion and waste.

## Reproduction

In most fishes, as in all the land vertebrates, the sexes are separate. Female fish produce eggs, the males milt. In most species fertilization is external, the males discharging milt over the eggs as they are extruded by the female. However, in the sharks, skates, and rays, and in some killifishes and gambusia, fertilization is internal. Sharks and rays bring forth their young alive. Skates lay eggs. In these three groups the pelvic fins of the male are modified into claspers that serve as organs of intromission. In males of the viviparous killifishes and gambusia, the anal fin is so modified.

In the fishes that practice external fertilization, the eggs vary in size from $\frac{1}{50}$ of an inch (0.05 cm) in diameter in the Naked Goby to $\frac{7}{8}$ of an inch (2.2 cm) in the Gafftopsail Catfish. Eggs of different species may variously float or sink, be free in the water, or become attached to weeds, snags, or rocks. They may be separate or they may adhere to each other in globular or stringy masses. Noteworthy is the floating egg mass of the Goosefish, which may be 2 feet (61 cm) wide and 40 feet (12.2 m) long and weigh up to 30 pounds (13.6 kg).

The number of eggs of any one species has little or nothing to do with the number of individuals of that species in the sea. Population sizes are controlled by the environment, in particular by the availability of food. Given ample food supplies and an increase in the amount of suitable territory, any species, no matter how few eggs it may lay, will soon have occupied the additional territory and stocked it to the limit.

Only a few saltwater fishes pay any attention to their eggs after they are laid. Among those that do, the most famous are the sea catfishes, whose males for as long as 65 days incubate the eggs and shelter the young in their mouths. Many freshwater species, on the other hand, such as the Largemouth and Smallmouth Bass, build nests of sticks or stones or scooped-out sand. The males often assist incubation by stirring the water around the eggs. They zealously guard the eggs during incubation and sometimes the young after they are hatched. The male seahorse and pipefish brood the eggs in an abdominal pouch, where they are deposited by the female.

### Evolution
The notochord, regarded as the prototype of the spinal column and the distinguishing feature of the first vertebrates, is believed by some authorities to have arisen in animals that lived in flowing water. The first fishlike creatures presumably had only notochords; later ones developed spinal columns of cartilage or bone.

The earliest such creatures of which there is any record were the armored, jawless ostracoderms. They originated in the late Silurian period, flourished during the Devonian, and then died out, leaving as the sole relics of their class the hagfishes and lampreys. Then jaws were developed—one of the great steps in the history of evolution—by the placoderms, which are all now extinct but which gave rise to the sharks and the true fishes. The well-jawed, well-toothed sharks appeared in the Devonian period and for the rest of the Paleozoic era dominated the sea. They have continued, in an aggressive though subordinate capacity, to roam the saltwater regions ever since.

The bony fishes also appeared in the Ordovician or late Silurian period, but the more advanced forms did not become conspicuous until the Mesozoic era. From then on, their varieties and their numbers multiplied and they successfully withstood all competition, including that of the marine reptiles, the mosasaurs and ichthyosaurs, and of the marine mammals, the whales, all of which were members of more advanced groups that invaded the sea. Today, if success were to be reckoned by variety and numbers, the bony fishes would be judged the most successful backboned animals on the globe. They display a greater variety of species, and an astronomically greater number of individuals, than all other vertebrates combined.

## Conservation

Fishes are of great importance to man. We eat probably several billion of them a year; we use many for oil, fertilizer, and other purposes; and we catch them for sport. Angling for sport, in its various branches, has increased considerably in recent years. Some 25,000,000 North Americans buy licenses each year to fish the fresh waters.

Fishes need conservation just as land animals and birds do. Even though commercial fisheries often take only a relatively small harvest of the great bounty of the sea, there are many problems to consider, such as the overexploitation of fishery resources by the efficient use of large commercial trawls and factory ships, and the pollution caused by industrial wastes. These and many other conservation questions engage the attention of federal and state fishery authorities.

In fresh waters the situation is quite different. Entire river systems in the past have been polluted and the fish destroyed by the dumping of toxic factory wastes. Much of this still goes on. The East Coast run of shad is a pitiful fraction of what it was 100 years ago. The Atlantic Salmon has been extirpated from southern New England. Similar situations occur elsewhere. Anglers, too, can soon overfish a stream and do so in heavily fished areas. Most states have had to institute elaborate hatchery and restocking programs in order to offset the effects of pollution and of heavy angling pressure. Such programs, if wisely conducted, are beneficial; but much must still be done to stop pollution.

The question of conservation leads inevitably to the question of the sea as a new food frontier for man, because fish are notably rich in vitamins, minerals, and proteins and make an excellent human food. The resources of the oceans and even the topography of their floors are still most inadequately understood. Scientists recognize that there is a finite limit to fishery resources and that the situation is presently at its upper limit or even has exceeded the capacity of fish populations to renew themselves.

## Where to Look for Fishes

To begin knowing what kinds of fishes are available to find, exploration of local markets is a good start. Here one can become familiar with the species in a particular locale. During the warmer months, scuba and snorkeling can then prove productive in any body of water, whether fresh or salt water.

## Range and Scope

Included in this chapter are all species of food and game fishes that occur regularly in appreciable numbers, and many other readily identifiable in the field (a) that occur annually within a depth of 25 fathoms (45.7 m) off the coasts of North America, principally from southern Florida north and west to the delta of the Mackenzie River in Canada, and (b) that occur annually in the fresh waters of continental North America east of the 100th meridian between the Canadian border and southern Florida. Ranges apply to the species as a whole only within the area of geographic coverage of the chapter. Sizes given are the maximum average sizes. Occasionally the record size is given.

The Atlantic Coast from Cape Hatteras to Cape Cod is temperate and has, in general, temperate-water species of fishes. North of

Cape Cod, the fish fauna changes, in large part, to circumpolar Arctic forms. Few temperate species or individuals are found north of the elbow of Cape Cod; few boreal fishes are south of it.

**Classification and Nomenclature**

Within recent decades, numerous changes have been proposed for the scientific names of many fishes. There is as yet no firm agreement among scientists as to the proper generic placement for many fishes or even wholehearted agreement as to the relative order in which some fish families should appear.

In this chapter, the classification of fishes follows Nelson, *Fishes of the World,* 1976, with slight modifications. The scientific nomenclature has been brought up to date and the English names follow, with some modifications, the 1970 American Fisheries Society List of Common Names.

# USEFUL REFERENCES

Bigelow, H., and Schroeder, W. C. 1953. *Fishes of the Gulf of Maine.* Washington: U.S. Government Printing Office.

Breder, C. M., Jr. 1948. *Field Book of Marine Fishes of the Atlantic Coast.* New York: Putnam.

Clay, W. M. 1975. *The Fishes of Kentucky.* Frankfort: Kentucky Department of Fish and Wildlife Resources.

Dahne, R. A. 1950. *Saltwater Fishing.* New York: Holt.

Eddy, S. 1957. *Freshwater Fishes.* Dubuque: W. C. Brown.

Eddy, S., and Underhill, J. C. 1974. *Northern Fishes.* Minneapolis: University of Minnesota Press.

Harlan, J. R., and Speaker, E. B. 1956. *Iowa Fish and Fishing.* Des Moines: State Conservation Commission.

Hildebrand, S. F., and Schroeder, W. C. 1927. *Fishes of Chesapeake Bay.* Bulletin U.S. Bureau of Fisheries, Vol. 43.

Hubbs, C. L., and Lagler, K. F. 1964. *Fishes of the Great Lakes Region.* Ann Arbor: Univ. of Michigan Press.

LaMonte, F. 1950. *North American Game Fishes.* New York: Doubleday.

Leim, A. H., and Scott, W. B. 1966. *Fishes of the Atlantic Coast of Canada.* Ottawa: Fisheries Research Board of Canada, Bulletin 155.

Moore, G. A., 1968. Fishes. In Blair, Blair, Brodkorb, Cagle, and Moore, *Vertebrates of the United States.* New York: McGraw-Hill.

Nelson, J. S. 1976. *Fishes of the World.* New York: John Wiley & Sons.

Pflieger, W. L. 1975. *The Fishes of Missouri.* St Louis: Missouri Dept. of Conservation.

Scott, W. B., and Crossman, E. J. 1973. *Freshwater Fishes of Canada.* Ottawa: Fisheries Research Board of Canada, Bulletin 184.

Trautman, M. D. 1957. *The Fishes of Ohio.* Columbus: Ohio State Univ. Press.

# GLOSSARY

*Abdominal ridge* Belly ridge from region below gills to anus.

*Adipose fin* A fleshy fin, without rays, between dorsal and caudal fin.

*Anadromous* Ascending rivers from salt water to spawn.

*Anal fin* Fin situated between anus and caudal fin.

*Anterior* Toward the front of the body (usually toward the head or cranial end).

*Barbel* Fleshy appendage projecting from upper jaw, lower jaw, or chin region.

*Body depth* Greatest vertical distance through body, not including fins.

*Branchiostegal rays* Bony rays that support the membranes under the head below the opercular bones.

*Caudal fin* Tail fin.

*Caudal peduncle* Posterior end of body from last ray of anal fin to base of caudal fin.

*Cephalic fins* Detached part of the pectorals on the heads of certain rays.

*Circuli* A series of concentric ridges on the scales.

*Cirrus* (*pl.* cirri) Skin flap projecting outward from head or body.

*Claspers* Extension from the paired pelvic fins of male sharks, rays, and chimaeras that serve as copulatory organs.

*Compressed* Flattened from side to side (laterally compressed) or from top to bottom (dorsoventrally compressed).

*Ctenoid scales* Bony fish scales with small toothlike projections from the posterior edge.

*Cycloid scales* Bony fish scales with a smooth posterior edge.

*Dermal denticles* Spinelike "scales" of a cartilaginous fish; also termed placoid scales.

*Disk* The flat, circular or diamond-shaped forepart of skates and rays, formed by fusion of pectoral fins to head.

*Dorsal* Toward the upper back region of body.

*Dorsal fin* Fin situated on top of the back, not including the adipose fin.

*Epipelagic* The oceanic zone into which enough light penetrates for photosynthesis.

*Furcate* Forked.

*Gill arch* Bony or cartilaginous structure to which the gill filaments and rakers are attached.

*Gill cover* Bony or cartilaginous gill cover; synonomous with *operculum*.

*Gill filaments* Fleshy red protuberances on the outer sides of the gill arch that serve in respiration.

*Gill rakers* Cartilaginous (sometimes bony) protuberances on the inner sides of the gill arch that direct food into the gullet.

447

*Gill slits* Openings between gill arches for passage of water, visible externally in cartilaginous fishes, covered by an operculum in bony fishes.

*Heterocercal* Unequally lobed; said of caudal fin when upper lobe is larger than lower.

*Keel* A longitudinal fin extending along the center of the bottom of the body.

*Lateral line* Sensory organ composed of a canal which connects a series of openings along sides of body; or a posterior extension of the sensory canals on the head.

*Lobule* Either a small lobe or a subdivision of a lobe.

*Lunate* Between crescent and halfmoon in shape.

*Mandible* Lower jaw, composed of from four to seven bones in fishes.

*Maxilla* The bone lying on each side of the two halves of the upper jaw of fishes.

*Medial* Toward the center of the body.

*Mesopelagic* Relating to ocean depths from about 600 to 3000 feet.

*Nostrils* Paired, cuplike structures on the snout of fishes that do not connect with the mouth cavity for respiration but serve in the reception of chemical stimuli.

*Notochord* A flexible, rodlike structure that supports the body of vertebrate embryos and the adults of jawless fishes. (In most higher vertebrates, the notochord is later replaced by the vertebral column.)

*Operculum* Bony or cartilaginous gill cover.

*Palatines* Paired bones of the palate, lateral and posterior to the upper jaw bones.

*Parr marks* Dark vertical markings on sides of young fish, especially trout.

*Pectoral fins* Paired fins attached to the shoulder or pectoral girdle, located just posterior to gill openings. They may be fused with pelvic fins into a sucking disk (in the clingfish) or completely absent (as in eels).

*Pelvic fins* Paired fins attached to pelvic girdle, located on belly between throat and anus. Both fins may be united into a sucking disk (in gobies), be completely absent (in eels and eellike fishes), or form barbels (as in cusk-eels and brotulas).

*Peritoneum* The membrane lining the visceral or abdominal cavity.

*Pharyngeal teeth* Teeth attached to bones of the paired fifth gill (pharyngeal) arch, located immediately anterior to esophagus or gullet. (Pharyngeal tooth counts are listed in order from left to right.)

*Placoid scales* Spinelike scales of a cartilaginous fish; also termed dermal denticles.

*Posterior* Towards the rear of the body; usually toward caudal fin.

*Precaudal pit* Depression that is found immediately anterior to the base of the caudal fin in some sharks.

*Preopercle* Paired bones of the posterior cheek region, anterior to bones of gill cover (operculum).

*Premacilla* Bone at front of upper jaw, or forming entire upper jaw.

*Preorbital* Region between eye and tip of snout.

*Prickles* Small, fine, sometimes curved spines in place of scales.

*Ray* Flexible (cartilaginous) supports of the soft fins.

*Roe* Fish eggs.

*Scute* An external scale.

*Shagreen* Rough, hard-scaled skin of some sharks.

*Spinous rays* The fins of advanced bony fishes, which are sharp, bony, and usually inflexible, and contain spines.

*Spiny* Composed of sharp, inflexible spines.

*Spiracle* A vestigial gill slit present behind eyes of sharks, rays, and a few primitive bony fishes. In rays it is large and serves for intake of water during respiration.

*Striae* Small ridges or lines.

*Tubercle* Small rounded lump; a modified scale, hard or soft.

*Vent* Anus.

*Viviparous* Gives birth to living young.

*Vomer* Bone at the midline of the palate, immediately behind the upper jawbone.

# Saltwater Fishes

## LANCELETS
Class Leptocardii

## Lancelets
Order Amphioxiformes

## LANCELETS
Family Branchiostomidae

### VIRGINIA LANCELET
*Branchiostoma caribaeum*            *Fig. 98*

**Description**
Size, to 2¼ in. (5.7 cm). No jaws; mouth a longitudinal slit below.
Body long, tapering, compressed, colorless, naked; many small gill
slits, eyes rudimentary, fins only slightly developed.
**Habitat**
Shallow, sandy bottoms.
**Other name**
Amphioxus.
**Range**
Chesapeake Bay and s.

Virginia Lancelet

*Fig. 98*

Hagfish

## LAMPREYS
Class Agnatha

## Hagfishes
Order Myxiniformes

### HAGFISHES
Family Myxinidae

Members of this family are eel-shaped, jawless fishes lacking
paired fins and scales. They generally inhabit cool marine waters.
They are scavengers, feeding on dead or dying fish. When
disturbed or handled, hagfishes secrete a great quantity of slime.
Each egg is contained in a horny capsule.

## HAGFISH
*Myxine glutinosa*

*Fig. 98*

**Description**
Size, to 31 in. (78.7 cm). Eellike body; no jaws or eyes. Skeleton boneless, skin without scales; single finlike fold running around length of body. Mouth lipless and with tentacles, star-shaped when closed; tongue studded with horny teeth, mucous sacs along sides on abdomen.
**Habitat**
Soft mud, where it lies embedded with tip of snout projecting.
**Food**
Carrion; also parasitic.
**Range**
Arctic seas to N.C.

# SHARKS, SKATES, AND RAYS
Class Chondrichthyes

Members of this class are distinguished from the bony fishes by having the entire skeleton cartilaginous. Fertilization is internal, effected by a pair of rodlike copulating organs developed from the pelvic fins in the male. Most species give birth to live young.

Sharks are distinguished from the skates and rays by the free eyelids, at least partly lateral gill openings, and pectoral fins not attached to the sides of the head. Most sharks have distinctive teeth.

# Sharks
Order Lamniformes

Sharks of Order Lamniformes have two dorsal fins (with the exception of one genus) without spines, an anal fin, and five gill slits. Gill rakers are generally absent; spiracles are generally present.

# SAND SHARKS
Family Odontaspididae

## SAND TIGER
*Odontaspis taurus*

**78:13**

**Description**
Size, to 9 ft. (2.7 m) and 250 lb. (113.4 kg). Last gill opening in front of pectorals, base of first dorsal in front of pelvics; no keel on peduncle, upper tail lobe notched and twice the size of lower.
**Habitat**
Relatively shallow inshore waters.
**Food**
Fishes, squid, crustaceans.
**Range**
Maine and s.

# MACKEREL SHARKS
## Family Lamnidae

Members of this family are distinguished by the following characteristics: The body is spindle-shaped and the teeth are large and sharp. The last gill opening is set in front of pectorals, which are large and half as long as they are high. The first dorsal fin is high, and its base is wholly in front of the small pelvics; the second dorsal and anal are very small. The caudal peduncle is slender and keeled and the tail crescent-shaped, less than one-third the total length of the fish. This family goes back to the Cretaceous.

*Fig. 99*
Shark Teeth

| | | | | | |
|---|---|---|---|---|---|
| Sand Tiger | Porbeagle | Shortfin Mako | White | Basking | Thresher |

| | | | | | |
|---|---|---|---|---|---|
| Smooth Dogfish | Tiger | Great Blue | Bull | Sandbar | Smooth Hammerhead |

| | | | |
|---|---|---|---|
| Dusky | Spiny Dogfish | Greenland | Angel |

# PORBEAGLE SHARK
*Lamna nasus*                                    78:10

**Description**
Size, to 12 ft. (3.7 m); all-tackle record, 271 lb. (122.9 kg). Body stout and heavy-shouldered; first dorsal fin large, as high as long and originating slightly behind armpits or pectorals; furrow on upper and lower surface of root of tail. Tail fin has secondary longitudinal keel toward the front; second dorsal directly above anal.

**Habitat**
Surface waters offshore.

**Habits**
Surfaces on calm days.

**Food**
Fishes.

**Other name**
Mackerel Shark.

**Range**
Newfoundland to S.C.

## SHORTFIN MAKO SHARK
*Isurus oxyrhynchus*                                                78:11

**Description**
Size, to 11 ft. 5 in. (348.0 cm); all-tackle record, 1000 lb. (453.5 kg). Body moderately slender. First dorsal fin large and triangular, about as high as long; no secondary lateral keels on caudal. First dorsal originates above or behind rear base of the pectorals; and second dorsal originates in front of the anal.
**Habitat**
Surface of warm seas.
**Food**
Fishes.
**Range**
Newfoundland and s.

## WHITE SHARK
*Carcharodon carcharias*                                             78:9

**Description**
Size, to 21 ft. (6.4 m); all-tackle record, 2536 lb. (1150.1 kg). Body stout, lead-white in large individuals, darker in smaller ones. First dorsal originates over armpits of sickle-shaped pectorals; second dorsal slightly in front of anal.
**Habitat**
Temperate and tropical deep water; also in shallow waters inshore.
**Food**
Large fishes, porpoises.
**Other name**
Maneater Shark.
**Range**
Newfoundland to Brazil.

## BASKING SHARK
*Cetorhinus maximus*                                                 78:12

**Description**
Size, to 40 ft. (12.2 m). Gill openings so large they almost meet under the throat. Body thick, snout short, teeth very small. Base of first dorsal fin wholly in front of small pelvics; second dorsal and anal very small, former slightly in front of latter. Caudal peduncle keeled; tail crescent-shaped. Color dark metallic gray fading to light belly.
**Habitat**
Surface waters inshore in summer, presumably deeper offshore in winter.
**Habits**
Often basks in sun at surface.
**Food**
Plankton.
**Range**
Newfoundland to N.C.

# THRESHER SHARKS
Family Alopiidae

## THRESHER SHARK
*Alopias vulpinus*                                              **78:8**

**Description**
Size, to 20 ft. (6.1 m); all-tackle record, 922 lb. (418.1 kg); tail as long as its head and body. Body stout, snout blunt; first dorsal fin about as high as long, set midway between sickle-shaped pectorals and small pelvics; second dorsal and anal fins extremely small, caudal peduncle thick. Color blue-gray to purplish back fading to white belly; fins dark gray to purplish.

**Habitat**
Surface of temperate seas.

**Food**
Small schooling fishes.

**Range**
N.S. and s.

# REQUIEM SHARKS
Family Carcharhinidae

The members of this family have sharp, saw-edged or flat and hard teeth, and eyes with nictitating, or winking, membranes. The last gill openings are above or behind the base of the pectorals; the first dorsal fin lies above the similar-sized anal. The tail, notched in many species, is less than one-third total length, and the upper lobe is longer.

## GREAT BLUE SHARK
*Prionace glauca*                                              **78:3**

**Description**
Size, to 12 ft. 7 in. (3.8 m). First dorsal nearer the pelvics than the pectorals. Body color deep blue, slender, thickest at mid-length; snout very long. Pectorals narrow, curved, very long.

**Habitat**
Open sea.

**Food**
Small fishes.

**Range**
Warm seas, n. to Newfoundland.

## BULL SHARK
*Carcharhinus leucas*                                          **78:1**

**Description**
Size, to 10 ft. (3 m). Short, broadly rounded snout; smooth back with no ridge between the dorsals. Dorsal profile more convex than the ventral, head very wide, eyes small. First dorsal originates over armpit of long pectorals; anal is more concave than second dorsal; upper lobe of tail more than twice length of lower.

**Habitat**
Relatively shallow inshore waters.

**Range**
Brazil to N.C., rarely to N.Y.

## SMOOTH DOGFISH
*Mustelus canis*                                                 **77:14**

**Description**
Size, to 5 ft. (1.5 m). Body slender, flattened below; snout blunt,
spiracles present. Both dorsals spineless, first dorsal originating
over rear angle of pectorals; pectorals large, higher than long; anal
fin smaller than, and to rear of, second dorsal; no pit at root of tail.
**Habitat**
Inshore.
**Food**
Larger crustaceans.
**Remarks**
Edible.
**Range**
Bay of Fundy and s.

## TIGER SHARK
*Galeocerdo cuvieri*                                            **78:4**

**Description**
Size, to 18 ft. (5.5 m); all-tackle record, 1382 lb. (626.7 kg). First
dorsal above the armpit of the pectorals; caudal peduncle keeled,
notched above and below. Head large, snout blunt, spiracles
present; mouth broad, with long grooves along upper jaw. First
dorsal high, nearly as large as pectorals; second dorsal $1/3$ to $1/4$ size
of first; upper lobe of tail 3 times length of lower.
**Habitat**
Surface of warm seas.
**Food**
Sea turtles, other sharks, fishes, carrion.
**Range**
Tropical and temperate seas occasionally n. to N.S.

## SANDBAR SHARK
*Carcharhinus milberti*                                         **78:5**

**Description**
Size, to 8 ft. (2.4 m). Triangular snout; dorsals connected by a low
ridge on the back. First dorsal fin originates over armpit of
pectoral; length of front edge less than $1/2$ distance from base of fin
to snout. Body stout, heavy forward; broad first dorsal and
pectorals.
**Habitat**
Inshore waters.
**Food**
Bottom fishes and crustaceans.
**Range**
Cape Cod to Brazil; relatively common along the coasts of N.Y.
and N.J.

## DUSKY SHARK
*Carcharhinus obscurus*                                         **78:2**

**Description**
Size, to 14 ft. (4.3 m). Body slender, tapering; snout rounded;
dorsal fins connected by low ridge on back. First dorsal large and
originating over rear corner of pectoral; length of front edge at
least $1/2$ distance from base of fin to snout; anal fin slightly longer
than second dorsal.

**Habitat**
Inshore.
**Food**
Fishes, squid.
**Range**
Cape Cod to Brazil.

# HAMMERHEAD SHARKS
Family Sphyrnidae

## SMOOTH HAMMERHEAD
*Sphyrna zygaena*                                             **78:6**

**Description**
Size, to 11 ft. (3.4 m). Hammer-shaped head; widely separated eyes
at the far edges.
**Habitat**
Open seas near surface.
**Food**
Fishes.
**Range**
Worldwide in tropical and temperate waters n. to Cape Cod, rarely
to N.S.

# Dogfish and Angel Sharks
Order Squaliformes

These sharks have two dorsal fins, with or without spines, have no
anal fins, and have five or six gill slits.

# DOGFISH SHARKS
Family Squalidae

## SPINY DOGFISH
*Squalus acanthias*                                           **78:7**

**Description**
Size, to 4 ft. (1.2 m). Single spine in front of each of 2 dorsal fins;
no anal fin. Head flattened and tapering to blunt tip; first dorsal
nearer to pectorals than to pelvics, second dorsal originating well
behind pelvics; no notch on tail. Color gray to brownish back
fading to lighter belly.
**Habitat**
Cold water, inshore in winter, offshore in summer.
**Habits**
Travels in large schools.
**Food**
Fishes, squid, worms.
**Range**
N.S. to Cuba.

## GREENLAND SHARK
*Somniosus microcephalus*                                    78:15

**Description**
Size, to 21 ft. (6.4 m). Dorsal fins small, almost equal in size;
upper tail lobe not much larger than lower. Body tapering to rear
but cylindrical in front, snout rounded. Pectorals small, hardly
larger than pelvics; second dorsal originating to rear of origin of
pelvics; no spines on dorsals, no anal.
**Habitat**
Near the bottom of cold seas.
**Range**
Polar seas s. to Cape Cod.

# ANGEL SHARKS
Family Squatinidae

## ANGEL SHARK
*Squatina dumerili*                                         78:14

**Description**
Size, to 5 ft. (1.5 m). Head broadly rounded, eyes on top of head.
Both dorsal fins brush-shaped and similar; lower tail lobe larger
than upper.
**Similarities**
Distinguished from all other sharks by its flattened body.
Distinguished from the skates by the following: pectoral fins not
attached to sides of head; gill openings extending from lower
surface up onto sides of neck.
**Habitat**
Warm waters.
**Range**
Cape Cod and s.

# SKATES AND RAYS
Order Rajiformes

The members of this order have flattened bodies and spiracles.
They lack anal fins. If dorsal fins are present, they are located on
the tail. Skates and rays are distinguished from sharks in the
following respects: they do not have free eyelids, gill openings are
below only, and the edges of the pectoral fins are attached to the
sides of the head in front of the gill openings.

# ELECTRIC RAYS
Family Torpedinidae

## ATLANTIC TORPEDO
*Torpedo nobiliana*                                         79:11

**Description**
Size, to 5 ft. (1.5 m). Body almost completely rounded. Skin
smooth, eyes very small and set far forward. Pelvic fins somewhat
in front of first dorsal; 2 dorsal fins situated at forward end of
thick, spineless, triangular tail; electric organ, consisting of
hexagonal tubes between pectoral fins and head, capable of giving
a strong shock.

**Habitat**
Ocean bottom.
**Habits**
Bottom feeder.
**Food**
Fishes.
**Other name**
Electric Ray.
**Range**
N.S. to Fla. Keys.

# SKATES
Family Rajidae

Skates have the skin more or less covered with spines and a muscular tail usually supporting two dorsal fins and sometimes a caudal fin. Distinguish skates from rays by their concave, instead of convex, pelvic fins and by the absence of a tail spine. The family goes back to the Cretaceous.

## CLEARNOSE SKATE
*Raja eglanteria*                                                79:5

**Description**
Size, to 3 ft. (0.9 m). Brown above, white below; sides of snout form a right angle, outer corners of pectoral fins angular. Row of small, stout, star-shaped spines along back and tail; 1-4 rows of spines on each side of tail; groups of large spines opposite and behind eyes; 1-5 spines on each shoulder.
**Habitat**
Sandy bottoms.
**Other name**
Brier Skate.
**Range**
Fla. to Mass.

## LITTLE SKATE
*Raja erinacea*                                                79:3

**Description**
Size, to 21 in. (53.3 cm). No row of spines along midline of back (except in young); sides of snout meet at obtuse angle and bulge opposite eyes; about 50 series of teeth in each jaw.
**Habitat**
Smooth shallow bottoms.
**Other name**
Common Skate.
**Range**
N.S. to Va.

## BARN-DOOR SKATE
*Raja laevis*                                                79:2

**Description**
Size, to 5 ft. (1.5 m). Large, brown with some light spots, paler below. Skin nearly smooth, spines few and small; sides of snout meet at acute angle and bulge opposite eyes; row of spines along tail.

**Habitat**
Smooth bottoms, often in deep water. Tolerates or enjoys an unusually wide range of temperature; is found in waters from 32° to over 68°F (0° to over 20° C).
**Range**
N.S. to N.C.

## WINTER SKATE
*Raja ocellata*                                                         **79:4**

**Description**
Size, to 34 in. (86.4 cm). Large white spot with a black center near the rear corner of each pectoral fin. No row of spines along midline of back; sides of snout meet at obtuse angle and bulge opposite eyes; about 90 series of teeth in each jaw.
**Habitat**
Smooth bottoms.
**Range**
Newfoundland to N.C.

## THORNY SKATE
*Raja radiata*                                                         **79:1**

**Description**
Size, to 40 in. (1.0 m). One large spine in front of each eye and 1 behind, 3 on each shoulder; small spines scattered elsewhere above; sides of snout meet at obtuse angle, 30–50 series of teeth in each jaw.
**Habitat**
Smooth bottoms in cool waters.
**Range**
Hudson Bay to N.Y.

# STINGRAYS
Family Dasyatidae

Stingrays lack dorsal fins. They are distinguished from skates by their convex, instead of concave, pelvic fins and their long tails, often bearing venomous spines. They bear living young.

## SMOOTH BUTTERFLY RAY
*Gymnura micrura*                                                       **79:9**

**Description**
Size, to 2 ft. (61.0 m) in width. This ray and its relative the Spiny Butterfly are the only rays with tail shorter than body.
**Habitat**
Ocean bottom.
**Habits**
Bottom feeder.
**Food**
Crustaceans.
**Reproduction**
Usually 2 living young.
**Range**
Cape Cod to Brazil.

Note: The **SPINY BUTTERFLY,** *Gymnura altavela,* gets up to 6 ft. (1.8 m) in width.

## ROUGHTAIL STINGRAY
*Dasyatis centroura*                                    **79:8**

**Description**
Size, to 10 ft. (3.1 m). Snout forms obtuse angle, pectorals angled; tail about 2½ times length of body and keeled below only.
**Habitat**
Ocean bottom.
**Habits**
Bottom feeder.
**Range**
Cape Cod and s.

## ATLANTIC STINGRAY
*Dasyatis sabina*                                      **79:6**

**Description**
Size, to 3½ ft. (1.1 m). Sides of snout meet at right angle, pectorals rounded; tail less than twice as long as body and keeled both above and below.
**Habitat**
Ocean bottom.
**Habits**
Bottom feeder.
**Range**
Chesapeake Bay and s.

# EAGLE RAYS
Family Myliobatidae

The members of this family are distinguished by pectorals that stop on the sides of the head; cephalic, or head, fins that are not below the level of the body and are not long and earlike; a high-domed crown; eyes and spiracles on the sides of the head; and a long, whiplike tail, with one dorsal fin near its root. Eagle rays are swimmers as well as bottom feeders. They bear living young. The family goes back to the Cretaceous.

## BULLNOSE RAY
*Myliobatis freminvillei*                              **79:12**

**Description**
Size, to 5 ft. (1.5 m). Body not covered with small spots; teeth in several flat series.
**Habitat**
Ocean bottom.
**Habits**
Bottom feeder.
**Food**
Crabs, lobsters, larger mollusks.
**Range**
Cape Cod to Brazil.

## SPOTTED EAGLE RAY
*Aëtobatus narinari*                                   **79:10**

**Description**
Size, to 12 ft. (3.7 m) and 450 lb. (204.1 kg). Body covered with small spots; teeth in single series.

**Habitat**
Ocean bottom.
**Habits**
Bottom feeder.
**Food**
Shelled mollusks.
**Reproduction**
4 living young.
**Range**
Va. coast (common) to Brazil (rare).

## COWNOSE RAY
*Rhinoptera bonasus* 79:13

**Description**
Size, to 7 ft. (2.1 m). Pectoral fins stop on sides of head, 1–2
rounded head fins in front, crown high-domed with eyes and
spiracles on sides of head; teeth large flat grinding plates in mosaic
pattern. Tail armed with poisonous spines.
**Habitat**
Bottom and higher levels.
**Habits**
Bottom feeder.
**Food**
Mollusks, especially soft clams.
**Range**
Nantucket to Brazil.

# MANTAS
Family Mobulidae

## ATLANTIC MANTA
*Manta birostris* 79:7

**Description**
Size, to 22 ft. (6.7 m) wide and 3500 lb. (1587.3 kg). Head fins
curve out in front. Crown high-domed, eyes on sides instead of on
top; pectoral fins stop on sides of head.
**Habitat**
Not confined to the bottom.
**Habits**
A swimmer.
**Food**
Small pelagic animals.
**Range**
Brazil to Va., rarely to Cape Cod.

# BONY FISHES
Class Osteichthyes

These are the modern, bony fishes, distinguished from the sharks,
skates, and rays by the possession of true bones. Generally they
have erectile fins with stiff or soft rods between which stretch
membranes, often transparent. Normally, their bodies are covered
with thin overlapping scales, but sometimes the body is naked. The
gill opening is usually covered with a bony gill cover, or
operculum. An air bladder is generally present. We call them
modern, but actually they appeared before the sharks and go back
in time to the Lower Devonian, 340,000,000 years ago.

# Bonefishes
Order Elopiformes

These species lack spines in the vertical fins, have the ventral fins
in an abdominal position, have a single dorsal fin, and usually have
a well-forked tail. Gill openings are wide. The body is elongate
and more or less compressed.

## TARPONS
Family Elopidae

### LADYFISH
*Elops saurus*                                                    *Fig. 100*

**Description**
Size, to 3 ft. (0.9 m). Bluish above, silvery on sides, whitish
beneath; plate between branches of lower jaw. Dorsal lacks
prolonged ray of the tarpon and is depressible into a scaly sheath;
belly not compressed or covered with specialized scales.
**Habitat**
Warm waters.
**Food**
Crustaceans and smaller fishes.
**Other names**
Ten Pounder, Big-eyed Herring.
**Range**
Fla., n. infrequently to Cape Cod in fall.

*Fig. 100*

Bonefish          Ladyfish

### TARPON
*Megalops atlantica*                                              *Fig. 101*

**Description**
Size, to 8 ft. (2.4 m) and 315 lb. (142.9 kg); all-tackle record, 283
lb. (128.3 kg); scales to 3 in. (7.6 cm). Large scales, long dorsal
ray, curved anal. Plate between branches of lower jaw.
**Habitat**
Estuaries, coast.
**Food**
Fishes, crabs.
**Range**
Gulf of Maine to Brazil.

*Fig. 101*

Tarpon

# BONEFISHES
Family Albulidae

## BONEFISH
*Albula vulpes*                                           *Fig. 100*

**Description**
Size, to 31 in. (78.7 cm); record, 18 lb. (8.2 kg). Face unique; mouth small, teeth hard, flat in back of mouth; snout piglike; scales small.
**Habitat**
Relatively shallow water.
**Habits**
A solitary bottom feeder.
**Food**
Small mollusks.
**Other name**
Ladyfish.
**Range**
Warm seas; n. sparingly to Cape Cod.

# Herringlike Fishes
Order Clupeiformes

These species are silvery and have the ventral fins in an abdominal position. They have branchiostegal rays, usually fewer than fifteen, and the abdomen often has keeled scutes along the ventral midline. Most are plankton feeders with numerous long gill rakers.

# HERRINGS
Family Clupeidae

These are compressed, silvery, fork-tailed fishes without a lateral line, with very small teeth or none, and with very long gill rakers. The family goes back to the Cretaceous.

## MENHADEN
*Brevoortia tyrannus*                                      **76:13**

**Description**
Size, to 18 in. (45.7 cm) and 1 lb. 13 oz. (0.8 kg). Large scaleless head, nearly ⅓ total length; no teeth. Scales comb-toothed, mouth large, lower jaw projects; dorsal above or slightly behind very small ventrals; tail deeply forked.
**Habitat**
Near surface.
**Habits**
Occurs in vast compact schools.
**Food**
Plankton.
**Age**
To 10 years.
**Other names**
Pogy, Mossbunker, Fatback.
**Range**
N.S. to Fla., Mar. to Nov.

## ATLANTIC HERRING
*Clupea harengus*                                          85:3
**Description**
Size, to 17 in. (43.2 cm) and 1½ lb. (0.7 kg). Body deeper than
thick; scales large, loose, rounded; mouth large, lower jaw projects,
oval patch of small teeth in roof of mouth. Dorsal midway along
body and directly above smaller ventrals, belly sharp.
**Habitat**
Open sea.
**Habits**
Travels in vast schools.
**Food**
Plankton.
**Reproduction**
Spawned in spring, summer, and fall; spring and fall the chief
seasons; hatch in 22 days at 45°F (7°C).
**Age**
To 20 years.
**Remarks**
An important food fish, and one of the most abundant fishes in the
world. Young are marketed as "sardines."
**Other name**
Sea Herring.
**Range**
N. Atlantic; most abundant in winter.

## SPANISH SARDINE
*Sardinella anchovia*                                       85:5
**Description**
Size, to 6 in. (15.2 cm). Body long, sides of nape striped; dorsal
originates distinctly forward of ventral.
**Habitat**
Warm coastal waters.
**Food**
Copepods and other plankton.
**Range**
Cape Cod to Rio de Janeiro; erratic.

## ROUND HERRING
*Etrumeus teres*                                            85:2
**Description**
Size, to 15 in. (38.1 cm). Dorsal wholly in front of the ventrals.
Slender; depth only ⅙ length; belly rounded, not sharp.
**Habitat**
Offshore marine waters.
**Range**
Bay of Fundy to Gulf of Mexico, but rare n. of Cape Cod.

# ANCHOVIES
Family Engraulididae

These fishes have a small, compressed body, an undershot, cleft
mouth, and usually a silvery lateral band. They are found in large
schools often close to shore. The family goes back to the Tertiary.

## STRIPED ANCHOVY
*Anchoa hepsetus*                                    85:11

**Description**
Size, to 6 in. (15.2 cm). Lateral band well defined, as broad as eye.
**Habitat**
Shallow water.
**Habits**
Schools.
**Food**
Copepods and small mollusks.
**Range**
N.S. to Brazil; first of May to late fall.

## BAY ANCHOVY
*Anchoa mitchilli*                                    85:8

**Description**
Size, to 4 in. (10.2 cm). Lateral band diffuse or indistinct, scarcely wider than pupil of eye.
**Habitat**
Sandy shores, estuaries; usually abundant.
**Range**
Maine to Tex., but rare n. of Cape Cod.

# Eels
Order Anguilliformes

These are fishes with jaws and snakelike bodies. They are naked or have only tiny scales. Their small gill openings are on the sides. They have no ventral fins, sometimes no pectorals, no spines in the fins, and no distinct tail fin.

The amazing life story of the eel is a classic of natural history. The species spawns in deep water southwest of the Bermudas. The transparent young, called leptocephali and long thought to be a different species, slowly make their way to the mouth of the freshwater rivers whence their parents came. Still translucent, they become elvers and finally grow into dark-colored eels. The females then ascend these rivers. After several years they are ready to spawn. In the fall, they descend the rivers and are joined by the males in the estuaries (the males do not go up the rivers). The adult males and females seek their traditional breeding grounds, where they spawn once and die.

## TRUE EELS
Family Anguillidae

## AMERICAN EEL
*Anguilla rostrata*                                    80:1

**Description**
Size, to 5 ft. (1.5 m). Distinguished from the Conger by the fact that it is often found in fresh water and by its small, imbedded scales.
**Habitat**
Shallow water, fresh or salt.
**Habits**
Hibernates in mud.

**Food**
Shrimps, crabs, mollusks, worms, small fishes.
**Range**
Labrador to the Guianas; common.

## CONGER EELS
Family Congridae

### CONGER EEL
*Conger oceanicus*                                       **80:2**
**Description**
Size, to 12 ft. (3.7 m) and 8 lb. (3.6 kg). Distinguished from the
American Eel by an entire absence of scales.
**Habitat**
Salt water offshore, crevices and nooks in rocks.
**Food**
Fishes, worms.
**Range**
Gulf of Maine to Brazil.

# Catfishes
Order Siluriformes

## SEA CATFISHES
Family Ariidae

The sea catfishes, which are the saltwater family of this largely
freshwater order, have a transverse mouth surrounded by barbels.
Each dorsal and pectoral fin bears a stout spine. The family goes
back to the Eocene.

### GAFFTOPSAIL CATFISH
*Bagre marinus*                                       *Fig. 102*
**Description**
Size, to 2 ft. (0.6 m). Lower jaw with 2 barbels. Long streamers on
dorsal and pectoral fins and on barbel on the jaw.
**Habitat**
Coastal estuaries.
**Reproduction**
Eggs are incubated within mouth of the male.
**Range**
Cape Cod to Panama, but rare n. of Del.

*Fig. 102*

Sea Catfish

Gafftopsail Catfish

## SEA CATFISH
*Arius felis*                                                   *Fig. 102*

**Description**
Size, to 1 ft. (0.3 m). Lower jaw with 4 barbels, no streamers on pectoral or dorsal spines.
**Habitat**
Harbors of southern states.
**Habits**
Active mostly at night.
**Reproduction**
Males incubate eggs, numbering up to 55, in their mouths. The young may reach 4 in. (10.2 cm) in length before leaving paternal shelter.
**Range**
Cape Cod (rarely) to West Indies.

# Codfishes
Order Gadiformes

These fishes have an elongate body, usually with long dorsal and anal fins. The pelvics, if present, are below or in front of the pectorals. There are cycloid scales in most species.

# CODFISHES
Family Gadidae

These fishes have soft fins only. Their ventrals are under or in front of their pectorals, not behind them. The only fishes with three dorsal fins are found in this family. Most are bottom feeders. This family includes some of the most important food fishes.

## ATLANTIC TOMCOD
*Microgadus tomcod*                                              **84:2**

**Description**
Size, to 15 in. (38.1 cm) and 1¼ lb. (0.6 kg). Resembles a young Atlantic Cod except for its ventral fins, which have a filament as long as the rest of the fin.
**Habitat**
Less than 3 fathoms (5.5 m) deep; moves inshore in fall.
**Food**
Invertebrates, fishes.
**Other name**
Frostfish.
**Range**
Labrador to Va.

## ATLANTIC COD
*Gadus morhua*                                                  **76:12**

**Description**
Size, to 6 ft. (1.8 m) and 211 lb. (95.7 kg); all-tackle record, 57 lb. (25.9 kg). Speckled gray-green or reddish above, white below; head large, mouth wide, teeth small and numerous.
**Habitat**
Continental shelf and submarine banks, often lying a fathom (1.8 m) or so from the bottom.

# CODFISHES

**Habits**
Rapacious.
**Food**
Mollusks, crustaceans, squid, fishes.
**Age**
At least to 9 years.
**Other names**
Rock Cod, Scrod.
**Range**
N. Atlantic, particularly the Grand Banks, s. to N.Y. rarely to Va.

## POLLACK
*Pollachius virens*                                    **76:11**

**Description**
Size, to 3½ ft. (1.1 m) and 35 lb. (15.9 kg); all-tackle record, 36 lb. (16.3 kg). Dorsals, 3; projecting lower jaw. Greenish-blue; lateral line light; 2 anals, tail slightly forked.
**Habitat**
Cool water, any moderate level.
**Habits**
More active than Cod or Haddock.
**Reproduction**
Spawns Oct. to Dec., at depth of 15–50 fathoms (27.4–91.4 m); 200,000–400,000 eggs hatch in 9 days at 43° F (6.1° C).
**Age**
To 19 years.
**Remarks**
A very important commercial fish.
**Other name**
Boston Bluefish.
**Range**
N. Atlantic, rarely s. to Chesapeake Bay.

## HADDOCK
*Melanogrammus aeglefinus*                             **76:9**

**Description**
Size, to 44 in. (1.1 m) and 37 lb. (16.8 kg), but usually 3–4 lbs. (1.4–1.8 kg). Distinguished by 3 dorsal fins, a black lateral line, and a dusky blotch over each pectoral fin. Purplish-gray above; mouth smaller than in other codfishes and with shorter gape; 2 anals.
**Habitat**
Bottoms in cold water, 10–75 fathoms (18.3–137.2 m).
**Food**
Mollusks, worms, squid, fishes.
**Age**
To 14 years.
**Remarks**
A great food fish. The Georges Bank–South Channel area has often yielded 37,000,000 fish a year, but numbers are now greatly reduced.
**Other name**
Scrod.
**Range**
N. Atlantic s. to N.J.

# HAKES
Genus *Urophycis*

These hakes have two dorsal fins, the second much longer than the first, one anal fin, and long ventrals.

## SPOTTED HAKE
*Urophycis regius*                                                    84:1

**Description**
Size, to 16 in. (40.6 cm) and 1½ lb. (0.7 kg). No streamer on first dorsal fin.
**Habitat**
Shallow water.
**Habits**
Bottom feeder.
**Food**
Fishes, squid, crustaceans.
**Range**
Fla. to Cape Cod, rarely to Maine.

## WHITE HAKE
*Urophycis tenuis*                                                    84:3

**Description**
Size, to 4 ft. (1.2 m) and 40 lb. (18.1 kg), average 8 lb. (3.6 kg). Short streamer on first dorsal. Distinguish from Red Hake by ventrals that do not reach beyond vent, smaller scales, and mouth gaping to back of eye.
**Habitat**
Soft bottoms, to 300 fathoms (0.6 km).
**Habits**
Most active at night. Locates much of its food by swimming slowly over the bottom and trailing the thin tips of its ventrals, which act as feelers.
**Food**
Shrimps, squid, fishes.
**Other names**
Boston Hake, Ling, Mud Hake.
**Range**
Newfoundland to Cape Hatteras.

## RED HAKE
*Urophycis chuss*                                                    84:4

**Description**
Size, to 30 in. (76.2 cm) and 7 lb. (3.2 kg). Long streamer on first dorsal. Distinguish from White Hake by ventrals that reach beyond vent, larger scales, and mouth gaping only to pupil of eye.
**Habitat**
Soft bottoms, to 300 fathoms (0.6 km).
**Food**
Shrimps, squid, fishes.
**Other names**
Ling, Squirrel Hake.
**Range**
Newfoundland to Chesapeake Bay.

## CUSK
*Brosme brosme*      **84:5**

**Description**
Size, to 3½ ft. (1.1 m) and 27 lb. (12.2 kg). Only 1 dorsal fin,
extending entire length of back; dorsal and anal fins continuous
with tail fin, the boundaries marked merely by notches.
**Habitat**
Cool waters over rocks or gravel from 10 fathoms (18.3 m) down.
**Habits**
Solitary, a bottom feeder.
**Food**
Crustaceans, mollusks.
**Remarks**
Although Cusk is excellent eating, landings of this species have
decreased in recent years with the decrease in long-line fishing and
increase in otter trawling. This is because the cusk frequents rough
bottoms that are not reached by the trawl.
**Range**
N. Atlantic, s. to Cape Cod, rarely to N.J. in deeper water.

## SILVER HAKE
*Merluccius bilinearis*      **76:10**

**Description**
Size, to 2½ ft. (0.8 m) and 5 lb. (2.3 kg). Distinguished from other
hakes by the absence of chin barbels and by its ordinary and not
elongated ventrals.
**Habitat**
Surface to 300 fathoms (0.6 km), sandy or pebbly bottoms.
**Food**
Fishes.
**Remarks**
A fine food fish, but soft.
**Other name**
Whiting.
**Range**
Grand Banks to N.Y.

# Toadfishes
Order Batrachoidiformes

# TOADFISHES
Family Batrachoididae

## OYSTER TOADFISH
*Opsanus tau*      **80:10**

**Description**
Size, to 15 in. (38.1 cm). Body tapering, belly plump; head, mouth,
and eye large; head covered with fleshy flaps. Pectorals large and
fanlike; 2 dorsals separated by a notch, soft dorsal 5 times as long
as spiny; anal shorter than second dorsal, with outer ends of rays
free; tail rounded. Somewhat resembles the sculpins, but ventrals
are set under the throat in front of pectorals, not below or behind.
**Habitat**
Shallow inshore waters with hiding places.

**Habits**
Vocally active during the summer spawning season.
**Remarks**
This fish can severely pinch the fingers while one is trying to
extract the hook.
**Range**
Maine to Fla.

# Frogfishes
Order Lophiiformes

# ANGLERS
Family Lophiidae

## GOOSEFISH
*Lophius americanus*                                   *Fig. 103*

**Description**
Size, to 4 ft. (1.2 m) and 50 lb. (22.7 kg). Colossal mouth. Body
soft, flattened, and tadpolelike, tapering back from enormous head
to small, broom-shaped tail; teeth many, curved; lower jaw projects,
gill openings very small. Above 3 thin, stiff spines; dorsals
separated, second longer than first; pectorals with wristlike joint at
base.
**Habitat**
Any bottom, tide line to 365 fathoms (0.7 km).
**Habits**
Uses end tab on first dorsal spine as a fishing lure while it lies in
wait on the bottom.
**Food**
Almost anything edible and marine, including sea birds.
**Other names**
Angler, Allmouth.
**Range**
Newfoundland to Brazil.

*Fig. 103*                                              Goosefish

# Needlefishes, Halfbeaks, Killifishes, and Silversides
Order Atheriniformes

Fishes of this order are varied. Gill cover and preopercular margins are without spines or serrations. Generally surface feeding fishes, many species are found in fresh water.

# NEEDLEFISHES
Family Belonidae

## ATLANTIC NEEDLEFISH
*Strongylura marina*                                    80:12

**Description**
Size, to 4 ft. (1.2 m). Greenish above, silvery on sides, white below; both jaws extended into a long, toothed beak.
**Habitat**
Coastal, often ascends rivers part way.
**Habits**
Thrusts itself up through ocean surface like a spear; can be dangerous to man.
**Food**
Fishes.
**Other name**
Billfish.
**Range**
Maine to Fla., but rare n. of Cape Cod.

# HALFBEAKS
Family Exocoetidae

## HALFBEAK
*Hyporhamphus unifasciatus*                            80:11

**Description**
Size, to 1 ft. (0.3 m). Unique for its long lower jaw and short upper jaw.
**Habitat**
Tropic waters and Gulf Stream.
**Food**
Small crustaceans, mollusks, plant material.
**Range**
Maine to Argentina, more common in tropics.

# KILLIFISHES
Family Cyprinodontidae

These are small fishes with a very small mouth, ventrals on the abdomen, one dorsal fin set to the rear, and a thick peduncle. The brilliant colors of the spawning males rival the beauty of many tropical reef fishes.

## MUMMICHOG
*Fundulus heteroclitus*                                    80:3

**Description**
Size, to 6 in. (15.2 cm). Body stout and covered with large rounded
scales; snout blunt, teeth sharp, back and belly rounded. Anal fin
of male larger than in female and used as an intromittent organ
when spawning. Male assumes brilliant orange and blue colors in
breeding season.
**Habitat**
Quiet water, less than 100 yards (91.4 m) from shore; usually
associated with mud bottoms of tidal marsh creeks.
**Food**
Omnivorous to carnivorous.
**Remarks**
Provides good bait.
**Other names**
Killifish, Saltwater Minnow, Chub, Mummy.
**Range**
Gulf of St. Lawrence and s.

## STRIPED KILLIFISH
*Fundulus majalis*                                         80:4

**Description**
Size, to 7 in. (17.8 cm). Male takes on bright colors in breeding
season. Distinguished from Mummichog by black bars, transverse
on male, lengthwise on adult female; and by more pointed snout
and more slender body.
**Habitat**
Quiet water close to shore; usually associated with sandy bottoms.
**Habits**
Can flop several feet back into the water if stranded by outgoing
tide.
**Food**
Mollusks, crustaceans, fishes, insects.
**Other name**
Striped Mummichog.
**Range**
N.H. to Fla.

## SHEEPSHEAD MINNOW
*Cyprinodon variegatus*                                    80:9

**Description**
Size, to 3 in. (7.6 cm). Back high, arched; sides flat. Males assume
brilliant blue above and orange below in breeding season.
Distinguished from Mummichog by greater depth of body, large
wedge-shaped teeth, and square tail.
**Habitat**
Shallow waters of inlets.
**Other names**
Broad Killifish, Pupfish.
**Range**
Cape Cod to Mexico.

# SILVERSIDES
Family Atherinidae

## ATLANTIC SILVERSIDE
*Menidia menidia*                                              **76:2**

**Description**
Size, to 6 in. (15.2 cm). Two dorsals, long anal, and no adipose
fin. Head and body slender and eye large.
**Habitat**
Shallow water near shore.
**Food**
Small crustaceans, mollusks, plankton, bottom organisms.
**Range**
Va. to N.S. Abundant.

# Pipefishes and Cornetfishes
Order Syngnathiformes

This order includes species with long snouts ending in a small
mouth.

## CORNETFISHES
Family Fistulariidae

## BLUESPOTTED CORNETFISH
*Fistularia tabacaria*                                    *Fig. 104*

**Description**
Size, without lash, to 6 ft. (1.8 m). Long whiplike lash growing
from the center of forked tail. Body long, silvery; scales small,
inconspicuous.
**Habitat**
Drifts into eastern range in the Gulf Stream.
**Food**
Small fishes.
**Range**
Cape Cod to Brazil, but rare out of the tropics.

*Fig. 104*

Northern Pipefish                    Bluespotted Cornetfish

## PIPEFISHES
Family Syngnathidae

These fishes have segmented body plates; the gill opening is a
small pore and the gills are small, rounded tufts.

The mode of reproduction among these fishes suggests, remotely,
that of marsupial mammals, but here it is the male that has the

pouch on his abdomen. The female deposits her eggs in this pouch. There they are incubated and hatched and the young are protected until they can shift for themselves. The family goes back to the Eocene.

## NORTHERN PIPEFISH
*Syngnathus fuscus*                                      *Fig. 104*

**Description**
Size, usually 6 in. (15.2 cm), but up to 12 in. (30.5 cm). Axis of head in line with axis of body, tail not prehensile.
**Habitat**
Shallow water among seaweed.
**Habits**
Not very active; may sometimes be caught by hand.
**Food**
Small crustaceans.
**Range**
Halifax to N.C., most common near N.Y.

## LINED SEA HORSE
*Hippocampus erectus*                                      **80:7**

**Description**
Size, to 7½ in. (19.1 cm). Axis of head at angle to that of body, tail prehensile.
**Habitat**
Shallow water among seaweed.
**Habits**
Curls tail about stalk, swims vertically by undulating dorsal fin.
**Reproduction**
Spawns in summer; up to 150 young.
**Food**
Tiny organisms.
**Range**
Mass. to S.C., rarely to N.S.

# Perchlike Fishes
Order Perciformes

This large and varied order is the most diversified fish order today. Its members have spines on the first dorsal fin and the anal fin. The pelvic fins, if any, are usually located on the throat, not the abdomen. Scales are ctenoid or absent.

# TEMPERATE BASSES
Family Percichthyidae

## STRIPED BASS
*Marone saxatilis*                                      **77:7**

**Description**
Size, average 20–48 in. (50.8–121.9 cm), 3–40 lb. (1.4–18.1 kg); maximum 6 ft. (1.8 m), 125 lb. (56.7 kg); all-tackle record, 73 lb. (33.1 kg). Longitudinal stripes; 3 spines and 12 rays on anal fin. Body stout, mouth oblique and gapes back to eye; lower jaw projects. Dorsals about equal in length, anal similar to second dorsal and originates below it; tail wide, slightly forked.

**Habitat**
Coastal waters.
**Food**
Fishes and larger crustaceans.
**Age**
To 23 years.
**Remarks**
This species has occasioned much controversy. The anglers want it taken only on hook and line; commercial fishermen say it is hard to prevent a few from getting caught occasionally in nets at river mouths designed for other species.
**Other names**
Rockfish, Striper.
**Range**
Gulf of St. Lawrence to Fla.; most common from Cape Cod to N.C.

# SEA BASSES
Family Serranidae

These carnivorous fishes are highly valued as food. Both spiny and soft-rayed portions of the dorsal are well developed, either as separate fins or divided by a deep notch. The ventral fins are under the pectorals; the anal is about as long as the soft part of the dorsal. There is a large caudal peduncle and a broad tail. The family goes back to the Eocene.

## RED GROUPER
*Epinephelus morio*                                    80:6

**Description**
Size, to 3 ft. (0.9 m). Brownish, bases of jaws reddish, sides with pale blotches, dark dots around eye. Dorsals adjoin, second dorsal spine highest.
**Habitat**
Close to bottom of inlets and reefs.
**Remarks**
Large and commercially valuable as food fish; a well known frequenter of reefs.
**Range**
Brazil to Va., straggling to Cape Cod.

## BLACK SEA BASS
*Centropristes striata*                                77:3

**Description**
Size, to 2 ft. (0.6 m) and 7½ lb. (3.4 kg); all-tackle record, 8 lb. (3.6 kg). Dorsal fins continuous; in large fish, upper ray of the tail fin lengthened. Scales large but head scaleless; flat spine near rear angle of gill cover. Front dorsal saw-toothed in outline, pectorals broad and rounded, anal with 3 short spines at front, tail rounded.
**Habitat**
Shallow waters offshore over rocky bottoms.
**Habits**
Bottom feeder over rocky ground.
**Food**
Crustaceans, mollusks, fishes, squid.
**Range**
N. Fla. to Cape Ann, rarely in Maine; May to November.

# TILEFISHES
Family Branchiostegidae

## TILEFISH
*Lopholatilus chamaeleonticeps*                                76:14

**Description**
Size, to 3½ ft. (1.1 m) and 50 lb. (22.7 kg). Fleshy, finlike flap on
nape in front of dorsal fin. Mouth wide, both jaws with 2 rows of
teeth, eye set high in head. Spiny and soft portions of dorsal fin
continuous and of even height.
**Habitat**
Continental slope.
**Food**
Crabs, squid, shrimps, shelled mollusks, worms, fishes.
**Range**
Chesapeake Bay to Maine.

# BLUEFISHES
Family Pomatomidae

## BLUEFISH
*Pomatomus saltatrix*                                           77:8

**Description**
Size, to 3½ ft. (1.1 m) and 27 lb. (12.2 kg); all-tackle record, 24 lb.
(9 kg). Blue above, white below; mouth large, lower jaw projects,
teeth prominent. First dorsal low, spiny, and depressible in a
groove; second dorsal concave and twice as long as first. Anal fin,
originating to rear of second dorsal, shaped like it and preceded by
2 short spines; tail broad and forked.
**Habitat**
Warm water inshore or on offshore banks.
**Habits**
One of the most ferocious fishes of its size.
**Food**
Fishes and squid.
**Other names**
Snapper (for the young), Tailor.
**Range**
Gulf of Mexico to Cape Cod, straggling to Maine; arrives April to
May off N.Y.

# COBIAS
Family Rachycentridae

## COBIA
*Rachycentron canadum*                                         80:5

**Description**
Size, to 5 ft. (1.5 m); all-tackle record, 102 lb. (46.3 kg). Body
spindle-shaped, broadly striped; mouth large, lower jaw projects.
First dorsal made up of low, separate spines; second dorsal long
and concave; anal originates behind second dorsal, but is similar in
outline. Young Cobias up to 4 in. (10.2 cm) have convex tails; later
tails become concave.

**Habitat**
Offshore or near wharves in bays and inlets.
**Food**
Crabs and fishes.
**Other names**
Sergeantfish, Cabio, Crabeater.
**Range**
Cosmopolitan in warm waters n. to Chesapeake Bay, rarely to
Cape Cod.

# REMORAS
Family Echeneidae

## SHARKSUCKER
*Echeneis naucrates*                                    **80:8**

**Description**
Size, to 38 in. (96.5 cm). Unique in having on top of its head a flat
oval sucking disk with 20–28 cross-ridges; 3 other species with 14–
18 ridges occur far offshore. Body very slender, nearly round in
cross section, tapering to very slender caudal peduncle. Belly as
dark as back. Pectorals pointed and set high, ventrals pointed and
directly under pectorals, dorsal and anal fins long and tapering; tail
convex in young, concave in old.
**Habitat**
Open sea.
**Food**
Attaches itself by its sucking disk to the underside of sharks and
tarpons, and feeds on scraps from its host's meals.
**Other name**
Remora.
**Range**
Warm seas, rarely n. to Cape Cod.

# JACKS
Family Carangidae

In this family the first, spiny, dorsal is much shorter than the soft-
rayed second, and it may be reduced to a few short spines or be
lacking. The anal fin is preceded by three short spines, which may
become a permanent finlet or be lost in old age. The caudal
peduncle is very slender and the tail deeply forked. Most species
have a prominent hard keel on the rear portion of the lateral line,
which serves to strengthen the peduncle. These are largely oceanic
fishes. The family goes back to the Eocene.

## BIG-EYED SCAD
*Selar crumenophthalmus*                                **81:2**

**Description**
Size, to 2 ft. (0.6 m). Eyes large; lateral line slightly arched but
without spots, heavily scaled to rear only; 2 spines in front of anal.
**Habitat**
Warm seas.
**Other name**
Goggle-eyed Scad, Cigarfish.
**Range**
Atlantic Coast n. to Cape Cod, straggling to N.S.

## ROUND SCAD
*Decapterus punctatus*                                  81:5

**Description**
Size, to 1 ft. (0.3 m). First dorsal originates over middle of
pectorals; single finlet behind dorsal and anal. Distinguished from
Mackerel Scad by toothed jaws, longer second dorsal, and spots
along arched lateral line, rear half of which is covered with large
keeled scales.
**Habitat**
Open ocean.
**Range**
Warm Atlantic, n. to Cape Cod, rarely to N.S.

## MACKEREL SCAD
*Decapterus macarellus*                                 81:4

**Description**
Size, to 1 ft. (0.3 m). Distinguished from Round Scad by absence
of spots along lateral line, untoothed jaws, and less developed scales
on peduncle.
**Habitat**
Open ocean.
**Range**
Warm waters n. to Cape Cod, rarely to N.S.

## ROUGH SCAD
*Trachurus lathami*                                     81:1

**Description**
Size, to 1 ft. (0.3 m). Strongly arched lateral line heavily scaled
along its entire length; 2 small detached spines in front of anal fin.
No finlets behind dorsal or anal; tail deeply forked.
**Habitat**
Open ocean.
**Other name**
Saurel.
**Range**
Atlantic Coast n. to Cape Cod, rarely to Maine.

## YELLOW JACK
*Caranx bartholomaei*                                   81:16

**Description**
Size, to 15 in. (38.1 cm). Body deep; dorsal and anal fins completely
covered with small scales, last rays not notably enlarged.
**Range**
Brazil to Nantucket, rarely to N.S.

## HORSE-EYED JACK
*Caranx latus*                                          81:3

**Description**
Size, to 2 ft. (0.6 m). Body deep, eye large, lateral line strongly
arched. Dorsal and anal fins scaly only on elevated portions.
**Food**
Small fishes.
**Range**
West Indies n. to Va.

479

## BLUE RUNNER
*Caranx crysos*                                      **81:17**

**Description**
Size, to 22 in. (55.9 cm) and 4 lb. (1.8 kg). Breast scaly; lateral
line strongly arched and with bony plates increasing in size to rear.
Dorsal and anal fins scaly only on elevated portions; anal preceded
by finlet of 2 short spines.

**Food**
Shrimps, fishes.

**Other name**
Hard-tailed Jack.

**Range**
Brazil to Cape Cod, rarely to N.S.

## CREVELLE JACK
*Caranx hippos*                                      **81:15**

**Description**
Size, to 2½ ft. (0.8 m). Well-rounded forehead, black spot on gill
cover, scimitar-shaped pectoral fins. Body oblong; lateral line
arched, with strongly keeled scales to rear; 2 pairs of small canine
teeth in lower jaw. Breast naked, with only a small triangular
patch of scales in front of ventrals.

**Food**
Fishes.

**Range**
Warm seas n. to Cape Cod, rarely to Maine.

## AFRICAN POMPANO
*Alectis crinitus*                                   **81:8**

**Description**
Size, to 7 in. (17.8 cm). Unique for long fin streamers. Body short,
very deep, head convex.

**Other names**
Threadfin, Threadfish.

**Range**
Tropical waters, straying to Cape Cod.

## LOOKDOWN
*Selene vomer*                                       **81:7**

**Description**
Size, to 1 ft. (0.3 m). Oblique head; front rays of dorsal and anal
fins lengthened. Body deep, thin; mouth low; first dorsal reduced to
7 or 8 short spines, pectorals sickle-shaped.

**Habitat**
Warm waters.

**Range**
Argentina to N.S.

## ATLANTIC MOONFISH
*Vomer setapinnis*                                   **81:9**

**Description**
Size, to 1 ft. (0.3 m). Deep, thin body; concave head; low dorsal
and anal fins. Pectorals short; first dorsal reduced to 8 short spines;
second dorsal and anal low and even in height.

**Habitat**
Warm seas.
**Other names**
Horsefish, Bluntnose, Shiner, Dollarfish.
**Range**
Uruguay to Cape Cod, straggling to N.S. The young appear off
N.Y. in late summer, brought up by the Gulf Stream.

## PALOMETA
*Trachinotus goodei*                                           81:11

**Description**
Size, to 1 ft. (0.3 m). Body deep, thin; black crossbars on sides.
First dorsal and anal prolonged past middle of tail fin.
**Habitat**
Clear waters with sandy bottoms; rare.
**Range**
Argentina to Mass.

## PERMIT
*Trachinotus falcatus*                                         81:10

**Description**
Size, to 18 in. (45.7 cm). Profile arched; dorsal fin spear longer
than head. Lacks black crossbars of Palometa on sides, and dorsal-
and anal-fin spines rarely reach to tail.
**Habitat**
Shallow water over sandy bottoms in bays and near islands.
**Range**
Brazil to Cape Cod.

## FLORIDA POMPANO
*Trachinotus carolinus*                                        81:12

**Description**
Size, to 18 in. (45.7 cm) and 2 lb. (0.9 kg). Profile not strongly
arched; spines of dorsal and anal fins very short.
**Habitat**
Shallow water close to shore.
**Range**
Gulf and Atlantic coasts n. to Cape Cod; only young common
above Va.

## LEATHERJACKET
*Oligoplites saurus*                                           81:13

**Description**
Size, to 12 in. (30.5 cm). Rear parts of soft dorsal and anal fins
divided into 12 low brushlike finlets. Scales embedded in corrugated
skin giving fish a leathery look; lateral line nearly straight. First
dorsal reduced to 5 short separate spines, and preceded by 2 stout
spines; caudal peduncle slightly keeled.
**Habitat**
Clear, swift water, often in bays or inlets.
**Habits**
Often leaps from the water as it pursues its prey.
**Range**
Tropical waters n. to N.Y., sometimes to Cape Cod.

## BANDED RUDDERFISH
*Seriola zonata*                                          81:14

**Description**
Size, to 3 ft. (0.9 m). Body thin, nose pointed. Anal half as long as second dorsal and, in young, preceded by 1–2 short spines. Young have sides crossbarred with 5–6 broad dark bands which fade or disappear with age.
**Habits**
Travels in schools following drift or large fish.
**Other names**
Shark Pilot, Slender Amberjack.
**Range**
Gulf of Mexico to N.S.

## GREAT AMBERJACK
*Seriola dumerili*                                        81:6

**Description**
Size, to 6 ft. (1.8 m); all-tackle record, 120 lb. (54.5 kg). Resembles smaller Banded Rudderfish but has longer second dorsal (36–38 rays instead of 30–34).
**Habitat**
Warm waters over offshore flats and reefs, around wrecks.
**Remarks**
Both fish and roe make good eating.
**Range**
West Indies to N.J., rarely to Cape Cod.

# DOLPHINS
Family Coryphaenidae

## DOLPHIN
*Coryphaena hippurus*                                     77:9

**Description**
Size, to 6 ft. (1.8 m); all-tackle record, 76 lb. (34.5 kg). Distinguished by high arched head, long dorsal fin, brilliant color, swiftness, and spectacular leaps. Sides gold with blue polka dots, tail golden; changes color rapidly when dying. Body flattened sideways; head massive, blunt (forehead higher in male); small comblike teeth in jaws and roof of mouth. Ventrals below pectorals, and ½ height and ½ length of dorsal; tail deeply forked.
**Habitat**
Near drift in warm open seas, notably Gulf Stream.
**Habits**
Solitary or in schools; often feed under kelp beds.
**Food**
Fishes, particularly flying fishes.
**Remarks**
Both fish and roe make good eating.
**Range**
Warm seas n. to Cape Cod, occasionally to N.S.

# TRIPLETAILS
Family Lobitidae

## TRIPLETAIL
*Lobotes surinamensis*                                    *Fig. 105*

**Description**
Size, to 3 ft. (0.9 m). Rear dorsal and anal fins prominently
extended to the rear, suggesting a triple tail. Young light, often
mottled; adults almost black. Body large, deep, compressed; scales
small, rough; lower jaw projects, head concave above eye. First
dorsal of stout spines joins second dorsal; tail rounded.
**Habitat**
Drifting debris in channels; near wrecks, pilings.
**Food**
Mainly fishes.
**Range**
Warm waters, Argentina to Cape Cod.

*Fig. 105*

Pigfish

Tripletail

# GRUNTS
Family Pomadasyidae

## PIGFISH
*Orthopristis chrysoptera*                                *Fig. 105*

**Description**
Size, to 15 in. (38.1 cm). Noted for piglike mouth, continuous
dorsals, forked tail, and pointed pectorals. Body compressed, mouth
small with lengthwise groove on chin and projecting lower jaw;
teeth small, pointed. Fins more or less scaly, scaly flap at base of
ventrals.
**Habitat**
Sandy shores (most commonly); grassy flats.
**Food**
Annelids, crustaceans, fishes.
**Range**
Mass. to Fla.

# CROAKERS
Family Sciaenidae

In croakers the lateral line extends to the back edge of the caudal fin and the anal fin has only one or two spines. The spiny and soft-rayed dorsals are often continuous. Most members of the family have air bladders which can vibrate, producing the sounds from which the name is derived. Many species are valued for sport and food.

## BANDED DRUM
*Larimus fasciatus*                                    82:11

**Description**
Size, to 10 in. (25.4 cm). Body somewhat compressed, back arched; mouth large, oblique; lower jaw projects, teeth in both jaws; tail wedge-shaped.
**Habitat**
Offshore.
**Range**
Mass. to Fla., only as a straggler n. of Chesapeake Bay.

## SILVER PERCH
*Bairdiella chrysura*                                  82:13

**Description**
Size, to 1 ft. (0.3 m). Straight tail and medium-sized, only slightly oblique mouth; 2 anal spines and lateral line extending to rear edge of tail.
**Habitat**
Sandy shores.
**Range**
N.Y. to Tex.; season, N.Y., June to Dec.

## RED DRUM
*Sciaenops ocellata*                                   77:12

**Description**
Size, to 5 ft. (1.5 m); all-tackle record, 83 lb. (37.6 kg). Black spot at base of tail; lateral line extending to the back edge of tail. Reddish stripes, upper jaw projects.
**Habitat**
Sandy shores.
**Food**
Mostly crustaceans.
**Other names**
Channel Bass, Redfish.
**Range**
Mass. to Fla.

## ATLANTIC CROAKER
*Micropogon undulatus*                                 77:6

**Description**
Size, to 12 in. (30.5 cm). Distinguished from its relatives by row of small barbels on each side of lower jaw.
**Habitat**
Sandy bottoms.
**Food**
Mostly worms, crustaceans, fishes taken largely from the bottom.

**Reproduction**
Spawn in cold half of the year.
**Other name**
Hardhead.
**Range**
Mass. to Argentina, not common n. of N.J.; great fluctuations in population size.

# SPOT
*Leiostomus xanthurus* **76:4**

**Description**
Size, to 14 in. (35.6 cm). Black spot behind gill cover; lateral line extending to back edge of tail. Snout blunt, mouth small. Spiny dorsal triangular, with top rounded; second dorsal only ½ as high; tail concave.
**Habitat**
Inshore waters.
**Range**
Gulf of Maine to Tex. Quite variable in numbers; have some years of great runs.

# SOUTHERN KINGFISH
*Menticirrhus americanus* *Fig. 106*

**Description**
Size, to 15 in. (38.1 cm). Closely resembles Northern Kingfish, but normally light in color, and longest spine of first dorsal does not reach front of second dorsal.
**Habitat**
Shallow, sandy waters; surf; inlets.
**Food**
Invertebrates and small fishes.
**Other name**
Southern King Whiting.
**Range**
N.Y. to Argentina.

*Fig. 106*

Southern Kingfish

# NORTHERN KINGFISH
*Menticirrhus saxatilis* **76:1**

**Description**
Size, to 17 in. (43.2 cm) and 3 lb. (1.4 kg). Has chin barbel. Third spine of the first dorsal greatly extended and, in adults, threadlike; reaches past second dorsal. Dark in color. Lower half of tail fin rounded, upper half concave.
**Habitat**
Sandy shores.

**Habits**
Bottom feeder.
**Food**
Crabs, squid, shrimps, young fishes.
**Other names**
King Whiting or Northern Whiting, Minkfish.
**Range**
Maine to Fla.

## BLACK DRUM
*Pogonias cromis*                                         **82:14**

**Description**
Size, to 4 ft. (1.2 m) and 146 lb. (66.2 kg); all-tackle record, 94 lb. (42.6 kg). Deep body, high arched back, many barbels. Scales large, first dorsal triangular, second dorsal oblong, pectorals long and pointed, anal fin with 2 spines, tail almost square. Young characterized by 4–6 broad black bands on the sides.
**Habitat**
Inlets and surf of sandy shores.
**Food**
Shellfish.
**Range**
Cape Cod to Argentina.

## WEAKFISH
*Cynoscion regalis*                                       **77:2**

**Description**
Size, to 32 in. (81.3 cm); all-tackle record, 17½ lb. (7.9 kg). Snout pointed, mouth large, lower jaw projects, 2 canine teeth in upper jaw. First dorsal triangular, second longer and rectangular; soft rays of dorsal and anal scaled; tail broad and slightly concave. Lacks the spots on the fins and tail of the spotted sea trout.
**Habitat**
Shallow water.
**Habits**
Travels in schools.
**Food**
Crabs, shrimps, fishes.
**Reproduction**
Many thousands of eggs, hatching 48 hours at 65°F (18°C), in June and July.
**Age**
To 12 years.
**Other names**
Gray Weakfish, Sea Trout.
**Remarks**
This is a very popular and important species both commercially and for sport. Its jaws are weakly set in its head and are sometimes pulled out by the fisherman's hook, whence the name is derived.
**Range**
N.S. to Fla.

## SPOTTED SEA TROUT
*Cynoscion nebulosus*                                     **82:12**

**Description**
Size, to 7 lb. (3.2 kg); all-tackle record, 15 lb. (6.8 kg). Similar to weakfish in appearance and habits, but has black spots on the second dorsal and tail, and soft rays of dorsal anal fins are scaleless.

**Habitat**
Coastal bays and estuaries.
**Remarks**
One of the important food fishes of Chesapeake Bay.
**Other name**
Spotted Squeteague.
**Range**
N.Y. to Tex., but scarce n. of Del.; season, Chesapeake Bay, Mar. to Dec.

# PORGIES
Family Sparidae

## SCUP
*Stenotomus chrysops*         **76:6**

**Description**
Size, to 18 in. (45.7 cm) and 4 lb. (1.8 kg). Body thin, depth ½ length; scales large, thick; head short, forehead concave, eye set high. Pectorals long, dorsal long and preceded by a spine; anal and dorsal fins can be laid back in grooves; tail deeply concave.
**Habitat**
Sandy bottoms inshore and offshore.
**Habits**
Ground feeder.
**Food**
Crustaceans, worms, fishes, squid.
**Age**
To at least 5 years.
**Other name**
Northern Porgy.
**Range**
N.S. to e. Fla.

## PINFISH
*Lagodon rhomboides*         **83:8**

**Description**
Size, to 10½ in. (26.7 cm). Similar to Scup but with molars in each jaw, shorter pectorals, notched incisors, and a widely forked tail.
**Habitat**
Shallow water near pilings.
**Range**
Cape Cod to Cuba, but common only Del. and s.

## SHEEPSHEAD
*Archosargus probatocephalus*         **83:7**

**Description**
Size, to 30 in. (16.2 cm) and 20 lb. (9.1 kg). Body has 7 broad vertical bands. Lacks concave forehead of Scup.
**Habitat**
Near pilings and jetties close to the bottom; tidal streams and bays.
**Food**
Mollusks and crustaceans crushed by teeth.
**Reproduction**
Eggs hatch in 40 hours at 77°F (25°C).
**Range**
N.S. to Fla., rare n. of Va.

# GOATFISHES
Family Mullidae

## RED GOATFISH
*Mullus auratus* 82:10

**Description**
Size, to 8 in. (20.3 cm). Red with a long double barbel. Dorsal fins high and well separated, ventrals beneath pectorals, anal beneath second dorsal; tail well forked.
**Habitat**
Sandy bottoms.
**Range**
N.S. to West Indies.

# SEA CHUBS
Family Kyphosidae

## BERMUDA CHUB
*Kyphosus sectatrix* 83:9

**Description**
Size, to 18 in. (45.7 cm). Dorsals continuous, low, almost even height; 2 spines in front of low anal fin. Body deep, compressed, dull in color; both jaws with small incisor teeth; tail forked.
**Habitat**
Shoals and bars.
**Food**
Omnivorous, including offal.
**Other name**
Rudderfish.
**Range**
Mass. to Brazil.

# WRASSES
Family Labridae

These are chiefly tropical fishes with large scales. Their mouths are small and thick-lipped, with strong canine teeth. Their dorsal fins are continuous, usually with a long spiny portion. The family goes back to the Paleocene.

## CUNNER
*Tautogolabrus adspersus* 83:10

**Description**
Size, to 15 in. (38.1 cm) and 2½ lb. (1.1 kg). Distinguished from Tautog by extended snout, saw-toothed preopercle, and scaly gill cover.
**Habitat**
Rocky bottoms to 35 fathoms (64 m), and wharves.
**Food**
Omnivorous.
**Reproduction**
Spawns June to early July.
**Age**
To 7 years.

**Other name**
Bergall.
**Range**
Newfoundland to Chesapeake Bay.

# TAUTOG
*Tautoga onitis*                                                          76:8

**Description**
Size, to 36½ in. (92.7 cm) and 22½ lb. (10.2 kg); all-tackle record,
21 lb. (9.5 kg). Distinguished from Cunner by blunt snout,
untoothed preopercle, and naked gill cover.
**Habitat**
Shallow water over rocky bottoms near wrecks, pilings.
**Habits**
When not feeding, have the odd habit of often lying crowded
together on their sides in a hole amid rocks.
**Food**
Mollusks, crustaceans.
**Age**
Probably to 10 years.
**Other name**
Blackfish.
**Range**
N.B. to S.C.

# BARRACUDAS
Family Sphyraenidae

These fishes have strong jaws and teeth and a small first dorsal of
a few spines, set well forward of the second. The family goes back
to the Eocene.

## GREAT BARRACUDA
*Sphyraena barracuda*                                                     83:3

**Description**
Size, to 10 ft. (3 m); all-tackle record, 103 lb. (46.7 kg). Scales and
teeth large, lower jaw projects and has fleshy tip. First dorsal
arises behind root of ventrals.
**Habitat**
Bays, inlets, and surface water over reefs.
**Food**
Fishes.
**Range**
Brazil to Fla., straggling to Cape Cod.

## NORTHERN SENNET
*Sphyraena borealis*                                                      83:5

**Description**
Size, to 12 in. (30.5 cm). Tiny version of Great Barracuda, lacking
fleshy tip on jaw.
**Habitat**
Coastal inshore waters.
**Food**
Fishes.
**Range**
Cape Cod and s.; N.Y., June to Nov.; Woods Hole, July to Dec.

# MULLETS
Family Mugilidae

These are medium-sized bay fishes with silvery, cylindrical bodies, small toothless mouths, a small first dorsal of four slender spines, and a longer, soft-rayed second dorsal. The family goes back to the Oligocene.

## STRIPED MULLET
*Mugil cephalus*      83:4

**Description**
Size, to 2½ ft. (0.8 m). Sides striped; rear dorsal and anal fins almost scaleless.
**Habitat**
Warm and temperate seas.
**Food**
Young feed on plankton, adults on detritus and associated microscopic benthic animals.
**Remarks**
As diet changes from plankton to detritus, intestine gets longer and more convoluted in order to cope with new digestive problems.
**Range**
N. to N.Y., occasionally to N.S.

## WHITE MULLET
*Mugil curema*      83:6

**Description**
Size, to 3 ft. (0.9 m). Sides plain, unstriped; rear dorsal and anal fins scaled.
**Habitat**
Warm and temperate seas.
**Food**
Same as Striped Mullet.
**Range**
Cape Cod to Brazil, more abundant to s.

# STARGAZERS
Family Uranoscopidae

## NORTHERN STARGAZER
*Astroscopus guttatus*      *Fig. 107*

**Description**
Size, to 12 in. (30.5 cm). Eyes on top of head; large vertical mouth. Electric organ behind eye.
**Habitat**
Sand bottoms.

*Fig. 107*

Rock Gunnel            Northern Stargazer

**Habits**
Remains buried for long intervals with only eyes and nostrils protruding.
**Remarks**
Specimens from 6 in. (15.2 cm) up of this unusual-looking species can give an electric shock either to stun their prey or for self-defense.
**Range**
N.Y. to Va.

# WOLFFISHES
Family Anarhichadidae

## ATLANTIC WOLFFISH
*Anarchicas lupus*                                                  83:2
**Description**
Size, to 5 ft. (1.5 m) and 40 lb. (18.1 kg). Body tapers from savage-looking head to small tail; tusks large, projecting; teeth formidable. Dorsal, of flexible spines of uniform height, extends from nape to base of tail. Pectorals large and rounded; no ventrals; anal ½ height and length of dorsal, and with angular rear corner.
**Habitat**
Hard bottoms and seaweed.
**Habits**
Solitary bottom feeder.
**Food**
Hard-shelled echinoderms, mollusks, and crustaceans.
**Remarks**
This large fish is dangerous to handle when caught and is reported to have attacked waders in rock pools.
**Range**
S. Labrador to Nantucket.

# GUNNELS
Family Pholidae

## ROCK GUNNEL
*Pholis gunnellus*                                               Fig. 107
**Description**
Size, to 12 in. (30.5 cm). Unique for low, long spiny dorsal of uniform height, with similar half-length anal, both separated from small, rounded tail by only a notch. Body eellike, scales very small; head short, naked; mouth small and oblique, several rows of teeth in upper jaw, 1 in lower. Pectorals small, rounded; ventrals inconspicuous and reduced to 1 short spine and 1 ray apiece.
**Habitat**
Tide pools and seaweed down to 40 fathoms (73.2 m) with gravelly ground or shell beds.
**Habits**
Swims like an eel, curls about eggs.
**Food**
Worms, crustaceans, mollusks.
**Other name**
Rock Eel.
**Range**
Hudson Strait to Delaware Bay.

# WRYMOUTHS
Family Cryptacanthodidae

## WRYMOUTH
*Cryptacanthodes maculatus*                                    83:1

**Description**
Size, to 3 ft. (0.9 m). Oblique mouth, continuous vertical fins, oval
tail. Body naked, eellike, much flattened sidewise; head flat-topped,
eyes high, mouth oblique; dorsal, caudal, and anal fins continuous;
dorsal spiny, pectorals low; no ventrals.
**Habitat**
Bottom fish.
**Habits**
Burrows in mud.
**Food**
Fishes, mollusks, crustaceans.
**Other name**
Ghostfish.
**Range**
Labrador to N.J.

# SAND LANCES
Family Ammodytidae

## AMERICAN SAND LANCE
*Ammodytes americanus*                                    83:12

**Description**
Size, to 8½ in. (21.6 cm). Distinguished from other eel-shaped
fishes by forked tail. Head long, nose pointed; mouth large,
toothless, with lower jaw projecting. Dorsal long, low, soft-rayed,
and more than twice length of low anal; pectorals low, pointed; no
ventrals.
**Habitat**
Sandy shores.
**Habits**
Occurs in dense schools; swims like an eel; burrows with great
speed into sandy beach, where it remains at low tide.
**Food**
Small crustaceans.
**Other name**
Sand Eel.
**Range**
Labrador to Cape Hatteras.

# CUTLASSFISHES
Family Trichiuridae

## ATLANTIC CUTLASSFISH
*Trichiurus lepturus*                                    83:11

**Description**
Size, to 5 ft. (1.5 m). Pointed whiplike tail; no caudal fin. Body
naked, bandlike, tapering. Eye large; snout pointed; mouth with
long barbed fangs, 4 in upper jaw, 2 in lower jaw, which projects.

Dorsal soft-rayed, extending from nape to near tip of tail; no ventrals; long anal of very low spines.
**Habitat**
Surface waters of warm seas.
**Food**
Fishes.
**Other names**
Hairtail, Scabbardfish, Ribbandfish.
**Range**
Mass. to Argentina.

# MACKERELS AND TUNAS
Family Scombridae

These are swift, powerful, spindle-shaped fishes, frequently of large size. Distinctive characteristics are small finlets behind both dorsal and anal fins; and from one to three keels on the thin caudal peduncle. Mackerels have velvety skins, small scales, no bony scales on the rear lateral line, both soft and spiny dorsal fins, and deeply forked or crescent-shaped tails. The family goes back to the Eocene. All eastern mackerels are highly regarded as food for man.

## SPANISH MACKEREL
*Scomberomorus maculatus*                           **82:3**

**Description**
Size, to 36 in. (91.4 cm) and 10 lb. (4.5 kg). Yellow spots on upper sides; blue above, silvery below. Mouth oblique and with 32 large conical jaw teeth; lateral line bends in center. Second dorsal concave, pectorals scaleless, ventrals originating behind origin of first dorsal, anal symmetrical with second dorsal; tail deeply concave.
**Habitat**
Warm surface waters offshore; bays, inlets.
**Food**
Fishes.
**Range**
Maine to Brazil.

## CERO
*Scomberomorus regalis*                             **82:4**

**Description**
Size, average 5–10 lb. (2.3–4.5 kg). Gradually sloping lateral line; narrow brown stripe from behind pectoral fin to tail. Mouth oblique and with 40 large conical jaw teeth. Spots on sides in rows below lateral line. Upper half of first dorsal deep blue; pectorals scaled; ventrals below (instead of behind) origin of first dorsal, symmetrical with second dorsal.
**Habitat**
Warm waters offshore.
**Other name**
Painted Mackerel.
**Range**
Brazil to N.C., rarely to Cape Cod.

## KING MACKEREL
*Scomberomorus cavalla*                                        82:7

**Description**
Size, to 5 ft. (1.5 m) and 100 lb. (45.4 kg); all-tackle record, 77 lb.
(34.9 kg). Lateral line with abrupt bend downward in the middle;
mouth oblique and with 40 large conical jaw teeth; first dorsal
concave and uniform in color.

**Habitat**
Warm waters inshore or offshore.

**Remarks**
This is one of the fastest and most game of all fish, a furious
striker and a savage fighter. It can make long twisting underwater
runs and leap as high as 10 feet (3 m) in the air. It will often
follow a hooked fish to the boat and be caught in the process.

**Other names**
Cero, Kingfish.

**Range**
Brazil to N.C., rarely to Gulf of Maine.


## ATLANTIC MACKEREL
*Scomber scombrus*                                             76:3

**Description**
Size, to 22 in. (55.9 cm) and 4 lb. (1.8 kg). Widely separated
dorsal fins; sides uniformly silver below lateral line. Scales very
small; teeth small, sharp; 2 small longitudinal keels (but no median
lateral keel) on the caudal peduncle. First dorsal with 11 or more
spines, which can be laid down in a groove on the back; 4–6 finlets
behind second dorsal, 5 behind anal.

**Habitat**
Open sea.

**Habits**
Found in vast schools of many thousands. A swift swimmer; uses
so much oxygen that in warm water it must keep swimming all the
time to maintain the necessary flow of water to its gills.

**Food**
Plankton and small fishes.

**Age**
To at least 8 years.

**Remarks**
One of the greatest food fishes.

**Range**
N. Atlantic from Labrador to Cape Hatteras; season, N.Y., May to
July; Woods Hole, May to Dec.


## CHUB MACKEREL
*Scomber japonicus*                                            82:5

**Description**
Size, to 25 in. (63.5 cm). Fine black network of lines on back,
lower part of sides mottled with dusky blotches, eye larger, gap
between dorsals smaller, 9–10 spines in first dorsal.

**Habitat**
Temperate waters, inshore or offshore.

**Range**
N.S. to Brazil; season, N.Y.: June to Sept., but irregular. Varies
greatly in numbers.

## SKIPJACK TUNA
*Euthynnus pelamis*                                    82:2

**Description**
Size, to 34½ in. (87.6 cm); all-tackle record, 31 lb. (14.1 kg). Has
4 long stripes on sides. Body robust; lateral line curves and is above
side stripes. First dorsal high, slopes sharply down and back;
second dorsal symmetrical with anal; tail slightly concave.
**Habitat**
High seas.
**Food**
Flying fishes.
**Other name**
Oceanic Skipjack.
**Range**
Tropical waters n. to Gulf of Maine.

## LITTLE TUNA
*Euthynnus alleteratus*                                82:8

**Description**
Size, to 3 ft. (0.9 m) and 20 lb. (9.1 kg). Dorsals triangular and
concave; anal originates behind second dorsal; tail concave. Black
stripe above the gently sloping lateral line.
**Habitat**
Open sea.
**Food**
Fishes.
**Other name**
False Albacore.
**Range**
Gulf of Maine to Brazil.

## BLUEFIN TUNA
*Thunnus thynnus*                                      77:13

**Description**
Size, to 14 ft. (4.8 m) and 1600 lb. (725.6 kg); all-tackle record,
977 lb. (443.1 kg). Very large; plain colors; continuous dorsals, 9
or more dorsal finlets. Lower gill rakers, first arch, 24–28. Steel-
blue above, silver below, no markings; body robust, scaled; jaw
teeth small. First dorsal triangular; can be laid down. Second
dorsal higher than long, concave and with top pointed. Short
second dorsal and anal do not reach peduncle, which bears strong
median longitudinal keel; anal finlets 8–9.
**Habitat**
Open sea.
**Habits**
Frequently jumps or splashes about at surface; younger ones of
same size school together; very large ones often solitary.
**Food**
Menhaden, mackerel, other fishes and squid.
**Remarks**
One of the great food resources of the deep.
**Other names**
Horse Mackerel; small individuals are called School Tuna.
**Range**
N. to Newfoundland.

## YELLOWFIN TUNA
*Thunnus albacares*                                        **82:9**

**Description**
Size, to 6 ft. (1.8 m); all-tackle record, 265 lb. (120.2 kg). Lower
gill rakers, first arch, 18–22. Very long second dorsal and anal fins
which reach to the caudal peduncle. Pectorals long and slim; tail
concave.
**Habitat**
Offshore.
**Range**
N.S. to Md. and s.

## BLACKFIN TUNA
*Thunnus atlanticus*                                       *Fig. 108*

**Description**
Size, average to 25 lb. (9.3 kg); all-tackle record, 44 lb. (20 kg).
Pectoral more than ⅘ but less than 1⅕ length of head, usually
does not reach anal. Lower gill rakers, first arch, 15–18; upper, 3–
7; total, 20–24. Dorsal finlets dusky brown, anal finlets dusky steel;
rear edge of tail not white.
**Habitat**
Offshore.
**Range**
Mass. to Brazil.

*Fig. 108*

Blackfin Tuna

## ALBACORE
*Thunnus alalunga*                                         **82:1**

**Description**
Size, to 5 ft. (1.5 m), average 15–30 lb. (6.8–13.6 kg); all-tackle
record, 55 lb. (24.9 kg). Lower gill rakers, first arch, 15–22. Very
long pectorals reaching to the third dorsal finlet. Rear edge of tail
white. Body stout and with small scales; first dorsal long and
sloping; second dorsal and anal concave, pointed.
**Habitat**
Near surface offshore.
**Habits**
Migratory, erratic; schools.
**Food**
Small fishes, plankton, squid.
**Range**
N.J. to Lesser Antilles.

## ATLANTIC BONITO
*Sarda sarda*                                              **77:11**

**Description**
Size, to 3 ft. (0.9 m) and 12 lb. (5.4 kg). Faint, iridescent stripes
that slant upward to the rear. Body stout, scaly; mouth and jaw

teeth large; median longitudinal keel on peduncle. Dorsals join; first dorsal triangular, long, tapering. Second dorsal, anal, and caudal concave; dorsal finlets, 7–8; anal finlets, 7.

**Habitat**
Open sea, notably the Gulf Stream.

**Range**
Abundant n. to Cape Cod, rare to N.S.

# MARLINS
Family Istiophoridae

Marlins differ from the swordfishes in having a shorter, rounded (not flattened) sword; narrow, often embedded, scales; and a very long first dorsal, capable of being depressed into a groove. Their edibility varies from poor to good, depending upon locality and whether fresh or smoked. The family goes back to the Eocene.

## BLUE MARLIN
*Makaira nigricans*                                              82:6

**Description**
Size, to 15 ft. (4.6 m) and 1000 lb. (453.5 kg); all-tackle record, 756 lb. (342.9 kg). Scales thornlike; no lateral line; caudal peduncle with 2 small longitudinal keels. Dorsal and anal fins, 2 each; first dorsal very long, second dorsal and second anal low, short. Pectorals long, pointed; ventrals reduced to 2 long spines; tail concave, much broader than long. Differs from White Marlin in its thick body; darker coloration, blue above, silver below; and pointed dorsal fin that is brilliant cobalt.

**Habitat**
Gulf Stream and environs.

**Habits**
Generally solitary; often leaps from water.

**Food**
Mackerel, mullet, and other fishes.

**Remarks**
One of the great game fishes.

**Range**
N. to Mass.

## WHITE MARLIN
*Tetrapturus algidus*                                           77:1

**Description**
Size, to 8½ ft. (2.6 m); all-tackle record, 161 lb. (73.0 kg). Similar to larger Blue Marlin but body slender, belly white, lateral line conspicuous, and first dorsal rounded.

**Habitat**
1–25 miles (1.6–40.2 km) offshore.

**Habits**
Solitary or in schools.

**Food**
Other fishes.

**Range**
Warm Atlantic n. to N.S.

# SWORDFISHES
Family Xiphiidae

## SWORDFISH
*Xiphias gladius*                                                77:10

**Description**
Size, to 16 ft. (4.9 m); all-tackle record, 1182 lb. (536.0 kg). Lacks
scales and ventral fins. Sword flattened toward the tip; first dorsal
short, high, and sharklike. Sword nearly ⅓ total length of fish;
mouth large, toothless, and with pointed lower jaw; peduncle with
longitudinal keel. Dorsal and anal fins, 2 each; first anal similar to
first dorsal but far to rear, second dorsal and second anal very
small and near caudal peduncle; tail concave.
**Habitat**
Open sea.
**Habits**
Usually solitary, sometimes suns at surface.
**Food**
Squid, other fishes.
**Other name**
Broadbill.
**Range**
Warm Atlantic n. to Newfoundland; season, N.Y., June to Sept.;
N.S., July to Sept.

# BUTTERFISHES
Family Stromateidae

## BUTTERFISH
*Peprilus triacanthus*                                          *Fig. 109*

**Description**
Size, to 12 in. (30.5 cm) and 1½ lb. (0.7 kg). No pelvics; long, low
anal; a forked tail. Body deep and very thin, scales small, head
short, snout blunt, mouth small, teeth weak. Dorsal and anal fins
somewhat symmetrical.
**Habitat**
Near surface over sandy bottoms.
**Food**
Plankton, small fishes, squid, crustaceans.
**Remarks**
A good pan fish.
**Range**
N.S. to Cape Hatteras.

*Fig. 109*                                                      Butterfish

# GOBIES
Family Gobiidae

Most gobies occur abundantly on mud flats and in tidepools and some in freshwater streams. Several are important to fishermen as bait. Gobies are distinguished by their pelvic fins that are completely joined. These fishes are usually small-sized carnivorous bottom feeders.

## NAKED GOBY
*Gobiosoma bosci*

**Description**
Size, to 2 in. (5.1 cm). Pelvic fins form a sucking disk. Body naked, cylindrical, and without caudal peduncle. Dorsal fins short and high, first spiny, second soft and longer than first; anal short and high; tail rounded.
**Habitat**
Grassy flats.
**Food**
Small crustaceans.
**Reproduction**
Eggs become elliptical in shape; often attached to oyster shell.
**Range**
Cape Cod and s.; particularly abundant in Chesapeake Bay.

# SCORPIONFISHES
Family Scorpaenidae

## REDFISH
*Sebastes marinus*                                      **76:5**

**Description**
Size, to 2 ft. (0.6 m) and 13½ lb. (6.1 kg). All red; projecting lower jaw. Head concave; mouth large and oblique; bony knob at tip of lower jaw; gill covers large, pointed near top. Dorsal fin continuous, spiny part longer, soft part higher; pectorals large, lozenge-shaped; anal fin has 3 spines and 7 or 8 rays; tail small, concave.
**Habitat**
Cool water to 260 fathoms (0.5 km).
**Food**
Crustaceans, small fishes.
**Age**
To 20 years.
**Remarks**
Valuable commercially.
**Other name**
Ocean Perch.
**Range**
N. Atlantic, s. to Cape Cod, rarely to N.J.

# SEA ROBINS
Family Triglidae

The members of this family resemble sculpins, with their broad heads, slender bodies, and large fanlike pectoral fins. They differ, however, in that their entire heads are armored with rough bony and spiny plates and each pectoral fin is preceded by three detached fingerlike rays. These rays bend downward at the tips; as the fishes rest on the bottom they drag themselves forward by hooking the rays into the sand. The family goes back to the Eocene.

## NORTHERN SEA ROBIN
*Prionotus carolinus*       **77:4**

**Description**
Size, to 16 in. (40.6 cm). Snout flat and depressed; 2 short spines over each eye; 1 spine on each cheek, side of neck, and shoulder. Spiny and soft dorsal fins barely separated, first higher, second longer; pectorals long, rounded, and extending past origin of second dorsal; ventrals long; anal symmetrical with second dorsal. Distinguished from Striped Sea Robin by absence of longitudinal stripe on body and by concave tail.

**Habitat**
Bottoms.

**Food**
Crustaceans, squid, small fishes.

**Reproduction**
Spawns in summer at N.Y.

**Other name**
Carolina Sea Robin.

**Range**
Bay of Fundy to S.C., common Cape Cod to Hatteras.

## STRIPED SEA ROBIN
*Prionotus evolans*       *Fig. 110*

**Description**
Size, to 18 in. (45.7 cm). Pectorals very long, overlapping most of second dorsal. Distinguished from Northern Sea Robin by a dusky stripe on each side below the lateral line, and by a straight instead of a concave tail.

**Range**
Mass. to Fla.

*Fig. 110*

Striped Sea Robin

# SCULPINS
Family Cottidae

Sculpins resemble sea robins, having broad heads, slender bodies, and large fanlike pectoral fins. They differ, however, in that the

tops of their heads are naked, instead of armored, and in that they
have no detached rays preceding the pectoral fins. They are chiefly
Arctic fishes, and voraciously bite on any bait. The family goes
back to the Oligocene. They are sometimes used as bait for lobster
pots.

## LONGHORN SCULPIN
*Myoxocephalus octodecemspinosus*                                     84:8

**Description**
Size, to 18 in. (45.7 cm). Body elongate, slender; head blunt,
flattened; 3 spines on the preopercle, upper one at least 4 times as
long as the one below. First dorsal sooty with irregular markings,
second dorsal with 3 or 4 dark crossbars.
**Habitat**
Open ocean.
**Habits**
Bottom feeder.
**Food**
Crustaceans, mollusks, squid; scavenger.
**Range**
Coastal waters, Newfoundland to Va.

Note: The **SHORTHORN SCULPIN,** *Myoxocephalus scorpius* **(84:7),**
has a more rounded body and a short spine on the preopercle.

## GRUBBY
*Myoxocephalus aenaeus*                                               84:6

**Description**
Size, to 6½ in. (16.5 cm). Differs from other sculpins in the
following combination of characteristics: longitudinal ridge with 2
spines along top of head over each eye; 2 spines between nostrils; 6
short spines on each side of face. Spiny dorsal fin originates slightly
in front of upper corner of gill opening and is separated from soft
dorsal by deep notch. Soft dorsal lower than spiny and about equal
in length, ventral fins of 3 rays each.
**Habitat**
Open oceans.
**Habits**
Bottom feeder.
**Range**
Coastal waters, Gulf of St. Lawrence to N.J.

## SEA RAVEN
*Hemitripterus americanus*                                           84:9

**Description**
Size, to 25 in. (63.5 cm) and 7 lb. (3.2 kg). Fleshy tabs on head.
Brown or purplish, sometimes bright yellow; skin prickly, both
jaws with large teeth. Spiny dorsal highest at front, jagged in
outline, higher and longer than soft dorsal; ventrals have 3 rays;
anal longer than second dorsal; tail brush-shaped.
**Habitat**
Rock, pebble, or sand bottom, 2–50 fathoms (3.7–91.4 m).
**Food**
Invertebrates, small fishes.
**Range**
Newfoundland to Cape Cod, some to Chesapeake Bay.

# FLYING GURNARDS
Family Dactylopteridae

### FLYING GURNARD
*Dactylopterus volitans*                    *Fig. 111*

**Description**
Size, to 1 ft. (0.3 m). Can skim through the air on its pectorals or
walk in the water with its ventrals as if on stilts. Eye large and set
high; armor of head extends back in a spine to beyond origin of
spiny dorsal. First 2 spines of dorsal separate; pectorals enormous
and reach caudal peduncle when laid back; first 5–6 rays of each
pectoral, with membrane, form a separate fin; tail concave.
**Habitat**
Warm waters.
**Range**
N. to N.Y., rarely to Cape Cod.

*Fig. 111*

Flying Gurnard

# Clingfishes
Order Gobiesociformes

# CLINGFISHES
Family Gobiesocidae

### SKILLETFISH
*Gobiesox strumosus*                    *Fig. 112*

**Description**
Size, 4 in. (10.2 cm). Dark, tadpole-shaped body; naked skin; large
ventral sucking disk.
**Habitat**
Clings by sucking disk to submerged objects, such as piles, rocks,
and shells.
**Food**
Small crustaceans.
**Range**
N.J. to Brazil.

*Fig. 112*

Skilletfish

# Flatfishes
Order Pleuronectiformes

Members of this order are bottom fishes and have the body flattened so that the fish rests on one side on the ocean floor. The underside is white, and adults have both eyes on the upper side of the head. The young have one eye on each side and swim like other fishes; but in the course of a few weeks one eye travels over the top of the head to join its mate and the fish rests on one side. The order goes back to the Paleocene. All are edible, and several species are taken in great numbers each year by commercial fishermen.

## LEFTEYE FLOUNDERS
Family Bothidae

Members of this family are sinistral, that is, they have the eyes and coloring on the left side of the body in adults. Pelvic fins are asymmetrically located on abdominal ridge.

### SUMMER FLOUNDER
*Paralichthys dentatus*                    77:15

**Description**
Size, to 46 in. (1.2 m) and 26 lb. (11.8 kg); all-tackle record, 20 lb. (9.1 kg). Eyes on left; mouth large; single dorsal fin extends from above eyes to peduncle. Pectoral and ventrals small; anal extending from beneath origin of pectoral to peduncle; tail somewhat wedge-shaped. Differs from Fourspot Flounder in having many small dark spots.
**Habitat**
Any kind of bottom; moves offshore in winter.
**Food**
Crustaceans, mollusks, fishes.
**Remarks**
One of the tastiest flatfishes.
**Other names**
Fluke, Summer Fluke.
**Range**
Cape Cod to Fla.

### FOURSPOT FLOUNDER
*Paralichthys oblongus*                    84:14

**Description**
Size, to 15 in. (38.1 cm). Similar to Summer Flounder but with 4 large, oblong pink-edged black spots.
**Habitat**
7–150 fathoms (12.8–274.3 m).
**Food**
Mollusks, small crustaceans, fishes.
**Range**
Cape Cod to S.C.

### WINDOWPANE
*Scophthalmus aquosus*                    84:11

**Description**
Size, to 18 in. (45.7 cm) and 2 lb. (0.9 kg). First 10 or 12 rays of dorsal fins partially free and branched at the tips. Body almost

round, translucent; eyes on left. Ventral fins as wide at base as at tip; left (upper) ventral a continuation of anal; right (lower) ventral smaller and nearer throat; tail rounded.

**Habitat**
Sandy shores to 40 fathoms (73.2 m).

**Food**
Mollusks, small crustaceans, fishes.

**Age**
To at least 7 years.

**Range**
Cape Cod to S.C.

## EYED FLOUNDER
*Bothus ocellatus*                                                   *Fig. 113*

**Description**
Size, to 8 in. (20.3 cm). The most rounded flatfish in the range; eyespots on back. Eyes on left, lateral line arched in front. Dorsal and anal fins of virtually uniform height throughout; tail small and rounded.

**Habitat**
Sandy shores.

**Range**
Va. to Brazil, occasionally n. to N.Y.

*Fig. 113*

American Plaice                                    Eyed Flounder

## SMALLMOUTH FLOUNDER
*Etropus microstomus*

**Description**
Size, to 6 in. (15.2 cm). Small, elongate; small mouth; very close-set eyes. Eyes on left; lateral line without arch in front; 1 row of teeth in each jaw, tail rounded.

**Habitat**
Shallow coastal waters.

**Remarks**
Anglers often characterize small species such as this one, and the young of larger species, as "postage stamps." Large individuals and species they call "bathmats."

**Range**
N.Y. to Brazil.

# RIGHTEYED FLOUNDERS
Family Pleuronectidae

Members of this family usually have the eyes and coloration on the right side. A number of exceptions do occur, however. Pelvic fins are symmetrically located, one on each side of the abdominal ridge.

## GREENLAND HALIBUT
*Reinhardtius hippoglossoides*

**Description**
Size, to 40 in. (1 m) and 25 lb. (11.3 kg). Eyes on right; body long, lateral line almost straight; mouth large, with jaws and teeth nearly equally developed on both sides. Dorsal extends from nape to peduncle, highest at center; anal originates behind pectoral but otherwise symmetrical with dorsal; tail slightly concave.
**Habitat**
To 530 fathoms (1 km).
**Range**
Arctic and n. N. Atlantic; occurs regularly only n. of Maine.

## ATLANTIC HALIBUT
*Hippoglossus hippoglossus*                                   76:15

**Description**
Size, to 9 ft. (2.7 m) and 700 lb. (317.5 kg); rare over 450 lb. (167.9 kg). Largest and thickest flatfish in range. Eyes to right, well separated; body thick, long; mouth large, gaping back to eyes; teeth sharp, curved; lateral line arched above pectoral fin. Dorsal extends from above eye to caudal peduncle, rising to low angle at center; upper pectoral pointed, lower rounded; ventrals small, symmetrical, and in front of pectorals; anal originates just behind pectorals, similar to dorsal in shape; tail concave.
**Habitat**
Sand, clay, or gravel bottom down to 500 fathoms (0.9 km).
**Habits**
Occurs in schools; voracious.
**Age**
Probably to 50 years.
**Remarks**
A valuable food fish.
**Range**
Newfoundland to Cape Cod.

## AMERICAN PLAICE
*Hippoglossoides platessoides*                               Fig. 113

**Description**
Size, to 32½ in. (82.6 cm) and 14 lb. (6.4 kg). Dorsal and anal fins smoothly rounded; tail rounded. Body long, lateral line nearly straight, scales saw-toothed on eyed (upper) side; mouth large, gaping back to eyes; teeth sharp, conical. Pectoral on eyed side longer and more rounded than that on blind side; ventrals symmetrical and close in front of anal; anal originates in front of pectorals, shaped like dorsal, and preceded by short sharp spine.
**Habitat**
Sandy or muddy bottoms.
**Food**
Invertebrates.
**Reproduction**
To 60,000, spawned March to June, hatch in 12 days at 39°F (3.9° C).
**Age**
To 30 years.
**Other names**
Rough Dab, Rough Sole.
**Range**
Labrador to Cape Cod.

## YELLOWTAIL FLOUNDER
*Limanda ferruginea*                                              84:10

**Description**
Size, to 22 in. (55.9 cm). Eyes on right, barely separated; body
oval, top of head concave, snout pointed, mouth and teeth small;
lips thick, fleshy, no mucous pits on underside. Pectoral fin on eyed
side longer than that on blind side, ventrals symmetrical, anal
smoothly rounded, tail slightly convex. Resembles the American
Plaice, but dorsal fin comes to high point near the middle, and
lateral line is arched in front.
**Habitat**
Rocky bottoms.
**Other name**
Yellowtail.
**Range**
Labrador to N.J.

## WINTER FLOUNDER
*Pseudopleuronectes americanus*                                  84:12

**Description**
Size, to 22 in. (55.9 cm). Body thick, oval; scales rough above,
smooth below; eyes well separated and with rough scales between;
mouth small; lips thick, fleshy; lateral line nearly straight, no
mucous pits below. Dorsal fin almost uniform in height; tail broad,
rounded. Resembles the American Plaice, but anal fin comes to
apex just to rear of the middle.
**Habitat**
Sand and mud bottoms to 20 fathoms (36.6 m); ascends rivers
almost to fresh water.
**Food**
Shrimps, mollusks.
**Remarks**
This is the most common and meatiest shallow-water flounder in
the range.
**Other names**
Blackback, Sole, Lemon Sole.
**Range**
Labrador to Ga.

## SMOOTH FLOUNDER
*Liopsetta putnami*

**Description**
Size, to 1 ft. (0.3 m) and 1½ lb. (0.7 kg). Both dorsal and anal fins
come to an apex near the middle. Body thick, oval; no scales
between eyes; mouth small; lips thick, fleshy; lateral line straight; no
mucous pits on blind side of head.
**Habitat**
Soft mud bottom, 2–5 fathoms (3.7–9.1 m) deep.
**Reproduction**
Spawns in winter.
**Range**
Arctic Atlantic to R.I. The most plentiful flatfish along the Strait
of Belle Isle.

## WITCH FLOUNDER
*Glyptocephalus cynoglossus*

**Description**
Size, to 25 in. (63.5 cm). Right-eyed, small-mouthed; dorsal and
anal fins of even height, with many rays. Above brownish-gray.
Body thin, long; scales smooth; head small, with about 12 large
open mucous pits on blind side. Pectorals and ventrals symmetrical,
pectoral fin membrane dark on eyed side.
**Habitat**
Muddy sand or clay, 10–150 fathoms (18.3–274.3 m).
**Food**
Small invertebrates.
**Other name**
Gray Sole.
**Range**
Gulf of St. Lawrence to Cape Cod, offshore to Va.

# SOLES
Family Soleidae

## HOGCHOKER
*Trinectes maculatus*                                    84:13

**Description**
Size, to 8 in. (20.3 cm). Unique among eastern flatfishes in having
no pectorals. Eyes on right, small and set flat; body and head
rounded; scales very rough on both sides, slimy with mucous; no
snout or caudal peduncle; mouth small, gape shorter and more
crooked on blind side than on eyed side. Dorsal and anal highest to
rear and extending to base of tail; right ventral continuous with
anal; tail rounded.
**Habitat**
Brackish bays and estuaries.
**Food**
Worms, small crustaceans.
**Range**
Mass. to Panama.

# Triggerfishes, Filefishes, and Others
Order Tetrodontiformes

# TRIGGERFISHES AND FILEFISHES
Family Balistidae

## PLANEHEAD FILEFISH
*Monocanthus hispidus*

**Description**
Size, to 10 in. (25.4 cm). Distinguished by its 1-spined first dorsal
and the long thread of the first ray of the second dorsal. Body deep,
thin; scales small, eye set high, gill openings nearly vertical. Dorsal
spine with rear edge barbed; pectorals short, rounded, and set
below gill openings; external ventral spine conspicuous; soft dorsal,
anal, and tail fins rounded.

**Habitat**
Warm waters, particularly inshore and about pilings.
**Remarks**
Filefish skin is so hard and rough that the early settlers are said to have dried it and used it for sandpaper, whence the name.
**Range**
N.S. to Brazil.

## ORANGE FILEFISH
*Aluterus schoepfi*                                      77:5

**Description**
Size, to 2 ft. (0.6 m). Distinguished by its l-spined first dorsal and rounded second dorsal, without extended first ray. Body thin, oval; scales small, eye not set high, gill openings oblique. Pectorals short and rounded, no external ventral spine, tail narrower than in Planehead.
**Habitat**
Around wharves.
**Habits**
Often swims head down.
**Food**
Marine vegetation.
**Range**
N.S. to Brazil.

# BOXFISHES
Family Ostraciidae

These fishes are encased in hard, rigid shells with openings for fins, gills, eyes, mouth, and the thin, longish caudal peduncle. There are no spiny dorsal or ventral fins, but there are spines on the ventral ridges to the rear. Young are almost spherical in cross section, adults are triangular. The family goes back to the Eocene. Its members make excellent food.

## TRUNKFISH
*Lactophrys trigonus*                                   Fig. 114

**Description**
Size, to 9 in. (22.9 cm). Body deep, profile concave behind eye. Anal and soft dorsal small, rounded, and pointing rearward; tail slightly rounded. Distinguished from the Scrawled Cowfish by the lack of spines over its eyes.
**Habitat**
Shallow marine waters.
**Other name**
Boxfish.
**Range**
New England to Gulf of Mexico.

*Fig. 114*

Scrawled Cowfish            Trunkfish

## SCRAWLED COWFISH
*Lactophrys quadricornis*            *Fig. 114*

**Description**
Size, to 1 ft. (0.3 m). Forward spine over each eye. Body deep, profile convex behind eye. Soft dorsal, anal, and caudal with nearly straight rear margins.
**Habitat**
Warm waters.
**Other name**
Horned Trunkfish.
**Range**
Brazil to the Carolinas, rarely in the Gulf Stream to Mass.

# SWELLFISHES
Family Tetraodontidae

Like porcupine fishes, swellfishes can inflate themselves with air or water until they become almost spherical, and their teeth are fused into a strong parrotlike beak. They lack spiny dorsal or ventral fins. The family goes back to the Miocene.

## SMOOTH PUFFER
*Lagocephalus laevigatus*            *Fig. 115*

**Description**
Size, to 2 ft. (0.6 m). Body somewhat elongated, rounded in front, tapering to rear. Pectorals broad, dorsal and anal fins short and somewhat pointed, ventral surface with a few scattered prickles. Tail concave, with skin folds below. Distinguished from Northern Puffer by absence of dark spots along side below.
**Habitat**
Shallow sandy shores.
**Range**
N. to Cape Cod.

*Fig. 115*

Smooth Puffer          Striped Burrfish, p. 510

## NORTHERN PUFFER
*Sphoeroides maculatus*            **76:7**

**Description**
Size, to 10 in. (25.4 cm). Dark spots along side below; body somewhat elongated, tapering from above eye fairly steeply to front, gradually to rear. Pectorals broad, dorsal and anal fins small and angular; no skin folds below tail; tail rounded.
**Habitat**
Sandy shores, tide line to a few fathoms.
**Habits**
Partially buries itself in sand; a group will attack, apparently in concert, and dispose of a blue crab.
**Food**
Crustaceans, mollusks.
**Range**
Fla. to N.S.

# PORCUPINE FISHES
Family Diodontidae

### STRIPED BURRFISH
*Chilomycterus schoepfi*                           *Fig. 115*

**Description**
Size, to 10 in. (25.4 cm). Inflatable, spiny body. Body oval, covered
with short, stout spines; teeth fused into a strong parrotlike beak.
No spiny dorsal or ventrals; pectorals broad; soft dorsal and anal
broad and rounded; tail slender and rounded.
**Habitat**
Shallow inshore waters.
**Habits**
This fish, like the swellfishes, can inflate a saclike part of its
gullet with air or water, swelling the abdomen to such a degree
that the fish becomes almost spherical.
**Other name**
Spiny Boxfish.
**Range**
N. to Cape Cod.

# OCEAN SUNFISHES
Family Molidae

### OCEAN SUNFISH
*Mola mola*                                        *Fig. 116*

**Description**
Size, to 10 ft. 11 in. (3.3 m) and 2000 lb. (907 kg). Body oblong,
flattened sidewise; ½ as deep as long, ¼ as thick as deep; skin 1½
in. (3.8 cm) thick; mouth very small, teeth fused into bony beak;
eye small and in line with mouth; bones soft. No spiny dorsal or
ventrals; pectorals small; soft dorsal and anal fins high, pointed,
symmetrical; no caudal peduncle; tail very short, broad, scalloped.
**Habitat**
Open sea; all temperate and tropical seas.
**Food**
Jellyfish, small crustaceans.
**Other name**
Headfish.
**Range**
Atlantic Coast, drifting n. to Newfoundland; Pacific.

*Fig. 116*

Ocean Sunfish

# Freshwater Fishes

## LAMPREYS
Class Agnatha

Members of this class are eellike in shape and lack scales, jaws, gill covers, and paired fins. The mouth is circular and adapted for sucking. Behind the head are seven paired gill clefts. Eggs are deposited in spring in fresh water in or at the head of riffles, where they hatch into blind larvae which live buried in the bottom for four to six years. The larvae feed on microorganisms; the adults of some species are often parasitic on fish, to which they attach themselves by their suctorial mouth. The class goes back to the Silurian. There are seven freshwater lampreys in the range; five have an undivided dorsal fin; four are not parasitic.

## Lampreys
Order Petromyzoniformes

## LAMPREYS
Family Petromyzonidae

### SILVER LAMPREY
*Ichthyomyzon unicuspis*                        *Fig. 116a*

**Description**
Size, to 1 ft. (0.3 m). Silvery above, lighter below; teeth usually single-pointed; parasitic.
**Habitat**
Large rivers and lakes.
**Habits**
Lives only one year as an adult, then spawns and dies.
**Range**
Minn. to N.Y., Miss. R. Valley s. to Mo.

*Fig. 116a*

Sea Lamprey

Lamprey on Trout        Silver Lamprey

### SEA LAMPREY
*Petromyzon marinus*                        *Fig. 116a*

**Description**
Size, to 30 in. (76.2 cm). Dorsal fin divided in 2. Slate-blue to brown, speckled with blue or black; inner teeth have 2 points.
**Habitat**
Ocean, rivers, and large lakes. Ascends into fresh water to spawn and die; the larvae on maturing normally enter salt water.
**Range**
Atlantic Ocean to Great Lakes and Finger Lakes and s. to n. Fla.

# BONY FISHES
Class Osteichthyes

The fishes in this class have jaws, paired fins, and a single gill slit and gill cover. The class goes back to the Devonian.

# Sturgeons and Paddlefishes
Order Acipenseriformes

Members of this order have a long snout, inferior mouth, and two to four barbels, threadlike sense organs, on the undersurface of the snout. The tail lobes are unequal in size, the upper being the larger. The skeleton is largely of cartilage. The order goes back to the Cretaceous.

# STURGEONS
Family Acipenseridae

The bodies of these fishes are covered with five rows of bony plates. They reach 12 ft. (3.7 m) and 300 lb. (136.1 kg). Adults have four barbels but are toothless. They are bottom dwellers. When they achieve sexual maturity at about twenty years of age, they spawn in spring or early summer, ascending streams or moving to shallow water.

Members of the genus *Acipenser* have a cone-shaped snout, a small opening, or spiracle, between the eye and the upper corner of the gill cover, and a caudal peduncle without plates.

## SHOVELNOSE STURGEON
*Scaphirhynchus platorhynchus*                     88:3

**Description**
Size, to 3 ft. (0.9 m), average 4 lb. (1.8 kg); rarely to 8 lb. (3.6 kg). No spiracle; belly and slender caudal peduncle covered with small bony scalelike plates.
**Habitat**
Streams and lakes with clear bottoms.
**Range**
Miss. R. drainage.

Note: The similar **PALLID STURGEON**, *Scaphirhynchus albus* **(88:4)**, of the upper Mississippi drainage has the belly largely naked.

## ATLANTIC STURGEON
*Acipenser oxyrhynchus*                     88:1

**Description**
Size, to 12 ft. (3.7 m). Sharp snout; 7–14 dorsal plates. Olive-gray or brown, paler below. Snout almost as long as head; anal fin entirely below middle of dorsal fin. Barbels short, midway between snout tip and mouth; lateral plates, 26–30.
**Habitat**
Mud bottoms off coast.
**Food**
Mud-dwelling marine invertebrates and some fishes.
**Range**
Coastal waters of the Atlantic; ascends rivers.

## SHORTNOSE STURGEON
*Acipenser brevirostris*
88:2
**Description**
Size, to 30 in. (76.2 cm). Blunt snout; 11 dorsal plates. Color brownish, paler below. Snout about ¼ length of head; anal fin below front of dorsal; lateral plates, 32.
**Habitat**
Salt water; ascends streams to spawn.
**Range**
Cape Cod to Fla., rarer in n.

## LAKE STURGEON
*Acipenser fulvescens*
86:15
**Description**
Size, to 8 ft. (2.4 m). Very blunt snout in adult; 13 dorsal plates. Above dark olive; sides paler or reddish, especially in young, often spotted with black. Anal fin below middle of dorsal fin; lateral plates, 29–42.
**Habitat**
Lakes and large rivers.
**Food**
Mollusks, aquatic insect larvae, crayfish, some fishes, aquatic vegetation.
**Range**
Great Lakes, upper Miss. R., Sask. R., and Hudson Bay drainages, s. to Ala. and Mo.

# PADDLEFISHES
Family Polyodontidae

Paddlefishes, like sturgeons, are an ancient group represented by only two living species—one in North America and the other in the valley of the great Yangtze River in China. The fins, like those of sturgeons, are archaic and sharklike, and the scales have degenerated to a patch on the upper lobe of the caudal fin. The long, paddle-shaped snout, whose precise function is still unknown, sets this fish apart from all others in North America. Although of large size, the paddlefish feeds throughout life on microscopic plants and animals.

## PADDLEFISH
*Polyodon spathula*
87:15
**Description**
Size, to 6 ft. (1.8 m) and 180 lb. (87.6 kg). Large paddle-shaped snout. No scales except for small patch on tail; gill covers have long backward projecting tips. Very young lack the paddle.
**Habitat**
Large silty rivers; oxbow and flood-plain lakes.
**Food**
Feeds by straining plankton from water by its efficient gill rakers.
**Range**
Great Lakes and Miss. R. systems.

# Bowfins
Order Amiiformes

## BOWFINS
Family Amiidae

### BOWFIN
*Amia calva*                                                                     88:9

**Description**
Size, to 31 in. (78.7 cm) and 8 lb. (3.6 kg). Long dorsal fin.
Cycloid (rounded) scales; bony gular plate between angle of lower
jaws.
**Habitat**
Lakes and sluggish waters.
**Habits**
Sometimes gulps air at the surface.
**Food**
Fishes, crayfish.
**Reproduction**
In spring, male builds circular nest among weeds, where he guards
eggs and young.
**Range**
Minn. to Que. and, w. of Appalachians, s. to Tex. and Fla.

# Gars
Order Semionotiformes

## GARS
Family Lepisosteidae

Gars are long cigar-shaped fishes, olive above, gray below, and
with thick ganoid (diamond-shaped) scales. They are predators and
have beaklike jaws with sharp, pointed teeth. They frequent large
streams and rivers and shallow weedy lakes, where they spawn in
spring. They can use atmospheric oxygen and may bask on the
surface. The family goes back to the Cretaceous.

### LONGNOSE GAR
*Lepisosteus osseus*                                                             88:6

**Description**
Size, to 5 ft. (1.5 m); rod and reel record, 50 lb., 5 oz. (22.8 kg).
Thin snout twice the length of its head. In adult, black spots near
tail and on vertical fins; 1 row of teeth in upper jaw.
**Habitat**
Large streams and rivers; shallow, weedy lakes.
**Range**
Mont. and Vt. and s. to Mexico and Fla.

### SPOTTED GAR
*Lepisosteus oculatus*                                                           88:8

**Description**
Size, to 44 in. (111.8 cm). Large dark spots on top of head. Snout
broad, longer than rest of head but not twice as long.

**Habitat**
Clear, weedy lakes and bayous.
**Range**
Lakes Mich. and Erie, and Miss. R. drainage to Gulf of Mexico.

## SHORTNOSE GAR
*Lepisosteus platostomus*                                  88:7

**Description**
Size, to 30 in. (76.2 cm). Snout short, longer than rest of head but
not twice as long; large teeth in single series. Spots only on rear
fins and rear part of body; the spots are distinct.
**Range**
Long, muddy rivers of Miss. R. drainage.

## ALLIGATOR GAR
*Lepisosteus spatula*                                      88:5

**Description**
Size, to 93 in. (236.2 cm); rod and reel record, 279 lb. (126.5 kg).
In adult, 2 rows of large teeth in upper jaw. Snout relatively broad
and short; body mottled toward tail.
**Habitat**
Large streams and rivers.
**Range**
Miss. R. drainage and Gulf of Mexico tributaries.

# Mooneyes
Order Osteoglossiformes

# MOONEYES
Family Hiodontidae

This is a small family of silvery, deep, thin-bodied fishes, averaging
a foot in length, with a short round snout, a slanting, well-toothed
mouth, and a lateral line. They lack head scales, gular plate, spines
in fins, and an adipose fin. In females the lower edge of the anal
fin is almost straight; in the male it is lobed.

## GOLDEYE
*Hiodon alosoides*

**Description**
Size, to 20 in. (50.8 cm). Eye golden; dorsal fin rays 9–10; fleshy
keel on belly extends forward of pelvic fins; anal fin originates
before dorsal fin.
**Habitat**
Shallow, turbid lakes and rivers.
**Food**
Insects, mollusks, crayfish, fishes.
**Reproduction**
Moves upstream to quiet marshy shoals, where it spawns over
gravel beds.
**Range**
N. from Man. and Sask., throughout the Mo. R. system, s. to
Mont.

## MOONEYE
*Hiodon tergisus*                                                 85:12

**Description**
Size, to 18 in. (45.7 cm). Eye silvery; dorsal-fin rays 11–12; belly usually not keeled; anal fin originates behind dorsal fin.
**Habitat**
Clearest, largest waters, preferably swift.
**Food**
Insects, mollusks, crayfish, small fishes, plankton.
**Reproduction**
In May to July ascends rivers and streams to spawn in shallow water.
**Range**
Lake Champlain system w. to the Great Lakes; Miss. R. drainage as far s. as Ark.

# Herringlike Fishes
Order Clupeiformes

These fishes have soft fins, and pelvic fins are abdominal. The order goes back to the Triassic.

## SHADS AND HERRINGS
Family Clupeidae

Fishes in this family are compressed and have a naked head, weak teeth, and a keeled belly. They have no adipose fin, gular plate, or lateral line. Many of them are marine; of these we treat here only those that go up freshwater streams and rivers to spawn. The family goes back to the Cretaceous.

Fishes of the genus *Alosa* are silvery, with jaws about equal or the lower jaw projecting. The upper jaw has a deep or shallow notch to receive the lower jaw; jaw teeth are absent or very weak. The dorsal fin is short and the last ray is never greatly lengthened. All, except the Glut Herring, have a silvery or pale peritoneum.

## GIZZARD SHAD
*Dorosoma cepedianum*                                            85:9

**Description**
Size, to 15 in. (38.1 cm). Last ray of dorsal fin greatly lengthened. Mouth small, toothless, upper jaw projecting, dark spot behind gill cover.
**Habitat**
Mud-bottomed, shallow water. Its numbers have increased in reservoirs in recent years, perhaps in part because it has a greater tolerance for turbid water than other species.
**Food**
Plankton.
**Reproduction**
Eggs laid in shallow water during spring.
**Range**
Canada to Fla. excluding New England.

## SKIPJACK HERRING
*Alosa chrysochloris*                                      **85:6**

**Description**
Size, to 15 in. (38.1 cm). Pale lower jaw; no spots behind gill cover. Sides of upper jaw meet in an obtuse angle and form a shallow notch to receive lower jaw; always has tongue teeth; sides plain.
**Habitat**
Swift, deep waters.
**Habits**
Large, swift schools of this species force schools of shiners, its prey fish, to jump into the air. It then lightly skips into the air in pursuit, hence its name.
**Food**
Plankton and small fishes.
**Reproduction**
Eggs laid in early spring.
**Range**
Miss. Valley, s. to Gulf of Tex.

## HICKORY SHAD
*Alosa mediocris*                                      **85:10**

**Description**
Size, to 15 in. (38.1 cm). Dark spots, 5–6, behind gill cover. Sides of upper jaw meet in an obtuse angle and form a shallow notch to receive lower jaw; tongue teeth sometimes present; silvery patch on cheek longer than wide; faint stripes lengthwise on sides.
**Habitat**
Marine; anadromous.
**Range**
N.S. to Fla.

## BLUEBACK HERRING
*Alosa aestivalis*                                      **85:1**

**Description**
Size, to 12 in. (30.5 cm). Dark or black peritoneum. Somewhat similar to Alewife, but darker, thinner, and with smaller eyes; tongue with single row of teeth and no black spot behind operculum.
**Habitat**
Anadromous.
**Range**
N.S. to Fla.

## ALABAMA SHAD
*Alosa alabamae*

**Description**
Similar to American Shad, but much of lower jaw black; no dark spot behind gill cover; peritoneum pale.
**Habitat**
Freshwater rivers and streams.
**Range**
Ohio R. Valley and Miss. R. from Iowa to W.Va.; streams entering Gulf of Mexico from w. Fla. to Miss. R.; rare.

## ALEWIFE
*Alosa pseudoharengus* **85:4**
**Description**
Size, to 15 in. (38.1 cm). Small patch of teeth on tongue, single
black spot behind gill cover, and silvery patch on cheek longer than
wide.
**Habitat**
Anadromous; also landlocked in deep water of large lakes.
**Food**
Plankton and small insects.
**Reproduction**
Eggs laid over shallow, sandy bottoms in late spring, early
summer.
**Range**
Newfoundland to Fla.; is in the process of occupying the Great
Lakes via the Welland Canal. The first Lake Erie record was
1931; the first from Lake Huron, 1934; Lake Superior, 1954.

## AMERICAN SHAD
*Alosa sapidissima* **85:7**
**Description**
Size, to 3 ft. (0.9 m). Jaws about equal; sides of upper jaw join in
an acute angle and form a deep notch. Lower jaw light; 1 or more
dark spots behind gill cover; about 60 long, slender gill rakers on
lower arm of first arch; silvery patch on cheek wider than long;
peritoneum white.
**Habitat**
Anadromous.
**Food**
Plankton.
**Reproduction**
Eggs deposited in spring in upstream areas.
**Range**
Newfoundland to n. Fla.

# Salmons, Trouts, and Pikes
Order Salmoniformes

## SALMONS, TROUTS, AND WHITEFISHES
Family Salmonidae

These fishes have a long body with small cycloid scales and often
an adipose fin. The salmon and trout have large mouths with large
teeth. The lake herrings have a small mouth with weak or no
teeth. These fishes feed on insects, plankton, and bottom organisms.
They prefer water cooler than 70°F (21.1°C).

**SALMON AND TROUT COMPARISON CHART**

| Species | Body Spots | Tail Spots | Anal-Fin Rays |
|---|---|---|---|
| Sockeye | none | none | 13–17 |
| Atlantic | red, with halos | none | 9 |
| Brown | red, with halos | none | 8–12 |
| Rainbow | dark | yes | 10–13 |
| Cutthroat | dark | yes | 10–13 |
| Brook | red or pink | none | 10–13 |
| Lake | pale | yes | 10–13 |

## SOCKEYE SALMON
*Onchorhynchus nerka*                                        89:13

**Description**
Size, to 2 ft. (0.6 m) and 8 lb. (3.6 kg). Body silver or blue-black,
unspotted; turns bright red with a greenish head when spawning in
fall.
**Habitat**
Sea, rivers, and lakes; anadromous.
**Reproduction**
Eggs laid in streams or on shallow gravel shoals, after which adults
die.
**Range**
Introduced from American Pacific.

## ATLANTIC SALMON
*Salmo salar*                                               86:8

**Description**
Size, to 5 ft. (1.5 m), rod and reel record, 79 lb. 2 oz. (35.9 kg); for
landlocked variety, 28½ lb. (12.9 kg). Silvery to steel-blue; dark
spots large and diffuse, hardly visible on tail; orange or reddish
spots prominent and often with blue halos. Teeth on shaft of vomer
weakly developed or absent, maxilla extends barely behind eye; fins
pale to white; adipose fin small, olive in young; tail deeply forked.
**Habitat**
Anadromous.
**Reproduction**
Eggs laid in nest and covered during fall.
**Other name**
Ouananiche.
**Range**
Labrador to Maine; landlocked in cold lakes of e. Canada and
New England; introduced elsewhere, as N.Y.

## BROWN TROUT
*Salmo trutta*                                              86:6

**Description**
Size, to 2 ft. (0.6 m) and over; rod and reel record, 39½ lb. (17.9
kg). Spotting similar to Atlantic Salmon, but adipose fin relatively
large and orange in young, teeth on shaft of vomer well developed,
and maxilla extends well behind eye.
**Habitat**
Clear water with cover.
**Habits**
Feeds chiefly at dusk or night.
**Reproduction**
200–5000 eggs laid in nest during fall.
**Range**
All states except those of the extreme se. and middle s.; widely
introduced.

### RAINBOW TROUT
*Salmo gairdneri* 86:10

**Description**
Size, to 3 ft. (0.9 m); rod and reel record, 37 lb. (16.8 kg). Broad
pink or red band on the side; spotted straight tail. Body has dark
spots.
**Habitat**
Similar to Brown Trout, but prefers swifter current.
**Reproduction**
400–3000 eggs laid in nest on clear gravel in Apr. to May.
**Other name**
Steelhead (for anadromous specimens).
**Range**
100th meridian and e., as far s. as the mountainous areas of N.C.
and Tenn.; widely introduced.

### CUTTHROAT TROUT
*Salmo clarki* 89:10

**Description**
Size, commonly to 15 in. (38.1 cm); rod and reel record, 41 lb.
(18.6 kg). Red or pink streak on underside of lower jaw. Body
spotted.
**Habitat**
Cold-water streams and lakes.
**Reproduction**
Eggs left in nest made on gravel in May or June.
**Range**
Upper Mo., Ark., Platte, Colo., and Rio Grande R. drainages.

### BROOK TROUT
*Salvelinus fontinalis* 86:7

**Description**
Size, to 32 in. (81.3 cm), rod and reel record, 14½ lb. (6.6 kg).
Body dark with many blue-bordered red spots, dorsal fin with dark
spots; tail not forked.
**Habitat**
Cold, small streams and ponds with cover.
**Habits**
Cautious, easily frightened away.
**Reproduction**
200–500 eggs deposited in nest in shallow water over a gravel
bottom during late fall.
**Range**
Same as Brown Trout.

### LAKE TROUT
*Salvelinus namaycush* 86:9

**Description**
Size, to 5 ft. (1.5 m), rod and reel record, 63 lb. 2 oz. (28.6 kg).
Sides with red or pale spots, dorsal fin with pale spots; tail deeply
forked.
**Habitat**
Deep-water lakes; moves to shallow water in fall and winter.
**Reproduction**
Up to 20,000 eggs deposited over rocks at various depths in late fall.
**Range**
Canadian border states.

## ROUND WHITEFISH
*Prosopium cylindraceum*                                    *Fig. 117*

**Description**
Size, to 15 in. (38.1 cm). Single flap between the nostrils; 15–20 first-arch gill rakers. Body slender, head short, no jaw teeth.
**Habitat**
Shallow lake water.
**Habits**
A bottom feeder.
**Food**
Aquatic insects, mollusks, plankton.
**Reproduction**
Eggs laid in Nov. and Dec. in the mouths of streams and rivers.
**Range**
N. America from the Great Lakes and Maine n.

Note: The similar **MOUNTAIN WHITEFISH,** *Prosopium williamsoni,* of the upper Saskatchewan and Missouri rivers has 19 to 26 first-arch gill rakers.

*Fig. 117*

Cisco

Lake Whitefish, p. 522

Round Whitefish

## CISCO
*Coregonus artedii*                                          *Fig. 117*

**Description**
Size, 8–12 in. (20.3–30.5 cm); rarely to 2 ft. (0.6 m). Body thin, deepest at middle; mouth at end of pointed snout; fins black-tipped; first-arch gill rakers, variable from lake to lake, 38–64.
**Habitat**
Moderately deep lake water.
**Habits**
May school.
**Food**
Plankton, insects, small fishes.
**Reproduction**
Eggs laid in late fall in shallow water over sand or gravel.
**Range**
Yukon to N.S. and s. to n. Ind. and Vt.

Note: The **NIPIGON CISCO,** *Coregonus nipigon,* dark green, with 54 to 64 first-arch gill rakers and living in the shallow waters of Lakes Nipigon and Winnipeg and some smaller Ontario lakes, has been combined with the Cisco as a single species.

521

## SHORTNOSE CHUB
*Coregonus reighardi*

**Description**
Size, to 14 in. (35.6 cm). Body large, deepest at middle; head large, snout projects; mandible thick, blackish, and without knob at tip; first-arch gill rakers, 32–42; back, light green.
**Habitat**
13–50 fathoms (23.8–91.4 m).
**Range**
Lakes Ontario, Huron, Superior, Mich., and Nipigon.


## SHORTJAW CISCO
*Coregonus zenithicus*

**Description**
Size, to 12 in. (30.5 cm). Body thin, green, deepest at middle; jaws equal, mandible thick, not blackish, and without knob at tip. First-arch gill rakers, 32–46.
**Habitat**
11–30 fathoms (20–55 m).
**Habits**
May school.
**Range**
Great Lakes (except Erie), also Nipigon, Winnipeg, Athabasca, and other lakes in N.W.T.

Note: In the similar **LONGJAW CHUB**, *Coregonus aplenae,* of Lakes Michigan and Huron, the lower jaw projects.


## BLOATER
*Coregonus hoyi*

**Description**
Size, to 9 in. (22.9 cm). Body thin, deepest at middle; lower jaw projects, mandible thin and with knob at tip; fins black-tipped; first-arch gill rakers, usually 37–47.
**Habitat**
15–70 fathoms (27.4–128 m).
**Range**
Lakes Superior, Huron, Mich., Ontario, Nipigon, and others to n.


## LAKE WHITEFISH
*Coregonus clupeaformis*                                    *Fig. 117*

**Description**
Size, to 3 ft. (0.9 m), but average 15 in. (38.1 cm). Body thin, inferior mouth; snout rounded, mouth below and behind. Gill rakers 19–33, seldom less than 22.
**Habitat**
Shallow to moderate depths.
**Habits**
May school.
**Food**
Includes bottom organisms.
**Reproduction**
About 35,000 eggs laid in late fall over shallow shoals or streams.
**Range**
Great Lakes, Canada, and to n. New England.

## DEEPWATER CISCO
*Coregonus johannae*

**Description**
Size, to 1 ft. (0.3 m). Body deepest to front of middle; first-arch gill rakers 25–36, usually fewer than 33.
**Habitat**
16–100 fathoms (29.3–183 m).
**Range**
Lakes Mich. and Huron; rare, not seen in Lake Mich. since 1951.

## KIYI
*Coregonus kiyi*

**Description**
Size, to 1 ft. (0.3 m). Body small, thin, deepest to front of middle; mandible thin and with knob at tip; first-arch gill rakers, 34–47.
**Habitat**
30–100 fathoms (54.8–183 m).
**Range**
Lakes Superior, Mich., Ontario, Huron; numbers greatly reduced.

## BLACKFIN CISCO
*Coregonus nigripinnis*

**Description**
Size, average 13 in. (33 cm). Body large, thick, deepest to front of middle; mandible thick and without knob at tip; fins often black; first-arch gill rakers, 41–43.
**Habitat**
15–100 fathoms (27.4–183 m).
**Range**
Lakes Superior, Huron, and others to n.

## ARCTIC GRAYLING
*Thymallus arcticus*                                    **89:4**

**Description**
Size, to 24 in. (61 cm); and 5 lb. (2.3 kg). Body long, compressed, purplish-gray with black spots; mouth small, teeth in jaw and roof of mouth; no spines in fins. Head naked; dorsal fin flaglike; adipose fin and lateral line present; tail forked. Smells like a cucumber.
**Habitat**
Clear cold streams with pools.
**Habits**
May school.
**Food**
Insects and their larvae.
**Reproduction**
About 6000 eggs laid, March to June, over sand and gravel in headwaters.
**Range**
Mo. R. tributaries above Great Falls, n. to Arctic coast w. of Hudson Bay, w. to Siberia; widely introduced.

# SMELTS
Family Osmeridae

Members of this family are small, slender, silvery fishes inhabiting either marine, brackish or fresh water.

## RAINBOW SMELT
*Osmerus mordax*                                        **89:6**

**Description**
Size, average 7–8 in. (17.8–20.3 cm). Body slender; greenish with dark dots, silvery band on sides. Head naked, teeth in jaw; lateral line and adipose fin present, dorsal placed over pelvics. Smells like a cucumber.

**Habitat**
Coastal waters, ascending rivers and lakes in spring to breed; some landlocked.

**Habits**
Schools.

**Food**
Mainly plankton and insects, but some fishes.

**Range**
Atlantic Coast, Labrador to N.Y.; w. to Great Lakes. Introduced into Great Lakes drainage in 1912, this species is now well established in all the Great Lakes.

# MUDMINNOWS
Family Umbridae

These are small red-brown fishes with an oblong body, blunt snout, scaled head, and rounded tail. They have no lateral line or adipose fin. They live in soft-bottomed sluggish or stagnant water, burrow into the mud when alarmed, and are extremely resistant to adverse conditions. They feed on insects, crustaceans, and some vegetation. Their eggs are laid in early spring.

## CENTRAL MUDMINNOW
*Umbra limi*                                            **89:2**

**Description**
Size, usually 2–4 in. (5.1–10.2 cm). Body with faint vertical bars, lower jaw light, dark bar at base of tail.

**Habits**
Extremely hardy and tolerant. This species, it is said, can even be frozen more or less solidly in ice, provided its own flesh is not frozen, and revived upon thawing.

**Range**
Man. to mouth of St. Lawrence R.; s. through upper Miss. R. Valley.

Note: The **EASTERN MUDMINNOW**, *Umbra pygmaea*, size, 2–4 in. (5.1–10.2 cm), has longitudinal streaks, a dark lower jaw, and lacks the vertical bars of the Central Mudminnow. It swims in coastal streams from New York to Florida

# PIKES
Family Esocidae

These carnivorous fishes are distinguished by a duck-billed snout, pores on the head, sharp teeth, a long cylindrical body, a forked tail, and an incomplete lateral line. The dorsal and anal fins are set far back and are opposite each other; there is no adipose fin. Pikes live in vegetated waters and feed largely on fish and frogs. They spawn in the spring in shallow water and flooded marshes.

### PIKE COMPARISON CHART

| Species | Lower Half of Gill Cover | Cheek | Sides | Rod and Reel Record lb. oz. | |
|---|---|---|---|---|---|
| Redfin | scaled | scaled | dark bands | — | — |
| Chain | scaled | scaled | dark chains | 9 | 3 |
| Northern | unscaled | scaled | light spots | 46 | 2 |
| Muskellunge | unscaled | unscaled | dark markings | 69 | 15 |

## REDFIN PICKEREL
*Esox americanus*                                           **89:7**

**Description**
Size, to 1 ft. (0.3 m). Gill cover and whole cheek (usually) fully scaled; dark oblique bands on the sides.
**Habitat**
Ponds, lakes, and streams.
**Range**
*Esox americanus americanus,* coastal plain from St. Lawrence R. s. to Fla., w. to Miss. *Esox americanus vermiculatus,* Great Lakes and Miss. R. drainages.

## CHAIN PICKEREL
*Esox niger*                                               **87:1**

**Description**
Size, usually 15–20 in. (38.1–50.8 cm); rod and reel record, 9 lb. 3 oz. (4.2 kg). Gill cover and whole cheek fully scaled; dark chainlike markings on sides.
**Habitat**
Ponds, lakes, and streams.
**Range**
St. Lawrence R. and Lake Ont. s. to Fla., e. of Appalachians; also lower Miss. Valley.

## NORTHERN PIKE
*Esox lucius*                                              **86:4**

**Description**
Size, usually 18–30 in. (45.7–76.2 cm); rod and reel record, 46 lb. 2 oz. (20.9 kg). Fully scaled cheeks, but no scales on lower half of gill cover. Pores on underside of lower jaw, 10–11. Body olive, sides with light barlike spots on a dark background, dark spots on fins.
**Habitat**
Ponds, lakes, and streams.
**Habits**
Summers in shallows, winters in deep water.
**Range**
N. North America, s. to Nebr. and Mo., w. of Appalachians.

## MUSKELLUNGE
*Esox masquinongy* **86:5**

**Description**
Size, to 60 in. (152.4 cm); usually 28–48 in. (71.1–121.9 cm); rod and reel record, 69 lb. 15 oz. (31.7 kg). Lower half of both cheeks and gill covers scaleless. Greenish above, paler below; sides and fins with dark bars or spots on light background. Lower jaw with 6–9 pores on each side underneath.
**Habitat**
Ponds, lakes, and streams.
**Habits**
Summers in deep water, comes into shallow water in fall.
**Reproduction**
Eggs laid with much commotion, in flooded marshy areas.
**Range**
Great Lakes, St. Lawrence R., and n. Miss. R. systems.

# Minnowlike Fishes
Order Cypriniformes

These fishes, in the East, have scaleless heads and abdominal pelvic fins. Their anterior vertebrae are modified so as to form a hearing aid (Weberian apparatus), connecting the air bladder to the ear. They go back to the Cretaceous.

## SUCKERS
Family Catostomidae

In suckers the mouth is usually behind the point of the snout; the lips are thick and suckerlike. The teeth are located in the throat in a single comblike row. All fins lack spines. Suckers are bottom feeders and often move in large schools. They spawn in spring. The family goes back to the Eocene.

## BLUE SUCKER
*Cycleptus elongatus*

**Description**
Size, to 3 ft. (0.9 m). Eyes in the rear half of head; head abruptly more slender than body. Body bluish, long and slender.
**Habitat**
Rivers, impoundments.
**Range**
Miss. R. drainage, s. from s. Minn. and Wis., into n. Mexico.

## SPOTTED SUCKER
*Minytrema melanops* **90:3**

**Description**
Size, 9–15 in. (22.9–38.1 cm), rarely to 18 in. (45.7 cm), and 3 lb. (1.4 kg). Body silvery, lateral line incomplete; in adults, a black spot on each scale. Dorsal fin rays, usually 12; pelvic rays, 9.
**Habitat**
Clear lakes and ponds.
**Food**
Mollusks and insect larvae.
**Range**
Minn. to Pa., s. to Fla. and w. to Tex.

## LAKE CHUBSUCKER
*Erimyzon sucetta* 90:8

**Description**
Size, to 10 in. (25.4 cm) and 14 oz. (0.4 kg). Body deep, compressed, and with bands or blotches; mouth slanted; dorsal-fin rays, 11–12; lateral line missing.
**Habitat**
Sluggish streams, lakes and ponds.
**Food**
Insects and mollusks.
**Reproduction**
Eggs laid in spring.
**Range**
Minn. to Conn. and s. Apparently rare in midwest.

Note: The similar but smaller **CREEK CHUBSUCKER**, *Erimyzon oblongus,* has 9–10 dorsal-fin rays and inhabits clear streams.

## NORTHERN HOG SUCKER
*Hypentelium nigricans* 90:12

**Description**
Size, to 14 in. (35.6 cm). Head concave between eyes; lateral line complete; lower fins light orange or brown; dorsal-fin rays, 10–12; pectoral-fin rays, both sides, 34.
**Habitat**
Rocky riffles or clear, cool streams.
**Food**
Organic material, algae, insect larvae.
**Reproduction**
Eggs laid in small streams in April.
**Range**
Cen. U.S. and s. Canada, Great Lakes to Gulf of Mexico.

Note: The similar but smaller **ROANOKE HOG SUCKER**, *Hypentelium roanokense,* with 31 pectoral-fin rays (total), is confined to the upper Roanoke River system in Virginia.

## WHITE SUCKER
*Catostomus commersoni* 87:2

**Description**
Size, 12–20 in. (30.5–50.8 cm), rarely to 24 in. (61 cm), and 5 lb. (2.3 kg). Body long, slender; mouth large and behind end of snout; lateral line complete; dorsal-fin rays, 10–13.
**Habitat**
Bottoms of shallow lakes.
**Food**
Insects, mollusks, worms, plant material.
**Reproduction**
Some 50,000 eggs deposited on rock or gravel shoals in streams in spring.
**Range**
Throughout eastern range.

Note: The somewhat similar **LONGNOSE SUCKER**, *Catostomus catostomus,* has a long snout, a broad rosy band on its sides, prefers deeper and colder water, and ranges north from the Ohio River and Colorado.

## MOUNTAIN SUCKER
*Catostomus platyrhynchus*                                      **90:10**

**Description**
Size, usually 6–8 in. (15.2–20.3 cm). Head small; mouth
rectangular and underslung, distinct notch at mouth corners; lateral
line complete; scales smaller toward head; dorsal-fin rays, 9-11.
**Habitat**
Clear, cold streams.
**Food**
Slime, algae, some insects.
**Reproduction**
Eggs laid in late spring in shallow streams.
**Range**
Upper Mo. R. and Columbia R. drainages.

# BUFFALOFISHES
Genus *Ictiobus*

Buffalofishes are deep-bodied, heavy-headed, and usually golden or
reddish-brown. The subopercle, the area below the gill cover, is
almost semicircular. They feed on mollusks, crustaceans, insect
larvae, and plants. They are found in the Great Lakes and
Mississippi River drainage. Buffalofishes are prolific and will often
dominate a lake at the expense of other fishes. In Iowa, the State
Conservation Commission removes almost 500,000 pounds
(226,750 kg) a year to make room for more desirable species.

## SMALLMOUTH BUFFALO
*Ictiobus bubalus*

**Description**
Size, to 30 in. (76.2 cm) and 15 lb. (6.8 kg). Body thin, back
strongly arched; mouth small and horizontal, lips thick and
streaked by parallel lines, upper lip far below eye.
**Habitat**
Deep, clear, swift waters of large rivers.
**Reproduction**
Eggs spawned at random over mud or vegetation in May.
**Range**
Same as for Bigmouth Buffalo.

## BIGMOUTH BUFFALO
*Ictiobus cyprinellus*                                      *Fig. 118*

**Description**
Size, to 40 in. (101.6 cm) and 50 lb. (22.7 kg). Large, slanted
mouth at tip of snout; the upper jaw as long as snout. Tip of upper
lip about level with lower margin of eye; lips thin, faintly streaked
by parallel lines.

*Fig. 118*

Bigmouth Buffalo

**Habitat**
Large rivers, turbid lakes and sloughs.
**Reproduction**
A few hundred thousand eggs are spawned at random over muddy or vegetated bottoms in late April or early May.
**Range**
Miss. R. and Mo. R. drainages w. to Mont.; s. to La. and Tex.

## BLACK BUFFALO
*Ictiobus niger*                                        90:9
**Description**
Size, to 28 in. (71.1 cm), usually 14–20 in. (35.6 –50.8 cm), and 16 lb. (7.3 kg). Somewhat similar to Smallmouth Buffalo, but body more torpedo-shaped and back gently arched.
**Habitat**
Clear to moderately turbid waters, sluggish to swift.
**Range**
Throughout Miss. R. Valley, greater abundance in s.

# CARPSUCKERS
Genus *Carpiodes*

Carpsuckers are somewhat similar to buffalofishes, but they have silvery bodies, the mouth is ventral, and the subopercle is almost triangular. They live in large rivers, feeding on bottom material. The meat of these fishes is well flavored but bony, so they are little sought by fishermen; but if these species were adequately harvested, they could provide much valuable high-protein food.

**CARPSUCKER COMPARISON CHART**

| Species | Long Front Dorsal Ray | Knob at Tip of Front Jaw |
|---|---|---|
| Quillback | yes | no |
| River | no | yes |
| Highfin | yes | yes |

## QUILLBACK
*Carpiodes cyprinus*                                     90:6
**Description**
Size, to 2 ft. (0.6 m), usually 10–15 in. (25.4–38.1 cm) and 9 lb. (4.1 kg). No knob at tip of mandible; long front dorsal-fin ray.
**Habitat**
Lakes, creeks, rivers, especially those with sandy bottoms.
**Reproduction**
Eggs spawned at random over mud in streams and bayous in late April or May.
**Range**
St. Lawrence R. system, across Ont. to Lake of the Woods; s. from Minn., Iowa, and Mo. to Tenn. R. Valley, e. to Chesapeake Bay drainages.

Note: The **PLAINS CARPSUCKER,** *Carpiodes forbesi,* has been combined with the Quillback.

## RIVER CARPSUCKER
*Carpiodes carpio*                                        *Fig. 119*

**Description**
Size, 12–18 in. (30.5–45.7 cm), rarely 24 in. (61 cm). Small knob at tip of mandible; lacking long front dorsal-fin ray.
**Habitat**
Large silty streams and rivers.
**Reproduction**
Eggs deposited at random, April to May.
**Range**
E. Mont. to w. Pa., s. to Mexico and Tenn.

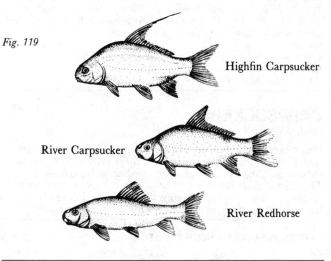

*Fig. 119*

Highfin Carpsucker

River Carpsucker

River Redhorse

## HIGHFIN CARPSUCKER
*Carpiodes velifer*                                       *Fig. 119*

**Description**
Size, 10–12 in. (25.4–30.5 cm), rarely to 15 in. (38.1 cm), and 2 lb. (0.9 kg). Small knob at tip of mandible; long front dorsal-fin ray.
**Habitat**
Clear rivers.
**Reproduction**
Eggs laid in shallows in May.
**Range**
Miss. R. Valley (but not Great Lakes).

# REDHORSES
Genus *Moxostoma*

This is a genus of silvery to reddish-brown fishes with a round body, a large head convex between the eyes, a short dorsal fin, and a mouth set behind the tip of the snout. These fishes feed principally on insect larvae and mollusks. Most spawn in spring in small streams. There are eighteen somewhat similar species in the East, of which we discuss four and illustrate two.

## RIVER REDHORSE
*Moxostoma carinatum*                                        *Fig. 119*

**Description**
Size, 12–22 in. (30.5–55.9 cm), rarely more than 1 lb. (0.45 kg).
Large, long head; large mouth; dark spots at bases of scales. Snout
blunt, tail pinkish.
**Habitat**
Deep waters of clear rivers.
**Range**
Upper St. Lawrence R., nw. Pa.; s. to Ala. and Ga., w. of
Appalachians; and e., through n. Mich. and Ohio.

## SILVER REDHORSE
*Moxostoma anisurum*

**Description**
Size, 10–16 in. (25.4–40.6 cm). Distinguished by convex dorsal fin
with 15–16 rays and absence of dark spots at bases of scales. Body
silvery, tail gray.
**Habitat**
Large, clear streams and lakes.
**Range**
Man. to St. Lawrence R. drainage; s. to n. Ala., w. of
Appalachians; w. to 100th meridian.

## SHORTHEAD REDHORSE
*Moxostoma macrolepidotum*                                   **86:3**

**Description**
Size, to 18 in. (45.7 cm) and 2 lb. (0.9 kg). Short head; 8–9 pelvic-
fin rays; front dorsal-fin rays do not reach end of last ray. Each
upper-body scale has a black crescent at its base.
**Habitat**
Clear, shallow waters of lakes and rivers over clean bottoms.
**Range**
Upper St. Lawrence R. and s., e. of Appalachians, to S.C.; w. to
100th meridian.

Note: The **OHIO REDHORSE**, *Moxostoma breviceps*, has been
combined with the Shorthead Redhorse.

## TORRENT SUCKER
*Moxostoma rhothoecum*                                       **90:11**

**Description**
Size, to 10 in. (25.4 cm). Head small with top convex, snout
slightly pointed, lower lips partly covered with small fleshy
projections.
**Habitat**
Mountain streams.
**Range**
James, Kanawha, and Potomac R. systems in Va. only.

Note: The similar **RUSTYSIDE SUCKER**, *Moxostoma hamiltoni*,
occupies the upper Roanoke River system in Virginia.

## GOLDEN REDHORSE
*Moxostoma erythrurum* 90:13

**Description**
Size, 1–18 in. (2.5–45.7 cm) and 5 lb. (2.3 kg). Concave or straight dorsal fin with 11–14 rays; no dark spots at bases of scales. Body bronze to golden, fins red or orange, tail gray.
**Habitat**
Large, fairly clear streams.
**Range**
Drainages of Lakes Mich. and Huron, and Miss. R. system s. to Ark.

# MINNOWS
Family Cyprinidae

Eastern minnows in general have thin lips, teeth in the throat, a forked tail, and a dorsal fin in the middle of the back with less than eleven rays. Only the Carp and Goldfish have any spines in the fins. Many species are hard to distinguish. The family goes back to the Paleocene.

## CARP
*Cyprinus carpio* 87:3

**Description**
Size, 1–2 ft. (30.5–61 cm), rarely to 3 ft. (0.9 m); rod and reel record, 55 lb. 5 oz. (25.1 kg). Strong saw-toothed spines at base of dorsal and anal fins; 2 barbels on each side of upper jaw. Scales large and with dark spot on base; back ruddy, belly lighter; mouth straight; lateral line present; Dorsal-fin rays, usually 18–20; anal-fin rays, 5–6.
**Habitat**
Warm, often very shallow waters, especially lakes.
**Food**
Insects, snails, crustaceans, vegetation.
**Reproduction**
Eggs laid June to July in shallow, weedy water.
**Remarks**
This is a native of Asia introduced into North America from Europe, where it is valued for food and raised in ponds for the market.
**Range**
S. Man. to N.S. and s.

## BITTERLING
*Rhodeus sericeus*

**Description**
Size, to 2–4 in. (5.1–10.2 cm). Body with lateral band; short lateral line, first 5–6 scales; mouth small; dorsal-fin rays, 10; anal-fin rays, 8–9.
**Habitat**
Clear waters, usually with vegetation.
**Reproduction**
By means of a long ovipositor, the female Bitterling lays her eggs in the gills of a fresh-water mussel, where in time they hatch.
**Range**
Introduced from e. Europe and cen. Asia.

## GOLDFISH
*Carassius auratus*                                    90:14

**Description**
Size, typically 5–10 in. (12.7–25.4 cm). Strong saw-toothed spine
at base of dorsal and anal fins; no barbels. Brown to dull gold; no
spot on base of scales; mouth slanted.
**Habitat**
Dense aquatic vegetation.
**Food**
Insects, snails, crustaceans, plant material.
**Reproduction**
Eggs laid in spring in weedy shallow.
**Range**
Introduced; Mont. to Maine and s. Sporadic in Canada. Native to
Asia.

## GOLDEN SHINER
*Notemigonus crysoleucas*                              91:10

**Description**
Size, usually 3–5 in. (7.6–12.7 cm), up to 12 in. (30.5 cm). Body
silvery or gold; scaleless belly keel; mouth very slanted; edge of
anal fin concave.
**Habitat**
Clear, quiet water with much vegetation.
**Food**
Plankton, insects, plant material.
**Reproduction**
Eggs adhesive, scattered over submerged plants, June to August.
**Range**
N.S. w. to Dakotas; s. to Gulf of Mexico.

## REDSIDE DACE
*Clinostomus elongatus*                                91:9

**Description**
Size, to 4 in. (10.2 cm). Sides red; body compressed, head pointed;
mouth large, upper jaw separated from snout by a groove, lower
lip fleshy; lateral line complete; anal-fin rays 8–9.
**Habitat**
Small, clear streams.
**Remarks**
A victim, too often, of wastes of factory, farm, and mine.
**Range**
E. Iowa and e. Mo. through Ill., s. Wis., Ind., Ohio, w. N.Y., and
all but se. Pa.

Note: The coarser-scaled **ROSY DACE**, *Clinostomus vandoisulus*,
with sides rosy, not red, ranges from New York to Georgia and
west in tributaries of the Ohio and Tennessee rivers.

## FALLFISH
*Semotilus corporalis*                                 90:7

**Description**
Size, average 4 in. (10.2 cm), grows to over 12 in. (30.5 cm). Back
bluish; scales large; head large, mouth at end of snout, upper jaw
separated from snout by a groove with a small flaplike barbel.
Differs from Creek Chub in having uncrowded scales and no black

spot on dorsal fin, which begins directly over pelvic fin; dorsal- and anal-fin rays, 8 each. The largest of native eastern minnows.
**Habitat**
Eddies of falls and rapids; also lakes.
**Reproduction**
Eggs laid in large pile of small stones.
**Range**
Drainages of James Bay, n. St. Lawrence R., e. Lake Ontario; Atlantic Coast to Va.

## CREEK CHUB
*Semotilus atromaculatus*                                              90:5

**Description**
Size, average 4 in. (10.2 cm) to 1 ft. (0.3 m). Somewhat resembles Fallfish, but scales crowded forward, black spot at base of dorsal fin, which begins behind forward edge of pelvics; dorsal-fin rays, 8.
**Habitat**
Clear, small streams with clean bottoms.
**Food**
Insects, crustaceans, mollusks, fishes, plant material.
**Reproduction**
Male builds a nest of stones in spring and later guards the eggs.
**Remarks**
Much used as a bait minnow for Bass, Walleye, and Channel Catfish.
**Range**
Mont. to Que. and s. to Gulf of Mexico.

Note: The related **PEARL DACE,** *Semotilus margarita,* of northern North America (south to Virginia) is duskier.

## SOUTHERN REDBELLY DACE
*Phoxinus erythrogaster*                                              87:10

**Description**
Size, to 3 in. (7.6 cm). Has 2 dark bands on each side with a red streak between. Scales tiny, lower lips fleshy, lateral line incomplete.
**Habitat**
Clear, cool gravelly creeks.
**Habits**
Easily frightened; may school.
**Range**
N. Ark. and n. through Iowa; e. to cen. Pa.; s. to Tenn.

## RIVER CHUB
*Nocomis micropogon*                                                  90:1

**Description**
Size, 4–6 in. (10.2–15.2 cm), rarely to 10 in. (25.4 cm). Slender body; upper jaw completely separated from snout by a groove; slender round barbel at rear of maxilla. Color brownish.
**Habitat**
Clear streams.
**Range**
Tenn. and Va., n. to s. Mich. and Lake Ont. drainage.

# SILVER CHUB
*Hybopsis storeriana* **90:4**

**Description**
Size, to 9 in. (22.9 cm). Resembles River Chub, but silvery; lower 3–4 rays of caudal fin unpigmented and immaculate white.
**Habitat**
Silty streams.
**Range**
Wyo., N.Dak., and from Lake Ont. s. to Red R. and Ala.

# BLACKNOSE DACE
*Rhinichthys atratulus* **91:14**

**Description**
Size, to 4 in. (10.2 cm). Mouth at tip of snout; upper jaw not separated from snout by a groove; barbel at rear of maxilla. Dorsal fin begins behind origin of pelvics; dorsal-fin rays, 8; anal-fin rays, 7.
**Habitat**
Clear brooks with clean bottoms.
**Range**
N.S. and w. to s. Ont. and N.Dak.; s. to the lower Miss. R. drainages.

# LONGNOSE DACE
*Rhinichthys cataractae* **91:15**

**Description**
Size, to 5 in. (12.7 cm). Somewhat similar to Blacknose Dace, but snout long and overhangs mouth.
**Habitat**
Fast water.
**Range**
Coast to coast; s. to Mexico and N.C.

# CUTLIPS MINNOW
*Exoglossum maxillingua* *Fig. 120*

**Description**
Size, 3½–4½ in. (8.9–11.4 cm). Snout overhangs mouth; mandible 3-lobed, no groove between snout and mouth; dorsal fin origin over pelvic fin origin; lateral line complete; anal-fin rays, 7.
**Habitat**
Clear, gravelly creeks.
**Range**
Watersheds of St. Lawrence R. and e. Lake Ont., s. to Va., e. of Appalachians.

Note: The somewhat similar **TONGUETIED CHUB,** *Exoglossum laurae,* has a maxillary barbel. It occurs in isolated areas in the western parts of New York, Virginia, and West Virginia.

*Fig. 120*

Cutlips Minnow

## SUCKERMOUTH MINNOW
*Phenacobius mirabilis*                                    **90:2**

**Description**
Size, to 4 in. (10.2 cm). Body slender, silvery; snout overhangs
mouth; no maxillary barbels; lateral band black; dorsal fin starts in
front of pelvics; peritoneum silvery.
**Habitat**
Riffles of large, turbid streams and rivers.
**Range**
S. Minn. and Wis., w. to include Okla., Kans., and Nebr.; e. to
Ohio, w. Ky., and w. Tenn.; s. to the lower Miss.
drainages.Originally a plains species, this fish has now become
established quite far eastward with the spread of agriculture and
muddy water. There are three other similar members of this genus
in the range, in the Kanawha, Tennessee, and Alabama river
systems.

## SILVERJAW MINNOW
*Ericymba buccata*                                        **91:1**

**Description**
Size, to 3 in. (7.6 cm). Olive back, silver sides; upper jaw separated
from snout by a groove; no maxillary barbels, lower lip fleshy,
underside of head with readily visible cavernous spaces; first ray of
dorsal fin thin, small, and attached to first full ray; anal- and
dorsal-fin rays, 8 each.
**Habitat**
Sandy brooks and small streams.
**Range**
Se. Mich. to w. Pa., s. to La. and w. Fla.; also Potomac R. system
above Great Falls.

## SILVERY MINNOW
*Hybognathus nuchalis*                                    **91:2**

**Description**
Size, to 6 in. (15.2 cm). Dark back, silver sides; upper jaw
separated from snout by a groove; no maxillary barbels, lower lip
fleshy, underside of head without cavernous spaces; first ray of
dorsal fin thin, small, and attached to first full ray; dorsal fin
begins forward of pelvic. Peritoneum black, intestine long and
coiled.
**Habitat**
Large, quiet waters.
**Reproduction**
Eggs laid in water with much vegetation.
**Range**
Sw. Que. and se. Ont., s. to Ga., La., Miss.; n. and w. in Miss. R.
and Mo. R. systems, excluding the Great Lakes except e. Lake
Ont., to Mont. and s. Ala.

## FATHEAD MINNOW
*Pimephales promelas*                                     **91:11**

**Description**
Size, to 3 in. (7.6 cm). Somewhat similar to Bluntnose Minnow,
but body stout, mouth at end of snout and slanted, lateral line
short and incomplete; tail spot absent or indistinct.

**Habitat**
Common; muddy brooks, small creeks, ponds, and small lakes.
**Remarks**
Highly adaptable; widely eaten by larger fish, and used by man for bait and for fish propagation.
**Range**
Appalachians and w. to Rocky Mts.

## BLUNTNOSE MINNOW
*Pimephales notatus*                                          91:3

**Description**
Size, to 3 in. (7.6 cm). Back dusky green; body slender; upper jaw separated from snout by a groove; no maxillary barbels, mouth below end of snout and almost straight, fleshy lower lip. First small ray of dorsal fin stout, blunt, and separated from first full ray; lateral line complete; dark tail spot conspicuous.
**Habitat**
Streams and lakes rich in plankton.
**Reproduction**
Eggs laid Apr. to Sept. on the underside of stones or boards; male guards and cleans eggs.
**Range**
S. Que. to Man. and Minn., s. to Iowa and e. of Miss. R. as far s. as Tenn., Ky., Va.

## STONEROLLER
*Campostoma anomalum*                                        91:13

**Description**
Size, to 7 in. (17.8 cm). Horny ridge extending to the edge of the lower lip, used to scrape food from the bottom. Upper jaw separated from snout by a groove, mouth below end of snout, no maxillary barbels; dorsal-fin rays, 8; anal-fin rays, 7; peritoneum black. In spring males become reddish or orange and grow large nuptial tubercles on the head and the body.
**Habitat**
Clear creeks to small rivers.
**Range**
Wyo. to w. N.Y., s. to Mexico and w. Fla.

# SHINERS
Genus *Notropis*

This is the largest genus of American minnows. Its members in the East have the upper jaw separated from the snout by a groove. They have fleshy lower lips but no maxillary barbels. The first dorsal-fin half-ray is thin and tightly attached to the first of the eight full rays. Shiners feed mainly on plankton and have relatively short intestines. Members of this genus furnish bait for the angler and natural food for many of the larger game species, such as pike, bass, perch, and crappie. Of 101 species, 5 are illustrated, as follows, on Plate 91:

**EMERALD SHINER,** *Notropis atherinoides* **(91:12),** 2–3 in. (5.1–7.6 cm). Back emerald, sides silvery. Occurs in large clear open lakes and rivers, southern Canada and Great Lakes to Gulf Coast.

**COMMON SHINER,** *Notropis cornutus* **(91:6),** 2½–4 in. (6.4–10.2 cm). Silvery, dark blotches; no lateral band. Frequents clear streams, southern Alberta to Nova Scotia and south to Colorado and Virginia.

**SATINFIN SHINER,** *Notropis analostanus* **(91:4),** to 4 in. (10.2 cm). Found in coastal lowlands, St. Lawrence R. to North Carolina.

**SPOTFIN SHINER,** *Notropis spilopterus* **(91:8),** to 3 in. (7.6 cm). Spot on last rays of dorsal fin. Range extends from North Dakota to Montreal and south to northeast Oklahoma, Alabama, and Delaware.

**SPOTTAIL SHINER,** *Notropis hudsonius* **(91:5),** to 5 in. (12.7 cm). Silvery; lateral band; black spot on tail. Occurs in Canada and the Great Lakes, south to Iowa and Missouri (not in Kansas); also Georgia.

# Catfishes
Order Siluriformes

## FRESHWATER CATFISHES
Family Ictaluridae

Catfishes have scaleless skins, a lateral line, a single strong spine in both dorsal and pectoral fins, an adipose fin, and eight barbels—two on the snout, two on the jaw, and four on the chin. They have bristlelike teeth in bands in the upper jaw. Catfishes are virtually omnivorous and are principally active after dark. Caution should be exercised in handling them, as a poison gland at the base of the pectoral spines in some species can cause a painful wound. The poison, however, is no more dangerous to man than a wasp sting, and it does not affect the edibility of the fish. Spawning takes place in the spring and summer. The family goes back to the Miocene.

**CATFISH COMPARISON CHART**

| Species | Tail | Tail Lobes | Anal-Fin Rays |
|---|---|---|---|
| Channel Catfish | forked | pointed | 24–29 |
| White Catfish | slightly forked | rounded | 18–22 |
| Blue Catfish | forked | pointed | 30–36 |
| Yellow Bullhead | rounded | | 25–26 |
| Black Bullhead | straight* | | 17–24 |
| Brown Bullhead | straight* | | 17–24 |
| Flathead Catfish | long, straight | | under 16 |
| Stonecat | long, straight | | 16 |

* or slightly notched

## CHANNEL CATFISH
*Ictalurus punctatus*                                        **87:14**

**Description**
Size, usually 14–21 in. (35.6–53.3 cm), 2–4 lb. (0.9–1.8 kg); rod and reel record, 55 lb. (24.9 kg). Deeply forked tail with pointed

lobes; 24–29 anal-fin rays. Body bluish or silvery, often with black spots; eyes large; anal fin slightly convex.

**Habitat**
Mainly large waters.

**Reproduction**
In summer, up to 20,000 eggs are deposited in a gelatinous mass in a nest which the male guards; the young travel in schools for several weeks.

**Remarks**
During the early 1950s, commercial fishermen took nearly 270,000 lb. (122,445 kg) annually from the Mississippi River.

**Range**
Widely distributed throughout the range.

## WHITE CATFISH
*Ictalurus catus*                                                    **92:10**

**Description**
Size, to 2 ft. (0.6 m). Deeply forked tail with rounded lobes; 22–23 anal-fin rays. Bluish above, silvery below, unspotted (sometimes mottled). Lower jaw shorter than upper; bony ridge between head and dorsal fin incomplete; tips of anal-fin rays form an arc.

**Habitat**
Fresh to brackish streams, ponds, and bayous.

**Remarks**
Owners of private lakes often introduce this species. It grows to a good size and bites well.

**Range**
Coastal plains, s. N.Y. to Fla.; widely introduced elsewhere.

## BLUE CATFISH
*Ictalurus furcatus*                                                 **92:12**

**Description**
Size, usually to 30 in. (76.2 cm), rarely to 40 in. (101.6 cm); rod and reel record, 94½ lb. (42.9 kg). Deeply forked tail with pointed tips; 30–36 anal-fin rays. Milky white to bluish, unspotted. Lower jaw shorter than upper, eyes small; back highest at front of dorsal fin; base of anal very long, tips of anal-fin rays almost in straight line.

**Range**
Large rivers of Miss. R. Valley and impoundments.

## YELLOW BULLHEAD
*Ictalurus natalis*                                                  **92:13**

**Description**
Size, usually 8–12 in. (20.3–30.5 cm). Body brown, chunky; belly yellow; jaws about equal, chin barbels white. Rear edges of pectoral spines sharply barbed, anal fin usually with dark band parallel to base; anal-fin rays, 22–26.

**Habitat**
Shallow waters of large ponds, lakes, and streams.

**Reproduction**
In May and June, up to 3000 eggs laid in nests in shallow water; parents guard fry.

**Range**
Se. N.Dak. to Vt., s. to Mexico and Fla.

## BLACK BULLHEAD
*Ictalurus melas*                                                    87:13

**Description**
Size, 5–10 in. (12.7–25.4 cm); rod and reel record, 8 lb. (3.6 kg).
Slightly notched tail; unmottled body. Back dark, belly yellow to
milky white, chin barbels black. Rear edges of pectoral spines
smooth; bar at base of tail white; edge of anal fin dark; anal-fin
rays, 15–19.
**Habitat**
Mud-bottomed lakes, ponds, and oxbows; also large rivers.
**Reproduction**
Nest saucer-shaped, in sand in shallow water; about 6000 eggs laid
in May or June; the parents guard the eggs and fry.
**Range**
Mont. to w. N.Y., s. to Mexico and Fla.

## BROWN BULLHEAD
*Ictalurus nebulosus*                                                92:14

**Description**
Size, to 8–12 in. (20.3–30.5 cm). Square to slightly notched tail;
mottled body. Jaws about equal; chin barbels black. Rear edges of
pectoral spines sharply barbed; anal-fin rays, 17–24; no white bar
at base of tail.
**Habitat**
Quiet, weedy, mud-bottomed lakes and ponds; also large rivers.
**Reproduction**
Nest saucerlike, protrudes out of mud in spring; about 10,000 eggs;
the parents guard the eggs and young.
**Range**
Se. Man. to N.B., s. to Ark. and Fla.

## FLATHEAD CATFISH
*Pylodictis olivaris*                                                92:15

**Description**
Size, 15–35 in. (38.1–88.9 cm), rarely to 48 in. (121.9 cm); weight
to 25 lb. (11.3 kg), uncommon to 50 lb. (22.7 kg). Long, greatly
flattened head. Body brown; lower jaw longer than upper, mouth
large and at end of snout; eyes very small. Dorsal fin begins
forward of origin of pelvic fin; adipose fin as large as head and free
at rear; tail rounded.
**Habitat**
Deep pools of large, slow-moving streams.
**Reproduction**
Eggs laid in June or July in a secluded nest; parents guard eggs
and young.
**Range**
S.Dak. to w. Pa., s. in Miss. Valley to Gulf coastal plain.

# MADTOMS
Genus *Noturus*

These small catfishes have a short lower jaw, horizontal mouth,
poison glands at the base of their pectoral spines, a complete lateral
line, and an adipose fin attached to the back and separated from
the tail only by a notch. There are twelve to twenty-three anal
rays.

## STONECAT
*Noturus flavus*                                                    **92:7**

**Description**
Size, 6–8 in. (15.2–20.3 cm). Body yellow-olive, fins yellow-edged.
Lower jaw shorter than upper; dorsal spine short, back edges
separated from tail only by a notch; tail rectangular and with light
border.
**Habitat**
Fast-water streams and riffles; also weedy lake-shore waters.
**Reproduction**
Eggs laid in June under stones; the parents guard the nest.
**Range**
S. Que. to n. New England, w. to Mont., s. to Miss. R. and Ohio
R. valleys.

## TADPOLE MADTOM
*Noturus gyrinus*                                                   **92:11**

**Description**
Size, to 3½ in. (8.9 cm). Body brown to yellowish, usually mottled;
sides with a thin black streak. Jaws equal; eyes small; back
humped at front of dorsal fin; tail rounded.
**Habitat**
Slow, clear streams, oxbows, and ponds with vegetation.
**Range**
S. Que. to n. New England, w. to Mont., s. to Fla. and Tex.

## MARGINED MADTOM
*Noturus insignis*                                                  **92:8**

**Description**
Size, to 4 in. (10.2 cm). Body dark, sides with no black streak;
lower jaw shorter than upper; fins usually black-edged; tail
rounded.
**Habitat**
Under stones in riffles of creeks and rivers.
**Range**
Atlantic Coast from s. tributaries of Lake Ont. and N.H. to n. Ga.

# Percopsiform Fishes
Order Percopsiformes

These fishes have ctenoid scales, spines in some of their fins,
pectorals usually below the mid-body line, and a small mouth not
separated from the snout by a groove. They are exclusively
freshwater fishes and are found only in North America. The order
goes back to the Eocene.

# CAVEFISHES
Family Amblyopsidae

These unusual fishes are live-bearing and, in the adults, the anus
is located under the throat. Many, such as the Southern Cavefish
and Northern Cavefish, are blind; the power of sight is not needed
in the darkness of an underground stream. Scales are small and
embedded, the head is naked; no groove separates the snout from
the mouth. The dorsal fin is positioned above the anal.

## SWAMPFISH
*Chologaster cornuta*                                           93:6

**Description**
Size, to 2 in. (5.1 cm). Brown body; no pelvic fins. Body with 3
dark longitudinal stripes and sprinkled with black specks. Eyes
present but covered by transparent skin. Dorsal fin white, often
black-edged; tail with black and white blotches; rays in tail
branched, 9–11.
**Habitat**
Swamps and weedy streams.
**Range**
Atlantic coastal plain from s. Va. to n. Fla.

Note: The similar **SPRINGFISH,** *Chologaster agassizi,* of western
Kentucky and southern Illinois, lacks the side stripes.

## SOUTHERN CAVEFISH
*Typhlichthys subterraneus*                                    93:1

**Description**
Size, to 2 in. (5.1 cm). No pelvic fins; body usually unpigmented.
Blind.
**Habitat**
Streams of cave regions.
**Range**
Ky., Tenn., n. Ala., and the Ozarks.

## NORTHERN CAVEFISH
*Amblyopsis spelaea*                                           93:2

**Description**
Size, to 5 in. (12.7 cm). Small pelvic fins; body unpigmented.
Blind.
**Habitat**
Subterranean cave streams.
**Range**
Mammoth Cave region of Ky. and s. Ind.

Note: The similar **ROSA'S BLINDFISH,** *Amblyopsis rosae,* without
pelvic fins, is confined to caves in southwestern Missouri.

# TROUTPERCHES
Family Percopsidae

## TROUTPERCH
*Percopsis omiscomaycus*                                       89:3

**Description**
Size, average 5 in. (12.7 cm). Noted for the peculiar translucence
of its body, the adipose fin, and the fine saw-toothed edges to its
scales. Body pale olive, sides with a silvery stripe. Head naked,
mouth small and straight, teeth in jaw; 2 weak dorsal spines;
adipose fin; pelvic fins and anus in rear half of body; 1 frail anal
spine.
**Habitat**
Large lakes at medium depth.
**Habits**
Nocturnal.

**Food**
Insects and plankton.
**Reproduction**
Eggs laid in spring in tributary streams and lake shallows.
**Range**
Yukon to Labrador, s. to Kan. and the Potomac R.

# PIRATE PERCHES
Family Aphredoderidae

## PIRATE PERCH
*Aphredoderus sayanus*                                              89:5

**Description**
Size, to 4 in. (10.2 cm). Dark vertical tail bar; no adipose fin. Body
dark to pinkish; sides of head scaly, mouth large and slanted; pelvic
fins in front half of body; anus (in adult) under throat. Spines
short and weak; dorsal, 3; anal 2.
**Habitat**
Vegetated creeks with thick mud bottoms; rarely in large rivers and
lakes.
**Food**
Small fishes, insects, and their larvae.
**Reproduction**
Nest built in spring; parents guard eggs and young.
**Remarks**
The odd position of the anus in the throat region distinguishes this
species and the cavefishes from almost all other species.
**Range**
Se. Minn. to w. Ohio, s. to e. Tex. and the Gulf states; also
Atlantic Coast from N.Y. to Fla.

# Codlike Fishes
Order Gadiformes

# CODFISHES
Family Gadidae

## BURBOT
*Lota lota*                                                         89:1

**Description**
Size seldom exceeds 30 in. (76.2 cm) and 10 lb. (4.5 kg). Long
second dorsal and anal; 1 chin barbel. Body olive or gray, marbled
with black. Scales tiny, embedded; no spines in fins; pelvics under
pectorals.
**Habitat**
Deep, cool lake water.
**Habits**
Retires to deeper water in summer.
**Food**
Other fishes, insects.
**Reproduction**
Eggs laid January to March in rivers or rocky lake bottoms.
**Range**
N. North America, s. to Kans., Mo., New England, and N.Y.

# Cyprinodont and Atherine Fishes
Order Atheriniformes

Members of this order lack spines in their fins. The head as well as the body is scaled. The pelvic fins, when present, are small and positioned near the middle of the belly. There is no lateral line and the caudal fin is either straight or rounded (never forked).

## KILLIFISHES
Family Cyprinodontidae

The fishes of this shallow-water family are frequently barred or striped and have large scales, extending onto their heads. The mouth is small, the snout often flattened and with a groove. Many are surface feeders. Of the thirteen species in the range, we treat four.

## KILLIFISHES
Genus *Fundulus*

This is a genus of slender fishes with pelvic fins, three to six pores on the lower jaw, and no lateral line.

### BANDED KILLIFISH
*Fundulus diaphanus*                                                    **93:5**

**Description**
Size, to 4 in. (10.2 cm). Dorsal fin starts in front of anal fin; dorsal-fin rays, 13–15; anal-fin rays, 10–12.
**Habitat**
Quiet, weedy waters of lakes, rivers, and estuaries.
**Habits**
Surface feeder; often travels in schools at surface.
**Food**
Insects, tiny crustaceans, vegetation.
**Reproduction**
Eggs laid in summer in weedy shallows.
**Range**
E. U.S. and Canada, w. to the Plains.

### NORTHERN STUDFISH
*Fundulus catenatus*

**Description**
Size, to 7 in. (17.8 cm). Bluish body; orange spots. Dorsal fin starts above anal; dorsal-fin rays, 13–16; anal-fin rays, 15–18; female with short brown bar.
**Habitat**
Clear streams.
**Range**
Kans. and Ark., e. to Va. and Ala.

## STARHEAD TOPMINNOW
*Fundulus notti*                                          **93:3**

**Description**
Size, to 3 in. (7.6 cm). Dorsal fin starts well behind origin of anal;
dorsal-fin rays, 6–8; anal-fin rays, 8–10.
**Habitat**
Clear, weedy backwaters.
**Habits**
Surface feeder.
**Range**
Lake Mich. drainage and Miss. R. Valley from Iowa to w. Tenn.
and ne. Ark.

## BLACKSTRIPE TOPMINNOW
*Fundulus notatus*                                        **93:4**

**Description**
Size, to 3 in. (7.6 cm). Big black stripe on side. Dorsal fin starts
well behind start of anal; dorsal-fin rays, 9; anal-fin rays, 11.
**Habitat**
Clear ponds, lakes, and streams; canals with some vegetation.
**Range**
E. Iowa to s. Mich. and Ohio, s. to e. Tex., Miss., and Tenn.

# LIVEBEARERS
Family Poeciliidae

## MOSQUITOFISH
*Gambusia affinis*                                        **93:7**

**Description**
Size, to 2 in. (5.1 cm). Back dusky, sides silvery, body and fins
with small black spots. Scales large; mouth separated from snout
by a groove; dorsal fin starts behind origin of anal fin; dorsal-fin
rays, 6–8.
**Habitat**
Clear vegetated water in ponds, pools, ditches, and marshes.
**Food**
Insects and crustaceans.
**Reproduction**
Bears live young. The modified anal fin of the male serves as an
intromittent organ to introduce sperm into the body of the female,
where the eggs are fertilized.
**Range**
E. from the Rio Grande to Atlantic; s. from Del. to Fla.

# SILVERSIDES
Family Atherinidae

## BROOK SILVERSIDE
*Labidesthes sicculus*                                    **89:12**

**Description**
Size, to 4 in. (10.2 cm). Silver sides; 2 dorsal fins, concave anal fin.
Scales cycloid and extending onto head; mouth forms a short beak;
2 dorsal fins separated; anal fin with 1 weak spine.

**Habitat**
Clear lakes and quiet parts of streams.
**Habits**
Surface swimmer, often seen in large schools; jumps and skips in air.
**Food**
Insects and plankton.
**Reproduction**
Eggs adhesive, laid in spring. Spawns after its first winter; few live through a second.
**Range**
Minn. to Appalachians, s. to e. Tex. and Fla.

# Sticklebacks
Order Gasterosteiformes

## STICKLEBACKS
Family Gasterosteidae

Sticklebacks are unique in the East for their several separate dorsal spines. The males build elaborate nests of plants and sticks and guard the eggs and young. Of the five eastern species, we discuss and illustrate one. The others, quite widely distributed in the range, have self-identifying names: Fourspine Stickleback, *Apeltes quadracus;* Ninespine Stickleback, *Pungitius pungitius;* Threespine Stickleback, *Gasterosteus aculeatus;* Blackspotted Stickleback, *Gasterosteus wheatlandi.*

### BROOK STICKLEBACK
*Eucalia inconstans*                                    **91:7**

**Description**
Size, to 2 in. (5.1 cm). Has 5 separate dorsal spines. Back dark, sides mottled, belly speckled.
**Habitat**
Clear, cold vegetated springs and brooks with mud bottoms.
**Food**
Small insects and crustaceans.
**Range**
Most of eastern range; s. Canada and s. to Iowa and N.Y.

# Perchlike Fishes
Order Perciformes

Members of this large order have the head scaled, the pelvic fins forward, and spines in both dorsal and anal fins. The order goes back to the Cretaceous.

## TEMPERATE BASSES
Family Percichthyidae

Eastern members of this family have pointed gill covers, two dorsal fins, three anal spines, and a lateral line which does not extend onto the slightly forked tail. Each pectoral has one spine and five rays.

# WHITE BASS
*Morone chrysops*                                    86:13

**Description**
Size, to 1 ft. (0.3 m) and 3 lb. (1.4 kg). Separate dorsal fins; lower jaw longer than upper. Silvery with 5–7 bold stripes on each side; dorsal spines graduated; anal rays, 11–12.
**Habitat**
Lakes and large rivers.
**Habits**
Usually travels in schools.
**Food**
Small fishes, insects, plankton, crayfish.
**Reproduction**
In spring, up to a million eggs spawned in gravelly shoals.
**Remarks**
The populations of this species and of the Yellow Bass fluctuate much more widely in numbers than do those of most other fishes.
**Range**
S. Minn. to w. N.Y., s. to Mexico and Gulf of Mexico; widely introduced.

# WHITE PERCH
*Morone americana*                                   89:9

**Description**
Size, usually 5–7 in. (12.7–17.8 cm); rod and reel record 4¾ lb. (2.2 kg). Dorsal fins joined; jaws almost equal; faint streaks on sides. Back olive, sides silvery and unmarked; mouth small, forehead long and sloping; anal spines not graduated; anal-fin rays, usually 8–10.
**Habitat**
Lakes and brackish water.
**Food**
Crustaceans, fishes, insects.
**Reproduction**
Eggs adhesive, laid in spring in freshwater shallows.
**Remarks**
A good pan fish.
**Range**
Atlantic Coast, N.S. to Ga.; widely introduced. May have started an invasion of the Great Lakes above Niagara; first Lake Erie specimen caught in 1953.

# YELLOW BASS
*Morone mississippiensis*                            87:8

**Description**
Size, to 1 ft. (0.3 m). Similar to White Perch, but sides with 6–7 bold, discontinuous stripes.
**Habitat**
Lakes, oxbows, large rivers.
**Food**
Crustaceans, small fishes, insects, insect larvae.
**Reproduction**
In May, 1–2 million tiny eggs deposited over gravel or rocky reefs.
**Range**
Minn. to Ind., s. to e. Tex. and Ala.; widely introduced.

# SUNFISHES
Family Centrarchidae

Sunfishes are carnivorous and have thin oblong or circular bodies, dorsal fins that are completely joined (except in *Micropterus*), a lateral line, and three or more anal spines. The males make a shallow depression for a nest and guard the eggs and fry.

## BANDED SUNFISH
*Enneacanthus obesus*                                            **92:2**

**Description**
Size, to 3 in. (7.6 cm). Olive with purple or golden spots; 5–8 broad vertical crossbars; gill cover notched and with a large dark spot. Dorsal spines, usually 9; tail rounded.
**Range**
Coastal plain from Mass. to Fla.

Note: The related **BLUESPOT SUNFISH,** *Enneacanthus gloriosus,* ranging from New Jersey to Florida, has bright blue spots on sides and fins.

## BLACKBANDED SUNFISH
*Enneacanthus chaetodon*                                        **92:5**

**Description**
Size, to 4 in. (10.2 cm). Has 4 vertical black bands and black forepart of first dorsal fin. Body yellow; mouth small, gill cover notched, tail rounded; dorsal-fin spines, 10.
**Range**
Coastal plain, N.J. to Fla.

## ROCKBASS
*Ambloplites rupestris*                                          **87:7**

**Description**
Size, 6–10 in. (15.2–25.4 cm) and 1 lb. (0.4 kg). Dark markings, red eye, 6 anal-fin spines. Body brassy, scales each with a dark spot; mouth large, lower jaw protrudes, gill cover notched. Lateral line complete; dorsal fin with 10–12 spines and much longer than anal.
**Habitat**
Rocky lake shallows, clear streams with rock bottoms.
**Habits**
Often schools; hibernates in winter.
**Food**
Insects, crayfish, fishes.
**Reproduction**
Nest built by male in gravelly shallows; about 5000 eggs laid in May or June; male guards nest and young.
**Range**
Great Lakes, St. Lawrence R. and Atlantic drainages, and Miss. R. Valley to Gulf of Mexico; widely introduced.

# BLACK BASSES
Genus *Micropterus*

This is the genus of sunfishes in which the dorsals are separated by a deep notch.

## SMALLMOUTH BASS
*Micropterus dolomieui*                                    87:6

**Description**
Size, usually 8–15 in. (20.3–38.1 cm); rod and reel record, 11 lb.
14 oz. (5.4 kg). Shallow notch in dorsal fin; upper jaw rarely
reaches rear of eye. Back olive, belly dusky silver, no stripe of dark
spots on sides (but young have dark vertical bar on sides.). Shortest
dorsal spine more than ½ length of longest; dorsal-fin rays, 12–15,
pectoral-fin rays, 16–18; anal-fin rays, 11.

**Habitat**
Clear, rocky lakes and rivers.

**Food**
Fishes.

**Reproduction**
Male fans out nest in gravelly bottom, where up to five females
may deposit total of 2000 eggs; male guards eggs and young;
respawning within several days is frequent.

**Range**
Minn. to Que., s. to Ark. and n. Ala.; widely introduced elsewhere.

## SPOTTED BASS
*Micropterus punctulatus*                                 92:9

**Description**
Size, 8–17 in. (20.3–43.2 cm) and 3 lb. (1.4 kg). Shallow notch in
dorsal fin; dark spot at base of tail; upper jaw rarely reaches rear
of eye. Back yellow, sides white, prominent dark spot on gill cover,
band of connected black blotches on side. Shortest dorsal spine
more than ½ size of longest; dorsal-fin rays, 11–13; pectoral-fin
rays, 15–16; anal-fin rays, 9–11.

**Habitat**
Slow streams and deep pools.

**Food**
Fishes.

**Range**
Most of s. U.S. from Fla. to Tex. and n. to Mo.

## LARGEMOUTH BASS
*Micropterus salmoides*                                   87:12

**Description**
Usually 10–18 in. (25.4–45.7 cm), rarely to 24 in. (61 cm), and 10
lb. (4.5 kg); rod and reel record, 22¼ lb. (10.1 kg). Dorsal fins
almost divided by notch; upper jaw extending beyond eye. Back
olive, sides and belly silver, side with a dark band (which becomes
broken with age). Shortest dorsal spine less than ½ longest; dorsal-
fin rays, 12–15; anal-fin rays, 11.

**Habitat**
Ponds, small lakes, oxbows; fairly tolerant of turbidity.

**Food**
Fishes, crayfish, frogs, insects.

**Reproduction**
Nest often on mud bottom with vegetation; eggs similar to
Smallmouth Bass.

**Range**
S. Canada from Que. to Man., entire Great Lakes area and Miss.
R. Valley to Gulf of Mexico.

# SUNFISHES
Genus *Lepomis*

These sunfishes have deep, thin bodies, ten dorsal and three anal spines, a notched tail, and no teeth on the tongue. They are colorful and popular game fishes, and some species have been widely introduced.

## REDBREAST SUNFISH
*Lepomis auritus*                    92:4

**Description**
Size, 5–7 in. (12.7–17.8 cm). Gill cover long, black, narrow, flexible. Back olive, belly orange-red, no black spot on hind dorsal rays. Teeth on roof of mouth; pectoral fins short, rounded.
**Habitat**
Rivers.
**Food**
Insects, mollusks, fishes.
**Range**
N.B. to Fla.; introduced w. to s. Okla.

## LONGEAR SUNFISH
*Lepomis megalotis*                    92:6

**Description**
Size, to 9 in. (22.9 cm). Color variable, often brilliant; sides blue-spotted, belly yellow, iris red. Mouth small, but reaching to front edge of eye. Ear flap large, rear edge red or bluish-white and with thin, flexible projection. No teeth on roof of mouth; pectoral fins rounded; pectoral-fin rays, 10–13.
**Food**
Insects.
**Range**
Miss. R. Valley, from Minn. to Gulf states, n. to N.C.

## REDEAR SUNFISH
*Lepomis microlophus*                    *Fig. 121*

**Description**
Size, usually 6–7 in. (15.2–17.8 cm), rarely to 10 in. (25.4 cm). Back olive, belly brassy, gill cover with red rear edge. Mouth barely reaching front edge of eye. Pectoral fin long and pointed; no distinct spots on rear of dorsal and anal fins.
**Habitat**
Clear, still waters with some vegetation.
**Range**
Miss. R. Valley, Iowa to Ind., and s. to Gulf of Mexico.

*Fig. 121*

Redear Sunfish

# WARMOUTH
*Lepomis gulosus*                                                  **92:3**

**Description**
Size, rarely to 10 in. (25.4 cm) and 1 lb. (0.4 kg). Anal spines, 3;
notched tail. Mouth large, upper jaw extending to eye. Color, olive;
3–5 dark spokes around eye; tip of gill cover with white border.
Pectoral fins short, rounded; dorsal-fin spines, usually 10; second
dorsal fin with dark wavy lines.
**Habitat**
Vegetated ponds, oxbows, marshes, sluggish streams, preferably
with muddy bottoms.
**Food**
Insects, snails, crustaceans, fishes.
**Reproduction**
Nest built on sand or gravel bottom.
**Range**
Kans., Mich., and N.Y., s. to Gulf of Mexico.

# GREEN SUNFISH
*Lepomis cyanellus*                                                **92:1**

**Description**
Size, to 8 in. (20.3 cm) and 15 oz. (0.4 kg). Body green; mouth
large and extending past front edge of eye; gill cover stiff to smooth
rear edge; lateral line complete. Pectoral fin short and rounded;
dark spot on gill cover and on bases of dorsal and anal fins.
**Habitat**
Warm streams and ponds.
**Remarks**
A good pan fish.
**Range**
S.Dak. to N.Y. and s. to N.Mex. and Ala.; introduced elsewhere.

Note: This species frequently hybridizes with the Bluegill. The
**BANTAM SUNFISH,** *Lepomis symmetricus,* of the Mississippi Valley
from southern Illinois south, has the lateral line incomplete. The
**SPOTTED SUNFISH,** *Lepomis punctatus,* has many bronze specks
on an olive background, and occurs in Mississippi basin from
southern Indiana south.

# BLUEGILL
*Lepomis macrochirus*                                              **87:4**

**Description**
Size, usually 4–9 in. (10.2–22.9 cm), rarely to 11 in. (27.9 cm); rod
and reel record, 4¾ lb. (2.2 kg). Back olive with blue and orange
on sides; mouth small and rarely reaching front edge of eye; 2
bluish bars extend back from mouth and chin. Rear edge of gill
cover thin, lower edge usually bluish; dusky spot on last 3 dorsal
rays. No teeth on roof of mouth; pectoral fins long and pointed;
anal-fin rays, 10–12.
**Habitat**
Warm, weedy waters of bays, ponds, and lakes.
**Food**
Insects, fishes, crayfish.
**Range**
Drainages of Great Lakes, Miss. R., Gulf Coast, and Atlantic
Coast, n. to N.J.

## PUMPKINSEED
*Lepomis gibbosus*      **87:5**

**Description**
Size, to 9 in. (22.9 cm) and 11 oz. (0.3 kg). Long, pointed pectoral fins; small red spot near edge of gill cover. Body very deep; mouth small, not reaching eye. Color orange to olive, sometimes with blue markings on cheek. No teeth on roof of mouth; gill cover stiff to smooth rear edge; rear of dorsal fin mottled.

**Habitat**
Clear, still water with many submerged plants.

**Food**
Insects, mollusks, fishes.

**Range**
N.Dak. to N.B., s. to Kans., Ky., and Ga.

## ORANGE SPOTTED SUNFISH
*Lepomis humilis*      **86:14**

**Description**
Size, to 4 in. (10.2 cm). Conspicuous orange spots on silvery sides. Back green; lobe of gill cover with broad red or orange margin; edges of gill cover white; rear edge thin with flexible projection. Mouth extending past front edge of eye; teeth on roof of mouth. Pectoral fins pointed; anal-fin rays, 9–10. The colors of the breeding male are exceptionally brilliant.

**Habitat**
Turbid lakes, ponds, and slow streams with silt-covered bottoms.

**Food**
Insects, small crustaceans, fishes.

**Range**
Dakotas to Ohio, s. to N.Mex. and Miss.

# CRAPPIES
Genus *Pomoxis*

These members of the sunfish family have long heads with concave foreheads.

## BLACK CRAPPIE
*Pomoxis nigromaculatus*      **87:9**

**Description**
Size, usually 7–12 in. (17.8–30.5 cm); rod and reel record, 5 lb. (2.3 kg). Body greenish with black blotches; base of dorsal fin as long as distance from eye to dorsal fin; dorsal-fin spines, 6–10 (usually 7–8).

**Habitat**
Clear, quiet water with mud or sand bottoms and submerged vegetation.

**Food**
Small fishes, insects.

**Reproduction**
Nest in 3–6 ft. (0.9–1.8 m) of water on sandy or mud bottom among plants; eggs laid in late spring.

**Range**
Across s. Canada from upper St. Lawrence R. to Man.; s. through e. and cen. U.S. to Fla. and Tex. Widely stocked by state conservation departments; but it often suffers in competition with the White Crappie.

## WHITE CRAPPIE
*Pomoxis annularis*                                     **87:11**

**Description**
Size, to 1 ft. (0.3 m); rod and reel record, 5 lb. 3 oz. (2.4 kg). Body greenish or silvery, with indistinct vertical bars; mouth large, lower jaw heavy and protruding, gill cover notched. Base of dorsal fin shorter than distance from eye to dorsal fin; dorsal-fin spines, 5–7 (usually 6).

**Habitat**
Ponds, lakes, bayous, oxbows, and sluggish pools, preferably with vegetation; very tolerant of turbidity.

**Habits**
Often travels in schools.

**Food**
Insects, crustaceans, small fishes.

**Reproduction**
Eggs laid in late spring.

**Remarks**
An abundant, tolerant, and popular pan fish, much stocked by state conservation departments.

**Range**
E. S.Dak. to sw. N.Y., s. to Gulf of Mexico.

# PERCHES
Family Percidae

These predaceous fishes have ctenoid scales, two completely separate dorsal fins, and one or two anal-fin spines.

## YELLOW PERCH
*Perca flavescens*                                      **86:1**

**Description**
Size, average 4–10 in. (10.2–25.4 cm), to 14 in. (35.6 cm); rod and reel record, 4 lb. 3½ oz. (1.9 kg). Prominent broad vertical bars on yellow sides. Body deep, compressed; no canine teeth; head slightly concave above eyes, making humped outline before first dorsal. Pelvics close together; first dorsal with 13–15 spines and well separated from second; anal-fin rays, 6–8.

**Habitat**
Vegetated lakes and streams.

**Habits**
Usually travels in schools.

**Food**
Plankton, insects, fishes.

**Reproduction**
Eggs laid in spring in long strands of gelatinous material.

**Range**
Alta. to N.S.; s. to ne. Kans., sw. Pa., and Atlantic Coast to S.C.

## SAUGER
*Stizostedion canadense*                                **89:11**

**Description**
Size, commonly to 15 in. (38.1 cm); rod and reel record, 8 lb. 3 oz. (3.7 kg). Many round black spots (but no single large black blotch) on the dorsal fin. Body elongated, cheeks scaled, free edge of preoperculum toothed; dorsal-fin spines, 13–15; dorsal-fin rays, 17–20; anal-fin rays, 11–14.

**Habitat**
Shallow, turbid lakes and large rivers.
**Food**
Fishes.
**Reproduction**
Eggs laid in early spring in shallow water.
**Range**
Cen. Alta. to N.B.; s. to Colo. and Ala.

## WALLEYE
*Stizostedion vitreum*                                    **86:11**

**Description**
Size, average 13–20 in. (33–50.8 cm); rod and reel record, 22¼ lb.
(10.1 kg). Distinct black blotch on the rear end of the first dorsal
fin. Strong canine teeth; dorsal-fin rays, 19–22.
**Habitat**
Clear and moderately deep lakes; large rivers.
**Food**
Fishes, crayfish.
**Reproduction**
Eggs laid in early spring, in lake shoals or tributary systems.
**Range**
Athabaska Lake (Alta.–Sask. border), e. to Labrador and s. to
N.C.; Great Lakes, Miss. R. Valley to Tenn. R.

# DARTERS
Genus *Percina*

These are small elongate fishes with small mouths, complete lateral
lines, and unforked tails. Their dorsal fins are well separated, with
ten to seventeen spines in the first fin, and their pelvic fins are
separated by a width at least equal to the base of a pelvic fin.
There are usually two anal spines. The midline of the belly either
is naked or has large spiny scales.

## BLACKSIDE DARTER, *Percina maculata* (**93:9**), to 2–3 in.
(5.1–7.6 cm). Mouth at end of snout, 7 oblong side blotches. Its
range extends from southern Manitoba and Ontario south to
northeast Texas and Alabama.

## LOGPERCH, *Percina caprodes* (**93:15**), usual size, 3–4 in. (7.6–
10.2 cm), to 7 in. (17.8 cm). Mouth below snout, dark vertical
stripes, spot at base of tail. Occurs from central Alberta to southern
Quebec south to east Texas, west of Appalachians.

# SAND DARTER
Genus *Ammocrypta*

## EASTERN SAND DARTER
*Ammocrypta pellucida*                                    **93:10**

**Description**
Size, to 3 in. (7.6 cm). Body long, slender, pellucid, lower half
almost scaleless; sides with 12–19 dusky blotches. Mouth small and
separated from upper jaw by groove; dorsal fins well separated;
dorsal-fin rays, 9–12; anal-fin rays, 8–10.

**Habitat**
Sandy bottoms.
**Habits**
Buries itself in sand up to its eyes, which scan the surrounding water for its prey; darts out after prey, then quickly reburies itself in the sand.
**Food**
Insects.
**Reproduction**
Eggs laid in spring.
**Range**
Se. Mich., Que., and Lake Champlain, s. to Ky. and W.Va.

# OTHER DARTERS
Genus *Etheostoma*

Members of this genus are small; their bodies, including the midline of the belly, are well scaled. The lateral line may be absent, incomplete, or complete. The anal fin is smaller than the second, or soft, dorsal. Eighty species are recognized.

**JOHNNY DARTER,** *Etheostoma nigrum* (**93:8**), to 3 in. (7.6 cm). Occurs in streams of eastern North America west to the Plains.

**BANDED DARTER,** *Etheostoma zonale* (**93:13**), to 3 in. (7.6 cm). Its range includes clear streams east of Plains, Lake Michigan drainage to Pennsylvania, and south to Gulf of Mexico.

**GREENSIDE DARTER,** *Etheostoma blennioides* (**93:11**), to 4 in. (10.2 cm). This species is found in riffles of Mississippi River system and Ozark region, east to New York and Georgia.

**FANTAIL DARTER,** *Etheostoma flabellare* (**93:12**), to 3 in. (7.6 cm). Found in riffles from Minnesota and Quebec south to Oklahoma and Alabama.

**NORTHERN SWAMP DARTER,** *Etheostoma fusiforme* (**93:14**), to 2 in. (5.1 cm). Range extends from Maine to North Carolina.

**RAINBOW DARTER,** *Etheostoma caeruleum* (**86:2**), to 2½ in. (6.4 cm). Found from southern Minnesota and eastern Ontario to Arkansas and Alabama.

# CROAKERS
Family Sciaenidae

## FRESHWATER DRUM
*Aplodinotus grunniens*                                     86:12

**Description**
Size, average 18–20 in. (45.7–50.8 cm). High back; long, continuous dorsal fins; short anal fin; lateral line extending onto the tail. Dusky above, sides silvery; body and head scaled, throat teeth coarse. Dorsal spines, 8–9; anal spines, 2, second long and heavy.
**Habitat**
Lake shallows, large rivers.
**Habits**
Produces a grunting sound.
**Food**
Mollusks, crayfish, insects.
**Reproduction**
Eggs laid in spring.
**Remarks**
An important commercial fish.
**Range**
Lakes and rivers as far n. as the St. Lawrence Basin to Man. and Mont.; s. to Tex.; peak numbers in Miss. R. and lower Mo. R. systems.

# SCULPINS
Family Cottidae

## MOTTLED SCULPIN
*Cottus bairdi*                                            89:8

**Description**
Size, average 3 in. (7.6 cm). Body dark, mottled; head and mouth large, eyes high. Dorsal-fin spines, 7–9; dorsal rays, 16–18; pectoral-fin spines, 1; pectoral rays, 13–17; anal-fin rays, 10–14.
**Habitat**
Cool streams.
**Food**
Insects, crustaceans, small fishes.
**Range**
S. Canada to s. Appalachians; also Ozarks. Discontinuous into parts of Mo. R. and Columbia R. drainages.

# Mollusks

**Consulting Editor**
Harald A. Rehder
Zoologist Emeritus
National Museum of Natural History
Smithsonian Institution

**Illustrations**
Plates 94–98 Klarie Phipps
Plates 99–107 Jennifer Emry-Perrott
Plate 108 Nancy Lou Gahan

# Mollusks
Phylum Mollusca

The enormously successful phylum of mollusks, sometimes termed the "soft ones" from translation of the Latin name Mollusca, constitutes one of the most advanced groups of animals in the invertebrate world. To be sure, an oyster may not seem very advanced, but the squid that can remove the lid from a box with its tentacles to get the food within is certainly more intelligent than some of the lower vertebrates. Moreover, its eye is the most highly organized visual organ known among the invertebrates.

## Nature of Mollusks
Mollusks, in general, are soft-bodied, unsegmented, shell-bearing invertebrates. Five classes are treated in this chapter: the gastropods, class Gastropoda; the bivalves, class Bivalvia; the scaphopods, class Scaphopoda; the chitons, class Polyplacophora; and the cephalopods, class Cephalopoda. The first four classes are the seashells, which make up two-thirds of all mollusk species. In these, the shell is external and conspicuous; in the cephalopods, it is internal and inconspicuous. In a few groups, such as the sea slugs, or nudibranchs, and a few chitons, the shell has been lost or never developed.

These five classes are, basically, as different from one another as are the five classes of the phylum Chordata, that were dealt with earlier in the book. In other words, in fundamental structure, a clam is as different from a conch shell as a mammal is from a bird, and a squid and chiton are as far removed from each other as man is from a mud turtle.

## Classification
In general terms, mollusks are soft-bodied, unsegmented, and usually hard-shelled invertebrate animals. The living mollusks comprise seven classes; the five major ones discussed in this chapter are:

## Class Polyplacophora
The chitons, also known as coat-of-mail shells, are primitive, ancestral-type mollusks of very sluggish habits. Elongated and flattened, the chitons bear a shelly armor of eight saddle-shaped plates (imbricating valves) arranged in an overlapping series along the back and held together by a surrounding girdle often bearing spiny hairs. On the underside is the broad, flat foot. Although most chitons inhabit shallow water close to shore, others live in depths as great as 2000 fathoms (3.7 km).

## Class Gastropoda
The gastropods are the snails, most of which have shells. Most gastropods are univalves; that is, they have a single shell, or valve, which is usually spiral. A few exhibit two valves. Gastropods have a low order of sight, hearing, and possibly taste but well-developed senses of smell and touch. Most are carnivorous and predatory. They move over the ocean floor, and some are lifelong wanderers on the open sea.

## Class Scaphopoda
The scaphopods comprise a small class of burrowing univalve mollusks having a tapering conical shell open at both ends and slightly arched, like a miniature elephant's tusk. The members of this class are commonly known as tooth or tusk shells, after their shape. From the larger end of the shell projects the foot and several

slender tentacles. The animals live partially buried in sand in clear water at various depths, feeding on minute marine organisms.

## Class Bivalvia

The bivalves, also known as pelecypods, are the clams, which have two valves joined along a hinge line by a tough ligament and held together by one or two strong muscles. Except in a short larval stage, most clams are sedentary throughout their lives.

## Class Cephalopoda

The cephalopods are highly specialized, carnivorous mollusks with relatively keen sight and a well-developed nervous system. The shell, if present, is generally internal and lengthened. The head supports two large eyes and is armed with a powerful parrotlike beak; it is surrounded by long flexible tentacles, or arms, studded with powerful sucking disks. Most cephalopods can discharge a caustic, inky fluid in defense. They move rapidly through the water with a jet-propelled action, the tentacles streaming behind.

*Fig. 122*

Parts of a Typical Snail and Clam

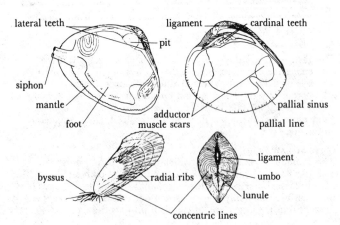

### Habitat

There are some 80,000 to 100,000 species of mollusks in the world, a little over half of them marine. Marine species live from three miles below the surface of the ocean to the top of the springtide splash line on sea cliffs. In their different varieties they occur on sandy, muddy, pebbly, or rocky coasts; in clear, muddy, or brackish water; in rocky crevices and tide pools; or afloat and adrift on the high seas. They are found from the Arctic to the tropics. Some mollusks not included in this chapter inhabit freshwater lakes and streams, live underground, or are found on trees.

### Habits

Many, perhaps most, mollusks are slow-moving or sedentary. A few, however, such as the squids, are among the swiftest travelers. Various species can burrow in mud, wood, stone, or even lead cables; some are parasitic. A number can swim; others can crawl or jump; and the sea arrow, or flying squid, can even travel in the air. As in most great aggregations of living forms, some are carnivorous, some are vegetarians, and some are scavengers.

### Reproduction

The phylum exhibits great diversity in reproduction. In some species the sexes are separate; some species are hermaphroditic; in others the same individual changes, sometimes even alternates, its sex. An interesting example is the Virginia Oyster, which indulges in an alternation of the sexes. The Quahog practices sex reversal, changing from male to female as it grows older.

Some species, such as the oyster, cast their eggs and sperm adrift on the surrounding sea. Higher forms are more careful. The male octopus has a tentacle designed to gather and deliver the sperm into the mantle cavity of the female. To make sure it arrives safely, the male argonaut takes the precaution of detaching and leaving the tentacle along with the sperm.

The shelled mollusks are essentially sedentary, although some scallops are active swimmers. Breeding seasons vary among different species; the custom of eating oysters only in months with an *r* coincides roughly with protection for the bivalve during the breeding season.

### Evolution

Geologically speaking, mollusks go back to the Cambrian period. Because of their calcareous shells they are commonly preserved as fossils, and the rocks of former ages often abound with remains. Because of their abundance and the ease with which shallow-water forms have spread throughout the world in past ages of widespread equable temperatures, many have become index fossils and are used by the paleontologist to correlate geologic formations in different regions.

Under conditions of a relatively stable environment, such as the sea, the rate of evolutionary change slows down and is influenced more by competition with other animals than by the environment. So slowly does change proceed among certain well-adapted species living in such a near-stable environment that the same species of seashells living today may often be found in deposits 50,000,000 years old. This is a situation unknown among birds or mammals, which are subject to the wide variations of rapidly changing conditions on land. Molluscan genera living today may often be recognized in Mesozoic and occasionally even in Paleozoic strata.

### Conservation

Mollusks which are eaten, such as clams and oysters, often suffer from pollution and overharvesting. However, much government research and legislation have gone into the problem of helping ten clams live and grow where only one did before. Commercial interest in the production of these shellfish is so great that noncommercial conservationists are more disposed to worry about other mollusks.

Most marine mollusks are not at the moment in danger of extinction or even significant depletion. However, this may change with excessive collecting. Even now, along some favored Florida beaches, the near-shore flats have been so well combed by amateurs that few individuals of the most sought-after species remain. The student, naturalist, and collector should remember to put back in its original position every rock or underwater shelter under which he looks. Any one of these may be a mollusk habitat, and without shelter the mollusks may go elsewhere or disappear.

### Shell Collecting Techniques

The most productive place to collect shells is the seashore at low tide. Strolling along a sandy stretch of beach can prove fruitful in finding shells that have been washed up by a late tide. Much of such shelly material, however, will be wave-eroded, broken, or damaged. The valves of a bivalve shell are commonly found separated, and the lips of a univalve usually show broken edges and worn shoulders.

When no longer content with beach-worn specimens, a collector should search for living invertebrates in their natural habitats. This opens up unlimited possibilities for acquiring a worthwhile shell collection. It is thus important to pay attention to tides, seasonal moods of the ocean, and the effects of wind and current.

Some states may require possession of a regular collecting license before collection of living seashells is permitted. Additionally, for certain species, there are stringent protective regulations. The collector should understand these laws before collecting living mollusks.

The best time to collect and look for living shellfish is at low tide, when exploration of the exposed tidal zone is easiest. Because many marine creatures hide under objects to await the return of the tide, any stone, plank, or piece of driftwood should be turned over, and then replaced as originally found. Bivalves can often be detected in sand or on mud flats by small holes which reveal the siphon; and weight above may cause the animal to withdraw its siphon with a sharp squirt of water. Snails, which usually bury themselves just beneath the surface, reveal their locations by small telltale mounds or by their trails.

Shell collecting, of course, is only part of the fascination of malacology. The study of the living animal offers much more. Many of the life histories of almost all of the 50,000 marine mollusks still await investigation. In this field science has taken only its first steps.

### Nomenclature

In this chapter every effort has been made to use the latest acceptable scientific nomenclature. The scientific name (genus and species) tells much that is known about a particular mollusk from its most distant ancestry to the present. Common or popular names are an attempt at standardization, and to many scientists translations seem inappropriate. It is suggested that the student concentrate on mastering the scientific name. However, the common names are included here, as an aid to the beginning malacologist.

### Range and Scope

Of the more than 50,000 marine mollusks living in the world today, there are 6000 known in North America, most of them found in deep water. In this chapter 171 shallow-water species are discussed. They represent all species above ¼ inch (0.64 cm) in size that occur regularly to a depth of 10 fathoms (18.3 m) from the southern Canadian border south along the Atlantic Coast of the United States to southern Florida. Land and freshwater mollusks are not included; nor are those normally found in waters less than 10 fathoms deep, which are under ¼ inch in size.

### Illustrations

Virtually every species discussed in this chapter is illustrated. The illustrations are captioned on the plates by common names of the individual species. The name is followed by a number indicating the size of the illustration relative to the actual life size of the species.

## USEFUL REFERENCES

Abbott, R. T., 1968. *Seashells of North America.* New York: Golden Press.

————.1974. *American Seashells.* 2nd ed. New York: Van Nostrand Reinhold Co.

Emerson, W. K., and Jacobson, M. K. 1976. *The American Museum of Natural History Guide to Shells, Land, Freshwater and Marine, from Nova Scotia to Florida.* New York: Alfred A. Knopf.

Warmke, G. L., and Abbott, R. T. 1975. *Caribbean Seashells.* (Reprint of 1961 edition.) New York: Dover Publishing Co.

## GLOSSARY

*Aperture* Opening or mouth of a snail.

*Beak* Noselike angle near hinge of bivalve shell, where growth started.

*Byssus* Strong threads that attach some clams to their supports.

*Callus* Shelly deposit sometimes covering umbilicus.

*Canal* A narrow extension of the lip of the aperture containing the siphons in many snails.

*Columella* Central pillar of snail shell.

*Cord* A band, found on shells with whorls.

*Frondose* Thin, flattened, often wavy, leaflike projection on a rib.

*Ganglion* (*pl.* ganglia) Knot or collection of nerve cells.

*Hinge* Area where the valves of a bivalve join.

*Ligament* Horny, elastic structure joining two shells of a bivalve.

*Lunule* Impressed or modified area in front of the beak on the outside of many bivalve shells.

*Mantle* A membranous flap or outer covering of the soft parts of a mollusk; it secretes the shelly material.

*Operculum* In some snails, a "trap door" that closes the aperture when animal's foot is withdrawn into shell.

*Pallial line* A mark on the inner surface of a bivalve shell, more or less parallel with the margin, caused by attachment of the mantle.

*Parietal wall* The inner lip area.

*Periostracum* Thin, noncalcareous outer layer of shells, often worn off those found on beach.

*Radula* Ribbonlike, saw-edged "tongue" of snails, used in feeding.

*Sinus* A deep cut.

*Siphon* Organ for admitting and expelling water from body cavity of bivalves.

*Spire* Upper whorls from the apex to the body whorl.

*Suture* Spiral line of spire in snails, separating the whorls.

*Turbinate* Top-shaped.

*Umbilicus* Hole in center of snail shell, at base of body whorl.

*Umbo* (*pl.* umbones) Swollen part of bivalve shell, near beak.

*Varix* (*pl.* varices) One of the prominent ridges across each whorl of certain gastropods showing a former position of the outer lip of the aperture; often called a "rib."

*Whorl* One of the turns of a snail shell.

# CHITONS
Class Polyplacophora

Chitons, or coat-of-mail shells, are sluggish, primitive saltwater mollusks, believed to be close in structure to the common ancestor of all mollusks. Typically they are flat and oblong and have eight armorlike overlapping plates held together by a surrounding girdle, often bearing spiny hairs. The animal uses the foot beneath to move about on or to hold fast to a rock by suction. Most chitons prefer shallow water and darkness. They are often found on the underside of rocks, where they feed on small algae. Some 600 species are distributed worldwide except in polar seas.

# True Chitons
Order Chitonida

## CHAETOPLEURID CHITONS
Family Chaetopleuridae

### EASTERN CHITON
*Chaetopleura apiculata*                                               94:4

**Description**
Size, ⅓–¾ in. (0.9–1.9 cm). Shape oval to oblong; plates raised, covered with beads, which concentrate in rows toward the center; girdle narrow, mottled, slightly grained, and with scattered short hairs. Color gray or pale yellow, sometimes tinged with red.
**Habitat**
Low tide to 15 fathoms (27.4 m).
**Range**
Cape Cod to Fla.

## COMMON CHITONS
Family Chitonidae

### TUBERCULATE CHITON
*Chiton tuberculatus*                                                  94:1

**Description**
Size, 2–3 in. (5.1–7.6 cm). Oval; valves with strong radial riblets on lateral areas; girdle with small overlapping scales. Color of valves various shades of grayish green, with girdle showing alternate patches of pale gray-green and black.
**Habitat**
Common on rocky shores.
**Range**
S. Fla. and W. Indies.

### FUZZY CHITON
*Acanthopleura granulata*                                              94:3

**Description**
Size, 2–3 in. (5.1–7.6 cm). Rather broadly ovate, granular sculpture; valves usually lack color because of erosion; girdle rather broad, with coarse hairlike spines, pale grayish color, sometimes with black splotches.

**Habitat**
Common on rocks.
**Range**
S. Fla. and W. Indies.

# SLENDER CHITONS
Family Ischnochitonidae

The shells of this family have comparatively large plates which are not covered by the girdle. The front and back plates are attached in the same manner and the back plate has no radiating ribs. These chitons are found on rocks between tide levels.

## WHITE CHITON
*Ischnochiton albus*                                              94:5

**Description**
Size, ½ in. (1.3 cm). Plates smooth with some growth ridges; girdle rough, scaly. Outside white, cream, or pale orange, occasionally with brown marks; inside white.
**Similarities**
Red Chiton is slightly larger and has a convex anterior-valve front slope.
**Habitat**
Common in cold, shallow water.
**Range**
Arctic seas to Mass.

## MOTTLED CHITON
*Tonicella marmorea*                                              94:6

**Description**
Size, 1 in. (2.5 cm). Oval to oblong, without hairs. Color creamy brown, mottled heavily with deep red.
**Habitat**
Common in 1–50 fathoms (1.8–91.5 m).
**Range**
Greenland to Mass.

## RED CHITON
*Tonicella ruber*                                                94:2

**Description**
Size, ½–1 in. (1.3–2.5 cm). Oblong, valves smooth; girdle with minute, elevated scales. Color pale brown with orange-red blotches or overall red.
**Similarities**
White Chiton is slightly smaller, with front slope of anterior valve straight to slightly concave.
**Habitat**
Common in 1–80 fathoms (1.8–146.3 m).
**Range**
Arctic seas to Conn.

# SNAILS
Class Gastropoda

There are worldwide some 40,000 species of marine snails. All those that are treated in this book have a single, usually spiral, shell. In the limpets this spiral feature is more apparent in the early stages. The nudibranchs, or sea hares, which are not discussed, are naked sea slugs without external shells.

As with other mollusks, the soft mantle secretes the shell, which grows around the mantle and increases in size as the animal grows. Truly marine snails breathe with gills and move by means of an extensible foot. Most snails are quite mobile, traveling about at will over the ocean floor. Some, such as the floating violet snail, are lifelong wanderers on the open sea.

Snails may be carnivorous, herbivorous, or scavengers. The Flat Slipper is typical of those that feed on tiny organisms suspended in sea water; ceriths are typical of the detritus feeders that strain their nourishment out of the mud.

Snails have a low order of sight, hearing, and presumably taste; but the senses of smell and touch are well developed. Various collections of ganglia control the movements of the animal and correspond to a brain in higher forms. In attacking their prey most snails use a long ribbon-shaped "tongue," or radula, well toothed and remotely resembling a file. This is protruded through the opening and is worked back and forth over the shell of the snail's prey, such as a clam. Finally, a hole is bored into the interior of the prey and the snail sucks out its meal. Most snails are nocturnal.

# Ancient Stomach-footed Gastropods
Order Archaeogastropoda

## KEYHOLE LIMPETS
Family Fissurellidae

These are limpets with a hole or slit near the summit. The shell is cone-shaped and has an oval base and many radiating ribs.

### LINNAEUS' KEYHOLE LIMPET
*Puncturella noachina*                                    **94:8**

**Description**
Size, ½ in. (1.3 cm). Shell white, glossy within; 21–26 main ribs with smaller secondary ribs between them, part way down; small slit in front of apex.
**Habitat**
Low-tide mark to depth of 1 mi. (1.6 km).
**Range**
Arctic and s. to Cape Cod.

### LITTLE KEYHOLE LIMPET
*Diodora cayenensis* 94:9

**Description**
Size, 1–2 in. (2.5–5.1 cm). Shell whitish to dark gray and pinkish; inside white to bluish gray; every fourth radiating rib larger; hole keyhole-shaped.
**Habitat**
Intertidal to moderate depths on stones, shells, or seaweeds; common.
**Range**
Chesapeake Bay to Brazil.

### LISTER'S KEYHOLE LIMPET
*Diadora listeri* 94:7

**Description**
Size, 1–2 in. (2.5–5.1 cm). Shell whitish to gray with every second radiating rib larger and the concentric ribs stronger, forming a more distinctly cancellate (crisscrossed ridges) sculpture than in *D. cayenensis*; intersections of ribs often nodulose or scaly.
**Habitat**
Intertidal; common.
**Range**
S. Fla. to Brazil.

# LIMPETS
Family Acmaeidae

### ATLANTIC PLATE LIMPET
*Collisella testudinalis* 94:10

**Description**
Size, to 1½ in. (3.8 cm). Shell oval, conical, depressed, smoothish. Outside blue-gray to cream with irregular axial bars checkered brown; inside rich brown with wide light gray border narrowly checkered with brown at the outer margin. Top of spire usually nearly central. Unlike Family Fissurellidae, has no opening at top of shell.
**Habitat**
Tide pools and shallow-water rocks.
**Remarks**
By developing the power of suction on rocks limpets have learned to survive and flourish in the zone between the tides.
**Range**
Arctic seas to Long Island.

# TOP SHELLS
Family Trochidae

The top shells are snails with pearly insides. They are generally cone-shaped and have a horny operculum. These common snails feed on vegetation.

## GREENLAND TOP SHELL
*Margarites groenlandicus*                                    **94:11**

**Description**
Size, to ¾ in. (1.9 cm). Shell glossy cream to brown; 4–5 whorls
flattened above, wavy at sutures; smooth, or with about 12 low,
smooth spiral ridges; initial whorl smooth, shiny; umbilicus broad,
funnel-shaped.
**Habitat**
Dredged to 150 fathoms (274 m).
**Range**
Arctic seas to Mass.

## MOTTLED TOP SHELL
*Calliostoma jujubinum*                                       **94:12**

**Description**
Size, ½–1¼ in. (1.3–3.2 cm). Shell pyramidal, brownish or reddish,
usually with a pale zone below the suture and often with irregular
darker or lighter splotches near lower part of whorls, which are
weakly beaded; last whorl with swollen, rounded periphery.
**Habitat**
Under rocks on reefs.
**Range**
N.C. to Brazil.

# TURBAN SHELLS
Family Turbinidae

These are solid shells with rounded, usually sculptured whorls, and
with a shelly operculum.

## KNOBBY TURBAN
*Turbo castaneus*                                             **95:3**

**Description**
Size, 1–1½ in. (2.5–3.8 cm). Shell nodulose, usually brownish,
grayish, or orange, and often with flamelike markings; operculum
gray-greenish with a white border.
**Habitat**
Common in shallow water.
**Range**
N.C. to Fla. and Tex., and s. to Brazil.

## LONG-SPINED STAR SHELL
*Astraea phoebia*                                            **95:2**

**Description**
Size, 1½–2½ in. (3.8–6.4 cm). Shell depressed, almost flattened on
its underside, whorls strongly sculptured, with flattened triangular
spines on the periphery; aperture silvery. Color grayish to
yellowish green; young shells with long spines and with iridescent
green and yellow color.
**Habitat**
Common in shallow water.
**Range**
Fla. to W. Indies and Brazil.

## AMERICAN STAR SHELL
*Astraea americanum* 95:1

**Description**
Size, 1–1½ in. (2.5–3.8 cm). Shell with high conical spire, flat sides, and swollen, rounded periphery; sculptured with numerous fine axial riblets; color grayish to yellowish white.
**Habitat**
Common in shallow water.
**Range**
Fla. to Mexico and W. Indies.

# NERITES
Family Neritidae

These snails have solid, usually broadly oval to flattened, smooth to spirally ribbed shells with a broad columella which may be smooth, tuberculate, or ribbed and has teeth at the inner edge. The operculum is shelly, with an internal peglike projection at one end.

## BLEEDING TOOTH
*Nerita peloronta* 95:4

**Description**
Size, ¾–1½ in. (1.9–3.8 cm). Shell broadly oval, weakly spirally sculptured, yellowish white with zigzag black and red markings. Rather broad white callus in columellar area marked with an orange-red splotch near the 2 teeth at the margin.
**Habitat**
Abundant on rocky ocean shores.
**Range**
S. Fla. through W. Indies.

## VARIEGATED NERITE
*Nerita versicolor* 95:7

**Description**
Size, ¾–1 in. (1.9–2.5 cm). Shell spirally ribbed, grayish with irregular spiral rows of black and red spots. Columellar callus with usually 4 strong teeth.
**Habitat**
Found with Bleeding Tooth, on rocky ocean shores.
**Range**
Fla. through W. Indies.

## TESSELLATED NERITE
*Nerita tessellata* 95:6

**Description**
Size, ½–¾ in. (1.3–1.9 cm). Shell sculptured with strong spiral cords of varying width separated by narrow grooves, the ribs irregularly marked by alternate spots of black and white; the broad, concave columellar area white, with 2 small teeth at apertural edge.
**Habitat**
Common under rocks in shallow water.
**Range**
Fla., W. Indies, to Brazil.

## ZEBRA NERITE
*Puperita pupa*                                                    **95:5**

**Description**
Size, ⅓–½ in. (0.9–1.3 cm). Shell smooth, white with irregular
rather narrow black axial lines; inside of outer lip and columellar
callus yellow; operculum smooth.
**Habitat**
In pools above tide line.
**Range**
Fla. through W. Indies.

# Middle Stomach-footed Gastropods
Order Mesogastropoda

# PERIWINKLES
Family Littorinidae

These are shore dwellers, some living in brackish and some in salt
water. Several species are amphibious and can survive long periods
without water. The shell is spiral, turbinate, or globular, and it has
few whorls and no umbilicus. Eggs are hatched inside or outside
the body.

## COMMON PERIWINKLE
*Littorina littorea*                                              **95:13**

**Description**
Size, ½–1 in. (1.3–2.5 cm). Shell thick, smoothish; 6–7 whorls
(seem fewer in beach-worn specimens) with very fine threads and
wrinkles. Brownish to dark gray; base of columella white; inside
chocolate-brown; white border of outer lip often checkered with
brown.
**Habitat**
Common on algae and seaweed clinging to rocks.
**Range**
Labrador to N.J.

## SMOOTH PERIWINKLE
*Littorina obtusata*                                              **95:9**

**Description**
Size, ⅓–½ in. (0.9–1.3 cm). Shell stout, smooth, globular; 3–4
whorls; point low, dull. Color variable, usually bright yellowish
brown; sometimes with bold brown spiral band; columella white.
**Habitat**
Common on rocks and rockweed.
**Remarks**
In the conquest of new habitats this periwinkle has ventured
farthest from its ancestral home, the sea. It gives birth to living
young and thus, independent of the water for reproduction, is able
to fight drought and the hazards of land life at the very edge of the
spring-tide saltwater splash line.
**Range**
Labrador to Cape May.

## SALT MARSH PERIWINKLE
*Littorina irrorata*           **95:10**

**Description**
Size, 1 in. (2.5 cm). Shell heavy, plump; top strongly pointed; 5 whorls with fine ridges. Grayish white with streaks of brownish dots on spiral ridges; opening oval, whitish; outer lip strong, sharp, slightly flaring.
**Habitat**
Marine vegetation between tides.
**Range**
N.Y. to cen. Fla. and Tex.

## ROUGH PERIWINKLE
*Littorina saxatilis*           **95:8**

**Description**
Size, ½ in. (1.3 cm) or less. Shell ovate, coarse; spire moderately elevated; 4–5 well-defined convex whorls. Color ashy or yellowish gray; surface markings show growth lines, revolving grooves. Opening oval, operculum horny.
**Habitat**
On rocks, often above high-tide mark.
**Range**
New England and n.

## KNOBBY PERIWINKLE
*Tectarius muricatus*           **95:11**

**Description**
Size, ½–1 in. (1.3–2.5 cm). Shell gray, rather broadly conical, whorls with spiral rows of regularly spaced beads.
**Habitat**
Common on rocks well above the water line.
**Range**
S. Fla. through W. Indies.

# CHINK SHELLS
Family Lacunidae

The stoutly conical chink shells are smooth, rather fragile periwinkles with thin shells, a half-moon aperture, and a shelflike columella alongside which is the characteristic lengthened groove, or chinklike umbilicus.

## LITTLE CHINK SHELL
*Lacuna vincta*           **95:12**

**Description**
Size, to ⅜ in. (0.95 cm). Typical of its family, this shell is fragile, plump, and conical, with a half-moon–shaped aperture. Identify it by the lengthened groove or chink beside the columella. Shell smooth, with small spiral scratches; dingy white to tan, often banded with purple; groove of columella (central pillar of spiral whorls) deep, umbilicus very small.
**Habitat**
On marine growths; often found washed ashore after storms.
**Range**
Arctic Ocean to R.I.

# SUNDIAL SHELLS
Family Architectonicidae

## COMMON SUNDIAL
*Architectonica nobilis*                                  **96:1**

**Description**
Size, 1–2 in. (2.5–5.1 cm) dia. Shell solid, flattened, almost discus-shaped, with strongly sculptured whorls above and below; deep, moderately narrow umbilicus with a strongly nodulose edge. Color whitish yellow with spiral rows of brown spots particularly strong below the suture.
**Habitat**
In sand to 20 fathoms (36.6 m).
**Range**
N.C. to W. Indies.

# SCREW SHELLS
Family Turritellidae

These are long, slender shells of many whorls, with a generally rounded aperture. In some forms the later whorls become uncoiled and very irregular, as in the following species.

## COMMON WORM SHELL
*Vermicularia spirata*                                    **96:7**

**Description**
Size, 2–7+ in. (5.1–17.8+ cm). Early part of shell coiled for about ¼–1 in. (0.6–2.5 cm) in regular elongate spiral; then growth becomes highly irregular and usually marked by irregular axial ridges; color grayish to yellowish brown.
**Habitat**
In shallow water, often partially embedded in sponges or ascidians.
**Range**
S. Fla. through W. Indies.

# HORN SHELLS
Family Cerithiidae

The horn shells are distinguished by their long shape, numerous whorls, and oblique opening. They are generally found on grass and seaweed in fairly shallow water.

## FLORIDA HORN SHELL
*Cerithium atratum*                                       **96:3**

**Description**
Size, 1–2 in. (2.5–5.1 cm). Shell with variable sculpture, whorls more or less angulate, sometimes with strong nodes, or else with spiral beaded sculpture dominant. Color dark brown to pale yellowish white with brown markings.
**Habitat**
In shallow water in bays or coral reefs.
**Range**
N.C. to Brazil; W. Africa.

### DWARF HORN SHELL
*Cerithium lutosum*          **96:6**

**Description**
Size, ⅓–¾ in. (0.9–1.9 cm). Shell small, whorls sculptured with
spiral rows of even-sized beads and fine spiral threads. Color pale
to dark brown, grayish white with brown markings or pure white.
**Habitat**
Common and often in great numbers in intertidal zones.
**Range**
N.C. through W. Indies and Tex. to Venezuela.

### VARIABLE BITTIUM
*Bittium varium*          **96:4**

**Description**
Size, ⅛ in. (0.3 cm). Shell with spiral rows of beads, occasionally
with nodulose axial riblets; a thickened varix is usually present.
**Habitat**
In eelgrass below low tide.
**Range**
Md. to Brazil.

# AWL-SHAPED CERITHS
Family Cerithiopsidae

### EMERSON'S CERITH
*Cerithiopsis emersoni*          **96:5**

**Description**
Size, ½–¾ in. (1.3–1.9 cm). Shell strong, slim; chocolate-brown but
may become ash-gray from erosion; 14–15 whorls with flattish
sides, each having 3 rows of rounded beads of lighter brown.
Aperture small, inner lip twisted.
**Habitat**
1–33 fathoms (1.8–60.4 m).
**Range**
Mass. to W. Indies and Brazil.

### ADAMS' MINIATURE CERITH
*Seila adamsi*          **96:2**

**Description**
Size, ¼–½ in. (0.6–1.3 cm). Shell long, slim; somewhat smooth and
concave at base; about 12 flat whorls each with 3 spiral ridges,
giving impression of threads on a screw; last whorl has 4 spiral
ridges. No beads as in Emerson's Cerith. Dark brown to orange
brown; aperture small; lip sharp, fragile, undulating.
**Habitat**
Common to 40 fathoms (73.2 m).
**Range**
Mass. to W. Indies.

# MODULUS
Family Modulidae

## ATLANTIC MODULUS
*Modulus modulus*                                              **96:10**

**Description**
Size, ½–¾ in. (1.3–1.9 cm). Shell with a low, conical, flat-sided
spire; low axial ribs; and a rounded base, with spiral cords, last
whorl with a sharp peripheral keel; small toothlike projection at
base of columella. Color grayish white.
**Habitat**
In algae and eelgrass in shallow water.
**Range**
N.C. to Brazil.

# WENTLETRAPS
Family Epitoniidae

Wentletraps are easily recognized by their high-spired, white shells
consisting of many ribbed whorls increasing in size from top to
bottom. The shells of these carnivorous snails are often called
"staircase shells."

## ANGLED WENTLETRAP
*Epitonium angulatum*                                          **96:8**

**Description**
Size, ¾–1 in. (1.9–2.5 cm). Shell stout, glossy; 8 convex whorls
with 9–10 strong, thin, somewhat angled ribs. China-white; outer
lip thick, flaring backward; no umbilicus.
**Habitat**
Common in shallow water to 25 fathoms (45.7 m).
**Range**
N.Y. to Tex.

## BROWN-BANDED WENTLETRAP
*Epitonium rupicolum*                                          **96:9**

**Description**
Size, ½–1 in. (1.3–2.5 cm). The only wentletrap in the range with
1 or 2 brown spiral bands on the body whorl. Shell has 8–11
rounded whorls with 12–18 low ribs; white, pinkish, or yellow to
purplish brown.
**Habitat**
Common from low water to 20 fathoms (36.6 m) and found
deeper.
**Range**
Cape Cod to Tex.

# PURPLE SEA SNAILS
Family Janthinidae

This is a small group of pelagic snails, usually purple, sometimes brown, that float on the surface of the sea by means of a raft of bubbles. Shells often cast up on beaches after an onshore blow.

## COMMON PURPLE SEA SNAIL
*Janthina janthina*                                                    **96:11**

**Description**
Size, 1–1½ in. (2.5–3.8 cm). Shell thin, whitish above, purple below, last whorl bluntly angular; outer lip sinuate.
**Habitat**
Normally pelagic but often cast up on shore in great numbers.
**Range**
Worldwide tropical.

# BOAT SHELLS
Family Calyptraeidae

These stationary snails are distinguished by their round or oval shape and horizontal or cuplike platform inside, which helps keep the animal in the shell.

## STRIATE CUP-AND-SAUCER
*Crucibulum striatum*                                                  **97:1**

**Description**
Size, 1 in. (2.5 cm). Caplike without; has only ⅓ of the cuplike platform attached to the main shell. Conical; top smooth, somewhat twisted, may or may not be central; rest of outside has light raised radial lines. Outside, yellowish or pinkish white, streaked with brown; inside glossy, yellowish, translucent.
**Habitat**
Common on rocks below low-water line.
**Range**
N.S. to S.C.

## FLAT SLIPPER SHELL
*Crepidula plana*                                                      **97:4**

**Description**
Size, ½–1½ in. (1.3–3.8 cm). Shell flat; platform less than ½ length of shell, top pointed and slightly off-center. Shell translucent, glossy within.
**Habitat**
Shallow water; on coastal mud flats many are found attached to large empty shells.
**Range**
Canada to Tex.

## COMMON BOAT SHELL
*Crepidula fornicata*                                                  **97:2**

**Description**
Size, ¾–2 in. (1.9–5.1 cm). Shell varies from nearly flat to high and arched, often corrugated from attachment to a ribbed surface;

platform concave, wavy at edge; apex drawn to one side. Color dingy white to tan, often flecked with brown; inside, white to purplish brown, glossy; platform whitish.
**Habitat**
Common in shallow water, often adhering to each other in clusters or to other shells.
**Range**
Canada to Tex.

## LITTLE BOAT SHELL
*Crepidula convexa*          **97:3**

**Description**
Size, ¼–½ in. (0.6–1.3 cm). Shell varies from sturdy to fragile, sometimes lightly wrinkled; small muscle scar under right outer corner or deck. Similar to Common Boat Shell but much smaller and has a straight edge to the platform. Color reddish to purple brown, sometimes spotted; inside including platform, pale bluish to chestnut.
**Habitat**
Attached to stones, shells, and seaweeds.
**Range**
Mass. to W. Indies.

# COFFEE BEAN SHELLS
Family Eratoidae

## COFFEE BEAN TRIVIA
*Trivia pediculus*          **97:9**

**Description**
Size, ½ in. (1.3 cm). Shell small, broadly oval, brownish pink to rosy pink with three pairs of irregular brown spots on back. Sculpture consists of strong transverse ribs on back and base, with a depression in the center of back.
**Habitat**
Intertidal to 25 fathoms (45.7 m).
**Range**
N.C. to Brazil.

## FOUR-SPOTTED TRIVIA
*Trivia quadripunctata*          **97:13**

**Description**
Size, ⅛–¼ in. (0.3–0.6 cm). Shell small, oval, pink; four small irregular brown spots on back; fine riblets crossing back, narrow shallow furrow in center of back.
**Habitat**
In shallow water.
**Range**
S. Fla. through W. Indies.

# CONCHS
Family Strombidae

## FLORIDA FIGHTING CONCH
*Strombus alatus*                                      **97:10**

**Description**
Size, 3–4 in. (7.6–10.2 cm). Shell reddish brown or yellowish brown, spire usually lighter in color, inner lip darker. Last whorl has pointed knobs on shoulder; outer lip winglike with deep notch on lower portion.
**Habitat**
In sand in shallow water.
**Range**
N.C. to Fla. and Tex.

# MOON SHELLS
Family Naticidae

These carnivorous snails have rounded shells, a sharp-edged aperture, and a large foot which in some species covers the whole shell when extended. The sand dollars so often seen on beaches are the egg cases of these handsome and ferocious snails.

## SHARK EYE
*Polinices duplicatus*                                  **97:5**

**Description**
Size, 1–2½ in. (2.5–6.4 cm). Shell generally broader than high, giving a flattened appearance; 4–5 whorls. Outside chestnut-brown tinged with blue, sometimes slate-gray or tan; inside brownish. Brown lobe wholly or partially covering umbilicus.
**Habitat**
Mud and sand flats at low tide.
**Remarks**
The thin, brown operculum of this species (and of the Common Moon Shell) may be known as the "dead sailor's toenail."
**Range**
Cape Cod to the Gulf of Mexico.

## COMMON MOON SHELL
*Lunatia heros*                                         **97:6**

**Description**
Size, 3–4 in. (7.6–10.2 cm). The largest moon shell in the range. Shell dirty white to ashy brown, 4–5 whorls, body whorl broadly expanded behind; aperture large, oval, glossy white to brownish; umbilicus large, round, extending almost to top of shell; periostracum yellowish brown, thin.
**Habitat**
Moist sand at low tide.
**Range**
Gulf of St. Lawrence to off N.C.

## COLORFUL MOON SHELL
*Natica canrena*                                          **97:8**

**Description**
Size, 1–2½ in. (2.5–6.4 cm). Shell smooth except for wrinkles near suture, with wavy axial brown lines and four rows of small squarish or arrow-shaped brown spots on spiral white bands usually on a light brown background. Umbilicus rather large. Calcareous operculum with spiral grooves.
**Habitat**
In sand in shallow water.
**Range**
N.C. through W. Indies to Brazil.

## MINIATURE NATICA
*Natica pusilla*                                              **97:7**

**Description**
Size, ¼–⅓ in. (0.6–0.9 cm). Shell glossy, yellow-white, occasionally with faint bands; oval in shape, and often with an open chink next to the umbilical callus.
**Habitat**
2–18 fathoms (3.7–32.9 m).
**Range**
Cape Cod to Fla., the Gulf states, and W. Indies.

## COMMON BABY'S EAR
*Sinum perspectivum*                                       **97:11**

**Description**
Size, 1–1½ in. (2.5–3.8 cm); height less than ½ in. (1.3 cm). The flattest moon shell in the range. Shell white; 3–4 whorls with many spiral lines and a few growth lines. Aperture large, wide, flaring; inside glossy; no umbilicus; periostracum yellowish, thin.
**Habitat**
Sandy beaches and shallow water.
**Range**
Va. to W. Indies.

# COWRIES
Family Cypraeidae

This is a large family. The shells are usually shiny, brightly colored, and oval. Generally, the last whorl envelops most of the spire, and the opening is long and narrow with teeth on both lips. It is a favorite with collectors.

## DEER COWRY
*Cypraea cervus*                                            **97:14**

**Description**
Size, 3–7 in. (7.6–17.8 cm). A relatively large shell, brown with white spots on the back. It is large, inflated, with numerous small spots.
**Habitat**
In water at or near low tide.
**Range**
Off N.C. to Bermuda and Cuba.

## GRAY COWRY
*Cypraea cinerea*                                    97:12

**Description**
Size, ¾–1½ in. (1.9–3.8 cm). Shell broadly ovate; color pinkish brown when fresh, yellowish brown when faded, often with small black spots, or showing obscure pale transverse bands; base creamy white.
**Habitat**
In shallow water under rocks on reefs.
**Range**
N.C. through W. Indies to Brazil.

# EGG SHELLS
Family Ovulidae

## SINGLE-TOOTHED SIMNIA
*Neosimnia uniplicata*                               98:2

**Description**
Size, ½–¾ in. (1.3–1.9 cm). A long, slender, cylindrical shell with a narrow, slitlike aperture. Shell smooth, glossy; 1 twisted columellar fold, which folds under upon itself at the opening. Color pink or purplish, sometimes yellowish white.
**Habitat**
In shallow water, generally attached to sea fans and sea whips.
**Habits**
Color adapts to color of host.
**Range**
Va. through W. Indies to Brazil.

## FLAMINGO TONGUE
*Cyphoma gibbosum*                                   98:1

**Description**
Size, ¾–1¾ in. (1.9–4.5 cm). Shell elongate, glossy, somewhat flattened; strong transverse ridge in middle of back; aperture without teeth on lips. Color pinkish or yellowish orange, white area in center of back. The living animal pale pinkish with numerous black squarish rings.
**Habitat**
On sea fans and gorgonians in shallow to moderately deep water.
**Range**
N.C. to Brazil.

# HELMET SHELLS
Family Cassidae

## SCOTCH BONNET
*Phalium granulatum*                                 98:6

**Description**
Size, 1½–3 in. (3.8–7.6 cm). Shell usually with numerous grooves which may be weakly beaded on body whorl, but sometimes smooth; outer lip often thickened with many teeth on inner side; inner lip with a white shieldlike callus which is usually more or less granulose. Color pinkish yellow to whitish with irregular pale brown spots spirally arranged.

**Habitat**
In shallow to moderately deep water.
**Range**
N.C. to Brazil.

# TRITON SHELLS
Family Cymatiidae

## HAIRY TRITON
*Cymatium pileare*                    98:8

**Description**
Size, 1½–4 in. (3.8–10.2 cm). Shell elongate-ovate; moderately
high, broad spire; whorls sculptured with weakly beaded spiral
cords of irregular size; usually has a periostracum with flattened
hairlike projections; varices prominent. Aperture pale brown with
inner lip dark brown between the white teeth.
**Habitat**
Common in shallow water.
**Range**
S.C. to Brazil.

## WHITE-MOUTHED TRITON
*Cymatium muricinum*                    98:7

**Description**
Size, 2 in. (5.1 cm). Shell strong, with 5 whorls, sharp apex. Color
spotty, may be banded with brown. Opening large, notched at
upper end; canal long and open. Surface ribbed.
**Habitat**
Common in shallow water.
**Range**
S. Fla. to W. Indies.

# FROG SHELLS
Family Bursidae

The frog shells are ovate, laterally somewhat compressed, and bear
two rows of continuous varices, one row on each side.

## GRANULAR FROG SHELL
*Bursa granularis*                    98:9

**Description**
Size, ¾–2 in. (1.9–5.1 cm). Shell flattened laterally, sides marked
by varices on each whorl; whorls sculptured by spiral cords of
varying size which are adorned with uniformly sized beads and
nodules. Color brownish with the larger nodes appearing white.
**Habitat**
In shallow water.
**Range**
S. Fla. to Brazil.

# New Stomach-footed Gastropods

Order Neogastropoda

## ROCK SHELLS

Family Muricidae

These shells are strong and thick, usually spiny, and with three varices per whorl. The snails are carnivorous and active and prefer shallow rocky or pebbly bottoms.

### APPLE MUREX
*Phyllonotus pomum*                                         **98:4**

**Description**
Size, 2–4½ in. (5.1–11.4 cm). Shell stout, grayish or whitish to yellowish white, often irregularly spotted with brown; 3 nodose varices per whorl, 1–3 axial ribs between the varices, all crossed by spiral cords of varying strength; siphonal canal narrow, almost covered by a broad rim, giving strength; inner lip with broad, marginate callus; outer lip usually spotted with brown within.
**Habitat**
Common in shallow water.
**Range**
N.C. to Brazil.

### LACE MUREX
*Chicoreus florifer dilectus*                               **98:3**

**Description**
Size, 1–3 in. (2.5–7.6 cm). Shell light brown or whitish with 3 stout and strongly and densely frondose varices per whorl; varix at outer lip reaches to end of rather long siphonal canal; strong knob between the varices; aperture small and oval.
**Habitat**
Common in shallow water.
**Range**
S.C. to s. Fla.

### GIANT EASTERN MUREX
*Muricanthus fulvescens*                                    **98:5**

**Description**
Length, 5–7 in. (12.7–17.8 cm). Shell stout, solid, white to grayish; body whorl broad with moderately long siphonal canal and 7–9 varices with separated stout triangular spines.
**Habitat**
Common seasonally in 5–40 fathoms (9.1–73.2 m).
**Range**
N.C. to Fla. and Tex.

### THICK-LIPPED DRILL
*Eupleura caudata*                                          **98:12**

**Description**
Size, ½–1 in. (1.3–2.5 cm). The rock shell with the most prominent ridges and knobs. Shell has 5–6 whorls, flattened at shoulders, with

9–11 raised ridges; the 2 diametrically opposite ridges have knobs. Color red-tinged to grayish white; outer lip thick, with 6 small teeth; edge of lip often bluish; canal straight, narrow.
**Habitat**
Rocks at low tide.
**Remarks**
An enemy of the oyster.
**Range**
Cape Cod to Fla.

## FLORIDA DRILL
*Calotrophon ostrearum* 98:10

**Description**
Size, ¾–1 in. (1.9–2.5 cm). Shell grayish to grayish brown in color with strongly angulate ribs crossed by spiral cords. Aperture pale yellowish orange to purplish brown, with outer lip with fine incised lines within.
**Habitat**
Common in shallow water to 30 fathoms (54.9 m).
**Range**
W. coast of Fla.

## OYSTER DRILL
*Urosalpinx cinerea* 98:11

**Description**
Size, ½–1 in. (1.3–2.5 cm). Shell dirty gray or yellowish, often with brownish spiral bands; 5–6 whorls with many raised lines and wavy folds. Opening oval, purple to tan; lip slightly thickened, sharp, sometimes with 2–6 small teeth; canal short.
**Habitat**
Oyster beds; very common.
**Remarks**
An enemy of the oyster.
**Range**
N.S. to Fla.

## ROCK PURPLE
*Nucella lapillus* 98:13

**Description**
Size, 1–2 in. (2.5–5.1 cm). The most variously colored rock shell in the range. Spire short, sharp; 5 whorls; body whorl large, furrowed, wrinkled; spines and varices not prominent. Color ranges from dull white to yellowish orange and purplish brown, with or without dark spiral bands; aperture oval, lip arched.
**Habitat**
Crevices in rocks between tide levels.
**Range**
Labrador to N.Y.

## FLORIDA ROCK SHELL
*Thais haemastoma floridana* 98:14

**Description**
Size, 2–3 in. (5.1–7.6 cm). Shell stout with angulate whorls, sometimes knobbed on the angle, usually rather finely spirally grooved or almost smooth. Color grayish to yellowish with

irregular markings of brown, the spiral threads often whitish, aperture orange-pink.
**Habitat**
Common in shallow water.
**Food**
Bivalves and barnacles.
**Range**
N.C. to Brazil.

# DOVE SHELLS
Family Columbellidae

These small shells are somewhat long and narrow, colorful and glossy, with rather narrow apertures and thickened outer lips.

## MOTTLED DOVE SHELL
*Columbella mercatoria*                                          **99:6**

**Description**
Size, ½–¾ in. (1.3–1.9 cm). Shell solid, broad, with shallow spiral grooves; outer lip thick with strong teeth on inner edge. Color variable; whitish, yellow, or orange, variously spotted and blotched with brown or white.
**Habitat**
Common in shallow water under rocks.
**Range**
E. Fla. to Brazil.

## GREEDY DOVE SHELL
*Anachis avara*                                                 **99:8**

**Description**
Size, ⅜–½ in. (0.95–1.3 cm). Shell with 6–7 whorls; upper half of body whorl has 12 smooth ribs interrupting the revolving lines found on entire shell. Color straw, sometimes dull gray, with irregularly placed large white spots; aperture small, oval.
**Habitat**
Low tide; common.
**Range**
N.J. to Tex.

# WHELKS
Family Buccinidae

The shells of this northern carnivorous family are large and thick and have few whorls. They are usually pear-shaped and have a pointed spire. The aperture is large.

## WAVED WHELK
*Buccinum undatum*                                              **99:4**

**Description**
Size, 2–4 in. (5.1–10.2 cm). Shell solid; about 6 whorls form a sharp spire with 12 vertical ribs on each, becoming weaker on the body whorl. Many raised revolving lines, together with the ribs, give the shell a wavy appearance. Color gray to reddish or yellowish brown, aperture white to yellow. Aperture oval, extends to ½ length of shell; lip sharp, canal short.

**Habitat**
Offshore to several fathoms.
**Food**
A scavenger.
**Remarks**
The egg capsules of this snail once were used for soap by sailors, who called them "sea wash balls." The snail is edible.
**Range**
Arctic seas to N.J.

## STIMPSON'S WHELK
*Colus stimpsoni*                                               **99:14**

**Description**
Size, 3–5 in. (7.6–12.7 cm). The most spindle-shaped whelk in the range. Shell gray to chalk white; 6–8 whorls with many slightly raised spiral lines; aperture oval, about half length of shell; lip not flaring; periostracum thin, grayish green.
**Habitat**
1–640 fathoms (1.8–1170 m).
**Range**
Labrador to off N.C.

## BROWN-CORDED NEPTUNE
*Neptunea decemcostata*                                        **99:10**

**Description**
Size, 3–4½ in. (7.6–10.2 cm). Shell heavy, pointed, grayish white; 6–7 whorls; 7–10 red-brown spiral cords on body whorl, 2–3 on upper whorls; aperture white, canal short.
**Habitat**
Cold water offshore.
**Range**
N.S. to Mass.

# LARGE WHELKS
Family Melongenidae

## CHANNELED WHELK
*Busycon canaliculatum*                                         **99:2**

**Description**
Size, 5–7½ in. (12.7–19.1 cm). Has 5–6 whorls; is distinguished by deep channel at the spiral seam of each whorl, forming a terrace. Shell pear-shaped; body whorl large, narrowing below, ending in a long straight canal. Outside creamy gray, inside yellowish; periostracum thick, gray, velvety.
**Habitat**
Common in sandy shallow water.
**Food**
Clams, oysters.
**Reproduction**
Egg ribbon contains capsules with sharp edges; often found on beach.
**Remarks**
Edible.
**Range**
Cape Cod to St. Augustine.

## KNOBBED WHELK
*Busycon carica*                                    **99:3**

**Description**
Size, 4–9 in. (10.2–22.9 cm). Shell pear-shaped, usually right-handed; spire low, knobbed; 6 whorls; large, broad body whorl has blunt knobs on shoulder. Color gray to yellowish, young with purplish streaks; aperture oval, yellow or orange-red within; outer lip sharp, inner lip arched and twisted; canal long, open.
**Habitat**
Common in shallow water with pebbly bottom.
**Food**
Clams, oysters.
**Reproduction**
Egg ribbon contains capsules with flat edges; often found on beach.
**Remarks**
Edible.
**Range**
Cape Cod to cen. Fla.

## LIGHTNING WHELK
*Busycon contrarium*                                **99:1**

**Description**
Size, 4–16 in. (10.2–40.6 cm). Shell left-handed; short, triangular knobs at shoulder; long, tapering siphonal canal. Color grayish white with rather widely spaced axial brown lines and spirally arranged large splotches of pale brown.
**Habitat**
Rather abundant in shallow to moderately deep waters.
**Range**
N.J. to Fla. and Gulf of Mexico.

## CROWN CONCH
*Melongena corona*                                  **99:11**

**Description**
Size, 2–5¼ in. (5.1–13.3 cm). Shell with 1, sometimes 2, rows of hollow spines below the flattened shoulder, and a smaller row of spines near base whorl with more or less obscure spiral threads. Color whitish with dark brownish spiral bands of varying width, the surface covered when fresh with a yellowish brown periostracum.
**Habitat**
Common in shallow water, especially on shores of Gulf of Mexico.
**Range**
Fla. and Gulf Coast.

# DOG WHELKS
Family Nassariidae

These scavenging snails have small, plump shells with pointed spires, an oval aperture, a short notchlike canal, and usually a distinct columellar callus. They inhabit all seas.

## MOTTLED DOG WHELK
*Nassarius vibex*                                   **99:9**

**Description**
Size, ½ in. (1.3 cm). Shell heavy; about 6 whorls with many folds, beaded ridges, and distinct spiral bands; body whorl large, with 12

folds, 10 lines. Color grayish brown to white, often mottled or with broken bands of dark brown; aperture notched at both ends; outer lip thick, toothed; heavy callus on inner lip. Shell shorter, chunkier than in New England Dog Whelk.

**Habitat**
Common on sandy shores, mud flats.

**Range**
Cape Cod to W. Indies.

## NEW ENGLAND DOG WHELK
*Nassarius trivittatus*                                          99:7

**Description**
Size, ½–¾ in. (1.3–1.9 cm). Shell with 8–9 whorls, slightly flattened at shoulder and lightly grooved; 4–5 rows of beading on whorls; spire sharp. Color gray to yellowish white, sometimes with a few brown bands; aperture oval; outer lip thin, scalloped. Shell larger, taller than in Mottled Dog Whelk.

**Habitat**
Common between tides to 45 fathoms (82.3 m).

**Range**
N.S. to S.C.

## MUD SNAIL
*Nassarius obsoletus*                                          99:5

**Description**
Size, ¾–1 in. (1.9–2.5 cm). The commonest and the darkest dog whelk in the range. Shell has 6 whorls with uneven lines crossed by light growth lines and oblique fold lines toward base; spire usually worn away. Color dark brown to purplish black; aperture oval; outer lip thin; inner lip arched, folded.

**Habitat**
Mud flats.

**Remarks**
Their keen sense of smell makes these snails move actively after their food. In so doing they leave groovelike trails in the mud, often seen by the bay-shore beachcomber.

**Range**
Gulf of St. Lawrence to ne. Fla.

# TULIP SHELLS AND HORSE CONCHS
Family Fasciolariidae

## TULIP SHELL
*Fasciolaria tulipa*                                          99:13

**Description**
Size, 3–8 in. (7.6–20.3 cm). Shell almost smooth or with obscure spiral threads but always with a few strong, somewhat crinkled spiral threads below the suture. Color whitish, grayish, or pinkish white with irregular orange, brown, or greenish brown spots and with many interrupted spiral chestnut lines; sometimes shell is wholly suffused with orange or brown.

**Habitat**
Common intertidally in bays and estuaries.

**Range**
N.C. to Brazil.

## FLORIDA BANDED TULIP
*Fasciolaria lilium hunteria*                                      **99:12**

**Description**
Size, 2½–3 in. (6.4–7.6 cm). Shell yellowish white or grayish with
5–6 red-brown spiral lines on last whorl, and with irregular
orange or bluish gray axial flammules.
**Habitat**
Common in quiet shallow waters.
**Range**
N.C. to Fla. and Ala.

## FLORIDA HORSE CONCH
*Pleuroploca gigantea*                                             **99:15**

**Description**
Size, 1–2 ft. (0.3–0.6 m). Shell with high spire and moderately
long siphonal canal; strong nodose ribs on whorls, crossed by rather
strong spiral cords. Color whitish to pale reddish orange, usually
covered by a thick periostracum which flakes off when dry.
**Habitat**
On sandy bottom in 3–20 ft. (0.9–6.1 m).
**Range**
N.C. to Fla. and Mexico.

# OLIVE SHELLS
Family Olividae

The shells of this family are more or less cylindrical in shape, since
the much-inflated body whorl tends to conceal all the earlier
whorls. Widely distributed in warm seas, the shells are often
brightly colored and usually smoothly polished.

## LETTERED OLIVE
*Oliva sayana*                                                     **100:1**

**Description**
Size, 2–2½ in. (5.1–6.4 cm). Shell glossy, elongate, sides
moderately convex. Color gray to grayish yellow with irregular
brownish zigzag or tentlike axial markings of light and dark
brown, the latter usually occurring in transverse bands.
**Habitat**
In sand in shallow water.
**Range**
N.C. to Brazil.

## MINUTE DWARF OLIVE
*Olivella mutica*                                                  **100:2**

**Description**
Size, ¼–⅝ in. (0.6–1.6 cm). Shell glossy, variously banded with
gray and brown, sometimes with irregular spots below suture or
completely brown. Parietal and columellar callus rather heavy,
extending above top of aperture to next whorl; lower part of
columella with obliquely entering folds on a raised callus.
**Habitat**
In sand or mud in shallow water.
**Range**
N.C. to s. Fla. and Bahamas.

## RICE DWARF OLIVE
*Olivella floralia*                                    100:3

**Description**
Size, ¼–½ in. (0.6–1.3 cm). Shell slender, glossy, white; spire often colored yellow, grayish, brown, or rosy; body whorl sometimes mottled with brown. Parietal callus thin, extending beyond the aperture.
**Habitat**
Common in shallow water.
**Range**
N.C. through Fla. and W. Indies to Brazil.

# MITER SHELLS
Family Mitridae

## BEADED MITER
*Mitra nodulosa*                                       100:8

**Description**
Size, ¾–1½ in. (1.9–3.8 cm). Shell rather elongate, with narrow axial ribs crossed by spiral grooves resulting in a beaded sculpture; columella with 4 folds, the upper one strongest. Color yellowish to orange-brown, often with the top of the whorls whitish.
**Habitat**
Under rocks at or below low tide.
**Range**
N.C. to Brazil.

# VOLUTE SHELLS
Family Volutidae

## JUNONIA
*Scaphella junonia*                                    100:14

**Description**
Size, 5–6 in. (12.7–15.2 cm). Shell solid, smooth, whitish to pinkish with spiral rows of squarish dark brown spots; 4 rather strong folds on columella.
**Habitat**
1–30 fathoms (1.8–54.9 m).
**Remarks**
A striking shell, much prized by Florida collectors and rarely washed ashore. There are several varieties differing in coloring and spotting.
**Range**
N.C. to Fla., Gulf states, and Mexico.

# NUTMEG SHELLS
Family Cancellariidae

## COMMON NUTMEG
*Cancellaria reticulata*                               100:7

**Description**
Size, 1–1¾ in. (2.5–4.5 cm). Solid, ovate shell, sculptured by axial and spiral threads giving a reticulate appearance; strong fold on

columella, with a weaker one below it. Color grayish to yellowish white irregularly mottled with brown.
**Habitat**
Common in sand in shallow water.
**Range**
N.C. to Brazil.

# MARGINELLA SHELLS
Family Marginellidae

## COMMON ATLANTIC MARGINELLA
*Prunum apicinum*                                            100:9

**Description**
Size, ¼–⅓ in. (0.6–0.9 cm). Shell glossy, smooth, subtriangular; rather thick, smooth outer lip, 4 folds at base of inner lip. Color yellowish or grayish, with usually 3 obscure dark bands on body whorl and 2 or 3 small orange-brown spots on outer margin of outer lip.
**Habitat**
Common in shallow water.
**Range**
N.C. to W. Indies.

# CONE SHELLS
Family Conidae

This is a large family of conical tropical shells with a generally narrow aperture and thin outer lip. Cone shells are carnivorous and are provided with retractile harpoonlike teeth that can be venomous in some species.

## ALPHABET CONE
*Conus spurius atlanticus*                                   100:12

**Description**
Size, 2–3 in. (5.1–7.6 cm). Shell smooth except for lines of growth. Color whitish or yellowish with spiral rows of irregular squarish brownish orange blotches.
**Habitat**
Common in shallow water.
**Range**
Fla. and Gulf of Mexico.

## JASPER CONE
*Conus jaspideus*                                            100:13

**Description**
Size, ½–¾ in. (1.3–1.9 cm). Spire of shell conically elevated, last whorl sometimes smooth but usually with incised lines, and some forms (i.e., form *verrucosus*) with spiral rows of nodules. Color variable, generally mottled with reddish brown or yellowish orange, occasionally pinkish.
**Habitat**
In shallow to moderately deep water, generally under stones and coral.
**Range**
S. Fla. to Brazil.

# AUGER SHELLS
Family Terebridae

Members of this family typically are very long and slender and have many whorls.

## ATLANTIC AUGER
*Terebra dislocata*                                          100:4

**Description**
Size, 1½–2 in. (3.8–5.1 cm). Shell long and slender; white, pinkish, or brown. Small aperture; lip notched below. Has 12–15 whorls, each divided by a deep spiral groove, giving the impression of double this number of whorls; some 25 ribs per whorl; columella twisted.
**Habitat**
Sandy bottoms in shallow water.
**Range**
N.C. to W. Indies.

## SALLÉ'S AUGER
*Impages salleana*                                          100:5

**Description**
Size, 1–1⅓ in. (2.5–3.4 cm). A slender, pointed, straight-sided shell; strong axial folds on whorls below the suture. Color bluish or brownish gray, white band below suture, dark brown spots between the folds.
**Habitat**
Burrowing in intertidal sand.
**Range**
Fla. to Mexico.

# Bubble Shells and Sea Hares
Order Cephalaspidea

In this order the shell may or may not be present. If present, it is in a reduced condition and becomes thinner as the size of the snail increases and envelops more of the shell.

# SMALL BUBBLE SHELLS
Family Haminoeidae

These slugs have shells which are nearly cylindrical, and the large body whorl almost engulfs the spire. The spiral line is channeled and forms a continuous groove. The shell gives the impression of having been turned on a lathe, hence its other name of "lathe shell."

## EASTERN PAPER BUBBLE
*Haminoea solitaria*                                        100:6

**Description**
Size, ½ in. (1.3 cm). Shell white or yellowish, translucent, oval; faint, fine growth lines and revolving ridges.
**Habitat**
In shallow water with muddy or sandy bottoms.
**Range**
Cape Cod to N.C.

# Notch-banded Gastropods
Order Pyramidellida

## PYRAMS
Family Pyramidellidae

Members of this family of tiny augerlike shells are parasites on worms or bivalves. They hold on by suction, pierce the host's body with a tiny dagger, and suck its juices.

### INTERRUPTED TURBONILLE
*Turbonilla interrupta*                                    100:10
**Description**
Size, ¼ in. (0.6 cm). Tall, slender, with 8 broad whorls and many ribs; color pale yellow, luster waxy; aperture roundish.
**Habitat**
2–107 fathoms (3.7–195.7 m).
**Range**
Maine to W. Indies.

### IMPRESSED ODOSTOME
*Odostomia impressa*                                       100:11
**Description**
Size, ¼ in. (0.6 cm). Shell elongate, pear-shaped, with about 8 whorls carrying flattened revolving ridges; body whorl surrounded by ridges; color milky white; aperture oval.
**Habitat**
Shallow water.
**Range**
Mass. Bay to Gulf of Mexico.

## TUSK SHELLS
Class Scaphopoda

This small class of sluggish univalves is distinguished by a single hollow tusklike shell with an opening at each end. The animals live in the sand and feed on minute marine organisms in the sea water, which is sucked in through the small rear end and expelled through the front. Tiny filaments surrounding the internal mouth extract this food from the water. Tusk shells have an extensible wedge-shaped foot. They breathe through the mantle, not through the gills, and they live in clean water at various depths.

## TUSK SHELLS
Family Dentaliidae

The shells of this family are elongate, curved somewhat like a miniature elephant's tusk, open at both ends with the greatest diameter at the aperture.

## STIMPSON'S TUSK
*Dentalium entale stimpsoni*                                    101:3

**Description**
Size, 1–2 in. (2.5–5.1 cm). Shell long, slender, tubular, and
moderately curved; round in cross section; top usually worn. Color
ivory-white with buffy or tan tones.
**Habitat**
Common in 8–1200 fathoms (14.6–2,194.6 m).
**Range**
N.S. to Cape Cod.

## IVORY TUSK
*Dentalium eboreum*                                            101:4

**Description**
Size, 1–2½ in. (2.5–6.4 cm). Shell slender, glossy, gently curved,
smooth except for concentric growth lines and obscure microscopic
longitudinal ridges at apex, which has a narrow slit; color whitish.
**Habitat**
In sand in moderately deep water.
**Range**
N.C. to Tex. and W. Indies.

# CLAMS
Class Bivalvia

There are worldwide some 12,000 species of marine clams. These
are bivalves; that is, they have two shells, or valves, which are
joined at the hinge by a ligament and held together by one or two
strong muscles. Except in a short larval stage, most are sedentary
throughout their lives. Some, however, have a fleshy and extensible
foot and can move about, and the scallops can swim by clapping
their valves together.

Clams breathe both through their gills and through their mantle,
which, as in the snails, secretes the calcium carbonate which makes
the shell. Growth lines on the shell show seasons of relative quiet,
such as winter. Clams, in general, possess no sense of taste or smell
and are apparently limited to identifying edible bits of food and to
closing their valves if a starfish goes by. They have no head; three
somewhat enlarged sets of ganglia take the place of a brain. Some
species, however, such as scallops, have little eyespots and can see.

Most clams are vegetarians. They live by extracting minute
particles of food from the sea water that passes over their gills or is
pumped over them by their siphons. These clams are called
suspension feeders, as they get their food from particles suspended
in the water. Others are deposit feeders and suck up food from the
mud with their siphons. Certain clams and oysters are the source
of pearls, but few of commercial importance are produced by the
species in the East.

# Awning Clams
Order Solemyoida

## VEILED CLAMS
Family Solemyacidae

These clams are primitive in form and are fairly uncommon in the eastern range. Their two valves are equal in size, somewhat long and fragile, and are covered by a shiny, horny brown periostracum which extends beyond the shell. There is a gap between the valves both in front and behind.

### ATLANTIC AWNING CLAM
*Solemya velum*                                                101:2

**Description**
Size, ½–1 in. (1.3–2.5 cm). Shell oblong, delicate, smooth, yellowish brown; 15 slightly indented wide radiating lines; inside grayish blue. Periostracum hangs in scallops beyond edges of valves.
**Habitat**
Mud and sand beyond low water.
**Range**
N.S. to Fla.

# Nut Clams
Order Nuculoida

## NUT SHELLS
Family Nuculidae

These are small, three-cornered or ovate shells with pearly interiors, finely denticulated ventral margins, and a row of teeth on each side of the beak cavity but no ligamental pit between them.

### ATLANTIC NUT CLAM
*Nucula proxima*                                               101:7

**Description**
Size, ⅓ in. (0.9 cm). Shell whitish, inflated, thin, translucent; inside grayish; periostracum light olive.
**Habitat**
Mud just offshore.
**Range**
N.S. to Tex.

## YOLDIAS
Family Nuculanidae

This family has small, oval valves that are equal in size and have teeth at the hinge, with a pit for the ligament. The shell is usually prominently angled behind and rounded in front.

## POINTED NUT CLAM
*Nuculana acuta*                                            **101:6**

**Description**
Size, ¼–⅜ in. (0.6–0.95 cm). Shell white, rounded in front, sharply pointed behind, strongly and evenly grooved; many teeth at hinge, pit central; periostracum thin, yellowish to brown.
**Habitat**
Common in shallow waters with sandy bottoms.
**Range**
Cape Cod to W. Indies.

## FILE YOLDIA
*Yoldia limatula*                                          **101:5**

**Description**
Size, 1–2½ in. (2.5–6.4 cm). Shell shiny green to light brown, faintly lined; long, oval, narrowing toward back; 20–22 prominent teeth in filelike arrangement; umbo nearly central; inside bluish white.
**Habitat**
Common below low-tide mark.
**Habits**
Active; can swim and jump.
**Range**
Maine to Cape May.

# Arks and Bittersweets
Order Arcoida

# ARK SHELLS
Family Arcidae

Most arks are found in warm seas, but some are found in cooler waters. The shells are somewhat box-shaped and heavily ribbed, with a narrow hinge line bearing many teeth on both sides. The umbones are toward the back of the shell, which usually carries a heavy periostracum.

## TURKEY WING
*Arca zebra*                                               **101:12**

**Description**
Size, 2–3¼ in. (5.1–8.3 cm). Shell with a broad ligamental area between the beaks; ribs of varying widths; surface covered with a coarse brown matted periostracum, generally more or less worn off.
**Habitat**
Attached to rocks and corals by byssus in shallow water.
**Range**
N.C. to Brazil.

## MOSSY ARK
*Arca imbricata*                                           **101:9**

**Description**
Size, 1½–2½ in. (3.8–6.4 cm). Shell smaller and relatively broader than Turkey Wing, with a fine beaded sculpture and a larger byssus opening; suffused with dark brown.

**Habitat**
Attached by byssus to underside of rocks.
**Range**
N.C. to Brazil.

## TRANSVERSE ARK
*Anadara transversa*                                101:10

**Description**
Size, ½–1½ in. (1.3–3.8 cm). The smallest of the ark shells in the range. Shell oblong, dull white, with heavy gray-brown periostracum; 30–35 ribs on each valve; left valve overlaps right valve, beaks curve inward.
**Habitat**
Sandy bottoms offshore.
**Range**
Cape Cod to Tex.

## BLOOD ARK
*Anadara ovalis*                                    101:8

**Description**
Size, 1½–2½ in. (3.8–6.4 cm). One of the very few mollusks that have red blood. Shell white, thick, solid, oval; umbones knobby. Each valve has 26–35 broad radiating ribs separated by narrow grooves; lower part of valves covered with thick brown periostracum.
**Habitat**
Sandy bottoms in shallow water.
**Range**
Cape Cod to W. Indies.

## HEAVY ARK
*Noetia ponderosa*                                  101:11

**Description**
Size, 2–2½ in. (5.1–6.4 cm). Shell creamy white, heavy, inflated; valves equal in size, symmetrically curved with about 32 radiating ribs on each; prominent ridge from beak to hind margin with valve falling off almost straight behind. Umbones placed toward front of valves and bend backward; periostracum blackish, heavy.
**Habitat**
Sandy bottoms.
**Remarks**
These are often found as fossils on Wildwood and Cape May beaches in New Jersey. They come from an interglacial deposit laid down when the sea there was as warm as it is in Virginia today.
**Range**
Va. to Gulf of Mexico.

# BITTERSWEET CLAMS
Family Glycymerididae

This family consists of a small group of heavy, usually orbicular, equivalve, porcellaneous shells, generally with a soft velvety periostracum; the beaks are incurved, the hinge heavy and with many small teeth, and the ligament external with grooves diverging from the area. The largest muscle scar is at the anterior end.

## COMB BITTERSWEET
*Glycymeris pectinata*                                           101:1

**Description**
Size, ½–1¼ in. (1.3–3.2 cm). Shell orbicular in outline, rather
flattened, with numerous smooth, flattened radial ribs crossed by
fine crowded concentric threads. Color grayish or yellowish gray,
irregularly spotted with brown.
**Habitat**
Common in shallow water.
**Range**
N.C. to Brazil.

# Mussels
Order Mytiloida

# MUSSELS
Family Mytilidae

The shells of this family are long, narrow, and oval, with valves of
equal size and shape and often with a pearly lining. The umbo is
close to the front of the shell. The periostracum is heavy, dark, and
sometimes hairy. Some species burrow; others attach themselves to
rocks or pilings with a byssus, or set of threads, which they spin.

## BLUE MUSSEL
*Mytilus edulis*                                              102:5

**Description**
Size, 1–3 in. (2.5–7.6 cm). Shell long, a flattened oval, blue-black,
often with pale violet patches; no ribs, but often with fine growth
lines. Umbo at extreme top, 4 teeth; inside white and blue-violet;
periostracum thin, shiny.
**Habitat**
Very common clustered on rocks and pilings at low tide.
**Range**
Arctic Ocean to S.C.

## HOOKED MUSSEL
*Ischadium recurvus*                                          102:4

**Description**
Size, 1–2½ in. (2.5–6.4 cm). Shell green-brown to gray-black;
rather flat, wide, strongly hooked at front end; many wavy ribs,
edge of shell at umbo has 3–4 small, longish teeth; inside pink to
purplish brown with narrow grayish blue border.
**Habitat**
Clustered on rocks and shells in shallow water.
**Range**
Cape Cod to W. Indies.

## PAPER MUSSEL
*Amygdalum papyrium*                                          102:3

**Description**
Size, 1–1½ in. (2.5–3.8 cm). Shell smooth, long, oval, very fragile,
shiny; green-blue to yellowish brown; inside white, shiny.

**Habitat**
Common in bays in mud at roots of grass.
**Reproduction**
This species builds a nest with its byssal threads.
**Range**
Md. to Tex.

## HORSE MUSSEL
*Modiolus modiolus*                                    102:1

**Description**
Size, 2–6 in. (5.1–15.2 cm). Shell coarse, oval, dark blue to mauve-white, with concentric lines; beaks slightly to one side of short, narrow front end. Hind end broad and round, inside white; periostracum thick, dark.
**Habitat**
Burrows in sand or rock crevices in cool, moderately deep water; common.
**Range**
Arctic seas to ne. Fla.

## RIBBED MUSSEL
*Geukensia demissus*                                   102:2

**Description**
Size, 2–4 in. (5.1–10.2 cm). Shell shiny, pinkish or black-brown to greenish yellow; coarse radiating ribs at back, becoming finer toward front; bottom margin straight in young, curved in old; old often have top outside of shell eroded away toward front. Umbo to one side of top; inside bluish white, often purplish white toward back; periostracum dark brown.
**Habitat**
Common in mud and sand at low tide.
**Range**
Gulf of St. Lawrence to S.C.

# PEN SHELLS
Family Pinnidae

The members of this family are generally elongately triangular with rather thin shells, the pointed end embedded in sandy mud where it anchors itself to a hard object by means of a strong byssus composed of many fine threads.

## FLESH PEN SHELL
*Pinna carnea*                                         102:10

**Description**
Size, 4–11 in. (10.2–27.9 cm). Shell rather thin, more or less translucent, with 8–12 ribs which may or may not bear erect, hollow spines. Color light brown or pinkish to orange-brown.
**Habitat**
In shallow water in sandy mud.
**Range**
Fla. to Brazil.

## STIFF PEN SHELL
*Atrina rigida*                                                                102:8

**Description**
Size, 5–11 in. (12.7–27.9 cm). Shell moderately stout with usually
10–25 radial ribs generally having rather long tubular spines.
Color light brown to dark grayish brown.
**Habitat**
In shallow water.
**Range**
N.C. to W. Indies.

# Scallops and Oysters
Order Pterioida

## TREE OYSTERS
Family Isognomonidae

Some species in this family of flattened bivalves are commonly
found attached by their byssus to the roots of mangrove trees,
others to rocks and coral reefs.

## FLAT TREE OYSTER
*Isognomon alatus*                                                             102:6

**Description**
Size, 2–3 in. (5.1–7.6 cm). Compressed shells, purplish brown to
grayish, irregularly oval in outline.
**Habitat**
Attached in clumps by byssus to mangrove roots.
**Range**
Fla. to Brazil.

## SCALLOPS
Family Pectinidae

The valves of scallops may or may not be equal. The shells have
radiating ribs and scalloped edges; they lack teeth. There is an ear-
shaped projection on either side of the umbones. Many of the
animals are edible.

Scallops are one of the relatively few mollusks that swim. They
have eyespots on the edge of their body, which give them some
power of sight. On favored beaches or after violent storms, thousands
of their shells will strew the strand. Almost no two are alike.

## DEEP SEA SCALLOP
*Placopecten magellanicus*                                                     103:9

**Description**
Size, 5–8 in. (12.7–20.3 cm). Shell almost round; upper valve
yellow to red-brown, slightly convex; lower valve flat, pinkish white.
Ears symmetrical; valves gap at front and have many fine, raised
threads and grooves, crossed by growth lines; inside shiny white.
**Habitat**
10–100 fathoms (18.3–182.9 m).
**Range**
Labrador to N.C.

## LION'S PAW
*Nodipecten nodosus*                                    **103:10**

**Description**
Size, 3–6 in. (7.6–15.2 cm). Shell rather stout, with about 9 broad, nodose ribs, which are finely sculptured with whitish radial threads or riblets and separated by deep interspaces; nodes often protuberant and hollow. Color orange or dark red, often irregularly spotted with white.
**Habitat**
Rather abundant in moderate depths.
**Range**
N.C. to Brazil.

## BAY SCALLOP
*Argopecten irradians*                                  **103:8**

**Description**
Size, 2–3 in. (5.1–7.6 cm). The well-known edible scallop. Shell varies greatly in color; white, gray, or bluish to yellowish or reddish brown, shading to deep violet at hinge; weathered shells slaty. Valves equal in size, each with 17–21 strongly rounded ribs with narrower grooves; ears slightly unequal and ridged; edge evenly scalloped; inside purplish white or mottled.
**Habitat**
Common among eelgrass in shallow water of bays.
**Range**
Cape Cod to N.J.

## CALICO SCALLOP
*Argopecten gibbus*                                     **103:7**

**Description**
Size, 1–2½ in. (2.5–6.4 cm). Shell smaller than Bay Scallop, more inflated, with ribs closer together and rather squarish in cross section. Lower valve usually white; upper valve variously and brightly colored in rose or red, or white with red or purplish mottlings.
**Habitat**
Common in shallow to moderately deep water.
**Range**
Md. to Brazil.

# SPINY OYSTERS
Family Spondylidae

This is a family of usually rather large, thick-shelled, radially ribbed, and generally more or less spiny shells with a ball-and-socket hinge and a central internal ligament. The margin of the interior is often distinctly colored. The shells are always attached to the substrate by the lower valve.

## SPINY OYSTER
*Spondylus americanus*                                  **102:7**

**Description**
Size, 3–4 in. (7.6–10.2 cm). Shell often quite thick and heavy; many irregular and scabrous radial riblets; several larger ribs often with spines which may be long, slender, and flattened. Color white, reddish orange, or yellow.

**Habitat**
Attached to hard substrate in shallow to rather deep water.
**Range**
N.C. to Brazil.

# CAT'S PAW OYSTERS
Family Plicatulidae

### KITTEN'S PAW
*Plicatula gibbosa*                                          102:9

**Description**
Size, ¾–1 in. (1.9–2.5 cm). Shell rather compressed with 6–9
strong, occasionally divergent ribs, resulting in a wavy margin;
hinge with 2 strong interlocking teeth in each valve. Color whitish
to gray with short, fine reddish lines on the ribs.
**Habitat**
Attached to rocks or coral in shallow to moderately deep water.
**Range**
N.C. to Brazil.

# JINGLE SHELLS
Family Anomiidae

This family is characterized by the hole in the lower valve, which
gives the anchoring threads an outlet to attach themselves to a rock,
shell, or other support. The shells are fragile, roundish, waxy in
luster, and have unequal-sized valves. The upper valve is left-
handed and dome-shaped; the lower one with the hole is right-
handed, smaller, and concave. The upper one has three muscle
scars inside and is the "jingle shell" most commonly found washed
up on shore.

### COMMON JINGLE SHELL
*Anomia simplex*                                          103:4

**Description**
Size, 1–2 in. (2.5–5.1 cm). Shell varies from silvery to yellow, red,
or blue-black, often translucent; shape varies with habitat, usually
unevenly rounded, thin, smooth; edges irregular, sometimes jagged;
upper valve plump, lower valve flat, hole near top.
**Habitat**
Common; attached to shells, rocks, boats, in moderately deep water.
**Range**
Maine to W. Indies.

### PRICKLY JINGLE SHELL
*Anomia squamula*

**Description**
Size, ½–¾ in. (1.3–1.9 cm). Shell opaque, yellowish white,
unevenly oval. Upper valve rough, convex, with rows of small
prickly scales. Lower valve flat, small hole near hinge; inside
whitish.
**Habitat**
Attached to rocks, shells, and seaweed in cold water.
**Range**
N.S. to N.C.

# OYSTERS
Family Ostreidae

Oyster shells are all asymmetrical and are greatly varied, depending largely upon the bottom to which they are affixed.

---

## VIRGINIA OYSTER
*Crassostrea virginica*                                                103:5

**Description**
Size, 2–6 in. (5.1–15.2 cm). Shell dingy gray, coarse-textured, thick, somewhat elongated and curved. Upper valve smaller and less rounded than lower; beaks long, curved, and grooved by growth of the ligament. Inside white; muscle scar almost central and deep purple.
**Habitat**
Attached to hard surfaces.
**Reproduction**
Up to 50,000,000 eggs are released by 1 animal at 1 spawning.
**Remarks**
Perhaps the most valuable commercial mollusk in the range; widely cultivated. "Blue Points" are round, "Lynnhavens" broad and long.
**Range**
Gulf of St. Lawrence to W. Indies.

---

## COON OYSTER
*Lopha frons*                                                          103:6

**Description**
Size, 1–2½ in. (2.5–6.4 cm). Shell oval to elongate with numerous irregular ridged ribs. Has shelly projections or claspers for attaching to support.
**Habitat**
Attached to corals in shallow to moderately deep water.
**Range**
Fla. to Brazil.

---

# Lucines, Clams, and Razors
Order Veneroida

---

# CARDITAS
Family Carditidae

---

## BROAD-RIBBED CARDITA
*Carditamera floridana*                                                104:2

**Description**
Size, 1–1½ in. (2.5–3.8 cm). Shell elongate, solid, with about 20 strong, flattened, radial ribs; elongate beads on earlier part of ribs, becoming fine, crowded threads near margin. Color whitish gray with oblong reddish brown spots on ribs arranged in concentric rows.
**Habitat**
Common on sandy bottom, moderate depths.
**Range**
Fla. to Mexico.

---

# LUCINES
Family Lucinidae

The equivalve shells of this family are orbicular, strong, and laterally compressed and have small but definite beaks. Most species are white.

## PENNSYLVANIA LUCINE
*Lucina pensylvanica*                                103:2

**Description**
Size, 1–2 in. (2.5–5.1 cm). Shell ovate, rather inflated, stout; conspicuous, fine, concentric ridges; strong furrow running from beak to posterior ventral edge. Color white, occasionally orange.
**Habitat**
May be locally common in shallow water.
**Range**
N.C. to W. Indies.

## GREAT WHITE LUCINE
*Codakia orbicularis*                                103:3

**Description**
Size, 1½–3¾ in. (3.8–9.5 cm). Shell slightly broader than high; stout, compressed, with numerous coarse radial riblets crossed by fine, concentric threads that make the riblets beaded. Color white inside and outside; occasionally tinged with yellow, often suffused with rose near hinge or along margins.
**Habitat**
Common in sandy, shallow water.
**Range**
Fla. to Brazil.

## CROSSHATCHED LUCINE
*Divaricella quadrisulcata*                          103:1

**Description**
Size, ½–1 in. (1.3–2.5 cm). Shell white, glossy, solid, and circular; valves with 5 or more concentric grooves and many fine radial lines, which form a unique pattern running obliquely toward the end of each valve; right valve has 1 small tooth, left has 2; inside margin with fine scalloping.
**Habitat**
Common in 10–50 fathoms (18.3–91.4 m).
**Range**
Mass. to W. Indies.

# JEWEL BOXES
Family Chamidae

These thick-shelled bivalves are attached by one valve to the substrate, possess a strong hinge with few large teeth, and are sculptured with frondose, concentric ridges or spiny radial ribs.

## LEAFY JEWEL BOX
*Chama macerophylla*                                                    104:3

**Description**
Size, 1½–3 in. (3.8–7.6 cm). Shell sculptured with concentric rows of leafy ridges; ridges in specimens from quiet deep water with delicate scaly spines. Color may be white, yellow, orange, to purplish or combinations; internally often more or less suffused with purple.
**Habitat**
In shallow to deeper water.
**Range**
N.C. to Brazil.

# ASTARTES
Family Astartidae

The shells of this family have prominent umbones which are almost central. They are small, usually brownish, triangular, and with well-developed cardinal teeth at the hinge. They all have conspicuous concentric grooves and growth lines.

## BOREAL ASTARTE
*Astarte borealis*                                                     104:1

**Description**
Size, 1–2 in. (2.5–5.1 cm). Shell white, oval, only slightly raised, smooth, with low concentric growth lines; valves equal; inside smooth, margin not scalloped, periostracum brownish.
**Habitat**
Common offshore to 30 fathoms (54.9 m).
**Range**
Arctic seas to Mass. Bay.

# COCKLES
Family Cardiidae

The shells of this family have valves of equal size, which often gape behind; the prominent beaks are almost central. The margins are toothed or scalloped; the hinge teeth are arched; the pallial line is wavy behind; and the general shape of the shell, viewed from the side, is that of a heart. These cockles are mobile and have no byssus.

## GREAT HEART COCKLE
*Dinocardium robustum*                                                 104:5

**Description**
Size, 3–6 in. (7.6–15.2 cm). The largest and most colorful cockle in the range. Shell yellowish brown, streaked and spotted with chestnut and purplish brown; plump, abruptly flattening toward

the back; 30–35 ribs radiating from the high, rounded beak. Valve margins toothed, inside salmon-pink.
**Habitat**
Sandy bottoms at moderate depth.
**Range**
Va. to Mexico.

## COMMON COCKLE
*Trachycardium muricatum*                                       **104:4**

**Description**
Size, 1¼–2 in. (3.2–5.1 cm). Shell nearly circular in outline with many radial ribs, more or less spiny except at the center. Color whitish or yellowish; interior white or tinged with yellow.
**Habitat**
Common in shallow water.
**Range**
N.C. to Brazil.

# VENUS CLAMS
Family Veneridae

These clams are distinguished by an unusual symmetry. The valves are equal, the teeth interlocking, the colors often arresting. The lunule is clear and deep. The shell is thick and strong, the pallial line is wavy, and the muscle scars are oval. The inside edge is often ridged or scalloped.

## NORTHERN QUAHOG
*Mercenaria mercenaria*                                          **106:8**

**Description**
Size, 3–5 in. (7.6–12.7 cm). Shell dingy white, thick, rounded; prominent concentric lines fairly widely spaced at the beaks, becoming very close toward margins. Valves have smooth central area; beaks are toward the short front; back long with an oval sweep; lunule ¾ as wide as long. Inside chalky white, violet muscle scars, inside margin lightly ridged.
**Habitat**
Mud or sand at low tide.
**Remarks**
This is the most common hard-shelled clam.
**Range**
Gulf of St. Lawrence to Gulf of Mexico.

## DISK DOSINIA
*Dosinia discus*                                                **106:4**

**Description**
Size, 2–3 in. (5.1–7.6 cm). The flattest Venus clam in the range; has finer and closer concentric lines than the other species. Shell snowy white, valves round, beaks shallow and small, hinge strong and thick; periostracum thin, yellowish.
**Habitat**
In sand, moderate depths.
**Range**
Va. to the Bahamas.

## CROSS-BARRED CHIONE
*Chione cancellata* 106:3
**Description**
Size, 1–1¾ in. (2.5–4.5 cm). Shell solid, rather stout, roughly
triangular in outline, with widely spaced lamellar concentric ribs
and rather regular axial riblets. Color whitish or grayish, generally
maculated with reddish brown to brown, often in axial rays;
interior often tinged with bluish purple.
**Habitat**
Common in shallow water.
**Range**
N.C. to Brazil.

## SUNRAY SHELL
*Macrocallista nimbosa* 106:9
**Description**
Size, 4–6 in. (10.2–15.2 cm). Shell elongately oval, rather stout,
smooth, glossy; pale grayish yellow with grayish purple radial rays,
generally more or less interrupted; surface covered by a thin
brownish periostracum.
**Habitat**
Rather common in sandy areas in shallow water.
**Range**
N.C. to Fla. and Tex.

## CHECKERBOARD CLAM
*Macrocallista maculata* 106:10
**Description**
Size, 1½–3 in. (3.8–7.6 cm). Shell smaller, more broadly oval than
in Sunray Shell and covered with irregular spots arranged in
concentric and often radial patterns; otherwise similar.
**Habitat**
Moderately common in shallow water.
**Range**
N.C. to Brazil.

# ROCK DWELLERS
Family Petricolidae

These bivalves are long, are rounded in front and narrowing
behind, and have a weak hinge, which is almost toothless.

## FALSE ANGEL WING
*Petricola pholadiformis* 106:1
**Description**
Size, 2–2¼ in. (5.1–5.7 cm). Shell chalky white, fragile, rectangular.
Umbones raised, placed close to short, sharply rounded front end.
Valves have 2 teeth, many radiating ribs crowded and weak behind,
strong and widely spaced in front; growth lines emphasized at intervals.
**Habitat**
Burrows in clay between tides.
**Remarks**
Usually found with the valves attached together in pairs.
**Range**
Gulf of St. Lawrence to Gulf of Mexico.

# RAZOR CLAMS
Family Solenidae

The shells of this family are very long and narrow, and gape at each end. The valves are equal in size and shape. Distributed worldwide in the sandy bottoms of shallow coastal waters, all species are considered edible.

## RIBBED POD
*Siliqua costata*                                                      **104:8**

**Description**
Size, 2–2½ in. (5.1–6.4 cm). Shell oblong, greenish purple, smooth, fragile, somewhat flat; beaks very small. Inside purplish white, glossy, with white raised rib from hinge to halfway across valve; periostracum olive-brown, iridescent, smooth.
**Habitat**
Common in sandy bottoms of shallow water.
**Range**
Gulf of St. Lawrence to N.J.

## COMMON RAZOR
*Ensis directus*                                                      **104:10**

**Description**
Size, 6–10 in. (15.2–25.4 cm). Shell white, thin, long, narrow, slightly curved; hinge, with sharp plate of 1 tooth on 1 valve between other valve which has 2 teeth and a double plate. Ends sharp, squarish, and gaping, inside white, stained purple; periostracum dark greenish brown, thin, and suggesting varnish.
**Habitat**
Common in sand at low water.
**Remarks**
These active animals burrow rapidly in the sand if disturbed. If captured they will attempt to jump away with amazing speed.
**Range**
Labrador to S.C.

# TELLINS
Family Tellinidae

These bivalves are usually polished and colorful. The valves are equal, rounded in front, but sharp and slightly folded behind. They are fairly flat, and the edges close evenly. Tellins are among the most beautiful shells.

## LINED TELLIN
*Tellina alternata*                                                    **105:1**

**Description**
Size, 2–3 in. (5.1–7.6 cm). Shell compressed, rather solid, surface sculptured with evenly spaced grooves. Color whitish or yellowish or tinged with pink.
**Habitat**
In sand in shallow to moderately deep water.
**Range**
N.C. to Fla. and Tex.

## NORTHERN TELLIN
*Tellina agilis*                                                    105:3

**Description**
Size, ⅓–¾ in. (0.9–1.9 cm). Shell iridescent white or rose, inside
and out, fragile, flat, translucent, with many very fine growth lines;
front rounded, back slopes to a rounded point. Valves have 2
cardinal teeth; glossy but not enameled; no radial grooves; pallial
sinus large, rounded; ventral margin slightly curved.
**Habitat**
Sandy bottoms below tides.
**Range**
Gulf of St. Lawrence to N.C.

## DeKAY'S TELLIN
*Tellina versicolor*                                                105:5

**Description**
Size, ½ in. (1.3 cm). Shell white, opalescent, with pink rays
widening toward edge. Pallial sinus close to muscle scar; ventral
margin straight, not curved. Similar to Northern Tellin but slightly
longer.
**Habitat**
In sand in shallow water to 50 ft. (15.2 m).
**Range**
N.Y. to W. Indies.

## CHALKY TELLIN
*Macoma calcarea*                                                   105:6

**Description**
Size, 1½–2 in. (3.8–5.1 cm). Shell chalky white, thin, and with fine
concentric lines; front evenly rounded, back narrow and slightly
twisted; valves have 2 hinge teeth; periostracum olive-gray.
**Habitat**
Common in cold water, 5–40 fathoms (9.1–73.2 m).
**Range**
Greenland to Long Island.

## BALTIC MACOMA
*Macoma balthica*                                                   105:4

**Description**
Size, ½–1½ in. (1.3–3.8 cm). Shell thin and white or rosy in a
sandy location; thick, bluish red in mud; slightly triangular,
rounded in front; back angular and with a rounded tip. Beaks
nearly central, fairly prominent; inside white or rose. Remnants of
thin grayish periostracum often seen on margins.
**Habitat**
Common between tides and up creeks and rivers.
**Range**
Arctic seas to off Ga.

## TENTA MACOMA
*Macoma tenta*                                        **105:2**

**Description**
Size, ½–¾ in. (1.3–1.9 cm). Shell shiny, thin, oval, elongate, pinkish white; many sharp, fine growth lines; front long, rounded; back narrow, twisted, and gaping. Inside white, tinged with yellow, with fine radiating lines; 2 teeth on right valve, 1 on left.
**Habitat**
Muddy or sandy bottoms in shallow water; common.
**Range**
Cape Cod to W. Indies.

# WEDGE SHELLS
Family Donacidae

These small wedge-shaped clams are long and rounded at the front, short and straight in back. The beaks are toward the back. The edges of the valves are usually ridged.

## SOUTHERN COQUINA
*Donax variabilis*                                    **105:10**

**Description**
Size, ½–1 in. (1.3–2.5 cm). Color varies ranging from shades of white, yellow, pink, and blue to mauve, with darker rays sometimes crossed with bands of other colors as in a plaid. Front end of shell smooth, occasionally minutely lined; small radial lines toward the center, becoming larger toward blunt back end. Inside glossy, varying in color, with finely toothed edges.
**Habitat**
Sandy beaches between tides.
**Range**
Va. to Tex.

# SANGUIN CLAMS
Family Psammobiidae

## FALSE COQUINA
*Heterodonax bimaculatus*                             **105:8**

**Description**
Size, ½–1 in. (1.3–2.5 cm). Shell broadly oval, smooth except for fine growth lines; 2 small teeth in each valve. Color variable; pure white, purplish, or whitish, suffused with rose or orange, or with internal rose rays, or maculated with purplish gray.
**Habitat**
Common intertidally on sandy beaches.
**Range**
S. Fla. to W. Indies.

# SEMELES
Family Semelidae

Chiefly warm-seas animals, the Semeles have rounded oval shells that are little inflated, with the posterior end characterized by relatively obscure folds.

## WHITE SEMELE
*Semele proficua* 105:7

**Description**
Size, ¾–1½ in. (1.9–3.8 cm). Shell subcircular, rather stout, with fine concentric ridges and minute radial striations. Color whitish to yellowish white, internally white or yellowish often spotted with purple or pink.
**Habitat**
Common offshore.
**Range**
N.C. to Brazil.

# LONG SIPHON CLAMS
Family Solecurtidae

These clams are somewhat similar to the tellins but larger. The sides of the shell are more nearly parallel and are often marked with fine concentric lines.

## PURPLISH RAZOR
*Tagelus divisus* 105:9

**Description**
Size, 1½ in. (3.8 cm). Shell fragile, smooth, translucent, purplish; oblong, with round end; back narrower than front. Beaks nearly central; purplish rib stripe runs from beaks toward margin; 2 cardinal teeth in each valve; inside purplish, glossy; periostracum thin glossy brown.
**Habitat**
Sandy bottoms in shallow water.
**Range**
Cape Cod to the Caribbean.

# BLACK CLAMS
Family Arcticidae

## BLACK CLAM
*Arctica islandica* 106:5

**Description**
Size, 3–5 in. (7.6–12.7 cm). Shell almost round, dull, chalky, smoothish, with low, concentric growth lines; inside white; periostracum thick, blackish.
**Habitat**
Sandy mud, 5–80 fathoms (9.1–146.3 m).
**Range**
Newfoundland to N.C.

# SURF CLAMS
Family Mactridae

These clams have equal valves, usually tightly closed, but sometimes with a very small gap. The hinge has two cardinal teeth.

## ATLANTIC SURF CLAM
*Spisula solidissima*                                              104:6

**Description**
Size, 4–7 in. (10.2–17.8 cm). Shell yellowish white, thick, oval; smoothish with many growth lines. Beaks large, nearly central, hinge strong, ligament internal with large, fairly straight spoon-shaped pit, cardinal tooth small; lateral teeth long, with notched edges.

**Habitat**
Under sand in surf and farther out; abundant just offshore.

**Remarks**
This is probably the best-known single shell along the North Atlantic Coast. The shells are conspicuous and common, so the first-time visitor to the ocean often takes some home as souvenirs. The meat makes a good chowder. Since the animals are abundant just offshore, a real northeaster will pile them up on the beach.

**Range**
N.S. to S.C.

## DWARF SURF CLAM
*Mulinia lateralis*                                               104:7

**Description**
Size, ⅓–¾ in. (0.9–1.9 cm). Shell whitish, inflated, smoothish, triangular and convex, with concentric wrinkles and a radial ridge toward the back. Beaks prominent, nearly central, pointing forward; hinge strong; marginal teeth V-shaped; periostracum thin, yellowish brown.

**Similarities**
Resembles young Atlantic Surf Clam, which has a larger spoon at hinge and tiny saw teeth along the larger side hinge teeth.

**Habitat**
Abundant in sand in warm, shallow water.

**Range**
Maine to Tex.

## WEDGE RANGIA
*Rangia cuneata*                                                  104:9

**Description**
Size, 1–2½ in. (2.5–6.4 cm). Shell heavy, stout, smooth; high, strong beaks; whitish but usually covered by a grayish brown periostracum; interior white, glossy.

**Habitat**
Common in mud in fresh to brackish water of coastal rivers.

**Range**
Md. to Tex.

# Clams, Rock Borers, Piddocks, and Shipworms
Order Myoida

## SOFT-SHELLED CLAMS
Family Myidae

A distinctive feature of these shells is the spoon-shaped tooth in the left valve which fits into a corresponding pit in the right valve. The valves are usually unequal and gape at both ends.

### COMMON SOFT-SHELLED CLAM
*Mya arenaria*                                                    106:7

**Description**
Size, 1–6 in. (2.5–15.2 cm). Shell gray or chalky white, thick, wrinkled by growth lines; elliptical. Spoon-shaped tooth long, shallow; pallial sinus somewhat V-shaped; periostracum thin, gray to straw-colored.
**Habitat**
Common in shallow bottoms between tides.
**Habits**
Reveals presence by abrupt, vertical squirt of water from suddenly indrawn siphon.
**Remarks**
Highly edible, especially when steamed.
**Range**
Labrador to off N.C.

## BASKET CLAMS
Family Corbulidae

Formerly the Family Aloididae, the shells are small, solid, inequivalve with one valve usually overlapping the other. Commonly ribbed centrally, the valves may gape slightly at the anterior end. Each valve has one upright conical tooth. They occur worldwide in temperate waters.

### COMMON BASKET CLAM
*Corbula contracta*                                              106:6

**Description**
Size, ½ in. (1.3 cm). Shell white, plump, irregularly oval with smooth, regular, concentric ridges. Valves unequal, rounded in front, pointed behind, lower back of right valve overlaps left. Angular fold from beak to margin; left valve has V-shaped notch in front of beak; hinge tooth upright, slender; large conical tooth fits into corresponding pit in the other valve; periostracum thin, brownish.
**Habitat**
Common in sandy or muddy shallows.
**Range**
Cape Cod to W. Indies.

# ROCK BORERS
Family Hiatellidae

The nestling and burrowing habits of these clams cause irregularities and considerable variation in their shells. They are chalky, are toothless, and have an interrupted pallial line.

## ARCTIC ROCK BORER
*Hiatella arctica*                                          106:2

**Description**
Size, 1–3 in. (2.5–7.6 cm). Shell chalky white, oblong, oval, or distorted, with heavy, uneven growth lines. Valve margins almost parallel in front and behind, occasionally gaping behind; radial rib toward rear; beaks close somewhat behind top of valves; inside whitish; periostracum grayish, thin, flaky.
**Habitat**
Common, burrowing in mud or rock in cold water.
**Range**
Arctic seas to deep water in W. Indies.

# PIDDOCKS
Family Pholadidae

These clams are able to bore through hard rocks. The long, narrowed shells have sharp toothlike ridges in front. They gape at both ends and lack both ligament and hinge teeth.

## COMMON ANGEL WING
*Cyrtopleura costata*                                       107:9

**Description**
Size, 4–8 in. (10.2–20.3 cm). Shell snowy white, rather fragile, widely gaping; valves wing-shaped, each with about 30 strong radial ribs crossed by growth lines, forming cuplike scales; beaks close to front with internal spoon-shaped brace; on the inside, sculpture is reversed; periostracum thin, gray.
**Habitat**
Burrows to depth of 12 in. (30.5 cm) in sandy mud.
**Range**
Mass. to W. Indies.

## FALLEN ANGEL WING
*Barnea truncata*                                           107:2

**Description**
Size, 2–2½ in. (5.1–6.4 cm). Shell white, delicate, oblong; triangular, scaly, and pointed in front; smoothish, bluntly cut off behind; both ends gaping. Growth lines prominent, thin; broken into scales at radial ribs, giving beaded appearance; long, shelly plate reinforces hinge. Beaks prominent, leaning toward back, siphons can extend 3 times length of shell.
**Habitat**
Burrows in clay, rock, or wood between tides.
**Range**
Mass. Bay to Fla.

## GREAT PIDDOCK
*Zirfaea crispata*                                          **107:5**

**Description**
Size, 1–2 in. (2.5–5.1 cm). Shell grayish, pointed in front, rounded
behind; valves widely gaping at ends, touching at center. Surface
equally divided by deep furrow from beak through middle of
valves; front area has tooth ridges, entire valve has coarse
concentric growth lines; inside whitish.
**Habitat**
Bores into soft rock in cold water.
**Range**
Newfoundland to N.J.

## STRIATE WOOD PIDDOCK
*Martesia striata*                                          **107:7**

**Description**
Size, ¾–1½ in. (1.9–3.8 cm). Shell elongately ovate, front end
broadly rounded, gaping at hind end; area of beaks covered by a
roundish shieldlike plate. Front end with fine, curved, denticulate,
concentric riblets; shallow radial groove separating it from posterior
end, which has smoothish concentric ridges that at very end become
irregular wrinkles.
**Habitat**
Bores into wood.
**Range**
N.C. to Brazil.

## WEDGE-SHAPED PIDDOCK
*Martesia cuneiformis*                                      **107:6**

**Description**
Size, ½–¾ in. (1.3–1.9 cm). Shell smaller, less elongate, more oval
than in Striate Wood Piddock; the plate on top triangular with a
groove in the center.
**Habitat**
Bores into wood.
**Range**
N.C. to Brazil.

# SHIPWORMS
Family Teredinidae

Shipworms are an aberrant group of clams in which the shell is
used for boring into saltwater-soaked wood, and the body, which
maintains the connection with the outside world of sea water,
becomes greatly lengthened, sometimes to forty times the length of
the shell. Numerous species are recognized throughout the world.
They are distinguished by differences in the featherlike pallets,
composed of a series of cones, at the far end of the body, which
control the admission of sea water into the siphons.

One female lays millions of eggs. The free-swimming young soon
attach themselves to anything wooden and start boring in. They
never leave this hole but keep on boring and lengthening their
bodies the rest of their lives.

## NAVAL SHIPWORM
*Teredo navalis*                                        107:8

**Description**
Size of shell, ½ in. (1.3 cm); length of pallets, to 2 in. (5.1 cm).
Valves thin, sharp, globular; pallets pointed, slightly compressed,
and symmetrical, with shallow cones, widening from a slender stalk
and tapering to a somewhat hollow tip. Extreme ⅓ of pallet
covered by yellowish brown periostracum.
**Habitat**
Common in wood soaked by sea water.
**Food**
Plankton from sea water; cellulose from the wood it excavates.
**Range**
Arctic to tropic seas.

Note: The **GOULD'S SHIPWORM,** *Bankia gouldi*, has ½-inch (1.2-
cm) pallets of deep-cupped cones.

# Paper Shells, Slender Clams, and Spoon Shells
Order Pholadomyoida

# PAPER SHELLS
Family Lyonsiidae

These shells have small fragile valves that are unequal in size and
somewhat rectangular.

## GLASSY LYONSIA
*Lyonsia hyalina*                                        107:3

**Description**
Size, ½–¾ in. (1.3–1.9 cm). Shell very fragile, white or beige,
pearly translucent; rounded in front; long, narrow somewhat cut
off behind and gaping. Hinge weak, toothless; beaks small, pointed
toward front. Valves covered with thin periostracum having radial
wrinkles bearing minute fringes, which catch sand with which to
coat shell.
**Habitat**
Low water to 34 fathoms (62.2 m); common.
**Range**
N.S. to S.C.

# SLENDER CLAMS
Family Pandoridae

These small shells are very thin and flat. The valves are white, unequal, and pearly inside. The beaks are not noticeable, and there are two teeth at the hinge which fit into matching grooves.

## GOULD'S PANDORA
*Pandora gouldiana*                                                107:4

**Description**
Size, 1 in. (2.5 cm). Shell chalky white, roughly oval, rounded in front; back cut off almost straight, curving strongly at lower edge. Right valve flat, left valve convex; hinge line concave, curving up toward back; inside pearly; brownish periostracum on edges of valves.
**Habitat**
Rocky bottoms to 20 fathoms (36.6 m).
**Range**
Gulf of St. Lawrence to Cape May.

# SPOON SHELLS
Family Periplomatidae

These shells are oval, gape slightly, and have a faint pearly glow. A spoon-shaped tooth at the hinge is supported by a small triangular process beneath, which is often lost.

## UNEQUAL SPOON SHELL
*Periploma margaritaceum*                                          107:1

**Description**
Size, ¾–1¾ in. (1.9–4.5 cm). Shell with left valve more inflated than and slightly overlapping the right; surface smooth, except front end minutely granulose and separated from rest of shell by a strong ridge and groove from beak to lower margin.
**Habitat**
Common on certain beaches.
**Range**
S.C. to Tex.

# CEPHALOPODS
Class Cephalopoda

Cephalopods are carnivorous mollusks with the shell, when present, lengthened and usually inside, not outside, the animal. The region of the head is surrounded by tentacles or arms, which have prehensile suckers. The mouth is round with a parrotlike beak and there are two eyes, often large. The body is either cylindrical or rounded and frequently has terminal fins. The outside has colored spots which, when contracted or expanded, change the color of the surface of the animal. With the exception of the Paper Nautilus, all cephalopods can discharge an inky fluid as a means of defense.

This class includes some of the most highly organized, swift-moving, and intelligent of the invertebrates. Some octopi in tests have shown an intelligence greater than that of many vertebrates.

# Cuttlefish
Order Sepioidea

# RAM'S HORNS
Family Spirulidae

## RAM'S HORN
*Spirula spirula*                                                    **108:2**

**Description**
Size, ¾–1 in. (1.9–2.5 cm). Shell fragile, in the form of a flat, few-whorled spiral, the coils not touching; internally divided into chambers by concave partitions; the pearly concave surface of last partition, visible at aperture, has a tiny hole near inner edge. Color yellowish white with the internal partitions showing through as opaque white lines.
**Habitat**
Pelagic, living in depths of 60–500 fathoms (109.8–914.5 m).
**Range**
Cast up on beaches, Mass. to Brazil.

# BOB-TAILED SQUIDS
Family Sepiolidae

## ATLANTIC BOB-TAILED SQUID
*Rossia tenera*                                                     **108:1**

**Description**
Size, including arms, 3–4 in. (7.6–10.2 cm). The smallest squid in the range; characterized by the comparatively large suckers at the middle of the side arms. Body short, soft; pinkish with small brownish spots (papillae). Short arms about ½ length of long arms; fins about ½ body length. Internal shell small, soft, slender.
**Habitat**
7–250 fathoms (12.8–457.2 m).
**Range**
N.S. to se. Fla.

# Squids
Order Teuthoidea

## COMMON SQUIDS
Family Loliginidae

These animals have a long, tapering, cylindrical body with ten
arms and large triangular terminal fins. The arms bear two rows
of suckers; each sucker is surrounded by a horny, dentated ring.
The tentacular arms bear four rows of suckers on small clubs. The
horny internal pen is slightly broader at its base and tapers to a
point, with a keel on the underside.

### COMMON SQUID
*Loligo pealei*                                                   **108:3**

**Description**
Size, 1–1½ ft. (30.5–45.7 cm). Distinguished by long triangular
fins over ½ length of body, with their tips meeting at the end. Body
long, narrow; gray with reddish to purple spots; large; 8 short arms
with 2 rows of suckers; 2 long arms with 4 rows on clublike ends.
**Habitat**
Common from low water to 50 fathoms (91.4 m).
**Remarks**
This species includes mackerel in its diet. If some of them attack a
school of mackerel of the same size, a rugged battle may ensue.
The squid try to fasten on to the neck of the mackerel; the
mackerel try to sink their jaws in the squid. Sometimes the
mackerel win. These squid are used commercially for food and
bait.
**Range**
S. Labrador to Fla.

### DWARF SQUID
*Lolliguncula brevis*                                            **108:5**

**Description**
Size, 5–10 in. (12.7–25.4 cm). The short, rounded fins which meet
to form an oval at the base of the body are characteristic of this
animal. Body short, with large purplish spots. Upper arms very
short; long arms equal to length of body, with 4 rows of suckers at
ends; fins white underneath.
**Habitat**
Warm waters.
**Range**
Del. Bay to Brazil.

## SEA ARROWS
Family Ommastrephidae

### SEA ARROW
*Illex illecebrosus*                                             **108:4**

**Description**
Size, 12–18 in. (30.5–45.7 cm). Body long, cylindrical. Notch in
front of eyes; head ridged transversely behind the eyes, narrows
abruptly to neck, and has smooth, deep depression underneath to

hold siphon. Superficially similar to the Common Squid, but easily distinguished by shorter fins and small eyes.

**Habitat**
Common in open sea; during summer in schools close to shore in New England.

**Habits**
Can skim through air.

**Remarks**
Used commercially for food and bait.

**Range**
Greenland to Gulf of Mexico.

# Octopods
Order Octopoda

The octopods are characterized by having only eight arms, lacking the tentacular arms of the squids. The arms bear suckers which, unlike those of the squids, are not on stalks and lack horny rings. There is no internal pen or shell.

# OCTOPI
Family Octopodidae

## ATLANTIC OCTOPUS
*Octopus vulgaris*                                    108:7

**Description**
Size, with longest arm, 1–2½ ft. (30.5–76.2 cm). Body short, thick, depressed, rounded behind. Skin smoothish, translucent, blue-white with brown dots; 8 arms connected partway by a web, each has 2 rows of suckers; third right arm in males serves as sexual intromittent organ. Head broad, enlarged, and warty around eyes; depressed between.

**Habitat**
Rocks and niches close to shore.

**Range**
Gulf of Maine to W. Indies.

# ARGONAUTS
Family Argonautidae

## PAPER NAUTILUS
*Argonauta argo*                                      108:6

**Description**
Size, 4–8 in. (10.2–20.3 cm). Dorsal arms of the female secrete a thin, spiral, few-whorled papery shell, which is used to contain the eggs and into which the female can retreat when disturbed. Shell symmetrical, compressed, with branching radial ridges on surface terminating in sharp nodes on twin median keels. Keels tinged with brownish black in the early parts of shell; remainder of shell white.

**Habitat**
Pelagic, in tropical seas.

**Range**
Found on beaches in tropical Fla.

# Other Marine Invertebrates

**Consulting Editor**
Robert D. Barnes
Department of Invertebrate Zoology, Gettysburg College

**Illustrations**
Plates 109–110 Klarie Phipps
Text Illustrations by John Cameron Yrizarry and
Nancy Lou Gahan

# Other Marine Invertebrates

Marine invertebrates are sea animals which possess no backbone. In this book they have been divided into two groups. The phylum Mollusca, or mollusks, includes the study of the chapter preceding this one. Included in this chapter are the most conspicuous species of eleven other phyla of marine invertebrates. Because of the great number of species this includes, it is not possible to treat them with the completeness given to the vertebrates in other sections of this book. However, the bare essentials of habits and life histories given below reveal at least a little of the fascinating world that lies between the tides and beneath the waves.

## Habitat

The diversity of animal life in the oceans is enormous. Many species live in deep water and are therefore not easily seen. However, there is a rich fauna at the edge of the sea, between the tide marks, and in water no deeper than knee level, which can be studied and enjoyed by all who are interested. Unfortunately, surf beaches, which are frequented by the largest number of beachcombers and visitors to the shore, are one of the most inhospitable and impoverished marine habitats. Few animals can live in sand pounded by waves.

The naturalist who wants to see marine animals must explore other kinds of habitats. At low tides, one should comb the protected beaches of inlets, bays, and sounds, looking beneath driftwood and algae cast up by spring tides. Tide pools on rocky shores are menageries; ways to seek out the inhabitants include poking beneath and between clumps of shells on oyster reefs (wearing gloves), examining wharf pilings and rock jetties at low tide, rambling over exposed sand and mud flats at low tide, and digging beneath surface markings.

## Conservation

Many of these habitats and species are protected because they are in state parks, and collecting or damaging specimens is prohibited by law. In some states, all marine life is protected and may not be collected without a permit. As with most wildlife, the formerly abundant supply of marine invertebrates has been reduced by vandals, collectors, students, overfishing, insecticides, and loss of habitat.

## Range and Scope

This chapter describes the most common, typical, or spectacular marine invertebrates (aside from the mollusks) that occur along the Atlantic coastline, principally from southern Florida to Canada, seaward to ten fathoms. The selection includes most of the species that the amateur field naturalist is likely to encounter. All of the species included are marine, or saltwater, invertebrates. The size given is usually the maximum growth of the species.

## Illustrations

Each species illustrated is shown as it would be seen by the naked eye, just as it would be found on the seashore, rather than with the use of a high-powered microscope. Common names, for those species for which they are accepted, have been included. When there is no widely used common name, the common name of the group to which the species belongs has been captioned in parentheses on the illustrated plates, numbers 94 and 95, and in the text illustrations to facilitate species identification.

## USEFUL REFERENCES

Barnes, R. D. 1980. *Invertebrate Zoology*, 4th ed. Philadelphia: Saunders.

Gosner, K. L. 1971. *Guide to Identification of Marine and Estuarine Invertebrates*. New York: Interscience Publishers.

Kaestner, A. 1967–1970. *Invertebrate Zoology*. New York: Interscience Publishers.

MacGinitie, G. E., and MacGinitie, N. 1968. *Natural History of Marine Animals*. New York: McGraw-Hill.

Pennak, R. W. 1978. 2nd ed. *Freshwater Invertebrates of the United States*. New York: John Wiley and Sons.

Smith, F. G. W. 1971. *Atlantic Reef Corals*. Coral Gables, Fla.: Univ. of Miami Press.

Smith, R. I., ed. 1964. *Keys to Marine Invertebrates of the Woods Hole Region*. Contribution No. 11. Systematics–Ecology Program. Woods Hole: Marine Biology Lab.

## GLOSSARY

*Aboral* The side away from the mouth.

*Ampulla (pl.* ampullae) A membranous sac.

*Anterior* Front; forward part.

*Branchiae* Gills.

*Capitulum* Barnacle plate.

*Carapace* Shieldlike plate covering above and on sides.

*Chelipeds* Claw-bearing appendages.

*Cilia* Short, movable hairlike processes of cell surface.

*Cirri* Tendrils or fleshy appendages.

*Cloaca* Common duct into which digestive, excretory, and reproductive systems empty.

*Detritus* Fine particles of organic or inorganic origin.

*Dorsal* Pertaining to the upper side of the body.

*Flagellum* Long, movable hairlike process of cell surface.

*Ganglion (pl.* ganglia) A mass of nerve cell bodies.

*Gnathopods* Crustacean appendages which manipulate food.

*Gonads* Organs producing the sex cells (ovary and testis).

*Hermaphroditic* Having both male and female reproductive organs in one individual.

*Hydranth* Individual polyps in a cnidarian colony.

*Interradius* Space between radii.

*Interray* Space between rays.

*Lobule* Part of a lobe.

*Madreporic plate* Water intake disk.

*Mantle* An enveloping external covering of the body.

*Medusa* Jellyfish.

*Mesenteries* Folds of tissue in the body cavity.

*Nephridia* Excretory and internal water regulatory systems.

*Palp* A small, often fleshy, feeler.

*Papilla (pl.* papillae) A small, nipplelike projection.

*Parapodia* Side feet.

*Peduncle* Stalk.

*Peristome* Region around the mouth.

*Polyp* Attached tentaculate animal like hydra and sea anemone. Some species colonial.

*Prehensile* Grasping; capable of suspending body.

*Proboscis* Anterior body extension or tubular organ for feeding.

*Radius (pl.* radii) An imaginary radial plane dividing the body of a radially symmetrical animal into similar parts.

*Ray* Any of the radiating divisions of the body of an echinoderm with all its included parts.

*Rostrum* Beak or beaklike part of an animal.

*Scutum* Barnacle plate.

*Sessile* Attached directly by its base.

*Setae* Bristles.

*Spicule* Rod or needlelike skeletal unit.

*Spongin* A horny substance.

*Stolon* Extension of body wall.

*Telson* Posterior projection of last body segment.

*Test* Hard internal skeleton.

*Thoracic* Pertaining to the thorax, or middle division of the body.

*Umbo* Prominence above hinge in bivalves.

*Ventral* Pertaining to the underside of the body.

*Zooid* Individual animal in a colony.

# SPONGES
Phylum Porifera

Sponges are very different from most other animals. They are attached, and there is little discernible movement of any part of the body. There is no front end, no head, no mouth or gut. Essentially a sponge is a mass of tissue organized around a system of water canals. The in-flowing canals are provided with microscopic openings, from which the name *Porifera*, or "pore bearer," is derived. The out-flowing canals open to the surface through one or more large openings, or oscula. Seawater is driven through the canals by internal flagellated cells, the water stream providing the animal with food and oxygen and removing wastes.

Sponges range in size from species that can just be seen with the naked eye to others that would fill a bushel basket. The growth form differs depending upon the species—encrusting sheets, rounded masses, erect branches, tubes, and vases—and there may be some variation even within a species. The body is supported by a skeleton of minute mineral rods (spicules) or organic fibers (spongin) or both.

Sponges reproduce sexually to form a free-swimming ciliated larva, which after a few days attaches permanently to the bottom or to some object underwater and develops into a mature sponge.

There are a few freshwater sponges, but the vast majority of some 5,000 species live in the sea, from the intertidal zone to deep ocean basins.

# CALCAREOUS SPONGES
Class Calcarea

## *SCYPHA CILIATA*                                    *Fig. 123*

**Description**
Length, to 1 in. (2.54 cm). Cigar-shaped, covered with thin surface projections; excurrent canal surrounded by palisade of long, slender, needlelike limy spicules. Color yellowish white. Budded individuals separate completely, or almost completely, from the parent, giving the colony the appearance of a bunch of tiny bottles.
**Habitat**
Seaweed, shells, stones, and dock piles, low water to 300 ft. (91.4 m).
**Range**
Greenland to Long Island.

# SILICEOUS AND HORNY SPONGES
Class Demospongiae

## BORING SPONGE
*Cliona celata*                                         *Fig. 123*

**Description**
Sponge fills channels excavated in old mollusk shells and appears at the surface as yellow plugs ⅛ in. (0.32 cm) in dia., or less. In shells which have been extensively riddled, the sponge may overgrow the surface as a thick irregular crust as much as 8 in.

(20.3 cm) across. Excavation begins when the larva settles to the bottom. Signs of this sponge are the finely perforated clam and oyster shells it has attacked.

**Habitat**
Gravel or shell bottoms, low tide to 100 ft. (30.5 m).

**Remarks**
Boring sponge is an important agent in the breakdown of shell and coral.

**Range**
Arctic to the Carolinas.

*Fig. 123*

*Haliclona oculata*
Deadman's Fingers

*Cliona celata*
Boring Sponge

*Scypha ciliata*
(Sponge)

## RED SPONGE
*Microciona prolifera*                                    **109:6**

**Description**
Size, irregular masses to 1 ft. (30.48 cm) high and 6 in. (15.2 cm) square at base. Commonly limited to small, thin encrusting sheets. Surface has scattered oscula, numerous pores. Color bright red.

**Habitat**
Grows on rocks, oyster shells, piles, old bottles, from low tide to 60 ft. (18.28 m).

**Range**
Maine to the West Indies.

## DEADMAN'S FINGERS
*Haliclona oculata*                                    *Fig. 123*

**Description**
Size, fingers 3–6 in. (7.6–15.2 cm) long, oscula $\frac{1}{10}$ in. (0.25 cm) dia. Colonies compact, large; fingers close-set on slender stem; oscula numerous. Color pale orange-red.

**Habitat**
From low tide to 500 ft. (152.4 m); common.

**Remarks**
The fingerlike branchings from the central stem give this sponge its name. Bleached skeletons are often found on eastern beaches.

**Range**
Arctic to Cape May.

## *HYMENIACIDON HELIOPHILA*

**Description**
Size, erect peaked tufts up to several inches across. Color bright orange.

**Habitat**
From low tide to 150 ft. (45.7 m) on shells and stones; common in some areas.

**Range**
Bay of Fundy through the Carolinas.

# CNIDARIANS
Phylum Cnidaria

This numerous and largely marine phylum is distinguished by one or more rings of tentacles encircling the mouth, which is the only opening into the interior of the body. The symmetry is thus radial. Some cnidarians (polyps) have a tubular, attached body, with the oral tentaculate end directed upward; others (medusae, or jellyfish) are bowl- or bell-shaped and free-swimming, with the mouth directed downward. Many polypoid cnidarians are colonial, with the individuals connected together by living tissue. Most cnidarians feed on other animals, which they capture with the aid of unique stinging cells located on their tentacles and other parts of the body.

# HYDROIDS
Class Hydrozoa

Members of the class Hydrozoa exhibit polypoid, medusoid, or both body forms in the course of their life cycle. Most marine polypoid species are colonial and are known as hydroids. They are common on pilings, jetties, and rocks, where their attached plantlike growth form may cause them to be dismissed as "seaweed." There are also species which grow on the surfaces of the larger seaweeds. The medusae, or jellyfish, of this class are quite small, usually less than one inch in diameter, and are therefore less conspicuous than the larger, more familiar jellyfish of the next class. Hydrozoan jellyfish are usually a reproductive stage arising from the hydroid colony. Hydroids may produce a painful sting. Meat tenderizer quickly rubbed on jellyfish stings is very effective in reducing painful reaction.

Only a few of the many species of hydroids are described below.

## Uncupped Hydroids
Suborder Gymnoblastea

In this group the polyps do not form a protective cup (hydrotheca) around themselves, as those of the suborder Calyptoblastea. However, the stems of the colony may have a firm protective covering.

### *TUBULARIA CROCEA*                          109:5, *Fig. 124*
**Description**
Size, to 4 in. (10.2 cm) high. Straight, stalked clumps with tangled and spreading bases, growing in colonies. Stalks gray, tentaculate heads (hydranths) pink.
**Habitat**
Submerged rocks, pilings, cables.
**Range**
N.B. to Fla.

### *HYDRACTINIA ECHINATA*                          *Fig. 124*
**Description**
Lives on surface of old snail shells that are inhabited by hermit crabs. Size depends upon size of shell. Hydroid colony gives shell a furry appearance; color brown to cream.

**Habitat**
Shallow water, on shells inhabited by hermit crabs.
**Range**
Labrador to the Carolinas.

*Fig. 124*  **Hydroids**

*Hydractinia echinata*      *Tubularia crocea*   *Pennaria tiarella*

## *PENNARIA TIARELLA*                    *Fig. 124*

**Description**
Size, up to 6 in. (15.2 cm) long. Featherlike or fernlike growth
form. Pale orange.
**Habitat**
Common on pilings and rocks in shallow water.
**Range**
Cape Cod to Fla.

## *EUDENDRIUM RAMOSUM*

**Description**
Size, colony to 4 in. (10.2 cm) long. Colonies form irregular
branching tufts. Color bright orange.
**Habitat**
On pilings, jetties, rocks, from shallow to deep water.
**Range**
Bay of Fundy to Fla.

# Cupped Hydroids
Suborder Calyptoblastea

In this group of hydroids the transparent protective covering not
only occurs on the stems but also extends up around the polyps to
form a protective cup, or hydrotheca, into which the polyp can
withdraw.

## *OBELIA* (Species)

**Description**
Size, colony to 1 in. (2.5 cm) long. Small, delicate, irregular
branching colonies. Oral end of polyps within a protective cup but
difficult to see without magnification. Color pink or translucent gray.
**Habitat**
On pilings, rocks, and seaweeds to 200 ft. (60.9 m).
**Remarks**
There are many other hydroids with growth form similar to
*Obelia*. They can be distinguished only with the aid of a
microscope and knowledge of structural details.
**Range**
Bay of Fundy to Fla.

## SERTULARIA PUMILA

**Description**
Size, up to 1½ in. (3.8 cm) long. Branching stalks resembling
jigsaw blades placed back to back, the "teeth" being the protective
houses of the polyps. Color silvery.

**Habitat**
Common on seaweed.

**Remarks**
All of the members of this family (Sertulariidae) have a similar
growth pattern, although size differs. Many different species are
distributed along the entire East Coast.

**Range**
Bay of Fundy to N.J.

## PORTUGUESE MAN-OF-WAR
*Physalia physalis*                                    109:3

**Description**
Floating colony up to 1 ft. (30.5 cm) in length, surmounted by an
elongate purplish-blue gas-filled sac. Long tentacles, 2 to many feet
in length.

**Habitat**
Warm tropical waters.

**Habits**
Tentacles sting small animals and convey captured food to mouth
by slow contraction. Some small fish, immune to sting, live and
travel among tentacles.

**Remarks**
Has a painful sting, which can be extremely dangerous and should
be carefully avoided by swimmers.

**Range**
Gulf of Mexico, West Indies, and Gulf Stream. Blown ashore in
Fla. and the tropics and occasionally in N.J.

# JELLYFISHES
Class Scyphozoa

These are the large jellyfishes that are seen swimming near the
shore or washed up on beaches. They are found in all seas and
vary in diameter from about one inch (2.5 cm) to seven and one-
half feet (2.3 m). Scyphozoan jellyfish lack the flaplike shelf, or
velum, within the bell margin that is found in the little hydrozoan
jellyfish, and four long flaps, or oral arms, hang down from around
the mouth. The eggs of the sexually reproducing adult jellyfish
develop into an inconspicuous polypoid stage, the scyphistoma,
which attaches to seaweeds or to hard bottoms. It is not colonial
and is usually less than one-half in. (1.3 cm) high. When budding,
it may look like a stack of tiny plates. The top plate, showing
tentacles, detaches at regular intervals and swims away as a tiny
juvenile jellyfish.

## SEA NETTLE
*Dactylometra quinquecirrha*                           109:1

**Description**
Size, bell to 8 in. (20.3 cm), tentacles to 3 ft. (91.4 cm). Oral arms
long, ruffled. Tentacles of bell margin, 40, located between lobes
on the bell margin; every sixth position between lobes lacks a

tentacle, the area being occupied by an inconspicuous sense organ.
Stinging cells on both bell and tentacles. Bell milky white or rosy
pink to tan, sometimes with rosy stripes; oral arms milky white or
pink.

**Habitat**
Common in less saline waters of bays and estuaries.

**Range**
Cape Cod to Fla.; especially common in the Chesapeake Bay and
certain parts of the Carolina sounds.

## CYANEA CAPILLATA                                    109:2

**Description**
Size, bell dia. to 3 ft. (91.4 cm), rarely to 8 ft. (24.4 m). Bell
margin notched, forming 32 lobes. Long tentacles arranged in 8
clusters around bell margin; number of tentacles increases with bell
diameter, reaching as many as 800 in large specimens. Bell
brownish to purplish-red; margin translucent.

**Habitat**
Cool coastal waters.

**Range**
Bay of Fundy to Cape Hatteras.

## MOON JELLY
*Aurelia aurita*                                      109:4

**Description**
Size, bell dia. 6–10 in. (15.2–25.4 cm). Marginal lobes of bell, 8,
broad, flattened. Tentacles numerous, short, having the appearance
of a fringe around the bell margin. Color bluish-white; male
gonads yellow, female gonads pink, having a four-leaf-clover
pattern when viewed through bell.

**Habitat**
Coastal waters.

**Range**
Bay of Fundy to Cape Hatteras.

## STOMOLOPHUS MELEAGRIS                        *Fig. 125*

**Description**
Size, bell dia. 6 in. (15.2 cm). Bell rigid, shaped like mushroom cap.
No tentacles on bell margin; oral arms form short, wide, frilly mass.

**Habitat**
Coastal waters.

**Habits**
Strong swimmers, moving horizontally through water when
swimming rapidly.

**Range**
Va. and s.

*Fig. 125*

*Stomolophus meleagris*          *Leptogorgia virgulata*
(Jellyfish)                      Whip Coral, p. 632

# SEA ANEMONES AND CORALS

Class Anthozoa

Members of the class Anthozoa are entirely polypoid; there is no free-swimming medusa stage in the life cycle. The mouth leads through a short gullet to the internal gut cavity, which is partially compartment by vertical partitions (mesenteries).

In contrast to most other anthozoans, sea anemones are solitary; that is, individuals are not attached together in colonies. They also attain the largest size of any anthozoan polyps. (The colonies of other groups may be quite large, but the individual polyps are generally small.) The heavy cylindrical body of sea anemones, called the column, is surmounted by the oral disc bearing the tentacles and mouth. In many sea anemones a collarlike fold around the top of the column closes over the oral disc when the sea anemone is contracted.

The name *coral* most commonly refers to those anthozoans known as stony, madreporarian, or scleractinian corals. They are closely related to sea anemones, but differ in being colonial and in secreting a skeleton of calcium carbonate. The living colony covers or rests on top of the skeleton. The position of the polyps is indicated in the skeleton of most corals by the presence of cuplike depressions.

The *Astrangia danae* is the only common shallow-water coral along the East Coast, but in the Florida Keys and West Indies there are many species of reef corals.

## WHIP CORAL
*Leptogorgia virgulata*

109:13
*Fig. 125*

**Description**
Size, colony 1–3 ft. (30.5–91.4 cm) long. Colonies have form of whiplike rods the diameter of drinking straws. Small polyps embedded in the common tissue of the rod. Colony supported by an internal organic core and microscopic calcareous spicules. Color yellow, orange, or purple.
**Habitat**
Attached to rocks and shells in shallow water.
**Range**
Va. to Brazil.

## SEA PANSY
*Renilla reniformis*

**Description**
Size, colony about 2 in. (5.1 cm) dia. Colony rounded, leaflike, attached by a short stem, fleshy but containing microscopic spicules; small polyps embedded on upper side of "leaf." Color purple.
**Habitat**
Attached on sandy bottoms in shallow water to 40 ft. (12.2 m).
**Range**
Va. and s.

## BURROWING OR TUBE-DWELLING SEA ANEMONE
*Ceriantheopsis americanus*                                    *Fig. 126*

**Description**
Size, extended body 6 in. (15.2 cm) long, ½ in. (1.3 cm) dia.
Mouth surrounded by two circlets of 125 or more tentacles. Color
brown.
**Habitat**
Soft bottoms of sand and mud, from intertidal zone to 60 ft.
(18.3 m).
**Habits**
Lives within a soft tube buried in mud or sand. Only oral end
emerges to surface.
**Range**
Cape Cod and s.

*Fig. 126*                                    **Sea Anemones**

*Diadumene
leucolena*

*Ceriantheopsis americanus*
Burrowing Sea Anemone

*Metridium senile*
Brown Sea Anemone, p. 634

## *AIPTASIA PALLIDA AND ERUPTAURANTIA*

**Description**
Size, *A. pallida* to 1½ in. (3.8 cm) dia., *A. eruptaurantia* smaller.
Short sea anemones with one circlet of about 50 tentacles. Color
highly variable: *A. pallida* white to dark brown with white stripes
on tentacles; *A. eruptaurantia* may be orange, also with white
stripes on tentacles.
**Habitat**
On rocks and shells in shallow water.
**Range**
Cape Hatteras and s.

## *DIADUMENE LEUCOLENA*                                    *Fig. 126*

**Description**
Size, to 1 in. (2.5 cm) long. Body long, cylindrical, translucent;
mesenteries visible. Tentacles, 40–60 in 4 circlets, delicate,
tapering; inner ones are "catch tentacles," longer, thicker, and
more opaque (a distinguishing feature); central series longest.
Color white, pinkish, or greenish.
**Habitat**
Oysters, dark rock overhangs, ships' bottoms, under stones on rocky
beaches; common.
**Range**
Cape Cod to Cape Hatteras.

## BROWN SEA ANEMONE                    *Fig. 126*
*Metridium senile*

**Description**
Size, column to 4 in. (10.2 cm) high, to 3 in. (7.6 cm) dia. Adult, column dark brown blotched with white, orange, or straw, and streaked with brown. Disc, when expanded, is convex, very broad, with up to 1000 short, white, tapered tentacles, which distinguish it from *Diadumene*.

**Habitat**
Docks or rocks from low water to 300 ft. (91.4 m); common.

**Range**
Arctic seas to N.J.

## *ASTRANGIA DANAE*                   **109:14**

**Description**
Size, colony up to 4 in. (10.2 cm) wide. Skeleton with distinct cups about $\frac{3}{16}$ in. (0.48 cm) dia. Colony more or less encrusting; few tentacles. Living colony pinkish or white, translucent.

**Habitat**
Attached to stones and shells in shallow water. Broken and wave-worn skeletons often found on beaches.

**Range**
Cape Cod to Fla.

# COMB JELLIES
Phylum Ctenophora

Members of the phylum Ctenophora are somewhat similar to jellyfish in having transparent jellylike bodies. The distinguishing feature is the eight bands of comb plates composed of fused cilia, which are used in locomotion. Comb jellies are usually globular in shape with a bilateral symmetry, rather than the radial symmetry of the Cnidaria. Some have retractile tentacles but without stinging cells. The digestive system starts with a slitlike mouth that leads to a tubular stomach. This branches and leads to tubes which lie directly beneath the comb plates.

Almost all comb jellies are free-swimming or -floating and occur in vast numbers when conditions are favorable. They are carnivorous, feeding on plankton. Some species feed on other comb jellies.

## SEA WALNUT
*Pleurobrachia pileus*                 *Fig. 127*

**Description**
Size, body ¾ in. (1.8 cm), tentacles to 15 in. (38.1 cm). Oval, plump; 2 branched tentacles. Color pink, yellow, or clear.

*Fig. 127*

*Mnemiopsis leidyi*
(Comb Jelly)

*Beroe cucumis*
(Comb Jelly)

*Pleurobrachia pileus*
Sea Walnut

**Habitat**
Coastal water to deepest ocean; frequent in winter.
**Range**
Arctic seas to Long Island.

## *MNEMIOPSIS LEIDYI*                                  *Fig. 127*

**Description**
Size, to 6 in. (15.2 cm) long. Oval, with oral end divided into two
large lobes. Combs produce blue-green luminescence when animal
is disturbed at night.
**Habitat**
Shallow water to deep sea; often extremely abundant on bathing
beaches.
**Range**
Cape Cod and s.

## *BEROE CUCUMIS AND OVATA*                     *Fig. 127*

**Description**
Size, to 4 in. (10.2 cm) long, to 3 in. (7.6 cm) wide. Body bell-
shaped with neither tentacles nor lobes. Color pinkish or
translucent; comb plates iridescent.
**Habitat**
Shallow water to open ocean.
**Range**
*B. cucumis,* Arctic seas to S.C.; *B. ovata,* Cape Cod to Cape
Hatteras.

# RIBBON WORMS
Phylum Nemertea

Most ribbon worms are small, inconspicuous animals. They are
greatly flattened, but the body is elongate and more wormlike than
flatworms. They possess a long eversible proboscis that can be shot
out of the anterior to capture prey. In addition to many small
species which occupy the same sorts of habitats as the flatworms,
there are a few very large ribbon worms, such as the one described
below, that live beneath stones or burrow in sand.

## RIBBON WORM                                          *Fig. 128*
*Cerebratulus lacteus*

**Description**
Size, rarely to 6 ft. (1.8 m) long, to 1 in. (2.5 cm) wide. Head
lance-shaped; front end of worm thicker, narrower than rear end,
which has a small, slender projection, often lost. Color pink or
creamy, with internal organs showing as a brownish band.

*Fig. 128*                                          Ribbon Worm

**Habitat**
Sandy or muddy bottoms in shallow water. May be found under
flat rocks on beaches at low tide.
**Remarks**
Popular bait for hand-line fishing.
**Range**
Maine to Fla.

# MOSS ANIMALS

Phylum Bryozoa

The phylum Bryozoa contains many species of common colonial
marine animals, but they are unknown to most beachcombers
because of their minute size. The attached erect or encrusting
colonies may reach several inches high or across, but the
individuals composing the colony are less than $\frac{1}{16}$ in. (0.16 cm) in
length. Each is enclosed within a horny or calcareous exoskeleton.
Bryozoans feed on plankton and detritus, which is collected by
means of a crown of tentacles protruded through an opening in the
skeletal encasement. However, the tentacles and other body
structures can be seen only with magnification.

A few of the more common conspicuous species are described
below. The description is based on aspects of the colony which may
be seen with the naked eye.

## *CRISIA EBURNEA*                                          *Fig. 129*

**Description**
Size, colony ½–¾ in. (1.3–1.8 cm) high. Erect branching colony;
each branch composed of minute, white, outward-curving cylinders
arranged in 2 alternating rows.
**Habitat**
Attached to solid objects in shallow water to 100 ft. (30.5 m).
**Range**
Arctic to Long Island.

Note: The related species *C. denticulata* ranges from Bay of Fundy
to Florida and reaches one in. (2.5 cm) in height. It possesses a
fluffier colony with black jointed basal sections.

*Fig. 129*
**Moss Animals**

*Bugula turrita*

*Crisia eburnea*

## BUGULA TURRITA and NERITINA                    Fig. 129

**Description**
Size, colony 1–6 in. (2.5–15.2 cm) high, sometimes higher. Colony
erect, composed of branching tufts; superficially looks like seaweed,
but finely sectioned structure of each branch and brown or
purplish-brown color are distinctive.
**Habitat**
Species of *Bugula* are very common on pilings, jetties, buoys, and
other objects in shallow water.
**Range**
*B. turrita* mostly n. of Cape Hatteras and *B. neritina* s.

*Fig. 130*
**Moss Animals**

*Electra* species            *Membranipora* species

## ELECTRA AND MEMBRANIPORA (Species)      Fig. 130

**Description**
Size, colony usually less than 1 in. (2.5 cm) long. Species of these
two genera form white, lacy encrusting colonies. Separation of
species requires observation of microscopic characteristics.
**Habitat**
On algae, stones, and shells from the intertidal zone to over 1000
ft. (304.8 m).
**Range**
Entire East Coast. *Membranipora tuberculata* abundant on floating
*sargassum* (gulf weed), which is sometimes washed up on southern
beaches, more rarely in the north.

## SCHIZOPORELLA UNICORNIS                      Fig. 131

**Description**
Size, varies from small to very large since species is colonial. Heavy
encrusting colonies, wide spreading, sometimes vertically expanded.
Color whitish-orange or brick-red.
**Habitat**
On stones, shells, pilings, from low-tide mark to 100 ft. (30.5 m).
**Range**
Bay of Fundy to Fla.

*Fig. 131*

top view

side view

*Schizoporella unicornis*            *Terebratulina septentrionalis*
(Moss Animal)                  (Lamp Shell), p. 638

# LAMP SHELLS
Phylum Brachiopoda

The bivalved animals of this phylum bear a superficial resemblance to clams. However, the valves cover the upper and lower surfaces of the animal instead of the sides as in clams; thus the valves are bilaterally symmetrical, not symmetrical to each other. In articulate brachiopods (the type represented below), the upper, dorsal valve is smaller and usually flatter and the ventral valve extends behind the interlocking hinge joint. Most brachiopods are attached to the bottom by a stalk, which distinguishes them from any mollusk. A passage for the stalk is provided either by a notch or circular opening in the ventral valve. They usually are attached upside down.

## *TEREBRATULINA SEPTENTRIONALIS*    *Fig. 131*

**Description**
Size, ½ in. (1.3 cm) long. Shell oval, thin, semitransparent. Ventral valve points toward rear; dorsal valve with calcareous loop inside. Hinge line curves up in middle. Color yellowish white.
**Habitat**
Rocky bottoms, shallow water to 120 ft. (36.6 m).
**Range**
Arctic seas to Cape Cod.

# ACORN WORMS
Phylum Hemichordata

Acorn worms, which are the most conspicuous members of this small phylum of exclusively marine animals, burrow in sand and mud or live beneath stones or algae. The soft body is divided into three primary regions: the proboscis, the collar, and the trunk. The proboscis is a muscular structure attached to the collar by a narrow neck. The mouth opens widely at the anterior ventral margin of the collar, which is partly overlapping the proboscis. The trunk is the longest region of the body. It has paired gill slits in rows on either side in its anterior part and tapers toward the rear, ending with the terminal anus.

## *SACCOGLOSSUS KOWALEVSKII*    110:13

**Description**
Size, 6 in. (15.2 cm) long. Body slender, proboscis carrot-shaped, collar narrow. Odor faintly as of iodine. Color orange-yellow to brown, proboscis white to pink, collar red-orange.
**Habitat**
Fine, muddy sand in shallow water.
**Remarks**
Burrow may be marked in the intertidal zone by small coils of castings on the sand surface.
**Range**
Cape Cod and s.

## *BALANOGLOSSUS AURANTIACUS*  *Fig. 132*

**Description**
Similar to *Saccoglossus kowalevskii*, but proboscis short and
rounded; color often paler and size larger. Smells very strongly of
iodine.
**Habitat**
In sand in the intertidal zone and shallow water.
**Remarks**
Large piles of castings may be very conspicuous on a sand flat at
low tide.
**Range**
Va. and s.

*Fig. 132*    Castings of
*Balanoglossus aurantiacus*

# ECHINODERMS
Phylum Echinodermata

The members of this exclusively marine phylum are all relatively
large invertebrates and include such familiar animals as starfish,
sea urchins, and sand dollars. A five-part radial symmetry is a
distinguishing feature of echinoderms, although there are some
species of starfish in which there are more than five arms and some
other echinoderms in which the original radial symmetry has
become less apparent. All echinoderms contain an internal skeleton
of calcareous pieces, which usually includes surface spines of
various shapes and sizes. The name *echinoderm* means "spiny
skin." Also peculiar to echinoderms is the water vascular system, a
system of internal canals connected to small, hollow surface
appendages called tube feet. The tube feet are used for crawling,
feeding, or other functions and are located in five long areas
(ambulacral areas), one for each of the radial divisions of the body.

# STARFISH
Class Asteroidea

Starfish, or sea stars, have bodies composed of five tapered arms, or
sometimes more, arranged around a central disk. The lower side of
each arm bears a longitudinal ambulacral groove that contains the
tube feet. The five ambulacral grooves converge on the mouth,
located in the center of the underside of the disk. A buttonlike
structure, the madreporite, is usually conspicuous in an interradial
position to one side of the upper surface of the disk.

Starfish crawl by means of their tube feet. Some species live in
rocky habitats; others are adapted for life on sandy bottoms. Most
are carnivorous, eating clams, snails, and other echinoderms. They
feed either by swallowing their prey whole or by everting their
stomachs out of their mouths over the prey. Some species, such as
the common *Asterias* of the Atlantic coast, are able to slither their
everted stomachs between the valves of oysters and clams.

## ASTROPECTEN ARTICULATUS          Fig. 133

**Description**
Size, 4–12 in. (10.2–30.5 cm) dia. Distinguished by large, pale
marginal plates that border the arms and bear fringelike spines.
Surface smooth. Color variable, often orange to mauve or blue
above, pale below.
**Habitat**
Sandy bottoms from 20–400 ft. (6.1–121.9 m).
**Range**
N.J. to Mexico.

Fig. 133
**Starfish**

*Astropecten
articulatus*

*Luidia clathrata*

## LUIDIA CLATHRATA          Fig. 133

**Description**
Size, 4–10 in. (10.2–25.4 cm) dia. Like *Astropecten,* has a smooth
surface, but lacks the conspicuous marginal plates. Color rosy
orange to gray.
**Habitat**
Sandy bottoms, shallow water to 500 ft. (152.4 m).
**Range**
N.J. to Brazil.

## HENRICIA SANGUINOLENTA          109:9

**Description**
Size, 2½–4 in. (6.5–10.2 cm) dia. Radius of disk ⅙ length of arms.
Arms slender, smoothly rounded. Upper parts usually blood-red or
purple, often mottled, sometimes orange or creamy; underparts
yellow.
**Habitat**
Rock crevices from shore to 600 ft. (182.9 m).
**Range**
Greenland to Cape Hatteras.

## PURPLE SUN STAR
*Solaster endeca*          Fig. 134

**Description**
Size, 8½–16 in. (21.6–40.6 cm) dia. Arms, 7–13, usually 10,
slender, as long as disk is wide. Upper surface red-purple;
madreporic plate and margins yellow; underside white.
**Habitat**
Shallow water to 600 ft. (182.9 m).
**Range**
Greenland to Cape Cod.

## COMMON SUN STAR
*Solaster papposus*                                              *Fig. 134*

**Description**
Size, 8–15 in. (20.3–38.1 cm) dia. Arms, 8–14, shorter than disk
dia. Disk large; dorsal spines large, tufted. Single series of
marginal spines very prominent. Disk with concentric color bands
often of scarlet, crimson, or orange, sometimes with purple tones;
arms banded pink or white, tips crimson; underside whitish.
**Habitat**
Shoreline to depth of 1 mi. (1.609 km).
**Range**
Arctic seas to N.J.

*Fig. 134*

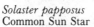
*Solaster papposus*       *Solaster endeca*
Common Sun Star          Purple Sun Star

## *ASTERIAS FORBESII AND VULGARIS*                    **109:10**
**Description**
Size, 6–12 in. (15.2–30.5 cm) dia. Arms, 4–7, about twice as long
as disk dia.; tuberculate spines give rough surface. Color brown to
purple, green, or orange; madreporite of *A. forbesi* red-orange, of
*A. vulgaris* light yellow.
**Habitat**
Rocky and shelly bottoms, shallow water to over 150 ft. (45.7 m).
**Range**
*A. forbesi* n. of Cape Cod, *A. vulgaris* s.

# BRITTLE STARS
Class Ophiuroidea

Brittle stars, or serpent stars, are similar to starfish in possessing
arms, but the arms are long, slender, and jointed and sharply
distinct from the small, round central disk. In the basket stars,
which are not included here, the arms are branched. There are no
ambulacral grooves on the underside of the arms, and the small,
inconspicuous tube feet are used in feeding and respiration rather
than in locomotion. The madreporite is on the lower side of the
disk and is not easily seen.

Brittle stars move by pushing with their flexible arms and are the
most agile of all echinoderms. They will readily sever one or more
arms if seized. Although common shallow-water animals, they hide
beneath stones, shells, and other objects and are thus not as
conspicuous as many other echinoderms. Brittle stars are
scavengers or feed on fine bottom detritus.

Species of brittle stars are differentiated by fine structural details,
most of which cannot be seen with the naked eye. The following
four species are commonly encountered in shallow water.

## DAISY BRITTLE STAR
*Ophiopholis aculeata*                                        *Fig. 135*

**Description**
Size, disk ½–1 in. (1.3–2.5 cm) wide, arms 3–3½ in. (7.6–8.9 cm)
long. Disk bulges between the arms, giving it an almost circular
outline. Spines on arms conspicuous, stand out from the arm.
Colors vary from rusty-red to blue, green, or brown. Disk usually
mottled; arms banded.
**Habitat**
Low water to depth of 1 mi. (1.6 km).
**Range**
Arctic seas to Long Island.

*Fig. 135*                     **Brittle Stars**

*Ophiopholis aculeata*     *Amphipholis*        *Ophiothrix*
Daisy Brittle Star         *squamata*           *angulata*

## AMPHIPHOLIS SQUAMATA                                *Fig. 135*

**Description**
Size, disk ⅛ in. (0.5 cm) wide, arms 1¼ in. (3.1 cm) long. Arms
with prominent lateral spines. Color gray; white spot on base of
each arm.
**Habitat**
Shoreline to 350 ft. (106.7 m).
**Range**
Arctic seas to Long Island.

## GREEN BRITTLE STAR
*Ophioderma brevispinum*                              **109:12**

**Description**
Size, disk ½ in. (1.3 cm) dia., arms 3 in. (7.6 cm) long. Arms, 4–6,
usually 5; disk 5-sided; spines very short, usually lying flat against
arm. Color dark greenish-brown to clear green, often banded.
**Habitat**
Tide pools, sand flats; to depth of 600 ft. (182.9 m).
**Range**
Cape Cod to Brazil.

## OPHIOTHRIX ANGULATA                        **109:11**, *Fig. 135*

**Description**
Size, disk ½ in. (1.3 cm) dia. Arms with long spines. Color highly
variable; brown, red, green, mottled and banded; white stripe on
upper surface of each arm, especially toward tip.
**Habitat**
Beneath stones and shell clumps from shallow to deep water.
**Range**
Va. and s.

# SEA URCHINS AND SAND DOLLARS

Class Echinoidea

The body of echinoids is not drawn out into arms but rather is globular and disk-shaped. The skeletal pieces are fused together to form a rigid internal shell, or test, parts of which are commonly picked up on beaches or sand flats. The surface of the test bears tubercles upon which are mounted movable spines, long and heavy in sea urchins and minute in sand dollars. The thousands of tiny spines give sand dollars a furry look. The five ambulacral areas containing the tube feet run over the body of echinoids as five meridians, converging on the mouth in the center below and on the anus above. Each tube foot connects to the interior parts of the water vascular system by a pair of canals through the test, which can be seen in a dried clean sea urchin test as paired perforations.

Most sea urchins are adapted for life on hard bottoms and, like starfish, move by means of their tube feet. They may also push with their spines and erect them in defense. Sea urchins are scavengers, scraping the bottom with their complex jaw apparatus, called Aristotle's lantern, which can be seen projecting slightly from the mouth.

Sand dollars are adapted for life on soft bottoms and use their minute spines to burrow partially or completely below the sand surface. The tube feet are used in respiration and in moving food. The five conspicuous petallike areas on the upper surface of the test mark the position of the respiratory tube feet. Sand dollars feed on fine detritus sifted out between the spines and transported by cilia and the tube feet to the mouth. Grooves for food transport can be seen on the underside of the test. The calcareous piece found in the center of the sand dollar when it is broken open is the vestigial jaw apparatus.

## PURPLE SEA URCHIN
*Arbacia punctulata*                                              **109:8**

**Description**
Size, test, 2 in. (5.1 cm) wide, spines, 1 in. (2.5 cm) high. Color purple to dark purplish brown. Test domed; center of dome and surrounding area bare; rest of test almost hidden by sturdy spines.
**Habitat**
Seaweed in tide pools; rocky and shelly bottoms in shallow water.
**Range**
Cape Cod to Gulf of Mexico.

## KEYHOLE URCHIN
*Mellita quinquiesperforata*                                     **109:7**

**Description**
Size, test to 4½ in. (11.4 cm) wide. Round, flattened on rear margin. Pattern of tube feet clear, somewhat irregular. Test has 5 slots (from above); 4 stretch from petal tips to margin; 1 is between petals.
**Habitat**
Sandy bottoms in shallow coastal waters.
**Range**
Md. to the West Indies.

## GREEN SEA URCHIN
*Strongylocentrotus droebachiensis*                    *Fig. 136*

**Description**
Size, test to 3½ in. (8.9 cm) wide, spines ½ in. (1.3 cm) long. Test
somewhat flattened; spines equal in length, crowded, completely
covering test. Test greenish brown, spines bright green.
**Habitat**
Tide pools and rocky bottoms to 4000 ft. (1219.2 m); common.
**Range**
N.J. (deep water); Cape Cod (shallow water) and n.

*Fig. 136*

alive                    dead

*Strongylocentrotus droebachiensis*      *Echinarachnius parma*
Green Sea Urchin                          Sand Dollar

## WHITE SEA URCHIN
*Lytechinus variegatus*

**Description**
Size, test to 2 in. (5.1 cm) wide, spines about ½ in. (1.3 cm) or
less. Spines shorter and more slender than those of *Arbacia*. Color
white, sometimes greenish (in the West Indies, very dark purple).
**Habitat**
Shelly and rocky bottoms, shallow water to 150 ft. (45.8 m).
**Range**
Cape Hatteras to the West Indies.

## SAND DOLLAR
*Echinarachnius parma*                    *Fig. 136*

**Description**
Similar to the sand dollar above but lacks slots.
**Habitat**
Sandy bottoms in shallow water.
**Range**
Bay of Fundy to Cape Hatteras.

## HEART URCHIN
*Moira atropos*                    *Fig. 137*

**Description**
Size, 2¼ in. (5.7 cm) long. Spines short and numerous as in sand
dollars but body not flattened and slightly longer than wide. Color
brown.
**Habitat**
Burrows beneath sand surface from low-tide mark to 450 ft.
(137.2 m).
**Range**
Va. (s. of Chesapeake Bay) to Fla.

# SEA CUCUMBERS
Class Holothurioidea

Sea cucumbers are elongated along the radial axis, which runs
between the mouth and anus, giving the body a cucumber- or
worm-like shape. These animals lie on their side rather than on
the oral end as do sea urchins. The body has a leathery texture, for
the skeleton is reduced to microscopic pieces.

Sea cucumbers creep over the bottom using the rows of tube feet,
which run down the length of the body, or they burrow in sand
and mud. They feed on bottom detritus or plankton collected by
means of a crown of tentacles representing modified tube feet,
which can be extended from around the mouth. The tentacles are
usually contracted and not easily seen when the animal is collected.

## *PENTAMERA PULCHERRIMA*                  109:16

**Description**
Size, 1½–2 in. (3.8–5.1 cm) long. Gherkin-shaped, narrowing at
ends. Tentacles, 10, branched; lower 2 shorter. Tube feet in 2 rows
on each of 5 tracts. Color yellowish or white.
**Habitat**
Shallow, muddy bottom with eelgrass.
**Range**
Cape Cod to the Carolinas.

## *SCLERODACTYLA BRIAREUS*                  *Fig. 137*

**Description**
Size, 6 in. (15.2 cm) long. Tentacles, 10, treelike. Body saclike,
with tube feet scattered all over the surface, giving it a furry
appearance. Color greenish, brown, or black.
**Habitat**
Buried in muddy sand in shallow water.
**Range**
Cape Cod and s.

*Fig. 137*

*Leptosynapta inhaerens*
(Sea Cucumber)

*Moira atropos*
Heart Urchin

*Sclerodactyla briareus*
(Sea Cucumber)

*Caudina arenata,* p.646
(Sea Cucumber)

## *LEPTOSYNAPTA INHAERENS*                  *Fig. 137*

**Description**
Size, 5–9 in. (12.7–22.8 cm) long, and to ⅜ in. (0.96 cm) dia.
Delicate wormlike body. Tube feet absent; ambulacral areas
marked by five longitudinal lines extending length of body.
Tentacles, 12, each having 5–7 side branches. Color transparent to
whitish.
**Habitat**
Within muddy sand, shallow water to 600 ft. (182.9 m).
**Range**
Bay of Fundy to the Carolinas.

### CAUDINA ARENATA                                    *Fig. 137*

**Description**
Size, 4–7 in. (10.2–17.8 cm) long, ½–1 in. (1.3–2.5 cm) dia.
Forward portion of body chunky; terminal ⅓ slender, taillike. Skin
smooth. Mouth surrounded by 15 short, stubby tentacles, split
crosswise at tips, each forming 4 fingerlike projections. Tube feet
absent; 5 clusters of papillae around anus. Color translucent
pinkish to purple.
**Habitat**
Burrowing, with tail protruding, in mud or sand from shoreline to
100 ft. (30.5 m).
**Range**
Bay of Fundy to Cape Cod.

# SEGMENTED WORMS
Phylum Annelida

Members of the phylum Annelida have the body divided into a
linear series of similar parts, or segments. At the anterior end of
the body is the nonsegmental prostomium, which makes up most of
the head and lies just in front of the ventral mouth. Behind the
prostomium lies the peristomium (surrounding the mouth), which
is in turn followed by the trunk segments. The trunk segments
usually carry similar external and internal structures. Annelid
worms are found in the sea, in fresh water, and in soil.
Earthworms are familiar examples of terrestrial species.

# MARINE SEGMENTED WORMS
Class Polychaeta

Of approximately 9000 described species of segmented worms, two-
thirds live in the sea. Most of these marine annelids belong to the
class Polychaeta, which not only is the largest class but contains
the greatest diversity of forms. Polychaetes are distinguished from
other annelids by the pair of fleshy lateral projections, or
parapodia, carried by each trunk segment. The parapodia contain
bundles of fine horny rods, or bristles, called setae, from which the
name *polychaeta,* or "many setae," is derived. A head is often well
developed in polychaetes and may bear eyes, antennae, and other
sensory projections. In some polychaetes the head carries elaborate
feeding appendages, such as tentacles or featherlike projections.

Many different life styles are found among polychaetes. Some
species live beneath stones, in rock crevices, on algae, or among
other sessile animals; some burrow in sand or mud; and others live
within tubes either secreted by the worm or composed of sand
grains or other materials cemented together. The tubes are planted
in mud or sand or attached to various objects, and they are usually
much more conspicuous than their inhabitants. Polychaetes use
their parapodia for crawling or gripping the sides of their burrow

The feeding habits of polychaetes vary greatly. Some are carnivores
or algae eaters and possess jaws for seizing and tearing. Many
consume detritus or even mud and sand from which the detritus is
digested. Some are filter feeders, removing plankton by passing a
current of water over some part of the body (the filter).

The following species are a few of the larger, more conspicuous
polychaetes that might be encountered. Identification has been
limited to family where that category is more easily recognized
than species.

## SEA MOUSE
*Aphrodita hastata*                                          *Fig. 138*

**Description**
Size, to 7 in. (17.8 cm) long, 2 in. (5.1 cm) wide. Body elliptical.
Head end bears 1 tentacle and 2 long palps. Back slightly arched,
with dense coat of long setae. Setae gray on top of back, golden
green and iridescent on sides.
**Habitat**
In mud.
**Range**
Gulf of St. Lawrence to Chesapeake Bay.

*Fig. 138*

*Aphrodita hastata*
Sea Mouse

*Lepidonotus squamatus*
(Scale Worm)

## SCALE WORMS
Families Polynoidae and Sigalionidae                        *Fig. 138*

**Description**
Size, ¼–3 in. (0.64–7.6 cm) long. Upper surface of body covered by
flat overlapping scales. Color varies; black, grays, browns common.
**Habitat**
Beneath stones, in rock crevices, among oyster or mussel shells, in
algae.
**Range**
Entire E. Coast.

## CLAM WORMS
Family Nereidae
**110:12**
*Fig. 139*

**Description**
Size, 2 in. (5.1 cm) to many inches long. Distinguished by large,
flat parapodia and a distinct head structure. (The latter, however,
may be difficult to discern without a hand lens.)
**Habitat**
Burrows in muddy sand in shallow water.
**Range**
Bay of Fundy to Va.

Note: The **GREEN CLAM WORM,** *Nereis virens,* is a large greenish
species reaching a length of eighteen inches, and is commonly used
as bait.

retracted pharynx

everted proboscis

*Fig. 139*
**Clam Worms**

*Nereis virens*        *Glycera dibranchiata,* p. 648

## MARINE SEGMENTED WORMS

### *DIOPATRA CUPREA*                                    110:14

**Description**
Size, 12 in. (30.5 cm.) long, ½ in. (1.3 cm) wide. Body with 5
tentacles in front, tapering rows of feathery gills on front ⅓ of
body. Color bluish green with metallic iridescence, tentacles pale,
gills bright red.
**Habitat**
Shallow water or intertidal zone.
**Remarks**
The tube of this worm is commonly abundant and conspicuous in
the intertidal zone or protected beaches or sand flats. About the
size of a pencil, it is shaped like a ship's funnel and is adorned
with bits of algae and shell. About an inch or so of the tube sticks
up above the sand surface.
**Range**
Cape Cod and s.

## BLOOD WORMS
*Glycera dibranchiata* and *americana*                  *Fig. 139*

**Description**
Size, 6–8 in. (15.3–20.3 cm) long. Cylindrical body, pointed at
anterior end. Easily recognized by long proboscis that is
periodically shot out of the mouth when the worm is handled.
Proboscis armed at tip with 4 black jaws.
**Habitat**
Muddy sand from intertidal zone to deep water; live in a relatively
complex burrow system.
**Remarks**
Widely used for bait.
**Range**
Cape Cod to the Carolinas.

## LUGWORMS
*Arenicola* (species)                                   *Fig. 140*

**Description**
Size, 6–12 in. (15.2–30.5 cm) long, dia. up to that of a finger.
Body large, cylindrical; parapodia reduced to ridges; 11–13 pairs of
lateral gills. Color light tan, green, greenish black.
**Habitat**
Intertidal zone. Live in L-shaped burrows in sand; top of burrow
often marked by castings. Large transparent gelatinous egg masses,
oval or elongate, commonly attached to top of burrow.
**Range**
Entire E. Coast.

*Fig. 140*

Exposed end of
sand grain tube

*Arenicola marinus*
Lugworm

*Clymenella torquata*
Bamboo Worm

648

## *CHAETOPTERUS VARIOPEDATUS*     *Fig. 141*

**Description**
Size, 6 in. (15.2 cm) long. Strange-looking, flaccid worm; body
with 3 sections, flattened anterior region, middle region with
semicircular parapodia, posterior section with rather elongate
parapodia. Most easily recognized by secreted parchmentlike tube,
U-shaped, 1–1½ ft. (30.5–45.7 cm) long, about 1½ in. (3.8 cm)
dia. at middle; ends, about size of large drinking straws, stick ½–1
in. (1.3–2.5 cm) above sand surface, about 1 ft. (0.3 m) apart.
**Habitat**
Sandy protected bottoms from the low-tide mark to 20 ft. (6.1 m).
**Range**
Cape Cod to Fla.

Note: Living in the tube with *Chaetopterus* are often found both
sexes of the little commensal crabs *Pinnixa chaetopterana* or
*Polyonyx gibbesi*. The latter species can be recognized by the very
small fifth pair of legs.

**Marine Segmented Worms**

*Chaetopterus variopedatus*

*Fig. 141*

*Pectinaria gouldii*
Trumpet Worm, p. 650

*Sabellaria vulgaris*, p. 650

## BAMBOO WORMS
*Clymenella torquata* and *Axiothella mucosa*     *Fig. 140*

**Description**
Size, to 4 in. (10.2 cm) long. Elongated segments give worm the
appearance of a bamboo cane. Straight sand-grain tubes up to size
of large matchsticks planted upright in sandy bottoms; end of tube
projects about ½ in. (1.3 cm) above sand surface.
**Habitat**
Sandy bottoms in intertidal zone and shallow water. *C. torquata*
found to 300 ft. (91.4 m).
**Range**
*C. torquata*, Bay of Fundy to the Carolinas; *A. mucosa*, Cape
Hatteras and s.

## *AMPHITRITE ORNATA*     110:15

**Description**
Size, 12–15 in. (30.5–38.1 cm) long. Body bears great mass of
coiling tentacles at anterior end and 3 pairs of branched gills on
dorsal side just behind tentacles. Color pinkish, gills red.
**Habitat**
Lives in mucus-lined burrow in muddy bottoms from shallow
water to intertidal zone.

**Remarks**
*Amphitrite* belongs to the family Terebellidae, whose members can easily be recognized by the mass of coiling tentacles at the anterior end.
**Range**
Cape Cod and s.

Note: The similar *johnstoni* has a brownish body with about 24 pairs of setae clusters in anterior body region. It occurs from the Arctic to Cape Cod.

## TRUMPET WORM
*Pectinaria gouldii*                                    *Fig. 141*

**Description**
Size, tube to 1½ in. (3.8 cm) long, ¼ in. (0.64 cm) wide; worm slightly longer than tube in which it lives. Worm with 2 feathery gills on each side of head. Head with oblique double coronet of stiff bristles for digging; when withdrawn form a defensive phalanx. Tentacles numerous and in a cluster; each flattened, folded lengthwise to form a groove underneath. Body flesh-colored, mottled; gills bright red; bristles golden; tentacles flesh-colored. Tube trumpet-shaped, formed of single layer of large and small sand grains.
**Habitat**
Buried in sand and mud from low water to 60 ft. (18.3 m).
**Range**
Maine to Fla.

## *SABELLARIA VULGARIS*                              *Fig. 141*

**Description**
Size, less than 1 in. (2.5 cm) long. Worm with concentric rows of yellow setae at anterior end. Firm sand-grain tubes plastered on shells and stones, frequently massed in colonies.
**Habitat**
Intertidal zone and shallow water.
**Range**
Cape Cod to Va.

## FEATHER DUSTER WORMS, FAN WORMS, PEACOCK WORMS
Families Sabellidae and Serpulidae

**Description**
Two families of tube-dwelling polychaetes in which the anterior end bears a crown of long featherlike projections, or radioles. (The radioles are arranged in a funnel in many species and as two spirals in others.) Most easily distinguished by their tubes.
**Habitat**
Most members of both families attach their tubes to firm objects in shallow water.
**Habits**
The sabellids secrete membranous tubes or ones composed of foreign particles; serpulids secrete white tubes of calcium carbonate, and one radiole is modified as a stopper.
**Range**
Both worldwide.

Note: Two common serpulids are *Hydroides dianthus* (*Fig. 142*), which builds twisted chalky tubes on shells or stones, and *Spirorbis borealis,* which constructs little spiral tubes resembling snail shells attached to seaweeds. *S. borealis* (*Fig. 142*) occurs from Bay of Fundy to Cape Cod, but other species of *Spirorbis* are found beyond that range.

*Fig. 142*

*Spirorbis borealis*
(Segmented Worm)

*Hydroides dianthus*
(Segmented Worm)

*Echiurus pallasii*
(Echiuran Worm)

# ECHIURANS
Phylum Echiura

The Echiura is a small phylum of worms that differ from their relatives the annelids in lacking segmentation and in possessing a fleshy, shovel-shaped anterior end, called the proboscis. The trunk is somewhat sausage-shaped. Echiurans burrow in mud and sand or inhabit rock crevices. One species along the east coast of the United States lives within old sand-dollar tests.

## *ECHIURUS PALLASII*                                        *Fig. 142*

**Description**
Length, 1–12 in. (2.5–30.5 cm). Proboscis fleshy, spoon-shaped. Body gray, yellow, or orange; proboscis light yellow with orange grooves.
**Habitat**
Mud, sand, and rock bottoms in shallow water.
**Range**
Arctic to the Carolinas.

# ARTHROPODS
Phylum Arthropoda

The phylum Arthropoda contains the spiders, mites, crabs, shrimps, insects, centipedes, and millipedes and is the largest phylum of animals. There are more species of arthropods than of all other animals combined. The body of arthropods is covered by an external skeleton containing a hard substance called chitin. This exoskeleton, or carapace, is the so-called shell of crabs and shrimps. Periodic molting, or shedding, of the exoskeleton permits growth to occur. The phylum is characterized by jointed legs, from which the name *arthropod*, or "jointed feet," is derived. Arthropods are segmented, but the segmentation is often modified and not as distinct as in annelids.

# HORSESHOE CRABS
Class Merostomata

The members of this little group of arthropods are not crabs at all but are related to the terrestrial scorpions and spiders. Like their relatives, they lack antennae, and the head and anterior part of the trunk are fused together as a cephalothorax. The cephalothorax has the outline of a horseshoe and bears a pair of large lateral eyes. Behind the cephalothorax is the large triangular abdomen, which carries short lateral spines and a large, terminal, spikelike telson. On the underside of the cephalothorax there are six pairs of leglike appendages, the first five with claws and the sixth modified for clearing the body undersurface of sand. On the underside of the abdomen are leaflike gills.

Horseshoe crabs live on sandy bottoms in shallow water, where they feed on soft-bodied burrowing animals. They crawl with their legs but may use the telson for pushing or righting themselves if turned over. Horseshoe crabs are harmless animals, and the telson makes a convenient handle with which to pick them up or carry them.

## HORSESHOE CRAB
*Limulus polyphemus*                                        *Fig. 143*

**Description**
Size, to 28 in. (71.1 cm) long or more. Adults dark brown, young yellowish.
**Habitat**
Sandy bottoms in shallow water.
**Range**
N.S. to Fla.

*Fig. 143*                                        *Limulus polyphemus*
                                                  Horseshoe Crab

# CRUSTACEANS
Class Crustacea

Crustaceans are distinguished from other arthropods by their two pairs of antennae. They are largely aquatic animals and are the principal representatives of the phylum in the sea. The great diversity in form and size ranges from microscopic wormlike species which live between sand grains to certain crabs with 10-ft. (3.05-m) leg spans.

# BARNACLES
Subclass Cirripedia

Barnacles are the only crustaceans that are attached, living on stones, pilings, and other objects. The larva, which is free-swimming, attaches at the end of its planktonic life by means of a cement gland at the base of the antennae. The fixed animal then becomes encased within protective calcareous plates, another distinctive feature of the group. Two pairs of these plates are movable and guard the opening into the mantle cavity, which houses the legs and other parts of the body. The legs, or cirri, bear long bristles (setae) and can be extended out of the opening like scoops, collecting plankton as food.

Not surprisingly, there are species of barnacles which attach to other animals, such as crabs, turtles, and whales, and species which are parasitic. There are also some barnacles which bore into coralline rock or old shells.

# Stalked Barnacles
Suborder Lepadomorpha

## GOOSE BARNACLES
*Lepas* (species)                                                         110:9
**Description**
Size, about 2 in. (5.1 cm) long. Only 5 calcareous plates; encased part of body laterally compressed. Attach by a long, heavy, naked stalk.
**Habitat**
Attached to floating objects, such as timbers and bottles, often in dense clusters.
**Range**
Entire E. Coast; most species cosmopolitan.

# Acorn Barnacles
Suborder Balanomorpha

The barnacles below lack a stalk and are called sessile or acorn barnacles. The two pairs of movable plates are surrounded by a ring of vertical plates forming a rigid circular wall. The attachment surface is called the basis.

## CHTHAMALUS FRAGILIS
**Description**
Size, to ⅜ in. (0.96 cm) dia. Basis membranous. Plates brown or gray.
**Habitat**
Marsh grass, rocks, pilings in intertidal zone.
**Range**
Cape Cod to the Carolinas.

## BALANUS BALANOIDES                                          Fig. 144
**Description**
Size, to ⅜ in. (0.96 cm) dia. Basis membranous. Plates white, smooth.
**Habitat**
Rocks, pilings in intertidal zone.
**Range**
Bay of Fundy to Cape Hatteras.

## *BALANUS EBURNEUS*                    *Fig. 144*

**Description**
Size, to 1 in. (2.5 cm) dia. Basis calcareous. Plates white, smooth.
**Habitat**
Rocks, pilings in intertidal zone; will tolerate low salinities.
**Range**
Maine to West Indies.

*Fig. 144*
**Acorn Barnacles**

*Balanus eburneus*     *Balanus balanoides,*
                        p. 653

*Balanus improvisus*     *Balanus crenatus*     *Balanus amphitrite*

## *BALANUS CRENATUS*                    *Fig. 144*

**Description**
Size, to 1 in. (2.5 cm) dia. Basis calcareous. Wall plates white,
rough; edges of wall sharp, saw-toothed.
**Habitat**
Shells, rocks from tide mark to 300 ft. (91.4 m).
**Range**
Bay of Fundy to Long Island Sound.

## *BALANUS IMPROVISUS*                    *Fig. 144*

**Description**
Size, to ½ in. (1.3 cm) dia. Basis calcareous. Wall plates white,
surface and edges smooth.
**Habitat**
Rocks, pilings in intertidal zone; tolerates water of low salinity.
**Range**
N.S. to Patagonia.

## *BALANUS AMPHITRITE*                    *Fig. 144*

**Description**
Size, to ⅜ in. (0.96 cm) dia. (greater in south). Basis calcareous.
Wall plates smooth, tinged or striped with pink, red, or purple.
**Habitat**
Rocks, pilings in intertidal zone to 150 ft. (45.7 m).
**Range**
Cape Cod to the West Indies.

# Isopods
Order Isopoda

Members of this order have somewhat dorsoventrally flattened bodies. Viewed from above, the anterior half, or thoracic region, bearing the seven pairs of legs, is not sharply demarcated from the posterior half, or abdomen, which carries flattened respiratory appendages. The young are brooded in a marsupium beneath the thorax.

Sow bugs, pill bugs, and wood lice are familiar land forms, but the far greater number of isopods are marine, either free-living or parasitic. The latter are often twisted or distorted with one side shorter than the other.

The following species are a few common forms that may be encountered in the intertidal zone or in shallow water.

## FISH LOUSE
*Lironeca ovalis*                                              *Fig. 145*

**Description**
Size, 1 in. (2.5 cm) long, ½–¾ in. (1.3–1.8 cm) wide. Body oval, flattened, asymmetrical; foremost pair of antennae widely separated. Thoracic segments, 7; head compactly set in the first; legs, doubly curved, hooked, prehensile.
**Habitat**
Coastal waters where bluefish and other hosts abound.
**Habits**
Parasitic on the gills and in the mouths of fish.
**Range**
Cape Cod to Tex.

*Fig. 145*
**Isopods**

Gammarus locusta
Seaweed Hopper, p. 657

male

Talorchestia
longicornis, p. 657

Lironeca ovalis
Fish Louse

Limnoria lignorum
Gribble

Idotea baltica, p. 656

## GRIBBLES
*Limnoria lignorum* and *tripunctata*                          *Fig. 145*

**Description**
Size, ³⁄₁₆ in. (0.48 cm) long. Body 3 times longer than wide. Color gray.
**Habitat**
Wood borers, attacking the surface layers of pilings and other submerged timbers in shallow water.
**Range**
*L. lignorum,* Newfoundland to N.C.; *L. tripunctata,* se. U.S.

## *IDOTEA* (Species)         *Fig. 145*

**Description**
Size, to ¾ in. (1.8 cm) long. Abdomen appears from above to be composed of three segments counting terminal telson. Eyes round, large. Color gray, brown, greenish.
**Habitat**
In algae, eelgrass, gravel bottoms, swimming in shallow to deep water.
**Range**
Entire E. Coast.

## *SPHAEROMA QUADRIDENTATUM*      *Fig. 146*

**Description**
Size, ⅜ in. (0.96 cm) long. Body oval in outline; abdomen from above with only two visible segments counting terminal telson. Color gray.
**Habitat**
On pilings, in old barnacle shells, on algae, beneath stones, from intertidal zone to 3 ft. (91.4 cm).
**Habits**
Capable of rolling up into a ball.
**Range**
Cape Cod to Key West, Fla.

*Fig. 146*
**Isopods**

*Lygia oceanica*            *Sphaeroma quadridentatum*

## *LYGIA OCEANICA AND EXOTICA*     *Fig. 146*

**Description**
Size, to 1 in. (2.5 cm) long. Resemble large pill bugs. Color gray.
**Habitat**
Intertidal and supratidal zone, but animals always stay above water level.
**Habits**
Run rapidly over docking, pilings, and stones above the water line.
**Range**
*L. oceanica*, n. of Cape Cod; *L. exotica*, Del. and s.

# Amphipods
Order Amphipoda

In members of this order the body is usually laterally compressed and the axis strongly curved. They may be distinguished from small shrimps and prawns by the absence of a carapace, by the body that is jointed and flexible along its entire length, and by their smaller size (most are less than one-half inch long). Some of the legs (usually the first two, called gnathopods) are subchelate; that is, the last segment, the finger, closes against the end of the preceding segment. As in isopods, the young are carried in a marsupium beneath the thorax.

There are over 5500 species in this order, almost exclusively marine. Many are found in the beach wrack or climbing on hydroids and ascidians attached to rocks and pilings. Identification of species is difficult without magnification and knowledge of technical characters. The following groups can be recognized most easily by their habitat.

## SEAWEED HOPPERS, SCUDS
Family Gammaridae

*Fig. 145*

**Description**
Size, ¼–1 in. (0.64–2.5 cm) long. Color brown, olive, mottled.
**Habitat**
Common within algae and under stones in intertidal zone and very shallow water, i.e. *Gammarus locusta*.
**Habits**
Scurry rapidly, listing to one side.
**Range**
Entire E. Coast.

## BEACH FLEAS, SAND FLEAS, BEACH HOPPERS
Family Talitridae

**109:15**
*Fig 145*

**Description**
Size, ¼–1 in. (0.64–2.5 cm) long. Of the two genera, *Orchestia* and *Talorchestia*, the latter reaches the largest size; species of *Orchestia* are no longer than ½ in. (1.3 cm). Color waxy-white to brown or gray.
**Habitat**
High-tide mark; semiterrestrial, living beneath seaweed and driftwood washed up on beaches by extremely high tides.
**Habits**
Ability to jump is a distinctive feature (very noticeable when drift is turned over).
**Range**
Entire E. Coast.

## WOOD-BORING AMPHIPOD
*Chelura terebrans*

*Fig. 147*

**Description**
Size, ¼–⅜ in. (0.48–0.64 cm) long.
**Habitat**
Bores into pilings and other submerged marine timbers much like the isopod *Limnoria*.
**Range**
Cape Cod to Cape Hatteras.

*Fig. 147*

*Chelura terebrans*
Wood-boring Amphipod

*Aeginella longicornia,* p. 658
(Skeleton Shrimp)

## SKELETON SHRIMP
Family Caprellidae                                                    *Fig. 147*

**Description**
Size, to 1 in. (2.5 cm.) long, but usually less. Strange-looking
aberrant amphipods with long, narrow, rather cylindrical bodies
with a greatly reduced abdomen.

**Habitat**
On attached organisms in shallow and deep water; may be
especially abundant on hydroids and such erect bryozoans as
*Bugula.*

**Habits**
Climb about various attached organisms, using the grasping claws
at the ends of their legs.

**Range**
Entire E. Coast.

# Shrimps, Lobsters, and Crabs
Order Decapoda

These are the most highly organized crustaceans. The head and
thorax are rigidly fused to each other and form the cephalothorax,
which is covered by the carapace. The anterior tip of the head
often projects forward as a rostrum (snout), like the prow of a
ship. The ten pairs of legs give this order its name; although some
terminate in pincers, most of them are adapted for walking or
swimming. The majority of this group are marine, but some have
invaded fresh water, and a few live on land. Decapods include
familiar shrimps, prawns, lobsters, crayfishes and crabs. This
diverse order is divided into a number of sections and tribes,
encompassed by the suborders Natantia and Reptantia.

# Swimming Decapods
Suborder Natantia

## EDIBLE SHRIMP
*Penaeus aztecus, setiferus,* and *duorarum*

**Description**
Size, body to 6 in. (15.2 cm) long, antennae to 12 in. (30.5 cm).
The largest shrimps on the East Coast. Rostrum ⅔ length of
carapace, forms a low keel on the carapace with lateral grooves.
Color brownish green.

**Habitat**
Shallow water to about 300 ft. (91.4 m); tolerate low salinities.

**Range**
All three species occur from the Chesapeake to the Gulf of Mexico,
but *P. aztecus* extends to Cape May, N.J., and *P. setiferus* to Long
Island Sound.

## SAND SHRIMP
*Crangon septemspinosa*                                        **110:11**

**Description**
Size, to 2 in. (5.1 cm) long. Body somewhat tubular, tapers to tail;
gray-green, translucent on sand, darker on mud. Rostrum short,
front end blunt; this and the comparatively large size help identify

this animal when swimming, although its translucency makes it hard to see.

**Habitat**
Mud, sand, surf, drainage ditches.

**Remarks**
This species is the food of many fishes. Other members of the genus are eaten by man, more so on the West Coast and abroad than in eastern North America.

**Range**
Labrador to N.C.

## GRASS SHRIMP

*Palaemonetes* (species)                                    *Fig. 148*

**Description**
Size, 1 in. (2.5 cm) long. The three eastern species are difficult to separate and have similar ranges. Tip of rostrum forked or unforked, with teeth on underside; two sets of small claws, the second slightly larger; first segment of abdomen partly hidden by plate of second segment. Color more or less transparent, spotted with brown.

**Habitat**
Eelgrass, in ditches, salt marshes.

**Range**
Maine to Tex.

*Fig. 148*                    *Palaemonetes vulgaris*
                             Grass Shrimp

## SNAPPING SHRIMP

Family Alphaeidae

**Description**
Size, to 2 in. (5.1 cm) long, but usually smaller. Distinguished by the single greatly enlarged claw.

**Habitat**
Among oyster shells, within rock crevices.

**Remarks**
The structure and closing mechanism of the claw is such that it produces a popping sound. The claw functions in threat displays toward other snapping shrimp of the same species, apparently ensuring that pairs of shrimp are spaced out within available habitats.

**Range**
The family is largely tropical and semitropical, but some species extend as far north as Va. on the E. Coast.

# Creeping or Walking Decapods
Suborder Reptantia

## AMERICAN LOBSTER
*Homarus americanus*                                         *Fig. 149*

**Description**
Size, to 12 in. (30.5 cm) long, sometimes more; occasionally over 2
ft. (0.6 m) and up to 44 lb. (24 kg). The largest crustacean in the
range. Claws, 2, very large, asymmetrical; one has crushing edge,
other is a sharp grasper.
**Habitat**
Rocky or sandy bottoms from shallow to very deep water.
**Remarks**
The greatest commercial catch is made between Nova Scotia and
Long Island. The excellent flavor of the lobster is a perpetual
threat to its numbers, which in many areas have been sharply
reduced. A wise conservation program rigidly enforced is essential
in order to guard against the further depletion of this natural
resource.
**Range**
Labrador to the Carolinas.

*Fig. 149*                                         *Homarus americanus*
                                                   American Lobster

## MOLE CRAB, SAND BUG
*Emerita talpoida*                                         **110:2**

**Description**
Size, 1½ in. (3.8 cm) long. Carapace olive-shaped, shiny white
with faint purple markings. First antenna small, second long and
feathery. First pair of legs directed forward; second, third, and
fourth strong, leaflike at tips; fifth pair threadlike, hidden.
**Habitat**
Clean sand between tides on surf beaches.
**Habits**
Animals reburrow rapidly after being washed out by incoming
waves.
**Remarks**
One of the few conspicuous inhabitants of ocean-front beaches.
**Range**
Cape Cod to Fla.

## MUD SHRIMP, BURROWING SHRIMP
*Upogebia affinis*                                         *Fig. 150*

**Description**
Size, to 4 in. (10.2 cm) long. Shrimplike, but abdomen
dorsoventrally flattened as in lobsters and crayfish. First pair of
legs with equal-sized claws. Exoskeleton rather soft. Color pale
gray.

**Habitat**
Lives in burrows in muddy sand or shelly mud.
**Range**
Cape Cod and s.

Note: The similar *Callianassa atlantica* has claws of unequal size
and burrows in cleaner sand than does *Upogebia*. It occurs from
N.C. to La.

*Fig. 150*

*Upogebia affinis*
Mud Shrimp

# HERMIT CRABS AND MOLE CRABS
Section Anomura

Hermit crabs house themselves in the empty shells of snails,
whelks, conchs, or other gastropod mollusks. They cannot live
without a shell and must locate another empty shell when they
outgrow the old one. The shell is carried with the fourth or fifth
pair of legs, which is jammed against the inner shell wall, and with
a hook-shaped appendage at the end of the abdomen. The abdomen
is asymmetrical and soft. Mole crabs do not house themselves in
empty mollusk shells.

## *PAGURUS POLLICARIS*                    110:1

**Description**
Size, carapace to 1½ in. (3.8 cm) long. Claws covered with fine
tubercles; right much larger than left, wide and flat, acting as a
"door" to the shell when animal is withdrawn. Color reddish
brown or tan, claws white or gray.
**Habitat**
Shallow water over sand or mud bottoms.
**Range**
Mass. to Tex.

## *PAGURUS LONGICARPUS*

**Description**
Size, carapace ⅜ in. (0.96 cm) long. Larger right claw
subcylindrical. Color variable, upper surface of claws and legs may
be iridescent.
**Habitat**
Shallow water to 150 ft. (91.4 m).
**Range**
Mass. to Tex.

## STRIPED HERMIT CRAB
*Clibanarius vittatus*

**Description**
Size, carapace to 1¼ in. (5.2 cm) long. Claws equal. Color greenish
to dark brown; legs with white, gray, or light orange stripes.
**Habitat**
Low-tide mark to 6 ft. (1.8 m).
**Range**
Va. to Brazil.

# TRUE CRABS
Section Brachyura

A short body form has evolved in decapods as a consequence of reduction and folding of the abdomen beneath the cephalothorax. These decapods are called crabs. The cephalothorax is as wide as or wider than it is long, and a sideways gait is usually well developed.

## SPIDER CRAB
*Libinia emarginata*                                    Fig. 151

**Description**
Size, carapace to 4 in. (10.2 cm) long. Body somewhat heart-shaped, rostrum long and forked; carapace with tubercles and with 9 spines on middle line. Legs long, slender, giving crab a spiderlike appearance. Color brownish or whitish; algae growing on shell may vary this appearance.
**Habitat**
Mud, rocks, and oyster beds; common.
**Range**
N.S. to Fla.

Note: The similar species, *L. dubia,* has a longer, more deeply divided rostrum and only 6 spines on its middle line. It occurs from Cape Cod south.

female

Fig. 151

*Libinia emarginata*
Spider Crab

## GREEN CRAB
*Carcinus maenas*                                       Fig. 152

**Description**
Size, carapace 2¼ in. (5.7 cm) long. Claws short, sturdy. Last legs slightly flattened. Color orange, red, or dark green, mottled with black and yellow.
**Habitat**
Rocky bottoms in shallow water.
**Range**
Maine to N.J. Since 1900, has extended its range n. from Cape Cod to N.S.

Fig. 152

male

*Carcinus maenas*
Green Crab

*Pinnotheres ostreum*
Oyster Crab, p. 664

## LADY CRAB
*Ovalipes ocellatus*                                          110:3

**Description**
Size, 2 in. (5.1 cm) long, 2¼ in. (5.7 cm) wide. Last legs equipped with swimming paddles. Color bluish white with red or purple spots.
**Habitat**
Sandy bottoms; common.
**Range**
Cape Cod to Gulf of Mexico.

## BLUE CRAB
*Callinectes sapidus*                                         110:4

**Description**
Size, 3 in. (7.6 cm) long, 7 in. (17.8 cm) wide. Carapace wide, having front margin with spines and ending with a large spike on either side; 4 teeth between eye sockets; last pair of legs paddle-shaped. Carapace gray- or blue-green, legs blue.
**Habitat**
Grassy, sandy, and muddy bottoms from low-tide mark to 100 ft. (30.5 m). Can tolerate very low salinities. Common in bays, inlets, and estuaries.
**Remarks**
Members of the family Portunidae, to which the Lady Crab and Blue Crab belong, have the last pair of legs paddle-shaped and adapted for swimming. There are other species having the general form of the Blue Crab but differing in structural details and in coloration.
**Range**
Cape Cod to Tex.

## ROCK CRAB
*Cancer irroratus*                                           110:7

**Description**
Size, carapace 3 × 5 in. (7.6 × 12.7 cm). Carapace transversely oval, granulated but smooth, yellowish with bronzy-purple spots; sometimes with light spots in circular design, and light borders to front spines.
**Habitat**
Among rocks between tide marks to a depth of several hundred feet.
**Range**
Labrador to Fla.; in deeper water south of Cape Cod.

Note: The similar **JONAH CRAB,** *C. borealis,* has the large marginal teeth of the carapace subdivided into smaller teeth. It occurs from Nova Scotia to Florida.

## *SESARMA RETICULATUM*

**Description**
Size, ⅞ in. (2.2 cm) long, 1⅛ in. (2.9 cm) wide. Carapace a short transverse rectangle. Color dark olive to black or purple.
**Habitat**
Amphibious; lives in burrows in mud in salt marshes.
**Range**
Woods Hole to Tex.

## *SESARMA CINEREUM*

**Description**
Size, ¾ in. (1.8 cm) long. Carapace almost square. Color brown.
**Habitat**
Semiterrestrial; lives beneath drift at the high-tide mark on
protected beaches; sometimes found at the high-tide mark on
pilings and jetties.
**Range**
Chesapeake Bay to British Honduras.

## MUD CRABS
Family Xanthidae

**Description**
Size, usually less than 1 in. (2.5 cm) wide. Carapace transversely
oval with teeth on the laterofrontal margin. Legs and claws stout.
Color drab, mostly dark gray and brown; claws with black, white,
and brown.
**Habitat**
Beneath rocks and in oyster clumps and reefs on sandy and muddy
bottoms; very common. Some species tolerate very low salinities.
The six species which inhabit eastern shores are difficult to
distinguish.
**Remarks**
*Panopeus herbstii* is the easiest of the six species to distinguish. It
reaches a length of 1 in. (2.5 cm) and possesses a large, blunt tooth
near the base of the movable finger of the large claw. It may be
found as far north as Massachusetts.
**Range**
Entire E. Coast.

## OYSTER CRAB OR PEA CRAB
*Pinnotheres ostreum*                                    *Fig. 152*

**Description**
Size, to ¾ in. (1.8 cm). Carapace round, pinky white, smooth.
**Habitat**
The eastern crab that is found inside the shells of living oysters.
Females spend their lives commensally within oysters. Males are
more free-living.
**Range**
Cape Cod to Tex.

Note: The similar but hairy **MUSSEL CRAB**, *P. maculatus,* of
similar range, lives in the shells of mussels, clams, and scallops.

## GHOST CRAB
*Ocypode quadrata*                                    **110:5**

**Description**
Size, carapace 1½ in. (3.8 cm) long, 2 in. (5.1 cm) wide. Carapace
rectangular. Pale and ghostly looking, the color of dry sand. Eye
stalks very prominent.
**Habitat**
Terrestrial; lives in burrows in dunes above tide mark.
**Habits**
Comes out at night to feed.
**Range**
N.J. to Fla.

## MUD FIDDLER CRAB
*Uca pugnax*                                                   *Fig. 153*

**Description**
Size, carapace 1 in. (2.5 cm) wide. Carapace trapezoidal. Inner
surface of great claw ridged diagonally from base of fixed finger
toward hind margin of hand or palm. Color dark olive; great claw
paler than carapace. Distinguished from *U. minax* by lack of red
joints.
**Habitat**
Salt marshes.
**Range**
Cape Cod to Mexico.

Note: Members of the genus *Uca* are amphibious. They make their
home in burrows excavated in the intertidal zone, from which they
emerge at low tide to feed on fine organic matter mixed with
surface sand or mud. Male fiddlers, in contrast to the female, in
which both claws are small and of equal size, have one very large
claw.

*Fig. 153*      *Uca minax*
            Red-jointed Fiddler Crab

*Uca pugilator*          *Uca pugnax*
Sand Fiddler Crab       Mud Fiddler Crab

## SAND FIDDLER CRAB                                      **110:6**
*Uca pugilator*                                           *Fig. 153*

**Description**
Size, to 1¼ in. (3.2 cm). Similar to *U. pugnax* but slightly larger,
and rear of great claw more or less evenly pebbled on all sides,
with no strong oblique ridge as in other fiddlers; no red markings
at joints of great claw as in *U. minax*.
**Habitat**
On protected sand or sand-mud beaches where the salinity of the
water is relatively high.
**Habits**
Often occur in large aggregations.
**Range**
Boston to Tex.

## RED-JOINTED FIDDLER CRAB
*Uca minax*                                               *Fig. 153*

**Description**
Size, carapace 1½ in. (3.8 cm) across. The only fiddler in the range
with red joints on the great claw. Forward ⅓ light, rear ⅔ dark; H
in middle with red crossbar. Great claw diagonally ridged within.
**Habitat**
From beach inland almost to fresh water.
**Range**
Cape Cod to Tex.

# Mantis Shrimps
Order Stomatopoda

The members of this small order of crustaceans are not decapods, although they are about the same size as true shrimps. The distinctive feature of the group is the large, powerful second pair of legs, resembling those of the praying mantis. The terminal piece of this pair of legs folds into the penultimate segment like the blade of a pocket knife. These predatory crustaceans are most widely distributed in tropical and subtropical waters.

## MANTIS SHRIMP
*Squilla empusa*                                                  **110:10**

**Description**
Size, to 7 in. (17.8 cm) long. Thorax short, abdomen long. Second legs large, with terminal piece folding into penultimate segment. Color yellowish green, segments edged with yellow, tail rosy.
**Habitat**
Burrows in sandy bottoms in shallow water.
**Range**
Cape Cod to Fla.

# CHORDATES
Phylum Chordata

To this phylum belong the largest and most highly organized animals. Three structures distinguish them from the members of other phyla: (1) a notochord, a cylindrical stiff supporting rod of tissue dorsal to the gut; (2) a dorsal hollow nerve cord (the spinal cord), which lies above the notochord; and (3) lateral pharyngeal clefts, or gill slits.

The vertebrate chordates—fish, reptiles, birds, and mammals—are very familiar. The notochord in most of these chordates is found only in the embryo, and the vertebrae develop around the embryonic notochord. In higher terrestrial vertebrates the gill slits are formed in embryonic life but become modified or lost in the course of later development. The outer ear canal of humans is a modified part of a gill slit.

There are also some chordates without backbones. These invertebrate chordates are all marine and include some very common animals, but ones which are unfamiliar to most laymen.

# TUNICATES OR SEA SQUIRTS
Class Ascidiacea

Sea squirts are a very specialized group of marine chordates. In general they are baglike sedentary animals, with an incurrent and an excurrent siphon. The body is covered by a transparent or opaque envelope, or tunic, which contains tunicin, a substance very similar to the cellulose found in plants. Sea squirts may be either solitary or colonial. In colonial species, the zooids (individuals) are usually very small and are embedded in a common tunic, which may form a tough, flexible, attached mass.

The incurrent siphon brings water and food particles to the perforated pharynx, through which the water is filtered. The food particles are trapped in films of mucus and so carried to the mouth. When sea squirts contract at low tide, they often eject a stream of water from the siphons. The incurrent siphons of colonial forms are separate, but the excurrent siphons may open into a common cloaca.

The tiny free-swimming larva of these animals resembles a tadpole and possesses all of the chordate characteristics. Following attachment, the notochord and dorsal hollow nerve cord degenerate.

## STYELA PARTITA                                    110:8

**Description**
Size, to 1 in. (2.5 cm) long. Irregularly oval; attaches at posterior end or along entire ventral surface. Outer covering yellowish brown, tough, with an irregular surface. Openings of siphons rectangular, terminating prominent tubes marked with triangular spots of purple and white.
**Habitat**
On stones, shells, pilings.
**Range**
Maine to West Indies.

## SEA GRAPES
*Molgula manhattensis*                              *Fig. 154*

**Description**
Size, to 1 in. (2.5 cm) dia. The only eastern sea squirt with a round ball-shaped body, usually occurring in groups or colonies (hence its common name). Tunic ashy gray, rough, slightly translucent; often covered with sand, bits of seaweed, eelgrass, etc. Siphons long, incurrent with 6 lobes at margin, excurrent with 4.
**Habitat**
Piles, stones, and eelgrass.
**Range**
Maine to La.

*Fig. 154*
**Sea Squirts**

*Perophora viridis*, p. 668

*Ciona intestinalis*, p. 668
Sea Vase

*Ascidia callosa*,
p. 668

*Molgula manhattensis*
Sea Grapes

## *ASCIDIA CALLOSA AND INTERRUPTA*          *Fig. 154*

**Description**
Size, 2 × 1 in. (5.1 × 2.5 cm). Body oval, length twice width; flattened, usually attaches on one side. Outer covering translucent; greenish, yellowish, or gray; thick, fleshy, firm. Siphons short.
**Habitat**
Shallow to deep water on pilings, stones, and shells.
**Range**
*A. callosa*, Arctic to Cape Cod; *A. interrupta*, s. of Cape Hatteras.

## SEA VASE
*Ciona intestinalis*                                      *Fig. 154*

**Description**
Size, to 5 in. (12.7 cm) high. Body slender, attaches at rear. Outer tunic translucent, yellow, spotted with orange or red; spots vanish on death. Strong longitudinal muscle bands may be seen through the tunic; 2 siphons at upper end give animal a vaselike shape.
**Habitat**
Buoys, timbers, rocks.
**Range**
N.S. to Long Island.

## *PEROPHORA VIRIDIS*                        *Fig. 154*

**Description**
Size, zooids ⅛ in. (0.32 cm) across. Semicolonial; individual zooids transparent, soft green, or chartreuse; connected together by a vinelike stolon.
**Habitat**
Rocks, pilings, kelp, or other firm base.
**Range**
Cape Cod to Fla.

## SEA PORK
*Amaroucium stellatum*                                *Fig. 155*

**Description**
Size, colonies to 6 in. (15.2 cm) high, 1 in. (2.5 cm) thick. Colony in shape of long, erect plates; made up of tiny individuals enclosed in gelatinous tunic. Tunic not heavily sand-encrusted, color pink or bluish pink; color and texture similar to those of raw salt pork, hence the name. Individuals, usually 6–18, arranged in star-shaped clusters.

*Fig. 155*

*Amaroucium stellatum*
Sea Pork

**Habitat**
Coastal waters, 10–20 ft. (3.1–6.1 m) deep.
**Range**
Maine to Cape Hatteras.

Note: The similar *A. constellatum* forms pear-shaped masses 1½ to 2 inches (1.3–5.1 cm) high and has a cream-colored tunic through which the reddish-orange individuals may be clearly seen. It occurs from Cape Cod to Florida.

# Life Lists

## BIRDS

_____ Acadian Flycatcher
_____ Alder Flycatcher
_____ American Avocet
_____ American Bittern
_____ American Coot
_____ American Golden
Plover
_____ American Goldfinch
_____ American Kestrel
_____ American Redstart
_____ American Robin
_____ American White
Pelican
_____ American Wigeon
_____ American Woodcock
_____ Anhinga
_____ Arctic Loon
_____ Arctic Tern
_____ Audubon's Shearwater

_____ Bachman's Sparrow
_____ Bachman's Warbler
_____ Baird's Sandpiper
_____ Baird's Sparrow
_____ Bald Eagle
_____ Bank Swallow
_____ Barn Owl
_____ Barn Swallow
_____ Barred Owl
_____ Barrow's Goldeneye
_____ Bay-breasted Warbler
_____ Bell's Vireo
_____ Belted Kingfisher
_____ Bewick's Wren
_____ Black-and-White
Warbler
_____ Black-backed
Woodpecker
_____ Black-bellied Plover
_____ Black-billed Cuckoo
_____ Black-billed Magpie
_____ Blackburnian Warbler
_____ Black-capped
Chickadee
_____ Black-crowned Night
Heron
_____ Black Duck
_____ Black Guillemot
_____ Black-headed Grosbeak
_____ Black-headed Gull
_____ Black-legged Kittiwake
_____ Black-necked Stilt
_____ Blackpoll Warbler
_____ Black Rail
_____ Black Scoter
_____ Black Skimmer
_____ Black Tern
_____ Black-throated Blue
Warbler

_____ Black-throated Gray
Warbler
_____ Black-throated Green
Warbler
_____ Black-throated
Sparrow
_____ Black Vulture
_____ Black-whiskered Vireo
_____ Blue-gray Gnatcatcher
_____ Blue Grosbeak
_____ Blue Jay
_____ Blue-winged Teal
_____ Blue-winged Warbler
_____ Boat-tailed Grackle
_____ Bobolink
_____ Bobwhite
_____ Bohemian Waxwing
_____ Bonaparte's Gull
_____ Boreal Chickadee
_____ Boreal Owl
_____ Brant
_____ Brewer's Blackbird
_____ Brewer's Sparrow
_____ Brewster's Warbler
_____ Bridled Tern
_____ Broad-tailed
Hummingbird
_____ Broad-winged Hawk
_____ Brown Creeper
_____ Brown-headed
Cowbird
_____ Brown-headed
Nuthatch
_____ Brown Pelican
_____ Brown Thrasher
_____ Buff-breasted
Sandpiper
_____ Bufflehead
_____ Burrowing Owl

_____ Canada Goose
_____ Canada Warbler
_____ Canvasback
_____ Canyon Wren
_____ Cape May Warbler
_____ Cardinal
_____ Carolina Chickadee
_____ Carolina Wren
_____ Caspian Tern
_____ Cassin's Kingbird
_____ Cassin's Sparrow
_____ Cattle Egret
_____ Cedar Waxwing
_____ Cerulean Warbler
_____ Chestnut-collared
Longspur
_____ Chestnut-sided
Warbler
_____ Chimney Swift
_____ Chipping Sparrow

# BIRDS

_____ Chuck-will's Widow
_____ Clapper Rail
_____ Clark's Nutcracker
_____ Clay-colored Sparrow
_____ Cliff Swallow
_____ Common Crow
_____ Common Eider
_____ Common Flicker
_____ Common Gallinule
_____ Common Goldeneye
_____ Common Grackle
_____ Common Ground
Dove
_____ Common Loon
_____ Common Merganser
_____ Common Murre
_____ Common Nighthawk
_____ Common
Oystercatcher
_____ Common Puffin
_____ Common Raven
_____ Common Redpoll
_____ Common Snipe
_____ Common Teal
_____ Common Tern
_____ Common Wheatear
_____ Common Yellowthroat
_____ Connecticut Warbler
_____ Cooper's Hawk
_____ Cory's Shearwater
_____ Crested Caracara
_____ Curlew Sandpiper

_____ Dark-eyed Junco
_____ Dickcissel
_____ Dipper
_____ Double-crested
Cormorant
_____ Dovekie
_____ Downy Woodpecker
_____ Dunlin
_____ Dusky Flycatcher

_____ Eared Grebe
_____ Eastern Bluebird
_____ Eastern Kingbird
_____ Eastern Meadowlark
_____ Eastern Phoebe
_____ Eastern Wood Pewee
_____ Eurasian Wigeon
_____ European Tree
Sparrow
_____ Evening Grosbeak

_____ Ferruginous Hawk
_____ Field Sparrow
_____ Fish Crow
_____ Forster's Tern
_____ Fox Sparrow
_____ Franklin's Gull

_____ Fulvous Whistling-
Duck

_____ Gadwall
_____ Gannet
_____ Glaucous Gull
_____ Glossy Ibis
_____ Golden-crowned
Kinglet
_____ Golden Eagle
_____ Golden-winged
Warbler
_____ Grasshopper Sparrow
_____ Gray Catbird
_____ Gray-cheeked Thrush
_____ Gray-crowned Rosy
Finch
_____ Gray Jay
_____ Gray Kingbird
_____ Gray Partridge
_____ Great Black-backed
Gull
_____ Great Blue Heron
_____ Great Cormorant
_____ Great Crested
Flycatcher
_____ Great Egret
_____ Greater Prairie
Chicken
_____ Greater Scaup
_____ Greater Shearwater
_____ Greater Yellowlegs
_____ Great Gray Owl
_____ Great Horned Owl
_____ Great-tailed Grackle
_____ Green Heron
_____ Green-tailed Towhee
_____ Gull-billed Tern
_____ Gyrfalcon

_____ Hairy Woodpecker
_____ Harlequin Duck
_____ Harris' Sparrow
_____ Hawk Owl
_____ Henslow's Sparrow
_____ Hermit Thrush
_____ Herring Gull
_____ Hoary Redpoll
_____ Hooded Merganser
_____ Hooded Warbler
_____ Horned Grebe
_____ Horned Lark
_____ House Finch
_____ House Sparrow
_____ House Wren
_____ Hudsonian Godwit

_____ Iceland Gull
_____ Indigo Bunting
_____ Ivory-billed
Woodpecker

674

_____ Ivory Gull

_____ Kentucky Warbler
_____ Killdeer
_____ King Eider
_____ King Rail
_____ Kirtland's Warbler

_____ Ladder-backed
Woodpecker
_____ Lapland Longspur
_____ Lark Bunting
_____ Lark Sparrow
_____ Laughing Gull
_____ Lawrence's Warbler
_____ Lazuli Bunting
_____ Leach's Storm-Petrel
_____ Least Bittern
_____ Least Flycatcher
_____ Least Sandpiper
_____ Least Tern
_____ Le Conte's Sparrow
_____ Lesser Black-backed
Gull
_____ Lesser Goldfinch
_____ Lesser Prairie Chicken
_____ Lesser Scaup
_____ Lesser Yellowlegs
_____ Lewis' Woodpecker
_____ Limpkin
_____ Lincoln's Sparrow
_____ Little Blue Heron
_____ Little Gull
_____ Loggerhead Shrike
_____ Long-billed Curlew
_____ Long-billed Dowitcher
_____ Long-billed Marsh
Wren
_____ Long-eared Owl
_____ Long-tailed Jaeger
_____ Louisiana Heron
_____ Louisiana Waterthrush

_____ McCown's Longspur
_____ MacGillivray's Warbler
_____ Magnificent
Frigatebird
_____ Magnolia Warbler
_____ Mallard
_____ Mangrove Cuckoo
_____ Manx Shearwater
_____ Marbled Godwit
_____ Merlin
_____ Mew Gull
_____ Mississippi Kite
_____ Mountain Bluebird
_____ Mourning Dove
_____ Mourning Warbler
_____ Mute Swan

_____ Nashville Warbler

_____ Neotropical
(Olivaceous)
Cormorant
_____ Northern Fulmar
_____ Northern Goshawk
_____ Northern Harrier
_____ Northern Mockingbird
_____ Northern Oriole
_____ Northern Parula
_____ Northern Phalarope
_____ Northern Pintail
_____ Northern Shoveler
_____ Northern Shrike
_____ Northern Skua
_____ Northern Waterthrush

_____ Oldsquaw
_____ Olive-sided Flycatcher
_____ Orange-crowned
Warbler
_____ Orchard Oriole
_____ Osprey
_____ Ovenbird

_____ Painted Bunting
_____ Palm Warbler
_____ Parasitic Jaeger
_____ Pectoral Sandpiper
_____ Peregrine Falcon
_____ Philadelphia Vireo
_____ Pied-billed Grebe
_____ Pileated Woodpecker
_____ Pine Grosbeak
_____ Pine Siskin
_____ Pine Warbler
_____ Pinyon Jay
_____ Piping Plover
_____ Pomarine Jaeger
_____ Poor-will
_____ Prairie Warbler
_____ Prothonotary Warbler
_____ Purple Finch
_____ Purple Gallinule
_____ Purple Martin
_____ Purple Sandpiper
_____ Pygmy Nuthatch

_____ Razorbill
_____ Red-bellied
Woodpecker
_____ Red-breasted
Merganser
_____ Red-breasted Nuthatch
_____ Red-cockaded
Woodpecker
_____ Red Crossbill
_____ Reddish Egret
_____ Red-eyed Vireo
_____ Redhead
_____ Red-headed
Woodpecker

_____ Red Knot
_____ Red-necked Grebe
_____ Red Phalarope
_____ Red-shouldered Hawk
_____ Red-tailed Hawk
_____ Red-throated Loon
_____ Red-winged Blackbird
_____ Reeve
_____ Ring-billed Gull
_____ Ring-necked Duck
_____ Ring-necked Pheasant
_____ Roadrunner
_____ Rock Pigeon
_____ Rock Ptarmigan
_____ Rock Wren
_____ Roseate Spoonbill
_____ Roseate Tern
_____ Rose-breasted
Grosbeak
_____ Ross' Gull
_____ Rough-legged Hawk
_____ Rough-winged Swallow
_____ Royal Tern
_____ Ruby-crowned Kinglet
_____ Ruby-throated
Hummingbird
_____ Ruddy Duck
_____ Ruddy Turnstone
_____ Ruff
_____ Ruffed Grouse
_____ Rufous-crowned
Sparrow
_____ Rufous-sided Towhee
_____ Rusty Blackbird

_____ Sabine's Gull
_____ Sage Thrasher
_____ Sanderling
_____ Sandhill Crane
_____ Sandwich Tern
_____ Savannah Sparrow
_____ Saw-whet Owl
_____ Say's Phoebe
_____ Scarlet Tanager
_____ Scissor-tailed
Flycatcher
_____ Screech Owl
_____ Scrub Jay
_____ Seaside Sparrow
_____ Semipalmated Plover
_____ Semipalmated
Sandpiper
_____ Sharp-shinned Hawk
_____ Sharp-tailed Grouse
_____ Sharp-tailed Sparrow
_____ Short-billed Dowitcher
_____ Short-billed Marsh
Wren
_____ Short-eared Owl
_____ Short-tailed Hawk
_____ Smith's Longspur

_____ Smooth-billed Ani
_____ Snail Kite
_____ Snow Bunting
_____ Snow Goose
_____ Snowy Egret
_____ Snowy Owl
_____ Snowy Plover
_____ Solitary Sandpiper
_____ Solitary Vireo
_____ Song Sparrow
_____ Sooty Shearwater
_____ Sooty Tern
_____ Sora
_____ Spot-breasted Oriole
_____ Spotted Sandpiper
_____ Sprague's Pipit
_____ Spruce Grouse
_____ Starling
_____ Steller's Jay
_____ Stilt Sandpiper
_____ Summer Tanager
_____ Surf Scoter
_____ Swainson's Hawk
_____ Swainson's Thrush
_____ Swainson's Warbler
_____ Swallow-tailed Kite
_____ Swamp Sparrow

_____ Tennessee Warbler
_____ Thayer's Gull
_____ Thick-billed Murre
_____ Three-toed
Woodpecker
_____ Townsend's Solitaire
_____ Tree Sparrow
_____ Tree Swallow
_____ Tufted Titmouse
_____ Turkey
_____ Turkey Vulture

_____ Upland Sandpiper

_____ Varied Thrush
_____ Veery
_____ Vesper Sparrow
_____ Violet-green Swallow
_____ Virginia Rail

_____ Warbling Vireo
_____ Water Pipit
_____ Western Flycatcher
_____ Western Grebe
_____ Western Kingbird
_____ Western Meadowlark
_____ Western Sandpiper
_____ Western Tanager
_____ Western Wood Pewee
_____ Whimbrel
_____ Whip-poor-will
_____ Whistling Swan
_____ White-breasted
Nuthatch

_____ White-crowned Pigeon
_____ White-crowned Sparrow
_____ White-eyed Vireo
_____ White-faced Ibis
_____ White Ibis
_____ White-fronted Goose
_____ White-rumped Sandpiper
_____ White-tailed (Gray Sea) Eagle
_____ White-tailed Kite
_____ White-throated Sparrow
_____ White-throated Swift
_____ White-winged Crossbill
_____ White-winged Scoter
_____ Whooping Crane
_____ Willet
_____ Willow Flycatcher
_____ Willow Ptarmigan
_____ Wilson's Phalarope
_____ Wilson's Plover
_____ Wilson's Storm-Petrel
_____ Wilson's Warbler
_____ Winter Wren
_____ Wood Duck
_____ Wood Stork (Wood Ibis)
_____ Wood Thrush
_____ Worm-eating Warbler

_____ Yellow-bellied Flycatcher
_____ Yellow-bellied Sapsucker
_____ Yellow-billed Cuckoo
_____ Yellow-breasted Chat
_____ Yellow-crowned Night Heron
_____ Yellow-headed Blackbird
_____ Yellow Rail
_____ Yellow-rumped Warbler
_____ Yellow-throated Vireo
_____ Yellow-throated Warbler
_____ Yellow Warbler

## MAMMALS

_____ Arctic Fox
_____ Arctic Hare
_____ Arctic Shrew
_____ Atlantic Spotted Dolphin
_____ Atlantic White-sided Dolphin

_____ Badger

_____ Beach Vole
_____ Bearded Seal
_____ Beaver
_____ Big Brown Bat
_____ Big Free-tailed Bat
_____ Bison
_____ Black Bear
_____ Black-footed Ferret
_____ Black Rat
_____ Black Right Whale
_____ Black-tailed Jackrabbit
_____ Black-tailed Prairie Dog
_____ Blue Whale
_____ Bobcat
_____ Bottle-nosed Dolphin
_____ Bowhead Whale
_____ Brazilian Free-tailed Bat
_____ Brown Lemming

_____ Caribou
_____ Cave Myotis
_____ Coati
_____ Collared Lemming
_____ Common Dolphin
_____ Common Pilot Whale
_____ Cotton Mouse
_____ Coyote

_____ Deer Mouse
_____ Desert Shrew
_____ Dwarf Sperm Whale

_____ Eastern Chipmunk
_____ Eastern Cottontail
_____ Eastern Harvest Mouse
_____ Eastern Mole
_____ Eastern Pipistrelle
_____ Eastern Spotted Skunk
_____ Eastern Woodrat
_____ Elk
_____ Ermine
_____ European Hare
_____ Evening Bat

_____ False Killer Whale
_____ Fin Whale
_____ Fisher
_____ Florida Mouse
_____ Fox Squirrel
_____ Franklin's Ground Squirrel
_____ Fulvous Harvest Mouse

_____ Gaspé Shrew
_____ Gervais' Beaked Whale
_____ Golden Mouse
_____ Goose-beaked Whale
_____ Grampus
_____ Gray Fox

677

_____ Gray Myotis
_____ Gray Seal
_____ Gray Squirrel
_____ Gray Wolf

_____ Hairy-tailed Mole
_____ Harbor Porpoise
_____ Harbor Seal
_____ Harp Seal
_____ Heather Vole
_____ Hispid Cotton Rat
_____ Hispid Pocket Mouse
_____ Hoary Bat
_____ Hog-nosed Skunk
_____ Hooded Seal
_____ House Mouse
_____ Hump-backed Whale

_____ Indiana Myotis

_____ Keen's Myotis
_____ Killer Whale

_____ Labrador Collared
Lemming
_____ Least Chipmunk
_____ Least Shrew
_____ Least Weasel
_____ Little Brown Myotis
_____ Long-tailed Shrew
_____ Long-tailed Weasel
_____ Lynx

_____ Manatee
_____ Marsh Rabbit
_____ Marsh Rice Rat
_____ Marten
_____ Masked Shrew
_____ Meadow Jumping
Mouse
_____ Meadow Vole
_____ Mink
_____ Minke Whale
_____ Moose
_____ Mountain Lion
_____ Mule Deer
_____ Muskox
_____ Muskrat

_____ Narwhal
_____ New England
Cottontail
_____ Nine-banded
Armadillo
_____ North Atlantic Beaked
Whale
_____ North Atlantic Bottle-
nosed Whale
_____ Northern Bog
Lemming
_____ Northern Flying
Squirrel

_____ Northern Grasshopper
Mouse
_____ Northern Pocket
Gopher
_____ Northern Pygmy
Mouse
_____ Northern Yellow Bat
_____ Norway Rat
_____ Nutria

_____ Oldfield Mouse
_____ Olive-backed Pocket
Mouse
_____ Ord's Kangaroo Rat

_____ Plains Harvest Mouse
_____ Plains Pocket Gopher
_____ Plains Pocket Mouse
_____ Polar Bear
_____ Porcupine
_____ Prairie Vole
_____ Pronghorn
_____ Pygmy Shrew
_____ Pygmy Sperm Whale

_____ Raccoon
_____ Rafinesque's Big-eared
Bat
_____ Red Bat
_____ Red Fox
_____ Red Squirrel
_____ Red Wolf
_____ Richardson's Ground
Squirrel
_____ Ringed Seal
_____ Ringtail Cat
_____ River Otter
_____ Rock Squirrel
_____ Rock Vole
_____ Rough-toothed
Porpoise
_____ Round-tailed Muskrat

_____ Sei Whale
_____ Seminole Bat
_____ Short-finned Pilot
Whale
_____ Short-tailed Shrew
_____ Silver-haired Bat
_____ Small-footed Myotis
_____ Smoky Shrew
_____ Snowshoe Hare
_____ Southeastern Myotis
_____ Southeastern Pocket
Gopher
_____ Southeastern Shrew
_____ Southern Bog
Lemming
_____ Southern Flying
Squirrel

_____ Southern Plains Woodrat
_____ Southern Red-backed Vole
_____ Southern Short-tailed Shrew
_____ Sperm Whale
_____ Spotted Ground Squirrel
_____ Star-nosed Mole
_____ Striped Porpoise
_____ Striped Skunk
_____ Swamp Rabbit
_____ Swift Fox

_____ Texas Mouse
_____ Thirteen-lined Ground Squirrel
_____ Townsend's Big-eared Bat
_____ Tropical Beaked Whale
_____ True's Beaked Whale

_____ Virginia Opossum

_____ Walrus
_____ Water Shrew
_____ Western Harvest Mouse
_____ White-beaked Dolphin
_____ White-footed Mouse
_____ White-tailed Deer
_____ White-tailed Jackrabbit
_____ White Whale
_____ Wild Boar
_____ Wolverine
_____ Woodchuck
_____ Woodland Jumping Mouse
_____ Woodland Vole

## REPTILES

_____ Alabama Map Turtle
_____ Alabama Red-bellied Turtle
_____ Alligator Snapping Turtle
_____ American Alligator
_____ Atlantic Ridley

_____ Banded Water Snake
_____ Barbour's Map Turtle
_____ Black Kingsnake
_____ Black-knobbed Map Turtle
_____ Black Racer
_____ Black Rat Snake
_____ Black Swamp Snake
_____ Blanding's Turtle

_____ Blotched Water Snake
_____ Blue Racer
_____ Bog Turtle
_____ Broad-banded Copperhead
_____ Broad-banded Water Snake
_____ Broad-headed Skink
_____ Brown Snake
_____ Brown Water Snake
_____ Bullsnake
_____ Butler's Garter Snake
_____ Buttermilk Racer

_____ Cagle's Map Turtle
_____ Carolina Salt Marsh Snake
_____ Central Plains Milk Snake
_____ Checkered Garter Snake
_____ Chicken Turtle
_____ Coachwhip
_____ Coal Skink
_____ Collared Lizard
_____ Common Garter Snake
_____ Common Kingsnake
_____ Cooter
_____ Copper-bellied Water Snake
_____ Copperhead
_____ Corn Snake
_____ Cottonmouth
_____ Cumberland slider

_____ Desert Kingsnake
_____ Diamondback Terrapin
_____ Diamondback Water Snake

_____ Eastern Box Turtle
_____ Eastern Coral Snake
_____ Eastern Diamondback Rattlesnake
_____ Eastern Fence Lizard
_____ Eastern Garter Snake
_____ Eastern Glass Lizard
_____ Eastern Hognose Snake
_____ Eastern Kingsnake
_____ Eastern Milk Snake
_____ Eastern Mud Turtle
_____ Eastern Painted Turtle
_____ Eastern Ribbon Snake
_____ Eastern Yellow-bellied Racer

_____ False Map Turtle
_____ Five-lined Skink
_____ Flat-headed Snake
_____ Florida Box Turtle

_____ Florida Kingsnake
_____ Florida Red-bellied
Turtle
_____ Florida Scrub Lizard
_____ Florida Snapping
Turtle
_____ Florida Softshell
_____ Florida Water Snake
_____ Fox Snake

_____ Glossy Snake
_____ Glossy Crayfish Snake
_____ Gopher Tortoise
_____ Graham's Crayfish
Snake
_____ Gray Rat Snake
_____ Great Plains Rat Snake
_____ Great Plains Skink
_____ Green Anole
_____ Green Turtle
_____ Green Water Snake
_____ Ground Skink
_____ Ground Snake
_____ Gulf Coast Box Turtle
_____ Gulf Coast Salt Marsh
Snake

_____ Hawksbill

_____ Indigo Snake
_____ Island Glass Lizard

_____ Key Ringneck Snake
_____ Kirtland's Water Snake

_____ Lake Erie Water Snake
_____ Leatherback
_____ Lesser Earless Lizard
_____ Lined Snake
_____ Loggerhead
_____ Loggerhead Musk
Turtle
_____ Long-nosed Snake
_____ Louisiana Milk Snake

_____ Many-lined Skink
_____ Map Turtle
_____ Massasauga
_____ Mexican Milk Snake
_____ Midland Painted
Turtle
_____ Midland Water Snake
_____ Milk Snake
_____ Mississippi Ringneck
Snake
_____ Mole Kingsnake
_____ Mud Snake

_____ Night Snake

_____ Northern Copperhead
_____ Northern Pine Snake
_____ Northern Ringneck
Snake
_____ Northern Water Snake

_____ Osage Copperhead
_____ Ouachita Map Turtle

_____ Painted Turtle
_____ Pale Milk Snake
_____ Pigmy Rattlesnake
_____ Pine Snake
_____ Pine Woods Snake
_____ _Pituophis melanoleucus
sayi_
_____ Plain-bellied Water
Snake
_____ Plains Black-headed
Snake
_____ Plains Garter Snake
_____ Prairie Kingsnake
_____ Prairie Ringneck Snake
_____ Prairie Skink

_____ Queen Snake

_____ Racer
_____ Rainbow Snake
_____ Rat Snake
_____ Razor-backed Musk
Turtle
_____ Red Milk Snake
_____ Red-bellied Snake
_____ Red-bellied Turtle
_____ Red-bellied Water
Snake
_____ Red-eared Slider
_____ Red-sided Garter
Snake
_____ Ringed Map Turtle
_____ Ringneck Snake
_____ River Cooter
_____ Rough Earth Snake
_____ Rough Green Snake

_____ Scarlet Kingsnake
_____ Scarlet Snake
_____ Short-headed Garter
Snake
_____ Short-horned Lizard
_____ Short-tailed Snake
_____ Six-lined Racerunner
_____ Slender Glass Lizard
_____ Slider
_____ Smooth Earth Snake
_____ Smooth Green Snake
_____ Smooth Softshell
_____ Snapping Turtle

_____ Southeastern Crowned Snake
_____ Southeastern Five-lined Skink
_____ Southern Copperhead
_____ Southern Hognose Snake
_____ Southern Painted Turtle
_____ Southern Ringneck Snake
_____ Southern Water Snake
_____ Speckled Kingsnake
_____ Spiny Softshell
_____ Spotted Turtle
_____ *Sternotherus minor minor*
_____ Stinkpot
_____ Striped Mud Turtle
_____ Striped Crayfish Snake
_____ Stripe-necked Musk Turtle

_____ Texas Blind Snake
_____ Texas Horned Lizard
_____ Texas Map Turtle
_____ Texas Rat Snake
_____ Three-toed Box Turtle
_____ Timber Rattlesnake

_____ Western Box Turtle
_____ Western Hognose Snake
_____ Western Painted Turtle
_____ Western Rattlesnake
_____ Western Ribbon Snake
_____ Western Terrestrial Garter Snake
_____ Wood Turtle
_____ Worm Snake

_____ Yellow-bellied Slider
_____ Yellow-bellied Water Snake
_____ Yellow-blotched Map Turtle
_____ Yellow Mud Turtle
_____ Yellow Rat Snake

## AMPHIBIANS

_____ Alabama Waterdog
_____ American Toad

_____ Barking Treefrog
_____ Bird-voiced Treefrog
_____ Black-bellied Salamander

_____ Black Mountain Salamander
_____ Blue-spotted Salamander
_____ Brimley's Chorus Frog
_____ Bullfrog

_____ Canadian Toad
_____ Carpenter Frog
_____ Cave Salamander
_____ Cheat Mountain Salamander
_____ Couch's Spadefoot
_____ Crayfish Frog

_____ Dark-sided Salamander
_____ Dusky Salamander
_____ Dwarf Salamander
_____ Dwarf Siren
_____ Dwarf Waterdog

_____ Eastern Narrow-mouthed Toad
_____ Eastern Spadefoot

_____ Flatwoods Salamander
_____ Four-toed Salamander

_____ Gray Treefrog
_____ Greater Siren
_____ Great Plains Narrow-mouthed Toad
_____ Great Plains Toad
_____ Green Frog
_____ Green Salamander
_____ Green Toad
_____ Green Treefrog
_____ Grotto Salamander
_____ Gulf Coast Toad

_____ Hellbender

_____ Jefferson Salamander
_____ Jordan's Woodland Salamander

_____ Lesser Siren
_____ Little Grass Frog
_____ Long-tailed Salamander

_____ Mabee's Salamander
_____ Many-lined Salamander
_____ Many-ribbed Salamander
_____ Marbled Salamander
_____ Mink Frog
_____ Mole Salamander
_____ Mountain Chorus Frog

_____ Mountain Dusky Salamander
_____ Mudpuppy
_____ Mud Salamander

_____ Newt
_____ Northern Cricket Frog
_____ Northern Leopard Frog

_____ Oak Toad
_____ Oklahoma Salamander
_____ Ornate Chorus Frog

_____ Pickerel Frog
_____ Pig Frog
_____ Pigmy Salamander
_____ Pine Barrens Treefrog
_____ Pine Woods Treefrog
_____ Plains Leopard Frog
_____ Plains Spadefoot

_____ Ravine Salamander
_____ Red-backed Salamander
_____ Red Salamander
_____ Red-spotted Toad
_____ Ringed Salamander
_____ River Frog

_____ Seal Salamander
_____ Seepage Salamander
_____ Shovel-nosed Salamander
_____ Silvery Salamander
_____ Slimy Salamander
_____ Small-mouthed Salamander
_____ Southern Chorus Frog
_____ Southern Cricket Frog
_____ Southern Dusky Salamander
_____ Southern Leopard Frog
_____ Southern Red-backed Salamander
_____ Southern Toad
_____ Spotted Chorus Frog
_____ Spotted Salamander
_____ Spring Peeper
_____ Spring Salamander
_____ Squirrel Treefrog
_____ Strecker's Chorus Frog
_____ Striped Chorus Frog
_____ Striped Newt

_____ Tennessee Cave Salamander
_____ Texas Toad
_____ Three-toed Amphiuma
_____ Tiger Salamander
_____ Tremblay's Salamander

_____ Two-lined Salamander
_____ Two-toed Amphiuma

_____ Valley and Ridge Salamander

_____ Wehrle's Salamander
_____ Weller's Salamander
_____ Western Spadefoot
_____ Western Toad
_____ West Virginia Cave Salamander
_____ White-spotted Salamander
_____ Wood Frog
_____ Woodhouse's Toad

_____ Yonahlossee Salamander

_____ Zigzag Salamander

# FISHES

_____ African Pompano
_____ Alabama Shad
_____ Albacore
_____ Alewife
_____ Alligator Gar
_____ American Eel
_____ American Plaice
_____ American Sand Lance
_____ American Shad
_____ Angel Shark
_____ Arctic Grayling
_____ Atlantic Bonito
_____ Atlantic Cod
_____ Atlantic Croaker
_____ Atlantic Cutlassfish
_____ Atlantic Halibut
_____ Atlantic Herring
_____ Atlantic Mackerel
_____ Atlantic Manta
_____ Atlantic Moonfish
_____ Atlantic Needlefish
_____ Atlantic Salmon
_____ Atlantic Silverside
_____ Atlantic Stingray
_____ Atlantic Sturgeon
_____ Atlantic Tomcod
_____ Atlantic Torpedo
_____ Atlantic Wolffish

_____ Banded Darter
_____ Banded Drum
_____ Banded Killifish
_____ Banded Rudderfish
_____ Banded Sunfish
_____ Bantam Sunfish
_____ Barn-door Skate
_____ Basking Shark

_____ Bay Anchovy
_____ Bermuda Chub
_____ Big-eyed Scad
_____ Bigmouth Buffalo
_____ Bitterling
_____ Blackbanded Sunfish
_____ Black Buffalo
_____ Black Bullhead
_____ Black Crappie
_____ Black Drum
_____ Blackfin Cisco
_____ Blackfin Tuna
_____ Blacknose Dace
_____ Black Sea Bass
_____ Blackstripe
Topminnow
_____ Blackside Darter
_____ Bloater
_____ Blueback Herring
_____ Blue Catfish
_____ Bluefin Tuna
_____ Bluefish
_____ Bluegill
_____ Blue Marlin
_____ Blue Runner
_____ Bluespot Sunfish
_____ Bluespotted Cornetfish
_____ Blue Sucker
_____ Bluntnose Minnow
_____ Bonefish
_____ Bowfin
_____ Brook Silverside
_____ Brook Stickleback
_____ Brook Trout
_____ Brown Bullhead
_____ Brown Trout
_____ Bullnose Ray
_____ Bull Shark
_____ Burbot
_____ Butterfish

_____ Carp
_____ Central Mudminnow
_____ Cero
_____ Chain Pickerel
_____ Channel Catfish
_____ Chub Mackerel
_____ Cisco
_____ Clearnose Skate
_____ Cobia
_____ Common Shiner
_____ Conger Eel
_____ Cownose Ray
_____ Creek Chub
_____ Creek Chubsucker
_____ Crevelle Jack
_____ Cunner
_____ Cusk
_____ Cutlips Minnow
_____ Cutthroat Trout

_____ Deepwater Cisco
_____ Dolphin
_____ Dusky Shark

_____ Eastern Mudminnow
_____ Eastern Sand Darter
_____ Emerald Shiner
_____ Eyed Flounder

_____ Fallfish
_____ Fantail Darter
_____ Fathead Minnow
_____ Flathead Catfish
_____ Florida Pompano
_____ Flying Gurnard
_____ Fourspot Flounder
_____ Freshwater Drum

_____ Gafftopsail Catfish
_____ Gizzard Shad
_____ Golden Redhorse
_____ Golden Shiner
_____ Goldeye
_____ Goldfish
_____ Goosefish
_____ Great Amberjack
_____ Great Barracuda
_____ Great Blue Shark
_____ Greenland Halibut
_____ Greenland Shark
_____ Greenside Darter
_____ Green Sunfish
_____ Grubby

_____ Haddock
_____ Hagfish
_____ Halfbeak
_____ Hickory Shad
_____ Highfin Carpsucker
_____ Hogchoker
_____ Horse-eyed Jack

_____ Johnny Darter

_____ King Mackerel
_____ Kiyi

_____ Ladyfish
_____ Lake Chubsucker
_____ Lake Sturgeon
_____ Lake Trout
_____ Lake Whitefish
_____ Largemouth Bass
_____ Leatherjacket
_____ Lined Sea Horse
_____ Little Skate
_____ Little Tuna
_____ Logperch
_____ Longear Sunfish
_____ Longhorn Sculpin
_____ Longjaw Chub

# FISHES

_____ Longnose Dace
_____ Longnose Gar
_____ Longnose Sucker
_____ Lookdown

_____ Mackerel Scad
_____ Margined Madtom
_____ Menhaden
_____ Mooneye
_____ Mosquitofish
_____ Mottled Sculpin
_____ Mountain Sucker
_____ Mountain Whitefish
_____ Mummichog
_____ Muskellunge

_____ Naked Goby
_____ Nipigon Cisco
_____ Northern Cavefish
_____ Northern Hog Sucker
_____ Northern Kingfish
_____ Northern Pike
_____ Northern Pipefish
_____ Northern Puffer
_____ Northern Sea Robin
_____ Northern Sennet
_____ Northern Stargazer
_____ Northern Studfish
_____ Northern Swamp
        Darter

_____ Ocean Sunfish
_____ Ohio Redhorse
_____ Orange Filefish
_____ Orange Spotted
        Sunfish
_____ Oyster Toadfish

_____ Paddlefish
_____ Pallid Sturgeon
_____ Palometa
_____ Pearl Dace
_____ Permit
_____ Pigfish
_____ Pinfish
_____ Pirate Perch
_____ Plains Carpsucker
_____ Planehead Filefish
_____ Pollack
_____ Porbeagle Shark
_____ Pumpkinseed

_____ Quillback

_____ Rainbow Darter
_____ Rainbow Smelt
_____ Rainbow Trout
_____ Redbreast Sunfish
_____ Red Drum
_____ Redear Sunfish
_____ Redfin Pickerel

_____ Redfish
_____ Red Hake
_____ Red Goatfish
_____ Red Grouper
_____ Redside Dace
_____ River Carpsucker
_____ River Chub
_____ River Redhorse
_____ Roanoke Hog Sucker
_____ Rockbass
_____ Rock Gunnel
_____ Rosa's Blindfish
_____ Rosy Dace
_____ Rough Scad
_____ Roughtail Stingray
_____ Round Herring
_____ Round Scad
_____ Round Whitefish
_____ Rustyside Sucker

_____ Sandbar Shark
_____ Sand Tiger
_____ Satinfin Shiner
_____ Sauger
_____ Scrawled Cowfish
_____ Scup
_____ Sea Catfish
_____ Sea Lamprey
_____ Sea Raven
_____ Sharksucker
_____ Sheepshead
_____ Sheepshead Minnow
_____ Shortfin Mako Shark
_____ Shorthead Redhorse
_____ Shorthorn Sculpin
_____ Shortjaw Cisco
_____ Shortnose Chub
_____ Shortnose Gar
_____ Shortnose Sturgeon
_____ Shovelnose Sturgeon
_____ Silver Chub
_____ Silver Hake
_____ Silverjaw Minnow
_____ Silver Lamprey
_____ Silver Perch
_____ Silver Redhorse
_____ Silvery Minnow
_____ Skilletfish
_____ Skipjack Herring
_____ Skipjack Tuna
_____ Smallmouth Bass
_____ Smallmouth Buffalo
_____ Smallmouth Flounder
_____ Smooth Butterfly Ray
_____ Smooth Dogfish
_____ Smooth Flounder
_____ Smooth Hammerhead
_____ Smooth Puffer
_____ Sockeye Salmon
_____ Southern Cavefish
_____ Southern Kingfish

_____ Southern Redbelly
Dace
_____ Spanish Mackerel
_____ Spanish Sardine
_____ Spiny Butterfly
_____ Spiny Dogfish
_____ Spot
_____ Spotfin Shiner
_____ Spottail Shiner
_____ Spotted Bass
_____ Spotted Eagle Ray
_____ Spotted Gar
_____ Spotted Hake
_____ Spotted Sea Trout
_____ Spotted Sucker
_____ Spotted Sunfish
_____ Springfish
_____ Starhead Topminnow
_____ Stonecat
_____ Stoneroller
_____ Striped Anchovy
_____ Striped Bass
_____ Striped Burrfish
_____ Striped Killifish
_____ Striped Mullet
_____ Striped Sea Robin
_____ Suckermouth Minnow
_____ Summer Flounder
_____ Swampfish
_____ Swordfish

_____ Tadpole Madtom
_____ Tarpon
_____ Tautog
_____ Thorny Skate
_____ Thresher Shark
_____ Tiger Shark
_____ Tilefish
_____ Tonguetied Chub
_____ Torrent Sucker
_____ Tripletail
_____ Troutperch
_____ Trunkfish

_____ Virginia Lancelet

_____ Walleye
_____ Warmouth
_____ Weakfish
_____ White Bass
_____ White Catfish
_____ White Crappie
_____ White Hake
_____ White Marlin
_____ White Mullet
_____ White Perch
_____ White Shark
_____ White Sucker
_____ Windowpane
_____ Winter Flounder
_____ Winter Skate

_____ Witch Flounder
_____ Wrymouth

_____ Yellow Bass
_____ Yellow Bullhead
_____ Yellowfin Tuna
_____ Yellow Jack
_____ Yellow Perch
_____ Yellowtail Flounder

## MOLLUSKS

_____ Adams' Miniature
Cerith
_____ Alphabet Cone
_____ American Star Shell
_____ Angled Wentletrap
_____ Apple Murex
_____ Arctic Rock Borer
_____ Atlantic Auger
_____ Atlantic Awning Clam
_____ Atlantic Bob-tailed
Squid
_____ Atlantic Modulus
_____ Atlantic Nut Clam
_____ Atlantic Octopus
_____ Atlantic Plate Limpet
_____ Atlantic Surf Clam

_____ Baltic Macoma
_____ Bay Scallop
_____ Beaded Miter
_____ Black Clam
_____ Bleeding Tooth
_____ Blood Ark
_____ Blue Mussel
_____ Boreal Astarte
_____ Broad-ribbed Cardita
_____ Brown-banded
Wentletrap
_____ Brown-corded Neptune

_____ Calico Scallop
_____ Chalky Tellin
_____ Channeled Whelk
_____ Checkerboard Clam
_____ Coffee Bean Trivia
_____ Colorful Moon Shell
_____ Comb Bittersweet
_____ Common Angel Wing
_____ Common Atlantic
Marginella
_____ Common Baby's Ear
_____ Common Basket Clam
_____ Common Boat Shell
_____ Common Cockle
_____ Common Jingle Shell
_____ Common Moon Shell
_____ Common Nutmeg
_____ Common Periwinkle
_____ Common Purple Sea
Snail

_____ Common Razor
_____ Common Soft-shelled Clam
_____ Common Squid
_____ Common Sundial
_____ Common Worm Shell
_____ Coon Oyster
_____ Cross-barred Chione
_____ Crosshatched Lucine
_____ Crown Conch

_____ Deep Sea Scallop
_____ Deer Cowry
_____ DeKay's Tellin
_____ Disk Dosinia
_____ Dwarf Horn Shell
_____ Dwarf Squid
_____ Dwarf Surf Clam

_____ Emerson's Cerith
_____ Eastern Chiton
_____ Eastern Paper Bubble

_____ Fallen Angel Wing
_____ False Angel Wing
_____ False Coquina
_____ File Yoldia
_____ Flamingo Tongue
_____ Flat Slipper Shell
_____ Flat Tree Oyster
_____ Flesh Pen Shell
_____ Florida Banded Tulip
_____ Florida Drill
_____ Florida Fighting Conch
_____ Florida Horn Shell
_____ Florida Horse Conch
_____ Florida Rock Shell
_____ Four-spotted Trivia
_____ Fuzzy Chiton

_____ Giant Eastern Murex
_____ Glassy Lyonsia
_____ Gould's Pandora
_____ Gould's Shipworm
_____ Granular Frog Shell
_____ Gray Cowry
_____ Great Heart Cockle
_____ Great Piddock
_____ Great White Lucine
_____ Greedy Dove Shell
_____ Greenland Top Shell

_____ Hairy Triton
_____ Heavy Ark
_____ Hooked Mussel
_____ Horse Mussel

_____ Impressed Odostome
_____ Interrupted Turbonille
_____ Ivory Tusk

_____ Jasper Cone
_____ Junonia

_____ Kitten's Paw
_____ Knobbed Whelk
_____ Knobby Periwinkle
_____ Knobby Turban

_____ Lace Murex
_____ Leafy Jewel Box
_____ Lettered Olive
_____ Lightning Whelk
_____ Lined Tellin
_____ Linnaeus' Keyhole Limpet
_____ Lion's Paw
_____ Lister's Keyhole Limpet
_____ Little Boat Shell
_____ Little Chink Shell
_____ Little Keyhole Limpet
_____ Long-spined Star Shell

_____ Miniature Natica
_____ Minute Dwarf Olive
_____ Mossy Ark
_____ Mottled Chiton
_____ Mottled Dog Whelk
_____ Mottled Dove Shell
_____ Mottled Top Shell
_____ Mud Snail

_____ Naval Shipworm
_____ New England Dog Whelk
_____ Northern Quahog
_____ Northern Tellin

_____ Oyster Drill

_____ Paper Mussel
_____ Paper Nautilus
_____ Pennsylvania Lucine
_____ Pointed Nut Clam
_____ Prickly Jingle Shell
_____ Purplish Razor

_____ Ram's Horn
_____ Red Chiton
_____ Ribbed Mussel
_____ Ribbed Pod
_____ Rice Dwarf Olive
_____ Rock Purple
_____ Rough Periwinkle

_____ Sallé's Auger
_____ Salt Marsh Periwinkle
_____ Scotch Bonnet
_____ Sea Arrow
_____ Shark Eye
_____ Single-toothed Simnia

_____ Smooth Periwinkle
_____ Southern Coquina
_____ Spiny Oyster
_____ Stiff Pen Shell
_____ Stimpson's Tusk
_____ Stimpson's Whelk
_____ Striate Cup-and-saucer
_____ Striate Wood Piddock
_____ Sunray Shell

_____ Tenta Macoma
_____ Tesselated Nerite
_____ Thick-lipped Drill
_____ Transverse Ark
_____ Tuberculate Chiton
_____ Tulip Shell
_____ Turkey Wing

_____ Unequal Spoon Shell

_____ Variable Bittium
_____ Variegated Nerite
_____ Virginia Oyster

_____ Waved Whelk
_____ Wedge Rangia
_____ Wedge-shaped Piddock
_____ White Chiton
_____ White-mouthed Triton
_____ White Semele

_____ Zebra Nerite

## OTHER MARINE INVERTEBRATES

_____ *Aiptasia eruptaurantia*
_____ *Aiptasia pallida*
_____ American Lobster
_____ *Amphipholis squamata*
_____ *Amphitrite johnstoni*
_____ *Amphitrite ornata*
_____ *Ascidia callosa*
_____ *Ascidia interrupta*
_____ *Asterias forbesii*
_____ *Asterias vulgaris*
_____ *Astrangia danae*
_____ *Astropecten articulatus*

_____ *Balanoglossus aurantiacus*
_____ *Balanus amphitrite*
_____ *Balanus balanoides*
_____ *Balanus crenatus*
_____ *Balanus eburneus*
_____ *Balanus improvisus*
_____ Bamboo Worms
_____ Beach Fleas
_____ Beach Hoppers
_____ *Beroe cucumis*
_____ *Berod ovata*

_____ Blood Worms
_____ Blue Crab
_____ Boring Sponge
_____ Brown Sea Anemone
_____ *Bugula neritina*
_____ *Bugula turrita*
_____ Burrowing Sea Anemone
_____ Burrowing Shrimp

_____ *Callianassa atlantica*
_____ *Caudina arenata*
_____ *Chaetopterus variopedatus*
_____ *Chthamalus fragilis*
_____ Clam Worms
_____ Common Sun Star
_____ *Crisia denticulata*
_____ *Crisia eburnea*
_____ *Cyanea capillata*

_____ Daisy Brittle Star
_____ Deadman's Fingers
_____ *Diadumene leucolena*
_____ *Diopatra cuprea*

_____ *Echiurus pallasii*
_____ Edible Shrimp
_____ *Electra* (species)
_____ *Eudendrium ramosum*

_____ Fish Louse

_____ Ghost Crab
_____ Goose Barnacles
_____ Grass Shrimp
_____ Green Brittle Star
_____ Green Clam Worm
_____ Green Crab
_____ Green Sea Urchin
_____ Gribbles

_____ Heart Urchin
_____ *Henricia sanguinolenta*
_____ Horseshoe Crab
_____ *Hydractinia echinata*
_____ *Hydroides dianthus*
_____ *Hymeniacidon heliophila*

_____ *Idotea* (species)

_____ Jonah Crab

_____ Keyhole Urchin

_____ Lady Crab
_____ *Leptosynapta inhaerens*
_____ Lugworms
_____ *Luidia clathrata*
_____ *Lygia exotica*

687

# OTHER MARINE INVERTEBRATES

_____ *Lygia oceanica*

_____ Mantis Shrimp
_____ *Membranipora* (species)
_____ *Mnemiopsis leidyi*
_____ Mole Crab
_____ Moon Jelly
_____ Mud Crabs
_____ Mud Fiddler Crab
_____ Mud Shrimp
_____ Mussel Crab

_____ *Obelia* (species)
_____ *Ophiothrix angulata*
_____ Oyster Crab

_____ *Pagurus longicarpus*
_____ *Pagurus pollicaris*
_____ *Panopeus herbstii*
_____ Peacock Worms
_____ Pea Crab
_____ *Pennaria tiarella*
_____ *Pentamera pulcherrima*
_____ *Perophora viridis*
_____ *Pinnixa chaetopterana*
_____ *Polyonyx gibbesi*
_____ Portuguese Man-of-War
_____ Purple Sea Urchin
_____ Purple Sun Star

_____ Red-jointed Fiddler Crab
_____ Red Sponge
_____ Ribbon Worm
_____ Rock Crab

_____ *Sabellaria vulgaris*
_____ *Saccoglossus kowalevskii*
_____ Sand Bug
_____ Sand Dollar

_____ Sand Fiddler Crab
_____ Sand Fleas
_____ Sand Shrimp
_____ Scale Worms
_____ *Schizoporella unicornis*
_____ *Sclerodactyla briareus*
_____ Scuds
_____ *Scypha ciliata*
_____ Sea Grapes
_____ Sea Mouse
_____ Sea Nettle
_____ Sea Pansy
_____ Sea Pork
_____ Sea Vase
_____ Sea Walnut
_____ Seaweed Hoppers
_____ *Sertularia pumila*
_____ *Sesarma cinereum*
_____ *Sesarma reticulatum*
_____ Skeleton Shrimp
_____ Snapping Shrimp
_____ *Sphaeroma quadridentatum*
_____ Spider Crab
_____ *Spirorbis borealis*
_____ *Stomolophus meleagris*
_____ Striped Hermit Crab
_____ *Styela partita*

_____ *Terebratulina septentrionalis*
_____ Trumpet Worm
_____ Tube-dwelling Sea Anemone
_____ *Tubularia crocea*

_____ Whip Coral
_____ White Sea Urchin
_____ Wood-boring Amphipod

# Index

Animal species are indexed in general by common name. A number of species, however, especially the marine invertebrates, do not have common names, and these species are indexed by their Latin (genus-species) nomenclature. Higher taxa, such as classes, orders and families, are indexed by both common and scientific names. A number in *italics* indicates a text page on which the species is illustrated; a reference in **bold** type indicates the number of the color plate on which the species is illustrated and the position of the species on the color plate.

# INDEX

# INDEX